Dictionary of Environmental Science and Technology

Fourth Edition

By the rubbish in our wake, and the noble noise we make,
Be sure, be sure, we're going to do some splendid things!

Rudyard Kipling, *'The Road Song of the Bandar-Log'*

Dictionary of Environmental Science and Technology

Fourth Edition

ANDREW PORTEOUS

Emeritus Professor of Environmental Engineering,
the Open University, UK

John Wiley & Sons, Ltd

Fourth edition published 2008 by John Wiley & Sons Ltd,
Baffins Lane, Chichester,
West Sussex PO19 1UD, England
National (+44) 1243 779777
International (+44) 1243 779777

Third edition published 2000
Second edition published 1996
Revised edition published 1992
First edition published 1991 by Open University Press

Other Wiley Editorial Offices

John Wiley & Sons Inc., 111 River Street, Hoboken, NJ 07030, USA

Jossey-Bass, 989 Market Street, San Francisco, CA 94103-1741, USA

Wiley-VCH Verlag GmbH, Boschstr. 12, D-69469 Weinheim, Germany

John Wiley & Sons Australia Ltd, 42 McDougall Street, Milton, Queensland 4064, Australia

John Wiley & Sons (Asia) Pte Ltd, 2 Clementi Loop #02-01, Jin Xing Distripark, Singapore 129809

John Wiley & Sons Canada Ltd, 6045 Freemont Blvd. Mississauga, Ontario, L5R 4J3 Canada

Wiley also publishes its books in a variety of electronic formats. Some content that appears in print may not be available in electronic books.

Library of Congress Cataloging-in-Publication Data

British Library Cataloguing in Publication Data

A catalogue record for this book is available from the British Library

ISBN 978-0-470-06194-7 (cloth)
      978-0-470-06195-4 (paper)

Typeset in 10/12pt Times by SNP Best-set Typesetter Ltd., Hong Kong
Printed and bound in Great Britain by TJ International Ltd., Padstow, Cornwall

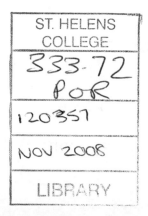

To Margaret, for forty-five years of support,
care and love. And to Neil for the future

Contents

Preface to Fourth Edition

This book is in effect, my '*apologia pro vita mea*,' thus all errors, omissions and commissions are down to me.

I have lived and breathed responsible environmental management all of my professional life, which commenced at the Thayer School of Engineering, Dartmouth College, USA, in 1964. The School instilled in me a thorough grounding in rational resource management, engineering rigour, and the necessity for accountable professional conduct.

It was with this background and in the belief that too much environmental illiteracy abounded that I essayed my first work *The Environment – A Dictionary of the World around Us*, published by Arrow in 1976. This was written in association with my friend and colleague Geoffrey Hollister, the visionary founding Dean of The Open University's Technology Faculty. Subsequently, this oeuvre metamorphosed into the current series of Dictionaries.

This edition would not have been possible without the unstinting help of my colleague and friend Dr Suresh Nesaratnam, Senior Lecturer in Environmental Engineering at The Open University, allied with the dedicated secretarial assistance of Mrs Rozy Carleton.

There is little left to say that has not been covered in the previous Prefaces except to bemoan (again) the levels of environmental ignorance and posturing of our political masters – *plus ça change.*

I conclude with the Persian proverb which has guided me since I discovered it in Dartmouth College Library in 1965.

'God will not seek thy race, nor will He ask thy birth. Alone, He will demand of thee "What hast thou done with the land that I gave thee?"'

Persian Proverb
Discovered in Dartmouth (USA) College Library (1965)

Preface to Third Edition

The need to consider the environmental impacts of any industrial activity is now taken for granted.

Advances in environmental literacy have led to much more questioning of, and accountability from, environmental professionals. Both are to be greatly welcomed.

This text has yet again been expanded to aid the above goals. I am gratified that it still is found to be useful.

My grateful thanks to all my colleagues (listed in the preface to the first edition) plus new colleagues, Dr Suresh T. Nesaratnam and Dr Stephen Burnley.

My secretary Mrs Morine Gordon has, as ever, helped the gestation immensely.

The holistic view of Lavoisier (1743–1794):

Rien ne se perd, rien ne se creé, tout se transforme

[Nothing is lost, nothing is created, everything is transformed] still holds true. It is how the transformation is effected that counts.

Andrew Porteous
Professor of Environmental Engineering
Faculty of Technology
The Open University
May 2000

Preface to Second Edition

Time marches on as they say, and this second edition is my way of documenting advances in environmental practices, knowledge and perceptions.

The opportunity has been taken to greatly expand the contents. I hope it meets with your approval.

My grateful thanks to my colleagues and my secretary Mrs Morine Gordon for typing this second edition.

Balaam (Numbers, Chapter 24), 'I came to curse [this task] but stayed to bless it!'

Andrew Porteous
Professor of Environmental Engineering
Faculty of Technology
The Open University

Preface

This text springs from an earlier attempt when, along with my colleague Geoffrey Holister, we endeavoured to lay the foundations for widespread environmental literacy in *The Environment – a dictionary of the world around us* published in 1976. This new work is not quite so ambitious; it is focused on the science and technology of environmental protection and resource management, as this is where the environmental payoffs are greatest.

The text has been principally written in basic SI units of kg, m, s, but in Tables and Figures, the units may be sub-values of the basic SI units, e.g. g/m^3 or kg/h. The SI system is admirable for setting out theory and equations but can become cumbersome if numerical values have to be written frequently in terms of powers of ten. For example, volume flows have been written as litre/s which is much more familiar than using 10^{-3} m/s. Equations are written for temperatures in K, but Tables and Figures are given in °C because this is the common unit of practical measurement. On occasions, concentrations have been given in non-SI units as these are often enshrined in current legislation or codes of practice.

I should like to put on record my thanks to my colleagues Judy Anderson and Rod Barratt, who have reviewed the text with diligence, and to Lesley Booth who had the trying task of typing it. My colleagues Keith Attenborough, David Cooke, and David Yeoman have kindly commented on specific entries. Caryl Hunter-Brown, OU liaison librarian, compiled the invaluable directory of environmental organizations.

Andrew Porteous
Professor of Environmental Engineering
Faculty of Technology
The Open University

Introduction

Stanley Clinton Davis in his 1987 Royal Society of Arts Lecture 'The European Year of the Environment' gave two reasons why the public wanted action by government on environmental issues now rather than later. These are given below.

> All governmental decisions tend to be taken under pressure from particular special interest groups. And it is a sad fact that the producers of pollution are on the whole better at exerting such pressure than those who have to live with its consequences. The manufacturers of nitrate fertilizers are well organized to ensure that their views are known in official circles. Those who worry about the fish that subsequently die are not. There are, of course, exceptions to this rule. There are well organized campaigns in which the view of the man and woman in the street is fully brought home to the man (and sometimes woman) in the Ministerial Office – I vividly recall the recent example of baby seals – but they are rare. I have direct explanations of why they should continue to do whatever it is they are doing. And I receive a steady stream of letters from the public inevitably tending to be more personal, and less well argued. I find that I have to be constantly on the alert not to let myself be misled by the different levels of presentation.

> The second reason why people are better than governments at spotting environmental needs goes still deeper. It is because environmental policy is fundamentally about the *future*, while governmental decision making is often too exclusively focused on the present.

These were brought home to me when I heard two senior British politicians respectively declaim on a television discussion programme 'PCBs cause toxic algae blooms in the North Sea' and 'Fitting catalytic converters to cars may reduce sulphur dioxide but will still produce gases that deplete the ozone layer'.

More recently, the December 2007 political announcement that the UK will install 33 GW-worth of wind turbines around the coast by 2020, without regard to the feasibility of such an undertaking in terms of its cost, rate of construction and installation, and ensuring the National Grid can cope, let alone the ecological consequences, beggars belief. This is equivalent to the construction of 15–20 thermal power stations.

This book is written in the hope that it may contribute to environmental literacy, by providing basic definitions and data plus demonstrate the nature of the issues. Where appropriate, the technology and/or measures that are already available to prevent pollution and aid resource conservation are outlined. As our understanding of these becomes clearer, we may at least begin to appreciate a number of things that we must obviously *not* do and that is a start of sorts.

In writing this book I have had to select information from an almost infinite source of data, much of it conflicting. In making this selection, I have naturally had to apply my own value judgements as to which data are relevant and important and which are not. In doing this, I have attempted to be as objective as possible, but under such circumstances perfect objectivity is quite impossible. We cannot assess the value of data without passing a tacit judgement on the character of the source. Objectivity does not consist in giving equal weight to all statements.

Environmental problems are essentially multi-faceted and demand at least a nodding acquaintance with many previously separate specialisms – ecology, economics, sociology, technology, physics, chemistry, and so on. The world is an enormously complex system and it is in the nature of complex systems that the characteristics of the connections between the constituent parts are often more important than the nature of the separate parts themselves.

This book is designed for a multi-access approach on the part of the reader – it can be dipped into, as well as read straight through. The reason for this format is the obvious diversity of backgrounds and interests of the readers. Most of you who read this book will have some specialist knowledge of some aspect of our technological society, and are likely to be interested in one particular aspect of our environment more than another. The format of this book allows you to select those areas of interest, although because of the complex nature of environmental issues, you will find that *wherever* you start in this book, you will be led inexorably by the references to other areas of the problem that are probably new to you. But that is the nature of environment; everything is related, in some way or another, to everything else.

As Chief Seattle of the Suquamish Indians said of the earth over 130 years ago:

> *...The earth does not belong to man; man belongs to the earth...all things are connected...*
> *Man did not weave the web of life, he is merely a strand in it. Whatever he does to the web, he does to himself.*

Acknowledgements

Grateful acknowledgement is made to the following sources for permission to reproduce material in this book:

Figure 1: Based on material supplied courtesy of CEGB Research, with permission of the Editor

Figures 6, 7, and 53: Based on material supplied by Enpure, Kidderminster, England

Figure 8: Based on material in *Renewable Energy*, Issue 6, January 1989, Department of the Environment

Figures 13 and 14: Based on material supplied courtesy of Keep Britain Tidy Schools Research Project, Brighton Polytechnic, Science Unit 2, Glass

Figure 17: From *Managing Water Abstraction*, supplied courtesy of the Environment Agency

Figure 26: S. H. Schneider and R. D. Dennett, 'Climatic barriers to long-term energy growth', *Ambio*, vol. 4, no. 2, 1975, reproduced by permission of the Royal Swedish Academy of Sciences

Figure 27: From material supplied courtesy of the World Coal Institute, Richmond-upon-Thames, England

Figures 28–32: Based on material in 'Focus on Combined Heat & Power', *Energy Management*, Issue 7, 1988

Figure 51: Based on material in *Renewable Energy*, Issue 6, January 1989, Department of the Environment

Figure 52: Reproduced from Open University, Course PT272, 'Environmental Control and Public Health', with permission

Figure 54: The Royal Commission on Environmental Pollution, *Report No. 1,* HMSO 1971, reproduced by permission of HMSO

Figure 55: *Coal and Energy Quarterly*, no.7, Winter 1975, reproduced with permission

Figure 58: R. Ashman, 'Air pollution control' in A. Porteous (ed.), *Developments in Environmental Control and Public Health,* Applied Science Publishers, 1979, reproduced by permission of the Publisher

Figures 60–62: From material supplied courtesy of WWF-UK, Godalming, England

Figure 64: L. Pringle, *Ecology: Science of Survival*, Macmillan Co., 1971, reproduced by permission of the Publisher

Figures 66 and 67: Open University, Course PT272, 'Environmental Control and Public Health', reproduced by permission of the Open University

Figure 79: Energy Conservation, HMSO, 1974, reproduced by permission of HMSO

Figure 92: Based on material supplied courtesy of CEGB Research, September 1988, with permission of the Editor

Figure 93: Plant portrait of Allington Quarry. From material supplied courtesy of Lentjes GmbH, Germany

Figure 100: A. Tucker, *The Toxic Metals*, Pan/Ballantine, 1972 reproduced by permission of the Publisher

Figure 103: W. Burns, *Noise and Man*, 2nd edition, John Murray, 1973, reproduced by permission of the Publisher

Figure 104: Based on data from G. Borgstrom, *Too Many,* Macmillan, 1967

Figure 107: Courtesy of Motherwell Bridge Engineering, PO Box 4, Logans Road, Motherwell, UK

Figure 115: Courtesy of Warwickshire County Council, Coping with Landfill Gas report, no. 4/4, November 1988, County Surveyors Society

Figure 116: Courtesy of Warwickshire County Council, Coping with Landfill Gas report, no. 4/4, November 1988, County Surveyors Society

Figure 119: Waste Management Paper 26B, produced by The Stationery Office, reproduced by permission of TSO

Figure 128: From material supplied courtesy of Bollegraaf UK Ltd, West Bromwich, England

Figure 129: From material supplied courtesy of Jonathan Clarke, of TiTech Visionsort Ltd

Figure 142: H. R. Warman, 'World energy prospects and North-Sea oil', *Coal and Energy Quarterly*, no. 6, Autumn 1975 reproduced with permission

Figure on page 494: Based on P. Chapman, *Fuel's Paradise*, Penguin, 1975, is reprinted by permission of Edward Arnold (Publishers) Ltd

Figure 145: From material supplied courtesy of INCPEN, Reading, England

Figure 154: J. Claerbout, 'Material and Energy recycling of PVC: Case Studies', in Karl J. Thomé-Kosmeinsky (ed.), *Recycling International*, vol. 3, reproduced by permission of EF-Verlag

Figures 180 and 197: From material supplied courtesy of the Institution of Chemical Engineers, UK

Figure 184: Land, Air and Sea – Research in Man's Environment, NERC, 1975, reproduced by permission of the Natural Environment Research Council

Figure 199: Courtesy of Blue Circle Group

Figures 201 and 203: From material supplied courtesy of Steve Eminton, of letsrecycle.com

Figure 208: Reproduced from K. Knox, Chapter 4, in A. Porteous (ed.), *Hazardous Wastes Management Handbook,* Applied Science Publishers, 1985, reproduced by permission of the Publisher

Figure 210: Reproduced from Metropolitan Water Division, *A Brief Description of the Undertaking*, 1973, reproduced by permission of the Thames Water Authority

Figure on page 763: From material supplied courtesy of Wavegen (a Voith and Siemens Company), Inverness, Scotland

Phillip Russell and Hannah Rogers, of West Sussex County Council Waste Management Services, for their valued contribution to the entries for Hazardous Household Wastes, and Waste Electrical and Electronic Equipment.

Veolia Environmental Services, United Kingdom, for permission granted to publish emission data from the Energy-from-Waste Plants in Sheffield and Portsmouth.

Whilst every effort has been made to trace the owners of copyright material, in a few cases this has proved impossible and we take this opportunity to offer our apologies to any copyright holders whose rights we may have unwittingly infringed.

Abbreviations

AC	Activated carbon
ADAS	Agricultural Development and Advisory Service
ADI	Acceptable daily intake
A/G ratio	Arithmetic–geometric ratio
AOX	Adsorbable organically-bound halogens
ATU	Allythiourea
BAF	Biological aerated filter
BATNEEC	Best Available Technique Not Entailing Excessive Cost
BHC	Benzene hexachloride
BMWP	Biological Monitoring Working Party
BOD	Biochemical oxygen demand (BOD_x signifies measurement over x days)
BP	British Petroleum
BPEO	Best Practicable Environmental Option
BPM	Best Practicable Means
BS	British Standard (i.e. a standard set by the British Standards Institute (BSI))
BSE	Bovine spongiform encephalopathy
BSI	British Standards Institute
CEFIC	European Chemical Industry Council
CFRR	Catalytic flow reversal reactor
Ci	Curie
CNG	Compressed natural gas
CNL	Corrected noise level
COC	Committee on the Carcinogenity of Chemicals in Food, Consumer Products and the Environment
COE	Catalytic oxidation of effluents
COMEAP	Committee on the Medical Effects of Air Pollutants
COP	Coefficients of Performance
CV	Calorific Value (units $MJ/kg : MJ/Nm^3$)
CWAO	Catalytic wet air oxidation
DANI	Department of Agriculture for Northern Ireland
dB	Decibel

dB(A)	Decibels A-scale
DDE	Dichlorodiphenyldichloroethylene
DDT	Dichlorodiphenyltrichloroethane
DDVP	Dichlorvos 2,2-dichlorovinyl dimethyl phosphate
DES	Diethylostilboestrol
DETR	Department of the Environment, Transport and the Regions
DIPNs	Diisopropylnaphthalenes
DMT	Dimethyl terephthate
DNA	Deoxyribonucleic acid
DO	Dissolved oxygen
DOA	Dioctyladipate
DRE	Destruction and removal efficiency
EA	Environmental Assessment/Environment Agency
EALs	Environmental assessment levels
EB	Environment burden
EC	European Commission
ECF	Elemental chlorine free
EDTA	Ethylenediamine tetraacetate disodium salt or ethylenediamine tetraacetic acid
EEZs	Economic exclusion zones
EfW	Energy from Waste
EG	Ethylene glycol
EMAS	Eco-Management and audit scheme
EMS	Environmental Management System
EPA	Environmental Protection Act (1990)/Environmental Protection Agency (USA)
EPAQS	Expert Panel on Air Quality Standards
EPNdB	Effective perceived noise level
ER	Electroremediation
ERTS	Earth Resources Technology Satellite
ES	Environmental statement
ESA	Environmental Services Association
EU	European Union
FBR	Fluidized bed biofilm reactor
FID	Flame ionization detector
FMA	Forest Management Association
FMC	Field moisture capacity
GAC	Granulated activated carbon
GC	Gas chromatography
GHG	Greenhouse gas
GMO	Genetically modified organisms
GNP	Gross National Product
HACCP	Hazard Analysis and Critical Control Point

HAS	Hygiene Assessment System
HGV	Heavy goods vehicle
HS&E	Health, safety and environmental
HSE	Health and Safety Executive
Hz	Hertz
ICRCL	Interdepartmental Committee on the Redevelopment of Contaminated Land
ICRP	International Committee on Radiation Protection
IMO	International Maritime Organization
IPC	Integrated Pollution Control
IPPC	Integrated Pollution Prevention and Control
IWC	International Whaling Commission
JVC	Joint Venture Company
K	Kelvin
L_{10}	Ten per cent level (road traffic noise index)
LAWDC	Local Authority Waste Disposal Company
LCA	Life Cycle Analysis
LC_{50}	Lethal concentration (50 per cent survival)
LD_{50}	Lethal dose (50 per cent survival)
LEL	Lower explosive limit
LFG	Landfill gas
LPG	Liquefied petroleum gas
MAC	Maximum allowable concentration
MAFF	Ministry of Agriculture, Fisheries and Food
MHD	Magnohydrodynamic generator
MHS	Meat Hygiene Service
MLVSS	Mixed Liquor Volatile Suspended Solids
MOX	Mixed oxide fuel
MPN	Most probable number
MSW	Municipal solid waste
MUF	Material unaccounted for
NFFO	Non-Fossil Fuel Obligation
NIC	National insurance charges
NNI	Noise and number index
NRA	National Rivers Authority
NRCan	Natural Resources Canada
OECD	Organization for Economic Co-operation and Development
OEL	Occupational exposure levels
OTMS	Over Thirty Month Slaughter
OVS	Official Veterinary Surgeon/Official Veterinary Services
PAH	Polynuclear aromatic hydrocarbon
PAN	Peroxyacetylnitrate
PCBs	Polychlorinated biphenyls

PCTs	Polychlorinated terphenyls
PET	Polyethylene terephthalate
PNAs	Polynuclear aromatic hydrocarbons
PNdB	Perceived noise decibels
POHC	Principal organic hazardous constituents
pphm	Parts per hundred million
ppm	Parts per million
PRNs	Packaging Recovery Notes
PTFE	Polytetrafluoroethylene (Teflon)
PVC	Polyvinyl chloride
PWR	Pressurised water reactor
RBE	Relative biological effectiveness
REC	Regional Electricity Company
RCEP	Royal Commission on Environmental Pollution
RIVPACS	River Invertebrate Prediction and Classification Scheme
RME	Rape methyl ester
R/P ratio	Reserves–production ratio
SBM	Specified bovine material
SCP	Single-cell protein
SCWO	Supercritical water oxidation
SD	Sustainable development
SEPA	Scottish Environment Protection Agency
SERM	Safety and Environmental Risk Management
SI	Système international d'unités (International System of Units)
SIDT	Solid insulation distribution transformer
SMD	Soil moisture deficit
SOAEFD	Scottish Office Agriculture, Environment and Fisheries Department
SRC	Short rotation coppicing
STP	Standard temperature and pressure
SWORD	Surveillance of Work Related and Occupational Respiratory Disease
TCD	Thermal conductivity detector
TCDD	Tetrachlorodibenzo-p-dioxin
TCF	Totally chlorine free
TDI	Total daily ingestion
TEQ	Toxic equivalent
TLV	Threshold limit value
TN	Total nitrogen
TNI	Traffic noise index
TSE	Transmissible spongiform encephalopathy

TUPE	The Transfer of Undertakings and (Protection of Employment) Regulations (1981)
UDC	Under-developed country
UEL	Upper explosive limit
UN	United Nations
UNCLOS	United Nations Convention on the Law of the Sea
WDA	Waste Disposal Authority
WDC	Waste Disposal Company
WHO	World Health Organization
WLM	Working level month
WML	Waste Management Licence
WOAD	Welsh Office Agriculture Department

A

Abatement. Reduction or lessening (of pollution) or doing away with a nuisance, by legislative or technical means, or both.

Absolute temperature scale. ◊ TEMPERATURE.

Absorption[1]. The taking up, usually, of a liquid or gas into the body of another material (the absorbent). Thus, for instance, an air pollutant may be removed by absorption in a suitable solvent. Not to be confused with ADSORPTION. This definition is most appropriate to sound absorption, for example by a porous material, and should be distinguished from sound insulation.

Absorption[2]. The taking up of radiant energy by a material it encounters or passes through. This can be the basis of measuring the concentration of a substance or identifying a number of substances. This definition is more like attenuation of energy, at least with regard to the transmission of energy. ◊ ATTENUATION.

Absorption chillers. District heating (DH) can also be used as the energy source for absorption chillers. The benefits of encouraging potential consumers to install absorption chillers for their chilled water and air conditioning requirements are two-fold. Firstly absorption systems use no CFCs or HFCs, so are environmentally friendly; secondly they have very few moving parts and are consequently relatively easy to maintain.

The chilling effect is produced in the absorption cycle by the evaporation of a fluid – the refrigerant – in the lower pressure side of a closed system. The refrigerant vapour is recovered by the absorbing action of a second fluid – the absorbent. In the higher pressure side of the system the absorbent containing the dissolved refrigerant is boiled, the refrigerant vapour and the absorbent being returned to the lower pressure side for further use.

Indirectly fired machines can use medium temperature hot water or steam in the range 104 °C to 130 °C, making them ideal for connection to the DH network.

Absorption chillers also produce a large benefit to the DH operator, in that they increase the summer demand for heat, thus providing a balanced loading over as much of the year as possible, to utilize the available energy efficiently. If a building's winter heating and summer cooling requirements can be met using the district heating network, it will greatly enhance the efficiency of the scheme and increase annual revenues.

Abstraction. The permanent or temporary withdrawal of water from any source of supply, so that it is no longer part of the resources of the locality.

Acceptable daily intake (ADI). The acceptable daily intake of a chemical is the daily intake which, during an entire lifetime, appears to be without risk on the basis of all known facts at the time. It is expressed in milligrams of chemical per kilogram of body weight (mg/kg). ADIs may be unconditional, conditional, or temporary.

Accuracy. The nearness of a measurement to the true value of the thing being measured. cf. PRECISION.

Acid. A substance which, in solution in water, splits up to a greater or lesser extent into ions, including hydrogen IONS. Strong acids are those which split up to a large degree; they are usually corrosive. Acids neutralize ALKALIS with the formation of SALTS. Acids have a multitude of industrial uses, e.g. the 'pickling' (cleaning) of steel, preparation of copper components (e.g. copper printed circuit boards), polishing/etching of glass, selective extraction of minerals or recovery from waste, descaling, the manufacture of 'nylon' through the medium of polymeric AMIDES (e.g. acetic acid, CH_3COOH, is used in the production of cellulose acetate). They are also used to adjust the pH value of solutions.

Acid dewpoint. As commonly used, this term applies to the temperature at which dilute acid (e.g. sulphuric acid) appears as a condensate (liquid droplets) when a flue gas containing sulphur trioxide and water vapour is cooled below saturation temperature. This value is related to the moisture and sulphur trioxide content of the flue gas. Therefore, gases must be released to atmosphere well above this temperature otherwise substantial corrosion will take place in the stack. The corrosion of many combustion installations especially if fired on heavy fuel oil is due to the flue gases being cooled below the acid dewpoint.

Acid mine drainage. Many mining operations, particularly those that work sulphide ores, e.g. nickel, copper or coal mining where pyrites (iron sulphide) are present, can, through a combination of air and moisture, form acidic and metal-bearing solutions. This combination of acids and metals can have severe local effects on the ecology of streams and rivers, and the metals can enter food chains and further affect life. If, as is likely, iron sulphates are formed, this will result in a brown coating on rocks and stream beds, further despoiling the appearance of the area.

In the UK the problems of acid mine drainage are not severe, but in nickel and copper mining areas of the world (e.g. Sudbury, Ontario) the devastation has to be seen to be believed.

The only way of controlling or preventing acid mine drainage damage other than not working the ores is to control the pH by raising it with lime and neutralizing it. (◊ PH)

Acid rain. Most rainfall is slightly acidic due to carbonic acid forming from the carbon dioxide content of the atmosphere, but 'acid rain' in the pollution sense is produced by the conversion of the primary pollutants sulphur dioxide and nitrogen oxides to sulphuric acid and nitric acid respectively. These processes are complex, depending on the physical dispersion processes and the rates of the chemical conversions. The basic cycle is shown in Figure 1.

However, the term 'acid rain' is misleading because the phenomenon and associated effects are broader than it suggests. Pure water has a neutral pH value of 7. Carbon dioxide in the atmosphere dissolves in rain reducing its pH to 5.6 and naturally occurring oxides of sulphur and nitrogen are responsible for unpolluted rain having a pH of about 5.0. Lower values of pH may result from strong acids produced from fuel use. An alternative unit of acidity is the 'microequivalent of hydrogen ion per litre', written as μeq H$^+$/l, as defined by

$$pH = -\log_{10}(\mu eq\, H^+/1 \times 10^{-6})$$

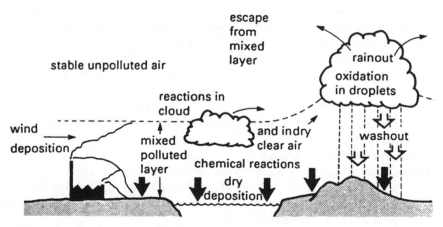

Figure 1 Acid rain cycle.

Hence, a value of $10\,\mu\text{eq}\,H^+/l$ equates to a pH $= -\log_{10}(10^{-5}) = 5$.

The following definitions are in use:

Acid precipitation: rainfall or snow with acidity greater than $10\,\mu\text{eq}\,H^+/l$ or pH lower than 5.

Acid mists: fog, mist or low cloud in which the water has an acidity greater than $10\,\mu\text{eq}\,H^+/l$ or pH lower than 5.

Acid deposition: total deposition of acid (hydrogen ions) or acid-forming compounds, e.g. SO_2 and NOX_x by both wet and dry deposition.

Acid rain: Precipitation and other deposition pathways which are more acidic than pH 5.0.

Source: 'The down-side of cleaner air', *Financial Times*, 28 February 1995.

The solutions to acid rain are as a consequence not simple. DESULPHURIZA-TION of flue gases alone will not necessarily effect a cure. The high levels of NITROGEN OXIDE and UNBURNT HYDROCARBONS from motor vehicles are also implicated.

Damage to trees is usually worst on hills where mists collect and deposition is greater. The soils are often poor and acidic to begin with, hence the additional stresses imposed by acid rain alone or in conjunction with ozone can prove too much as has been found in the German mountain forests and increasingly so in the UK as well.

To cause soil acidification it is necessary to have:

(a) A source of hydrogen ions to exchange for the exchangeable BASES.
(b) Some means of removing the displaced base CATIONS.

Acid precipitation supplies both. The source of hydrogen IONS is externally applied ACID (sulphuric or nitric acid). Cations are removed by leaching of the displaced base cation by the mobile sulphate or nitrate anion.

(It has been estimated that about 75 per cent of the UK now receives under 20 kg of atmospheric sulphur per hectare per year which is less than half the amount needed to produce a full crop of oilseed rape, insufficient for either silage grass or cereals and barely enough for potatoes or sugar beet.)

A short-term remedial measure for water sources and lakes consists of dosing with lime to restore the pH balance. ($\lozenge$ DEPOSITION PROCESSES; FLUE GAS DESULPHURIZATION; pH)

Acoustic. Pertaining to SOUND.

Actinides. The group of elements, with ATOMIC NUMBERS from 89 to 103, including URANIUM and PLUTONIUM. All members of this group are produced in nuclear reactions or by radioactive decay. Many of the more long lived are RADIATION emitters.

Action level. The level which triggers remedial/ameliorative action to safeguard public health. This is based on current knowledge and can be revised (usually downwards) ($\diamond$ DIOXINS, NITROGEN OXIDES, RADON). Also refers to noise exposure in the work place. As an example the first Action Level for noise is 80 dBA.

Activated carbon (AC). Carbon obtained from vegetable or animal matter by roasting in a vacuum furnace or by heat treatment. Its porous nature gives it a very high surface area per unit mass – as much as 1000 square metres per gram, which is 10 million times the surface area of one gram of water in an open container. The substance is a very good adsorbent for aromatic and unsaturated aliphatic compounds. It is extensively used for odour control and air-freshening applications, and can adsorb large quantities of gases. It is used in potable water treatment to adsorb trace organics. It is often used with regeneration. For effluent treatment purposes, it can be used to adsorb refractory, non-biodegradable compounds.

Powdered activated carbon, suspended in water, can adsorb large quantities of OXYGEN. In an ACTIVATED SLUDGE effluent treatment system, this allows for a higher aeration efficiency. The adsorbed oxygen is desorbed into the SLUDGE once the oxygen concentration around the carbon particles is lowered by microbiological activity. The addition of powdered carbon increases the rate of sedimentation of the sludge by a nucleation process giving denser, more compact floccules.

This allows greater concentrations of sludge to be maintained in the aeration basin. The sludge itself drains better, allowing discharge transport costs to be reduced. It can also adsorb ORGANIC and INORGANIC substances that are not, or only partly, decomposed by the activated sludge system. This applies to most toxic compounds.

When granulated activated carbon (GAC) is used, it is commonly regenerated, with the method chosen dependent on the nature of the adsorbed impurity and the ease with which it can be removed. The three most commonly used methods are:

Steam – This can be used if the adsorbed products are volatile, i.e. if they can be steam distilled.

Chemical Regeneration – It is sometimes the case that the adsorbed impurity can be desorbed from the surface by treatment with a chemical. Caustic soda has been used successfully with certain organic acid purification systems.

Thermal Regeneration – This is the most widely used method particularly in the large tonnage outlets in the sugar, glucose and water industries. This involves the burning off of the impurities on the carbon under controlled conditions in a regeneration kiln.

Exhausted AC can be dangerous as it tends to take on the properties of the material adsorbed and this factor should be considered when handling, reactivating or disposing of it.

AC case study – Entek International Ltd, of Newcastle, produces specialized POLYETHYLENE products for use in the manufacture of lead/acid batteries.

The polyethylene film from which these products are made is produced by heating POLYMER granules, extruding them through a die and forcing the polymer between two rollers. Significant quantities of lubricating oil are added to the polymer during the extrusion process to reduce friction and prevent adhesion to the rollers. The resulting membrane may, typically, contain 60% by weight of oil immediately after forming.

The oil is removed from the membrane surface after extrusion and before further processing. This involves passing the polymer film through a tank of warm trichloroethylene SOLVENT. As the membrane passes over guide rolles in the tank, it is washed by solvent circulating in the opposite direction. The solvent passes through a distillation unit for cleaning before being recirculated.

The film emerging from the washing tank passes through an air drier where the solvent evaporates. The dry film is cut to the required width and rolled onto spools before being packaged for despatch to the customer.

Because of the large quantities of solvent used in the washing process, Entek installed a GAC extraction and recovery plant to recover solvent from the solvent-laden air from the drier for reuse, and ensure that the solvent content of the workplace atmosphere remains within safe limits.

Reuse of the recovered solvent in the product cleaning process significantly reduces the quantities of solvent that would otherwise have to be purchased each year. It is estimated that the company is achieving net annual cost savings of more than £1.6 million, with a payback period on the capital outlay of 1.25 years (see table below).

Other benefits include reduced solvent emissions and improved workplace safety, plus compliance with ISO 14001.

Economic analysis	Annual value
Estimated solvent purchases without solvent recovery	£1752000
Solvent purchases with solvent recovery	£96000
Estimated reduction in solvent purchases	£1656000
Percentage saving	94.5%
Estimated costs of plant operation*	(£50000)
Estimated net savings	£1606000
Estimated capital expenditure on solvent recovery plant	£2000000
Estimated payback period	1.25 years

* Conservative estimate based on experience at other plants.
Sources: Environmental Technology Best Practice Programme GC100 Guide, DETR, 1998. IChemE '*Avert*', Issue no. 13, Winter 1998/99.

Activated sludge. The sludge removed from the activated sludge sewage treatment process. It consists of BACTERIA and PROTOZOA which can live and multiply on the sewage. Because of this multiplication, the excess organisms require continuous removal. Part of the surplus active sludge is returned to the raw sewage (hence the term 'activated sludge'), and part (approximately 90 per cent) is sent for disposal to land, or incineration. (◊ SEWAGE TREATMENT)

Activated sludge process. ◊ SEWAGE TREATMENT.

Activation product. Material made radioactive as a result of irradiation, particularly by neutrons in a nuclear reactor.

Active packaging. Food packaging which has an extra function, in addition to that of providing a protective barrier against external influence. Active packaging is intended to change the condition of the packed food, to extend shelf-life or improve sensory properties while maintaining the freshness and the quality of the food. This is done by the use of absorbers of food-related chemicals, such as scavengers of oxygen, water, carbon dioxide, ethylene, and off-flavours like amines and aldehydes, or by using emitters of food additives, which release preservatives, antioxidants, flavourings, or colours. Thus, for active packaging, the packaging is intended to influence the food.

Source: CSL Report FD 03/28 for the Food Standards Agency: Active Packaging – Current Trends & Potential for Migration, April 2004.

Acute. Of a disease or medical condition, coming to a rapid crisis, the opposite of CHRONIC.

Adiabatic. Without loss or gain of heat. When air rises, it experiences lower air pressure and it expands adiabatically in the atmosphere; since it can neither gain nor lose heat, its temperature will fall as it expands to fill a larger volume. This gives rise to the theoretical adiabatic temperature profile or lapse rate, which is used as a basis for comparison of actual temperature profiles (from ground level) and hence predictions of stack gas dispersion characteristics.

Adiabatic lapse rate (ALR). The rate of adiabatic cooling with height above the surface for a rising parcel of air in the troposphere. For a parcel of dry air, the rate of cooling is about 1 °C for each 100 m of ascent. The value is somewhat lower for moist air because the heat released by condensing water vapour slows the rate of cooling.

Adipates. ◊ PHTHALATES.

Adsorption. A phenomenon in which molecules of a substance (the adsorbate) are taken up and held on the surface of a material (the adsorbent) (compare ABSORPTION). (◊ ACTIVATED CARBON)

Aeration zone. The area in the ground above the SATURATION ZONE where WATER/LEACHATE infiltration percolates, partially filling PORE spaces and voids. The aeration zone can be split into three: (i) the soil water zone, (ii) the intermediate zone and (iii) the capillary fringe.

Aerobic processes. Many bio-technology production and effluent treatment processes are dependent on micro-organisms which require oxygen for their metabolism. These are termed aerobic processes. For example, water in an aerobic stream contains DISSOLVED OXYGEN and organisms can utilize it to oxidize organic wastes to simple compounds as below:

$$\text{organic materials} + O_2 \xrightarrow{\text{micro-organisms}} CO_2 + H_2O + NH_3$$
$$+ \text{micro-organisms}$$

Ammonia (NH_3) can then be further oxidized to a compound containing the nitrate ion (NO_3^-) by bacteria and the organic waste such as sewage is thereby totally decomposed. ($\diamond$ SEWAGE TREATMENT)

Aerobic respiration. The net effect of the reactions by which an organism uses oxygen to bring about the complete oxidation of simple organic compounds (carbohydrates) to carbon dioxide and water, thereby liberating the energy needed to live and grow.

Aerosol. Minute liquid droplets or solids of particle size up to $100\,\mu m$ suspended in a gaseous flow or the atmosphere. Due to their small size, they can be readily dispersed. There are two types: *condensation aerosols* formed when moisture-laden gases are cooled and *dispersion aerosols* formed from the break up of solids or atomization of liquids.

Aerosol propellant. Colloquially, the pressurized system used to disperse hair spray, deodorant, etc. More accurately, an inert liquid with a low boiling point, from the CHLOROFLUOROCARBONS or HYDROCARBONS, which vaporizes instantaneously at room temperatures on release of pressure. When the pressure in the aerosol canister is released, the vapour carries the AEROSOL of the desired substance to its target. The propellant then disperses into the atmosphere. Compounds containing chlorine are a hazard to the earth's OZONE SHIELD because of the free CHLORINE liberated in the upper atmosphere as a result of ultra-violet radiation.

The use of fluorocarbon propellants is intended to be virtually terminated except for medical purposes.

Other propellants are available, such as butane, which is highly flammable and therefore only suitable for spraying water-based products. CARBON DIOXIDE, nitrous oxide and nitrogen are possibly suitable for most spray purposes and are inert, but there is no commercial incentive for their adoption, and, as yet, no legal requirement that this be done to protect the environment.

After-burner. In COMBUSTION equipment, an after-burner may be fitted following the main zone of combustion in order to obtain virtually complete combustion, thus decreasing emissions of certain air pollutants.

After-burners may also be used in non-combustion plants to control pollutant emissions.

Agent Orange. ◊ DIOXINS.

Aggregate tax. A tax designed to:

1. Conserve virgin AGGREGATES and encourage RECYCLING and/or less wasteful uses.
2. Reduce the environmental impact of virgin/primary aggregate production and internalize the associated environmental costs.

An example of how the concept can be expanded is provided by Friends of the Earth, who have stated that such a tax can be designed to:

- reduce the inefficient use of this important non-renewable resource;
- increase efficiency by encouraging recycling, reuse and more durable designs;
- support other policy measures with similar aims through the price mechanism by internalizing many of the external costs of aggregate extraction that are not fully captured by regulation;
- allow a shifting of the burden off labour in the form of employer's National Insurance Contribution (NIC) and onto another production factor in the economy (resource use) that has significant external costs in the short and long term;
- increase employment both in its own right (as shown by research carried out by Cambridge Econometrics for Friends of the Earth and Forum for the Future) and even more effectively when recycled through a reduction in employer's NIC;
- reduce the environmental impacts of primary aggregates extraction by cutting output as recycling increases its share of total demand;
- complement rather than duplicate the landfill tax by increasing the comparative value of recycled aggregates and by reducing the production of potential landfill sites though quarrying.

Source: Friends of the Earth, 1998.

Aggregates. Inert tough materials used for concrete production (in conjunction with cement and sand). Also used for road foundations or top coating when covered with a bitumen binder. Typically crushed rock, gravel or graded ash from coal-fired power generation stations or municipal waste incinerators are used for these purposes. Fine aggregates, building sand, glass-making raw materials and foundry sands for moulding purposes are typical examples.

Aggregate extraction is a major source of contention in the UK due to the often permanent changes in topography in the form of lakes in river valleys, many of which are used for private recreation purposes that may bring their

own brands of NUISANCE in tow. Where the WATER TABLE permits, LAND-
FILL of waste is not uncommon ($\lozenge$ AGGREGATE TAX).

Agricultural economics. The application of traditional economic criteria to the
production of food. It is unfortunate that government policies are usually
designed by specialists whose experiences are essentially urban and industrial, and who have little or no understanding of the complex ecological balances required for a stable agricultural system.

Agriculture cannot be regarded simply as an industry. The majority of our
industries are concerned with the conversion of raw materials – which are
often regarded as virtually inexhaustible – into goods. The raw material of
agriculture is the soil, which should be husbanded – not mined.

Agricultural waste. Now that agricultural waste is 'controlled' an increase in the
number of exemptions (up from some 80 000 to about 1 million) is expected.
As the EA will only visit the 'average' farm about once in 20 years, effective
risk-based targeting will be essential. The Agency is adopting an approach
based on the OPRA (Operator and Pollution Risk Appraisal) concept and
using agriculturally focused and trained Environment Officers. At the heart
of the approach is the 'Whole Farm Appraisal' which allows self-assessment
of risk by farmers.

Source: Environmental Services Association News, 26 January 2007.
($\lozenge$ OPERATOR AND POLLUTION RISK APPRAISAL)

Agriculture, energy and efficiency aspects. The effectiveness of modern agriculture is invariably measured in traditional economic terms – percentage return
on invested capital, yields per acre, yields per man-year, etc. It is in these
terms that claims for the high efficiency of modern industrialized agricultural
techniques are made. At first sight, these claims are persuasive: in Britain,
average wheat yields are up from 0.95 tonne per acre in 1946 to 1.4 tonnes
in 1968–9; barley up from 0.9 tonne per acre in 1946 to 1.4 tonne per acre in
1968–9; and with similar increases in most other field crops. Cereal yields
approaching 3 tonnes per acre have been obtained. ($\lozenge$ AGRICULTURAL
ECONOMICS)

Unfortunately, close investigation shows that when such figures are appor-
tioned and averaged per agricultural worker, they misrepresent the true facts
because of the nature of the oversimplifications that are made. For instance,
figures showing increased yields per farm worker ignore all of those workers
who have moved from farms into agriculture-related industries, and are
engaged in the production of the machines, fuel and chemicals which give
the remaining farm workers an artificially greater efficiency.

A more realistic measure of agricultural efficiency is given by the relative
inputs and outputs of energy to the total system. This calculation is difficult
when applied to a modern industrial-type agricultural system; nevertheless,
the total energy budgets for British agriculture have been calculated and

compared with the efficiency of other agricultural systems. The general conclusions were that, where a fossil-fuel subsidy *does not exist*, human energy can be used very efficiently. For example, maize growers in Yucatan, Mexico, produce 13 to 29 units of food energy for each unit of (predominantly human) energy expended; primitive gardeners in Tsembaga, New Guinea, produce 20 units of energy for each unit expended. British wheat growers, on the other hand, produce 2.2 energy units for each unit expended; potato growers gain 1.1 units per unit expended; sugar beet, by the time it has been refined to white sugar, represents 0.49 units gained per unit expended; a battery egg, including the food value of the hen at the end of her laying life, represents 0.16 units for each unit expended; and a broiler chicken, 0.11 units gained.

Air. The mixture of gases which constitutes the earth's ATMOSPHERE. The approximate composition of dry air is, by volume at sea level, NITROGEN 78.0 per cent, OXYGEN 20.95 per cent, argon 0.93 per cent, and CARBON DIOXIDE 0.03 per cent, together with very small amounts of numerous other constituents. The water vapour content is highly variable and depends on atmospheric conditions. Air is said to be pure when none of the minor constituents is present in sufficient concentration to be injurious to the health of human beings or animals, to damage vegetation, or to cause loss of amenity (e.g. through the presence of dust, dirt, or odours or by diminution of sunshine).

Air, primary. ◊ COMBUSTION.

Air, secondary. ◊ COMBUSTION.

Air classifier (or air separator). Dry separation device used to separate shredded domestic REFUSE by DENSITY difference methods. It can be used in conjunction with ferrous-magnet and revolving screens to sort municipal waste or commercial wastes for recycling and WASTE-DERIVED FUEL production.

Figure 2 shows a vertical air classifier. The shredded refuse is fed through the airlock into a counter-current stream of air which entrains the light fractions (paper and plastics) while allowing the dense fraction (glass, tin cans, stones) to drop down onto the conveyor belt. The light fraction can then be further processed to a pelletized fuel or fed loose to a furnace as a fuel supplement.

Air pollutants and health. The system of AIR QUALITY BANDING used in the UK since 1990 was revised to take into account new AIR QUALITY STANDARDS, recommended by the DOE Expert Panel on Air Quality Standards (EPAQS). The banding system is intended to provide guidance as to the effects of air pollutants on health.

At concentrations of OZONE of less than 90 parts per billion (ppb) it is very unlikely that anyone will notice any adverse effects although effects are detectable at a population level. As concentrations rise towards 180 ppb

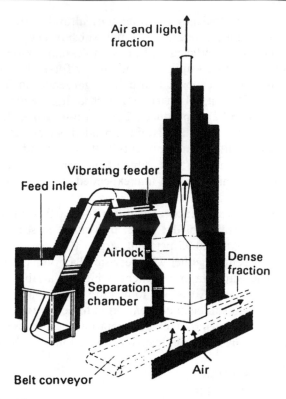

Figure 2   Air classifier.

some individuals, particularly those exercising out of doors, may experience eye irritation, coughing and discomfort on breathing deeply. At more than 180 ppb these effects may become more severe. Individuals suffering from asthma and other respiratory disorders associated with a reduction in respiratory reserve may experience earlier and more marked effects.

SULPHUR DIOXIDE is a respiratory irritant and causes tightening of the airways when inhaled at high concentrations. Those suffering from asthma are significantly more sensitive to sulphur dioxide than other people. There is little evidence to suggest that those suffering from asthma would be significantly affected by exposure to concentrations of sulphur dioxide of less than 200 ppb. The World Health Organization has suggested that exposure to 400 ppb sulphur dioxide may lead to significant narrowing of the airways in those suffering from asthma. For most people the effects expected would not be large although some individuals may be clinically affected. Exposure to such concentrations may add to the effects of exposure to other pollutants and allergens. Consequently, asthmatics may need to increase their medication since the effects can be reversed by use of inhalers. As concentrations rise above 400 ppb, more asthmatic individuals may experience adverse effects.

Evidence of effects of NITROGEN DIOXIDE at lower levels (200–300 ppb) is inconsistent and studies of its effects on those suffering from asthma and other individuals are more difficult to interpret. EPAQS recommended a standard of 150 ppb (1 hour average) representing 'low' levels of air pollution at which it is very unlikely that anyone will experience any adverse effects. Studies of volunteers, including those with asthma, exposed to concentrations of up to 300 ppb for 1 hour do not provide convincing evidence that significant effects on health are likely. Some increase in the response of the lung to substances which produce narrowing of the airways has been recorded on exposure to nitrogen dioxide at these concentrations. Again, the studies are inconsistent and the effects are small.

A few studies have shown small direct effects on indices of lung function at levels around 300 ppb. In addition, there is evidence from epidemiological studies of the effects of mixtures of pollutants characterized by concentrations of nitrogen dioxide in this range that adverse effects on health may occur.

Above 400 ppb (1 hour average) epidemiological studies have provided evidence of effects. These included increased admissions to hospital and consultations with General Practitioners. Individuals suffering from asthma do not appear to be at such increased risk on exposure to nitrogen dioxide as they are on exposure to sulphur dioxide.

PARTICULATE MATTER is monitored in the UK as PM_{10}, i.e. particles generally less than 10 micrometres (μm) in diameter. A large number of epidemiological studies have shown that day-to-day variations in concentrations of particles are associated with adverse effects on health. These include increased daily deaths, increased admissions to hospital of patients suffering from heart and lung disorders, and a worsening of the condition of those with asthma. Even at low particle concentrations, effects remain.

The most significant exposure to CARBON MONOXIDE occurs in the general population because of cigarette smoking. Carbon monoxide interferes with the transport of oxygen by the blood and at high concentrations produces unconsciousness and death. Some 60 accidental deaths occur in the UK each year because of exposure to carbon monoxide indoors. Concentrations of carbon monoxide in the air are related to concentrations of carbon monoxide in the blood in a well-understood and predictable way. In describing levels of exposure, it is usual to speak in terms of that percentage of haemoglobin, the essential oxygen-transporting protein of the blood, which is saturated with carbon monoxide. The effects of different percentage saturations have been studied in volunteers and EPAQS reviewed the results of these investigations closely in recommending a standard of 10 parts per million (ppm) (8 hour average concentration). At this concentration, the blood would reach a level of less than 2 per cent and effects on health would be unlikely even in those suffering from heart disease. However, epidemiological studies have reported associations between outdoor concentrations of carbon monoxide and admissions to hospital for treatment of heart disease.

Moderate levels of exposure (10–15 ppm as a running 8 hour average) lead to a level in haemoglobin of about 2.5 per cent. At this level, there is some evidence to show that those with angina and other heart diseases may experience a more rapid onset of chest pain on exercise.

Exposure to 20 ppm of carbon monoxide leads to a level of 4.5 per cent, which reduces peak exercise capacity of healthy subjects and the time needed for anginal pain to appear on exercise in those with heart disease.

As concentrations of carbon monoxide rise above 20 ppm ('very high') so the percentage saturation of haemoglobin will increase. Effects on those with heart disease become more likely. Sufficient outdoor exposure to reach these levels of saturation is unlikely.

Air pollution index. An arbitrary function of the concentration of one or more pollutants used to scale the severity of air pollution. For example, the following calculation has been used in the USA: 10 times the SO_2 concentration plus the CO concentration (both in ppm by volume) plus twice the coefficient of HAZE. The 'alarm' level on this scale is 50 (or more). The average value is 12. Measures of this nature are quite empirical and great caution is required in interpretation.

Air quality. Air quality usually refers to the CONCENTRATION in air of one or more pollutants. For many pollutants, air quality is expressed as an average concentration over a certain period of time, e.g. mg/m^3 over an 8 hour mean.

Air Quality Act. An Act passed by the US Government in 1967 requiring the states to establish regional AIR QUALITY STANDARDS (and setting up a timetable for their doing so) and to control emissions in accordance with national criteria where these existed. It was amended by the Clean Air Act of 1970. (◊ CLEAN AIR ACT (USA))

Air quality banding. The system of air quality banding used in the UK since 1990 and revised in 1998 to take into account the Air Quality Standards. The revised system, like the old, is intended to provide guidance as to the effects of air pollutants on health. The bands are briefly indicated in the facing table.

Air quality criteria. These describe the effects that may be expected to occur whenever and wherever the ambient air concentration of a pollutant reaches or exceeds certain values for specific time periods. Air quality criteria are descriptive.

Air quality banding in the UK.

of air pollution	Ozone	Sulphur dioxide	Nitrogen dioxide	Particulate matter*	Carbon monoxide
Air Quality standard (AQS) 'Low air pollution'	<50 ppb (running 8 h av.)	<100 ppb (15 min av.)	<150 ppb (1 h av.)	<50 μg/m^3 (running 24 h av.)	<10 ppm (running 8 h av.)
'Moderate'	50 ppb (running 8 h av.)–89 ppb (1 h av.)	100–200 ppb (15 min av.)	150–300 ppb (1 h av.)	50–75 μg/m^3 (running 24 h av.)	10–15 ppm (running 8 h av.)
'High'	90–179 ppb (1 h av.)	200–400 ppb (15 min av.)	300–400 ppb (1 h av.)	75–100 μg/m^3 (running 24 h av.)	15–20 ppm (running 8 h av.)
'Very High'	>180 ppb (1 h av.)	>400 ppb (15 min av.)	>400 ppb (1 h av.)	>100 μg/m^3	>20 ppm (running 8 h av.)

* Because of the continuous relationship between concentrations of particles and effects on health, a different approach to devising bands of air quality was advised by COMEAP.

Air quality management. The key principles of the UK Government air quality management strategy are:

- the setting of national standards and reduction targets for all main pollutants;
- supplementing national policies with new systems for local air quality management, focused on designated areas at risk;
- integrating air quality considerations with planning, transport and other policies;
- promoting a balanced approach to emission control designed to secure the most cost-effective improvement process; and including maintaining control of domestic emissions, pressing for the continuing improvement of industrial emissions on the basis of BATNEEC and securing early and sub-stantial improvement in VEHICLE EMISSIONS.

Source: Department of the Environment, Press Release 14, 19 January 1995.

Air quality standards. The Ambient Air Quality Assessment and Management Directive was formally adopted in September 1996 (96/62/EC). The Directive sets out the framework for air quality policy in Europe, giving the general monitoring strategy and the concerns to be addressed when setting limit values. Pollutants to be covered are: sulphur dioxide, nitrogen dioxide, fine particles, lead, ozone, benzene, carbon monoxide, polyaromatic hydrocar-bons, cadmium, arsenic, nickel and mercury.

The actual limit value, specific monitoring requirements and other issues particular to each pollutant are to be covered by a 'daughter' directive. The limits are set out in tables (a) to (d), with World Health Organization Guidelines shown in table (e) for comparison. Like the UK standards and those from other countries, these all provide benchmarks for acceptable air quality.

(a) Limit values for sulphur dioxide.

	Averaging period	Limit value	Date by which limit value is to be met
1. Hourly limit value for the protection of human health	1 hour	$350\,\mu g/m^{-3}$ not to be exceeded more than 24 times per calendar year*	1 January 2005
2. Daily limit value for the protection of human health	24 hours	$125\,\mu g/m^{-3}$ not to be exceeded more than 3 times per calendar year	1 January 2005
3. Limit value for the protection of ecosystems	calendar year and winter (1 October to 31 March)	$20\,\mu g/m^{-3}$	Two years from entry into force of the Directive

* Designed to protect against exceedances of the WHO 1996 10 minute guideline to protect health.

(b) Limit values for nitrogen dioxide and nitric oxide.

	Averaging period	Limit value	Date by which limit value is to be met
1. Hourly limit value for the protection of human health	1 hour	$200\,\mu g/m^{-3}$ NO_2 not to be exceeded more than 8 times per calendar year	1 January 2010
2. Annual limit value for the protection of human health	calendar year	$40\,\mu g/m^{-3}$ NO_2	1 January 2010
3. Annual limit value for the protection of vegetation	calendar year	$30\,\mu g/m^{-3}$ $NO + NO_2$	Two years from entry into force of the Directive

(c) Limit values for PM_{10}.

	Averaging period	Limit value	Date by which limit value is to be met
Stage 1			
1. 24 hour limit value for the protection of human health	24 hours	$50 \mu g/m^{-3}$ PM_{10} not to be exceeded more than 25 times per year	1 January 2005
2. Annual limit value for the protection of human health	calendar year	$30 \mu g/m^{-3}$ PM_{10}	1 January 2005
Stage 2			
1. 24 hour limit value for the protection of human health	24 hours	$50 \mu g/m^{-3}$ PM_{10} not to be exceeded more than 7 times per year	1 January 2010
2. Annual limit value for the protection of human health	calendar year	$20 \mu g/m^{-3}$ PM_{10}	1 January 2010

(d) Limit value for lead.

	Averaging period	Limit value	Date by which limit value is to be met
1. Annual limit value for the protection of human health	calendar year	$0.5 \mu g/m^{-3}$	1 January 2005

(e) WHO 1996 Air Quality Guidelines for Europe.

	Averaging period	Concentration ($\mu g/m^3$)
Sulphur dioxide: health	10 minutes	500
	24 hours	125
	one year	50
Sulphur dioxide: ecotoxic effects	annual and winter mean	10–30 depending on type of vegetation
Nitrogen dioxide: health	1 hour	200
	one year	40
Nitrogen dioxide and nitric oxide: ecotoxic effects	one year	30
PM10	24 hours	dose/response
	one year	dose/response
Lead	one year	0.5

($\lozenge$ UK AMBIENT AIR QUALITY STANDARDS AND OBJECTIVES; ENVIRON-MENTAL ASSESSMENT LEVELS.)

Airborne particulate matter. ◊ DUST.

Aircraft noise. Aircraft noise consists of a build-up to a peak level and then a fall-off, occurring at intervals, as opposed to the continuous but fluctuating noise from heavy road traffic.

NNI	Typical air-traffic conditions
60	This occurs only close to airports where there are many overflights at low altitude. Noise levels can interfere with sleep and conversation in ordinary houses and may also interfere even within sound-insulated houses.
45	This occurs mostly near busy routes from airports. Many aircraft are heard at noise levels which can interfere with conversation in ordinary houses.
35	The overflying is typically irregular at noise levels which are noticeable and occasionally will be intrusive within ordinary houses.

Aircraft noise is measured and assessed in terms of the EQUIVALENT CONTINOUS A-WEIGHTED SOUND PRESSURE LEVEL (L_{Aeq} 166) over the period 0700–2300 GMT. Planning Policy Guidance 24 (1992, Planning and Noise) suggests the following categories:

$L_{Aeq, 16}$	Category
<57	Where measured or predicted noise levels are in this category, noise need not be considered as a determining factor in granting planning permission although noise at the upper end of the suggested range for this category should not be taken as a desirable level.
57–66	Where measured or predicted noise levels are in this category, noise should be taken into account when determining planning applications and noise control measures should be required.
66–72	Where measured or predicted noise levels are in this category, there should be a strong presumption against granting planning permission. Where permission is given for example because there are no quieter alternative sites available, conditions should ensure an adequate level of insulation against external noise.
>72	Normally, where measured or predicted noise levels are in this category, planning permission should be refused.

(◊ NOISE; SOUND; NOISE INDICES; HEARING; ROAD-TRAFFIC NOISE; INDUSTRIAL NOISE MEASUREMENT)

Airline emissions. The reduced emissions from the new Boeing 747–8 Intercontinental are claimed to result in the CO_2 generation of only 75 g per passenger km. (*Source*: Boeing Advertisement, *The Times*, 30 July 2007.)

The True Cost of Flying:

Outbound and return flights of UK citizens = 7.4 per cent of UK CO_2 Emissions

Effect of aviation emissions at high altitude multiplies the damage by 2.5 times

Thus, the effect of these emissions is $7.4 \times 2.5 = 18.5$ per cent.

By 2050, the UK's aviation emissions are set to have increased by 547 per cent (from 1990 levels). This growth would wipe out any other CO_2 reductions made by the rest of us, destroying our chance to combat global warming.

What you can do:

• Profit from video-conferencing in your business
• Fly less
• Take trains when you can

Source: Advertisement by www.spurt-aviation.com in *The Times*, 22 March 2007, following the UK Chancellor's refusal to tax airline fuel (for their emissions) in his Budget of 21 March 2007.

Albedo. The ratio of light reflected from a particle, planet or satellite to that falling on it. Therefore, it always has a value less than or equal to 1. It is also used in nuclear physics.

The albedo of the earth plays an important part in the earth's radiation balance and influences the MEAN ANNUAL TEMPERATURE, and therefore the CLIMATE, on both a local and global scale.

Aldehydes. ORGANIC compounds containing the group –CHO attached to a HYDROCARBON. As air pollutants, a number of them have an unpleasant smell, e.g. in DIESEL exhaust, and can be an irritant to the nose and eyes; many can be poisonous. (◊ AUTOMOBILE EMISSIONS)

Aldrin. An agricultural insecticide which, together with other CHLORINATED HYDROCARBONS such as endrin, DDT, DIELDRIN, and BENZENE HEXACHLORIDE, are major and serious pollutants. They have all been found in significant quantities in the milk of human mothers in the United States (in the USA 99.5 per cent of human tissues taken during post-mortems in a 1971 study contained an average of 0.29 parts per million (ppm) of dieldrin). Aldrin is converted to dieldrin in the environment.

In October 1974 both aldrin and dieldrin were banned by the US government because of strong evidence that they are powerful CARCINOGENS. The use of both aldrin and dieldrin has been restricted since 1960. The UK Ministry of Agriculture, Fisheries and Food has allowed the use of aldrin for bulb crop protection against narcissus fly, but on 18 May 1989 ordered an immediate ban because of 'massive' concentrations of dieldrin in eels in the

Newlyn River in Penzance. Rising concentrations had been noticed for several years and it took strong conservation group protests to have the ban implemented. However, there is concern that stockpiles have been accumulated and its use may continue in some areas unless these are completely recalled. (A court case in June 1990 revealed that a gamekeeper and fruit grower respectively had stockpiled the toxic agricultural chemical, endrin, since it was banned in 1983.) (◊ RED LIST)

Alert level. A concentration of gaseous pollutants that has been defined by a competent authority as indicating an approaching, or constituting a potential or actual, hazard to health. Several different alert levels may be defined, ranging from a concentration at which a preliminary warning is issued to one that necessitates emergency action.

Algae. Simple unicellular or complex multicellular organisms that utilize the process of PHOTOSYNTHESIS for life. Most of them thrive in a wet environment (freshwater or marine) such as in lakes and rivers, on damp walls and in the sea. They can cause problems in lakes and reservoirs if there is an excess of NUTRIENTS, as their excessive multiplication can result in an algal 'bloom' (or population explosion) which, when it dies, can be detrimental to the environment due to the OXYGEN demand created by the decaying algae. Toxins may also be released by algal blooms. Free-floating algae are referred to as PHYTOPLANKTON. Lakes that are rich in plant nutrients and as a consequence are highly productive are said to be eutrophic.

A novel proposal is that algal blooms resulting from EUTROPHICATION should be harvested and fed to organisms higher up the FOOD CHAIN that have economic value, such as oysters. Others have suggested that algae should be cultivated in suitable nutrient liquids to produce vegetable protein.

Algae also form one of the constituent symbiotic partners in LICHENS. (◊ SYMBIOSIS)

Algal bloom. ◊ ALGAE.

Algicide. A chemical used for destroying algae.

Alkali. A substance which, on dissolving in water, splits up to a greater or lesser extent into IONS and results in an excess of hydroxyl ions over hydrogen ions. Alkalis neutralize ACIDS to form salts. The adjective is 'alkaline'.

Alkali and Clean Air Inspectorate. Until 1987, the Inspectorate (formerly known as the Alkali Inspectorate) was responsible for enforcement of the Alkali Works Regulation Act 1906, and the Health and Safety at Work etc. Act 1974 in England and Wales. The Inspectorate was incorporated into HMIP which is now subsumed into the ENVIRONMENT AGENCY. (◊ BEST PRACTICABLE MEANS; BEST AVAILABLE TECHNOLOGY)

Alkalinity. The capacity of a water to neutralize acids due to its bicarbonate, carbonate or hydroxide content. It is usually expressed in milligrams per litre of calcium carbonate equivalent.

Alkane(s). HYDROCARBONS of the METHANE series, e.g. methane CH_4, ethane $H_3C \cdot CH_3$.

Alkyl. A HYDROCARBON RADICAL derived from ALKANES by removing a hydrogen atom, e.g. methyl CH_3^-, ethyl $C_2 H_5^-$.

Alpha particle (α-particle). An alpha particle has a positive charge, consists of two protons and two neutrons (in effect the nucleus of a helium atom) and is emitted from the nucleus of an atom. A nucleus can spontaneously emit an alpha particle when its mass is greater than the combined masses of the product or daughter nucleus and the alpha particle. Spontaneous alpha emission takes place with many of the nuclei heavier than lead.

Alpha particles cause high ionization and are large in comparison with other radiating particles. They therefore lose energy very quickly. They have little penetrating force, 0.001–0.007 centimetres in soft tissue, but once inside the body, either by inhalation or through a wound, they are biologically very damaging. ◊ POLONIUM

Alpha radiation. A stream of fast-moving alpha particles emitted from the nuclei of radioactive elements. They are easily absorbed by matter. (◊ RADIONUCLIDE)

Alumina. Intermediary in the production of ALUMINIUM from BAUXITE which is dissolved in caustic soda with alumina as a precipitate. (◊ HALL PROCESS)

Aluminium (Al). The most abundant element on earth next to iron, it is the most commonly used metal by virtue of its excellent strength to weight ratio, lightness, electrical conductivity and ease of working.

It is made from BAUXITE which is used to manufacture the intermediate ALUMINA which is then electrolysed to produce aluminium (with gaseous emissions of CO and fluorine). Spent anodes are also produced as a solid waste stream (see Figure 3).

Aluminium manufacture has a high ENERGY requirement, e.g. to manufacture a 1/3 l soft drinks can from steel requires 0.9 MJ and a similar aluminium can requires 2.9 MJ (tonne for tonne, bulk Al consumes five times more energy than steel).

This means that very high RECYCLING rates are necessary if aluminium is to compete with steel on an energy per fill basis.

The UK steel industry has stated:

> Similar energy savings can be found when looking at the full life cycle of steel cans. This means that if all cans were made from steel, large savings could be achieved. Currently, 50 per cent of drinks cans are steel and 50 per cent are aluminium. If all drinks cans used in the UK were made from steel, the total energy savings would be the equivalent of the average household lighting requirements for every house in the UK – all 22 million of them – for at least four weeks!

This helps explain high profile aluminium can recycling campaigns (and high prices paid for the empty containers due to the inbuilt energy).

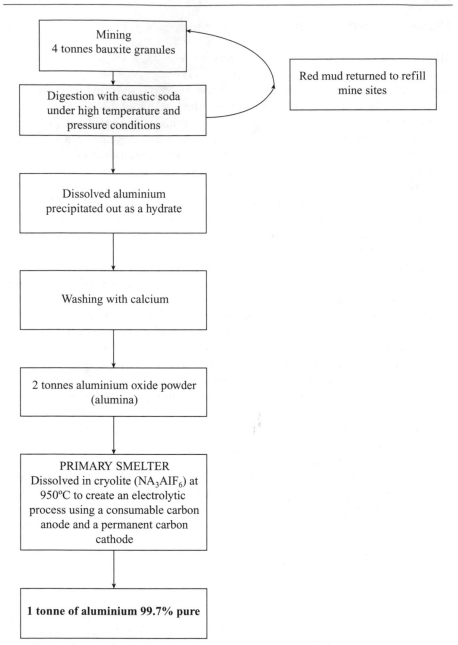

Figure 3 Aluminium production from bauxite.
Source: *Steel Cans and our Environment*, Steel Can Recycling Information Bureau leaflet, Aluminium Can Recycling Association, '*Alu*', No. 1 (no date).

Ninety per cent of aluminium refining energy can be saved by recycling and 75 per cent for steel. The arithmetic shows that aluminium can recycling will never outstrip the energy/fill of steel when steel can recycling is also taken into consideration.

There are other considerations to the equation, including the fact that social responsibility is encouraged by recycling.

Aluminium sulphate ($Al_2(SO_4)_3$). An additive used in water treatment as a coagulant to remove suspended solids. It may be implicated in excess aluminium levels in humans, although aluminium cooking vessels, food containers, etc. may be alternative routes. In addition, ACID RAIN allows aluminium to be leached from soils in water catchment areas and hence into drinking water.

The EC maximum admissible concentration level for aluminium in drinking water is 200 μg/l. It is estimated that more than 2 million people in the UK drink water with aluminium levels in excess of this limit.

Attention was focused on this topic when a 20 tonne load of 8 per cent solution of aluminium sulphate accidentally entered the water supply of the CAMELFORD area of North Cornwall on 6 July 1988. A population of ca. 20 000 was put at risk, with some people and animals becoming violently ill plus the death of an estimated 60 000 fish in the rivers Camel and Allen. Exposure levels of up to 2000 times the EC limit were postulated for the initial period of this event. The after-effects of the chemicals dosed into the water supply to neutralize the aluminium sulphate reacted with water pipes which subsequently produced high levels of copper, lead and zinc in the drinking water. The ill effects *attributed* to this incident are skin troubles, arthritis, nausea and kidney complaints. This is a classic case of (i) solving a problem in isolation and (ii) how easy it is for a seemingly minor act of carelessness to grossly pollute and consequently endanger a large group of people. (◊ CHERNOBYL). As a result of the Camelford incident other coagulants such as $FeCl_3$ are being used more. Also, the security of water & chemical tanks in water treatment plants is more carefully controlled.

Ambient Air Quality Standards and Objectives, UK. The framework for improving air quality in the UK is set by Part IV of the Environment Act 1995. Two main elements of the Act are:

1. a requirement to produce a National Air Quality Strategy containing standards, objectives and measures to achieve objectives;
2. a Local Air Quality Management system giving statutory duties for local authorities to review their air quality, assess it against air quality objectives set in regulations and, where it is likely the objectives will not be met, declare air quality management areas and draw up and implement appropriate action plans.

The National Air Quality Strategy was published in March 1997. The Strategy contains standards (see table overleaf) and policy objectives for eight pollutants, based on advice from the independent Expert Panel on Air Quality Standards (EPAQS).

Air Quality Standards and Objectives.

Pollutant	Standard concentration	Measured as	Objective (to be achieved by 2005)
Benzene	5 ppb	running annual mean	5 ppb
1,3-Butadiene	1 ppb	running 8 hour mean	1 ppb
Carbon monoxide	10 ppm	running annual mean	10 ppm
Lead	$0.5 \mu g/m^3$	annual mean	$0.5 \mu g/m^3$
Nitrogen dioxide	150 ppb	1 hour mean	150 ppb, hourly mean
	21 ppb	annual mean	21 ppb, annual mean
Ozone	50 ppb	running 8 hour mean	50 ppb, measured as the 97th percentile
Fine particles (PM10)	$50 \mu g/m^3$	running 24 hour mean	$50 \mu g/m^3$ measured as the 99th percentile
Sulphur dioxide	100 ppb	15 minute mean	100 ppb measured as the 99.9th percentile

ppm = parts per million; ppb = parts per billion; $\mu g/m^3$ = micrograms per cubic metre.

The Air Quality Regulations 1997 give legal status to seven air quality objectives so that local authorities can undertake their statutory duties under Part IV of the Environment Act 1995. The Strategy's objective for ozone has not been prescribed in the regulations because its international/transboundary nature means that action by a single local authority to control ozone is unlikely to be effective.

Amide. A class of compounds manufactured by replacing one, two or three of the HYDROGEN ATOMS of AMMONIA (NH_3) by an ACID RADICAL, e.g. acetamide $CH_3 CONH_2$ is manufactured from acetic acid ($CH_3 COOH$) plus ammonia and is widely used as a SOLVENT.

Amine. A class of organic compounds of nitrogen in which one or more of the hydrogen atoms in AMMONIA (NH_3) has been replaced with an ALKYL group, e.g. methylamine $CH_3 NH_2$. Used as a SOLVENT, FUEL additive and in dyeing and staining.

Amino acid. Essential component of PROTEINS, consisting of amino ($-NH_2$) and acidic carboxyl ($-COOH$) groups. There are some 20 different amino acids normally present in proteins and a balanced diet must consist of the right intakes of these acids. There are eight essential amino acids for humans that can only be obtained from the environment by heterotrophic means. The other non-essential amino acids can be manufactured by the human organism usually from the essential ones. 'First class' protein is the name given to a protein that contains these eight essential amino acids, and a human's daily needs are estimated at 30–70 grams, some of which should preferably come from animal sources.

Ammonia (NH₃). An alkaline compound which can cause both toxicity to fish and an OXYGEN demand in receiving waters. A high CHLORINE demand is also created if the water is to be used for drinking water. The presence of AMMONIACAL NITROGEN is taken as an indicator of sewage pollution in a stream.

Ammonia is present in very small amounts in exhaust gases from engines without CATALYTIC CONVERTERS and there is also evidence that emissions from vehicles fitted with three-way catalytic converters are very much higher than from non-catalyst vehicles. If all petrol cars were fitted with catalysts, they could contribute as much as 10 per cent of total UK emissions of ammonia.

Ammonia injection can also be used as a means of NITROGEN OXIDE control in COMBUSTION processes, producing nitrogen and water vapour as the products. (⟡ SCR)

Source: Royal Commission on Environmental Pollution 18th Report, published 1994.

Ammonia slip. The excess AMMONIA (NH₃) discharged in STACK GASES from SELECTIVE NON-CATALYTIC REDUCTION (SNCR) systems used for NITROGEN OXIDE (NO$_x$) control which make use of ammonia or UREA injection to achieve NO$_x$ reduction.

A balance has to be struck between NO$_x$ emissions and NH₃ slip. See Figure 4.

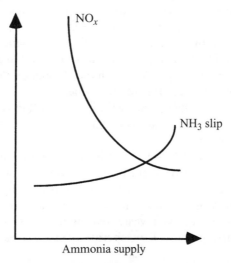

Figure 4 NO$_x$ emissions in relation to the ammonia slip in an SNCR system based either on ammonia or urea.

Ammoniacal nitrogen. Nitrogen present as AMMONIA and ammonium ion in liquid effluents. It has a high CHLORINE demand in water treatment and is toxic to fish.

Ammonium nitrate. Formula NH_4NO_3, made by reacting ammonia and nitric acid. Used as an explosive and (mainly) as a booster fertilizer, plants can absorb nitrogen as ammonium ions or as nitrate ions when they are reduced to ammonia.

Anaerobic. Oxygen-free conditions, typically found in waterlogged soils (bogs, swamps, marshes, etc.) and in the intestines of animals. The decomposition of organic matter (biomass) by bacteria adapted to these conditions (anaerobes) is the main natural source of atmospheric methane.

Anaerobic (organisms) – organisms that can only survive in the absence of oxygen.

Anaerobic digestion. This is the biological conversion of organic materials to methane & carbon dioxide in the absence of oxygen. Although AEROBIC PROCESSES are almost always used for the reduction of BIOCHEMICAL OXYGEN DEMAND (BOD) in organic effluents, interest in ANAEROBIC PROCESSES is growing, particularly for high BOD/COD wastes from industry. It is also used as a SLUDGE treatment process in conventional sewage treatment plants and for animal effluent on farms.

The overall process of anaerobic digestion occurs through the combined action of a consortium of four different types of micro-organisms. Hydrolytic and fermentative micro-organisms including common food spoilage bacteria break down complex wastes into their component sub-units and then ferment them to short-chain fatty ACIDS and carbon dioxide and hydrogen gases. Bacteria convert the complex mixture of short-chain fatty acids to acetic acid with release of more CARBON DIOXIDE and HYDROGEN gases to provide the main substrates for the METHANE BACTERIA.

Methane bacteria produce large quantities of methane and carbon dioxide from this acetic acid and also combine all of the available hydrogen gas with carbon dioxide to produce more methane. Sulphate-reducing bacteria are also present. They reduce sulphates and other sulphur compounds to hydrogen sulphide, most of which reacts with iron and other HEAVY METAL SALTS to form insoluble sulphides, but there will always be some hydrogen sulphide remaining in the biogas.

The ultimate yield of BIOGAS depends on the composition and biodegradability of the waste feedstock but its rate of production will depend on the population of bacteria, their growth conditions and the temperature of the fermentation. Microbial growth and natural biogas production are very slow at ambient temperatures. As a waste treatment process, the rate of anaerobic digestion is greatly increased by operating in the mesophilic temperature range (35–40 °C).

As this form of processing is expected to grow, several applications are summarized below to show the potential of this method.

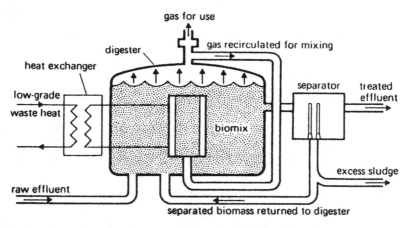

Figure 5 Anerobic effluent treatment.

Application 1. On waste containing a BOD of 12 000 milligrams per litre resulting from a starch and sugar content (*Process Engineering*, May 1975, p. 6). The process is shown in Figure 5 and uses a digester which must be kept heated to around 35 °C to allow sufficient bacterial activity. The result is methane, which is a source of light and heat. Between 85 and 95 per cent BOD is removed in this way, which means that the effluent from the digester would normally require aerobic treatment before final discharge.

The advantages that are claimed for the anaerobic process are that the plant is much smaller than that for an aerobic process, and the volume of sludge produced can be one-tenth of that from aerobic plants; it is said to be less objectionable to handle. It is extremely suitable for high BOD effluents which could overwhelm conventional sewage effluent plants. (◊ EFFLUENT, PHYSICO-CHEMICAL EFFLUENT TREATMENT)

Application 2. Animal excreta. The Bethlehem Abbey of the Cistercian monastery in Co. Antrim, Ireland, is heated completely by the biogas produced by the anaerobic digestion of a daily input of 12 tonnes of cow slurry and manure from its 300 head beef flattening farm and 22 000 broiler hens, respectively. The biogas saves the Abbey an estimated £1000 per month in fuel costs plus the fibre from the digested slurry is composted for 3 weeks at 60 °C and bagged into 25 litre sacks which sell for £3.85. The compost is marketed under the brand name of 'Dungstead'.

Application 3. Whey treatment (Hamworthy Engineering Report, *Whey Processing for Profit*, Poole, Dorset, 1987).

WHEY is notoriously difficult to treat as $1 m^3$ is equivalent in BOD to sewage effluent from 600 people. A typical comparison of before and after treatment is shown in the following table.

	Before	*After*	*Percentage reduction*
Total solids (mg/l)	66 000	9400	85
Volatile solids (mg/l)	65 000	6500	90
Chemical oxygen demand (mg/l)	65 000	6500	90
Biochemical oxygen demand (mg/l)	45 000	2250	95
Total nitrogen (mg/l)	1 500	1000	33.3

A waste of this strength would be highly uneconomic to treat by conventional aerobic means due to the high power requirement for aeration. Hence anaerobic treatment can be chosen. The individual components of whey, lactose, PROTEIN and fat are broken down by the digestion process to form BIOGAS. One cubic metre of whey yields $38 \, m^3$ of biogas at NORMAL TEMPERATURE AND PRESSURE (NTP).

The effluent from the digester has 90 per cent reductions in the COD/BOD and can, in some instances, be sent direct to sewer. If it is to be discharged to a watercourse, further treatment will be necessary to reduce the BOD, and total nitrogen and ammonia contents with the aid of an aerobic nitrification/denitrification system to meet ROYAL COMMISSION STANDARDS.

The BTA Process combines a pre-treatment system with anaerobic digestion and dewatering to provide an integrated solution for solid wastes. In the pre-treatment process (Figure 6), waste and process water are fed to the

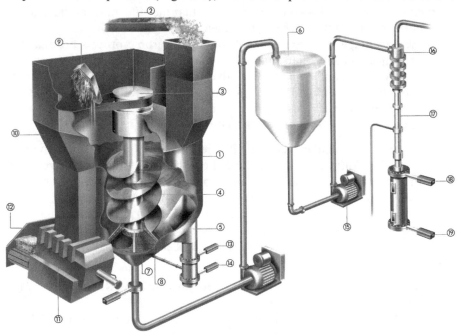

Figure 6   Schematic of the BTA Wet Pre-Treatment Process.

pulper vessel (1), already partly filled with process water remaining from the previous batch.

Pulping is preformed to facilitate three objectives:

1. Disintegration of biodegradable waste to enhance the subsequent digestion process.
2. Removal of non-biodegradable contaminants as a 'heavy' fraction (stones, large bones, batteries and metallic objects).
3. Removal of non-biodegradable contaminants as a 'light' fraction (textiles, wood, plastic film, string, etc.).

Material (up to 200 mm in size) is fed to the pulper by operation of the feed conveyors (2). When sufficient material has been loaded into the pulper the conveyor will stop. Loading is carried out automatically. During the feed stage the pulper drive (3) operates at low speed.

The pulper impellor (4) is designed to impart high hydraulic shear to the vessel contents without causing mechanical damage to biologically stable materials such as plastics, textiles, metal and glass. The pulper power and suspension concentration is closely controlled throughout the cycle in order to minimize energy usage and disintegration of inert materials and maximize the breakdown of biologically inert materials. At the same time it optimizes the release of digestible substrate. The pulper uses hydrodynamic forces to separate biodegradable materials into a thick pulp. It also separates and removes light materials such as plastics, textiles and dense materials such as glass and stones. The grit removal system removes and separates small stones and broken glass down to below 2 mm. The biodegradable pulp is hygenized, then digested to produce biogas and a stable digestate. The digestate can be aerobically matured to produce high quality compost.

At the start of the pulping cycle, stone, glass, large bones, batteries and metallic objects fall to the vessel floor and under the influence of centripetal force, drop into a peripheral heavy fraction pocket (5) built into the vessel. The impellor is operated at low speed during this part of the cycle to minimize wear and mechanical damage to non-biodegradable materials.

The impellor then increases speed to pulp the pulpable materials and continues until the correct pulp concentration is reached.

On completion of pulping, the vessel contents are pumped into the grit removal system surge tank (6). Pulp at approximately 8–9 per cent Total Solids leaves the pulper via a perforated plate screen (7) with approximately 10 mm openings. The screen retains the 'light' fraction (textiles, plastic film, string etc.) which is not damaged by the impeller. Separation is facilitated by a set of screen wiper blades (8) that rotate with the pulper impeller which continues to run at low speed during pulp discharge.

Once the pulp discharge cycle is complete, process water is pumped to the vessel and the light fraction is re-suspended. The pulper drive (3) is stopped and the raking cycle commences. An hydraulically driven mechanical rake arm (9) dips repeatedly into the vessel and transfers the light fraction into an adjacent chute (10), whence it descends into a press (11). The light fraction is compressed to reduce the water content before being discharged via a collection conveyor arrangement (12) to a storage bay.

At the base of the heavy fraction pocket a slide valve (13) opens to allow the heavy material to drop into a small chamber where it is washed with process water before being discharged through a second slide valve (14) onto a collection conveyor arrangement and to a storage bay.

Grit Removal. Grit removal is carried out batchwise. Screened pulp from the pulper is re-circulated from the Grit Removal System surge tanks (6), by means of a pump (15), through a dedicated hydrocyclone (16) and back to the surge tank. The pulp can be re-circulated to the feed tank and back through the hydrocyclone to improve the removal efficiency, Grit falls to the bottom of the hydrocyclone into an hydraulic classifier or elutriator (17) where process water flows counter-current to the falling grit. At the base of the classifier a slide valve (18) opens intermittently to allow the grit to drop into a small chamber. The rate of accumulation is monitored and the grit removal cycle is terminated when the rate falls below a preset value. Grit held in the chamber is washed with process water before being discharged through a second slide valve (19) onto a collection conveyor arrangement and to a storage bay.

On completion of the grit removal cycle, the re-circulation is stopped and the de-gritted pulp is transferred to the digester feed balance tank.

The system generates a strong, positive energy balance. The biogas produced is used in a CHP system to generate electricity for export and the waste heat produced is re-used in the process.

Aside from the biogas and digestate, the BTA process produces a washed heavy fraction of stones, broken glass and bones, a washed grit fraction of small stones and fine grit and a dewatered light fraction, consisting largely of film plastics and textiles.

The Anaerobic Digestion Process is shown in Figure 7.

1. Organic material such as sewage sludge, pulped solid waste, agricultural or industrial waste is pumped into the pasteurization unit through the inlet heat exchanges.
2. The first inlet heat exchanger raises the temperature of the feed using heat recovered from cooling the pasteurized feed before it is fed to the digester.
3. The second inlet heat exchanger raises the temperature of the feed to the pasteurization unit to 72 °C using waste heat from the CHP system.
4. The pasteurization unit comprises three stirred and insulated tanks that are operated on a batch basis. First a tank is filled over a period of an hour. Next the material is held in the tank for 60 minutes with the stirrer operating to eliminate cold spots, for example near the walls, and the temperature is monitored. Finally once the required temperature has

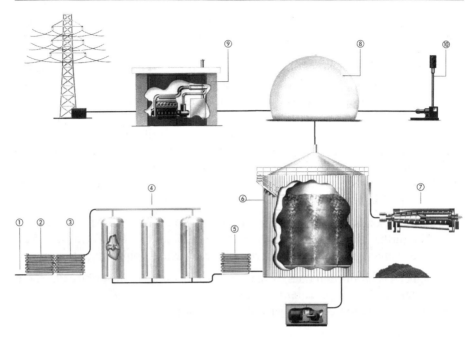

Figure 7 Schematic of the Anaerobic Digestion Process.

been held for the required time and validated by the control system, the tank contents are fed forward to the next process. Keeping the feed at 70 °C for 60 minutes ensures compliance with Animal By-Products Regulations. A significant pathogen kill results, allowing the treated material to be applied to farm land.

If the required time and temperature are not validated for any reason, the pasteurization tank contents are returned to the feed holding tank and are re-processed.

5. The pasteurized material passes through a cooling heat exchanger which reduces the temperature to that required by the anaerobic digestion process and recovers heat which is re-circulated to the first inlet heat exchanger. The cooled material is then fed to the anaerobic digester.

6. The anaerobic digester is a completely mixed stirred reactor where the biodegradable organic components of the organic material are converted largely to methane gas and carbon dioxide by hydrolytic, acidogenic and methanogenic bacteria. The reactor is insulated to minimize heat loss and continuously mixed using biogas recirculation. The reactor is fitted with pressure and vacuum relief systems, temperature monitoring and control, and a foam suppression system. The processing time in the anaerobic digester is typically in the range of 15 to 18 days.

7. The digested materials, or digestate, is fed to a centrifuge, which separates the digestate into a high solids cake for further processing or dis-

posal, and a liquid centrate which is recirculated for waste pulping in the case of solid waste digestion, or is treated prior to disposal, according to the application.

8. Biogas produced in the anaerobic digester is fed via flame traps and condensate drains to a low pressure gas holder for balancing storage prior to use.

9. The main use for the biogas is to generate electricity and heat in a combined heat and power (CHP) system. The electricity can be exported to the grid and can earn Renewable Obligation Certificates (ROC) from suitable materials. The waste heat is used to heat the pasteurization and digestion process and any surplus waste heat can be used for space heating.

10. A safety waste gas destructor is fitted. This is used to safely combust any surplus biogas, for example if the CHP system is being serviced, etc.

For high solids feedstocks such as MUNICIPAL SOLID WASTES (MSW) efficiency can be further increased by operation at thermophilic (50–60 °C) temperatures. The claimed advantages of thermophilic over mesophilic digestion are:

- higher gas production rates for similar retention times to mesophilic digestion;
- better separation of water from digestate;
- higher reduction of pathogens (disease-producing organisms).

The disadvantages are:

- higher energy consumption;
- higher fatty acids concentration in effluent;
- higher total solids (mg/l);
- more sensitivity to changes in process conditions and toxic substances.

The bacteria require mineral nutrients; a ratio of C:N of 10–16:1 is typical and N:P should be 7:1. Also, wastepaper can only be digested in low concentrations due to its effect on the C:N ratio.

Typical digester residence times are 25–35 days with biogas yields in the range 100–150 m^3/tonne of input waste.

The table below gives typical yields and calorific values.

		Mixed MSW	Vegetable wastes
Gas production	(Nm^3/ton)	105	95
Net calorific value (55–60% CH$_4$)	(MJ/Nm^3)	19.7–21.5	19.7–21.5
	(MJ/tonne)	2069–2258	1872–2043
Density	(kg/Nm^3)	1.284–1.221	1.284–1.221
Steam production	(ton/ton)	0.66–0.72	0.60–0.65

Sources: *Biogas from Municipal Solid Waste: Overview of Systems and Markets for Anaerobic Digestion of MSW*, IEA Bioenergy Brochure, published by Ministry of Energy/Danish Energy Agency, Copenhagen, Denmark, July 1994. *Feasibility Study for Anaerobic Digestion of Solid Wastes in Leicestershire*, ETSU B/EW/00389/12/REP, Contractor Tebodin PLT, Slough, prepared by A. Snuverink, J. van Seeters and E. C. Griffiths, 1994.

Anaerobic digestion systems for partial MUNICIPAL SOLID WASTE (MSW) disposal are detailed below.

Wet single-step. MSW is slurried with water, giving an input material with 10–15 per cent solids, and is fed to a complete mix tank digestor. This approach is amenable for co-digesting MSW with more dilute feedstocks.

Wet multi-step. Several stages are involved, in which slurried MSW is first fermented by hydrolytic bacteria to release volatile fatty acids. These are then converted to BIOGAS in a high-rate anaerobic digestor. This technique is useful for biowaste and food processing wastes.

Dry continuous. This system uses relatively dry (20–40 per cent solids) material which is batch-loaded into a continuously fed digestion vessel. Minimal water addition makes this technique suitable for thermophilic bacteria.

Dry batch. Here the vessel is charged with feedstock, inoculated with digestate from another reactor and sealed. LEACHATE is recirculated to maintain uniform conditions.

Sequencing batch. A variant of the dry batch process, leachate is exchanged between established and new batches. This helps start-up, inoculation and removal of volatile matter.

Bio-reactor landfills. With a LANDFILL divided into cells with leachate recirculation, the site can act as a large-scale high solids anaerobic digestion plant.

Source: The World Resource Foundation, Information Sheet, November 1998.

Anaerobic process. Any biological process that is carried out without the presence of air or oxygen, e.g. in a watercourse that is heavily polluted with no dissolved oxygen present. The anaerobic decay processes produce METHANE (CH_4) and HYDROGEN SULPHIDE (H_2S) which is an evil-smelling toxic gas.

Some anaerobic organisms, e.g. the denitrifying bacteria, are poisoned by the presence of oxygen. (◊ AEROBIC PROCESS)

Anaerobic respiration. Respiration in the absence of oxygen.

Anchovy fisheries. The anchovy is a small fish of the herring family which can be easily converted to fish-meal. The major source of anchovy is the Peruvian fishing industry, which became the world's largest producer in the early 1960s. Approximately 96 per cent of the anchoveta catch from the Peruvian (Humboldt) Current is converted into fish-meal. It represents the world's largest ocean harvest of a single species, and its export effectively deprives the South American continent of 50 per cent more PROTEIN than it is producing as meat, including Argentinian deliveries of beef to Europe and North America. (◊ MAXIMUM SUSTAINABLE YIELD; WHALE HARVESTS)

Angiosarcoma. A rare form of cancer commonly associated with the liver. (◊ POLYVINYL CHLORIDE)

Anion. Negatively charged ION.

Anode. An electrode which is at a positive potential.

Antagonism. A state in which the presence of two or more substances diminishes or decreases the toxic effects of the substances acting independently. It is the opposite of SYNERGISM.

Anthropogenic. Produced as a result of human activity.

Antibiotic. A chemical substance, produced by micro-organisms which has the capacity, in dilute solution, to inhibit the growth of or destroy other micro-organisms. Originally derived from moulds or bacteria but now produced synthetically as well, antibiotics are used to treat and control infectious diseases in humans, animals and food crops, to stimulate the growth of animals, and to preserve foods.

    With the large-scale use of antibiotics, particularly in farming, the emergence of resistant bacteria has become an increasing problem. As resistance develops, an antibiotic becomes less effective and dosages must be increased or mixtures of antibiotics must be used. Particularly disturbing is the mechanism of transferable drug resistance, by means of which one species of bacteria previously susceptible to a certain antibiotic suddenly becomes resistant to it by virtue of cell-to-cell contact with a species of bacteria which is already resistant to the antibiotic. Thus resistance to several drugs can be acquired simultaneously. Contaminated animal antibiotics from non-UK sources are held to be responsible for ca 5% of UK meat exceeding safe residue limits.

Anticyclonic blocking. A winter occurrence when high pressure periodically builds over Russia and Scandinavia which diverts unsettled and rainy systems into the western Mediterranean. When it does not occur, persistent low pressure systems develop across northern Europe, especially Scandinavia, leaving high pressure across southern Europe and an absence of the rainfall commonly associated with low pressure.

    The Andalucia area of Spain was without significant rainfall during 1991–95 due to this phenomenon.

Source: *Daily Telegraph*, 1 May 1995.

AOX. AOX is defined as organic halogens subject to ABSORPTION. This is a measure of the amount of CHLORINE (and other halogens) combined with ORGANIC compounds.

Aquifer. An underground water-bearing layer of porous rock, e.g., sandstone, in which water can be stored and through which it can flow, to wells and springs, after it has infiltrated from either the surface or another underground source. The London Chalk Basin is a typical example of such an aquifer; it supplies three million cubic metres of water per day. Water may also be extracted from the chalk under the Cotswolds to meet the expected demand from London. (◊ INFILTRATION)

Polluted surface water can enter the SATURATION ZONE of an aquifer and lead to its contamination. Typical pollutants can be LEACHATE, NITRATE, and SOLVENTS which can permeate via fissures, fractures and boreholes.

Aquifer management and RIVER REGULATION are now practised conjunctively in the UK to maximize the amount of water available for consumption from rainfall over a particular catchment area, e.g. the December 1988 rainfall in S.E. England was 17 mm against an average of 76 mm. Hence the water extracted from the chalk aquifers in early 1989 had to be cut back substantially to conserve this source of supply.

Arithmetic–geometric ratio (A/G ratio). A principle used by geologists for estimating the abundance of certain types of ore deposits, the idea being that as the grade of the ore decreases arithmetically its abundance increases geometrically until the average abundance in the earth's crust is reached. This has led certain economists to assert that the problem of non-renewable mineral resources is a non-problem, since as demand increases mining will simply move to poorer and poorer ores which are assumed to be progressively more and more abundant.

As one would expect, the facts are against such a simplistically reassuring view of the situation. The A/G ratio is applicable only to a very limited number of ores, and only within certain limits. In addition, although the economic costs of working lower and lower grade ores might be absorbed in some cases, the energy costs could not. (◊ ENERGY DEMAND)

Artificial recharge. The artificial replenishment of an AQUIFER by flooding the surface, ditches, excavations, or wells, etc. This augments the natural contribution from precipitation & infiltration. The UNSATURATED ZONE often acts as a giant free treatment plant, hence slightly polluted water may be used for recharge purposes.

Artificial reef. Reefs are rock chains near water-level which are often very useful for shore or wild-life protection.

Artificial reefs are constructed of various materials (e.g. tyres, building rubble, rocks, old ship hulks, etc) for the purpose of promoting marine life. They provide hard surfaces to which algae, & invertebrates such as barnacles, corals, & oysters attach. This community provides food for fish.

Source: Lloyd's List, 5 April 1995.

Asbestos. A collective term for a group of naturally-occurring magnesium silicate materials ($MgSiO_4$). It is a fibrous mineral of variable length with excellent resistance to fire, heat and chemical attack being indestructible up to 800 °C. It is widely used in building products, gaskets, brake linings, and roofing materials.

There are three main forms of asbestos: blue, white and brown.

Blue asbestos or *crocidolite*. A survey of male asbestos workers by the TUC Centenary Institute of Occupational Health suggests that 30 years after first

exposure about one in 200 workers will be found to have died from mesothelioma (malignant tumours) associated with blue asbestos. Blue asbestos
is still present in old buildings, boiler plant, ships, etc., and its presence is a
substantial threat to demolition workers.

White asbestos, or *chrysotile*, which can be readily spun or woven into
tape.

The third important variety is *amosite* (*brown asbestos*). Other varieties of minor commercial importance are *tremolite*, *actinolite* and
anthophyllite.

The dangers associated with asbestos are grave. Asbestosis, a scarring of
the lung, cancers of the bronchii, pleura and peritoneum may result from
breathing in the minute fibres. Asbestosis may result from exposures as short
as six weeks in heavy dust concentrations. Brief exposure to blue asbestos
can manifest itself later in life as mesothelioma, a specific and invariably fatal
form of cancer.

In the UK, the 1931 Asbestos Regulations put the onus squarely on
employers to ensure that asbestos dust was removed at source, respirators
being regarded as only secondary protection. Astonishingly, these regulations did not address certain important uses of asbestos, and were apparently
disregarded by employers, with a resulting high record of sickness and death
in the industry. The Factory Inspectorate, (now superseded by HSE) which
had the responsibility for ensuring that the Factory Acts relating to asbestos
were properly enforced, appears to have been completely ineffective, with a
total record of three prosecutions in 30 years.

In terms of legislation, the Control of Asbestos Regulations came into
force on 13 November 2006 (Asbestos Regulations – SI 2006/2739). These
Regulations bring together the three previous sets of Regulations covering
the prohibition of asbestos, the control of asbestos at work and asbestos
licensing. The Regulations prohibit the importation, supply and use of all
forms of asbestos. They continue the ban introduced for blue and brown
asbestos in 1985 and for white asbestos in 1999. They also continue the ban
in the second-hand use of asbestos products such as asbestos cement sheets
and asbestos boards and tiles; including panels which have been covered with
paint or textured plaster containing asbestos. The ban applies to new use of
asbestos. If existing asbestos-containing materials are in good condition, they
may be left in place; their condition monitored and managed to ensure they
are not disturbed.

The Asbestos Regulations also include the 'duty to manage asbestos' in
non-domestic premises. The Regulations require mandatory training for
anyone liable to be exposed to asbestos fibres at work. This includes maintenance workers and others who may come into contact with or who may
disturb asbestos (e.g. cable installers) as well as those involved in asbestos
removal work.

When work with asbestos or work which may disturb asbestos is being carried out, the Asbestos Regulations require employers and the self-employed to prevent exposure to asbestos fibres. Where this is not reasonably practicable, they must make sure that exposure is kept as low as reasonably practicable by measures other than the use of respiratory protective equipment. The spread of asbestos must be prevented. The Regulations specify the work methods and controls that should be used to prevent exposure and spread.

Worker exposure must be below the airborne exposure limit (Control Limit). The Asbestos Regulations have a single Control Limit for all types of asbestos of 0.1 fibres per cm^3. A Control Limit is a maximum concentration of asbestos fibres in the air (averaged over any continuous 4-hour period) that must not be exceeded. In addition, short-term exposures must be strictly controlled and worker exposure should not exceed 0.6 fibres per cm^3 of air averaged over any continuous 10-minute period using respiratory protective equipment if exposure cannot be reduced sufficiently using other means.

Respiratory protective equipment is an important part of the control regime but it must not be the sole measure used to reduce exposure and should only be used to supplement other measures.

Respiratory problems caused by asbestos include:

Asbestosis. This causes scarring and shrinkage of the lungs, and is thought to increase the risk of lung cancer by five times.
Mesothelioma. A cancer of the inner lining of the chest wall, associated primarily with exposure to asbestos (in the US it is called 'asbestos cancer'). There is no cure and patients usually die within a year of diagnosis.

Suspicions of a link between asbestos and lung disease were aired in the 1920s; the term 'asbestos' appeared in the *British Medical Journal* in 1927. The first full study of the effects of asbestos began the following year and the first legislation regulating its use was introduced in the UK in 1931.

Figures compiled in 1995 put asbestos-related deaths at an estimated 3000 a year in the UK. One in 40 men in their fifties who have been exposed to asbestos are expected to die of mesothelioma. Projected deaths from mesothelioma are expected to continue to rise until at least 2010, possibly 2025, and will peak at 5000 to 10000 deaths a year in the UK (Health and Safety Executive data).

Askarel. An insulating non-flammable liquid used in electrical applications such as transformers, capacitors and special electrical power cables. Most well known are the polychlorinated biphenyls (PCBs).

Aspergillus niger (AN). A fungus capable of breaking down vegetable matter (e.g. carob pods, which come from the locust tree which is extensively grown in many developing countries) to produce SINGLE CELL PROTEIN (SCP). The strain used is called M1 and is capable of doubling its weight in 5 to 10 hours, provided inorganic nutrients are added. It is also used for the industrial production of citric acid, gluconic acid, and enzymes such as glucoamylase and α-galactosidase.

Asphyxiating pollutants. Asphyxiation is deprivation of oxygen, either through obstruction of the air passages or, as in the case of CARBON MONOXIDE, the inability of the blood to carry oxygen. HYDROGEN SULPHIDE (H_2S) is an irritant at very low concentrations, but it can also paralyse the respiratory system at slightly higher concentrations causing death by asphyxiation.

Source: G. L. Waldbott, *Health Effects of Environmental Pollutants*, 2nd edition, C. V. Mosby, St. Louis, 1978.

Asset Management Plan (AMP). An AMP is a water company's detailed description of its investment plans for its underground assets. AMP is often used as a shorthand name for the companies' business plans.

Asthma and outdoor air pollution. The incidence of asthma has increased in the UK over the past 30 years but this is unlikely to be the result of changes in air pollution. The Committee on the Medical Effects of Air Pollutants (COMEAP) concluded that most of the available evidence does not support a causative role for outdoor air pollution (excluding possible effects of biological pollutants such as pollen and fungal spores). Most asthmatic patients should not suffer a worsening of symptoms or provocation of asthmatic attacks through exposure to such levels of non-biological air pollutants as commonly occur in the UK. A small proportion of patients may experience clinically significant effects that may require an increase in medication or attention by a doctor. Factors other than air pollution were considered much more important than air pollution.

Aswan Dam. ◊ DAM PROJECTS.

ATF. Authorized treatment facility, e.g. licensed facility for End-of-Life Vehicles (ELVs).

Atmosphere[1]. The envelope of air around the earth. The composition of our present atmosphere is the result of the CARBON CYCLE and the NITROGEN CYCLE and the atmosphere is renewed and maintained by these processes.

The atmosphere, like any other natural resource, is finite and 99 per cent of its mass is within 20 miles of the earth's surface. It is in contact at its inner edge with land and water and changes in the atmosphere can induce changes in land or water such as the amount of rainfall on a particular area. The changes can be direct or they can be complex chain reactions.

The role of the atmosphere in the earth's radiation balance is unique. The incoming solar radiation is absorbed by the earth and reradiated into outer space as long-wave radiation, but the two processes are not in balance except over, say, a year when the total incoming radiation may balance the total outgoing. A very small change in either the outgoing or incoming radiation can have very large effects. Hence, the concern for the CARBON DIOXIDE

released by combustion of fossil fuels, the DUST particles released by COM-
BUSTION processes, the effects of AEROSOL PROPELLANTS on the ozone layer,
or supersonic aircraft. (◊ ALBEDO; CLIMATE; CONCORDE; GREENHOUSE
EFFECT; THERMAL POLLUTION)

The atmosphere has two broad bands. The lower part, the TROPOSPHERE,
decreases in TEMPERATURE and DENSITY as altitude increases. The other
band is the upper atmosphere, the STRATOSPHERE, where temperature
increases with altitude.

Because of its greater density, the lower troposphere contains roughly 80
per cent of the mass of the atmosphere. The depth of the troposphere varies
widely, depending on latitude; it has an average depth of 8 km over the poles
and 16 km over the equator. In the troposphere there is considerable mixing
as the warmer air rises and the cooler air falls.

The stratosphere has the opposite temperature characteristic: the temperature rises as altitude increases, and so there is very little mixing.

NITROGEN OXIDES (NO_x) released by jet engines in the stratosphere destroy
ozone, but in the troposphere they contribute to a build-up of ozone, where
it has a greater GREENHOUSE EFFECT. The residence time of NO_x is 1–4 days
in the troposphere but over 1 year in the stratosphere. Hence, the emissions
from jet engines at high altitudes may have a disproportionate contribution
to global warming compared to the same mass of emissions at low levels
(below 5 km).

Atmosphere[2]. A unit of pressure equating to 760 mm of mercury. 1 normal
atmosphere $= 101\,325\,N/m^2$ or $14.72\,lb/in^2$. (◊ STANDARD TEMPERATURE
AND PRESSURE)

Atom. The smallest part of an ELEMENT that can take part in a chemical reaction. It consists of a positively-charged core, called the NUCLEUS, surrounded
by negatively-charged ELECTRONS.

Atomic mass (relative). A number that gives the mass of an atom of an element
relative to that of an isotope of carbon ^{12}C, which has an assigned atomic
mass of 12.

Atomic mass unit (AMU). One-twelfth of the mass of an ATOM of carbon-12.

Atomic number. The number of ELECTRONS around the NUCLEUS of an ATOM.
This determines the chemical behaviour of an atom. The atomic number is
constant for each ELEMENT and its ISOTOPES.

Attenuation. General term for a reduction in magnitude/intensity/concentration
of a substance dispersed in a gaseous or liquid medium, e.g. LEACHATE from
a LANDFILL SITE may be attenuated in its passage through underlying strata
by the action of ABSORPTION, ADSORPTION, and biological processes in any
UNSATURATED ZONE. Sound is attenuated by atmospheric absorption and
by interaction with absorbing surfaces.

Automobile emissions. Generic name for the emissions from car exhausts, the
principal components of which are NITROGEN OXIDES (NO_x), UNBURNT

HYDROCARBONS, water vapour, CARBON MONOXIDE (CO), ALDEHYDES, and CARBON DIOXIDE (CO_2).

Carbon monoxide and nitrogen oxides are major pollutants in our cities, due to increased automobile emissions resulting from an increase in the number of cars on the road and the use of high-powered engines. They have significant health effects, particularly under SMOG and PHOTOCHEMICAL SMOG conditions.

The high compression ratios required for high-performance engines need fuels with 'anti-knock' properties. This was been achieved since the 1920s by adding tetraethyl- and tetramethyl-lead to the fuel, but now is being limited by legal and fiscal incentives. A large part of the lead – which is toxic – is emitted from the car exhaust in a fine particulate form that can be readily inhaled. Lead also poisons the catalytic reactors that are used in car exhaust systems to minimize emissions.

Photochemical smog is triggered by the action of sunlight on exhaust emissions of nitrogen oxides, and the amounts emitted from cars have increased as compression ratios, and hence engine temperatures, have increased, particularly in urban areas.

In California legislation has been enacted to effect reductions of 87 per cent for carbon monoxide, 95 per cent for hydrocarbons, and 75 per cent for oxides of nitrogen. This can be achieved by means of a CATALYTIC REACTOR in the car exhaust system which converts the unburnt hydrocarbons to carbon dioxide and water, and the carbon monoxide to carbon dioxide. The nitrogen oxides can also be reduced. Recirculation of exhaust and crankcase vapours back into the combustion chamber can also be carried out to further reduce emissions. The use of lead in petrol is now banned. ($\Diamond$ LEAN COMBUSTION)

Autonomous house; also known as the *Ecohouse*. The self-contained dwelling which simulates the ways of ECOSYSTEMS, i.e. wastes are converted to fuel by ANAEROBIC DIGESTION for METHANE production and the residues from the digestion are used for growing food. The food residues are composted and/or used for methane production. Solar energy is trapped by the GREENHOUSE EFFECT and used for house, crop and water heating. A windmill would be used for electricity. Thus, given sufficient space, sunshine, rainfall and wind, the autonomous house is in theory a self-contained system recycling its own wastes and using the sun as its ENERGY input.

Much research and development needs to be done before this can be achieved. The practitioners still require an initial energy input to build the house, provide the raw materials such as glass, plastics, paint, etc., but the concept can lead to much more energy-efficient housing.

Autotrophic organism. An organism that does not require ORGANIC material for food from the environment but which can manufacture food from inorganic chemicals. For example, chlorophyll-containing plants or ALGAE can manufacture their food from WATER, CARBON DIOXIDE and nitrates using the sunlight for an ENERGY source. HETEROTROPHIC ORGANISMS depend on the activities of the autotrophic ones for food. ($\lozenge$ PHOTOSYNTHESIS)

Availability. A thermodynamic measure of the maximum useful work that can be done by a system interaction which is at constant pressure and temperature (Absolute). A useful concept in comparing heat or thermal energy conversion processes. The greater the source temperature compared with the environment temperature, the greater the availability ($\lozenge$ SECOND LAW OF THERMODYNAMICS)

Note: Not to be confused with availability factor (US); ($\lozenge$ AVERAGE ACHIEVED OUTPUT)

Availability factor. Availability factor is a power plant performance indicator widely used in the United States and Japan. It stipulates the amount of time over a given period during which the unit was available, whether synchronized to the grid or not.

$$\text{Availability factor} = \frac{\text{Hours operated or operable}}{\text{Total hours in period}}$$

($\lozenge$ CAPACITY FACTOR)

Average achieved output. A useful concept for comparing RENEWABLE ENERGY installations such as WIND ENERGY where the wind may be either too much or too little, and output(s) fluctuate accordingly. It is customary to quote performance for overall ELECTRICITY generation as a percentage of the maximum rated capacity of the installation. For wind energy in the UK, this can vary between 25 and 30 per cent of the maximum rated capacity. ($\lozenge$ PAYBACK)

Azodrin. $\lozenge$ ORGANOPHOSPHORUS COMPOUNDS.

B

Background concentration of pollutants. If the atmosphere in a particular area is polluted by some substance from a particular local source, then the background level of pollution is that concentration which would exist without the local source being present. Measurement would then be required to detect how much pollution the local source is responsible for.

Sometimes the word 'background' is used to mean the concentration of the substance some distance from the particular source and therefore largely uninfluenced by it.

The term is also used in radiation work to mean the background level of radiation from natural sources or from sources other than that being measured and is commonly used as a reference datum against which emissions are scaled for public information purposes. This is not quite the complete story, as background radiation sources are not the same as those from the nuclear power industry and hence attempts to equate both could be misleading. (◊ IONIZING RADIATION, EFFECTS)

Background level of noise. This is used when assessing NOISE nuisance, e.g. in BS4142. The difference between the offending level and the background (defined as L_{A90}) is used as an indication of the likelihood of complaints.

Backwashing. Cleaning of a filter or ion-exchange column by reversing the fluid flow through the bed so that the filter medium or ion-exchange resin is reclassified and dirt removed.

Bacteria. Class of simple, single-celled organisms which do not possess CHLOROPHYLL. They usually multiply rapidly, by division. They are usually unicellular rods or rounded cells ca. $1\,\mu m$ in diameter.

Bacteria occur virtually everywhere in the BIOSPHERE and are responsible, for example, for the souring of milk or the decay of dead animals. They participate in the CARBON CYCLE, can fix nitrogen in the NITROGEN CYCLE, and decompose sewage. As DECOMPOSERS they are responsible for the decomposition of organic matter into simple substances which can then be reincorporated into the biological cycles. They can also cause and transmit diseases, e.g. TYPHOID, tuberculosis, etc. (◊ SEWAGE TREATMENT)

Bacteria, use in mining. ◊ SOLUTION MINING.

Baffle chamber. A settling chamber containing a system of baffles, which enables coarse PARTICULATE matter (e.g. large particles of fly ash) to be removed from stack gases. This is done by changes of gas flow direction and reduction of its velocity.

Bag filter. A closely woven bag for removing DUST from dust-laden gas streams. The fabric allows passage of the gas with retention of the dust. Separation EFFICIENCIES greater than 99 per cent can be obtained on DUSTS >1 μm in diameter. Bag filters are cleaned either by mechanical shaking or reverse pressure by means of a reversed gas jet. The former method requires that the plant be built in sections so that a section is isolated in sequence for cleaning. The latter method allows continuous operation.

Bag house. An installation with the function of abating the particulate content of gas streams by filtering these streams through large fabric bags.

Ballast voyage. The journey an oil tanker or other vessel makes with sea water carried in the oil-tanks as ballast to provide essential stability on the empty leg of its journey. The ballast is usually discharged at sea after treatment to remove oil. Several ports have facilities to receive ballast water for treatment.

Band screen. An endless moving wire mesh band used for solids removal from liquid effluents.

Bar. Unit of PRESSURE equal to 10^5 N/m^2 or 750.7 mm of mercury. Used for high steam pressure measurements. 1 millibar = 100 N/m^2 which is used for barometric measurements. Standard atmospheric pressure = 1.01325 bar.

Bar screen. A set of regularly spaced bars used for removing large solids from sewage effluent. The bars may be mechanically raked for automatic solids removal.

Barnes' formula. Empirical relationship first proposed by A. G. Barnes (1916) for calculating the flow velocity in a sewer, i.e.

$$V = 107 \ R^{0.7} \ S^{0.5}$$

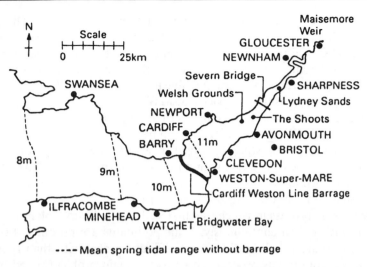

Figure 8 Location of preferred barrage line, showing tidal ranges in the Severn Estuary.

where V = flow velocity (m/s), R = hydraulic mean depth (m) and S is slope of sewer.

Sic gloria transit merde.

Barrage. A controlling dam or barrier placed in a stream or tidal flow. For renewable ENERGY purposes, a tidal barrage would have reversible turbines built into it so that both the ebbing and flowing tides can generate power.

One of the favoured UK sites for a tidal barrage is the Severn Estuary from Cardiff to Weston-Super-Mare as shown in Figure 8 which details the tidal ranges available. An ENVIRONMENTAL ASSESSMENT of such an undertaking must be made before implementation, as there could well be major changes in sediments, water quality, ECOSYSTEMS, effects on bird life, etc.

The Severn Estuary Scheme has been intensively studied and it is concluded that an installed turbine capacity of 8400 MW could produce a total annual energy output of 17 GWh at a 1989 cost of 3.7 p/kWh which is comparable to that of coal-fired power stations but without the environmental impacts associated with coal burning.

There are now (2008) calls for a reassessment of the potential of a Severn Barrage, with claims that up to 5 per cent of the UK's electricity demand could be met. The costs could be £15 bn, but could be much greater, given the UK's track record on major construction projects (e.g. The Dome, Wembley Stadium, and the Scottish Parliament).

Other UK tidal power sites have been identified where similar generating costs could be anticipated. These include the Mersey Estuary which has a mean spring tide range of 8.4m, possible installed capacity 600 MW; also the Conway Estuary, installed capacity 35 MW.

Barrel. Standard volumetric measure of crude oil production. Equivalent to 42 US gallons or 35 Imperial gallons or 159 litres.

Base. A substance which when dissolved in water generates hydroxide (OH⁻) IONS or is capable of reacting with an ACID to form a SALT. (◊ ALKALI)

Base exchange. An ion-exchange water softening process in which calcium and magnesium ions are exchanged for sodium ions.

Batch process. A process in which raw materials are fed into a plant in discrete batches rather than continuously. If air pollutants are produced by such a process they are usually rather more difficult to deal with than those produced by a continuous process. This was the main problem posed by early batch-fed refuse incinerators resulting in substantial combustion temperature reductions, incomplete combustion, DIOXIN emissions and gross air pollution. (◊ INCINERATION)

Battery directive. This requires manufacturers and importers to find accessible collection points for consumers. Battery distributors will have to take back used batteries irrespective of when they were marketed. Twenty-five per cent of household batteries have to be collected by 2012 and 45 per cent by 2016. All industrial/commercial batteries are to be collected by 2016.

Bauxite. Chemical formula $Al_2O_3xH_2O$. Major source of ALUMINIUM formed under tropical or subtropical conditions by weathering (and leaching) of aluminium-containing rocks. Between 4 and 6 tonnes of bauxite are needed to produce 1 tonne aluminium. A waste product, 'red mud' (iron oxide silica SLUDGE), is produced which can pose disposal difficulties. (◊ HALL PROCESS)

Beaufort Scale. The Beaufort Scale is an empirical measure for describing wind speed based mainly on observed sea conditions. Its full name is the Beaufort Wind Force Scale. The scale was created in 1805 by Irishman Sir Francis Beaufort, a naval officer and hydrographer. Initially the scale ranged from 0 to 12 but was extended in 1944 to 17. Wind speed on the 1946 Beaufort scale is defined by the empirical formula:

$$V = 1.87 \, B^{3/2} \text{ miles per hour}$$

where V is the equivalent wind speed in miles per hour, 30 feet above the surface of the sea, and B is Beaufort number. The table below describes the effects at each level of the scale.

The Beaufort Scale: Specifications And Equivalent Speeds

Force	Description	Specifications for use on land	Specifications for use at sea	Equivalent speed at 10m above ground				Description in forecasts	State of sea	Probable height of waves* metres
				Knots		Miles per hour				
				Mean	Limits	Mean	Limits			
0	Calm	Calm; smoke rises vertically.	Sea like a mirror.	0	1	0	1	Calm	Calm	00
1	Light air	Direction of wind shown by smoke drift, but not by wind vanes.	Ripples with the appearance of scales are formed, but without foam crests.	2	1–3	2	1–3	Light	Calm	0.1 (0.1)
2	Light breeze	Wind felt on face, leaves rustle, ordinary vane moved by wind	Small wavelets, still short but more pronounced. Crests have a glassy appearance and do not break.	5	4–6	5	4–7	Light	Smooth	0.2 (0.3)
3	Gentle breeze	Leaves and small twigs in constant motion, wind extends light flag.	Large wavelets, crests begin to break. Foam of glassy appearance. Perhaps scattered white horses.	9	7–10	10	8–12	Light	Smooth	0.6 (1.0)
4	Moderate breeze	Raises dust and loose paper, small branches are moved.	Small waves, becoming longer; fairly frequent white horses.	13	11–16	15	13–18	Moderate	Slight	1.0 (1.5)

The Beaufort Scale: Specifications And Equivalent Speeds (Continued)

Force	Description	Specifications for use on land	Specifications for use at sea	Equivalent speed at 10m above ground				Description in forecasts	State of sea	Probable height of waves* metres
				Knots		Miles per hour				
				Mean	Limits	Mean	Limits			
5	Fresh breeze	Small trees in leaf begin to sway, crested wavelets form on inland waters.	Moderate waves, taking a more pronounced long form; many white horses are formed. Chance of some spray.	19	17–21	21	19–24	Fresh	Moderate	2.0 (2.5)
6	Strong breeze	Large branches in motion; whistling heard in telegraph wires; umbrellas used with difficulty.	Large waves begin to form; the white foam crests are more extensive everywhere. Probably some spray.	24	22–27	28	25–31	Strong	Rough	3.0 (4.0)
7	Near gale	Whole trees in motion; inconvenience felt when walking against wind.	Some heaps up and white foam from breaking waves begins to be blown in streaks along the direction of the wind.	30	28–33	35	32–38	Strong	Very rough	4.0 (5.5)

8	Gale	Breaks twigs off trees; generally impedes progress.	Moderately high waves of greater length; edges of crests begin to break into spindrift. The foam is blown in well-marked streaks along the direction of the wind.	37	34–40	42	39–46	Gale	High	5.5 (7.5)
9	Strong gale	Slight structural damage occurs (chimney pots and slates removed).	High waves. Dense streaks of foam along the direction of the wind. Crests of waves begin to topple, tumble and roll over. Spray may affect visibility.	44	41–47	50	47–54	Severe Gale	Very high	7.0 (10.0)

The Beaufort Scale: Specifications And Equivalent Speeds (Continued)

Force	Description	Specifications for use on land	Specifications for use at sea	Equivalent speed at 10m above ground				Description in forecasts	State of sea	Probable height of waves* metres
				Knots		Miles per hour				
				Mean	Limits	Mean	Limits			
10	Storm	Seldom experienced inland; trees uprooted; considerable structural damage occurs.	Very high waves with long overhanging crests. The resulting foam, in great patches, is blown in dense white streaks along the direction of the wind. On the whole the surface of the sea takes a white appearance. The 'tumbling' of the sea becomes heavy and shock-like. Visibility affected.	52	48–55	59	55–63	Storm	Very high	9.0 (12.5)

11	Violent storm	Very rarely experienced: accompanied by widespread damage.	Exceptionally high waves (small and medium-sized ships might be for a time lost to view behind the waves). The sea is completely covered with long white patches of foam lying along the direction of the wind. Everywhere the edges of the wave crests are blown into froth. Visibility affected.	60	56–63	68	64–72	Violent storm	Phenomenal	11.5 (16.0)
12	Hurricane		The air is filled with foam and spray. Sea completely white with driving spray; visibility very seriously affected.	—	>64	—	>73	Hurricane force	Phenomenal	14.0 (–)

* These columns are a guide to show roughly what may be expected in the open sea, remote from land. Figures in brackets indicate the probable maximum height of waves. In enclosed waters, or when near land with an offshore wind, wave heights will be smaller and the waves steeper.

Source: The Meteorological Office, National Meteorological Library and Archive, Exeter, UK.

Becquerel (Bq). A measure of the intensity (activity) of a radioactive material, which has replaced the CURIE (Ci):

1 Bq is a rate of one disintegration per second. Hence

$$1\,Bq = 2.7 \times 10^{-11}\,Ci$$

or

$$1\,Ci = 3.7 \times 10^{10}\,Bq$$

Note: Both the intensity of the source (Bq) and the isotope must be known, e.g. STRONTIUM-90 decays by emission of a single β 'particle'. Hence a 1 Bq source will emit one particle per second. However, COBALT-60 emits a β particle and 2 γ rays per disintegration; such a source of strength 1 Bq will therefore emit 3 'particles' per second.

Furthermore, should the source disintegrate to produce an unstable DAUGHTER PRODUCT, the overall intensity of the source at any time is the sum of the disintegration of the parent and daughter products.

BedZED (Beddington Zero Energy Development). A pioneering UK attempt at providing low energy-usage housing, coupled with conservation of materials and the environment in its construction. The following extracts from BedZED sustainable resouce use give a flavour of the approach used in the housing construction.

BedZED is a mixed-use scheme in South London initiated by BioRegional Development Group and Bill Dunster Architects. BedZED has been developed by London's largest housing association, the Peabody Trust. The scheme comprises 82 homes and 3000 m^2 of commercial or living/working space. The first units were completed in March 2002 with total completion and occupation in September 2002.

BedZED employs state-of-the-art energy efficiency, with super-insulation, double and triple glazing and high levels of thermal mass. BedZED meets all its energy demands from renewable, carbon-neutral sources, generated on site, and so eliminates the 29% contribution that domestic household energy consumption makes to CO_2 emissions and global warming.

The total embodied CO_2 of BedZED is 675 kg/m^2 whilst typical volume house builders build to 600–800 kg/m^2. Despite the increased quantities of construction materials, the procurement of local, low-impact materials has reduced the embodied impact of the scheme by 20–30%.

The BedZED project has shown that in selecting construction materials, major environmental savings can be made without any additional cost. In many cases, the environmental option is cheaper than the more conventional material. For example, highly durable timber-framed windows are cheaper than uPVC and saved some 6% of the total environmental impact of the BedZED scheme and 12.5% of the total embodied CO_2. Recycled aggregate and sand are cheaper than virgin equivalents and are available as off-the-shelf products. Pre-stressed concrete floor slabs save time and costs on site, and by using fewer materials, saved some 7% of the BedZED's environmental impact compared with concrete cast *in-situ*. New FSC softwood from certified, sustainably managed woodlands is available at no cost premium, while local FSC green oak weatherboarding is cheaper than brick and shows a life cycle cost saving over imported preserved softwood. Reclaimed structural steel and timber are available cheaper than new, and offer 96% and 83% savings in environmental impact, respectively.

What you can do

1. Specify high quality timber window frames in preference to uPVC or aluminium
2. For any structural concrete, consider using a pre-stressed option
3. Introduce a local sourcing policy
4. For polystyrene-based insulation, specify HCFC- and HFC-free products
5. Specify recycled aggregates
6. Insist on FSC certified timber
7. Look into reclaimed materials. Always build in extra lead times and, if possible, extra storage space for these.

Source: BioRegional Development Group – BEDZED Report (funded by Biffaward), www.bioregional.com/retail/customer/home-ph-p.

Benefit/cost. The benefit/cost ratio seeks to compare the benefits of a particular action with its cost. In the case of anti-pollution expenditure the benefits of the effect of reduced pollution have to be assessed. A benefit/cost ratio of 10:1 implies that for every £1 spent on improving the pollution situation, the benefits can be valued at £10. The problem often arises in quantifying the benefits. If the polluter is divorced from the effects of the pollution, agreement may be hard to obtain.

Bentonite. A CLAY which can contain a large proportion of the mineral montmorillonite. This has the property of swelling when it comes in contact with water and thus if laid in sufficient thickness can make a very good LANDFILL SITE liner. The swelling bentonites are also used for drilling mud and dam sealing purposes.

Benzene. A clear, colourless, volatile, flammable liquid; an aromatic hydrocarbon, formula C_6H_6. Benzene is widely used in the chemical industries and is a minor constituent of petrol. It is highly toxic, and carcinogenic and is strongly implicated in LEUKAEMIA.

One estimate (US Govt.) is that there are 10 excess deaths per 1000 employees exposed to a working lifetime of 1 part per million benzene (US occupational exposure standard). The US EPA estimates leukaemia death RISK as 7×10^{-6} over a lifetime exposure to $1\,\mu g/m^3$ (0.3 ppb).

Wolffe (see footnote for reference) has stated:

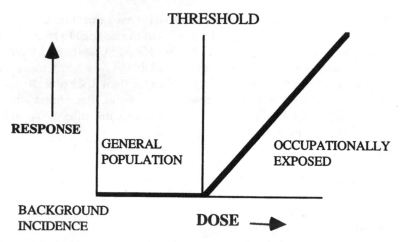

Figure 9   Dose–response curve for relationship between benzene exposure and leukae-mia assuming that background incidence of leukaemia is 'spontaneous' and unrelated to non-occupational exposure.

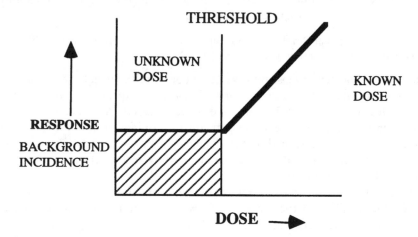

Figure 10   Revised dose–response curve for the relationship between benzene exposure and leukaemia suggesting that background incidence is related to non-occupational benzene exposure and that threshold dose is an arbitrary dividing line between known and unknown doses.

Certainly, although very high (occupational) levels of benzene are linked with leukaemia any risks of cancer at lower levels of benzene appear to disappear into the background of 'spontaneous' leukaemia incidence [see Figures 9 and 10]. The existence of the background rate, and the shape of the graph in the second Figure is used as an important justification for the concept of the 'threshold' dose which, translated into biological terms, argues for the presence of detoxification and

DNA repair/maintenance mechanisms which operate efficiently at lower concentrations of carcinogen but which can be saturated at higher doses. However, such an interpretation depends very much on how one views the revised dose response graph. The hypothesis that the background 'spontaneous' leukaemia incidence may also be related to non-occupational benzene exposure casts the threshold dose in a new light. In this instance, the threshold becomes the artificial dividing line between the high (and fairly well-defined) levels of occupational exposure, to which thousands of adults are exposed, and the less-defined and generally lower levels to which millions of people are exposed in the course of their day-to-day activities. This background level of exposure to benzene is suggested to contribute to the background level of leukaemia and lymphoma in the non-occupationally exposed population.

Unleaded petrol contains more benzene than leaded petrol. It is estimated (*Royal Commission on Environmental Pollution*, 18th Report Transport & the Environment) that 78 per cent of atmospheric benzene is from petrol engine exhausts, 10 per cent from DIESEL engine exhausts and 10 per cent from fuel evaporation and oil refineries. CATALYTIC CONVERTERS, if used properly, can reduce benzene by up to 90 per cent.

Wolffe has commented that:

detailed measurements of benzene and hydrocarbon exposure would, however, allow support for, or refutation of the hypothesis that internal combustion-powered vehicles cause cancer. There is a pressing need for biological and epidemiological investigation of non-occupational benzene exposure.

The EC maximum allowable concentration in drinking water is $1\,\mu g\,l^{-1}$.

Source: Wolffe, S.P. Correlation between car ownership and leukaemia: Is non-occupational exposure to benzene from petrol and motor vehicle exhaust a causative factor in leukaemia and lymphoma?, 1991.

Benzene hexachloride (BHC). The former name for the insecticide HCH (hexachlorocyclohexane). (◊ CHLORINATED HYDROCARBONS, LINDANE)

Benzo [alpha] pyrene. A polynuclear aromatic HYDROCARBON that occurs in coal tar, SOOT and tobacco smoke. It has been shown to be carcinogenic in laboratory animals when administered by a number of different routes.

Berylliosis. ◊ BERYLLIUM.

Beryllium (Be). A silver-white metal with very good resistance to corrosion, heat and stress. Its lightness and hardness find a wide variety of applications, especially in rocketry, aircraft, nuclear reactors. The main environmental sources are beryllium refining, alloying and fabricating operations and in the burning of coal.

Acute poisoning can result from exposure to airborne concentrations as low as 20 micrograms per cubic metre for less than 50 days. The lung tissue is inflamed and fevers, chills and shortness of breath can become chronic from such short exposures. Exposure leads to berylliosis.

In the UK the maximum allowable concentration at discharge has now been set at 0.1 micrograms per cubic metre. It has been classified as a HAZARDOUS POLLUTANT in the USA.

Best Available Techniques Not Entailing Excessive Costs (BATNEEC). This is essentially an updating and refinement of BEST PRACTICABLE MEANS (BPM) whereby a greater degree of control over emissions to land, air and water may be exercised. The techniques cover not only the hardware but also the way in which a particular process is operated, in fact, as operational management and staff training.

'Best available' techniques is taken to mean demonstrably the most effective techniques for an operation at the appropriate scale and commercial availability even if they are not in general use or have not been used in the UK – this is a most important development from the previous BPM philosophy. 'Not entailing excessive cost' means that the benefits gained by using the best available techniques should bear a reasonable relationship to the cost of obtaining them, otherwise it may not be justifiable to obtain very small reductions in emissions at very high costs unless emissions are very toxic, in which case very high specifications may be imposed, e.g. as in hazardous waste incineration.

Judgement about what constitutes BATNEEC can differ between new and existing plants. For example, the cost of adopting given techniques at an existing plant can depend upon its technical characteristics and could be much higher than those at a new plant which is purpose-designed to incorporate these techniques from the beginning.

The benefits from adopting these techniques could also be less at an existing plant, depending on the remaining life and likely utilization, in other words a degree of flexibility is allowed to the regulating authority.

Best Practicable Environmental Option (BPEO). This term was first used by the *Royal Commission on Environmental Pollution*, 5th Report (1976) in order to take account of the total pollution from a process and the technical possibilities for dealing with it. It is seen as a successor to 'BEST PRACTICABLE MEANS' which was primarily aimed at emissions to air. BPEO, in addition to controlling atmospheric emissions, includes

appropriate measures to deal with any harmful discharges to water and for the treatment or disposal of other solid and liquid wastes to land. A BPEO should take into account the risk of transfer of pollutants from one medium to another.
RCEP, 12th Report, BPEO, 1988.

Thus, BPEO specifically requires CROSS-MEDIA considerations which were often lacking under BPM. It does not imply that the best technology is to be used irrespective of cost. Quite the contrary, the costs and the benefits of action to deal with each problem will still need to be assessed. It establishes the option which provides the least damage to the environment as a whole at an acceptable cost. BPEO was included in Part I of the Environmental

Protection Act 1990 as a basis for the authorisation of processes under the Integrated Pollution Control (IPC) regime.

Summary of steps in selection of a BPEO

Step 1: *Define the objective.* State the objective of the project or proposal at the outset, in terms which do not prejudge the means by which that objective is to be achieved.

Step 2: *Generate options.* Identify all feasible options for achieving the objective: the aim is to find those which are both practicable and environmentally acceptable.

Step 3: *Evaluate the options.* Analyse these options, particularly to expose advantages and disadvantages for the environment. Use quantitative methods when these are appropriate. Qualitative evaluation will also be needed.

Step 4: *Summarize and present the evaluation.* Present the results of the evaluation concisely and objectively, and in a format which can highlight the advantages and disadvantages of each option. Do not combine the results of different measurements and forecasts if this would obscure information which is important to the decision.

Step 5: *Select the preferred option.* Select the BPEO from the feasible options. The choice will depend on the weight given to environmental impacts and associated risks, and to the costs involved. Decision-makers should be able to demonstrate that the preferred option does not involve unacceptable consequences for the environment.

Step 6: *Review the preferred option.* Scrutinize closely the proposed detailed design and the operating procedures to ensure that no pollution risks or hazards have been overlooked. It is good practice to have the scrutiny done by individuals who are independent of the original team.

Step 7: *Implement and monitor.* Monitor the achieved performance against the desired targets, especially those for environmental quality. Do this to establish whether the assumptions in the design are correct and to provide feedback for future developments of proposals and designs.

Throughout Steps 1 to 7: *Maintain an audit trail.* Record the basis for any choices or decisions through all of these stages, i.e. the assumptions used, details of evaluation procedures, the reliability and origins of the data, the affiliations of those involved in the analytical work and a record of those who have taken the decisions.

Note: the boundaries between each of the steps will not always be clear-cut: some may proceed in parallel or may need to be repeated.

Source: *Royal Commission on Environmental Pollution*, 12th Report, 1988.

Best Practicable Environmental Option (BPEO) Index. A numerical estimate of the overall environmental impact of a range of feasible pollution control options across the media of air, land, and water. The options can then be compared.

Best Practicable Means (BPM). An essentially pragmatic philosophy for the control of emissions from scheduled processes. The term was first used in the 1906 Alkali Act and was implemented by the now superseded UK ALKALI AND CLEAN AIR INSPECTORATE who were responsible for controlling emissions from industrial premises (oil refineries, chemical works, cement plant, etc.) listed in the schedule to the Act.

The ideology behind best practicable means caused much controversy as it was considered to be a loophole which allowed industry to emit noxious or injurious substances in greater amounts than would have been allowed had absolute standards been imposed. Basically, it implied that, while better emission standards may be obtainable, industry must not be unduly penalized in its operations as the costs of pollution are offset by the social and economic benefits of a thriving industrial sector. The air pollution emitted by the Bedfordshire brick kilns was a case in point. Thus, the effective emission standards may have surprisingly wide tolerances depending on local environmental factors and in practice often lagged behind those that were technologically quite feasible. (◊ ENVIRONMENT AGENCY)

Best Value. An innovative concept in UK local government whereby a balance is struck between cost and quality that local people desire, and there is economical, efficient and effective service delivery and continuous improvement.

The key to Best Value is a framework which enables:

- a corporate approach to service provision
- community consultation
- an ongoing programme of fundamental service reviews
- published targets for service improvement and reports back on performance
- a pragmatic view on who delivers service
- an external scrutiny by audit and inspection, and
- action to tackle failure.

The major claim is that it will drive up standards and drive down cost. Time will tell.

Source: Aldworth, 'The Route to Best Value', Institute of Waste Management annual conference, June 1998.

Beta particles (β particles). Beta particles are electrons and therefore negatively-charged. They are sparsely ionizing and have little penetrating ability, although more so than an ALPHA PARTICLE. Many atoms heavier than lead decay spontaneously by the emission of beta particles; however, a few emit positive particles (positrons) and these are also included under the term 'beta particle'. Positrons are equal in mass and charge to an electron.

In negative beta-particle emission, a neutron in the original nucleus breaks up into a proton and an electron; the latter is emitted. This decay will take place in nuclei which have too many neutrons for complete stability. Conversely, in positive beta-particle emission there are too many original protons and these break up into neutrons and positrons.

Also emitted in beta radiation are neutrinos, which have a zero charge and virtually zero mass.

Beta radiation. A stream of beta particles, i.e. electrons or positrons emitted from radioactive nuclei, possessing greater penetrating power than ALPHA RADIATION. (◊ RADIONUCLIDE)

Beverage cans. A container for beverages or juices, usually made of steel or ALUMINIUM.

Bilharziasis (bilharzia). ◊ SCHISTOSOMIASIS.

Bi-metal container. Container made out of two metals, with the body usually steel and the lid aluminium.

Bioaccumulation. ◊ BIOLOGICAL CONCENTRATION.

Biochemical oxygen demand (BOD). A standard water-treatment test for the presence of organic pollutants. It is an indicator of the ability of the pollutants to use up the oxygen which is dissolved in the water. This biochemical test depends on the activities of BACTERIA and other microscopic organisms which in the presence of OXYGEN feed upon ORGANIC matter. The test indicates the amount of dissolved oxygen in grams per cubic metre used up by the sample when incubated in darkness at 20 °C for five days. The notation for this is BOD_5 (Some 'difficult' effluents may be incubated over 7 days. This is denoted by BOD_7.) The 20 °C & 5 days are conditions used in the UK. They may be different in other countries.

The factors which influence the test are:

1. The presence of toxic substances which may inhibit the growth of micro-organisms during the test.
2. The chemical uptake of oxygen by substances such as AMMONIA or nitrites.
3. A departure from the specified incubation temperature of 20 °C.
4. The availability of dissolved oxygen and nutrients.

It is standard practice to suppress nitrification by the addition of Allythiourea (ATU). The BOD test may be supplemented by CHEMICAL OXYGEN DEMAND (COD) tests, but these do not solely measure the biodegradable matter which is of central interest. Both tests are commonly run when a 'new' effluent is encountered. (◊ DISSOLVED OXYGEN)

Typical BOD and COD values of organic wastes are shown in the table overleaf. The COD column is the chemical oxygen demand, which uses potassium dichromate to oxidize the organic matter completely and so includes the oxygen demand from materials that cannot be treated biologically. Both the BOD and COD tests are important for effluent monitoring. The BOD test requires five days whereas the COD test can take only two hours.

Once an effluent BOD:COD ratio has been determined, the monitoring of the effluent by the COD test automatically establishes the BOD, as BOD:COD ratios are relatively constant for particular effluents, but of course vary from one waste to another.

Type of waste	BOD (5 days, gm^{-3})	COD (gm^{-3})
Abattoir	2 600	4 150
Brewery	550	–
Distillery	7 000	10 000
Domestic sewage	350	300
Pulpmill	25 000	76 000
Petroleum refinery	850	1 500
Tannery	2 300	5 100

Typical BOD levels are given in the table below:

Type of waste	BOD (mg per litre)
Treated domestic sewage	20–60
Raw domestic sewage	300–400
Liquid sewage sludge	10 000–20 000
Pig slurry	20 000–30 000
Sludge effluent	30 000–80 000
Milk	140 000

Biodegradable material. Material which is capable of being broken down, usually by micro-organisms, into basic elements. Most ORGANIC wastes, such as food and paper, are biodegradable.

Biodegradation. The ability of natural decay processes to break down human-made and natural compounds to their constituent elements and compounds, for assimilation in, and by, the biological renewal cycles, e.g. wood is decomposed to CARBON DIOXIDE and water. Similarly, PAPER can be decomposed. The problem is that many human-made materials such as plastics are not biodegradable and their quantity has increased substantially.

Since the 1950s the chemicals industry worldwide has increased its yearly production of bulk chemicals (from plastics to antifreeze). By their very nature, most of those chemicals are still around, in some shape or form, lingering and polluting long after their original job is done.

Action needs to be taken in both developing biodegradable environment-friendly alternatives and also restricting the use of those that are not strictly necessary for human or animal requirements.

A note of caution needs to be entered on the biodegradable PLASTICS front, as the industry has reported that sales from some degradable material producers have been reported as virtually nil, with several companies having lost hundreds of millions since the 1980s, when biodegradable plastics were first commercially marketed.

Economically, biodegradable plastics made from starch can cost almost 10 times as much per kilogram as traditional petroleum-based plastics as they are made from virgin materials and require significantly more ENERGY to produce. To produce a degradable refuse sack would consume 90 MJ/kg whereas to produce a post-use recycled sack would consume 30 MJ/kg (including waste collection, transport and reprocessing). The current degradable market is around 14–18 million kg, with the potential to increase to about 112 million kg. With the current PETROLEUM-based market at around 112 billion kg, it is unlikely that biodegradable plastics will ever become an economically viable option. (◊ ENERGY ANALYSIS)

Clearly, there is no free lunch and if environmental protection is required, it has to be paid for by the consumer.

Note: Biodegradable bags break down in compost in ca: 12 weeks. Degradable plastics may take up to 5 years & are not compostable CAVEAT EMPTOR.

Source: *British Polythene Industries Review*, No. 59, February 1999.

Biodiesel. Biodiesel refers to a diesel-equivalent, processed fuel derived from biological sources (such as vegetable oils), which can be used in unmodified diesel-engined vehicles. It is thus distinguished from the straight vegetable oils (SVO) or waste vegetable oils (WVO) used as fuels in some modified diesel vehicles.

Biodiesel consists of alkyl esters made from the transesterification of vegetable oils or animal fats. In transesterification, vegetable and/or animal oils, fats or greases are combined with methanol in the presence of sodium or potassium hydroxide (as a catalyst). The process leaves behind two products – methyl esters (the chemical name for biodiesel) and glycerine (a valuable by-product usually sold to be used in soaps, sweets and other products). Popular raw sources for biodiesel processing include rapeseed oil (Europe), soybean oil (USA), and palm oil (SE Asia). Various cooking oils and fats are suitable (new and recovered).

Biodiesel refers to the pure fuel before blending with diesel fuel. Biodiesel blends are denoted as, 'BXX' with 'XX' representing the percentage of biodiesel contained in the blend (i.e. B20 is 20 per cent biodiesel, 80 per cent petroleum diesel).

Biodiesel is biodegradable and non-toxic, and produces significantly fewer emissions than petroleum-based diesel when burned, as it is essentially free of sulfur and aromatics.

There is much debate about the extent to which biodiesel can safely be used in conventional diesel engines without modification. Using biodiesel in unmodified engines may lead to problems: since biodiesel is a better solvent than standard diesel, it 'cleans' the engine, removing deposits in the fuel lines, and thus may cause blockages in the fuel injectors.

Biodiesel can also be used as a heating fuel in domestic and commercial boilers. Existing oil boilers may require conversion to run on biodiesel, but the conversion process is believed to be relatively simple.

Biodiesel is a light to dark yellow liquid. It is practically immiscible with water, has a high boiling point and low vapour pressure. Typical methyl ester biodiesel has a flash point of ~150°C (300°F), making it rather non-flammable. Biodiesel has a density of ~0.86 g/cm^3, less than that of water.

Biodiesel is used by millions of car owners in Europe (particularly Germany).

The energy content of biodiesel is about 90 per cent that of petroleum diesel. Recent studies using a species of algae with up to 50 per cent oil content have concluded that only 28 000 km^2 or 0.3 per cent of the land area of the USA could be utilized to produce enough biodiesel to replace all transportation fuel the country currently utilizes. Furthermore, otherwise unused desert land (which receives high solar radiation) could be most effective for growing the algae, and the algae could utilize farm waste and excess CO_2 from factories to help speed the growth of the algae. In tropical regions, such as Malaysia and Indonesia, oil palm is being planted at a rapid pace to supply growing biodiesel demand in Europe and other markets. It has been estimated in Germany that palm oil biodiesel has less than a third the production costs of rapeseed biodiesel. The direct source of the energy content of biodiesel is solar energy captured by plants during photosynthesis. Although on combustion it releases greenhouse gases (mainly, but not only, CO_2) it is described as carbon-neutral, since it is derived indirectly from living plant sources. It is argued that the plants absorb essentially the same amount of these gases during the growing stage.

Biodiesel reduces tailpipe particulate matter, hydrocarbon, and carbon monoxide emissions from most modern four-stroke compression ignition engines. The reason is mainly associated with the presence of oxygen in the fuel which allows for complete combustion. Particulates are mainly carbon dust, commonly described as soot, and contain very fine particles that can be dangerous in the atmosphere since they can get into the lungs by bypassing the body's natural filter mechanisms. Together with the unburned hydrocarbons, particulates are suspected of being carcinogens and can cause other serious toxic effects. Carbon monoxide is, of course, well known for its lethal action in the bloodstream. Fairly obviously, the higher the proportion of biodiesel in a mixture, the more these noxious exhausts are reduced.

Biodiesel has a very low concentration of sulphur, which can, for example, lead to SO_2 and acidic precipitation. The amount is very much less than traditional diesel oils and is comparable with the amount in Ultra Low Sulphur Diesel (ULSD).

But it has some disadvantages too. Biodiesel has a lower energy density than petro-diesel and because it is a denser fuel, the statistics vary slightly depending on whether you want a figure of energy per gallon or per pound. Roughly speaking the reductions are in the vicinity of 10 per cent. Viscosity at low temperatures is important with diesel, sometimes making cold-weather starting difficult; biodiesel can be at a disadvantage in this respect. There are reports of occasional use of dual tank vehicles where petro-diesel is used to start and shut down but switched to biodiesel for main running. This is a technique for the over zealous, not for the typical motorist. A more practical solution is to introduce some special additive or a proportion of petro-diesel.

The properties of biodiesel mean that rubber pipes may have to be replaced by plastic ones and it may remove engine deposits, hence filters may need changing more frequently.

If biodiesel-from-crops was widely adopted as a motor fuel then there would be a knock-on effect with respect to reduced biodiversity and the disrupted supply of other vital agricultural products. This will not happen overnight but could be a serious and complex problem requiring attention in the event.

Environmental concerns. When used in standard compression-ignition engines biodiesel can increase nitrogen oxide emissions. This phenomenon is not due to nitrogen in the fuel but comes from the intake air as a result of the particular combustion process. The order of the increase appears to be fairly linearly related to the proportion, reaching a maximum of around 10 per cent with B100. This phenomenon is not replicated in some other applications such as oil fired boilers and it is believed that the effect can be reduced in vehicle engines by refining the tuning.

The locations where oil-producing plants are grown is of increasing concern to environmentalists, one of the prime worries being that countries will clear out large areas of tropical forest in order to grow such lucrative crops, in particular oil palm. This has already occurred in the Philippines and Indonesia; both countries plan to increase their biodiesel production levels significantly, which will lead to the deforestation of tens of millions of acres if these plans materialize. Loss of habitat on such a scale could endanger numerous species of plants and animals. A particular concern which has received considerable attention is the threat to the already-shrinking populations of orangutans on the Indonesian islands of Borneo and Sumatra, which face possible extinction.

Source: wikipedia.org/wiki/Biodiesel. Accessed 3 January 2007.

Biodiversity (biological diversity). This is an area of study in biology, which focuses on the differences between organisms, as opposed to the study of cellular, physiological and biochemical biology. The term particularly emphasizes the huge numbers of different species of living organisms, their adaptations to diverse environments and the spectrum of ways in which humans interact with organisms.

Nobody knows exactly how many species there are in the animal and plant kingdoms in the world. The number of species already identified is about 1.4 million, and it is speculated, although opinions differ, that there may be 5–30 million in all. However, types and numbers of species of various organisms are not evenly dispersed throughout the world. For example, there only 12 species of frogs and toads (Anura) in northern Europe, of which there are only two types of toad in the UK, but there are more than 3500 Anura species worldwide. This biodiversity is the result of 4 billion years of evolution affecting all the different forms of life on earth, in the air, and in freshwater and seas.

There is a natural balance, within a region, governed by the activities of the components of the food web. The balance is between autotrophs and heterotrophs, producers and consumers, and herbivores and omnivores. Depending on changes in local conditions there may be species loss or dominance, but the real concern for the biodiversity of species is the possible acceleration of the rate of species loss by human activities.

The interactions between humans and organisms range from exploitation in hunting and logging, through management of agricultural systems to the efforts at conservation for the future. The diversity of life on earth is an irreplaceable asset to humans: food and clothes are obtained from crops and livestock; medicines from micro-organisms, plants and animals; building materials and fuel from forest products; and there are many other examples. Also affected by the existence of such biodiversity is soil generation and productivity, the biological cycles such as the water cycle, climate change and control, and our enjoyment of scenery and wildlife. In effect, the earth's biological resources are vital to economic and social development.

Following on from the setting up, by the United Nations Environment Programme (UNEP), of a working group, the Convention on Biological Diversity was opened for signature at the 1992 Earth Summit in Rio de Janeiro. In all, 168 countries signed and the Convention came into force at the end of December 1993. Its main objectives are 'the conservation of biological diversity, the sustainable use of its components and the fair and equitable sharing of the benefits arising out of the utilisation of genetic resources'.

Bioethanol. Bioethanol is ethanol produced from crops, and is the principal fuel used as a petrol substitute for road transport vehicles. It is mainly produced by the sugar fermentation process, although it can also be manufactured by the chemical process of reacting ethylene with steam. The main sources of sugar required to produce ethanol come from fuel or energy crops. These crops are grown specifically for energy use and include corn, maize and wheat crops, waste straw, willow and poplar trees, sawdust, reed canary grass, cord grasses, Jerusalem artichoke, myscanthus and sorghum plants. There is also ongoing research and development into the use of municipal solid wastes to produce ethanol fuel.

Ethanol is a high octane fuel and has replaced lead as an octane enhancer in petrol. Blending ethanol with gasoline also oxygenates the fuel mixture so it burns more completely and reduces polluting emissions. Ethanol fuel blends are widely sold in the USA. The most common blend is 10 per cent ethanol and 90 per cent petrol (E10). Vehicle engines require no modifications to run on E10 and vehicle warranties are unaffected also. Only flexible fuel vehicles can run on up to 85 per cent ethanol and 15 per cent petrol blends (E85). The first E85 pump in the UK opened in Norwich in 2006. The first UK bioethanol production facility will produce 70 million litres of biofuel per year from sugar beet grown by farmers in Norfolk, Cambridgeshire and Suffolk. The introduction of bioethanol is expected to create up to 10 000 jobs in agriculture and fuel production in the UK.

Russia plans to build its first bioethanol plant in Mikhailovka, in the Volgograd region, a major grain-growing area. It will take in 900 000 tonnes of wheat a year, and produce about 300 000 tonnes of ethanol. The plant is expected to cost $20–250 million, and take two years to build. Up to 12 huge pig farms with around 100 000 animals each, will be set up nearby to consume the spent grain (a waste product of the distillation process). The bioethanol from the plant will most likely be exported, as local consumption is discouraged by high taxes.

Bioethanol has a number of advantages over conventional fuels. It comes from a renewable resource, that is, crops and not from a finite resource, and the crops it derives from can grow well in the UK (like cereals, sugar beet and maize). Another benefit over fossil fuels is the greenhouse gas emissions. The road transport network accounts for 22 per cent of all greenhouse gas emissions and through the use of bioethanol, some of these emissions will be reduced as the fuel crops absorb the CO_2 they emit through growing. Also, blending bioethanol with petrol will help extend the life of the UK's diminishing oil supplies and ensure greater fuel security, avoiding heavy reliance on oil-producing nations. By encouraging bioethanol use, the rural economy would also receive a boost from growing the necessary crops. Bioethanol is also biodegradable and far less toxic than fossil fuels. In addition, using bioethanol in older engines can help reduce the amount of carbon monoxide produced by the vehicle, thus improving air quality. Another advantage of bioethanol is the ease with which it can be easily integrated into the existing road transport fuel system. Bioethanol is produced using familiar methods, such as fermentation, and it can be distributed using the same petrol forecourts and transportation systems as before.

Biofuel. A FUEL which is biological in origin, such as METHANE, BIOGAS and BIODIESEL, all of which can contribute to SUSTAINABILITY.

The smell of rancid cooking oil or vinegar from the exhaust fumes of buses running on biofuels are, at least for some people, a negative feature of these alternatives to petroleum-derived fuels (although to a committed user, they may be akin to silage odours to a farmer or wind turbine noise to an afficionado).

Drax, Britain's biggest coal-fired electricity generator, is planning to cover an area one-fifth the size of Wales with biofuel crops to meet its ambitious targets for cutting its carbon emissions. The company, which generates 8 per cent of the UK's electricity, said that 10 per cent of its output would come from burning energy crops such as rape seed and elephant grass by 2009, saving 2 million tonnes of CO_2 a year. This would involve planting between 200 000 and 400 000 hectares of land, or between 3 and 5 per cent of the UK's cropped land.

Source: *The Independent*, 9 March 2007.

Plant	Litres of oil obtained per hectare
Oil palm	5950
Coconut	2689
Jatropha	1892
Rapeseed	1190
Peanuts	1059
Sunflowers	952
Soyabean	446
Maize	172

Source: USDA, Foreign Agricultural Service GAIN Report ID 7019, cited by Wendy Larsen in tce September 2007.

Source: *Pollution: Nuisance or Nemesis*, HMSO, pp. 36–37.

Biogas. Gas formed by ANAEROBIC DIGESTION of organic materials, e.g. whey or sewage sludge, with a typical composition of 62 per cent methane, 38 per cent carbon dioxide. Can be used for heat and power purposes as spark ignition engines can be modified to use it as a fuel.

The table on the facing page gives the 1990 economics of processing slurry from 120 head equivalent dairy cattle (110 milking cows plus 50 followers), with a gravity/manual loading system (values are pounds sterling except where otherwise indicated).

The potential pollution reduction by biogas production from cattle, pig and poultry is highlighted by their respective population equivalent (in BOD terms) of 10.2, 3.5 and 0.1 times that of humans, respectively.

Biogas: 1990 economics of processing slurry from dairy cattle.

5 lid digester, fitted control house, stirring, heating, PLC controller, blower and separator, loading pump	42 000
Civil engineering at £88 per cubic metre of tank	8 800
Separator	5 200
Separator tower and electrical work	2 800
Rainwater control	200
Electrical engineering	500
Gas bags, $25\,m^3 \times 2$	3 000
Total	62 500
Department of Agriculture grant 50% of digester, gas regulator bag and slurry separator but not gas use equipment	31 250
Remainder to find for digester	31 250
Domestic boilers to burn gas at home	900
Total cost to farmer	32 150

INCOME

Fibre in 250 days/year during housing of cattle	150 tonnes
Plus imported slurries	50 tonnes
Total for the year	200 tonnes

Gas $100\,m^3$ gas per day with a digester heating requirement of $34\,m^3$ biogas per day, the available gas is valued at $(100 - 34) \times 6.25\,kWh/m^3 \times £0.022/kWh \times 365$ days = £3312 per year

SUMMARY

Fibre income per year (200 t × £30/t)	6000
Gas income per year	3300
Total	9300

£400 per year maintenance = net input of £8900 per year.
System cost to farmer £32 150 if all work is done by outside contractors.
Payback time 3.6 years at zero inflation, less with inflation.
Other benefits include an estimated 0.5% continuous annual improvement in grass production, odour control, weed seed destruction, pollution reduction by 85%, improved pasture land and ability to graze pastures soon after slurry spreading, the possibility of organic pastures fed with clover nitrogen and large scale organic farming.

Source: Reproduced by permission of Green Land Systems Ltd.

Biological aerated filter (BAF). A biological waste-water treatment unit comprising plastic media with a film of micro-organisms submerged in a tank and aerated.

Biological concentration. The mechanism whereby filter feeders such as limpets, oysters and other shellfish concentrate HEAVY METALS or other stable compounds present in dilute concentrations in sea or fresh water. One extreme example is that of limpets collected from rocks in the Severn Estuary which contained 550 parts per million of CADMIUM of which the ingestion of 50 milligrams could be lethal. The limpets also contained 900 ppm zinc. The Severn Estuary is near a large zinc smelter and it is not improbable that the high concentrations resulted from industrial activity.

Another example is the 'Irish Seabird Wreck 1969' in which 10 000 sea birds were affected by PCBs and pesticides. Analyses showed the following PCB concentrations: the PCB level in the Firth of Clyde, 23–30 October 1989, was 0.01 micrograms per litre yet the concentration in mussels (less shell) ranged from 200 to 800 micrograms per kilogram, a concentration of up to 80 000 times. The pesticide concentration of the Firth of Clyde in the same period (DIELDRIN and DDT) was ca. 0.001 to 0.005 micrograms per litre. The corresponding concentration range in shellfish was 100–300 micrograms per kilogram, a concentration factor of up to 300 000 times. (◊ PCBs)

This illustrates how minute traces of toxic compounds can be concentrated biologically and enter the food chain of human beings or in this case, be strongly implicated in the Irish Seabird Wreck, Autumn 1969 (NERC Series C No. 4, 1971). (◊ MINAMATA DISEASE)

Source: B. Silcok, 'Limpets reveal metal poison danger', *The Sunday Times*, 6 June 1971.

Biological filter. A bed of inert granular material, e.g. PLASTIC rings, clinker, etc., on which settled SEWAGE effluent is distributed and allowed to percolate. In doing so, the effluent is aerated thus allowing aerobic BACTERIA and FUNGI to reduce its BIOCHEMICAL OXYGEN DEMAND. It is in effect an AEROBIC PROCESS and not a filter *per se*.

Biological indicators of water quality. The differential response of river fauna and flora to varying water quality has been recognized since the early 20th century. Macroinvertebrates have been found to be particularly suitable as indicators of water quality. There are essentially a size-class of aquatic invertebrates comprising several taxa within the animal kingdom. All are relatively sedentary in habit and live in close proximity to the river bed, either amongst the river bed material or in association with attached benthic flora. In very clean waters invertebrates such as nymphs from the mayfly family may be found, while in extremely polluted water these will be absent and an abundance of worms found instead. A variety of species will be found in intermediate water quality levels. (◊ BMWP SCORE)

Biological systems. ◊ FOOD CHAIN; CARBON CYCLE; NITROGEN CYCLE.

Biomass. The mass of living organisms forming a prescribed population in a given area of the earth's surface. It is usually expressed in grams per square metre (g/m^2).

Biomass (Renewable Energy). Biomass is organic matter of recent origin. It does not include fossil fuels, which have taken millions of years to evolve. The CO_2 released when energy is generated from biomass is balanced by that absorbed during the fuel's production. This is thus a carbon neutral process. People have been producing energy from biomass for centuries.

Biomass or biofuels are produced from organic materials, either directly from plants or indirectly from industrial, commercial, domestic or agricultural products. Biofuels fall into two main categories: woody and non-woody biomass. Woody biomass includes forest products, untreated wood products, energy crops, and short rotation coppice (SRC), e.g. willow or elephant grass. Non-woody biomass includes animal waste, industrial and biodegradable municipal products from food processing and high energy crops, e.g. rape, sugar cane, maize.

For small-scale domestic applications of biomass the fuel usually takes the form of wood pellets, wood chips and wood logs. Wood pellets are a compact form of wood, with a low moisture content and high energy density.

There are two main ways of using biomass to heat a domestic property. Stand-alone stoves providing space heating for a room: these can be fuelled by logs or pellets but only pellets are suitable for automatic feed. Generally they are 6–12 kW in output, and some models can be fitted with a back boiler to provide water heating. Boilers connected to central heating and hot water systems: these are suitable for pellets, logs or chips, and are generally larger than 15 kW.

Producing energy from biomass has both environmental and economic advantages. It is most cost-effective when a local fuel source is used, which results in local investment and employment. Furthermore, biomass can contribute to waste management by harnessing energy from products that are often disposed of at landfill sites.

Biome. A large, well-defined biological community, e.g. tundra, grassland, coral atoll. The biome has a particular form of vegetation and associated animals which have become adapted to the local conditions. In other words, it is a balanced ecological community.

Biometallurgy. The application of microbiological processes to mineral treatment, especially for the extraction of non-ferrous metals from sulphide ores. (◊ SOLUTION MINING)

Biophotolysis. The storage and use of electrons produced in the first stages of PHOTOSYNTHESIS which can then be used to make free HYDROGEN. This is another research route which investigates the use of ENERGY from the sun to produce electricity or FUELS such as hydrogen, in this case directly.

Experimental work on algal systems has shown that this can be done in the laboratory.

Bioreactor (aerobic). Many industrial processes rely on the use of gas–liquid bioreactors which require the intimate contact of oxygen with the biological broth or effluent as the case may be.

As the characteristics of biological broths are different from those of liquid effluents, two designs of aerobic bioreactors have evolved, namely, the external loop and the internal loop configurations, respectively. The gas injected region is termed the *riser* and the relatively gas-free region, being heavier, is the *downcomer*. In the external loop, the riser and the downcomer are usually two separate tubes linked by connecting zones near the top and the bottom of the reactor. The internal loop is often of 'split cylinder' construction with a vertical baffle which allows internal circulation.

The performance characteristics of the two designs are quite different. The external loop has low oxygen transfer rates and high turbulence which suits effluent treatment whereas biological broths are better processed by the high oxygen transfer rates and low shear of the internal loop which is used extensively for mammalian cell culture. (⟡ AEROBIC PROCESSES; DEEPSHAFT TREATMENT)

Source: Y. Christi, 'Airlift reactors design and diversity', *Chemical Engineering*, February 1989, pp. 41–43.

Bioremediation. The use of biological methods to remediate/restore CONTAMINATED LAND. Typical methods make use of tailored MICROBES and break down PHENOLS which are major contaminants in many sites, especially old town gasworks. (Figure 11)

Bioremediation promotes the microbial destruction of organic pollutants, permanently removing these contaminants from the environment; it can precipitate or accumulate inorganics; and it can possibly improve soil structure.

In situ methods such as bioventing, air sparging, and injection and recovery systems are available and widely used in western Europe but the UK is slow on the uptake owing to traditional reliance on containment LANDFILL. Times change and progress can be expected due to restrictions on landfilling.

Source: R. Swannell, *The Chemical Engineer*, 12 February 1998.
(⟡ MICROBES, CONTAMINATED LAND)

Biosolids. The title for processed SLUDGES from SEWAGE TREATMENT. The use of biosolids in agriculture by farmers and the general public is the subject of intense promotion as a means of avoiding INCINERATION of sewage SLUDGE.

Biosphere. The region of the earth and its atmosphere in which life exists. It is an envelope extending from up to 6000 metres above to 10 000 metres below

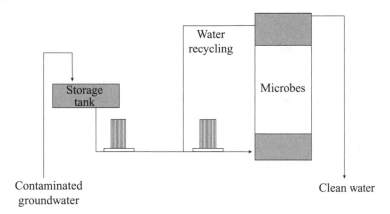

Figure 11 A flow diagram of the process of bioremediation of chlorophenol contaminated groundwater by a fluidized-bed biofilm reactor.
Source: *IEEE Bulletin*, 1/96. (Reproduced by permission of the IEEE. © 1996 IEEE)

sea level and embraces Alpine plant life and the ocean deeps. The special conditions which exist in the biosphere to support life are a supply of water; a supply of usable energy; the existence of interfaces, i.e. areas where the liquid, solid and gaseous states meet; the presence of nitrogen, phosphorus, potassium and other essential nutrients and trace elements; a suitable temperature range; and a supply of air (although there are anaerobic forms of life).

Biospheric cycles. To maintain the biosphere and thus life, essential materials must be recycled so that after use they are returned in a reusable form. In order to sustain this recycling a supply of energy is required – solar ENERGY.

The principal elements in the biosphere are nitrogen, OXYGEN, HYDROGEN and CARBON. As these elements constitute 99 per cent of all living things, including plants, they are obviously of prime importance. Other elements are also important, e.g. phosphorus, potassium, sulphur and calcium, and all have their own renewal cycles, so that they become available for reuse.

The major biospheric cycles are the CARBON CYCLE, the OXYGEN CYCLE, the NITROGEN CYCLE, and the water or HYDROLOGICAL CYCLE. They are all interlinked and operate simultaneously and maintain all life on earth.

Biotechnology. The process of using living organisms or components derived from living organisms, such as ENZYMES and biopolymers, to produce new materials or degrade existing substances at an enhanced rate. For instance, biopolymers can be produced by certain species of bacteria and these can be spun into fibres for conversion into textiles. In waste treatment, specific species of micro-organisms have been found to be effective against recalcitrant chemicals, e.g. white rot fungi can be used to treat pentachlorophenol in land remediation.

Biotechnology example. Engineering companies use large quantities of cutting fluid to cool and lubricate components while they are machined. In time, the fluid becomes contaminated with bacteria which makes it 'go off' and renders it useless. An estimated half a million tonnes of used cutting fluids and oily wastes are discharged to LANDFILL, sewers and incinerators each year. Industry pays to replace these fluids and to dispose of the waste.

A biotechnology solution to this problem is to pass used cutting fluids through an ANAEROBIC treatment tank (bioreactor) to digest the oil. The residue is then passed through an ultrafiltration system to extract any remaining contaminants. This results in purified water, which is then re-used in the process. (◊ BIOREMEDIATION)

Source: Department of Trade and Industry, *BIO-WISE* Information Leaflet, Issue 1, January 1999.

Bisphenol A. An intermediate compound in specialist plastics, coatings and adhesives, e.g. tin can linings. It is suspected of causing damage to male embryos, and producing low sperm counts as it is able to mimic naturally occurring hormones.

It is used to manufacture coatings for the inside of some cans, and to make polycarbonate food contact articles.

A study has conclusively demonstrated that there is no leaching of bisphenol A from polycarbonate baby bottles during use.

Source: *Food Safety Information Bulletin*, MAFF, no. 89, October 1997.

Bituminous coal. The most common form of British coal. It produces or contains a relatively high proportion of volatile HYDROCARBONS (over 13 per cent) and, burned in the traditional domestic grate, it will give rise to SMOKE. ◊ COAL

Black List. (◊ List I substances).

Blanket. 'Fertile' material (usually depleted URANIUM) placed round a fast reactor core to capture neutrons and create more fissile material (usually plutonium).

Blasting. A means of fragmentation to uncover coal, rock, etc. This is now strictly controlled. Figure 12 shows Northern Coal's 1993 blast monitoring results, using a fully computerized system to monitor properties located on the boundaries of their site operations.

A total of 3220 recordings have been made, not one of which has exceeded the authorized limits for their operations.

The following table lists commonly-accepted damage figures from blast overpressure. The duration of the pressure wave, however, has not been taken into account in the data.

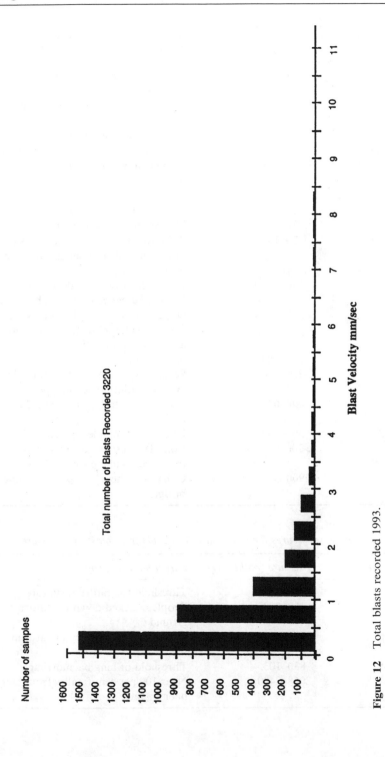

Figure 12 Total blasts recorded 1993.

Commonly-accepted damage figures from blast overpressure

Pressure (psi)	Pressure (millibar)	Damage
0.02	1.4	Annoying noise, if of low frequency
0.10	7	Breakage of small windows under strain
0.15	10	Typical pressure for glass failure
0.30	21	Damage to some ceilings
0.40	28	Minor structural damage
0.5–1.0	34–70	Large and small windows shattered, damage to window frames.
0.75	52	Minor damage to houses, 20–50 per cent of tiles displaced.
0.9	63	Roof damage to oil storage tanks
1.0	70	Houses made uninhabitable
1.0–2.0	70–140	Asbestos cladding shattered, fastenings of corrugated steel panels fail, tiled roof lifted and displaced
2.0	140	Partial collapse of walls and roofs of houses, 30 per cent of trees blown down
3.0	210	90 per cent of trees blown down, steel-framed buildings distorted and pulled away from foundations
3.0–4.0	210–280	Rupture of oil tanks
4.0–5.0	280–350	Severe displacement of motor vehicles
5.0	350	Wooden utility poles snapped
7.0–9.0	490–630	Collapse of steel girder-framed buildings
8.0–10	560–700	Brick walls completely demolished
>10	>700	Complete destruction of all un-reinforced buildings
70	4900	Collapse of heavy masonry or concrete bridges

Commonly-accepted figures for direct harm to people from blast overpressure

Pressure (psi)	Pressure (millibar)	Direct harm to people
2	138	Threshold for eardrum rupture
1.5–2.9	103–200	People knocked down or thrown to the ground
10–15	690–1035	90 per cent probability of eardrum rupture
12–15	830–1035	Threshold of lung haemorrhage
30–34	2068–2400	Near 100 per cent fatality from lung haemorrhage

HSE uses the following levels of overpressure for providing land use planning advice around hazardous installations. These are based on estimates of fatalities within occupied buildings.

Pressure (psi)	Pressure (millibar)	Direct harm to people
8.6	600	50 per cent fatalities for a normal population
2	140	Threshold of fatality (1–5 per cent) for a normal population
1	70	Threshold of fatality (1–5 per cent) for a vulnerable population

Source: Buncefield Major Incident Investigation Initial Report to the Health & Safety Commission and the Environment Agency of the investigation into the explosions and fires at the Buncefield oil storage and transfer depot, Hemel Hempstead, on 11 December 2005.

Source: RJB Mining, Gosforth, Newcastle upon Tyne.

Bloom. The sudden upsurge in numbers of ALGAE in a body of water, usually due to an excess of NUTRIENTS.

Blowdown[1]. Rapid depressurization, e.g. of a PWR primary coolant circuit.

Blowdown[2]. HYDROCARBONS purged during refinery start-up or shut-down.

Blowdown[3]. In boiler operation, a process for reducing the concentration of dissolved solids in the boiler water, or for removal of sludge from a boiler.

Blue-green algae. ALGAE which, in addition to chlorophyll, possess red and blue pigments. Some are capable of fixing nitrogen directly and so make it available for plants and animals; others convert nitrates to nitrites. They are, therefore, involved in the NITROGEN CYCLE. Many are capable of producing toxins which have caused death to animals, birds, and fish in many countries. These toxins have been linked to human illness through skin contact and ingestion.

BMWP Score. The Biological Monitoring Working Party Score. This is arrived at by first obtaining a sample of the macroinvertebrates (creatures longer than 1 mm) in a river. The macroinvertebrates are then classified into different taxa. Each taxon has a score between 1 and 10, with the species more sensitive to pollution having a higher score. The BMWP score is the sum of the scores for each taxa found. A BMWP score greater than 100 generally indicates good water quality.

Body burden. The amount of radioactive or toxic material in a human or animal body at any time.

Boreholes. ↻ WELLS AND BOREHOLES.

Bottle. GLASS or PLASTIC container, which can be returnable or non-returnable. The ratio of non-returnable to returnable bottles in the UK is ca. 25:1, whereas in Norway, all glass bottles are returnable.

Bottles come in different colours, shapes and sizes. Conservationists argue strongly against the throw-away philosophy, with some justification on ENERGY consumed. A strong case can be made for bottle standardization and the enforced use of returnables so that both energy and materials are conserved and waste minimized as non-returnable bottles account for around 8 per cent by weight of UK municipal wastes.

Figure 13 gives the energy required (kWh$_e$) to make a 0.3 litre refillable beer bottle and a non-refillable one but the former can last for many trips, requiring only the energy for transport, washing and refilling. Thus, as shown in Figure 14 the refillable bottle pays for itself in energy terms after two trips. The glass industry is promoting BOTTLE BANKS to recover bottles for remelt-

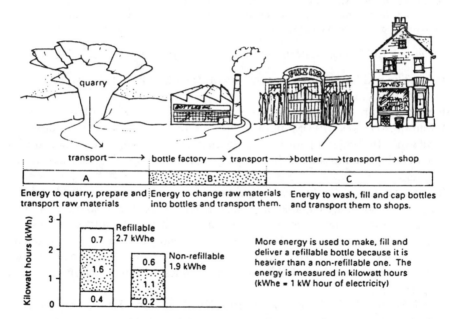

Figure 13 Comparison of energy required to manufacture half pint refillable and non-refillable bottles.

1st trip	0.4	1.6	0.7			Total energy used 2.7 kWh. Energy per trip 2.7 kWh.
2nd trip	0.4	1.6	0.7	0.7		Total energy used 3.4 kWh. Energy per trip 1.7 kWh.
3rd trip	0.4	1.6	0.7	0.7	0.7	Total energy used 4.1 kWh. Energy per trip 1.37 kWh.

Figure 14 Energy saved by the use of refillable bottles.

ing. However, the real pay-off in energy terms comes from the use of returnable refillable bottles.

A 1994 report for the US Department of Commerce concluded:

> Recycling of glass containers saves some energy but not a significant quantity compared to reuse. The energy saved is about 13% of the energy required to make glass containers from virgin raw materials. This estimate includes energy required for the entire product life cycle, starting with raw materials in the ground and ending with either final waste disposition in a landfill or recycled material collection, processing, and return to the primary manufacturing process. The actual savings depend on local factors, including population density; locations of landfills, recovery facilities and glass plants; and process efficiencies at the specific facilities available. The savings increase if wastes must be transported long distances to a landfill or if the containers are made in an inefficient furnace. They decrease if there is no local Municipal Recycling Facility (MRP) or glass plant, or if material losses in the recycling loop are high.

The options for disposal of used glass containers can be compared on the basis of the important decision factors, including energy saved, LANDFILL space required, and emissions. The table on page 79 gives a qualitative comparison of the options for the recycling or disposal of glass containers. In order of energy saved, the options for disposition of glass containers are reuse, recycling to the same product, recycling to a lower value product, and landfill.

The options with minimum energy use generally have the lowest other impacts as well. No real trade-off exists. The energy savings are small, and the balance can be altered by local or regional conditions. Although the analysis was performed for US conditions, they are not that dissimilar to conditions in the UK. (◊ ENERGY ANALYSIS)

A useful comparison of the energy requirements (kWh) for producing similar beverage containers has been performed by Scott (Reference shown below) The plastic container is a clear winner (0.11 kWh) and

Energy requirements for similar beverage containers

Container	Energy usage per container/kWh
Aluminium can	3.00
Returnable soft drink bottle	2.40
Returnable glass beer bottle	2.00
Steel can	0.70
Paper milk carton	0.18
Plastic beverage container	0.11

Source: Scott, G. *Polymers and the Environment*, p 21, Royal Society of Chemistry, London, 1999.

ALUMINIUM's 3.00 kWh should render it unacceptable. However, industry lobbying and recycling campaigns will keep it in business. Note that aluminium is 1 per cent household waste by weight but brings in 25 per cent of recycling income to local authorities by virtue of its ca £800/te value. The value of steel cans is perhaps £50/te, depending on the market.

Source: Argonne National Laboratory, *Energy Implications of Glass Container Recycling* report DE94000288, February 1994.

Bottle bank. A scheme promoted by the GLASS industry for the collection of glass bottles for CULLET. The percentage of domestic glass containers returned for remelting in the UK is ca. 34.0 per cent (2006), cf. rates of up to 60 per cent of container glass in Holland. The use of cullet effects an ENERGY saving compared with that required for melting the basic raw materials. (No account is taken of excess fuel consumption by the public in visiting bottle banks.)

Bottle Bill. A legislative (US) measure to encourage container recycling. 'Bottle Bills' are now in force in 12 states (including Massachusetts where a multi-million dollar industry-financed campaign was waged against such legislation). One such attempt is the Vermont Container Law 1973: it should be noted that no container is banned, but a deposit refund (in Vermont's case; other states require a plain refund) must be paid by the retail outlet on presentation of the empty container. By this means, a high degree of control over packaging entering the waste stream can be achieved and refillable bottles' use encouraged as it is cheaper for the bottler to refill empty bottles. The Vermont Deposit–Redemption Cycle is shown in Figure 15.

A 5 cent minimum deposit must be charged on a container sold at a retail **(3)**. It can be initiated at **(1)**, **(2)** or **(3)**. Glass containers destined for refilling usually have the deposit initiated at **(1)**. Most other deposits are initiated at **(2)**. Containers are redeemed for the refund value at any store or redemption

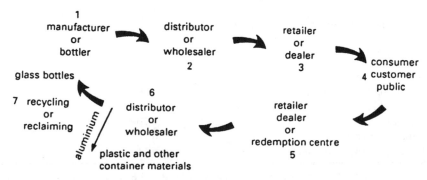

Figure 15   The deposit–redemption cycle – Vermont Container Law.

Comparison of options for disposal of glass containers.

Alternative	Energy impact	Environmental impact	Material sent to landfill	Comments
Reuse	Saves most of production energy	Avoids production emissions; possible impacts from water treatment	None	Bottles now made from ca. 35% (UK) old bottles; colour separation required
Recycle to containers	Small reduction in production energy	Small reduction in emissions[a]	Cullet processing losses	Inefficient furnaces used
Recycle to fiberglass	Small reduction in production energy	Small reduction in emissions[a]	Cullet processing losses	Can use clean, mixed cullet
Recycle to reflective beads	Small reduction in production energy	Small reduction in emissions[a]	Cullet processing losses	Can use clean, mixed cullet
Recycle to glassphalt	Saves no energy	Full production emissions	Cullet processing losses	Economic cost, no return
Landfill	Saves no energy	Full production emissions	Maximum	

[a] Dust and SO_2 reduced, but NO_x increased.

centres **(5)** which sells that kind, size and brand. Retailers and redemption centres **(5)** are reimbursed the refund value by the Distributor **(6)** and also are paid a handling fee (20 per cent of the refund value, or a minimum of 2 cents). Distributors **(6)** normally pick up containers from **(5)** on regular schedules or upon request. Glass containers are typically returned to bottlers for refilling. Other container materials are sold for recycling **(7)** along with the packaging materials such as corrugated cartons, etc.

Vermont (and other US states) have made great environmental gains through the use of Bottle Bills. California has followed suit with a novel twist in that redemption centres will repay the deposit, which can be increased, until satisfactory return rates are achieved. The California 'Bottle Bill' of 10 October 1987:

- Requires the cost of beverage containers to the consumer to include an initial charge of one cent per container.
- Enables consumers to receive a refund, a bonus and the scrap value of the container at redemption centres, not at retail establishments that sell beverage containers.
- Enables recycling operators to have a more active role in container redemption than in other states.
- Establishes a state agency to administer the Act, including conducting audits of recovery rates. If these fall below 65 per cent, the deposit will be increased.

Enlightened European practice regarding containers is, perhaps, typified by that of Denmark where their Act No. 297 of 8 June 1978, Reuse of Paper and Drink Packagings and the Reduction of Waste, has made significant contributions to resource conservation. The Danish National Agency for Environmental Protection has stated that without the present reuse of bottles and the ban on cans, the amounts of glass and metal waste would be much larger. They estimate that the percentage of glass in Danish municipal solid waste would rise by 46 per cent, were it not for these measures.

Metal beverage cans have been phased out from a level of 43.5 million cans in 1976 to zero in 1982. There are no mandatory arrangements for wines and spirits (beer and soft drinks are of course covered) but there is an excise duty of 0.97 DKr on all new wine and spirit bottles and a 50 per cent increase in the excise duty was followed by a more than 50 per cent increase in the collection and cleaning of wine and spirit bottles, which in 1982 reached a record high of 77 million bottles thus recycled. Container deposit legislation has the power to ensure much more effective recycling than BOTTLE BANKS. However do not expect the UK to implement this. Supermarkets want to sell, not store bottles. (◊ RECYCLING)

Bovine spongiform encephalopathy. ◊ SPONGIFORM ENCEPHALOPATHY.

Brackish water. Water (non-potable) containing between 100 and 10 000 ppm of total dissolved solids. (◊ BRINE, DESALINATION)

Breakpoint chlorination. The chlorination (disinfection) of drinking water until the chlorine demand of any organic impurities present is satisfied. Any further chlorine addition is then available as free chlorine.

Breeder reactor. ◊ NUCLEAR REACTOR DESIGNS.

Brine Water containing more than 50 000 ppm of salt (NaCl) which can yield salt after evaporation. Also the reject concentrate from DESALINATION plants. (◊ CHLORALKALI PROCESS)

BS4142. BS4142:1997 provides a technical means of assessing whether or not 'complaints are likely' where a noise is concerned. The result of an assessment carried out to BS4142 would normally be relevant to the deliberations of any court considering whether or not a nuisance exists.

Bring Recycling Site. This term refers to locations where the general public can bring waste items for subsequent recycling. The sites usually have permanent facilities where the user can deposit recyclable waste. Examples of bring sites include bottle banks, paper banks and other 'recycling banks'. Alternative terminology includes 'recycling centres', 'drop-off sites' and 'central-point recycling sites'.

BSE. ◊ SPONGIFORM ENCEPHALOPATHY.

Btu (British Thermal Unit). A unit of energy approximately equal to 1055 joules or 0.252 kilocalories. It is the amount of heat required to raise the temperature of 1 lb of water by 1 Fahrenheit Degree (60–61 °F). 3413 Btu = 1 kilowatt-hour, the common unit in which electricity is measured. Although Btu is superseded by the joule, it remains in very common UK and US usage.

Note: 100 000 Btu = 1 therm. – Still used in domestic boiler ratings.

Bubble. A concept of US origin which involves drawing a notional 'bubble' or dome-shaped boundary around a plant area, or even a state, and putting an upper limit on the total amount of a pollutant allowed to pass into the 'bubble'. Used in setting a limit or standard on total emissions. Hence an increase in pollution in one plant can be traded off against a decrease from one or more plants.

Buffer capacity. The capacity of certain solutions to oppose a change in properties, especially when ACID or ALKALI is added.

Buncefield. The location of the 'worst' fuel oil explosion in the UK, in Hemel Hempstead, on 11 December 2005. Briefly, Tank 912 was probably overfilled with unleaded petrol, resulting in a vapour cloud formation and subsequent ignition. Destruction of the tank farm followed, but fortunately there was no loss of life. Para 71 of the Initial Report to the Health and Safety Commission and the Environment Agency by the Buncefield Major Incident Investigations Board in 2006 stated:

In summary, the Investigation has revealed a number of matters concerning the design and operation of sites such as Buncefield where improvements to maintaining primary containment must be considered by the industry, working closely with the Competent Authority. These matters include:

The electric monitoring of tanks and pipework, and associated alarms that warn of abnormal conditions;
The detection of flammable vapours in the immediate vicinity of tanks and pipework;
The response to the detection of abnormal conditions, such as automatic closure of tank inlet valves and incoming pipeline valves;
The extent to which the exterior construction of tanks (e.g. tank top design) inhibits, or contributes to, flammable vapour formation;
The siting and/or means of protection of emergency response facilities; and
The recording of monitoring, detection and alarm systems and their availability (e.g. off-site) for periodic review of the effectiveness of the control measures by the operator and the Competent Authority, as well as in root cause analysis should there be an incident.

Source: Buncefield Major Incident Investigation Initial Report to the Health & Safety Commission and the Environment Agency of the investigation into the explosions and fires at the Buncefield oil storage and transfer depot, Hemel Hempstead, on 11 December 2005.

Buncefield by numbers:

33 million litres – of firewater and polluted surface water stored and later cleaned.
2.37 million – the number of tonnes of petrol, diesel and aviation fuel the depot used to handle each year.
2000 – the number of homes evacuated as the fires burned.
18 000 – the number of hours of Environment Agency effort during the first year after the explosion to protect the environment.
750 – the number of samples of water and soil taken to monitor pollution impact.
244 – the number of people treated at A&E, most of them emergency services personnel.

Source: Environment Agency, *Your Environment*, Issue 14, Feb 2007.

Buoyancy. The upward FORCE that acts on a body that is totally immersed in a fluid. It is equal to the weight of the fluid displaced by the body, so if the DENSITY of the immersed body is less than that of the fluid, there will be a net upward force on the body, e.g. hot gases discharged from a chimney into cooler air will rise due to buoyancy until they obtain the temperature of the surroundings. The effect of buoyancy of hot flue gases can be to give an EFFECTIVE CHIMNEY HEIGHT which can be two or three times greater than the actual chimney height, thus greatly aiding dispersion of any pollutants or ODOURS.

Bureaucracy. Leaving scatological definitions aside, it is relevant to give two examples of where it is perhaps excessive in the UK:

– the ruddy duck: It has a brown body and a blue bill with a healthy sex life, leading to miscegenation with the Spanish white headed duck. Such mating is a breach of Article 8(h) of the convention on Biological Diversity and Article II, 2 (b) of the Bern Convention on the Conservation of European Wildlife and Natural Habitats, not to mention the EU Birds Directive and the EU Habitats Directive. In order to implement the international agreements, a Working Group concluded that the Wildlife and Countryside Act 1981 should be modified to allow for the eradication of the ruddy duck. After a trial they then recommended that shooting ruddy ducks was the most effective control measure.

Source: *Times*, Letters, 21/06/06.

– this is from Defra the body which permits UK recovered waste to be recycled in China where environmental and health and safety standards are not at UK levels, yet will not permit incinerator bottom ash when used as a road base to be classified as recycling. QED?

Burn out[1]. A measure of the EFFICIENCY of INCINERATION which relies on determinations of fixed CARBON and putrescible matter in the residue after COMBUSTION.

Burn out[2]. The failure of a nuclear reactor (or industrial boiler) heat-exchange surface which can result in the loss of heat-exchange fluids and may, in very extreme circumstances, result in the release of radioactive products. It is equivalent to MELT-DOWN in US terminology.

Business Environmental Reporting. An example of responsible Environmental and Human Resource Business reporting:

Total water use by production volume (hl/hl)
2006   4.2
2005   4.5
Total energy use by production volume (kWh/hl)
2006   32.7
2005   37.6
Total CO_2 emissions by production volume (kg CO_2/hl)
2006   8.5
2005   9.5

Serious injury rate per 1000 employees
2006   22.9
2005   37.2
% of waste recycled
2006   94%
2005   93%

* All numbers relate to wholly-owned companies
Source: Scottish & Newcastle plc, Annual Report & Accounts 2006.

'Buy-back' programmes. Programmes to purchase recyclable materials from the public. (◊ RECYCLING)

By-product. An additional product which can be produced as an ancillary to the principal product, e.g. animal by-products from the offal remaining from meat processing or COMPOST from the fines of a waste derived fuel plant.

Byssinosis. ◊ FIBROSIS.

C

Caddie. The Government's preferred term for what used to be known as a 'slop bucket'. This item is to be used for the storage of wet kitchen waste (e.g. leftover chicken innards, fish heads, etc.). This is to be emptied weekly for recycling, presumably by ANAEROBIC DIGESTION.

Cadmium (Cd). A soft silvery HEAVY METAL, atomic weight 112.4. It is used in semiconductors, control rods for nuclear reactors, electroplating bases, PVC manufacture, and batteries. World production is around 18 000 tonnes per year produced in conjunction with zinc smelting.

Cadmium serves no biological function, is toxic to almost all systems, and is absorbed into the human organism without regard to the amount stored. Very small doses can cause severe vomiting, diarrhoea and colitis; pneumonia can develop as a consequence. Poisoning has developed from sources such as home-made punch stored in a cadmium-plated bowl. Changes in the heartbeat rate of persons in atmospheres containing as little as 0.002 micrograms per cubic metre (μg/m^3) have been reported (maximum allowable concentration is 100 μg/m^3). Hypertension is a common complaint of those who have ingested the metal at doses well below those regarded as toxic. Long exposure in lead and zinc smelting areas in Japan led to the so-called Ouch-ouch (Itai-itai) disease where the sufferers had severe joint pains and eventual immobility as a result of skeletal collapse. Cadmium leads to bone porosity and inhibition of bone-repair mechanisms.

Figure 16 shows how close co-operation by industry can significantly lower Cd levels in sewage SLUDGE. This enables it to be spread on land (not always the case in the UK due to high Cd levels). ($\lozenge$ POLYVINYL CHLORIDE).

g Cd/tonne Dry Matter

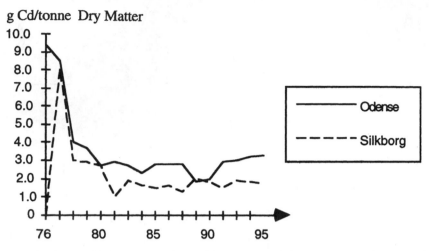

Figure 16   *Source*: J. A. Hansen, 'A new biowaste agenda', *ISWA Times*, issue number 2, 1995. (Reproduced by permission of ISWA)

However, caution on land use is necessary as abnormally high levels of the toxic metal can accumulate in the kidneys and livers of sheep grazing on pasture fertilized for years with sewage sludge.

Mike Wilkinson of the consultancy Chalcombe Agricultural Resources in Lincoln, working with Julian Hill of Writtle College in Chelmsford, Essex, measured levels of heavy metals accumulated by sheep when they grazed on one field that was treated with sewage sludge and another that was untreated. After 150 days, average levels of cadmium in sheep livers were 1.24 milligrams per kilogram (mg/kg) of dried tissue for the animals that had grazed on heavily treated pasture, eight times higher than the level found in sheep grazing on untreated land. In sheep kidneys, the levels in animals that had grazed on the treated land averaged 2.57 mg/kg of dried tissue, six times as high as in the sheep that had grazed on clean pasture.

Source: M. Wilkinson and J. Hill, 'Lamb's liver with cadmium garnish', *New Scientist*, 22 March 1997, p. 4.

Caesium-137. Caesium-137 is a radioactive isotope which is formed mainly by nuclear fission. It has a half-life of 30.23 years, and decays by pure beta decay to a metastable nulcear isomer of barium-137m which has a half-life of 2.55 minutes and is responsible for all of the gamma ray emission.

The photon energy of Ba-137m is 662 keV. These photons can be used in food irradiation, or in radiotherapy of cancer. Cs-137 is not widely used for industrial radiography as other isotopes offer higher gamma activities per given volume. It can be found in some moisture and density gauges, flow meters, and other sensor equipment.

The biological behavior of Cs-137 is similar to potassium. After entering the organism, all caesium gets more or less uniformly distributed through the body, with higher concentration in muscle tissue and lower in bone. The biological half-life of caesium isotopes is long (30.17 years for Cs-137).

Small amounts of Cs-134 and Cs-137 were released into the environment during nuclear weapon testing and nuclear accidents, most notably the Chernobyl disaster As of 2005, Cs-137 is the principal source of radiation in the exclusion zone around the Chernobyl power plant. Together with caesium-134, iodine-131, and strontium-90, it was among the most important isotopes regarding health impacts after the reactor explosion.

Improper handling of Cs-137 sources can lead to release of the isotope and radiation contamination and injuries. Perhaps the best known case is the Goiania accident, when a radiation therapy machine from an abandoned clinic in Goiania, Brazil, was scavenged and the glowing caesium salt sold to curious buyers. Metallic caesium sources can be also accidentally mixed with scrap metal, resulting in production of contaminated steel; a notable example is the case from 1998, when recycler Acerinox in Cadiz, Spain, accidentally melted a source. Many abandoned sources are scattered over the area of the former Soviet Union.

Source: en.wikipedia.org/wiki/Caesium-137.
Accessed 27 February 2007.

Calcium (Ca). Metal element found principally as CALCIUM CARBONATE.

Calcium carbonate. Limestone (also chalk), $CaCO_3$. Used in the production of CALCIUM OXIDE and CEMENT. $CaCO_3$ also causes temporary hardness in water if there is carbon dioxide in solution (as there invariably is in public supplies).

Calcium deficiency. DDT and other CHLORINATED HYDROCARBONS such as polychlorinated biphenyls severely interfere with a bird's ability to metabolize calcium. The result is the production of egg shells which are so thin that they are crushed by the weight of the nesting birds.

Calcium oxide. Also known as quicklime. Made by calcining (roasting) CALCIUM CARBONATE limestone $CaCO_3$ at 900 °C; this drives off carbon dioxide and produces calcium oxide:

$$CaCO_3 \rightarrow CaCO + CO_2$$

The addition of water produces SLAKED LIME:

$$CaO + H_2O \rightarrow Ca(OH)_2$$

The addition of water and sand and CO_2 from the atmosphere produces mortar, i.e. reversion or hardening:

$$Ca(OH)_2 + CO_2 \rightarrow CaCO_3 + H_2O$$

Calibration. All the operations for the purpose of determining the values of errors of a measuring instrument. Calibration of an instrument results in a correction factor or a series of correction factors that can subsequently be applied to readings given by the instrument.

Calorie (cal). Physicists define the calorie as the heat required to raise the temperature of 1 gram of water by 1 °C. This is a minute quantity of heat in practical terms. The term used on diet sheets refers to the kilocalorie (Cal), which is 1000 times larger.

Calorific value CV, gross. The number of heat units measured as being liberated when a unit mass of fuel is burned in oxygen saturated with water vapour under standardized conditions. (The remainder being gaseous oxygen, CARBON DIOXIDE, sulphur dioxide, and nitrogen, water and ash.) Used as a reference for fuel evaluation purposes.

Calorific value CV, net. The gross calorific value less the heat of evaporation of the water originally contained in the fuel *and* that formed during its combustion. It is important to clarify the basis on which CV is quoted.

Camelford. Area of N. Cornwall, UK where a 20 tonne load of ALUMINIUM SULPHATE solution was accidentally tipped into public drinking water supplies on 6 July 1988.

Various ailments were attributed to this incident and out-of-court settlements have been accepted by most claimants.

In 2002, a committee investigated claims of joint problem, brain damage & memory loss as a result of the incident but no conclusive links were found. The committee urged the government to study the issue better.

CAMP. An acronym for Continuous Air Monitoring Program (USA).

CAMS (Catchment Management Abstraction Strategies). A procedure devised by the Environment Agency for managing water resources effectively. The objectives for CAMS are: to make information publicly available on water resources availability and licensing within a catchment; to provide a consistent and structured approach to local water resources management, recognizing both abstractors' reasonable needs for water and environmental needs; to provide the opportunity for greater public involvement in the process of managing abstraction at a catchment level; to provide a framework for managing time-limited licences; and to facilitate licence trading.

The flow diagram in Figure 17 identifies the key stages of developing a CAMS. The key elements are: Resource assessment and resource availability status, Sustainability appraisal, Consultation, CAMS documents, and Implementation of the strategy and evaluation of the process.

Can[1]. The metal container in which nuclear fuel is sealed to prevent contact with the coolant or escape of radioactive fission products, etc. It may be made of MAGNOX, Zircaloy or stainless steel which is stripped off in nuclear fuel reprocessing to retrieve the spent fuel.

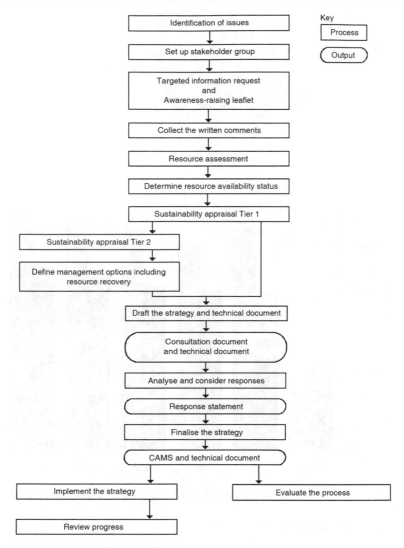

Figure 17 The CAMS Process (till 31 March 2008).
Source: Environment Agency, Managing Water Abstraction – The Catchment Abstraction Management Strategy Process, Environment Agency, Bristol, 2002.

Can². Tin plate, sheet aluminium or bimetallic (tin plate and aluminium) container for beverages and foodstuffs. (◊ RECYCLING)
    Figure 18 shows the can consumption in the UK for (a) 1994 and (b) 1980–1994.
Can banks. Containers used by public to deposit beverage and other cans, for later collection and recycling.

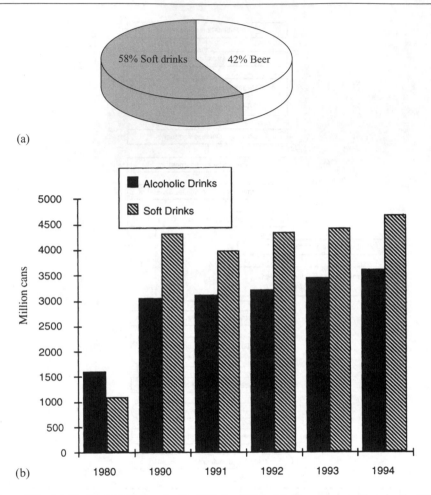

(a)

(b)

Figure 18 (a) Can consumption in the UK for 1994. (b) Can consumption in the UK for 1980–1994.
Source: Canmaker Report 1995.

According to the Steel Can Recycling Information Bureau, every tonne of steel cans recycled saves: 1.5 tonnes of iron ore; 0.5 tonnes of coke; 70 per cent of the ENERGY needed in production; and 40 per cent of the water needed in production.

It should be noted that recycling ALUMINIUM cans effects larger energy savings, but there is a much larger initial energy requirement in its manufacture.

Source: Steel Can Recycling Information Bureau.

Cancer. A disordered growth of cells which can invade and destroy body organs and thereby cause incapacity or death. It can be initiated by CARCINOGENS.

The many causes of cancer are subjects of major medical research. ($\lozenge$ ASBESTOS; HEAVY METALS; IONIZING RADIATION)

due to air pollution. $\lozenge$ CARCINOGEN.

due to asbestos. $\lozenge$ ASBESTOS.

due to chloroprene. $\lozenge$ CHLOROPRENE.

due to chrome wastes. $\lozenge$ CHROME WASTES.

due to DES. $\lozenge$ DES (DIETHYLSTILBOESTROL); FEED-CONVERSION RATIO.

due to hair dyes. $\lozenge$ MUTAGEN.

due to organophosphates. $\lozenge$ ORGANOPHOSPHORUS COMPOUNDS.

due to pesticides. $\lozenge$ ALDRIN.

due to PVC. $\lozenge$ PVC (POLYVINYL CHLORIDE).

Candle. Formerly used in air pollution measurement or studies of sulphur dioxide monitoring. It consisted of a porcelain cylinder covered with linen impregnated with lead dioxide. The amount of lead sulphate formed over, say, one month was measured and average concentrations of SULPHUR DIOXIDE determined.

Capacity factor. Capacity factor is a good indicator of the service supplied to the power grid by a power generating unit or group of units if operating on base load, i.e.

$$\text{Capacity factor} = \frac{\text{Output for period}}{(\text{Rated capacity})(\text{Period duration})}$$

($\lozenge$ AVAILABILITY FACTOR)

Carbohydrates. Group of ORGANIC compounds composed of CARBON, HYDROGEN and OXYGEN. Contains the sugars, i.e. monosaccharides and disaccharides and the polysaccharides (starch and CELLULOSE). General formula: $C_x (H_2O)_y$.

Carbon. A non-metallic element, ATOMIC NUMBER 6, relative atomic mass 12.011, symbol C. Carbon compounds make up all living matter and, consequently, fuels such as coal and petroleum.

Carbonation. The process by which atmospheric CARBON DIOXIDE penetrates the pores of concrete in humid conditions and depassivates (neutralizes) the natural ALKALINITY of concrete by reacting with calcium hydroxide to give calcium carbonate. As the pH-falls below 9, moisture can start to corrode the reinforcement steel. As this corrodes it expands (up to seven fold) and causes cracking and eventual destruction of the concrete. When SUSTAINABILITY is desired, measures need to be built in to ensure longevity of concrete structures.

Carbon, activated. $\lozenge$ ACTIVATED CARBON.

Carbon black. Finely divided carbon, usually produced from the COMBUSTION of gaseous and liquid HYDROCARBONS with restricted air supply. Since it is not easy to remove the last traces of the product from gases leaving the stack, the production of carbon black is often a source of intense air pollution.

Carbon–Capture and Storage (CCS). The claimed saviour of fossil fuel-fired power stations and large primary energy users as their CO_2 emissions will be separated from their stack gases using, possibly, an appropriate solvent followed by liberation and compression for appropriate transportation and storage in probably geological structures such depleted oil/gas fields. Other CO_2 separation methods embrace membrane filtration and adsorption/absorption. This is not cost free, but in common with all global warming mitigation technologies, a step change in costs is inevitable if the stop/go vagaries of wind power are to be avoided and reliable, fundamentally-sound power stations allowed to continue their humdrum existence.

Other proposals include:

(a) Carbon dioxide sequestration where CO_2 is separated and stored in underground reservoirs.
(b) A more readily realizable option is the retrofitting of supercritical boilers where steam is generated at supercritical temperature and pressures which leads to perhaps a 20 per cent efficiency improvement and consequently less CO_2. ($\Diamond$ CRITICAL POINT)

Carbon–Capture Power Station. The energy firm E.ON is planning to build a carbon–capture power station in Lincolnshire. The Project will go ahead if it gets government support. The 450 MW integrated gasification combined cycle (IGCC) coal-fired power station would convert coal into gases such as carbon dioxide and hydrogen. The CO_2 would be stripped out and stored, while the hydrogen would be used to generate electricity in a highly-efficient power station. The captured CO_2 would be converted into a liquid, pumped through pre-existing gas pipelines, and stored in depleted gas fields under the North Sea.

Carbon cycle. The atmosphere is a reservoir of gaseous CARBON DIOXIDE but to be of use to life this must be converted into suitable organic compounds, i.e. fixed, as in the production of plant stems. This is done by the process of PHOTOSYN-THESIS. The productivity of an area of vegetation is measured by the rate of carbon fixation. For example, tropical rain forest growth will fix 1–2 kilograms per square metre per year, whereas barren areas such as tundra will fix less than 1 per cent of this amount. Herbivores eat the plant material and carnivores eat the herbivores. The CARBON fixed by photosynthesis is eventually returned to the atmosphere as plants and animals die, and the dead organic matter is consumed by decomposer organisms. Thus the global carbon cycle on land (there is a complementary oceanic one as well) is as pictured in Figure 19.

In Figure 19, the atmosphere contains 700 thousand million (700×10^9) tonnes of carbon as carbon dioxide. The fixed amount of carbon in the biomass (plants and animals except the decomposers) is 550×10^9 tonnes and the dead organic matter reservoir is 1200×10^9 tonnes. The land-based system is in a steady state, i.e. the inputs and outputs balance. The input has been estimated as 110×10^9 tonnes carbon per year from the ATMOSPHERE as carbon dioxide, of which 60×10^9 tonnes per year are fixed by photosynthesis and 50×10^9 tonnes per year are returned to the atmosphere as carbon dioxide from respira-

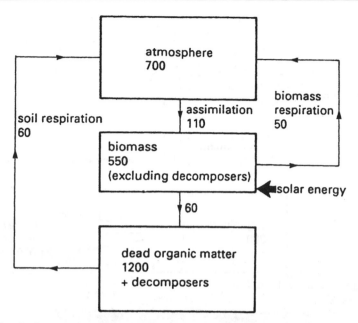

Figure 19 Carbon cycle on land. The quantities in Figures 19–21 are in thousand million (10^9) tonnes. The figures in the boxes show the amount of carbon either fixed or available for fixation. The figures with the arrows show the flow rates of carbon which are assimilated or respired per year.

tion processes after being used as an energy source to keep the plants alive. The dead organic matter reservoir has an input of 60×10^9 tonnes per year as FIXED CARBON which is decomposed to 60×10^9 tonnes per year as carbon dioxide. The system is in balance, powered by solar energy. In the process, oxygen is also produced so that the carbon and OXYGEN CYCLES interlink.

Complementing the CARBON CYCLE on land is the equivalent at sea. This differs from the land cycle in that carbon dioxide is soluble in sea water and is thus available for the phytoplankton. These are single-celled plants that take up carbon dioxide and, by photosynthesis, use it to produce carbohydrates and oxygen which also dissolves in the water, so the oceanic system is virtually closed. Further up the FOOD CHAIN, zooplankton and fish consume the CARBON fixed by the phytoplankton and in turn die and decompose, as do the uneaten phytoplankton, so that the carbon dioxide is replaced. Less than 1×10^9 tonnes carbon per year are deposited to sediments. Thus the oceanic system is also in a quasi-steady-state as shown in Figure 20. The inputs to the organic residues reservoir are 20×10^9 tonnes per year from fish and zooplankton and 20×10^9 tonnes per year from phytoplankton. This is balanced by an output of 35×10^9 tonnes per year as carbon dioxide to the surface layers and 5×10^9 tonnes per year carbon dioxide to the depths. Now,

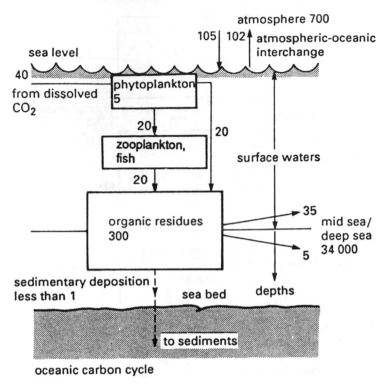

Figure 20   Oceanic carbon cycle.

the mixing time for exchange between surface layers and the depths is roughly 1000 years and it is this mixing time which controls the balance between carbon dioxide in the atmosphere and the oceans. Thus any extra carbon dioxide from fossil fuel combustion activities will increase the atmospheric carbon dioxide concentration and, since the natural system is a closed cycle, the assimilation of excess carbon dioxide will take 1000 years before equilibrium is attained through atmospheric–oceanic interchange. As we cannot control the carbon (or any other) cycle, we should be careful not to disturb natural systems too far.

The composite carbon cycle in the BIOSPHERE is shown in Figure 21 and includes the additional fossil fuel input and atmospheric–oceanic interchange fluxes which amount to an uptake of 105×10^9 tonnes of carbon per year and a release of 102×10^9 tonnes carbon per year, i.e. a net absorption of 3×10^9 tonnes of carbon per year which does not match the fossil fuel combustion and deforestation CO_2 contributions.

A controlling factor in the absorption of CO_2 in sea water is its ability to dissolve CO_2. This is influenced by sea water temperature. The greater the temperature, the less is the ability. Thus, the GREENHOUSE EFFECT could

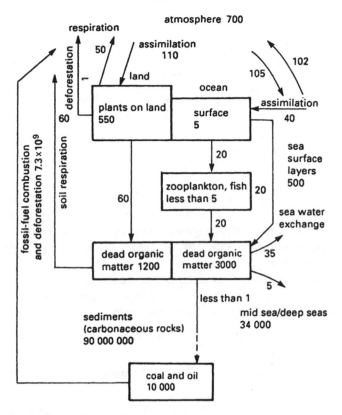

Figure 21 Composite carbon cycle.

accelerate the increase of the CO$_2$ content of the atmosphere by raising sea-water temperatures. This atmosphere–ocean interaction is the subject of detailed scientific scrutiny.

Attention is also being focused on the role of plankton as they absorb CO$_2$ during photosynthesis, most of which is returned by respiration in a very short time scale. However, some sink to the depths where the carbon may then be permanently locked in the sediments thus removing some CO$_2$ from the atmosphere.

Source: Oceans and the Global Carbon Cycle. *Biochemical Ocean Flux Study*, BOFS Office, Plymouth Marine Laboratory, May 1987.

Carbon dioxide (CO$_2$). Gas produced by the complete COMBUSTION of carbonaceous materials, by decay organisms such as aerobic DECOMPOSERS, by FERMENTATION, and by the action of ACID on limestone. It is exhaled by plants and animals and utilized in PHOTOSYNTHESIS in the CARBON CYCLE. Its atmospheric concentration is measured in ppm, as the concentrations are so low.

Reducing CO_2 Emissions

The Report 'The World in 2050: Implications of Global Growth for Carbon Emissions and Climate Change Policy' was released by PwC in September 2006. The Report assesses carbon emissions under six different scenarios:

Baseline Scenario – energy efficiency improves in line with trends of the past 25 years, with no change in fuel mix by country.
Scorched Earth – energy efficiency improvements are 1 per cent per annum lower than in the baseline scenario, with no change in fuel mix.
Constrained Growth – in which energy assumptions are as the baseline but GDP growth is lower, particularly in the E7 emerging economies.
Greener Fuel Mix Scenario – the baseline scenario except that there is a significant shift from fossil fuels to nuclear and renewable energy by 2050.
Green Growth – in which the green fuel mix assumptions in the previous scenario are combined with energy efficiency improvements 1 per cent per annum greater than in the baseline.
Green Growth + Carbon Capture and Storage (CCS) – a variant of the above scenario, which also incorporates possible emissions reductions due to the use of carbon capture and storage technologies.

Only the 'Green Growth and CCS' scenario offers the potential for reductions in CO_2 emissions.

Project Average Annual Global Economic & Primary Energy Consumption Growth in Alternative Scenarios

Growth (% pa except final column)	GDP	Primary energy	Non-fossil fuels*	Carbon emissions	
				Annual Average growth	Cumulative growth to 2050 (% change)
Baseline Scenario	3.2	1.6	1.6	1.6	112
Scorched Earth	3.2	2.6	2.6	2.6	233
Constrained Growth	2.6	1.0	1.0	1.0	61
Greener Fuel Mix	3.2	1.6	3.6	1.1	64
Green Growth	3.2	0.6	2.5	0.1	4
Green Growth + CCS	3.2	0.6	2.5	−0.4	−17

* Nuclear and renewables.

Average CO_2 emitted per person (tourist class) on a long-haul flight (round trip) Manchester to New York: 1.2 tonnes. CO_2 emitted by family car, 1000 miles: 1.2 tonnes. CO_2 emitted by round-trip, first-class long-haul flight to Barbados (per person), author's estimate: 4 tonnes.
Sustainable level of CO_2 per person per year: 1.2 tonnes.

Source: A. Monbiot, Radio 2 Interview, Jeremy Vine Show, 10/01/07.

The EU target for CO_2 reduction is 20 per cent by 2020, but preferably 30 per cent. The EU target for renewable energy production is 20 per cent by 2020. The time for 'carbon offset' trees to make a possible difference in CO_2 levels is ca. 10–20 years. Something doesn't add up.

This is illustrated in the Ryanair (Jan 07) claim to have halved their CO_2 emissions. This was analysed on BBC2 Newsnight (Jeff Gazard) on 11 January 2007. In 2002, Ryanair was estimated to have emitted 0.95 Mt CO_2. In 2006, this had risen to 2.89 Mt due to increase in passengers numbers. However, on a per passenger kilometre basis the calculated reduction was 28 per cent. It all depends! In defence of Ryanair, the 'Dutch Consumer Organisation' (Jan 2007) notes it is producing half the CO_2 emissions (in tonnes per passenger per 1500 km) compared with British Airways, viz. Ryanair 0.09 t, British Airways 0.17 t, Easy Jet 0.11 t.

If you must fly – minimize your carbon footprint!

The COMBUSTION of fossil fuels, slash and burn agriculture, and forest clearance, has raised the atmospheric carbon dioxide concentration from 290 ppm in the last century to 381 ppm (2005). Most of the increase in 2005 was particularly high at 2.6 ppm, as opposed to the average increase of 1.5 ppm over the last ten years. Estimates of the doubling of atmosphere CO_2 concentrations from the pre-industrial level of 280 ppm range from 30 years, if the oceans warm up and less CO_2 is dissolved, to ca. 50 years, if current growth rates are maintained.

CARBON DIOXIDE is partly responsible for the atmospheric GREENHOUSE EFFECT – estimates vary, but 50 per cent may be attributable to CO_2. This has led to the claim that nuclear power can reduce CO_2 emissions. The Rocky Mountains Institute has shown that constructing a nuclear plant every 3 days at a cost of $80 billion/year over 37 years would not reduce global CO_2 emissions.

The base measures that would help steady the atmospheric CO_2 concentration are:

1. The use of national and international programmes on energy conservation to reduce fossil fuel combustion, plus a tax on fossil fuel-based energy.
2. Phasing out CFCs and other substances which are depleting the OZONE layer.
3. Reversing the current level of deforestation, through the protection of the rainforests and large tree planting programmes.
4. More public transport and swingeing private vehicle taxes.
5. A switch to renewable sources of ENERGY such as SOLAR ENERGY, WATER and TIDAL POWER.

CO_2 capture and/or conversion is a very fruitful research area. Proposals include:

- CO$_2$ separation through inorganic membranes, gas adsorption membranes, and adsorption by sterically hindered AMINES;
- CO$_2$ hydrate formation as a selective removal process, also combining carbon dioxide with hydrogen dioxide recovery from OXYGEN blown GASIFICATION;
- conversion of carbon dioxide to METHANE using nickel CATALYSTS and conversion of CO$_2$ and CH$_4$ into synthesis gas, a mixture of carbon monoxide & hydrogen (this is achieved by PLASMA-assisted reforming followed by conversion to METHANOL).

Source: *Greenhouse Issues*, September 1998, p. 3.

Figure 22 shows how the use of alternative fuels can reduce vehicle emissions.

The table on page 100 shows commitments to limit or reduce emissions of equivalent CO$_2$ from 1990 to 2010 by Annex 1 parties as agreed to at Kyoto, compared with changes in CO$_2$ emissions from 1990 to 1995.

Annex 1 (developed countries) has specific targets, limitations or reductions of emissions, to be achieved by 2010. These are given as targets in terms of changes of equivalent CO$_2$ emissions. If emissions of other greenhouse gases are limited, restrictions on the use of fossil fuels will be correspondingly relaxed.

The Kyoto conference is a first step toward the introduction of economic instruments to achieve specific targets. Thus Article 6 of the protocol stipulates 'For the purpose of meeting its commitments under Article 3, any Party included in Annex 1 may transfer to, or acquire from, any other such Party emission reduction units resulting from projects aimed at reducing anthropogenic emissions by sources or enhancing anthropogenic removals by sinks of greenhouse gases in any sector of the economy. . . .'

Source: B. Bolin, 'The Kyoto Negotiations on Climate Change: A Science Perspective', *Science Journal*, Vol. 270, 16 January 1998, p. 330.

The contribution of the other GREENHOUSE GASES should not be overlooked. METHANE has a 'greenhouse factor' 65 (20-year time-scale, equal mass basis) times that of CO$_2$, hence measures should be directed towards its control, e.g. the reduction of LANDFILL GAS releases from the decomposition of DOMESTIC REFUSE in LANDFILL SITES could substantially reduce the greenhouse gas emissions from developed countries. Similarly, NATURAL GAS pipeline leakage should also be controlled. Most attention is, however, focused on CO$_2$ with coal fired power stations and automobiles as the main developed country targets. (◊ CLIMATE; GREENHOUSE EFFECT; GREENHOUSE GASES)

The table on page 101 shows the world's emissions of CO$_2$ from the consumption and flaring of fossil fuels in 2004.

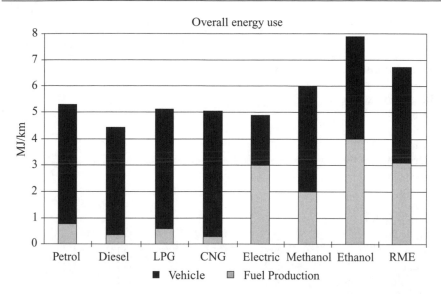

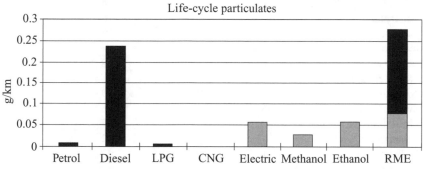

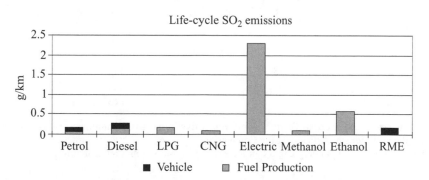

Figure 22 Alternative fuels as a means of reducing vehicle emissions.
Source: *The Engineer*, April 1996, p. 15.

CO_2 emissions: reductions agreed and actual emissions observed.

Party	Allowed 1990–2010 (%)	Observed 1990–1995 (%)
*European Union**	**–8**	**–1**
Austria	–8	–3
Belgium/Luxembourg	–8	+1
Denmark	–8	+8
Finland	–8	+3
France	–8	–4
Germany	–8	–9
Greece	–8	+7
Ireland	–8	–1
Italy	–8	–1
Netherlands	–8	+7
Portugal	–8	+49
Spain	–8	+14
Sweden	–8	+7
UK and N. Ireland	–8	–1
OECD, except EU	**(–6)**	**+8**
Australia	+8	+8
Canada	–6	+9
Iceland	+10	–4
Japan	–6	+8
New Zealand	0	+16
Norway	+1	+9
Switzerland	–8	–5
United States	–7	+7
*Countries in trans***	**(–1)**	**–2**
Bulgaria	–6	n.a.
Croatia	–5	n.a.
Czech Republic	–8	–23
Estonia	–8	n.a.
Hungary	–6	–15
Latvia	–8	n.a.
Poland	–6	n.a.
Romania	–8	n.a.
Russian Federation	0	n.a.
Slovakia	–8	n.a.
Ukraine	0	n.a.
Non-Annex 1 parties	—	**+25**

* Members of the European Union will implement their respective commitments in accordance with the provisions of Article 4 of the Convention.
** Countries that are undergoing the process of transition to a market economy.

World Carbon Dioxide Emissions from consumption and flaring of fossil fuels (2004 data, in million tonnes)

Region	CO_2 Emissions (million tonnes)
North America	6 886.88
Central & South America	1 041.45
Europe	4 653.43
Eurasia	2 550.75
Middle East	1 319.70
Africa	986.55
Asia & Oceania	9 604.81
World	27 043.57

Source: www.eia.doe.gov/pub/international/iealf/tablehlco2.xls
Accessed 16 February 2007.

Carbon, fixed. Not a precise constituent of a FUEL, but in the PROXIMATE ANALYSIS that proportion determined by deducting the sum of moisture, VOLATILE MATTER and ASH from 100 per cent.

Carbon footprint. A measure used to compare the CO_2 emissions from various alternatives towards the same goal, e.g., fruit supply to the UK. 'All of the following have about the same carbon footprint – one kg of fruit air freighted from the US, 30 kg of fruit sea freighted from the US, 200 kg of UK apples that have travelled 100 miles by road. So, the clear priority for the environmental shopper is to cut out air freight. One big problem that needs addressing is the difficulty for consumers of actually finding out whether their lettuce has come by plane or boat.' Mike Berners-Lee, Small World Consulting, letter to *Your Environment*, November 2007, no. 13.

Carbon monoxide (CO). A colourless odourless gas, lighter than air, formed as a result of incomplete COMBUSTION. It is a chemical poison when inhaled, as it can penetrate tissues and is absorbed into the bloodstream where it combines with the haemoglobin of blood cells, 300 times faster than OXYGEN and thus deprives the brain and heart tissues of oxygen. (◊ AUTOMOBILE EMISSIONS; SMOG; ASPHYXIATING POLLUTANTS; AIR QUALITY CRITERIA)

Carbon-neutral. A much-abused term implying that there is no nett addition to atmospheric CO_2 from whatever activity is being foistered upon the public. BIOFUELS, if made from grain (for example) require fertiliser inputs and energy-intensive cultivation, harvesting and processing. Organically-grown bio-fuels and wood-based ones which require fewer energy inputs may well be considered as carbon-neutral as the CO_2 from their combustion is reincorporated in new biomass. Note that most renewable energy systems require substantial energy investments upfront to make the concrete, steel, power cables, etc. It all depends!

Carbon offsetting[1]. The practice of levying a surcharge on air travellers to enable their CO_2 emissions to be offset by [usually] tree planting in Africa or India.

The trees take 30+ years to grow whereas the CO_2 emitted is immediately surcharged to the environment. This is not a quick fix, but a 'greenwash' gesture in too many instances.

Carbon offsetting[2]. Companies calculate their CARBON FOOTPRINT and offset this by doing something environmentally friendly, with a net CO_2 reduction e.g. wind farms. ($\lozenge$ RENEWABLE ENERGY)

Carbon taxes. Carbon taxes (based on FUEL CARBON content) work through the market mechanism, by changing the price incentives faced by Greenhouse Gas (GHG) emitters. The advantages of carbon taxes are that:

- they can be adopted on a national scale, without complex international agreements
- they are more cost-effective and impose fewer information requirements on policy-makers than mechanisms which rely on standards and norms
- they provide incentives for continual abatement efforts and technological improvement (ENERGY EFFICIENCY)
- they are easy to apply (technically) and offer a broad scope for implementation in several sectors, while possessing low administrative and transaction costs (e.g. from monitoring and enforcement)
- they raise substantial revenues that may be used to mitigate the negative impacts of carbon taxes (e.g. UK proposes to help underwrite National Insurance costs to employees) or to encourage use of renewable energy. ($\lozenge$ GREENHOUSE GASES)

Source: Adams Associates, International Academy of the Environment, Media Release, 22 October 1998.

Carbon taxes. A carbon tax is a tax on energy sources which emit carbon dioxide into the atmosphere. It is an example of a pollution tax, which has been proposed by economists as preferable because it taxes a 'bad' rather than a 'good' (such as income). A carbon tax, because of the link with global warming, is often associated with some kind of internationally administered scheme; however, this is not intrinsic to the principle, and politically improbable. The European Union has discussed a carbon tax covering its Member States, and has implemented a carbon emissions trading scheme.

The purpose of a carbon tax is both financial (like all taxes) and environmental. It can be implemented by taxes on gasoline and on certain types of energy production, such as coal-fired power plants.

On 1 January 1991, Sweden enacted a carbon tax, placing a tax of 0.25 SEK/kg ($100 per ton) on the use of oil, coal, natural gas, liquefied petroleum gas, petrol, and aviation fuel used in domestic travel. Industrial users paid half the rate (between 1993 and 1997, 25 per cent of the rate), and certain high-energy industries such as commercial horticulture, mining, manufacturing and the pulp and paper industry were fully exempted from these new taxes. In 1997 the rate was raised to 0.365 SEK/kg ($150 per ton) of CO_2 released.

Finland, the Netherlands, and Norway also introduced carbon taxes in the 1990s.

Carbon trading. The sale of CO_2 emission reduction capacity or purchase of the same. A European company failing to meet its target will be fined €40 per tonne. In 2005, 799 million tonnes of CO_2 (worth €9.4 billion) were traded in all carbon market segments. In the first half of 2006, 684 million tonnes of CO_2 (worth €12 billion) were traded.

Source: *The Times*, 12 September 2006.

(⟡ CARBON OFFSETTING, CARBON TAXES)

Carbon trading and offsetting. The European Commission sets emissions caps for CO_2 for each Member State. So far, 12000 energy-intensive plants have been included. These include fossil-fuel power generators, oil refineries, cement producers, iron and steel manufacturers, glass makers, and paper producers. Each is allocated an emissions limit. If exceeded, it must buy carbon credits from another company that has not utilized its allocation of CO_2. One credit is equivalent to one tonne of CO_2 reduced. Additional credits can be created from carbon offset projects, such as wind farms or solar power plants.

Carbon Trust. The Carbon Trust is an independent company funded by Government. Its role is to help the UK move to a low-carbon economy by helping business and the public sector reduce carbon emissions and capture the commercial opportunities of low carbon technologies. In short, the Carbon Trust works with UK business and the public sector to cut carbon emissions and develop commercial low carbon technologies.

Carbonate. A SALT of carbonic acid (H_2CO_3), e.g. $CaCO_3$, CALCIUM CARBONATE (limestone).

Carbone vs Clarkstown, New York. A landmark decision by the US Supreme Court ruled on 16 May 1994 that the states and local governments could not restrict interstate movement of solid wastes. This has permitted large-scale interstate movement of solid wastes and the development of 'mega fills' with consequent 'low' disposal costs. This has impacted on both RECYCLING and INCINERATION in the USA. (⟡ LANDFILL)

Carbonization. The treatment of coal or other solid fuels by heat in a closed vessel usually with the object of producing coke or gas. High-temperature carbonization is used for the production of coal gas and (at 1300 °C) blast furnace coke. Low-temperature carbonization (600 °C) is used to produce 'smokeless fuels'. Carbonization also gives many by-products (e.g. coal tar) of great importance to the chemical industries.

Carcinogen. Any compound or element which will induce CANCER in humans or other animals. It is now accepted that a large proportion of human cancers are directly associated with environmental agents, in particular chemical and physical agents, e.g. some chlorinated hydrocarbon PESTICIDES and ASBES-

TOS fibres. Carcinogens can be inhaled, e.g. tobacco smoke and asbestos fibres; ingested; or absorbed through the skin (insecticides, hair dyes).

A no-effect level cannot be assumed for carcinogens as the development of cancer is the result of an accumulation of irreversible cellular damage. In contrast, exposure to toxic substances can in many cases result in little or no damage below certain levels and they are eventually excreted from the human system.

Health Hazards of the Human Environment (World Health Organization, 1972) lists the following classes of environmental carcinogenic agents:

1. *Polynuclear compounds.* These occur in tobacco and coal smoke, exhaust gases; a major compound in BENZO (ALPHA) PYRENE (BP) found in all kinds of soot and smoke.
2. *Aromatic amines.* These are mainly found in industrial environments; they can also contaminate foodstuffs and plastics.
3. *Chlorinated hydrocarbons.* A major group which embraces industrial solvents, DDT, ALDRIN, DIELDRIN, and other pesticides. Some members of this family are notorious for their ubiquity, persistence and ability to accumulate in living systems. (◊ CHLORINATED HYDROCARBONS; RED LIST)
4. *N-nitroso compounds.* These can be found in industrial solvents of chemicals. They may be formed in the human intestine through bacterial action following the ingestion of nitrites.
5. *Inorganic substances.* Some toxic metals are included in this class, particularly beryllium, selenium, cadmium.
6. *Naturally occurring agents.* These are the so-called toxins that occur in spoiled food.
7. *Hormonal carcinogens.* Hormonal imbalances can induce cancer. The growth-promoting chemical DES (DIETHYLSTILBOESTROL) has been implicated as a carcinogen.

The known range is very extensive and any substance that acts as a MUTAGEN must be considered a potential carcinogen until proved otherwise.

Carnot efficiency. This is a direct outcome of the Second Law of Thermodynamics and places the upper limit on the maximum possible conversion of heat (thermal energy) to work in a heat engine. It is usually derived as:

$$\text{Carnot efficiency} = \frac{T_{\text{source}} - T_{\text{sink}}}{T_{\text{source}}}$$

where the TEMPERATURES are on the ABSOLUTE TEMPERATURE SCALE. Thus, comparing a heat engine supplied with steam at 600 °C (873 K) and the 'choice' of a sink in the Arctic at 5 °C (278 K) or a UK river at 30 °C (303 K) the respective Carnot efficiencies are:

$$\eta_{\text{Arctic}} = \frac{(873 - 278)}{873} \times 100 = 68\%$$

$$\eta_{\text{UK}} = \frac{(873 - 303)}{873} \times 100 = 65\%$$

This shows the influence on EFFICIENCY that sink temperatures can have, and why they should be as low as practicable in order to maximize efficiency. Conversely source temperatures should be as high as possible.

If a COMBINED HEAT AND POWER system is being considered, T_{sink} could typically be 130 °C in order to provide hot water at a useful temperature. In this case the Carnot efficiency becomes 54 per cent. This loss in conversion is of course made up by the use of the rejected heat in the associated heat network, thus leading to a higher overall efficiency.

In practice, the actual efficiencies are always lower than the Carnot efficiency due to the effects of friction and flow losses. Nevertheless, the Carnot efficiency is a useful basis for comparison (◊ LAWS OF THERMODYNAMICS; DIRECT ENERGY CONVERSION) and can filter out harebrained schemes such as using a minute difference in ocean temperature to drive a turbine.

Car tyre. Consumer product which presents a serious recycling problem. It is mainly composed of vulcanized rubbers, steel wire and fabric cord. The rubbers are chemically crosslinked with sulphur and form the bulk of the product comprising the carcass and tread. The tread is patterned with deep groves to assist water removal when driving in the wet and with use the tread is worn down in a more or less uniform manner. Current legislation imposes a minimum groove depth of 1.6 mm over the complete circumference, so a tyre with grooves less than this is illegal. Tyre replacement gives rise to a recycling problem, although the carcass may still be roadworthy. A small proportion may be retreaded, a practice more widely prevalent with truck tyres owing to their greater cost and complexity. The large proportion, however, are not fit for further use and are currently dumped as landfill or elsewhere. Since it is a hollow product, such tyre dumps easily catch fire, creating a serious air pollution problem (particularly from the CARBON BLACK soot and SULPHUR DIOXIDE gas emitted) as well as potential water run-off. A small proportion of waste tyres are incinerated to reclaim some heat energy in specially designed furnaces. Some *buffed crumb* is produced in retreading and can be recycled into sports surfaces, where the particles are bonded together with polyurethane adhesive (cryogenic crumbing is now a preferred route) (◊ PYROLYSIS)

Catalysis. The acceleration of chemical or biochemical reactions by a relatively small amount of a substance (the CATALYST), which itself undergoes no permanent chemical change, and which may be recovered when the reaction has finished.

Catalyst. A substance or compound that speeds up the rate of chemical or bio-chemical reactions. Catalysts used by the chemical industry are usually metals, such as vanadium which speeds up synthesis reactions. ENZYMES are organic catalysts.

One very interesting application of catalysis is in the combustion of natural gas without the presence of flames. The catalyst is a platinum-coated ceramic tube which radiates heat (thermal energy) and because combustion is at a lower temperature than normal, could be instrumental in reducing the production of NITROGEN OXIDES.

Catalytic converter. ◊ CATALYTIC REACTOR.

Catalytic oxidation of effluents (COE). A catalytic oxidation process which effects end-of-pipe destruction of aromatic hydrocarbons, typically from chemical manufacturing, pharmaceutical, petroleum refining, dyes/pigments, coal/coke, resin manufacture and explosives. Similar materials are found in contaminated groundwater and LANDFILL LEACHATE.

A COE process has been developed by waste treatment experts, Lanstar and High Force Research. The process uses hydrogen peroxide and a new catalyst. This new catalyst is non-hazardous and fully recyclable.

Source: *Envirotec*, August 1999, p. 34.

Catalytic reactor. (colloquially catalytic converter). An attachment fitted to the exhaust pipes of unleaded fuel internal combustion engines which converts UNBURNT HYDROCARBONS and CARBON MONOXIDE to carbon dioxide and water vapour by means of a platinum mesh CATALYST. Three-way catalytic reactors will in addition convert NITROGEN OXIDES to nitrogen using platinum, palladium and rhodium catalysts, supported on a ceramic shell to withstand and sustain the high temperatures required for the reactions to take place. Should the vehicle take only a short journey (less than 15 miles in winter), the reactor may not reach an adequate working temperature and therefore be rendered ineffective.

Diesel engines can also be fitted with a catalytic converter which is electrically heated for start-up purposes so that unburnt hydrocarbon and SOOT emissions can be reduced.

As an example of the technology in action, SÖDRA's new cargo vessel Cellus is the first of its kind in the world with catalytic conversion at a cost of SEK 2.2 million. The investment will be reimbursed thanks to a large extent to the fact that the vessel is to be on long-term charter to a Swedish company. Swedish environmental regulations mean that 40 per cent of the investment is repaid over a period of 5 years. As the vessel will sail mainly to Sweden, harbour dues are reduced for vessels that burn oil with low sulphur content and further reduced for vessels with catalytic converters.

Details of the vessel

- Diesel engine (3.8 MW) produces a speed of 14 knots with 85 per cent of the machine being used and 14 tonnes of oil being burned per 24 hours.

- The catalytic converter reduces the emission of nitrogen oxide by 98 per cent.
- Low sulphur oil reduces the emission of sulphur dioxide by 70 per cent.

Source: *Responze SÖDRA CELL*, Publication No. 2/98.

Catalytic wet air oxidation (CWAO). A method of removing organic contamination from process waters.

In the CWAO process the liquid phase and high-pressure air are passed co-currently over a stationary bed of CATALYST. The effect of the catalyst is to provide a higher degree of COD removal than is obtained by WAO at comparable conditions, or to reduce the residence time. CARBON MONOXIDE formation can also be eliminated. The process becomes autogenic at COD levels of about 5–10 g/l, at which the system will require external energy only at start-up.

The catalyst is a transition metal on supports such as alumina, zeolites and zirconia. (◊ CATALYTIC OXIDATION OF EFFLUENTS)

Source: IChemE '*Avert*', No. 13, Winter 1998/99.

Catchment. The natural drainage area for PRECIPITATION, the collection area for water supplies, or a river system. The area is defined by the notional line, or watershed, on surrounding high land. Typically a catchment is a valley, or series of valleys containing a river system.

Cathode. An electrode which is at a negative potential.

Cathodic protection. Cathodic protection works by applying a d.c. power source to reverse the natural flow of electrical current caused by galvanic corrosion. This stops the steel reinforcement in a structure from rusting. See Figure 23.

For it to work satisfactorily, there needs to be structural integrity, sound concrete, continuity of reinforcement, an ANODE system, a power source and monitoring and control systems. Should there be any gaps in the steel in a severely corroded structure, these need to be bridged first.

An electrode is applied to concrete adjacent to the reinforcement and connected to a positive source of direct current, around 100 mV, and the reinforcement is connected to the negative side of the current supply. This flow of current acts in direct opposition to that naturally occurring as a result of corrosion and provides the protection.

Anodes come in many shapes and forms and can be made from a variety of conductive materials, including meshes, plates, wires and sprays, made of zinc, titanium, graphite and conductive polymers.

Cation. A positively charged ION.

Caustic soda (NaOH). Sodium hydroxide, an alkali used, for example, in the manufacture of wood-free PAPER and PULP.

Caveat emptor. Latin for 'let the buyer beware'. This implies a buyer must ensure that goods about to be purchased are free from defects and that he/she bears the risk.

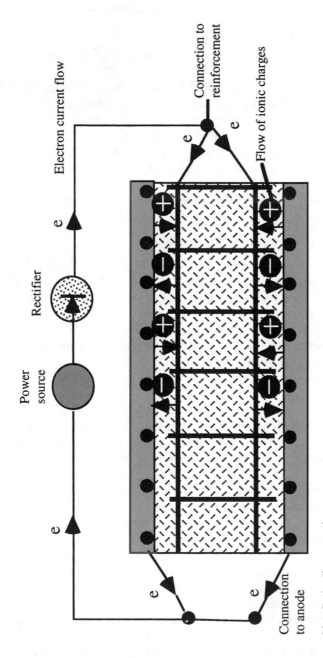

Figure 23   Cathodic protection.
Source: 'Cathodic protection', *Surveyor*, January 1994.

CEFIC. European Chemical Industry Council. Represents main chemical industry 'players'. Instrumental in raising environmental awareness and standards of accountability. For example, in the *CEFIC list of reportable parameters*, member companies report on the following: (1) Safety and occupational health (number of fatalities, lost time injuries frequency, occupational illness frequency); (2) environmental protection (HAZARDOUS WASTE for disposal, non-hazardous waste for disposal, SULPHUR DIOXIDE, NITROGEN OXIDES, CARBON DIOXIDE and other global warming gases, volatile ORGANIC COMPOUNDS, phosphorus compounds, nitrogen compounds, CHEMICAL OXYGEN DEMAND, HEAVY METALS, other substances with potential human health impacts, ENERGY consumption and EFFICIENCY); (3) distribution (distribution incidents).

Many of the 21 federation member countries already publish Responsible Care performance indicators which take the form of health, safety, and environmental aggregated performance data.

Note: these reflect national achievement and results are not intercomparable.

Source: 'Harmonic Bliss', *The Chemical Engineer*, 19 November 1998, p. 22.

Cellulose. A carbohydrate polymer constructed from the glucose building blocks formed by the action of PHOTOSYNTHESIS. It is the chief structural element and major constituent of the cell walls of trees and other higher plants where it is bound up with a glue (lignin) to form plant stems. The fixation of CARBON in the CARBON CYCLE is such that about 100 000 million tonnes of cellulose are produced annually. Many important derivatives stem from it, e.g. cellulose nitrate and cellulose acetate.

A major use is in PAPER manufacture when it has been freed from its matrix of LIGNIN and other ORGANIC matter by treatment with sodium sulphite or ALKALI to yield a pulp feedstock of cellulose fibres.

Cellulose is the principal chemical constituent of DOMESTIC REFUSE and can be processed to yield a variety of products such as glucose, SINGLE-CELL PROTEIN and ethyl alcohol. (◊ HYDROLYSIS; PYROLYSIS; COMPOSTING; RECYCLING)

Cellulose economy. The use of the sun's energy through the process of PHOTOSYNTHESIS and the subsequent production of CELLULOSE which is then processed, by enzymatic means or by chemical HYDROLYSIS, to yield glucose which can then be fermented to yield many industrial products, PROTEIN or FUELS.

The cellulose economy could use strains of plants which have a high ENERGY conversion EFFICIENCY, e.g. sugar-cane and sorghum in tropical areas which can fix 3 per cent of the incoming solar energy. These plants are then processed by the chemical or biochemical means already mentioned. This technique of energy cropping would also use marginal lands, swamps, etc., as productive areas. Species would be used that are capable of continuous cropping, such as fast-growing deciduous trees which re-sprout from stumps when cut and avoid the need for replanting. In the UK and USA, willows and hybrid poplars can be so cropped. One strain of hybrid poplar has been planted at 3700 trees per acre and produces eight harvests per planting with unit energy costs within an order of magnitude of those from fossil fuels.

Cellulose processing as a means of living off the sun's income is a fast-growing area of research and development. (◊ ENZYMES)

Cement – TX Active. TX Active ® is a photocatalytic cement which when used in building materials and coverings absorbs and eliminates 20 to 80 per cent of air pollutants, depending on atmospheric conditions and the level of sunlight available to trigger the process. Titanium dioxide is combined into wall facings and road surfaces which can remove pollutants highly effectively.

Sunlight starts a chemical reaction between the CO_2 and the titanium dioxide that removes the gas from the air and crystallizes it into a salt that sits on the surface of buildings and roads until washed away by rain.

Lab tests on photoactive cements indicate that three minutes of sunlight are sufficient to reduce pollutants by up to 75 per cent.

Source: Italcementi Group.

Cement kiln – waste combustion. Cement is made by heating a mixture of CLAY or shale plus chalk or lime in a ROTARY KILN up to 250 m long × 8 m diameter rotating at 1 rpm. The process can be wet, semi-dry or dry and the fuel can be pulverized coal, oil or gas. As the coal ash is similar in composition to the clay or shale, it can stay in the cement clinker.

As one of the kiln operator's major costs is fuel, and even a modest sized kiln can consume 8–10 tonnes of coal per hour, the cement kiln could, therefore, solve a waste disposal problem and also benefit the cement manufacturer by reducing fuel costs.

France used MSW and industrial waste as fuels in 12 cement kilns in 1986, and has subsequently increased capacity both to utilize the energy content in the waste and destroy hazardous wastes. This approach is also borne out by Dutch experience in that most inorganic materials bind very efficiently to cement clinker. Fuels with an 'appreciable' amount of cadmium, mercury or thallium should not be used. Regular analysis of the supplementary fuels and their binding efficiency is required. However, cement kilns do have potential as waste combustion outlets for PACKAGING wastes.

This development has not passed unnoticed by hazardous waste (merchant) incinerator operators one of whom has claimed that this 'will lead to unnecessary safety risks and POLLUTION levels in the vicinity of the operation . . .'. This is a space to watch as the formal authorization procedures to be followed for permanent use in the UK will set the parameters for any similar solvent-based fuel combustion. (The House of Commons Environment Committee, Second Report, *The Burning of Secondary Liquid Fuel in Cement Kilns*, 7 June 1995.)

In response to considerable public pressure, the House of Commons Environment Committee considered the matter and recommended:

- That the government formally classify Secondary Liquid Fuel (SLF) as waste (if necessary, by amending the law) thus applying the Duty of Care to all shipments of SLF to kilns and formally bringing the burning of SLF in kilns within the scope of the EU Hazardous Waste Incineration Directive.
- If the burning of SLF is to continue, Her Majesty's Inspectorate of Pollution (HMIP; now the ENVIRONMENT AGENCY) must enforce their own standards more effectively, and not be put off by commercial arguments such as the possible termination of contracts. For example, trials should be limited in duration and should not be allowed to continue unless the conditions of the HMIP Protocol can be complied with.
- That the adaptation of kilns to burn SLF should be designated a 'substantial change' under Part I of the Environmental Protection Act 1990, thus requiring a new authorization.
- That as a minimum HMIP apply additional controls to ensure that the emissions from burning coal plus SLF are not more polluting than those from burning coal alone. This would mean that:

  - the specification for SLF should be tightened up so that it is a cleaner fuel than coal, in terms of metals content, etc., and/or
  - effective abatement technology should be fitted, in order to trap the emissions which currently escape during kiln upsets.

- That HMIP carry out a more detailed analysis of the precipitator dust from kilns which burn SLF and its potential to contaminate soils and groundwater. If it is found to be an environmental hazard, it should be reclassified as Special Waste, which would bring it under a strict regulatory regime. It is also recommended that HMIP take immediate action to tackle the more general pollution problems associated with kilns.
- That the Government carry out a thorough survey of human and animal health in the vicinity of the kilns (both those burning SLF and those burning only coal) in order to establish or disprove the alleged links between kiln pollution and health effects.

CEMFUEL. ◊ SECONDARY LIQUID FUEL.

Ceramic filter. Ceramics are inert, inorganic, brittle materials that are heat bonded. Ceramic filtration media offer potential benefits: they can withstand high TEMPERATURES and corrosive atmospheres, so no pretreatment of the gas stream is necessary.

Particles are collected on the surface of the element and as filtration continues the layer of deposited particles becomes thicker, forming a cake which contributes to the pressure drop across the filter. At some point the cake must be removed for efficient filtration, usually by reverse pulse cleansing.

Figure 24 shows a typical schematic ceramic filter system.

Cetane number. A number on a conventional scale, indicating the ignition quality of a DIESEL FUEL under standardized conditions. It is expressed as the percentage by volume of cetane (i.e. hexadecane) in a reference mixture having the same ignition delay as the fuel for analysis. The higher the cetane number, the shorter the ignition delay. The lower the number, the longer the delay and the greater the noise levels produced in COMBUSTION.

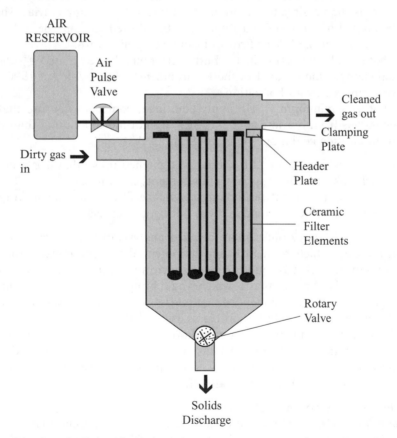

Figure 24   A typical ceramic filter system.
Source: 'Filter tips', *The Chemical Engineer*, 6 April 1995. (Reproduced by permission)

Chaos. Put simply, the flapping of a butterfly's wings in Africa could trigger a hurricane in Belize. Having had your credulity strained, the principle depends on determining those sensitive regions where accurate climatic measurements yield greater certainty in weather prediction for known geographical areas. The corollary of the massive (computerized) cross-checking also shows those locations on a 60 km global grid when additional data would make very little difference to forecast accuracy. In the UK attention has focused on cyclonic storms which form over the mid-Atlantic. Additional measurements of wind speed, barometric pressure, temperature and humidity have enabled markedly accurate predictions to be made for the UK.

Chaos theory. According to Kellert (1993), '*chaos theory*' is the qualitative study of unstable aperiodic behaviour in deterministic nonlinear dynamical systems' (italics in the original). A dynamical system is one that changes over time, so chaos theory looks at how things evolve. The systems studied are deterministic, that is, the state of the system at any stage depends on the state at the previous stage in a way that can be completely specified using simple mathematical operations on exact numbers. For instance, suppose that a system evolves by doubling at each iteration. If we start out with the number 10, we produce the sequence: 10, 20, 40, 80, 160, . . . This sequence is an example of a predictable system: we know that it will continue to infinity (note that this recursion formula is linear).

Non-linear recursion formulas, in contrast, often give rise to the complex, unpredictable behaviour that is the subject of chaos theory. For example, if we use the formula:

$$\times [\text{sub new}] = 4 \times [\text{sub old}](1 - \times [\text{sub old}])$$
$$\text{e.g. old} = 0.2; \text{ hence 'new'} = 4(0.2)(1 - 0.2) = 0.64.$$

And start with the number 0.2, we generate a sequence whose first few terms are 0.2, 0.64, 0.9216, 0.2890 . . . , 0.8219 . . . , 0.584 . . . , 0.9708 . . . , 0.1133 . . . , and so on. This sequence wanders all over the unit interval in an unpredictable manner. It is essentially a random-number generator.

Thus unpredictable behaviour may arise from the repeated application of a simple deterministic formula, not from the roll of a die or the intrusion of an unexpected new variable. The surprising insight of chaos theory is that extremely complex behaviour can arise from such a process.

The focus on sensitivity is one of the great contributions of chaos theory. It opens up a window on a whole class of problems that were considered intractable, problems in which the slightest error in initial information led to completely unreliable predictions.

Source: Bedford, Crayton W., *Mathematics Teacher*, Vol. 91, April 1998, p. 276. Kellert, Stephen H. In the Wake of *Chaos*: Unpredictable Order in Dynamic Systems. Chicago: University of Chicago Press, 1993.

Charcoal. A form of CARBON produced by the DESTRUCTIVE DISTILLATION of wood. Characterized by large surface area in a small volume.

Chelating agents. A GROUP of organic compounds which can incorporate metal IONS into their structure and so obtain a soluble, stable and readily excretable substance (known chemically as a chelate complex). Thus toxic metals such as LEAD (normally poorly excreted) can be removed from blood and bones of the human organism and excreted. The use of chelating agents is a standard treatment in such cases of poisoning.

   Chelating agents are, however, unable to penetrate the body's cellular material and to overcome this one method is to encapsulate them in substances known as liposomes which can penetrate cell walls. This is an important application as the liver, for example, is one of the vital organs where PLUTONIUM and other HEAVY METALS accumulate. Thus intracellular deposits may now be purged from the human organism.

   The medical use of chelating agents must be very carefully controlled as the sudden release of the burden of heavy metals can damage the liver and kidneys.

   Another extremely useful application of chelating agents is the use of disodium ethylenediamine-tetraacetate (EDTA) to remove (sequester) minute quantities of copper, iron, etc. in foods, thereby preventing discoloration. EDTA is also a preservative as it can inhibit the bacterial growth responsible for food poisoning.

Chemical oxygen demand (COD). The amount of OXYGEN consumed in the complete oxidation of carbonaceous matter in an effluent sample. This is done in a standard test which uses potassium dichromate as the oxidizing agent. (◊ DISSOLVED OXYGEN; BIOCHEMICAL OXYGEN DEMAND)

Chemical symbol. Either a single capital letter or a combination of a capital and lower case one which is used to represent an atom of a chemical element, e.g. Na for sodium (from the Latin name, *natrium*) and S for SULPHUR (from the English).

Chemosphere. Part of the lower section of the ATMOSPHERE in which chemical processes such as molecular dissociation and recombination take place during the day and night respectively, under the influence of ultraviolet radiation. (◊ IONOSPHERE)

Chernobyl. On 26 April 1986 the unit IV NUCLEAR REACTOR at the Chernobyl Nuclear Power Station in the Soviet Union caught fire and exploded. An estimated 100 million Ci (◊ CURIES) of radioactivity were emitted with up to half deposited within 30 kilometres of the plant. An official Soviet report estimated that the average external exposure to radiation in 1986 to Soviet citizens in the fallout path was doubled. The fallout will continue to contribute to background levels for many decades. Calculations indicate an increase in the death rate from cancer of up to 0.05 per cent, i.e. adding perhaps 5000 to the 9.5 million people who would normally be expected to die of cancer over the next 70 years.

However, there are also internal radiation aspects through the ingestion of contaminated materials. The two principal ISOTOPES are iodine-131 which has a HALF-LIFE of 8 days and CAESIUM-137, half-life 30 years. The iodine-131 uptake in plants is dependent on soil type. Poor soils allow much greater uptakes and it is for this reason and the prevailing deposition pattern that hill farmers' lambs in some parts of Wales and the Lake District are still (2008) contaminated with caesium-137 which in turn means that they cannot be sold. Indeed, it is now posited that the UK effects of the Chernobyl caesium-137 fallout will be felt for up to 150 years. In 1990, 400 000 acres of UK hill farming land were under sheep sale restrictions due to the effects of fallout.

Before the explosion, the annual incidence of childhood thyroid cancer in the Gomel region of Belarus was one in 2 million, about the same as in Britain. But by 1994 this had risen to one in 10 000 – a 200-fold increase. In some pockets of Belarus the incidence is ten times higher.

Over the same period, there was more than a 100-fold increase in the incidence of thyroid cancer in the most northerly parts of Ukraine. In the Bryansk and Kaluga regions of Russia, which lie to the north of Chernobyl, there has been an eight-fold increase since 1986.

The cause of the Chernobyl incident was an experimental reduction in power from 3200 MW to ca. 1000 MW. However, the reactor could not be stabilized quickly enough and power dropped to 30 MW. Neutrons absorbing xenon in the core prevented the raising of the reactor power level to above 200 MW. To counteract this, control rods were with-drawn and the reactor became unstable and a slow nuclear chain reaction ensued which blew the top off the reactor with the reactor's graphite moderator subsequently catching fire. This was eventually controlled by dropping 5000 tonnes of boron, limestone and sand into the reactor from helicopters.

Several points stand out from the story of Chernobyl:

(a) Human error and miscalculations can never be eliminated.
(b) The long-term effects of its radiation release, notably caesium-137, may be around for decades.
(c) A major nuclear reactor explosion was tamed and the core sealed off by the extraordinary bravery of the Soviet firefighters, some of whom knowingly and willingly gave their lives in the process.
(d) Chernobyl has made many people ponder on the need for nuclear power and question whether it is strictly necessary.

Source: USSR State Committee on the Utilization of Atomic Energy, *The Accident at Chernobyl Nuclear Power Plant and Its Consequences* (August 1986), p. 15.

The UK nuclear industry has stated (1992):

> The Chernobyl accident will cause a radiation dose of six millisieverts over the next 50 years to each of the 75 million people living nearest to it. But some British residents receive that each and every year from background radiation in their homes. The Chernobyl accident has rendered nearly 3,000 square kilometres of land sterile and useless. But major hydro schemes 'drown' areas up to three times as large all over the world.
>
> Each source of energy poses risks. Yet none of them pose risks which are unique or out of proportion to others. It is reckoned that the chance of another Chernobyl-scale accident anywhere in the world should be well below ten per cent during the next century, because of advances in plant technology and operation. At the same time, the world population is increasing and energy demand is likely to double by 2025. If the world is to meet this demand it will need to continue burning fossil fuel at the greatest rate consistent with a prudent attitude towards the threat of global climate change.
>
> It will need to develop all the practicable hydro schemes in the world.
>
> It will need nuclear.
>
> It cannot afford to abandon any of these options without jeopardising the ability to meet the demand. One might argue that the risks are trivial compared with the risks of having to do without any one of them.

From an industrial point of view, the above statements are not unreasonable. The looming energy crisis and/or global warming will need to be tackled on all fronts and the Chernobyl incident should not be used as a reason to impugn today's Western Nuclear Reactor Designs or plant operational procedures which are of the highest order.

Among the legacies of the Chernobyl disaster has been the proliferation of absurd claims. These achieved a surreal level at the time of the 1995 anniversary when Ukrainian officials reportedly claimed that 125 000 people had died as a result of the radiation released. It turned out that this was the total of all deaths in the affected area since the accident. But it was not the first time that the real victims of Chernobyl had been exploited to attract the sympathy of the West. *Network First* (ITV programme) told the tale of Igor Pavlovets, a child from Belarus born deformed after the accident. His case is a tragic one, but what evidence is there to link it to Chernobyl? Publicity for the programme asserts that 'over one million children' are deformed like Igor as a result of the accident. A sample survey of 500 children in Minsk, it says, has found only one to be completely healthy. But this may be a reflection of the poor health of Russian children in general and unconnected to the disaster at Chernobyl.

It may seem harsh to insist upon proper evidence in the face of misery, but only by doing so can the real causes of ill-health be properly addressed.

In an issue of *Nature*, Dr Valerie Beral of the Imperial Cancer Research Fund and a team from two Ukrainian scientific institutes examined the data on cancer of the thyroid, by far the most likely prompt response to a nuclear accident. Iodine-131 is a major part of the radioactive releases, contaminates grass, is eaten by cows and finds its way into milk. Since it is concentrated in the thyroid, that is where it is likely to cause damage, and children are at greater risk because their thyroids are smaller and they drink more milk.

The figures do indeed show an increase in cases. In the whole of the Ukraine, there were eight cases of thyroid cancer in children under 15 in 1986, 11 in 1989, and 42 in 1993. The data show that the risks are greatest among those who were living near the plant. There is no evidence of increased risk in children born after 1986, which is to be expected because iodine-131 decays relatively rapidly. Neither is there any evidence, Dr Beral says, of increases in any other forms of cancer.

Figures like these are no argument for complacency, but they do provide perspective. The actual evidence so far cannot possibly justify the wilder claims that have been made. Even a disaster needs to be kept in perspective.

Source: Nature Forum, June 1992.

Since 1977, Chernobyl has suffered a serious accident about once every 5 years. These events have left the site with only two of the original four units still in service today.

- In 1982, five years after first going critical, Chernobyl-1 experienced a core melt when coolant flow to a single fuel channel was inadvertently cut off. To this day, fuel fragments of the damaged sub-assemblies still remain inside the reactor.
- In 1986, an extreme power excursion completely destroyed Chernobyl-4. Lacking a containment, the reaction released great quantities of radioactive materials making it the world's worst nuclear accident.
- In 1991, an extensive turbine fire brought about the closure of Chernobyl-2. Serious core damage was narrowly averted as the fire disabled the systems essential for core cooling.

Source: USA Department of Energy, *Most Dangerous Reactors Report*, August 1995.

The UK does not have a blameless record in matters nuclear. For example, in 1999 BNFL admitted to having falsified fuel measurement records. A spokesman said workers who were supposed to check the measurements manually on a sample of 200 mixed-oxide (MOX) fuel pellets had filled out the inspection reports using figures from previous samples.

Source: *The Engineer*, 17 September 1999.

Chimney Height Memorandum. A formerly used UK guidance document, not binding in law, giving a sample method of determining the height appropriate for certain new chimneys serving COMBUSTION plant, in order to avoid unacceptable pollution (SO_2 or other gaseous POLLUTANTS). The 3rd edition, 1981, takes account of very low SULPHUR fuels. The document takes into account the type of district, the capacity of the plant and other buildings in the vicinity. Computer modelling is now used.

Chloracne. A sub-lethal effect of exposure to DIOXIN and other chlorinated chemicals, which manifests itself as a painful skin disorder.

Chlor-alkali process. A process used for the manufacture of CHLORINE by ELEC-TROLYSIS of brine. This process was formerly a source of mercurial discharge to estuaries. (◊ MINAMATA DISEASE)

Chlorinated hydrocarbons. One of three major groups of synthetic insecticides, others being ORGANOPHOSPHORUS COMPOUNDS and synthetic pyrethrins. The group includes DDT, ALDRIN, ENDRIN, BENZENE HEXACHLORIDE, DIELDRIN, and many others including ENDOSULFAN which is highly toxic to fish. An endosulfan spillage in the River Rhine in 1968 caused the death of millions of fish plus the cessation of drinking-water abstraction until the chemical contamination was thoroughly cleared.

In insects and other animals these compounds act primarily on the central nervous system. They also become concentrated in the fats of organisms and thus tend to produce fatty infiltration of the heart and fatty degeneration of the liver in vertebrates. In fishes they have the effect of preventing oxygen uptake, causing suffocation. They are also known to slow the rate of PHO-TOSYNTHESIS in plants.

Their danger to the ECOSYSTEM resides in their great stability and the fact that they are broad-spectrum poisons which are very mobile because of their propensity to stick to dust particles and evaporate *with* water into the atmosphere. (◊ PESTICIDES, ORGANOCHLORINES)

Chlorine (Cl). An element of the halogen group. A gas with an irritant smell which has severe effects on the lungs and respiratory system. Produced mainly by the CHLOR-ALKALI PROCESS, chlorine is widely used in the manufacture of organic chemicals, for example, carbon tetrachloride, HALOGE-NATED FLUOROCARBONS, AEROSOL PROPELLANTS, chloroform, PESTICIDES, PVC, and solvents, plastics and bleaching agents.

Euro Chlor estimates that about 12 tonnes of mercury a year are emitted through chlorine electrolysis. That compares with around 20 000 tonnes a year occurring naturally, says Barrie Gilliat, the chairman of Euro Chlor's technical committee. By contrast, it would cost about DM10 bn to replace mercury electrolysis with membrane technology. 'Does it make any sense to spend DM10 bn to eliminate 12 tonnes of mercury when nature and the rest of the world make 19 988 tonnes?' he asks.

Chlorine is often used as a disinfectant in water treatment to protect public health and indeed, for developing countries, there is no practicable substitute. Its use as a disinfectant has materially improved sanitary conditions throughout the world and saved or improved countless lives. ($\lozenge$ WATER SUPPLY)

Source: 'Economy v Ecology', *Financial Times*, 15 February 1995.

Chlorofluorocarbons (CFCs). A class of chemical compounds commonly used as SOLVENTS, AEROSOL PROPELLANTS, REFRIGERANTS, and in foam productions such as CFC-11 (CFCl$_3$ or TRICHLOROFLUOROMETHANE) or CFC-12 (CCl$_2$F$_2$ or dichlorodifluoromethane). The gases are chemically inert, but after release and entry into the STRATOSPHERE, where absorption of short-wave radiation takes place, each chlorofluorocarbon molecule decomposes with the release of the radical atomic CHLORINE which cannot exist independently and thus attacks OZONE (O$_3$) through catalytic chain reactions which lead to ozone depletion.

The decomposition of the trichlorofluoromethane (CFCl$_3$) under ultraviolet radiation is given by:

$$CFCl_3 \xrightarrow{ultraviolet} CFCl_2 + Cl$$

The chlorine atoms are extremely reactive and each chlorine atom can destroy up to 10^5 molecules of ozone by converting OZONE (O$_3$) molecules to ordinary OXYGEN (O$_2$) molecules before removal from the atmosphere by reaction with METHANE to form hydrochloric acid which is precipitated as given by:

$$Cl + CH_4 \rightarrow HCl + CH_3$$

The amount of chlorine in the atmosphere is estimated as 3 ppb. The target is to get it down to 2 ppb.

At the Antarctic, a different chemistry applies and inorganic forms of chlorine such as HCl and chlorine nitrate (ClONO$_2$) can be converted to the active forms of chlorine Cl and ClO respectively which then attack the ozone (O$_3$), as shown in the following equations:

$$HCl + ClONO_2 \xrightarrow{ice\ clouds} Cl_2 + HNO_3$$

$$Cl_2 \xrightarrow{ultraviolet} Cl + Cl$$

$$Cl + O_3 \rightarrow ClO + O_2$$

The virtual isolation of the polar atmosphere at certain times of the year is also a contributory factor in the ozone depletion process.

The gases are characterized by a long atmospheric lifetime because of their chemical and biological inertness. Their relative insolubility in water prevents rapid removal by cloud formation and precipitation.

The long atmospheric lifetime coupled with the delay period before the chlorofluorocarbons diffuse upwards and encounter the short-wave ultraviolet radiation mean that changes in the ozone layer will be observable through most of the next century even if production were held at current rates. Once again, the need for environmental acceptability tests for the products of the synthetic chemicals industry is demonstrated. (◊ BUTANE; GREENHOUSE EFFECT)

As CFCs are phased out, as a result of the Montreal Protocol many buildings using chilled water for air conditioning, where the refrigerant is currently a CFC, will have to use new plants or refrigerators. One environmentally-friendly alternative is provided by Citigen (London) which

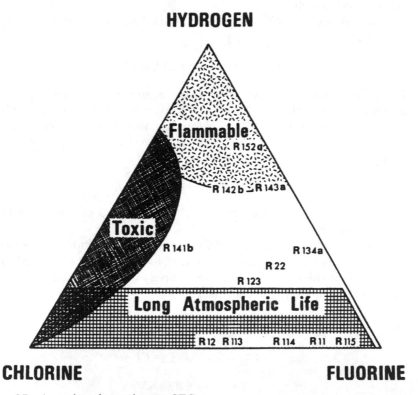

Figure 25 Assessing alternatives to CFCs.
Source: S. Pekbett, 'The changing atmosphere – ozone holes and greenhouses', *The Chemical Engineer*, August 1989, 14–18. (Reproduced by permission).

uses a COMBINED HEAT AND POWER (CHP) plant to supply electricity and both hot and cold water to key London buildings. The centralized system offers increased reliability and substantially lower running costs, allied with better utilization of the CHP plant which supplies chilled water in summertime when the heat demand is at its lowest, thus balancing the load.

Chlorophenols. Major group of CHLORINATED HYDROCARBONS, PESTICIDES and biocides which in the UK account for over 85 per cent of the non-agricultural pesticide use, such as anti-rotting agents in non-woollen textiles and wood preservatives.

The chlorophenols act as biocides by inhibiting the respiration and energy-conversion processes of the micro-organisms. They are toxic to humans above 40 parts per million (ppm), and to fish above 1 ppm, whilst concentrations as low a 1 part per thousand million can taint water. Little is known about their ecological effects, which is not to imply that they are harmless in this respect.

Chlorophyll. A combination of green and yellow pigments, present in all 'green' plants, which captures light energy and enables the plants to form carbohydrate material from CARBON DIOXIDE and water in the process known as PHOTOSYNTHESIS. It is found in all ALGAE, phytoplankton, and almost all higher plants.

Chloroprene. A chemical used for the production of synthetic rubber, Neoprene, which had been identified as a possible CARCINOGEN. Workers in a Russian Neoprene factory showed higher than normal levels of lung and skin cancer.

Chlorosis. A condition of reduced CHLOROPHYLL in green plants marked by yellowing or whitening which can be caused by a variety of diseases, pollutants and ACID RAIN.

Cholera. Bacterial infection spread by contamination of drinking-water supplies by sewage effluent and infected food. It is prevalent in Eastern countries and those areas where the growth in tourism has not been matched by the development of safe water supplies and adequate sewage disposal facilities. Tourist camping sites and recreation centres near lakes and rivers are particularly sensitive areas where precautions must be taken.

Prevention of infection is by the use of separate piped water and sewage systems and chlorination of water supplies before use. (◊ CHLORINE; WATER SUPPLY; SEWAGE TREATMENT)

Chromatography. An analytical technique used principally for separating and identifying the components of a sample by distributing them between a stationary and a moving phase. The stationary phase may be a solid, a liquid or a gas and may take the form of a column, a layer, or a film; the moving phase may be either a liquid or a gas. In gas chromatography (GC) the moving phase is a gas and the stationary phase is either a liquid (gas–liquid chromatography, GLC) or a solid (gas–solid chromatography, GSC). Gas chromatography, particularly with electron capture detection, can be used for the determination of certain CHLORINATED HYDROCARBON air pollutants; with flame ionization detection it can be used for the determination of CARBON MONOXIDE and other compounds.

Chrome wastes. Chromates and chromic acid are common wastes from certain industries, such as chromium plating and leather tanning. Chromates are soluble in water and are toxic to sewage treatment processes. (The hexavalent ion is the most toxic – it is present in chromic acid – and must be reduced to the trivalent state to form an insoluble product before being released into the environment.) Chromates act as irritants to eyes, nose and throat, and may cause dermatitis. On prolonged exposure there is liver and/or kidney damage and possible carcinogenic effects.

Hexavalent chromium may also cause abnormalities in chromosomes.

Chromium (Cr). A hard white metal. It is used as a steel-alloying element and in plating. The metal is stable and non-toxic, but there are water-soluble compounds which are extreme irritants and highly toxic. Chromium is a trace element essential for fat and sugar metabolism. Chromium aerosols can affect health in concentrations above 2.5 micrograms per cubic metre. Its presence as chromates can cause dermatitis at very low concentrations to individuals whose work has sensitized them, by contact with the compounds; for example, cement contains sufficient chromium (0.03 to 7.8 micrograms per gram) to cause dermatitis in sensitive people.

Source: G. L. Waldbott, *Health Effects of Environmental Pollutants*, 2nd edn, C. V. Mosby, St Louis, 1978.

Chronic. Medical term used in relation to the effects of certain pollutants, such as ASBESTOS, to describe a response which develops as a result of long and continuous exposure to low concentrations. It is contrasted with ACUTE. (◊ LEAD)

Civic amenity site. A civic amenity site (or CA site) is a waste treatment site where the public can dispose of household waste. Civic amenity sites are run by the municipality in a given area. Collection points for recyclable waste, such as green waste, metal and glass can also be located there. Items that cannot be collected by local waste collection schemes, such as bulky waste, can be deposited there.

Cladding.

1. The protective layer covering the fuel in a nuclear fuel element.
2. The stainless-steel lining of a PWR coolant circuit.

Clarifier. A mechanical device for removing solids from water. It is used for treating the effluent from paper mills to remove fibrous tissues and fillers. It is also used in water treatment for domestic or industrial use. (◊ DE-INKING; PULP; PAPER; WATER SUPPLY)

Clay. Fine-grained sedimentary rock of low PERMEABILITY which is capable of being shaped when moist. Consists of fine grains less than $4\,\mu m$ in diameter. (◊ MUD; SILT)

Clean Air Act (UK). The name given to two Acts passed by the United Kingdom Government. The 1956 Act dealt with the control of SMOKE from industrial and domestic sources. It was extended by the Act of 1968, particularly to control gas cleaning and heights of stacks of installations in which FUELS are burned and also to deal with smoke from industrial open bonfires. Subsequent legislation has amended the Acts.

Clean Air Act (USA). The name given to two Acts passed by the US Government. The Act of 1963 affirmed the authority of the Federal Government in dealing with interstate pollution situations, although it recognized air pollution to be primarily a state and local problem. The Act of 1970 empowered the Federal Government to set national air pollution standards (in place of state standards) and required the states to meet those standards (but to develop their own ways of doing so). It also set up air quality control regions immediately.

Clean coal technology. There is no real alternative to coal-based power generation in the long term other than nuclear power, and this may not necessarily be acceptable to many communities on grounds of both cost, perceived risk and the problems of NUCLEAR REACTOR WASTES.

Coal is commonly perceived as a dirty FUEL because of its emissions of CARBON DIOXIDE and relatively low thermal efficiencies as well. The way round this is to utilize clean coal technology which relies on GASIFICATION and the use of combined cycle gas and steam turbines. This lifts the EFFICIENCY of use from around 35 per cent to over 55 per cent, with the prospect of 60 per cent in the near future. The gasification process is a generic technology which can generate clean combustible gas from biomass and OIL as well as coal.

Clean Development Mechanism (CDM). Developing countries can use the Kyoto Treaty's Clean Development Mechanism (CDM) to sell carbon emissions to developed countries, e.g. one company (Climate Change Capital) uses Western technology to cut CO_2 emissions and then sell the spare tonnes to European companies. This enables the reduction of CO_2 emissions in say, China using advanced technology, to be traded. The environment is a net winner.

Source: *Times* 12 September 2006.

Climate. The earth's climate has gradually evolved as the ATMOSPHERE has stabilized. The controlling factor is latitude which governs the intensity of incident solar radiation. Atmospheric currents and oceans iron out large inequalities in TEMPERATURE. Of the incoming solar energy, an average of 30 per cent is reflected by the atmosphere back to space, 50 per cent is absorbed at the earth's surface, and the remainder is absorbed by the atmosphere, which in turn sets up atmospheric circulation patterns. Figure 26 shows on the left the incoming short-wave radiant ENERGY and on the right the outgoing long-wave radiant energy. The input of solar energy (100) equals the output of short-wave (28) plus long-wave (72) radiation. Thus the system is balanced. (◊ CARBON CYCLE)

A fraction of 1 per cent of the incoming solar energy is fixed by PHOTO-SYNTHESIS to form trees and plants, and this is the total energy fixation on which life on earth depends. The main anthropologied influences on global climatic patterns are:

1. Increased CARBON DIOXIDE content (◊ GREENHOUSE EFFECT; ENERGY, effect of conversion on CLIMATE).
2. A change in the ALBEDO (which can affect the MEAN ANNUAL TEMPERA-TURE) by clearing jungles, creating reservoirs, and laying vast areas under concrete and asphalt.
3. DUST/AEROSOL emission, which can reflect ionizing solar radiation.
4. Stratospheric properties may also be affected by SUPERSONIC FLIGHT (◊ CONCORDE) or possibly by CHLOROFLUOROCARBON propellant disso-ciation attacking the OZONE layer (◊ AEROSOL PROPELLANT).
5. OZONE SHIELD depletion.

Although these processes have some potential for changing the climate, it is unlikely that there will be drastic modification, but one caveat must be entered. *If the demand for fossil-fuel derived energy continues to grow, the heat balance of the earth could be upset as all energy used is ultimately degraded to heat* (thermal energy) which must be re-radiated into space to maintain the earth's radiation balance. This is linked with and to the GREENHOUSE EFFECT. The introduction and use of 'industrialized' energy sources such as fossil fuels and nuclear power mean that more energy must be re-radiated to space and the earth's radiation balance could be upset if the growth contin-ues. The calculations are such that if the 'industrialized' energy input reaches 1 per cent of the solar input, then major alterations in the earth's climate can be expected as atmospheric circulation patterns will be disturbed, with unknown consequences. Currently, we are not near this figure, but the expo-nential growth of energy consumption shows it is not entirely inconceivable that this could be reached. Thus, the earth's climate patterns place an ulti-mate restriction on long-term energy growth. While we are nowhere near the

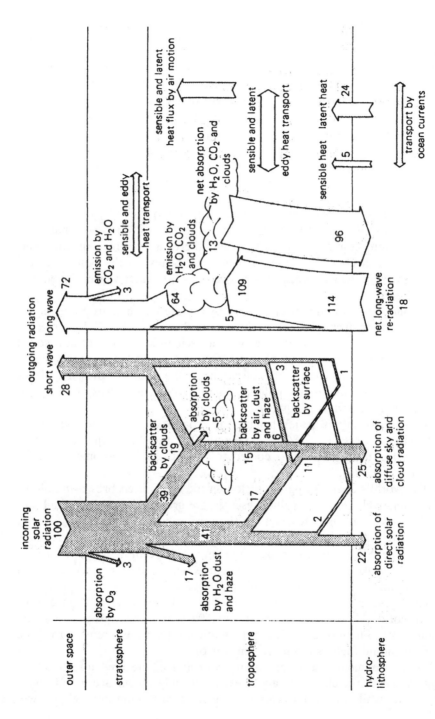

Figure 26 The earth's radiation balance between incoming (solar) radiant energy (on the left) and outgoing (terrestrial) radiant energy (on the right). The figure also shows the distribution of energy in the global system.

postulated limit, it is important to remember that future energy production may well be in very concentrated locations so that very high-energy densities will take place locally and could conceivably cause major regional climate upsets. The atmospheric system is finely tuned and there are limits to what it, or any other natural system, will tolerate. The current global warming trend shows humankind cannot really control atmospheric processes and should exercise much more care. (◊ EXPONENTIAL CURVE; THERMAL POLLUTION)

Source: S. H. Schneider and R. D. Dennett, 'Climate barriers to long-term energy growth', *Ambio*, vol. **4**, no. 2, 1975.

Clouds. A mass of water droplets formed in the atmosphere by condensation of water vapour around nuclei such as salt, dust and soil particles. Condensation commonly occurs when there is a drop in air temperature which cools the moist air mass to below its dewpoint (i.e. the temperature at which precipitation occurs in a water-vapour-laden gas stream).

Club of Rome, The. A body set up in 1968 by an international group of economists, scientists, technologists, politicians and others. The Club's object is the study of the interactions of economic, scientific, biological and social components of the present human situation, in the hope of eventually being able to predict, with some degree of certainty, the results of present policies and to formulate alternative policies where it is deemed necessary on environmental and survival grounds.

CNG (Compressed natural gas). When used as a road FUEL it is ca. 90% METHANE and emits 20% less CO_2, 70% less NO_x and 60% fewer PM_{10}s than an equivalent DIESEL fuelled vehicle. If cooled to −162°C, CNG becomes liquefied natural gas (LNG).

Coal. The world's coal and lignite (a low-grade coal, often brown in colour, with a relatively low heat value) reserves are greater than those of oil and gas. Two-thirds of these reserves are thought to be in Asia, but they are yet to be proven. Approximately 48 per cent of the world's known coal reserves are in the USSR, Eastern Europe, and China, 9 per cent in Western Europe, 6 per cent in Africa, 26 per cent in North America and 1 per cent in Latin America, and 9 per cent in Australia and Asia.

World resources of reasonably accessible coal are sufficient to meet projected world demands for the next 300–500 years, although this estimate would have to be severely reduced if significant quantities are used to make synthetic liquid fuels as a replacement for oil.

The environmental costs of mining coal and lignite are, unfortunately, high. Much of the cheapest can be extracted only by surface mining on a large scale while the social costs of conventional mining are well known.

Furthermore, when used as a fuel, coal is the dirtiest of the fossil fuels. It is possible that enlightened programmes of restoration, or remote mining, and of rigidly enforced controls on air pollution, may minimize such environmental impacts to acceptable levels, but the costs involved can be high. (◊ TAR SANDS; OIL SHALES; CLEAN COAL TECHNOLOGY)

Source: B.P. Statistical Review of World Energy, July 1989.

Atmospheric scientists at Texas A&M University, USA, recently reported that sooty particles from China's coal-fired power plants are fuelling cloud formation over the Pacific Ocean, which in turn causes more intense storms that could have a knock-on-effect on air-flow patterns around the world.

Source: tce, April 2007.

From World Energy Outlook 2006, published by the International Energy Agency:
As the second largest primary fuel, coal will continue to play a major role in the world energy mix. Demand for coal is projected to experience the biggest increase, rising by 32 per cent by 2015 and 59 per cent by 2030. Demand for coal will increase the most in developing countries (with 86 per cent coming from developing Asia).

Source: ECOAL, January 2007.

Coal production increased by over 7 per cent in 2005, to 4973 million tonnes – with China accounting for 2226 million tonnes of this total. Global coal consumption increased from 4646 Mt in 2004 to 4990 Mt in 2005.

Top Ten Hard Coal Producers (2005)

PR China	*2226 Mt*	*Russia*	*222 Mt*
USA	951 Mt	Indonesia	140 Mt
India	398 Mt	Poland	98 Mt
Australia	301 Mt	Kazakhstan	79 Mt
South Africa	240 Mt	Columbia	61 Mt

Source: E Coal, Oct 2006, Vol. 59.

The Table overleaf shows the different stages in the transformation of vegetable matter into coal.

Progressive Stages of Transformation of Vegetable Matter into Coal (increasing in rank, going down the list)

Fuel Classification by Rank	Locality	Moisture (as received)	Analysis on dry basis							
			Proximate			Ultimate				
			VM	FC	Ash	S	H_2	C	N_2	O_2
Wood		46.9	78.1	20.4	1.5		6.0	51.4	0.1	41.0
Peat	Minnesota	64.3	67.3	22.7	10.0	0.4	5.3	52.2	1.8	30.3
Lignite	North Dakota	36.0	49.8	38.1	12.1	1.8	4.0	64.7	1.9	15.5
Lignite	Texas	33.7	44.1	44.9	11.0	0.8	4.6	64.1	1.2	18.3
Subbituminous C	Wyoming	22.3	40.4	44.7	14.9	3.4	4.1	61.7	1.3	14.6
Subbituminous B	Wyoming	15.3	39.7	53.6	6.7	2.7	5.2	67.3	1.9	16.2
Subbituminous A	Wyoming	12.8	39.0	55.2	5.8	0.4	5.2	73.1	0.9	14.6
Bituminous High Volatile C	Colorado	12.0	38.9	53.9	7.2	0.6	5.0	73.1	1.5	12.6
Bituminous High Volatile B	Illinois	8.6	35.4	56.2	8.4	1.8	4.8	74.6	1.5	8.9
Bituminous High Volatile A	Pennsylvania	1.4	34.3	59.2	6.5	1.3	5.2	79.5	1.4	6.1
Bituminous Medium Volatile	West Virginia	3.4	22.2	74.9	2.9	0.6	4.9	86.4	1.6	3.6
Bituminous Low Volatile	West Virginia	3.6	16.0	79.1	4.9	0.8	4.8	85.4	1.5	2.6
Semianthracite	Arkansas	5.2	11.0	74.2	14.8	2.2	3.4	76.4	0.5	2.7
Anthracite	Pennsylvania	5.4	7.4	75.9	16.7	0.8	2.6	76.8	0.8	2.3

VM – volatile matter
FC – fixed carbon

Source: Combustion Engineering Inc. Reference Book, Published by Combustion Engineering, 277 Park Avenue, New York, 1966

Coal equivalent (ce). The amount of (standard) coal that would on combustion produce the same quantity of heat as a given mass of a given fuel. Coal equivalents can vary from one country to another.

Coal-fired power plant – CO_2 emissions. The EU has ranked 30 coal-fired power stations as being responsible for 10 per cent of the EU's CO_2 emissions. These include Greece's Agios Dimitrios (lignite fuel at 1350 g of CO_2/kWh, to Longannet (Scottish Power, low grade hard coal, at 970 g CO_2/kWh). Naturally, the UK's Drax (3945 MW) features on volume criteria at 22.8 Mt/y – context is all. Drax has one of the highest UK coal-fired thermal efficiencies and it just cannot be shut down. Once again, the problem is posed but the solutions are not provided. What do reports like this prove? Power stations are vital to our well-being.

Coal gasification. ◊ IGCC.

Coastal defences. Coastal erosion may be caused by the washing away of materials from the shore by waves or the current, or by the falling into the sea of cliffs and other structures because of geological instability. Other causes of erosion are harbour breakwaters and approach channels which disturb the natural coastal morphology, as do natural rivers and estuaries, and the action of groundwater or streams running through rock formations.

Since not all natural coastal changes are desirable, coastal defence works may be needed. Defence works are used to retard the natural erosion processes or, sometimes, to neutralize their effects.

Before taking any action to solve the problem, the material composition of the coastline must be ascertained. Knowledge of both the beach and the breaker zone materials is required before an assessment of the problem may be made. The composition may be SAND, CLAY (mud) or a mixture, or rock which is further classified as soft, or weak, brittle rock. The problem may be loss of sand from a beach or a change in shape, sand migrating inland, rock fall from cliffs or erosion at the base.

Depending on the circumstances, various remedies include seawalls, vertical or sloping, groynes and breakwaters, dykes, artificial beach formation, dunes plus grass planting, and shoring up of cliffs or building undercliff passes.

Cobalt-60. Radioactive ISOTOPE of cobalt. Cobalt-60 emits high-energy gamma radiation and has a HALF-LIFE of over five years, thus rendering it liable to accumulation in areas of discharge. It is particularly damaging to biological systems because of the RADIATION hazard, and therefore its discharge is subject to very strict licensing procedures.

Cobalt-60 is an inevitable waste product from nuclear reactors, which use high cobalt steel fuel cans which are, of course, irradiated in the fission process. (◊ NUCLEAR REACTOR DESIGNS; IONIZING RADIATION, EFFECTS)

Coefficient of Performance (COP). A measure of refrigeration plant or heat pump EFFICIENCY (sometimes termed reversed CARNOT) given by

$$COP = \frac{T_e}{T_e - T_c}$$

T_e = evaporation temperature (K).
T_c = condensing temperature (K).
A typical temperature range for a heat pump ($T_e - T_c$) is 55 K (55 °C), i.e. $T_c = 278$ K (5 °C), $T_e = 333$ K (60 °C). Hence

$$COP = \frac{333}{55} = 6.05.$$

This means that 6 times the quantity of electrical ENERGY supplied is discharged as 'heat' or thermal energy. In practice, the ideal (reversed carnot) COP cannot be achieved. Perhaps 60 per cent can be. HEAT PUMPS can make useful contributions to low grade heating application. (◊ CARNOT EFFICIENCY)

Co-generation. ◊ COMBINED HEAT AND POWER.

Coking coal. Also known as metallurgical coal – vital for steel production. Coking coal consumption was 633 Mt in 2005. Around 73 per cent of coking coal demand worldwide is from China, India, Japan, Russia and the Ukraine. See Figure 27 below.

Coliforms. A broad class of bacteria found in soil and faeces of mammals. Their presence in water may indicate the presence of disease-causing organisms.

Great concern has been raised about SEWAGE contamination of bathing waters. One well-known Cornwall beach had, in January 1993, levels of 2.6 million units per 100 ml water (legal level 10 000, and 500 for EC Blue Flag beaches).

Coliform count. A water-purity test: the number of presumptive coliform bacteria present in 100 millilitres of water. The organism *Escherichia coli* is used to indicate the presence of faecal matter which can spread enteric diseases. *Escherichia coli* is the main species of bacteria present in human excreta and while only 1 in 1 million of the bacteria may be pathogenic to

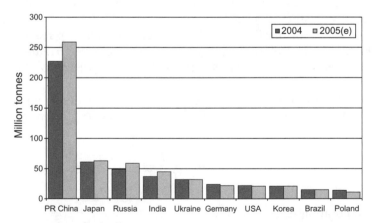

Figure 27 Top ten coking coal consumers.
Source: *IEA 2006*. e = estimated.
Source: World Coal Institute Newsletter, June 2007.

humans, the presence of any *E. coli* is an indication of polluted water. The water cannot, therefore, be distributed for consumption without proper sterilization.

Collection tax. This is a form of the polluter-must-pay philosophy. For example, the non-returnable BOTTLE is subsidized by the community who pay for the cost of refuse collection and disposal. A collection tax would make the purveyor/manufacturer of goods in non-returnable bottles pay for their collection and disposal and, of course, may influence an increase in the use of returnable bottles. (◊ RECYCLING, FINANCIAL INCENTIVES FOR)

Combined cycle power station. This type of plant is flexible in response and can be built in the 100–600 MW capacity range. It produces electrical power from both a gas turbine (ca. 1300 °C gas inlet temperature), fuelled by NATURAL GAS or OIL plus a steam turbine supplied with the steam generated by the 500 °C exhaust gases from the gas turbine. The THERMAL EFFICIENCY of these stations is ca. 50 per cent compared with a maximum of 40 per cent from steam turbine coal-fired power stations.

This type of plant can be built in two years compared with six years for a coal-fired station and 10–15 years for nuclear. The commonly used fuel in the UK is natural gas. (◊ COMBINED HEAT & POWER). However, Centrica is to build Britain's cleanest coal-fired (gasified) power station producing 0.15 tonnes of carbon per MW/ hour compared with 0.9 for a traditional power plant and 0.45 for a traditional gas-fired facility. The £1 bn plant could be operating by 2011 and would provide 800 MW of electricity. Most of the carbon dioxide would be removed and sent for storage. (◊ COMBINED HEAT & POWER)

Large COMBINED-CYCLE systems (with two ways of driving the same electric generator) are able to offer electrical efficiencies of 58 per cent, with 60 per cent in prospect within the next few years. Medium-sized combined-cycle plants are now capable of electrical efficiencies extended above 50 per cent. Such power stations, connected to industrial premises or district-heating networks, can achieve total efficiencies as high as 90 per cent. Merwedekanaal Station in Utrecht, when operating in district-heating mode, has an 87 per cent efficiency.

Combined-cycle plants burning natural gas can be the least polluting thermal generators of electric power since natural gas produces virtually no SO_2 and relatively little CO_2 and unburnt hydrocarbons during combustion.

Another environmental advantage is the reduced cooling-water requirement, amounting to one-third less than for a conventional thermal power station. This is because only a minor part of the power in a combined-cycle plant is generated by the steam turbine, which largely determines the cooling requirement. As a result, the cooling water system can be smaller and less water has to be treated before returning it to the environment.

Combined-cycle plants can burn a wide range of fuels at high efficiency. These include all types of natural gas, oils from diesel to crude and heavy oils, and also coal and residual oils in various states. (◊ IGCC).

Combined heat and power (CHP). Combined heat and power or co-generation are the terms used to describe joint production of heat (often in the form of steam) and power (usually in the form of electricity).

In the UK some 6 per cent of electricity is produced from CHP plants. Approximately 50 per cent is generated from steam condensing turbines with an overall efficiency of 32–38 per cent. Modern gas turbines achieve much higher electrical efficiencies especially when operating in the combined-cycle mode with the exhaust heat from the gas turbines used to raise steam and drive a steam turbine. Combined cycles convert 45–50 per cent of fuel into electricity and account for the remainder of UK output.

CHP plants can either be steam driven (Figure 28) where the output of electricity to steam is in the range of 1:2 to 1:6 or non-steam such as gas turbines where the high temperature products of combustion are used to produce heat in the form of steam or hot water (Figure 29), spark ignition engines or diesel engines (Figure 30). Non-steam plant can be combined with steam turbines to form a combined cycle and the residual heat used for the process applications (Figure 31).

In the case of a CHP plant, the electricity produced can be used onsite, sold privately to other consumers or alternatively fed into the national grid. The heat produced can be used onsite for process (industrial CHP) or for space and water heating (micro-CHP) or distributed through a mains network (district heating) (Figure 32).

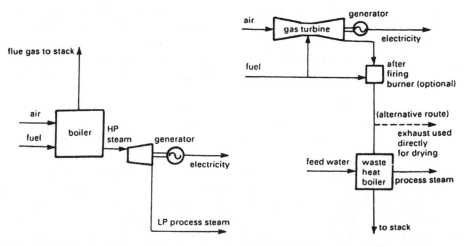

Figure 28   Steam turbine CHP plant.   **Figure 29**   Gas turbine CHP plant.

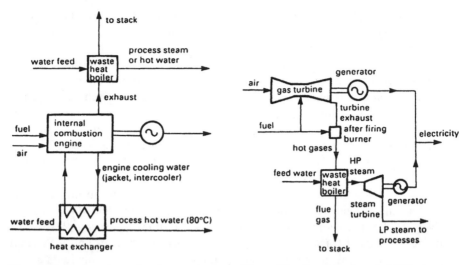

Figure 30 Internal combustion engine CHP plant.

Figure 31 Combined cycle CHP plant.

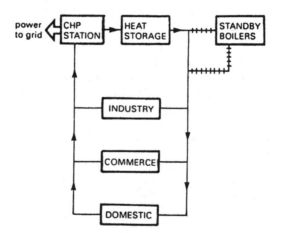

Figure 32 Combined heat and power with district heating.

The CHP market can be broadly sub-divided into three main categories of industrial CHP, micro-CHP in buildings and large-scale CHP/district heating (DH). The common factor is a higher energy conversion efficiency and consequent fuel savings. (◊ DISTRICT HEATING)

DH is widely adopted in many Western European cities and most Scandinavian towns. The UK lags badly in the DH field but there is widespread adoption of CHP in the process industries where a base heat load is guaranteed. Sewage SLUDGE digestion can also be assisted in a cost-effective manner by the adoption of CHP using the METHANE gas from the digestion as a fuel for modified internal combustion engines. The hot engine cooling jacket water is used to heat the digesters to enable optimum digestion to take place. Electrical power is also produced by the engines. Such a scheme cost the Northumbrian Water Authority £42800 in 1985 at their Cramplington treatment works. This generated £9000 per year in savings with a resulting PAY BACK of five years thus paving the way for the consideration of similar set-ups where populations in excess of 5000 are served. (◊ ANAEROBIC DIGESTION)

The (1997) 'Mitte' new combined cycle power plant built in the centre of Berlin, supplies electricity and heat rated at $430\,MW_e$ and $630\,MW_{th}$, respectively, to the civic centre, government offices and other major complexes. The plant is designed for an electrical efficiency of 47.4 per cent and a thermal efficiency of almost 90 per cent. The plant's central location made its architecture and environmental compatibility important planning issues. ABB is the lead contractor. When it is realised that common UK power plant thermal efficiency can be less than 40%, the scale of energy savings becomes apparent. Many UK wasted opportunities should be laid at the doors of successive governments.

Source: *Professional Engineering*, 6 April 1994.

Figure 33 shows reduction in emissions from the Vartaverket CHP plant run by Stockholm Energi. This plant heats most of Stockholm.

The following table shows data for the Holstebro/Struer CHP station.

Net power output	28 MW
Power rating	27%
Annual power production	$160 \times 10^6\,kWh$
Net heat output	67 MJ/s
Heat rating	61%
Heat production	1439000 GJ
Inlet temperature	75–90 °C
Return temperature	35–55 °C
Heat efficiency	88%
Steam temperature, turbine	520 °C
Steam pressure, turbine	65 bar
Station price (1989 prices include build.int)	515 million DKR

Source: Stockholm Energi, reported in *Local Government News*, September 1996.

The Roskilde (Denmark) waste incineration plant (see Figure 34) has a capacity of 20 tonnes of waste per hour, corresponding to 150000 to 160000

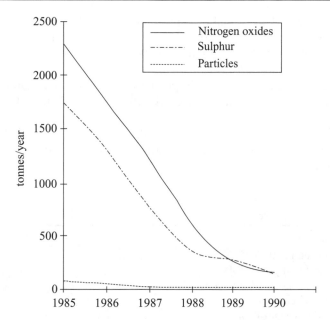

Figure 33 Reduction in emissions from the Vartaverket CHP plant.

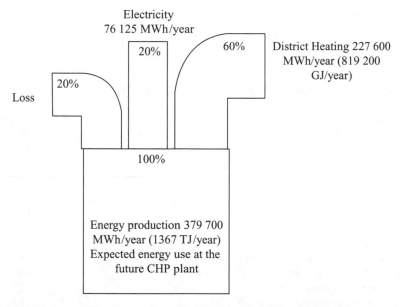

Figure 34 Energy utilization at the new CHP Roskilde (Denmark) plant.
Source: Kara (Denmark) Brochure, 1993.

tonnes per year. The annual heat production is 850 TJ and the annual production of electricity is 86×10^6 kWh. These figures correspond to the heat
consumption of 14 000 single-family houses and the electricity consumption
of 20 000 single-family houses.

This saves foreign currency for the purchase of fuels for CHP plants. And, most
importantly, the environment is spared pollution by CO_2 from burning fossil
fuels.

The performance parameters of the Berlin Mitte district heating power
plant (at the design point) and the main water–steam parameters are given
in the tables below.

Performance parameters.

Integrated plant	
Ambient temperature (°C)	0
Gross electrical power (MW)	380
Heat output, normal/maximum (MJ/s)	340/380
Thermal efficiency (%)	89.2
Electrical efficiency (%)	47.4
Boiler/steam system	
Pressure levels	2
HP steam: Mass flow (t/h)	210
Exit pressure (bar)	76.9
Temperature (°C)	525
Feedwater temperature (°C)	110
LP steam: Mass flow (t/h)	48
Exit pressure (bar)	5.3
Temperature (°C)	203
District heating water preheater – heat recovered (MW)	22.5
District heating	
Medium	Hot water
Temperature out (°C)	105
Temperature in (°C)	55
Mass flow (t/h)	5838

Water-steam parameters.

Water/steam section	HP
Steam generation rate$_{max}$ (kg/s)	67.49
Operating pressure max. (bar)	90
Steam drum pressure max. (bar)	83
Exit pressure max. (bar)	82
Exit temperature max. (°C)	525
Feedwater temperature, gas/oil (°C)	110/130

Source: *ABB Review*, January 1995.

Combined sewer overflow (CSO). CSOs operate in storm conditions to divert excess diluted sewage to a nearby watercourse, thus preventing a build-up of sewage within the wastewater collection system.

Combustion. A chemical reaction in which a FUEL combines with oxygen with the evolution of heat: 'burning'. The combustion of fuels containing CARBON and HYDROGEN is said to be *complete* when these two elements are oxidized to carbon dioxide and water, e.g. the combustion of carbon $C + O_2 = CO_2$, or hydrogen $2H_2 + O_2 = H_2O$. Incomplete combustion may lead to (1) appreciable amounts of carbon remaining in the ash; (2) emission of some of the carbon as CARBON MONOXIDE; and (3) reaction of the fuel molecules to give a range of products which are emitted as SMOKE. The combustion of VOLATILE products can be rendered more complete if secondary air is admitted over or beyond the fuel bed for coal, or flame for oil and gas (air supplied through the fuel bed is termed 'primary air'). Air in excess of the amount theoretically required for complete combustion ('excess air') is kept to the minimum compatible with complete combustion (and absence of smoke) in order to avoid undue loss of heat in the stack gases, poor heat transfer, and oxidation of sulphur dioxide to sulphur trioxide.

The proportions of the pollutants will vary according to the fuel analysis, e.g. CHLORINE is mainly present in coal, municipal solid waste and various spent solvents (which may be used as fuels) but not in petrol or DERV. A basic example calculation of combustion air requirements is given below.

A brown COAL containing 70 per cent carbon and the remainder ash requires sufficient air to provide oxygen to oxidize the carbon to carbon dioxide. The oxygen requirement is dictated by chemistry and is such that 1 kg C requires 2.67 kg O_2 for complete combustion (and produces 3.67 kg CO_2 in the process). Hence, 0.7 kg C requires 1.87 kg O_2 for complete combustion, producing 2.57 kg CO_2 in the process. However, excess air is always provided to ensure complete combustion (as far as possible). If 50 per cent excess air is provided, the actual amount of air provided to burn 1 kg brown coal is:

$$\left(\frac{\text{amount of oxygen required}}{\% \text{ oxygen in air}} \right) \times 1.5$$

$$= \frac{1.87}{0.23} \times 1.5 = 12.2 \text{ kg air}$$

The nitrogen in the air provided is 77 per cent by mass. Hence, in burning 1 kg brown coal, 9.39 kg nitrogen is contained in the 12.2 kg air supplied. Thus, some nitrogen oxide formation is a certainty. The only question is how much and this depends on the technology employed. As we do not know this, we can compute the total mass of flue gases. Thus, the mass products of combustion of 1 kg brown coal are (at 50 per cent excess air):

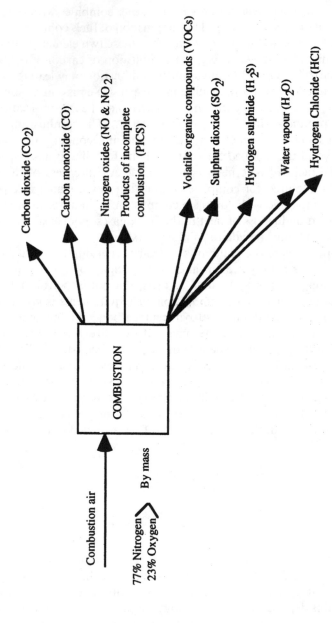

Figure 35   Possible pollutants emitted from a fuel containing carbon, hydrogen, sulphur and chlorine.

CO_2	2.57 kg
N_2	9.39 kg
O_2	0.94 kg
Total	12.9 kg gaseous products of combustion or flue gas.

It should be noted that 0.7 kg coal have been consumed in combustion, hence 0.7 kg carbon plus 12.2 kg air supplied, gives a total of 12.9 kg; this is known as a MASS BALANCE and accords with the 12.9 mass of flue gases discharged.

Also, as the volume of gas occupied by a KILOGRAM MOLE of gas at STANDARD TEMPERATURE AND PRESSURE (STP) is $22.4 \, m^3$, we can calculate the volume of flue gases emitted at STP from the combustion of 1 kg brown coal as follows, given that 1 kg mole $CO_2 = 44$ kg, 1 kg mole $N_2 = 28$ kg, 1 kg mole $O_2 = 32$.

Then from the combustion products table, the volume of flue gases discharged at STP from the combustion of 1 kg brown coal containing 70 per cent carbon in 50 per cent excess air is as follows:

$$CO_2 \text{ volume} = \frac{22.4 \times 2.57}{44} = 1.3 \, m^3$$

$$N_2 \text{ volume} = \frac{22.4 \times 9.39}{28} = 7.5 \, m^3$$

$$O_2 \text{ volume} = \frac{22.4 \times 0.94}{32} = 0.66 \, m^3$$

$$\text{Total} = 9.45 \, m^3$$

In addition, 0.3 kg of ash and particulates will be discharged as solids.

This simple calculation shows that very large quantities of combustion air are required to burn fuels and that commensurably large volumes of flue gases are emitted.

Theoretically, a $50 \, MW_T$ (thermal) output boiler plant will consume

$$\frac{50}{\text{calorific value of coal (MJ/kg)}} \text{ kg coal per second.}$$

As the CALORIFIC VALUE of 70 per cent brown coal is 27 MJ/kg, this means our $50 \, MW_T$ plant consumes

$$\frac{50}{27} \times \frac{60 \, (s)}{1000 \, (kg)} \times 60 \, (min) \times 24 \, (h) \text{ tonnes per day}$$

or 160 tonnes of coal per day (in practice, a greater amount will be consumed because of combustion losses). It emits (at 50 per cent excess air) $1\,513\,600 \, m^3$

of stack gases at STP per day, which contain 411.2 tonnes of CO_2, i.e. 411.2 tonnes of CO_2 are emitted per day from the $50\,MW_T$ boiler plant.

Note: If the combustion efficiency is 90 per cent the plant will consume 11 per cent more fuel and the pollutant outputs will be increased accordingly. If 80 per cent efficient it will consume 25 per cent more fuel with correspondingly increased emissions. This is a point to note well where sustainability is concerned – efficiency is important. Yet *CHP* adoption in the UK is abysmally low, eye catching wind turbines would appear to be preferred.

Corrections for combustion gas analysis

1. Excess O_2 correction

$$\text{Corrected value} = \text{measured value} \times \frac{21 - O_2 \text{ ref. value (\%)}}{21 - O_2 \text{ measured value (\%)}}$$

2. Wet to dry results

$$\text{Dry value} = \text{measured value} \times \frac{100}{100 - H_2O \text{ measured concentration (\%)}}$$

3. Temperature correction

$$\text{Corrected value} = \text{measured value} \times \frac{\text{temperature measured (K)}}{273\,(K)}$$

4. Corrected pressure

$$\text{Corrected value} = \frac{\text{measured value} \times 101.3\,(kPa)}{\text{measured pressure (kPa)}}$$

Source: ETI Group Ltd, Cheltenham, Glos., *The Little Blue Book* (Environmental Tables and Information 1997).

Commercial wastes. ◊ WASTES.

Common Inheritance, Government White Paper. A UK Government White Paper which looks at all levels of environmental concern and describes what the Government has done and proposes to do. It starts from general principles and objectives and it discusses the government's approach to the environmental problems affecting Britain, Europe and the world.

The main objectives are:

• protecting the physical environment through the planning system and other controls and incentives,
• using resources prudently, including increasing energy efficiency and recycling, and reducing waste,
• controlling pollution through effective inspectorates and clear standards, and

- encouraging greater public involvement, and making information available.

Commons, The. The concept of The Commons is one of common ownership of resources of value to the community. Originally the term referred to common land used for pasture, but it has recently been used to describe the common environmental resources: land, air, water and so on. The relevance of the analogy has been pointed out by Garrett Hardin, an American biologist, who has shown that the more an individual (or corporation) exploits The Commons, the greater the harm to the community as a whole. Hence, protective measures are required to control greed, otherwise we could all act as free enterprises putting little or nothing back for future generations.

It is sometimes said that market forces will take care of these problems. However, the price of a commodity depends on the difference between the rate of supply and the rate of demand. If supply slows down gradually, a pricing mechanism may evolve but too often the point of extinction is virtually reached before this happens. For example, fish can now be caught in much greater quantities by means of new technology and sold at a lower price because of this, yet this is a prime example of how to deplete a resource. The actual disappearance of blue whales, then the fin whales, etc., demonstrates this quite clearly. (Or the extinction of the dodo, the US passenger pigeon and virtually the buffalo.) In other words, market forces can lead to greater depletion rates; witness Saudi Arabia or Iran increasing oil production rates when revenues drop thus leading to faster resource depletion. (◊ WHALE HARVESTS)

Source: G. Hardin, 'The tragedy of The Commons', *Science (NY)*, vol. **162**, 1968, pp. 1243–8.

Community liaison group. A means of community involvement for large-scale industrial operations (e.g. INCINERATION). A community liaison group could comprise representatives of the development company (or of the owners and operators as appropriate), planning authority representatives and a cross-section of local community representatives. A third party facilitator may be appropriate. The frequency of the meetings and their remit should be agreed by all parties on a basis which is relevant to each site.

Compactor. Equipment for the compression and volume reduction of waste materials, usually into containers for onward transport.

Competent Authority. The Control of Major Accident Hazards Regulations (COMAH) are enforced by a joint Competent Authority comprising the Health and Safety Executive (HSE) and the Environment Agency in England and Wales, and HSE and the Scottish Environment Protection Agency in Scotland.

Composting. A process used to rapidly decompose ORGANIC materials and WASTES under controlled conditions. Biodegradable matter such as grasses, leaves, PAPER and kitchen waste may be broken down by aerobic micro-organisms to form a stable compost material which can be used for a variety of horticultural and agricultural applications. Efforts to reduce the amount of biodegradable waste sent to LANDFILL has encouraged the development of both domestic-scale and large-scale composting.

In the UK, home composting (i.e. composting on a domestic scale) is considered to be a waste minimization activity as it is designed to reduce the amount of biodegradable waste entering the waste stream. Home composting is strongly promoted by most local authorities working in partnership with WRAP (http://www.wrap.org.uk/) and together they have supplied over 1 million subsidized compost bins to householders. As well as traditional composting methods, home composting can take many forms, including VERMICOMPOSTING where earthworms are used to turn organic kitchen waste into compost, or the 'bokashi' method which maintains kitchen waste under anaerobic conditions to minimize foul odours during storage and to enhance subsequent home composting.

Treatment and disposal of municipal solid waste (MSW) which is not home composted becomes the responsibility of local authorities. As a consequence of the European Commission Landfill Directive, various forms of Landfill Allowance Scheme (LAS in Wales and LATS (Landfill Allowance Trading Scheme) in England, Scotland and N. Ireland where trading of allowances is permitted) have been introduced. Under this scheme, local authorities have been allocated decreasing annual tonnages of biodegradable waste which they are allowed to landfill until the year 2020. Diverting bio-degradable waste into large-scale composting is currently the preferred means of meeting these stringent landfill allowances and consequently this sector is growing rapidly. According to the Composting Association (http://www.compost.org.uk/), the amount of municipal waste composted in the UK in 2005/6 was 2.9 Mt, which was 4.7 times more than in 1999/2000. In 2005/6, the relatively simple open air, mechanically-turned windrow system of composting was the most frequently employed method (composting 81 per cent of waste) while more sophisticated enclosed facilities, often employing forced aeration systems, composted 14 per cent. Composting using enclosed systems is set to increase significantly, largely stimulated by the introduction of the Animal By Products Regulations in 2003, which imposed strict methods for composting kitchen waste due to the potential presence of meat products.

(◊ WRAP; VERMICOMPOSTING)

The first stage of composting is often to shred or pulverize the waste, after which the material is formed into large piles (see Figure 36). The process of decomposition generates heat and for composting to take place, the pile size must be sufficiently large to retain this heat, thereby raising pile temperature.

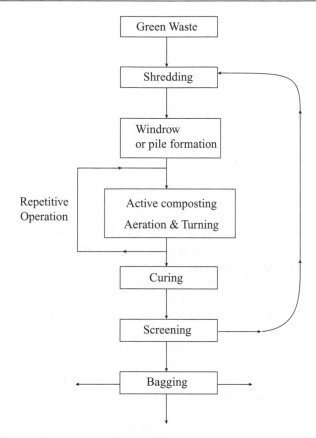

Figure 36 Composting system and operations.
Source: P. A. Wheeler and R. B. Bardos, *Preliminary Assessment of Heavy Metals in Anaerobic Digestion of Municipal Solid Waste*, ETSU Report, Harwell Laboratory, Oxon, 1991.

As pile TEMPERATURES enter the thermophilic range (45–70 °C), heat-loving micro-organisms proliferate, promoting rapid decomposition and killing most weed seeds and pathogens. After a number of weeks, the temperature drops back to ambient as decomposition slows down. On a large scale, most domestic, commercial and agricultural organic wastes can be composted within 12 weeks.

For effective composting, optimum conditions need to be maintained. In particular, sufficient moisture must be present (50–75 per cent) and the waste must contain a good balance of NUTRIENTS, with a CARBON to nitrogen ratio of around 24–35:1. Adequate levels of OXYGEN must be maintained (greater than 5 per cent). There are two main approaches to large-scale composting. The first involves open-air windrow systems where the waste is heaped into piles (windrows) and these are mixed and aerated by mechanically turning the piles. In the second approach, aeration and temperature regulation are achieved by pumping air through the piles and these composting operations are often enclosed and computer controlled.

Before waste-derived composts can be applied to land or sold they must first satisfy increasingly stringent standards. These standards typically regulate levels of potentially toxic elements (PTEs), human pathogens and physical contaminants. In order to market compost and to comply with national standards, there is a growing trend towards composting relatively uncontaminated municipal wastes, which have been segregated at source by householders. These include garden and kitchen wastes and cardboard. This is demonstrated in the table below where much lower levels of PTEs in compost derived from source segregated household waste can be seen compared with compost made from ordinary domestic refuse. For comparison, limit values for PTEs in compost are given for PAS 100 (2005), which is the voluntary specification for source segregated compost in the UK. In 2007, the UK also introduced the Quality Compost Protocol for composts which set out additional criteria to ensure that composts complying with PAS 100 can be classified as fully recycled and used without risk to the environment or harm to human health. In contrast to composting source segregated waste to produce high quality compost, un-segregated domestic waste can also be composted as part of a MECHANICAL AND BIOLOGICAL TREATMENT (MBT) process. Many countries in Europe have introduced MBT systems to pre-treat MSW prior to final disposal in order to greatly reduce methane emissions from landfill.

(◊ MECHANICAL AND BIOLOGICAL TREATMENT)

All values are in $mg\,kg^{-1}$ dry matter.

	Germany*	Witzenhausen† (average)	PAS100‡
Cadmium	5.5	1	1.5
Chromium	71	36	100
Copper	274	36	200
Lead	513	133	200
Mercury	2.4	—	1.0
Nickel	45	29	50
Zinc	1570	408	400

* Non-separated municipal solid waste.
† Source separated organic waste.
Source: L. R. Kuhlman, B. Dale, A. C. Groehof and T. F. E. Young, 'Windrow composting of Sewage Sludge, Garbage and Other Wastes', presented at the ReC, International Recycling Congress, Geneva, Switzerland, 19–22 January 1993.
‡ BSI, PAS100: British Standards Institution specification for composted materials.

Concentrate and contain. A means of waste management which relies on the wastes being concentrated and contained and kept separate from other environmental media. This is the opposite of DILUTE AND DISPERSE.

Concentration, units for. Many concentrations are expressed as parts per million (ppm) or parts per hundred million (pphm), where parts per million is based on proportional parts by volume. However, the International System of Units (SI) uses micrograms per cubic metre ($\mu g/m^3$) for air pollution, that is, weight per unit volume, and milligrams per litre (mg/l) or grams per cubic metre (g/m^3) for water pollution. (*Note*: 'gramme' is the Continental spelling.) The difference in scales for air and water is due to the density difference of a cubic metre of water and a cubic metre of air.

Now in this text, as in many books on the subject, micrograms per cubic metre values for air pollution are not used exclusively; data are often in parts per million (as in US and many UK texts). To covert from parts per million to micrograms per cubic metre, we make use of the fact that 1 mole (i.e. a mass of gas equivalent to its molecular weight in grams) of any gas at standard temperature and pressure occupies a volume of 0.0224 cubic metres (22.4 litres). Thus,

Concentration in micrograms per cubic metre

$$= \frac{(\text{concentration in ppm}) \times (\text{molecular weight of substance})}{0.0224}$$

For example, sulphur dioxide (SO_2) has a molecular weight of

$$S(32) + O_2(2 \times 16) = 64$$

Therefore, the concentration in micrograms per cubic metre under standard conditions of 1 part per million sulphur dioxide is

$$\frac{1 \times 64}{0.0224} = 2857 \text{ micrograms per cubic metre}$$

Note: for water, parts per million by volume is the same as grams per cubic metre, as 1 gram of water occupies 1 cubic centimetre, and 1 cubic metre equals 10^6 (one million) cubic centimetres. This relationship applies for dilute aqueous solutions. Where possible the units of parts per million should be avoided.

Note: The use of shorthand expressions of concentration such as ppb or ppm can be very misleading and should be used carefully. For example, the term ppb for parts per billion may mean one part in 10^9 or 10^{12} depending on whether one uses the American or European billion. Usually ppb refers to 1 in 10^9. Likewise the term ppt for parts per trillion is usually interpreted as 1 in 10^{12}, although as we have seen this can be expressed as ppb. A further confusion is that the term ppt has been used to express salinity, and in this context marine scientists refer to 1 part in 10^3 (i.e. 1 part per thousand). A further complication is that units such as ppb are ratios, but could be by weight to volume. This must be specified. Clearly a concentration in weight to volume cannot directly be expressed as ppm because the ratio must have both units the same. In such cases it is necessary to convert to common units. In the particular case of vapour concentrations in air, ppm and similar units are inadvisable because vapours do not always behave exactly as gases, and so where concentration conversions are necessary this is not always straightforward.

Further caution is necessary when, for example, concentrations in animals' tissues are quoted. While this is not always done, to be meaningful the concentration must specify whether it is in terms of wet weight, dry weight, ash weight or some other basis.

Concorde. A SUPERSONIC (faster than sound) aircraft (now withdrawn from service) which, in order to achieve the very large thrust needed for its flight speed, used turbo-jet engines which were inherently noisier than the turbo-fan types.

Two types of NOISE problem were associated with Concorde and other supersonic aircraft: the sonic boom and the extremely large noise 'footprint' (i.e. the area which receives a certain specific noise level). The sonic boom characterizes all supersonic transport (SST) and is produced as the result of a build-up of air pressure in front of the aircraft. From American studies, the pattern of the Concorde 100 EPNdB footprint ranges 54 square miles, some 41 times larger than the DC-10–30 footprint. EPNdB – effective perceived noise level – is the unit used in noise certification based upon perceived noise decibels (PNdB) corrected for particular pure-tone characteristics of jet noise that produce extra annoyance over and above that measured by PNdB (◊ DECIBEL; NOISE CONTROL; PNdB). Any substantial reduction will require major engine redesign.

Concorde's early flights broke Heathrow noise rules (measured in PNdB) on 26 take-offs out of 37. It repeatedly created noise levels above the threshold of pain (133 decibels) (*Observer*, 19 October 1975). Also, the low-frequency sound emitted by Concorde was five times as intense as that from a Boeing 707. This type of sound can cause structural damage to buildings.

In addition to noise problems, there were well-founded fears that the exhaust emissions in the stratosphere could have far-reaching and destructive effects on climatic conditions. The stratospheric ozone layer may be decreased, causing an increase in the amount of ultra-violet solar radiation reaching the earth's surface. The point really is, did we need to fly at this speed, given that quite acceptable and much more energy-efficient alternatives exist?

A former Concorde user's view is illuminating '. . . Others may consider my lifestyle – great food and drink; luxury hotels; Concorde; much travel, the best books and films – extravagance. To me they are the necessities of living a full life'. *Jonathan King, Weekend, The Guardian.* – No comment needed.

Source: The Times, 5 September 1996.

Concrete. A mixture of sand, cement, aggregate, water and additives to form a highly-durable structural material for a multitude of construction purposes. Its use can contribute to sustainability, as research has found that its high thermal mass can result in far less energy being used to provide cooling in the summer and heating in the winter, resulting in lower CO_2 emissions. Over a 60-year life cycle, a concrete and masonry home can emit up to 15 tonnes less CO_2 than a comparable lightweight alternative.

Operational CO_2 emissions have far more environmental impact than the embodied CO_2 of the materials used to construct a building. Some 50 per cent of the UK's CO_2 emissions are produced by the energy used to heat, cool and light buildings. This makes it essential that the energy consumption over a building's lifetime is taken into account when evaluating the environmental performance of construction materials.

Source: www.concretecentre.com/greenhomes. Accessed 20 March 2007.

Confidence interval (CI). In statistics, a confidence interval (CI) for a population parameter is an interval between two numbers with an associated probability *p* which is generated from a random sample of an underlying population, such that if the sampling was repeated numerous times and the confidence interval recalculated from each sample according to the same method, a proportion *p* of the confidence intervals would contain the population parameter in question. Confidence intervals are the most prevalent form of interval estimation.

Confined aquifer. An aquifer that is overlain by an impermeable bed.

Congestion charging. Congestion charging is a way of ensuring that those using valuable and congested road space make a financial contribution. It encourages the use of other modes of transport and is also intended to ensure that, for those who have to use the roads, journey times are quicker and more reliable.

Congestion charging was introduced in London because: London suffers the worst traffic congestion in the UK and amongst the worst in Europe; drivers in central London spend 50 per cent of their time in queues; every weekday morning, the equivalent of 25 busy motorway lanes of traffic tries to enter central London; and it has been estimated that London loses between £2–4 million every week in terms of lost time caused by congestion. The London scheme requires drivers to pay £8 per day if they wish to continue driving in central London during the scheme's hours of operation. The combined effects of charging and improved vehicle technology have led to NO_x

emissions within the charging zone falling by 13 per cent, and those of PM_{10} falling by 16 per cent.

The graphs on the facing page show the levels of NO_x and PM_{10} at various locations before and after the imposition of congestion charging.

Conservation of land. The maintenance of areas of the countryside for leisure and the preservation of threatened species of animals and plants is one of the main objectives of a conservation policy. Conservation is essentially the preservation of the environment in a condition to fulfil the needs for a healthy and satisfactory life as the pace of living and pressures of industrialized society increase. This is one aspect of THE COMMONS that requires external vigilance.

Construction and Demolition Waste (C & DW). Construction and Demolition Waste accounts for 15 per cent of the estimated annual waste arising in the UK. About 275 million tonnes of new construction aggregates are extracted annually. By 2012, if UK demands for aggregates increase by an expected 1 per cent per annum, an extra 20 million tonnes of aggregate will be needed each year. About 60 per cent of this is crushed rock and 40 per cent is sand and gravel. These are essential building materials but their extraction may cause significant environmental damage. Most C & DW are currently used for low-value purposes, such as road sub-base construction, or landfill engineering. Only 4 per cent is recycled for high-specification applications. The reason for this is that, while many C & DW materials could be used for higher-level uses, potential users are deterred by the perceived risks involved. However, work done by M.N. Soutsos, S.G. Millard and J.H. Bungey, at the University of Liverpool has demonstrated that concrete blocks which met exacting criteria can be made with appropriately processed C & DW.

Source: Soutsos, M.N., Millard, S.G., and Bungey, J.H., Concrete building blocks made with recycled demolition aggregate, *Concrete Plant International,* 2006, p70.

Construction (Design & Management) Regulations 1994. The UK Construction (Design & Management) (CDM) Regulations 1994 came into force on 31 March 1995.

The CDM Regulations seek to improve the overall management of health, safety and welfare and to ensure co-ordination between all the parties in construction and throughout the life of a structure. See Figure 39.

The CDM Regulations apply to:

- *construction work*
  - which is likely to last more that 30 days, or
  - which will involve more than 500 person days.
- *non-notifiable work* which involves 5 or more people on site at any one time.
- *any demolition work.*

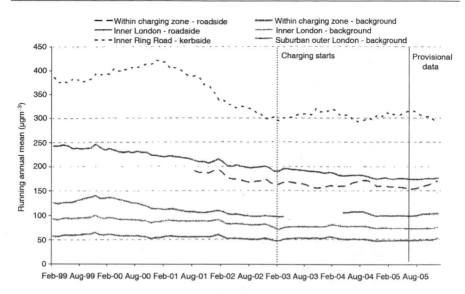

Figure 37 Running annual mean NO$_x$ concentrations.

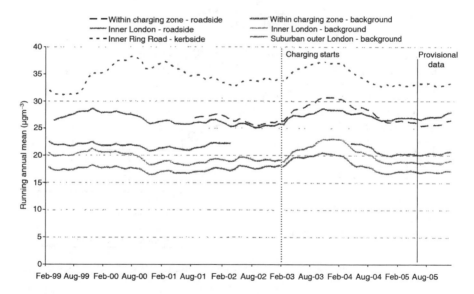

Figure 38 Running annual mean PM$_{10}$ concentrations.
Source: Transport for London, Congestion Charging – Impacts monitoring, Fourth Annual Report, June 2006.
Getting London Moving
© Transport for London

CDM Flowchart

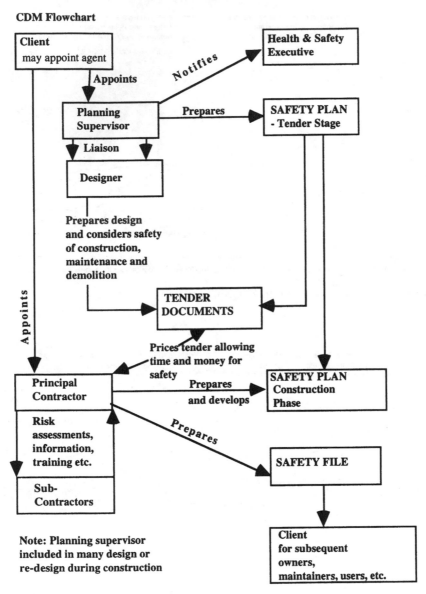

Figure 39 Construction design and management flowchart (Institution of Civil Engineers, 1995. Reproduced by permission).

Consumer Council for Water (CCW). The Consumer Council for Water was set up in 2005 in England and Wales, to provide a strong voice for water and sewerage customers in England and Wales.

Contaminated land. Land despoiled by previously poorly regulated uses such as gasworks, old landfill sites, foundries or tanneries, where high levels of heavy metals, phenols, ACIDS or ALKALIS may be found.

Initially (UK) a contaminated land register was proposed but political pressures have seen it off. That does not mean the problem will go away, only that more research has to be done by would-be purchasers. *Caveat emptor.* (◊ LIABILITY)

A suitable-for-use approach is now adopted for contaminated land management.

At the heart of the policy is the recognition that not all contaminated land needs to be treated with the same degree of urgency.

Instead, it focuses the force of statute law on requiring action in cases where 'the contamination poses unacceptable actual or potential risks to health or the environment; and there are appropriate and cost-effective means available [to carry out remedial action] taking into account the actual or intended use of the site'.

The trend is towards a RISK-based approach, finally laying to rest the idea of *carte blanche* contaminated land registers and confirming 'suitable for use' as a guiding principle (*CBI News*, April 1995).

Clause 54 in Part II of the Environment Act amends the Environmental Protection Act 1990, adding a new Part IIA comprising 16 sections which replace, in relation to contaminated land, the wider and more general definitions and procedures for statutory nuisance to be found in Part III of the Act. 'Contaminated land' is defined as follows: '. . . any land which appears to the local authority in whose area it is situated to be in such a condition, by reason of substances in, on or under the land, that (a) harm, or (b) pollution of controlled waters, is being, or is likely to be, caused'. A local authority (essentially a unitary authority or a district council) is obliged to have regard to guidance forthcoming from the ENVIRONMENT AGENCY in exercising this function. Each authority is obliged to identify contaminated land, closed landfill sites (defined as land in respect of which it appears that there is no waste disposal or waste management licence in force but where controlled waste has been deposited lawfully or unlawfully) and closed landfill sites which appear to be suitable for designation as 'special sites' by the Secretary of State. A 'special site' is a site in such a condition, by reason of substances in, on or under the land that (a) serious harm, or (b) serious pollution of controlled waters, is being, or is likely to be caused. It is the duty of the agency (in respect of a 'special site') and a local authority (in respect of a closed landfill site other than a special site) to prepare and publish a remediation statement. Such a statement must be served by the agency (in respect of a 'special site') or a local authority (in respect of any contaminated land other than a special site) or an 'appropriate person' who should be able thereby to ascertain what is required by way of remediation as well as the period for compliance. (The appropriate person is the person who caused or knowingly permitted the offending substances to be in, on or under the contaminated land.) This may not be the end of the matter as the problem of managing the effects of 200 years of heavy industry on a small island cannot be solved overnight.

As an example, there are an estimated 3000–4000 old town gas sites (where gas was manufactured using coal, before natural gas became available). These may be contaminated with coal-tar, phenols and sulphur compounds and asbestos.

Contaminated land remediation technologies

Several options exist to treat contaminated land, and these vary in complexity and cost. These are the main alternatives:

1. *Dilution to reduce hazard.* The contaminated material is mixed with clean soil or subsoil in order to reduce the maximum concentrations of contaminants to below the threshold trigger values.
2. *Excavation for off-site disposal.* Contaminated soil is excavated and deposited in a lined landfill site. The void is then filled with clean soil. This does not treat the problem, only shifts it.
3. *Containment and encapsulation.* The contaminated area is covered with layers of clay and then topsoil or hardstanding, to provide a physical barrier between users and the contaminant. It is usual to remove pockets of heavy contamination before containment. Membranes can also be used to cover the contaminated area.

   Encapsulation is a form of containment. This involves the provision of vertical and horizontal barriers around and below the contaminated soil.
4. *Physical treatment.* This class of treatment aims to remove the contaminants from the soil in a concentrated form suitable for detoxification by chemical, thermal or biological (BIOREMEDIATION) processes. *In-situ* processes include venting or vapour extraction, and washing by infiltration, whilst *ex-situ* processes might include size fractionation, gravity separation, froth flotation, magnetic separation, scrubbing and washing with appropriate reagents, air stripping and steam stripping. Physically treated soil can also be solidified or stabilized.
5. *Solidification/stabilization processes.* Solidification/stabilization processes are designed to reduce contaminant solubility, mobility and toxicity, to consolidate contaminated soils and/or to limit the potential migration of contaminants by reducing the exposed surface area-to-volume ratio. This

is often done by the addition of binders such as cement and quicklime. Solidification is the conversion of the contaminated soil to a more solid (water-impermeable) form while stabilization refers to the conversion of the contaminants to a more stable form.

6. *Chemical treatment*. The contaminants in the soil are either destroyed or converted to a less environmentally damaging form by *in-situ* processes such as neutralization, oxidation, reduction, HYDROLYSIS and ozonation. The disadvantages of chemical treatment include the possibility of chemical agents attacking natural compounds found in soil, which may damage soil structure or generate new toxic compounds.

7. *Thermal treatment*. Thermal processes include direct heating, indirect heating and incineration. Direct and indirect heating is used to volatilize organic compounds in the soil so as to remove them for subsequent treatment (e.g. incineration). INCINERATION is used to burn off organic compounds. The contaminated soil is heated to 800–2500°C *ex-situ*. In *in-situ* treatment, electric current can be passed through the soil, resulting in vitrification taking place.

Thermal processes are very costly, especially for wet soils, and leave a product that no longer looks like soil (although it can be used as fill material).

8. *Bioremediation*. In BIOREMEDIATION, micro-organisms are used to biodegrade the contaminants in the soil, either *in-situ* or in reactors off-site. The native micro-organisms already present can be harnessed, or 'tailor-made' species effective in degrading the pollutants of concern can be introduced into the soil. With bioremediation, a wide range of organic contaminants can be eliminated, given favourable conditions.

Microbial activity is enhanced by the supply of NUTRIENTS such as oxygen, N, P and trace elements. Oxygen may be supplied in a variety of ways – as air, pure oxygen, hydrogen peroxide, ozone and nitrate. Very often, for *in-situ* bioremediation, the groundwater at the site is pumped out and used as the transport medium for the nutrients. The nutrients are added to the abstracted groundwater which is then sent back into the ground through infiltration trenches. In many instances the groundwater is itself polluted with the hazardous waste and requires treatment first.

Figure 40 shows a possible treatment scheme for a site contaminated by FUEL which has leaked from an underground storage tank. Groundwater is drawn out through two extraction wells and passed to a skimmer tank where free oil rises to the surface and is separated. In the following stage, an air stripper removes volatile organic components from the groundwater. The separated volatile components are then degraded in a compost filter, or may be adsorbed onto ACTIVATED CARBON (if this is not done the pollutants will merely be transferred from the water into the air). After this stage, nutrients needed

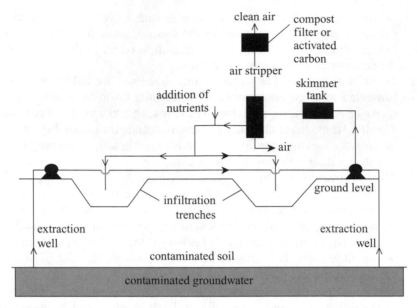

Figure 40   A possible treatment scheme for a site contaminated by fuel oil.
Source: Interdepartmental Committee on the Redevelopment of Contaminated Land, *Guidance on the assessment and redevelopment of contaminated land*, Department of the Environment, London, 1987.

for biodegradation are added and the water is put back into the ground, via infiltration trenches. As the water percolates through the ground, it takes with it the nutrients to the points at which the contaminant is located. The microbes present therein utilize the nutrients in biodegrading the contaminant.

Bioremediation *in-situ* is gaining popularity as a means of treating biodegradable hazardous waste. It eliminates the need for excavation and transportation of contaminated soil, which is a costly and disruptive operation. It also allows the treatment of both soil and groundwater at the same time. It has minimal equipment requirements and has generally been cost-effective. Treatment, however, may be slow or incomplete as there are organic compounds which are not amenable to biotreatment.

Continuous sampling. Uninterrupted sampling of, say, gas, usually at a fixed rate. Where the sample can be analysed continuously, the stream of gas may be passed through the measuring instrument continuously. Otherwise the sample is collected in an uninterrupted fashion for a given period and the total sample is finally analysed to give the mean composition of the gas over the whole period. (◊ COMBUSTION)

Continuously Regenerating Trap (CRT). A device for particulate removal from DIESEL engined vehicle exhausts which uses NITROGEN DIOXIDE (NO_2) to oxidize particulates (at temperatures of more than 250°C). The fumes from the engine pass through a catalyst designed to maximize NO_2 levels, into the trap system which is made from extruded ceramic containing millions of microscopic 'honeycomb' channels. The fumes react with the NO_2 to form oxides of NITROGEN and CARBON DIOXIDE. There is a small increase in CARBON DIOXIDE emissions of around 1 to 2 per cent.

Low sulphur diesel fuels (with a sulphur content less than 100 ppm) are necessary as SO_2 absorbs water which condenses on the filter and lowers its performance.

The CRT features a combination of catalyst and filter technology to remove particulates as well as CARBON MONOXIDE, HYDROCARBONS and the characteristic diesel smell. It is a significant advance on other technologies that have been used for cleaning up diesel emissions because it is able to continuously remove and eliminate soot at the vehicle's normal operating temperature, i.e. the trap continuously regenerates itself.

An oxidation catalyst and filter are combined in the CRT and, inside a stainless steel skin, can be engineered to provide sufficient sound attenuation to replace an ordinary silencer. The CRT is mounted on the vehicle in place of a conventional silencer, and provides a novel combination of filter and catalyst technology.

Exhaust gases containing carbon monoxide and hydrocarbons pass through the catalyst where they react with oxygen in the gas stream and are converted into carbon dioxide and water vapour. In addition, the specially developed catalyst increases the proportion of nitrogen dioxide to nitrogen monoxide, a key feature of the SOOT removal process. The exhaust stream then passes into the filter, a porous ceramic with cells about four times as large as those in the catalyst and in which channels are blocked at either end alternately, in a chequer-board style.

As the exhaust gas enters a channel it is forced through the wall into an adjacent channel to exit the filter. In so doing, the soot carried in the exhaust stream is deposited on the surface of the filter. Here, the soot (or carbon) is oxidized by nitrogen dioxide in the exhaust stream. The soot particles are gradually oxidized away and disappear completely as the exhaust gases continue to flow through the filter.

The continuous removal of the soot regenerates the filter so that it is working whenever the vehicle is in use. Importantly, the soot is removed at temperatures above 250°C, compared with 550–600°C necessary in earlier systems.

(New engines will have limited impact initially, because the durability of diesel engines means that trucks and buses remain in service for a long time. They typically have high mileages throughout their service lives. There will therefore be a large fleet of long-lived polluting diesel vehicles in our cities for ten years or more.)

Source: Johnson Matthey, Catalyst Systems Division, February 1995.

Contraction and Convergence. A shorthand term for the equitable distribution of carbon dioxide releases to atmosphere, so that each nation eventually has the same *per capita* emission level. This would achieve (in time) a stabilized global CO_2 ceiling level of say 450 ppmv. The Royal Commission on Environmental Pollution's 22nd Report 'Energy in a changing climate' of June 2000 has stated:

'The most promising, and just, basis for securing long-term agreement is to allocate emission rights to nations on a *per capita* basis – enshrining the idea that every human is entitled to release into the atmosphere the same quantity of greenhouse gases. But because of the very wide differences between *per capita* emission levels around the world, and because current global emissions are already above safe levels, there will have to be an adjustment period covering several decades in which nations' quotas converge on the same *per capita* level. This is the principle of contraction and convergence, which we support'.

The Royal Commission went on to say: 'For the UK, international agreement along these lines which prevented carbon dioxide concentrations in the atmosphere from exceeding 550 ppmv and achieved convergence by 2050 could imply a reduction of 60 per cent from current annual carbon dioxide emissions by 2050 and perhaps of 80 per cent by 2100. These are massive changes'.

The current global CO_2 concentration is about 370 ppmv. The UK government CO_2 reduction target is 20 per cent reduction by 2010 (cf. 1990 levels); the 'Kyoto' target is 12.5 per cent. Currently (2008) the UK is nowhere near its targeted reduction level. One may again conclude that carbon offsetting will not effect the necessary reduction in time, and major technological and societal changes will be required. ($\lozenge$ CARBON CAPTURE AND STORAGE. STERN REPORT)

Contraries. Materials which have a detrimental effect upon processing recycling, e.g. 'tramp' iron in quarry products, paper clips in waste paper.

Control of Major Accident Hazards Regulations 1999. The main aim of these Regulations is to prevent and mitigate the effects of those major accidents involving dangerous substances, such as chlorine, liquefied petroleum gas, and explosives which can cause serious damage/harm to people and/or the environment. The Regulations treat risks to the environment as seriously as those to people. They apply where threshold quantities of dangerous substances identified in the Regulations are kept or used.

Control limits. Control limits are occupational exposure limits relating to personal exposure. They are exposure limits contained in Regulations, Approved

Codes of Practice, in European Community Directives or have been adopted by the Health and Safety Commission, and should not normally be exceeded. They have been set following detailed consideration of the available scientific and medical evidence to be 'reasonably practicable' for the whole spectrum of work activities in Great Britain. Control limits apply to substances such as ASBESTOS, isocyanates, LEAD and chloroethylene. (◊ VINYL CHLORIDE)

Control limits, occupational. A means of controlling the DOSE received by a member of a workforce, of a hazardous substance. There are two limits, the LONG-TERM EXPOSURE LIMIT and the SHORT-TERM EXPOSURE LIMIT. The former is a time weighted average over 8 hours and the latter over 10 minutes.

As an example, the revised control limits for NICKEL and its inorganic compounds (except nickel carbonyl) adopted in the UK (*UK Health and Safety Commission Newsletter*, April 1989) from 1 June 1989, are that occupational exposure to these substances must be controlled so as not to exceed:

- 0.5 milligrams per cubic metre (mg/m^3) for nickel and water-insoluble inorganic nickel compounds
- 0.1 mg/m^3 for water-soluble inorganic nickel compounds

Both of these limits are expressed as 8-hour time weighted averages. As no specific short-term limits have been set, exposure during any 10-minute period should not exceed three times the 8-hour limit. (For the purpose of these limits a water-soluble nickel compound should be regarded as any single nickel salt or nickel complex which has a solubility greater than 10 per cent by weight in water at 20 °C.)

Because of its widespread use many thousands of employees may be exposed in processes such as engineering, electroplating and nickel manufacture.

A combination of methods may be necessary to control exposure to these substances, ranging from total enclosure of the process and the use of automatic handling techniques, to partial enclosure, local exhaust ventilation and the use of respiratory protective equipment (RPE). Engineering control methods can generally ensure that the occupational exposure limits are not exceeded. However, where it is not possible to use such methods, or the required degree of control is not achieved, then suitable RPE will be necessary. Such activities may include furnace cleaning and spray drying operations.

Control of Pollution Act 1974 (CoPA). CoPA 1974 covers a wide range of legislation on pollution issues: Part I – waste disposal; Part II – water pollution; Part III – noise; Part IV – atmospheric pollution; Parts V and VI – miscellaneous.

Control rod. A rod of neutron-absorbing material (e.g. cadmium, boron, hafnium) moved in or out of a nuclear reactor core to control the number of neutrons available for fission, and hence the power level of the reactor.

Controlled landfill. ◊ LANDFILL SITE.

Controlled tipping. The most common method of DOMESTIC REFUSE disposal where the refuse is tipped in layers, compacted and covered at the end of every working day with an inert layer of suitable material to form a seal. Controlled tipping avoids refuse being blown from the site, or tipping in static or running water, and ensures the general orderly appearance and running of the tip. The opposite of this practice is open tips which constitute public nuisances and can threaten public health by provided breeding grounds for rodents and flies. In practice, some sites are less than perfect and illegal tipping of hospital/hazardous wastes can take place.

Controlled wastes. Waste controlled under the Control of Pollution Act, 1974. These controlled waters comprise territorial waters, coastal waters, inland waters and ground water, which are subject to British pollution control legislation. (◊ WASTES)

Converter. A converter (packaging, waste, recycling, recovery) uses or modifies material supplied for the production of packaging. The converter recycling obligation is placed on the business which performs the last converting activity and supplies to the next stage (packer/filler) or another stage in the packaging chain. (◊ PRODUCER RESPONSIBILITY OBLIGATIONS)

Cooking. Pulp manufacture which involves treatment of the fibrous raw material at a minimum temperature of 100 °C with water and the addition of appropriate chemicals (◊ SULPHATE PULP).

Coolant. Gas, water, or liquid metal is circulated through a nuclear reactor core (primary coolant), to carry heat generated in it by fission and radioactive decay to boilers or heat exchangers where water (secondary coolant) is turned into steam for the turbines.

Cooling tower precipitation. The drizzle that occurs around water cooling towers that are not fitted with spray eliminators. If the circulating water within the tower contains salt, or other dissolved materials, such precipitation can be a source of POLLUTION, and a breeding ground for the bacteria that cause Legionnaires' disease. In addition, certain wind conditions that produce a down-draught inside the tower, or high tangential velocities, may result in water droplets being blown out of the base of the tower. Cooling-tower precipitation is also referred to as 'drift' and 'carry-over'.

Copolymerization. ◊ POLYMERIZATION.

Copper (Cu). A metal, commonly used for heat-transfer applications because of its high thermal conductivity. It is a very good electrical conductor and can be readily drawn or extruded into tubes and wire.

Copper is an essential constituent of living systems. However, copper ions (Cu^{2+}) are toxic to most forms of life, 0.5 parts per million being lethal to many algae. Most fish succumb to a few parts per million. In higher animals, brain damage is a characteristic feature of copper poisoning.

Copper pollution may arise from many sources. Soils receive high levels as a result of mining activities, intensive use of 'copper pellets' in pig rearing, or the application or copper fungicide. Effluents from factories and mines can cause serious water pollution.

Corporate Environmental Reporting. Corporate environmental reporting involves the voluntary publication of environmental information by a company, as a form of public accountability to all stakeholders. This usually includes details of environmental performance in areas such as greenhouse gas emissions, waste and water use. This information is expressed in quantified data and improvement targets and is published either in a company's annual report and/or self-standing reports. An environmental report is not a one off isolated activity but represents the final stage of a process of environmental management within a business.

Correlation. A statistical measure (usually denoted by 'r') of the closeness of the variations in the values of one VARIABLE to the variations in the values of another. Denoted by any value between +1 and −1.

Note that a value of r close to 1 does not imply a causative connection between the two variables (a common mistake).

However, if two variables move simultaneously in the same direction, they are positively correlated, e.g. sewage effluent discharge and population growth (& water consumption too).

Cosmic rays. Streams of atomic particles from space. They have been shown to induce cloud formation which has been posited as an (unproven) antidote/accelerant to global warming depending on cosmic ray activity.

Cost-benefit analysis. A technique which purports to evaluate the social costs and social benefits of investment projects in order to help decide whether or not such projects should be undertaken.

Cost-benefit methods have been offered to support the nuclear industry's claim that zero radiation release is too expensive. By using cost-benefit analysis, the minimum radiation levels that society can 'afford' to tolerate may be calculated. This approach is embodied in Figure 41, the thesis being that the dose that is 'as low as is readily achievable, economic considerations being taken into account', is thereby obtained.

One major objection to this approach is that exposure to radiation is an involuntary risk and is not one that an individual can opt out of. For example, a person driving a car takes a voluntary risk of an accident happening as opposed to completing the journey (benefit) and the costs and benefit apply to the same individual. In the case of nuclear power, costs and

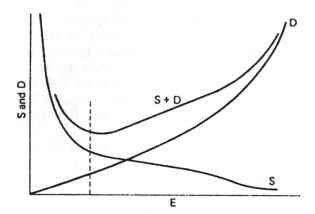

Figure 41 Differential cost-benefit analysis reveals the radiation level from nuclear operations, according to the considerations being taken into account. The two curves S and D represent the full cost of achieving a standard of safety (S) resulting in the collective dose (E), and the cost of the detriment (D) resulting from that dose. Combining these two curves gives the U-shaped curve representing the total cost of the collective dose. The minimum of the S + D curve is the dose that is 'as low as readily achievable'.

benefits do not apply to the same individuals and risks cannot be eliminated by voluntary action.

The questions which cost-benefit analyses have failed to answer in this and other areas are:

1. How do you put a financial cost on a risk when there is no mechanism for people to buy themselves out of the risk?
2. How can 'emotive issues' be taken out of cost-benefit analysis when ultimately all value is derived from emotion?
3. How can one evaluate either costs or benefits to non-direct participants in a system which does not yet exist?
4. On what evidence is the same discount rate applied to costs and benefits? In particular, why apply a discount rate to children? Most people seem to apply an interest rate to their children.

Among economists themselves, perhaps the most devastating indictment of cost-benefit analysis has come from Schumacher:

> It can therefore never serve to clarify the situation and lead to an enlightened decision. All it can do is lead to self-deception or the deception of others; for to undertake to measure the immeasurable is absurd and constitutes but an elaborate method of moving from preconceived notions to foregone conclusions; all one has to do to obtain the desired results is to impute suitable values to the immeasurable costs and benefits. The logical absurdity, however, is not the greatest fault of the

undertaking: what is worse, and destructive of civilization, is the pretence that everything has a price or, in other words, that money is the highest of all values.
E. F. Schumacher, *Small is Beautiful*, Blond & Briggs, 1973.

The controversy over cost-benefit analysis is central to the study of the environment.

Giving oral evidence to the European Communities Committee (Sub-Committee C) in January 1998, Dr Kramer (EC Environment Directorate) said:

> I am not aware that any Member State – either the United Kingdom or Denmark, Netherlands, Germany and so on–has been able to provide clear cost benefit analyses of the Landfill option in particular because of the long-term aspects that are extremely difficult to assess. We do not even know about all the impairments caused by landfills in relation to underground pollution. This is a policy question which is not capable of being fully supported by a cost benefit analysis for the moment.

You have been warned. 'Put not your trust in false gods'.

Source: House of Lords Inquiry into Sustainable Landfill, 1998, p. 121.

Cost function. The relationship between the degree to which a pollutant emission is reduced and the cost of attaining the reduction.

Cradle-to-grave analysis. ◊ LIFE CYCLE ANALYSIS.

Crank-case blow-by. A mixture of air, petrol/diesel vapour, and exhaust gases that escapes from the cylinders of an internal combustion engine by blowing past the piston rings into the crank case. If this mixture is not returned to the cylinders to be burned, it can account for some 20 per cent of the total emission of HYDROCARBONS from the engine. Positive crank-case ventilation, in which the mixture is returned to the cylinders by means of suction from the intake manifold, is now obligatory in many countries.

Critical group. A method of monitoring pollution emission which relies on the identification of the group of individuals most at risk to a particular discharge. If they are receiving doses of exposures below those recommended, it is assumed that the population at large is therefore not exposed or subjected to the same discharge. This technique has been used for the monitoring of IONIZING RADIATION emissions from nuclear reactors. It depends on the success with which the group most at risk can be identified and an accurate assessment of their exposure to the discharge.

It is also proposed as a method of controlling discharges of HEAVY METALS to estuaries. For example, crab meat (wet weight) from South Devon has been found to contain up to 21 parts per million of CADMIUM, and if the group of individuals were identified as, say, 'large' crab-meat eaters, and if their daily intake was below the Food and Agricultural Organization/World Health Organization limit of 60 micrograms for a 70-kilogram man, then the population at large could be assumed to be not at risk.

Children and old people are two distinct critical groups that require separate consideration. Children take in approximately twice as much food as an adult in relation to body weight. Thus their exposure to risk is greater. Old people generally have weaker internal organs, and this again may militate against bodily defence mechanisms functioning correctly.

Critical pathway. In planning a nuclear installation, the environmental pathway is found for every nuclide that is to be released on discharge through food chains or otherwise to any member of the population. Some member(s) of the population is thereby found to constitute the critical case of exposure, i.e. CRITICAL GROUP. If it is reasonably certain that the dose in the critical case is less than the International Commission for Radiological Protection (ICRP) limit, then no other persons will be at any greater risk. However, ICRP recommends that every exposure should be reduced as low as is readily achievable within the economic and social framework. Efforts are also made to exploit any opportunity to reduce the dose still further (and indeed discharges for Sellafield (Windscale) were drastically reduced in the 1980s as evidence accumulated that there were worrying local concentrations of radionuclides in estuary silt and parts of local beaches. (◊ CAESIUM-137; DERIVED WORKING LIMIT; IONIZING RADIATION, MAXIMUM PERMISSIBLE DOSE)

Critical point. At pressures above critical, there is no definite transition from liquid to vapour and the two phases cannot be identified. Physically the LATENT HEAT falls to zero.

SUPER CRITICAL WATER OXIDATION of toxic wastes makes use of these properties. For water the critical point is:

$p_c = 221.2$ bar
$T_c = 647.3\,°C$
$v_c = 0.00317\,m^3/kg$

Cross-media. Term used in pollution control to mean POLLUTION CONVERSION, e.g. a hazardous waste incinerator discharging toxic DUST from its dust collectors which requires further disposal would be assessed for cross-media effects.

Cryogenic scrap recovery. A form of scrap recovery which uses liquid nitrogen to freeze (at $-196\,°C$) the pot-pourri of metals and contaminants from used cars. The very low temperature causes mild steel to fracture like glass when put through an impactor. The steel can then be easily removed magnetically from its trapped contaminants and the result is a very high-grade scrap suitable for high-quality steel manufacture. The contaminants, which include copper and zinc, may be further processed for high-value scrap recovery.

Cryolite. Mineral (Na_3AlF_6) employed as a solvent in the ELECTROLYSIS of ALUMINA to produce ALUMINIUM. (◊ HALL PROCESS)

Cryptosporidium. A water-borne protozoan parasite which can cause severe stomach upsets. Agricultural run-off is a common cause.

An outbreak of cryptosporidiosis affected a large number of people in the Torbay area in August and September 1995 and, as a safeguard, the following were put in place.

- continuous monitoring of the treated water for cryptosporidium at water treatment plants where there is most risk of problems occurring;
- strengthening the power of the Drinking Water Inspectorate (UK) to prosecute water companies if this monitoring detects the presence of 'crypto' at an unacceptable level; and
- institution of a new criminal offence for failing to treat water adequately, and a penalty of an unlimited fine on conviction.

Cryptosporidium can be resistant to CHLORINE dosing and alternative control approaches are microfiltration which uses membrane filters with 0.2 micrometre pores which can remove 4–8 micrometre sized pathogens or photochemical systems which use a photocatalyst and titanium dioxide and ultraviolet light to produce hydroxyl RADICALS. They are powerful OXIDIZING AGENTS and can destroy pathogens such as cryptosporidium.

Source: *Wet News*, 20 May 1998.
M. Gibson, 'Something in the Water', *Financial Times*, 14/10/98.

Cube law. A shorthand method of stating that the power output of a wind turbine is proportional to the cube of the wind speed. Hence, if a 2 MW turbine has its maximum output at 13 m/s, and the wind speed drops to 2.5 m/s, the cube law tells us that there is a $(13/2.5)^3 = 140$-fold reduction. The output will now be 14.3 kW, enough to power 7.2 bar electric fires, to us a common analogy.

Cullet. Trade name for colour-graded glass fragments, offcuts, etc. suitable for remelting. Sources are mainly the glassworks themselves and bottling plants. DOMESTIC REFUSE is another source but the RECYCLING cost is often prohibitive when compared with the cost of the raw materials themselves as collection and transport costs outweigh the revenue from the sales.

Curie (Ci). Long established unit of radioactivity (now replaced by the BECQUEREL) defined as 37 000 million (3.7×10^{10}) disintegrations per second which is approximately equal to the activity of 1 gram of radium-226. Thus an amount of radioactivity was expressed effectively by stating the number of grams of radium that could provide the same number of disintegrations per second. In practice it was far too large and the picocurie was used, i.e. 1 million millionth of a curie (10^{-12}).

Cyanide. Generic name for a SALT of hydrocyanic acid (HCN), e.g. potassium cyanide (KCN) used for dissolving gold from crushed rock. The solution is then reduced and filtered. Considerable water and soil pollution can result from improperly conducted gold dissolution operations.

Sodium cyanide (NaCN) is extensively used in the specialist steels industry. Like the other cyanides, it is extremely toxic if swallowed, breathed in as a DUST or gas, or absorbed via eyes or mouth.

Cyanides form a highly toxic gas (hydrogen cyanide) when mixed with ACIDS, so they must be kept apart. This illustrates the risks of indiscriminate industrial waste disposal and highly toxic substances such as cyanides must be disposed of by approved methods such as oxidation by OZONE or HYDROGEN PEROXIDE or CHLORINE. The trouble is that most cyanide wastes also contain metals such as cobalt or NICKEL and hence oxidize very slowly. Again the need for specialist professional waste treatment is highlighted in handling dangerous materials.

Cycle. A process or series of operations performed by a system in which conditions at the end of the process are the same as the original state. Thus in the CARBON CYCLE the carbon dioxide in the air is fixed by photosynthesis into plant life which in turn decomposes or is consumed and eventually ends up as carbon dioxide again. In the STEAM power station cycle, water is heated in the boiler (energy supplied) to steam which expands in the turbine to produce power (workout). The exhaust steam is condensed in the condenser (energy rejection). The resulting condensate (water) is pumped back through the boiler for the cycle to recommence.

Cyclone dust separator. A device for removing DUST particles from air. The principle is shown in Figure 42. The dust-laden gas enters tangentially or axially and is spun in a helical path down the conical collector. The particles

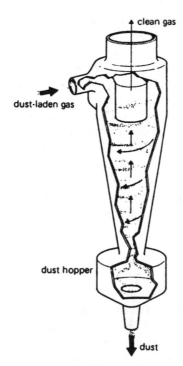

Figure 42   Cyclonic dust separation.

are flung to the wall by centrifugal force where they drop into the dust hopper while the clean gas leaves through a central 'core' tube at the top. Cyclones can remove 75–95 per cent of particulates in their range of applicability; for smaller particle sizes ELECTROSTATIC PRECIPITATORS, BAG FILTERS or other devices are often used, e.g. to remove FLY ASH.

D

Dam projects. Recent major dam projects have provided examples of high technology applied with the most benevolent of motives, but with inadequate consideration of any possible ecological side-effects. The most obvious example is the Aswan Dam. Designed to produce power and to store water for a permanent irrigation system, the effect of the irrigation ditches on the spread of a serious disease, SCHISTOSOMIASIS (bilharzia), was not considered. Similar unexpected side-effects occurred with the Kariba Dam, which helped spread a fly-borne disease which has disrupted the agriculture of people living along the river banks. Furthermore, for most water projects of this type, silting of the reservoir will eventually eliminate all the temporary gains.

In the case of the Aswan Dam, the changes produced in the flow of the Nile have had detrimental effects on the fisheries of the eastern Mediterranean. It will also probably have a deleterious effect on the soil fertility in the Nile Delta, since nutrients that were previously deposited annually by the flooding of the Nile will now be absent.

Damage function. The relationship between the degree to which a pollutant emission is reduced and the cost of the damage prevented by the reduction.

Darcy's law. A relationship that, in various forms, can be used to predict the rate of flow or the mass flux of a liquid through a porous medium. One version of Darcy's law states that the speed of water movement, v, down a slope through saturated soil between two points in a catchment that are a distance l metres apart, is partly determined by the difference in height, h, between them and the hydraulic conductivity of the soil, K: $v = Kh/l$.

Daughter product. The NUCLIDE which results from the radioactive disintegration of a 'parent' nuclide. Daughter products can be radioactive as well.

DDE (*d*ichloro*d*iphenyldichloro*e*thylene). A METABOLITE of DDT, it is formed by the action of some soil micro-organisms on DDT (in water DDT is refractory). It is more refractory and persistent than DDT and large quantities of both products could be accumulating in the biosphere. In the presence of ultra-violet light (i.e. sunlight) it can photodegrade to about nine other products, including several chlorinated biphenyls which can also have severe biological effects.

DDE is one of the most abundant organochlorine compounds in the biosphere.

DDT (*d*ichloro*d*iphenyl*t*richloroethane). An organic PESTICIDE developed during the Second World War as a delousing agent and later used to combat the insect carriers of malaria, yellow fever and typhus, thereby saving many lives. However, in the early 1960s certain immune strains of mosquitoes and other disease carriers developed. At the same time a decline in the reproduction rate of fish and bird life occurred.

The key problems are the persistence of DDT and its high biological activity. It is estimated that there are several million tonnes in circulation in the biosphere, spread by air and water currents, and most of this ends up in the ocean. Small concentrations (0.01 parts per million) reduce PHOTOSYNTHESIS in marine plankton by 20 per cent, and can be lethal to fish as they concentrate it. Such a low concentration would not be lethal to humans, but the use of DDT is now virtually banned world-wide because of its persistence. (◊ BIOACCUMULATION) despite its ability to help cheaply control malaria.

DDVP (Dichlorvos; 2,2-dichlorovinyl dimethyl phosphate). An insecticide often used in pest strips sold for domestic use. The US Environmental Protection Agency has cautioned against its use in rooms where food is prepared, or where infants or aged persons are confined – for example, in hospitals. Dietary deficiencies of protein, minerals, or vitamins may increase any ill effects. (◊ PESTICIDES; ORGANOPHOSPHORUS COMPOUNDS)

Decay. The disappearance of a pollutant as a result of absorption or chemical reaction at the earth's surface, or removal by rain, or transformation into some other substance. (◊ HALF-LIFE)

Decay heat. The heat produced by radioactive decay, especially of the fission products in irradiated fuel elements. This continues to be produced even after the reactor is shut down and the fuel removed.

Decibel (dB). A logarithmic measure used to compare the sound level of interest with a reference level. If we are concerned with sound power then reference is made to the smallest sound power that can be heard by someone with normal hearing at 1000 Hz. This reference power is 10^{-12} W. As an example we can deduce the sound power level of a jet aircraft on take-off. Now the noise of a jet aircraft at take-off (100 metres) is approximately 1 W which is

10^{12} as powerful as the reference power. Therefore, it is said to differ from the reference sound by

$$\log\left(\frac{\text{power}_2}{\text{power}_1}\right) = \log 10^{12} = 12 \text{ bels}$$

But bels are too large for convenience and so decibels are used instead, i.e. a factor of 10 is introduced. Hence, the jet aircraft at take-off has a sound power level of 120 dB with reference to a power of 10^{-12} W.

Sometimes it is necessary to compare sound pressures. Power is proportional to the mean square pressure under reflection-free conditions

$$dB = 10 \log\left(\frac{\text{power}_2}{\text{reference power}_1}\right) = 20 \log\left(\frac{\text{pressure}_2}{\text{reference pressure}_1}\right)$$

Reference pressure is 2×10^{-5} N/m^2 (1 N/m^2 is referred to as a pascal (Pa)).

Thus, an increase of 3 dB in the sound power level corresponds to a doubling of the sound power which corresponds to an increase of 6 dB in the sound *pressure* level.

Decibels are also used in telecommunications work as a measure of the system response (e.g. signal-to-noise ratio).

The specification of the scale is important. For most purposes loudness is quoted in DECIBELS A-SCALE (dB(A)) although other scales are used as well (for instance dB(C) gives emphasis to low frequency sounds) (◊ HEARING; LOGARITHMS; SOUND). The decibel A-scale corresponds to the frequency response of the human ear and therefore is correlated with the human perception of loudness.

Decibels A-scale (dB(A)). A frequency-weighted noise unit widely used for traffic and industrial noise measurement. The decibels A-scale corresponds approximately to the frequency response of the ear and thus correlates well with loudness. Other noise scales are used as well. (◊ HEARING; LOGARITHMS; SOUND; PNdB). For example, if the TIME-WEIGHTED AVERAGE industrial exposure for 8 hours is 90 dB(A), an increase in 3 dB doubles the dose and the exposure scale becomes:

> 90 dB(A) for 8 hours
> 93 dB(A) for 4 hours
> 96 dB(A) for 2 hours
> 99 dB(A) for 1 hour

Note: If someone is exposed to 99 dB(A) for 1 hour, he/she must be returned to very low noise levels for the remainder of the 8 hour period. (◊ DECIBEL; SOUND; HEARING; PNdB)

Decommissioning. The final closing down and putting into a state of safety of a nuclear reactor or other industrial plant or device when it has come to the end of its useful life.

Decommissioning of nuclear reactors. Decommissioning of nuclear reactors is costly in time and money. Sizewell A, one of Britain's oldest nuclear power stations is to be decommissioned at a cost of at least £870 million, and the work is expected to take 100 years! The reactor is one of 11 Magnox power stations in the UK, and on a typical day supplies more than 10 million kWh of electricity.

Decomposers. Organisms, usually BACTERIA or FUNGI, which use dead plants or animals as sources of food. They break down this material, obtaining the ENERGY needed for life and releasing minerals and NUTRIENTS back into the environment to be assimilated by other plant and animal life. They are an essential part of natural cycles and enable the components for life to be recycled. (◊ CARBON CYCLE)

Deepshaft treatment. ◊ SEWAGE TREATMENT.

Defoliants. ◊ HERBICIDES.

Degradable Packaging. The American Society for Testing and Materials and the International Standards Organization Definitions on Environmentally Degradable Plastics are as follows. They help to place 'biodegradability' into the context of 'degradability' in general:

Degradable plastic: a plastic designed to undergo a significant change in its chemical structure under specific environmental conditions resulting in a loss of some properties that may vary as measured by standard test methods appropriate to the plastic, and the application in a period of time that determines its classification.

Biodegradable plastic: a degradable plastic in which the degradation results from the action of naturally-occurring microorganisms such as bacteria, fungi and algae.

Photodegradable plastic: a degradable plastic in which the degradation results from the action of natural daylight.

Oxidatively degradable plastic: a degradable plastic in which the degradation results from oxidation.

Hydrolytically degradable plastic: a degradable plastic in which the degradation results from hydrolysis.

Compostable plastic: a plastic that undergoes degradation by biological processes during composting to yield carbon dioxide, water, inorganic compounds, and biomass at a rate consistent with other known, compostable materials and leaves no visually distinguishable or toxic residue.

Consumption of biodegradable plastic in the EU could rise to 1 Mt per year, compared with 30000t in 2001.

Source: R. Narayan & CA Pettigrew: Standards Help Define and Grow a new Biodegradable Plastics industry, *Standards News (ASTM)*, Feb 1999.

Degree-day. The difference, expressed in degrees, between the mean temperature of a given day and a reference temperature used as a predictor for fuel consumption or demand. One heating degree-day is counted for each degree that the daily mean temperature is lower than a base temperature, usually 15.5 °C in the UK. The total number of degree-days for the heating season is the sum of the degree-days for the different days of the season. Usually published monthly by area and used as a check on fuel consumption for heating purposes by large energy users.

De-inking. A process for removing ink from printed PAPER used in RECYCLING. De-inking takes place in four stages, as shown in Figure 43:

1. Pulping of the reclaimed paper with soda or other chemicals.
2. Centrifugal cleaning to remove paper clips, staples and dirt.
3. Screening to remove the freed fibres.
4. Washing to remove the printing ink solids either by mechanical or flotation methods.

The yield is ca. 75 per cent of the input paper.

PULP recovered from paper made from groundwood can be used for newsprint, magazines, and as a base for coated papers. Pulp recovered from non-groundwood paper can be used for the manufacture of writing/printing papers and tissues because of its longer fibre length.

The effluent from de-inking plants has a high BIOCHEMICAL OXYGEN DEMAND and requires considerable treatment before discharge.

Demand. The amount of a commodity that people are prepared to buy. If the price of a commodity varies, it would be reasonable to suppose that, at higher prices, less of the commodity will be bought. The fact that people still want the commodity does not mean that they demand it. If they cannot afford it, they exert no demand for it.

A graph showing the quantities of a commodity demanded at various prices is called a *demand curve* (see Figure 44). This shows, for each possible price, the total intended purchase of all buyers.

Demand management. If water companies and customers use water more efficiently, the need to invest in new water resources to meet increases in demand can be deferred. Demand management strategies, such as selective metering, appropriate tariff structures and leakage reduction, play an important role in maintaining a company's supply/demand balance.

'Deming' Circle. A graphical representation of a system for rational environmental improvement (see Figure 45).

Denitrification. A process in the nitrogen cycle carried out by microbes in anaerobic conditions, where nitrate acts as an oxidizing agent, resulting in the loss of nitrogen as it is converted to nitrous oxide (N_2O) and nitrogen gas (N_2).

Denitrifying bacteria. Bacteria (typically *Pseudomonas dentrificans* and *Hyphomicrobium* species) that carry out denitrification. ♦ NITROGEN CYCLE.

Density. The mass of a substance per unit volume (e.g. that of air at STP is 1.293 kg/m^3, water at 293 K is 998kg/m^3, and lead at 293 K is 11340kg/m^3).

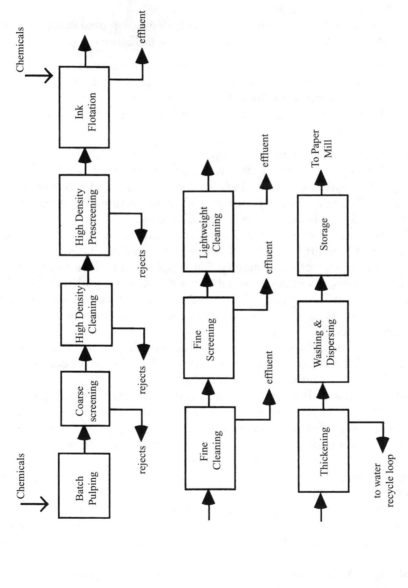

Figure 43 Recycled fibre mill – flow diagram.

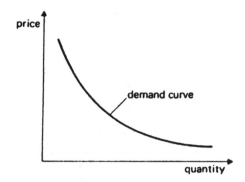

Figure 44 The demand curve.

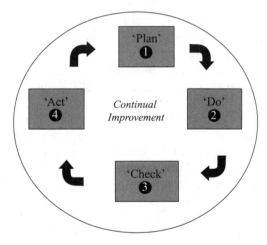

Figure 45 The 'Deming' Circle.

Density separation. The process by flotation in air or liquid to separate fractions of varying densities of certain types of material, e.g. PAPER and plastics from metals and other heavy contaminants in the waste stream for REFUSE DERIVED FUEL production.

Deposition processes. Air pollutants are removed from the atmosphere by two possible mechanisms – dry deposition and wet deposition. Dry deposition involves the pollutant impacting onto soil, water or vegetation at the earth's surface. Wet deposition involves the absorption of the pollutant into droplets either within clouds or below clouds, followed by removal by precipitation. Most pollutants are removed from the atmosphere by both processes. In dry climates, dry deposition will be the dominant process whereas in wet climates, wet deposition will dominate. In the UK, the amount of any pollutant

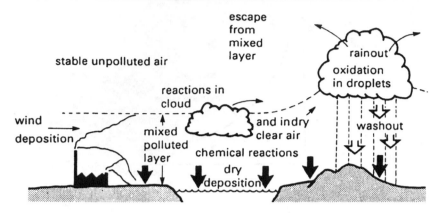

Figure 46   Acid rain deposition mechanisms.

removed by dry deposition approximately equals the amount removed by wet deposition averaged over a year.

Figure 46 shows the various deposition and transportation processes for the plumes from a power station stack. Typically, a plume from a 220 m stack first contacts the ground ca. 20 km downwind, by which time typical gas dilutions are 10 000 times. Dry deposition takes place from the point of plume impact and is assumed to be proportional to SULPHUR DIOXIDE concentration. Wet deposition rates can be much greater and are also influenced by the CONCENTRATION of photochemical oxidants in the atmosphere. The oxidants are also mainly anthropogenic pollutants thus ACID RAIN needs to be tackled by several routes and not just DESULPHURIZATION.

Derived working limit (DWL). This is a specific figure for each radioisotope released into the environment which is derived by determining both the CRITICAL PATHWAY for the radiation release and the group of people (the CRITICAL GROUP) who are most likely to be affected, i.e. given a radiation dose which is much larger than for the rest of the population. A derived working limit is then determined which limits the rate of discharge of the particular radioisotope to ensure that the ICRP dose limit would not be exceeded by the CRITICAL GROUP. A safety factor is always incorporated to guard against any uncertainties.

It should be noted that the determination of DWLs depends on a knowledge of the pathways, of the distribution of radioactive materials in the environment, and of their ability to accumulate in food chains. The critical pathway may change with time and so can the critical group. Hence, continuous monitoring is necessary, e.g. for the discharge of ruthenium-106 from Sellafield (named Windscale at the time), the critical group used were laver bread (fried seaweed) eaters who ate the seaweed *Porphyra*, collected near Sellafield. However, these collectors retired and the *Porphyra* was collected

Exposure pathways for discharges from the Sellafield reprocessing plant, 1970s.

Exposure pathway	Critical material	Critical group	Daily consumption rate or annual exposure	Critical organ	Radio-nuclides contributing to exposure	Derived working level	ICRP recommended dose limit (mrem/year)	Typical exposure of population group involved (mrem/year)
Internal	Seaweed	Laver bread consumers (100 persons)	160 g laver bread	Gastro-intestinal tract	^{106}Ru ^{144}Ce	300 pCi/g in the seaweed	1500	600
External	Estuarine silt	Fishermen (10 persons)	350 h	Whole body	^{95}Zr ^{95}Nb ^{106}Ru	1.4 mrem/h	500	50
Internal	Fish	Fishermen (100 persons)	25 g	Gastro-intestinal tract, whole body	^{106}Ru ^{137}Cs	900 pCi/g 1800 pCi/g	1500 500	2 0.5
External	Fishing gear	Fishermen (100 persons)	500 h	Hands	^{106}Ru ^{144}Ce	15 mrem/h	7500	20

elsewhere. The new critical group became salmon fisherman on the Raven-glass Estuary who received external exposure from radioactivity deposited on the shore.

The data opposite give an illustration of the 1970s (which ran at much less stringest levels c.f. today) exposure pathways for discharges from the Sell-afield reprocessing plant to show the critical groups, the basis for estimation of the dose and the levels of exposure theoretically permitted for each critical group.

Source: A. W. Kenny and N. T. Mitchell, 'United Kingdom waste-management policy', *Management of Low- and Intermediate-Level Radioactive Wastes*, Proc. Symposium, Aix-en-Provence 1970, IAEA, Vienna, 1970.

(◊ CAESIUM-137, IONIZING RADIATION EFFECTS; CRITICAL GROUP; CRITICAL PATHWAY; CURIE; REM)

DERV. Diesel-engined road vehicle. Also applied to a special grade of gas-oil with an extra-low sulphur content, which these vehicles use. (◊ PETROLEUM)

DES (diethylstilboestrol). A growth-promoting hormone used in the USA in the cattle industry to cut down on feed required to produce a given weight gain. It has been found to be carcinogenic in animals and is therefore a cause for concern in connection with humans. (◊ CARCINOGEN)

Desalination. The partial or complete removal of the dissolved solids in sea or saline water to make it suitable for domestic, agricultural or industrial purposes. The main techniques are ELECTRODIALYSIS, REVERSE OSMOSIS, freezing, and distillation. The processes and their applications are shown in Figure 47.

Electrodialysis: An electric current is passed through brackish or low-salinity water in a chamber in which many closely spaced ion-selective membranes are placed, thus dividing the chamber into compartments. The electric current causes the salts to be concentrated in alternate compartments, with reduced salt content in the remainder, thus producing a reduced salt product. A principal disadvantage of electrodialysis is that power consumption is proportional to total dissolved solids or salinity.

Reverse osmosis: The reverse application of osmotic pressure. When salt water and fresh water are separated by a semi-permeable membrane, osmotic pressure causes the fresh water to flow through the membrane to dilute the saline water until osmotic equilibrium is established. Now applying this in reverse, if a greater pressure is applied to the salt-water side of the membrane, then relatively pure water will pass through it leaving a concentrated brine to be disposed of. As with electrodialysis, power consumption is proportional to total dissolved solids. Seawater treatment is also possible

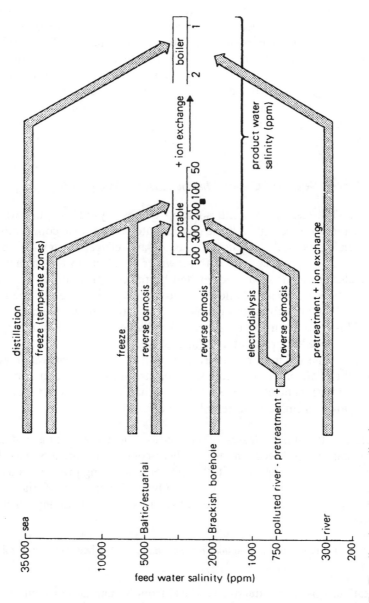

Figure 47 Desalination process application.

but both power capital costs do not allow for ready applicability. However, a novel development has been reported. Researchers at UCLA's Henry Samueli School of Engineering and Applied Science, USA, have made a breakthrough which aims to reduce desalination cost (and wastewater reclamation) by using a new membrane which contains a cross-linked matrix of polymers and nanoparticles designed to attract water ion, but repel most contaminants. Essentially it's a molecular tunnel capable of transporting water. It is claimed that water-loving nanoparticles embedded in the membrane repel organics and bacteria, which tend to clog up conventional membranes over time. As a result of the improvements, pumping the water across the membrane requires less energy. Tests suggests that the new membrane is twice as productive as the traditional method – or consumes 50 per cent less energy.

Source: Process News, *The Chemical Engineer*, Dec 06/Jan 07, p16.

Freezing: The freezing of a salt solution causes crystals of pure water to nucleate and grow, leaving a brine concentrate behind. One commonly proposed freezing method is the use of a secondary refrigerant in which butane is evaporated in direct contact with seawater, resulting in the formation of ice crystals. The ice is separated, melted by the compressed butane vapour, the two liquids are decanted and the product water obtained and the butane recycled. There are substantial problems in constructing large-scale freezing plants. This is an example of a process that looked good at laboratory scale but which proved impracticable at large scale.
Distillation: The boiling or evaporation of seawater to form water vapour which is then condensed to yield a salt-free stream. Energy requirements are virtually independent of the feed-water salinity, and product purity of less than 50 parts per million can readily be achieved.

There are two main classes of distillation: heat consuming and power (mechanical energy) consuming. A heat-consuming process has lower energy input costs compared with the power-consuming process, as dictated by the conversion of heat to power. Over 85 per cent of the world's installed desalting capacity is accomplished by heat-consuming distillation processes.

Source: A. Porteous (ed.) *Desalination Technology*, Applied Science Publishers, 1983.

Desertification. Loss of productive land use from human action such as deforestation, SOIL exhaustion, salinized irrigated fields, depleted groundwater resources, over-grazing, drought, wind and water erosion. The facing table gives an example of the Aral Sea basin desertification causes.

Types of land degradation	Classes of land degradation			
	Slight	Moderate	Severe	Total
Degradation of the vegetation cover	53.5	21.9	1.6	77.0
Wind erosion	1.0	0.2	0.3	1.5
Water erosion	3.8	2.1	—	5.9
Salinization of irrigated lands	0.9	7.4	0.8	9.1
Aral Sea level drop	0.4	7.4	0.8	9.1
Technogenic desertification	—	0.4	1.0	2.4
Waterlogging of desert rangelands	—	0.4	0.1	0.5
Total (per cent)	59.4	33.7	6.7	100.0

The table overleaf gives the extent of drylands on different continents. Pimentel *et al.* have stated:

> Of the world's agricultural land, about one-third is devoted to crops and the remaining two-thirds is devoted to pastures for livestock grazing. About 80% of the world's agricultural land suffers moderate to severe erosion, and 10% suffers slight to moderate erosion. Croplands are the most susceptible to erosion because their soil is repeatedly tilled and left without a protective cover of vegetation. However, soil erosion rates may exceed $100 \, \text{tons ha}^{-1} \text{year}^{-1}$ in severely overgrazed pastures. More than half of the world's pasturelands are overgrazed and subject to erosive degradation.
>
> Soil erosion rates are highest in Asia, Africa and South America, averaging 30 to $40 \, \text{tons ha}^{-1} \text{year}^{-1}$, and lowest in the United States and Europe, averaging about $17 \, \text{tons ha}^{-1} \text{year}^{-1}$. The relatively low rates in the United States and Europe, however, greatly exceed the average rate of soil formation of about $1 \, \text{ton ha}^{-1} \text{year}^{-1}$. Erosion rates in the undisturbed forests range from only 0.004 to $0.05 \, \text{ton ha}^{-1} \text{year}^{-1}$.

Sources: *Our Planet*, Vol. 6, No. 5, 1994.
Pimentel *et al.*, 'Environmental and economic costs of soil erosion and conservation benefits' *Science*, Vol. 267, p. 1117, 24 February 1995.

Destruction and Removal Efficiency (DRE). Used as an indicator of performance in HAZARDOUS WASTE INCINERATION. The levels of destruction and removal efficiency (DRE) for principal organic hazardous constituents (POHC) of the waste feed are given by:

$$\%DRE = 100 \times (W_{in} - W_{out})/W_{in}$$

where W = mass of the relevant POHC in the waste.

Source: *Hydrocarbon Processing*, May 1998, p. 120.

Destructive distillation. The heating of solid substances in closed retorts in the absence of air, and condensation of the ensuing volatile gases. The process

Drylands

EXTENT OF DRYLANDS IN DIFFERENT CONTINENTS OF THE WORLD (area numbers are in thousands of square km)

Continent	Hyper-arid	%	Arid	%	Semi-arid	%	Dry sub-humid	%	Total drylands (arid + semi-arid + dry sub-humid)	Moist sub-humid + humid	%	Cold	%	Grand total
Africa	8099	27	5052	17	5073	17	2808	9	12933	9171	30	0	0	30203
Americas & Caribbean	268	1	1201	3	7113	17	4556	11	12870	16926	41	11577	28	41641
Asia	2744	6	6164	13	7649	16	4558	9	18371	14997	31	12082	25	48223
Australia & Oceania	0	0	3488	39	3532	39	996	11	8016	1019	11	0	0	9035
Europe	0	0	5	0	373	7	961	17	1339	4059	71	289	5	5687
World Total	11110	8	15910	12	23740	18	13879	10	53529	46512	34	23948	18	134789

Source: www.sdnpbd.org/sdi/international_days/wed/2006/desertification/status_world.htm. Accessed 8 March 2007.

has been mooted for the utilization of DOMESTIC REFUSE for the production of METHANOL. It is akin to PYROLYSIS which is often used synonymously for the destructive distillation of refuse.

Desulphurization. ◊ FLUE GAS DESULPHURIZATION.

Detergents. Cleaning agents which include, as part of their chemical make-up, petrochemical or other synthetically derived wetting agents. They are made up of three main parts:

1. *Surfactant* – a wetting agent which permits water to penetrate fabric more easily. The surfactant molecules provide a link between the dirt molecules and the water molecules.
2. *Builder* – a sequestering agent. This ties up hard-water ions to form large water-soluble ions. The water becomes alkaline, which is necessary for removal of dirt.
3. *Miscellaneous* – brighteners, perfumes, anti-redeposition agents and enzymes.

A major drawback in the use of 'hard' detergents is that they are non-biodegradable. There is now a move to incorporate only biodegradable substances in detergents – although the use of hard detergents is considered necessary in certain industries such as wool scouring. The decomposition or breakdown of some detergents leads to PHOSPHORUS becoming available in aquatic systems where it can cause EUTROPHICATION. (◊ DETERGENTS, SUGAR-BASED; WASTE MINIMIZATION)

Development-induced displacement. The eviction of people because of national development is often hidden away. Yet between 90 and 100 million people have been involuntarily resettled over the past decade. In India alone some 23 million have been displaced since 1950. Hydroelectric dam projects each year lead to the involuntary relocation of between 1.2 and 2.1 million people.

In Africa, dams on the Volta (Ghana), Nile (Egypt), Zambesi (Zambia and Zimbabwe) and Bandama (Cote d'Ivoire) rivers have caused the forcible relocation of thousands of people, mainly farmers and herders, but also townspeople.

In China, water conservancy projects since the 1950s have created over 10 million displacees.

Source: C. McDonnell, *The Couriers*, No. 150, April 1995.

Dew-point. The temperature at which dew first appears on a solid surface whose temperature is steadily reduced below that of the surrounding moist air. (◊ ACID DEWPOINT)

Dichlorvos. ◊ DDVP.

Dieldrin. ◊ CHLORINATED HYDROCARBONS; ALDRIN.

Diesel. Common name for gas/oil which is a petroleum distillate used as boiler and compression ignition engine FUEL. Also called DERV (diesel engined road vehicle).

With the drive to control and reduce environmental emissions from diesel-engined vehicles, wide variations in key diesel fuel parameters are of concern to engine manufacturers and legislators. A diesel-engined vehicle travelling through Europe could find the following variations:

DENSITY	0.82 to 0.86 (5 per cent)
CETANE NUMBER	47 to 57 (20 per cent)
VISCOSITIES	1.8 to 3.2 (78 per cent)
SULPHUR CONTENTS	0.07 to 0.90 (1100 per cent)

which will potentially result in wide variations in HC, CO, NO_x, SO_x and particulate emissions, see Figure 48. (◊ BIODIESEL, DIMETHYLETHER)

Source: Ethyl European Diesel Fuel Survey, Winter 1988–89.

Diesel vehicle emissions. The diesel emissions standards for heavy duty vehicles, extant and proposed for the European Union, are shown in the table on the facing page.

Diethylene glycol. Produced from ETHYLENE, widely used as a SOLVENT and PLASTICIZER. Was responsible for up to 1080 infant deaths when it was found as a contaminant in the raw materials (glycerine) for cough medicine manufacture in Haiti (BBC2, *Newsnight*, 2 April 1997). This raised the issue of substandard medicines and practices in Third World countries.

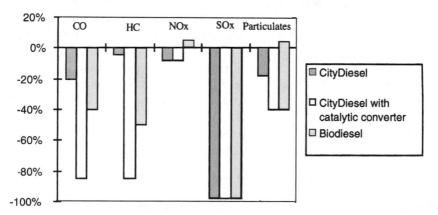

Figure 48 The effect of citydiesel and biodiesel on emissions c.f. conventional diesel engine.
Source: *IEA Renewable Energy Newsletter, Issue 3/94. Updated information is available from the International Energy Agency.*

EU Emission Standards for HD Diesel Engines, g/kWh (smoke in m^{-1})

Tier	Date	Emission test cycle	Carbon monoxide	Hydrocarbon	Nitrogen oxide	Particulate	Smoke
Euro I	1992, <85 kW	ECE	4.5	1.1	8.0	0.612	
	1992, >85 kW	R-49	4.5	1.1	8.0	0.36	
Euro II	Oct. 1996		4.0	1.1	7.0	0.25	
	Oct. 1998		4.0	1.1	7.0	0.15	
Euro III	*Oct. 1999 EEVs only*	*ESC & ELR*	*1.5*	*0.25*	*2.0*	*0.02*	*0.15*
	Oct. 2000	ESC & ELR	2.1	0.66	5.0	0.10 0.13*	0.8
Euro IV	Oct. 2005		1.5	0.46	3.5	0.02	0.5
Euro V	Oct. 2008		1.5	0.46	2.0	0.02	0.5

* for engines of less than 0.75l^3 swept volume per cylinder and a rated power speed of more than 3000 per minute. EEV is 'enhanced environmentally friendly vehicle'.
Source: www.en.wikipedia.org/wiki/European_emission_standards Accessed 16 February 2007.

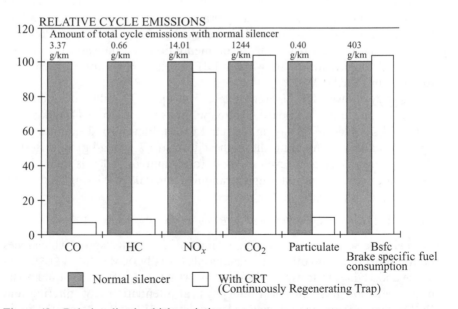

Figure 49 Relative diesel vehicle emissions.
Source: P.N. Hawker, Diesel emission connol technology, reprinted from Platinum Metals Review, Vol. 39, No. 1 (Reproduced by permission of Johnson Matthey PLC).

Diffusion tube sampling. In *passive sampling* no pump is used, but diffusion of the atmosphere occurs into the sampler. The collected sample is analysed later. These passive samplers for measuring ambient air quality are called diffusion tubes, although the term 'denuder' is sometimes seen. A typical diffusion tube for sampling nitrogen dioxide consists of a perspex tube about 7 cm long by 12 mm internal diameter, and fitted with a removable cap at one end. At the other end is a cap containing stainless steel meshes impregnated with triethanolamine. In use the removable cap is taken off the tube which is then placed in a suitable location where there is free movement of air around the open end. After a predetermined sampling period, the cap is replaced and the tube returned to a laboratory for analysis using the Griess-Saltzmann colorimetric method.

Digester[1]. A sealed container or vessel in which ANAEROBIC digestion of ORGANIC MATERIALS can take place under temperature controlled conditions.

Digester[2]. A heated pressure vessel in which wood chips are reacted with chemicals in the first stage of PULP production.

Diisopropylnaphthalenes (DIPNs). These are used as a SOLVENT for the colour former in carbonless copy-paper (also known as self-duplicating paper). Recycled PAPER used in making board may include carbonless copy-paper. Not all the DIPNs may be removed by the treatment of the recycled fibres and some may be present in the finished board and thus could migrate into food. Levels of DIPNs in food packaging made from recycled paper and board should be as low as reasonably practicable to minimize migration into food. The Committee on Toxicity of Chemicals in Food, Consumer Products and the Environment made this recommendation due the shortage of adequate toxicological information about DIPNs (e.g. there is not sufficient data to set a limit for DIPNs).

A survey showed DIPNs were detected at up to 33 milligrams/kilogram (mg/kg) in 51 samples of board from paper mills and at up to 44 mg/kg in 30 out of 34 samples of packaging for retail foods. Eleven food samples packaged in material containing the highest levels (6.7–44 mg/kg) were tested. DIPNs were detected in seven of these food samples at levels much lower than those present in the packaging, ranging from 0.06 to 0.89 mg/kg of food. Recycling is not a 'cure-all' for wastes.

Source: *MAFF Food Safety Information Bulletin*, No. 104, January 1999.

Dilute and disperse. A philosophy of liquid industrial WASTE disposal which relies on its absorption onto other waste materials (usually DOMESTIC REFUSE) and in the underlying strata especially any UNSATURATED ZONE. The liquid waste undergoes biological, chemical and physical attenuation (e.g. filtering and BOD reduction) during its downward percolation. When the attenuated liquid encounters an underlying AQUIFER, it is assumed that there has been both sufficient treatment before and subsequent dilution in the aquifer for any resulting pollution to be environmentally inconsequential.

The dilute-and-disperse philosophy has been strongly criticized by the EC on the grounds that there is no ultimate guarantee that environmental damage will not result. An alternative method is to seal the LANDFILL SITE with clay or a synthetic liner and contain the liquid waste within the receiving refuse mass and allow the attenuation processes to take place in the refuse mass alone. These processes may be aided by extraction of the liquids from the mass and spraying them on the landfill surface.

Following implementation of the LANDFILL DIRECTIVE, the co-disposal of industrial and household waste no longer continues in UK. The term may also apply to air pollution dispersal from a chimney stack. (◊ DISPERSION)

Dimethylether (DME). A claimed replacement for diesel fuel in urban locations because of low PM_{10} and nitrogen oxide emission levels.

The production process developed by the Danish company Haloror, Topsoe, uses STEAM reforming of NATURAL GAS, followed by conversion from the reformed gas into DME using a proprietary CATALYST.

Dioxins. A family of chlorinated organic compounds, a major member of which is 2,4,7,8-tetrachlorodibenzo-p-dioxin (TCDD), which is a manufacturing impurity in certain classes of HERBICIDES, disinfectants and bleaches. Both TCDD and PCDD (polychlorinated dibenzo-p-dioxin) are formed (along with variants including FURANS) when compounds containing chlorine are burnt at low temperature in improperly-operated/designed domestic refuse and industrial waste incinerators where the PCDDs and TCDDs can be found in both the flue gases and the FLY ASH. (◊ POLYVINYL CHLORIDE)

There are 210 dioxins and furans, 17 of which are toxicologically significant. 2,3,7,8-TCDD is taken as the reference standard with a maximum toxicity designation of 1. Each dioxin compound is given a Toxic Equivalence Factor (TEF) relative to that of TCDD. Dioxin/furan emissions are then expressed as Toxic Equivalents (TEQ). The sum of the individual TEQs is used in reporting dioxin/furan emissions.

Considerable public alarm on dioxin emissions (and near relatives) has been raised in many Western nations; usually these have been minute except in one or two cases of very poorly-designed industrial incinerators. One expert has stated that spontaneous (or deliberate) fires in rubbish tips emit much more dioxins than controlled incineration. 'A 20 kg piece of chipboard impregnated with CHLOROPHENOL (a "glue") creates as much dioxin when it burns as an entire incineration plant in a whole month.' (*Warmer Bulletin*, No. 9, March 1986.) The emission limit is now 0.1 ng/m³, i.e. 1 g dioxins in at least 10 000 000 000 m³ of flue gases emitted (STP conditions). This is the equivalent of a quarter of a standard 3 g sugar lump dissolved in Loch Ness (volume 7 000 000 000 m³). Sweden incinerates 70 per cent of its domestic refuse, with the energy used for DISTRICT HEATING and power generation. The UK ca 10% due to misguided opposition and a muddying of the water *vis-à-vis* RECYCLING with the claim that recycling is a better environmental option than incineration. The bald fact is both waste management options

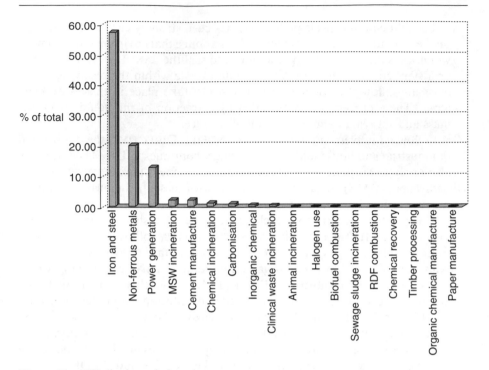

Figure 50   UK dioxin emissions by sector.

can work perfectly well together. Not in my back yard, or Nimbyism has a lot to answer for as far as have successive governments' pusillanimity in this area. We are all losers, financially and environmentally. (Does anyone really believe that shipping UK waste to China for reprocessing is environmentally sound?)

The dioxin contribution from municipal solid waste incineration is now less than 1 g per year. Figure 50 shows that the outputs from iron and steel and non-ferrous metals production and power generation dwarf that from Energy-from-Waste Incineration.

Many old, poorly-designed municipal solid waste (MSW) incinerators discharged dioxin at levels greater than 45 ng TEQ/Nm³ (Normal m³) and in 1989 all MSW incinerators then operating were estimated to contribute about one-fifth of the total known anthropological releases to the environment.

But the dioxin release levels from modern state-of-the-art energy-from-waste (EfW) plants are in the range 0.02–0.05 ng TEQ/Nm³. EC requirements are now <0.1 ng TEQ/Nm³ which all new plants can readily meet.

These standards imply a 98 per cent reduction in dioxin emissions from MSW combustion & in practice are negligible.

The RCEP (Royal Commission on Environmental Pollution) states that even with very large increases in the amount of MSW going to energy from waste, the total pollution load from modern plants under these new standards would not be a cause for concern. This reduction is made possible because the control of dioxins is now well understood.

Dioxin control. Dioxins (PCDD/PCDF) are principally formed in the 250–400 °C temperature range in various combustion processes. They can, however, be formed in improperly-controlled MSW incinerators as well.

However, compliance with EA requirements for combustion conditions of >850 °C for at least 2 seconds in at least 6 per cent O_2 with fabric filters, to clean the gases, ensures that discharge concentrations of <0.1 ng TEQ/μm^3 may be obtained. Still lower levels can be achieved by the injection of activated carbon in the flue gas stream, which is universally practiced in UK MSW incinerators.

Some dioxin reformation may take place in the 200–350 °C temperature range after combustion when the gases are cooled. This is the so-called 'de novo formation'. For this to take place, there must be unburnt carbon present and/or metal chlorides. As modern well-run incinerators produce approximately 2 per cent unburnt carbon and most of the heavy metals and particulates are trapped in the fabric filter, measured release rates of dioxins are approximately 0.02 ng TEQ/Nm^3.

A modern 400 000 tonnes per year MSW energy-from-waste plant emits less than a 20th of a gram of dioxin per year. The calculation is shown below.

Dioxin emissions
Plant refuse   Throughput 416 000 t/year
Availability   7446 h/year
Dioxin emission   0.02 ng/Nm^3 = 0.2 g × 10^{10} Nm^3
Flue gas flow   11% O_2 dry gas = 289 418 Nm^3/h

Plant dioxin emissions are therefore:

$$\frac{0.2 \times 289\,418 \times 7446}{10\,000\,000\,000} = 0.0431\,\text{g/year}$$

1/60th of the mass of a 3 g sugar lump per year!

Modern MSW incineration plants actually destroy 80 per cent of incoming dioxins in the waste, based on input–output measurements, i.e. state-of-the-art incinerators are dioxin sinks! (Yes, that is the case.)

Sampling work based on a 20 km square grid in lowland Scotland has shown that PCDDs are ubiquitous in the environment at low levels and that they may be found in trace quantities in snails, human and animal fat, milk and other biological tissues. A summary of the soil concentrations of PCDDs and related PCDFs (FURANS) found in the soil is given in Figure 51. The limit of detection is 0.5 parts per trillion (0.5 ng/kg).

The dioxin releases from the UK foot and mouth carcass pyres are worthy of note. To 23 April 2001, 63 g of dioxins were estimated to be released, whilst

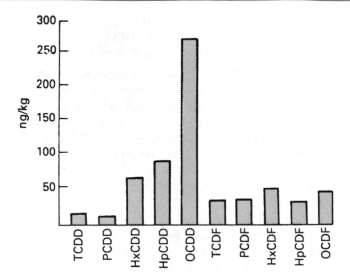

Figure 51   Mean dioxin soil concentrations, England and Wales.

all of the UK's Energy-from-Waste Incineration plants release less than 3 g per year.

Source: Porteous, A., 'Why energy from waste incineration is an essential component of environmentally responsible waste management', in *Biodegradable and Residual Waste Management* (Contemporary Waste Management Series), Edited by E.K. Papadimitriou and E.I.Stentiford, Published by Cal Recovery Europe Ltd., 2005.

Although MSW incineration has been perceived to be a major source of human exposure to dioxin, measurements in the US have shown that emissions from MSW incinerators at the national level account for less than 1 per cent of total current input into the environment.

Authors Travis and Blaylock (*Health Risks Associated with Air Emissions from MS Combustors*, American Chemical Society Conference, Washington DC, 21–26 August 1994) report that MSW combustion, with motor vehicles, hospital waste incinerators, residential wood burning, and pulp and paper mill effluent, together accounted for a maximum of 11 per cent of total 1993 US TCDD output. Concentrations of dioxin cannot therefore be linked to any one combustion source. Other potential sources include discharges from metal processing and treatment plants, pentachlorophenol production, and coal, forest and landfill fires.

TCDD is known as the Seveso poison, because of an incident in which 2 kg of TCDD equivalent dioxins were released over a 20 minute period from a runaway chemical reaction. This severely contaminated approximately 4 km^2 of soil. (It was also present as a contaminant in the US Vietnam War

Dioxin/furan	Equivalence factor
2,3,7,8-tetrachloridibenzodioxin (TCDD)	1
1,2,3,7,8-pentachloridibenzodioxin (PeCDD)	0.5
1,2,3,4,7,8-hexachloridibenzodioxin (HxCDD)	0.1
1,2,3,7,8,9-hexachloridibenzodioxin (HxCDD)	0.1
1,2,3,4,6,7,8-hexachloridibenzodioxin (HxCDD)	0.1
Octachloridibenzodioxin (OCDD)	0.1
2,3,7,8-tetrachloridibenzofuran (TCDF)	0.1
2,3,4,7,8-pentachloridibenzofuran (PeCDF)	0.5
1,2,3,7,8-pentachloridibenzofuran (PeCDF)	0.05
1,2,3,4,7,8-hexachloridibenzofuran (HxCDF)	0.1
1,2,3,7,8,9-hexachloridibenzofuran (HxCDF)	0.1
1,2,3,6,7,8-hexachloridibenzofuran (HxCDF)	0.1
2,3,4,6,7,8-hexachloridibenzofuran (HxCDF)	0.1
1,2,3,4,6,7,8-heptachloridibenzofuran (HpCDF)	0.01
1,2,3,4,7,8,9-heptachloridibenzofuran (HpCDF)	0.01
Octachloridibenzofuran (OCDF)	0.001

PCDD is shorthand for polychlorinated dibenzo-p-dioxin.
PCDF is shorthand for polychlorinated dibenzofuran.

herbicide Agent Orange.) There were no human fatalities from Seveso but several cases of CHLORACNE.

For the determination of the toxic equivalent (TEQ) value, stated as a release limit, the mass concentrations of the following dioxins and furans have to be multiplied with their equivalence factors before summing.

It is certainly true to say that not enough is known about dioxins and world-wide scientific research effort is continuing.

Both vinyl chloride monomer (VCM) manufacture and PVC waste incineration can give rise to the formation of dioxins. Modern VCM plants produce only very small quantities and emit much less to the environment. As an example, the Norsk Hydro plant in Norway makes 450 000 tonnes of VCM each year in the course of which 6 g of dioxins are made and 0.42 g escape to atmosphere or the fjord. Measurements in the USA by GEON detected no dioxins in PVC products. In the UK, the MAFF detected traces towards the limit of detection which the Ministry says presents no hazard to the public.

Source: *United Kingdom Comments on the United States Environmental Protection Agency's External Review Draft Reassessment of Dioxins*, Department of the Environment, Toxic Substances Division, January 1995.

Direct energy conversion. The conversion of chemical, solar or NUCLEAR ENERGY directly into electricity without producing mechanical work in the process (as in the conventional boiler-STEAM turbine-generator system). Direct energy conversion is a very desirable goal since it means that electricity could be made without intermediate equipment and perhaps with greater efficiency.

The classes of direct energy conversion devices are: FUEL CELLS, MAGNETOHYDRODYNAMIC GENERATORS, THERMIONIC CONVERTERS, and semiconductor THERMOELECTRIC CONVERTERS.

The First Law of Thermodynamics applies to all the above devices as does the Second Law in its general form, but the CARNOT EFFICIENCY restrictions do not apply to all direct energy converters. Thus, in theory, direct energy conversion could offer very attractive conversion efficiencies when realized.

Directive, EC. A form of binding European Community legislation either addressed to all Member States or only to specified ones. A directive states an objective and requires the addressee(s) to take such measures as are necessary to achieve the aim desired by the Community within a stated period. The method by which the aim is achieved depends on the addressee; it may mean new domestic law or the recasting or repealing of existing domestic legal and administrative rules.

Direct Toxicity Assessment. Direct Toxicity Assessment (DTA) is the use of whole effluent ecotoxicity testing to assist in the assessment and control of complex industrial effluents discharged directly to waters. Whole effluent ecotoxicity integrates the additive, synergistic and antagonistic effects of substances and their breakdown products within complex mixtures. The approach allows an integrated assessment of the combined biological effects of all constituents, including unknown substances.

Discharge consent. An authorization to discharge a liquid effluent to a receiving watercourse or to a sewer, provided by the Environment Agency or equivalent body. The consent will normally specify volumetric flow rate, BIOCHEMICAL OXYGEN DEMAND and/or CHEMICAL OXYGEN DEMAND, and SUSPENDED SOLIDS content. Other parameters such as pH or TEMPERATURE may also be specified. (◊ DISCHARGE STANDARDS)

Discharge Standards (Effluents). The Seminal Eighth Report of the Royal Commission on Sewage Disposal (1912) laid down discharge standards for SEWAGE effluents that are still used in the UK today. Basically these are 30 mg/l SUSPENDED SOLIDS (SS) and 20 mg/l BIOCHEMICAL OXYGEN DEMAND (BOD), the so-called ROYAL COMMISSION STANDARDS. This assumes that the receiving stream can provide a dilution of at least 8 times the volume of effluent. This standard is often adhered to today, except for discharges to non-tidal streams and/or where dilution is much greater or less than 8:1.

Where potable water is abstracted downstream of the effluent discharge, stricter standards may apply and a limit of not more than 10 mg/l ammoniacal nitrogen may also be imposed. The trend is for a 10:10:10 standard (SS:BOD:*ammoniacal nitrogen*) in areas where rivers are used as a source of potable water. It took the equally seminal Report 'Taken for Granted' Report of the Working Party on Sewage Disposal, HMSO. 1970. To kick start a clean up of Britains abysmally polluted rivers, estuaries and beaches. WORTH A READ!

Discounted cash flow (DCF). A common technique used to evaluate the relative costs of proposed schemes.

Knowing the rate of interest available, we can calculate what value any given sum will have at some future date. Conversely, if we know that, at some time in the future, we are going to have to spend a given sum, then we can easily work backwards to determine the present sum that must be put aside for such a future project. This process is called 'discounting'.

Interest rates gives us the amount by which an investment will grow annually. In discounting, the term 'discount rate' is used to signify the annual rate of interest at which the present value of the required future sum would have to grow over the intervening period in order to become that future sum.

In appraising projects, the term 'rate of return' is also used to indicate the return on investment that the project will generate.

The construction of any project involves the payment of cash at intervals, and its subsequent operation invariably involves further periodic payments (salaries, material, maintenance, etc.). Expressed on an annual basis, such a stream of payments represents a cash flow (outwards in this case, but inwards if you are calculating benefits, e.g. payments for goods or services resulting from the project).

Now, just as we can calculate the present value of a single future sum, so also can we obtain the present value of a series of future sums (costs of benefits), and such a procedure is called *discounted cash flow*. It is used to evaluate, on purely monetary grounds, the relative merits of various alternative projects or alternative ways of implementing a single project. A common application is in *least-cost analysis*. For example, in laying on a needed water supply, do we use a large pipe and let the water flow by gravity (high initial capital investment, low running costs) or a smaller pipe in association with pumps (lower capital investment, higher running costs)? Using discounted cash-flow methods, we can calculate which of the two schemes will cost less in terms of the present value of the sums to be committed.

Note: It is very difficult to incorporate environment matters into DCF calculations. (It is customary, in such calculations, to ignore costs that will be common to both schemes, such as pipeline maintenance, repairs, metering, etc.)

Dispersion. The dilution and reduction of concentration of pollutants in either air or water. Air pollution dispersion mechanisms are a function of the prevailing meteorological conditions. Figure 52 shows several modes of plume dispersion.

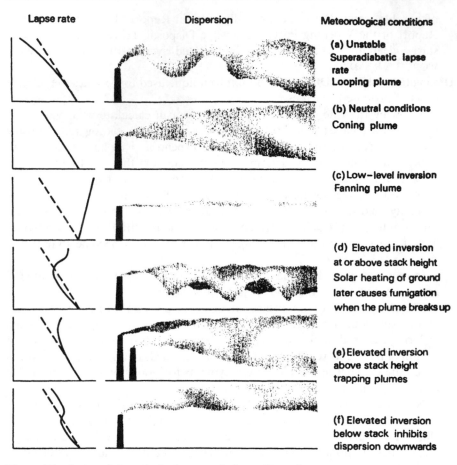

Lapse rate	Dispersion	Meteorological conditions

(a) Unstable
Superadiabatic lapse
rate
Looping plume

(b) Neutral conditions
Coning plume

(c) Low–level inversion
Fanning plume

(d) Elevated inversion
at or above stack height
Solar heating of ground
later causes fumigation
when the plume breaks up

(e) Elevated inversion
above stack height
trapping plumes

(f) Elevated inversion
below stack inhibits
dispersion downwards

Figure 52   Some of the principal types of plume dispersion.

Note that the ADIABATIC temperature profile is shown in solid for the saturated (i.e. wet air) adiabatic LAPSE RATE and dotted for the dry adiabatic lapse rate. The dry adiabatic lapse rate is used as a basis for comparison only. The saturated adiabatic lapse rate usually determines the pattern of dispersion. (◊ INVERSION LAYER; ADIABATIC; LAPSE RATE)

Disposal Levy. Levy charged on first purchase of an artefact to cover its eventual disposal. The levy is refundable only at accredited points of disposal where recycling or environmentally–sound disposal can be effected.

Dissociation. The breakdown of a substance into simpler substances owing to the addition of energy (e.g. heat) or the effect of a solvent.

Dissolved Air Flotation. Dissolved Air Flotation (DAF) is a well-established high-rate clarification process used in both water and waste water treatment. It differs from other clarification processes in that the separation mechanism

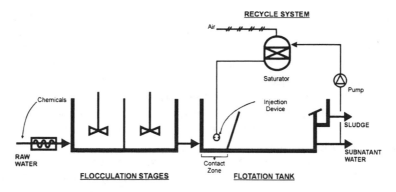

Figure 53   Schematic of the Dissolved Air Flotation Process.

is carried out by flotation rather than by sedimentation. The basic process is shown diagrammatically in Figure 53, with the main component parts identified.

Water that has been dosed with an appropriate quantity of a suitable coagulant and pH adjusted if needed, is fed to the flocculation tanks. Here it is gently agitated by low speed mixers to allow the solids gently to collide and form small accumulations of 'flocs'. The floc-bearing water enters the DAF tank where it is mixed in the 'Contact Zone' with a stream of recycled clarified water which has been super-saturated with air at high pressure. When the pressure of the recycled water is released, a stream of microbubbles is produced, which adhere to the floc particles. These bubbles decrease the bulk density of the flocs and they float to the surface in the DAF tank to form a sludge 'float'. This float is then removed by periodic de-sludging over a sludge beach located at the end of the tank. The clarified subnatant is collected from the bottom of the tank and passed on for subsequent treatment.

Advantages

Small footprint when compared with conventional sedimentation processes and comparable to other high rate methods.

Suitable for a range of raw water qualities, including problematical, highly coloured or algal-laden feeds.

Reduced water wastage as the floated sludge is typically between 2–3 per cent dry solids.

Capable of quick start ups and rapid changes of flows, making it suitable for plants with varying flows.

Dissolved oxygen (DO). The amount of oxygen dissolved in a stream, river or lake is an indication of the degree of health of the stream and its ability to support a balanced aquatic ecosystem. The OXYGEN comes from the ATMO-

SPHERE by solution and from PHOTOSYNTHESIS of water plants. The maximum amount of oxygen that can be held in solution in a stream is termed the *saturation concentration* and, as it is a function of temperature, the greater the temperature, the less the saturation amount. It is customary to express oxygen concentrations as percentages of saturation CONCENTRATION at any given temperature. The saturation concentration is also a function of atmospheric pressure and concentration of dissolved salts. As a result of the latter, sea water contains less oxygen than fresh water at the same temperature and pressure.

The discharge of an ORGANIC waste to a stream, e.g. sewage, imposes an oxygen demand on the stream. If there is an excessive amount of organic matter, the oxidation of the waste by micro-organisms will consume oxygen more rapidly than it can be replenished. When this happens, the dissolved oxygen is depleted and results in the death of the higher forms of life. In extreme cases, the river becomes anaerobic, the oxidation processes cease and noxious odours are given off. (◊ ANAEROBIC PROCESS)

The oxygen balance of a river is of great importance in controlling the effects of pollution. As soon as organic matter enters the water it exerts a BOD, the dissolved oxygen level falls, and as a result an oxygen deficit is created, which means that oxygen transfer from the air to the water takes place to try to restore saturation. The rate of oxygen transfer, i.e. the re-aeration rate, depends on the nature of the flow, the stream-bed characteristics, and the water temperature. A mountain stream will have a re-aeration rate many times greater than that of a stagnant quarry pond, and would thus be able to assimilate a much greater BOD load per unit volume without a serious oxygen deficit.

As the amount of oxygen consumed by organisms in the water increases, the oxygen deficit also increases and in turn the re-aeration rate increases until (if the pollution does not swamp the watercourse oxygen completely) the rates of deoxygenation and re-aeration eventually balance. From then on the re-aeration supplies more oxygen than is consumed by the BOD and the stream eventually recovers to saturation.

This process can be represented by an OXYGEN SAG CURVE, which is a graph of dissolved oxygen against distance (Figure 54). With moderate pollution, the dissolved oxygen levels are high enough to support fish life, while with high pollution the stream becomes anaerobic and the fish life disappears, and perhaps the only signs of life are tubificid worms (an indicator of low or zero dissolved oxygen). (◊ OXYGEN DEMAND)

Distillation. A process by which mixtures of liquids are heated and evaporated. Each vapour then condenses back to a liquid at a characteristic temperature and can be separated. Dissolved solids are left behind.

Distillation is a principal DESALINATION process for the production of pure water from saline water. Note that distillation is a two-stage process:

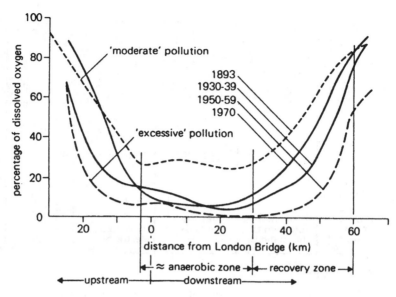

Figure 54   Oxygen sag curves for the river Thames.

first, evaporation and then condensation of the vapour. In the first stage the latent heat of evaporation has to be supplied to the liquid, and in the second stage the latent heat of condensation (virtually identical to the latent heat of evaporation) has to be removed from the vapour for it to condense.

District heating. The use of a large, efficient, centralized boiler plant or 'waste' steam from a power station to heat a district and/or supply steam to small industries. The heat is distributed by means of low-pressure STEAM or high-temperature water mains, to the consumers. (◊ ENERGY)

As the boiler plant is centralized, it combines greater utilization of FUEL compared with a myriad of small domestic or industrial appliances, together with a considerable reduction in low-level pollution compared with, say, open coal fires. If steam from a power station is used, it gives a much higher thermal EFFICIENCY for the system.

District heating has much to commend it in that it leads to much more efficient fuel use (see Figure 55). However, the cost per unit of energy supplied must be competitive with that of alternative fuels, unless subsidies are available (eg comparable to those for WIND ENERGY) to reflect the increased energy efficiency of district heating schemes. This requirement means that the customers must live very close to the boiler plant or power station, otherwise the cost of distribution makes the energy selling price high. (◊ COMBINED HEAT and POWER)

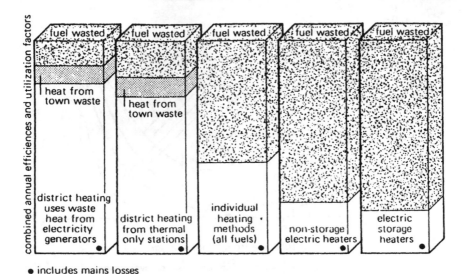

Figure 55   Primary fuel utilization factors for the alternative heating methods.

Diurnal. Active or happening during daylight hours.

DNA (deoxyribonucleic acid). An essential compound found in the nucleus of living cells. DNA is a long-chain molecule, which contains the genetic codes necessary for the development and functioning of the cell. It controls the formation of PROTEIN and ENZYMES. Carbon atoms make up 37 per cent of DNA, and in exposures to IONIZING RADIATION it is entirely possible that a RADIOACTIVE ISOTOPE known as carbon-14 may become incorporated in the DNA structure, thus causing a random change which may have genetic consequences.

Small changes in the complex DNA structure can lead to genetic disorders such as haemophilia, or mental and physical deformity. (◊ PHTHALATES – DOA)

Dobson unit (DU). 1 DU = 2.7×10^{16} molecules per square centimetre.

This unit is used to measure OZONE levels and is the average value of the total ozone column above $1\,cm^2$ of the earth's surface. It contains no information about the vertical distribution of ozone, only the total number of molecules in the column.

Dolomite. A natural calcium magnesium carbonate. It has been proposed as a source of ALKALI for various methods of removing SULPHUR OXIDES from stack gases.

Domestic refuse (UK); Garbage (USA). The generic name for waste emanating from households. In the UK, it has an average percentage composition shown below:

Material analysis of sample of municipal solid waste.

Material	% By weight
Dust and cinder	9.0
Vegetable matter	24.0
Paper and cardboard	31.0
Metals	7.8
Textiles	4.9
Glass	8.0
Plastics	11.0
Unclassified	4.7

Source: A. Porteous, 'Why energy from waste incineration is an essential component of environ-mentally-responsible waste management', *Waste Management*, 25 (2005), pp. 451–459.

In other words, it has approximately 65 per cent ORGANIC MATTER and 35 per cent INORGANIC MATTER. The organic portion is suitable for PYROLY-SIS, HYDROLYSIS, COMPOSTING, or INCINERATION. The inorganics are sources of metals and glass CULLET, while the ash and cinders may also be combustible. The UK **(2005/06)** production was **35.1 million** tonnes or approximately 1 tonne per household per year.

64 per cent of UK domestic refuse is disposed of by LANDFILL where steps must be taken to preserve public health by the prevention of pollution of groundwater, the breeding of flies and rats, and unsightly heaps or offensive odours. Wind-blown litter must also be controlled. Tip space is running out and INCINERATION is expected to gradually replace tipping. The adoption of RECYCLING via separate waste collection or at processing centres for materials recovery is expected to grow. The UK has a stated goal of 45 per cent recovery of recyclable materials by the year 2010. (◊ WASTE QUANTITIES)

Domestic waste. ◊ DOMESTIC REFUSE; WASTES.

Dongtan. A brand new city on Chong Ming Island, China, whose estimated ecological footprint (for life in the Dongtan start-up area) is 2.6 ha per person, compared with nearly 7 ha in the centre of Shanghai. The global earthshare is currently around 1.8 ha per person.

Source: Head, P., 'Responding to Climate Change by Rethinking Urban Infrastruc-ture, 19th Hugh Ford Management Lecture, A joint lecture by the Institution of Mechanical Engineers and the Institution of Civil Engineers, 2007.

Dose. The quantity of a substance or the amount of ENERGY either in a single application or experienced over an interval of time (i.e. the product of the dose rate × time) which produces some specific effect. If the effect is death, this gives rise to the expression LD_{50} (lethal dose), which is the dose large enough to kill 50 per cent of a sample of animals under test. LD_{50} values are normally quoted in mg/kg, and it should be realized that the kg refers to body weight of the animal concerned. (◊ HALF-LIFE; MERCURY; LETHAL

CONCENTRATION; TIME WEIGHTED AVERAGE). The toxicity of chemicals is usually classified as very toxic, toxic, harmful, and not classified. An LD_{50} greater than 7 g/kg weight is not classified as harmful.

'The right dose differentiates a poison and a remedy' (Paracelsus (1493–1541)).

Doubling time. ◊ EXPONENTIAL CURVE.

Downdraught. The low-pressure region created on the lee side when wind blows over a building. Emissions from stacks that terminate within the wind-generated turbulent region are drawn down to ground level in the low-pressure region, giving rise to intense pollution. A general rule for preventing this type of downdraught is that a stack should be at least 2.5 times the height, h, of any building situated within a circle of radius $2h$ around the stack.

Drinking Water Inspectorate. Set up in 1989 by the Government. Responsible for enforcing drinking water quality standards in England and Wales and ensuring that the water companies comply with requirements of the drinking water regulations.

Droplet. A particle of liquid substance of very small mass (up to $20 \mu m$ diameter), capable of remaining in suspension in a gas.

Dry solids. This is the common basis for measuring the dry matter content of sludges such as sewage sludge, and involves drying the sludge, driving off all water and weighing the dry matter left behind. (◊ DRY WEIGHT)

Dry weather flow. This is the rate of flow of sewage & trade waste, together with any infiltration, in a sewer in dry weather. This is measured as the average flow during seven consecutive days without rain, following seven days during which the rainfall did not exceed 0.25 mm on any one day. The winter and summer dry weather flows should preferably be obtained to arrive at an average dry weather flow. Storm overflows on sewers are constructed so that they come into operation when the flow exceeds $6 \times DWF$. Where trade effluents are discharged to sewers, the design DWF is sometimes set at twice the flow between 8 a.m. and 8 p.m. When the overflows come into operation during severe storms, in theory at least, the untreated sewage effluent is diluted six times and this can (presumably) be discharged untreated to a receiving watercourse.

Dry weight. The basis used for expressing the concentration of a compound in living organisms. For example, fish have a moisture content of 80 per cent, so that a concentration of 5 parts per million dry weight would only be 1 part per million wet weight in fish.

Duales System Deutschland (DSD). To reduce German recycling costs, a number of corporations from the retail, packing and filling industries, the producers of PACKAGING material, and the raw material suppliers for the packaging industry founded the Duales System Deutschland (DSD) GmbH in September 1990. The DSD organizes a private waste management network which ensures that all primary packaging waste is collected and sorted by local waste management firms and subsequently reprocessed by domestic or foreign RECYCLING firms.

Objectives of the German Packaging Decree include the following:

- A radical reduction in volume of packaging by avoiding use and stressing the value of reclamation.
- Manufacturers and retailers will be obliged to take responsibility for the packaging used as they are the originators.
- The local communities will be relieved from this part of waste disposal responsibility.
- Returnable packaging will be widely encouraged.
- Recycling will be given absolute priority over thermal recovery.

The DSD uses a 'green dot' to identify a product participating in the system. In order to meet the requirements stated by the Verpackungsveer-ordnung (VVO) the DSD grants the 'green dot' only to those packers and fillers (or respectively, packaging producers) who can present a reprocessing guarantee for the amount and type of packaging which they want to use. As the reprocessing capacities were insufficient, the producers of packaging and packaging materials set up their own – one for each type of packaging material. These reprocessing companies issue global reprocessing guarantees for the respective materials. Figure 56 illustrates the packaging waste flows in the DSD.

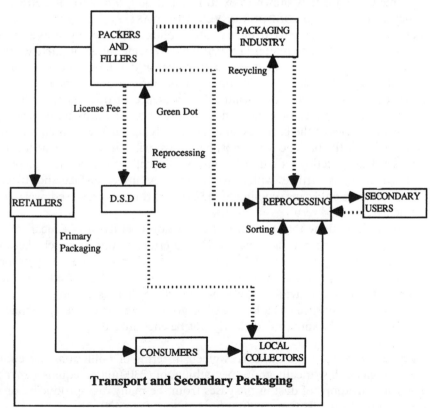

Figure 56 Packaging waste flows in the DSD.

The reprocessing as well as the collection and sorting activities are costly. Hence, the DSD finances collection and sorting costs through licence fees which have to be paid by each packer or filler who wants to use the 'green dot'. Initially, the licence fees varied according to the volume of the container, not according to the material in question, but in October 1993 the DSD introduced new charges based on the actual cost of collecting and sorting the different materials. This is something the UK could adopt.

Dumping at sea. The disposal of authorized wastes to sea such as FLY ASH from coal-fired power stations, and chemical industry wastes has a long history in the UK. Pre-1998, 30 per cent of UK sewage sludge, amounting to 6 million tonnes (95 per cent moisture content), plus 550 000 tonnes of fly ash annually was disposed of in this fashion. However, sludge dumping is now thankfully banned.

The UK was out of step, as it was the only signatory to the 1987 North Sea Conference which supported this method of waste disposal. The official reasoning was that scientific proof of damage to the environment was required before action was deemed to be necessary (PRECAUTIONARY PRINCIPLE). This is at odds with the precautionary policy of the EC, which states that by the time the damage is shown to exist, it could be too late to take effective remedial measures. (◊ LONDON CONVENTION, WASTES)

Dust. Often called airborne particulate matter; solids suspended in air as a result of the disintegration of matter. This embraces a wide range of particles but dust is normally taken as between 1 and $76\,\mu m$ in diameter (above $76\,\mu m$ is termed GRIT). Particles above $10\,\mu m$ in diameter are not carried far from their source except by strong winds. The presence of dust particles in the ATMOSPHERE could cause either a net cooling or a net warming, depending on the properties of the particles and the underlying surface. It is commonly supposed that the presence of atmospheric dust will cause the earth to cool, but as well as reflecting radiation from the sun, it will also reflect heat released from earth. Thus particles with a 'white' or 'grey' upper surface and 'black' lower surface would cause cooling of the earth; if this were reversed, warming of the earth would result.

Calculations show that the aerosol extinction coefficient – a measure of how much heat the particles absorb – has a critical value at which there is heating or cooling of the earth's surface. There is a balance in temperature on surfaces that have ALBEDOS in the range 0.35 to 0.60. For albedos greater than 0.6 there is heating; for values less than 0.35 there is cooling.

Atmospheric science has a long way to go, but the use of the atmosphere as a sink for pollutants should clearly not be encouraged.

Dust concentration in gases. This is given in terms of milligrams per cubic metre of gas or historically grains per cubic foot ($2300\,mg/m^3$ equals $1\,gr/ft^3$). The concentrations of dust in the gases from the many dust-producing processes range from less than $2000\,mg/m^3$ to more than $400\,g/m^3$.

The dust concentration in the cleaned gas from dust extraction equipment is, however, the really important figure and this can be predicted for any specific duty by relating the grade efficiency curve of the proposed equipment to the particle size distribution of the dust and the inlet concentration.

The tables below gives the size range of three typical industrial dusts classified as coarse, fine and ultra-fine respectively, and collection efficiencies for various types of gas-cleaning equipment.

British Standard sieve no.	Particle size (μm)	% by weight less than size*		
		Coarse (CD)	Fine (FD)	Ultra-fine (UFD)
100 (0.006 in.)	152	—	100	—
150 (0.004 in.)	104	—	97	—
200 (0.003 in.)	76	46	90	100
	60	40	80	99
	40	32	65	97
	30	27	55	96
	20	21	45	95
	15	16	38	94
	10	12	30	90
	7.5	9	26	85
	5.0	6	20	75
	2.5	3	12	56

* CD = coarse dust (21% below 20 μm); FD = fine dust (45% below 20 μm); UFD = ultra-fine dust (95% below 20 μm).
From R. Ashman, 'Air pollution control', in A. Porteous (ed.), *Developments in Environmental Control and Public Health*, Applied Science Publishers, 1979.

Type of equipment	Collection efficiency (%) for three typical industrial dusts			Approximate power requirements (kW) for a flow rate of $10^5 m^3/h$
	Coarse	Fine	Ultra-fine	
Medium efficiency cyclones	84.6	65.3	22.4	40
High efficiency cyclones	93.9	84.2	52.3	60
Self-induced spray scrubber	97.6	92.3	70.3	70
Pressure spray scrubber (medium to high energy)	99.9	99.7	99.5	150–300
Electrostatic precipitator	99.5	98.5	94.8	40
High efficiency electrostatic precipitator	99.96	99.85	99.35	50
Venturi-scrubber (medium to high energy)	99.97	99.9	99.6	200–370
Shaker-type filter	99.97	99.92	99.6	50
Reverse jet filter (blow ring)	99.98	99.95	99.8	130
Reverse jet filter (nozzle)	99.98	99.95	99.8	150
Reverse pressure filter	99.98	99.95	99.8	110

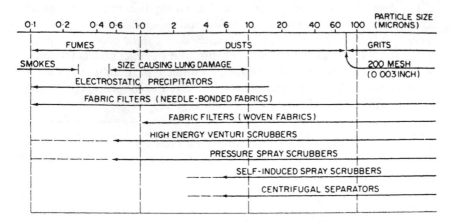

Figure 57 Characteristics of gas dispersoids and methods of control.

As an example of dust control, British Coal monitored the total dust deposit (including any potential coal content) at every operational location. Monitoring was undertaken using the British Standard directional dust gauge, 94.5 per cent of their results were below the level of $200\,mg/m^2/day$. The samples above $200\,mg/m^2/day$ were short-term incidents at Bates Staithes, Chester House and Keekle Ext. which have been addressed and remedial action taken (Fig. 58).

Duty of care[1]. Legal concept enshrined in pollution and Health & Safety legislation, e.g. Mines & Quarries Act 1954, which places a duty on, for example, a mine manager to take all possible steps to solve foreseeable problems and apply all the skill and care a manager should use to obviate the danger (*Mettam* v. *National Coal Board*, 12 July 1988).

Breach of duty of care can constitute negligence (e.g. *Gertsen* v. *Municipality of Metropolitan Toronto*, 1973) in which LANDFILL GAS escaped from a LANDFILL SITE onto adjoining land causing an explosion in garages in 1969. The authorities were found liable for damage and injury caused by the explosion by virtue of NUISANCE and negligence. (◊ *RYLANDS* v. *FLETCHER*)

Claims based on negligence may have substance if (Hawkins 'A duty not to be dumped', *Surveyor*, 15 September 1988):

1. the defendant has a duty of care towards the plaintiff;
2. there has been a breach of that standard of care imposed, e.g. a failure to act reasonably in a situation where reasonable care is required (this will depend, like a conviction for careless driving, on the facts of the case);
3. damage has been caused which was a reasonably foreseeable consequence of the breach of that duty of care.

In the *Gertsen* case, the defendants were held to have been negligent in:
(i) burying the garbage as they did when they knew, or ought to have known,

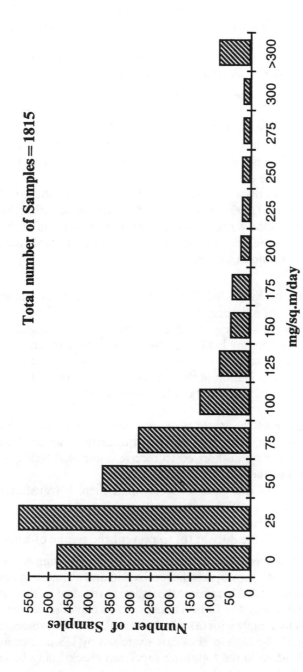

Figure 58 Monitoring of total dust deposits (1993).
Source: British Coal Opencast Northern, Environmental Review 1993.

that its decomposition resulted in the production of gas; (ii) in failing to take steps to prevent the escape of gas; and (iii) in failing to warn adjoining owners of the risk, i.e. they had a duty not to cause injury to persons or property.

Hawkins (Surveyor, 1988), states that in his opinion, a landfill operator in the UK would have a duty of care to adjoining owners. He also drew attention to *Miller* v. *South of Scotland Electricity Board* (1958):

> A person doing work on land owes a duty of care to all persons who are so closely and directly affected by his work that he ought reasonably to have them in contemplation when he is directing his mind to the task.

A recent UK law case has modified the above ruling. It appears that now the use of the land in question has to be 'non-natural' for strict LIABILITY to apply. Plenty of scope for argument here.

Duty of care[2]. This is applicable under Section 34 Part II of the UK Environmental Protection Act (1990) to 'those who import, produce, carry, keep, treat or dispose of controlled waste or, as brokers, have control of such waste'.

Those subject to the duty of care must take all reasonable measures to achieve the goals summarized below:

(a) to prevent
   - the deposit of controlled waste without a waste management licence or in breach of a licence;
   - the treatment, keeping, or disposal of controlled waste except as licensed in or on any land or by means of mobile plant;
   - the treatment, keeping or disposal of controlled waste in a manner likely to cause pollution of the environment or harm to human health;
(b) to prevent the escape of the waste from their control or that of any other person;
(c) on the transfer of the waste to secure that the transfer is only to an authorized person or to a person for authorized transit; there are six categories of person authorized to receive controlled waste, set out in Section 34(3) of the Act;
(d) on the transfer of the waste to secure that there is transferred such a written description of the waste as will enable each person
   - to avoid a contravention of Section 33 of the Act; and
   - to comply with the duty at (b) to prevent the escape of waste.

In addition to these aspects, those subject to the Duty must comply with regulations under Section 34(1) as respects the production and retention of documents, that is, they must make and keep records of waste as required by the regulations. They may also be required to furnish documents or copies of documents. Any person who fails to comply with (a)–(d) above is liable to be prosecuted.

An indication of the high level of care exercised by US producers of hazardous wastes is given in the following suggested checklists to be employed before disposing of their waste.

For a landfill:
1. Are there indications of improper drainage or ponding of water?
2. Is run-on diverted away from all portions of the landfill?
3. Is run-off from active portions of the landfill collected and treated?
4. Is waste that is subject to wind dispersal controlled?
5. Does the disposal operation occur within 200 ft of the property boundary?
6. Are untreated, ignitable or reactive wastes placed in the landfill?
7. Are incompatible wastes placed in the same landfill cell?
8. Are bulk and non-containerized liquid wastes landfilled?
9. Does the landfill have a liner that is chemically and physically resistant to liquids, and a functioning leachate-collection and -removal system with sufficient capacity to remove all leachate produced?
10. Are containers holding liquid waste, or waste containing free liquids, placed in this landfill?
11. Does the operator maintain a map with the exact location and dimensions, including depth, of each cell with respect to surveyed benchmarks?
12. Does the map show the contents of each cell and the approximate location of each hazardous waste type within the cell?

For an incinerator:
1. What type of incinerator is this (waterwall, boiler, fluidized bed, etc.)?
2. Is the residue from the incinerator a hazardous waste?
3. Does the operating record contain analyses listing the types of pollutants that might be emitted, including the heating value of the waste, halogen and sulphur content, and concentrations of lead and mercury?
4. Does the incinerator have instruments for measuring: waste feedrate, auxiliary-fuel feedrate, air flowrate, operating temperature, scrubber flowrate, and scrubber pH?
5. Are instruments relating to combustion and emission controls monitored every 15 minutes?
6. Is the stack plume observed visually at least hourly for opacity and colour?
7. Are there any signs of leaks, spills or fugitive emissions from pumps, valves, conveyors, pipes, etc.?
8. Are all emergency-shutdown controls and system alarms checked to ensure proper operation?
9. Is the incinerator inspected daily?
10. Is there open burning of hazardous waste?

For a chemical, physical or biological treatment facility:
1. Does the treatment system show any signs of ruptures, leaks or corrosion?
2. Is there a means to stop the inflow of continuously fed wastes?
3. Is the discharge-control safety equipment (e.g. waste-feed cutoff, bypass drainage, and pressure-relief systems) inspected daily?
4. Are data gathered for monitoring equipment at least once each day?
5. Are construction materials of the treatment process inspected at least weekly to detect corrosion or leaking of fixtures and seams?
6. Are the discharge confinement structures (e.g. dikes) immediately surrounding the treatment unit inspected at least weekly to detect erosion or obvious signs of leakage (e.g. wet spots or dead vegetation)?
7. Are ignitable or reactive wastes fed into the system treated or protected from any material or conditions that may cause them to ignite or react?
8. Are incompatible wastes placed in the same treatment process?

It is up to waste generators to ensure that the person to whom they consign their waste has the appropriate licence and meets the licensing conditions. It is advisable to audit the disposal route. Waste generators should also know exactly what they are disposing of and keep a record of this.

Breach of the duty of care is a criminal offence carrying, on summary conviction, a fine not exceeding the statutory maximum set by HMG or on indictment, an unlimited fine.

Source: Denise Braker Curtis, 'Waste treatment – better safe than sorry', *Chemical Engineering*, 23 May 1988.

(◊ CONTROLLED WASTE; WASTES)

E

E. coli. ◊ (ESCHERICHIA COLI)

Earth Resources Technology Satellites (ERTS). Orbiting laboratories launched by the National Aeronautics and Space Administration of the USA to monitor natural resources. The ERTS are designed to provide systematic repetitive global land coverage. They complete 14 orbits per day photographing three strips 185 kilometres wide in North America, and 11 strips in the rest of the world.

The orbit has been designed in conjunction with the photographic settings to pass over any location on the earth's surface once every 18 days. Thus, the repetition allows changes in the earth's surface features to be monitored over time. The possibilities for monitoring ocean dumping, changes in land use, jungle clearance, waste tips, etc. are endless. The ERTS also have great potential for monitoring crop health and identifying potentially hazardous situations before they develop, and so allow remedial action to be taken.

Ecobalance. A condition of equilibrium between ecosystems, thus allowing their coexistence.

Ecocide. Suicide by depletion of environmental resources to 'the point of no return' where regeneration is not possible and/or finite resources are squandered rather than conserved for future generations.

Eco-Emballages. Eco-Emballages is a private, non-profit company accredited by the French public authorities to install, organize and optimize sorting and selective collection of household packaging. Eco-Emballages S.A. is based on the Lalonde Decree No. 92-377 on Household Packaging Waste of 1 April 1992. It states that a company should be established to take over the collecting and recycling of household packaging. Moreover, the filling industry and importers should accept responsibility for their packaged products. At the same time, the local authorities retain their traditional responsibility for waste management. Eco-Emballages was created in 1992 on the initiative of industry and approved by the French Government. The shareholders of the company are product and packaging material manufacturers, importers and trading companies. Eco-Emballages provides financial and technical support to the local authorities which undertake the selective collection and valorization of household packaging waste. Eco-Emballages also provides a guaranteed recovery for all the secondary materials that conform to the contractual quality standards. The system is financed mainly by fillers, distributors and importers of household products who pay a license fee for the use of the Green Dot trademark. On average, companies contribute 0.6 eurocents per package. The system has 21 688 licensees' contracts representing 45 000 companies (2005).

With the mass of domestic packaging in 2005 being 4.7 Mt, the following were achieved:

Recovery rate (aggregates recycling and incineration with energy recovery):
    76 per cent of the total mass (3.6 Mt)
Recycling rate: 59.5 per cent of the total mass (2.8 Mt).
Recycling performance *by material* (as a percentage of the total volume):
Steel: 99 per cent
Aluminium: 25.2 per cent
Paper-card: 54 per cent
Plastic: 19.4 per cent
Glass: 73.3 per cent

Source: www.ecoemballages.fr/eco_emballages_in_brief.html. Accessed 16 February 2007.

Eco-cycle. The RECYCLING of products, residues and wastes to energy, raw materials and land reclamation by LANDFILLING. It is a philosophy that embodies:

- waste prevention and reduction, and less hazardous content in products and waste;
- producer responsibility;
- the establishment of environmentally and economically sound recycling, treatment and disposal using a combination of methods;
- the enforcement of environmental laws and standards;
- training and information; and
- functioning international networks.

The eco-cycle is illustrated in Figure 59.
Ecohouse. ◊ AUTONOMOUS HOUSE.
Eco-label. An EC scheme (with approved logo awarded by national bodies) to promote products with reduced environmental impacts during their life cycle.

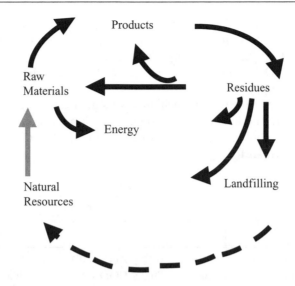

Figure 59 Eco-cycle society.
Source: R. Hakan, 'ISWA and the Global Community', ISWA International Waste Management Seminar, 10 June 1998, Paper Six, p. 2. (Reproduced by permission of ISWA)

The impacts are evaluated on a matrix similar to that above using LIFE CYCLE ANALYSIS.

Eco-label environmental impact matrix for complete product life cycle.

Environmental fields	Product life cycle			
	Production	Distribution	Utilization	Disposal
Waste relevance				
Soil pollution and degradation				
Water contamination				
Air contamination				
Noise				
Consumption of energy				
Consumption of natural resources				
Effects on ecosystems				

The devil is in the detail, e.g. PAPER products can be made from natural (managed forestry resources) or recycled paper, or both. The ENERGY used can be finite or RENEWABLE or both. Air pollutants may be restricted to SULPHUR OXIDES only. COD and AOX are the main water pollutant parameters to be considered. All these are still bones of contention since the scheme was mooted in 1992. Certainly, European paper mills use more finite energy than their Scandinavian counterparts and it is not unlikely that paper products will founder on this comparison. The Nordic (Scandinavian) weighting scale for AOX and COD is given below. European manufacturers have not yet signed up.

By comparison, a washing machine has its greatest impact during use (in energy and water) and is very easy to analyse. Paper products will remain a major bone of contention in the eco-label systems. (◊ ECO-MANAGEMENT and AUDIT SCHEME)

Weighting scale for parameters from paper manufacture.

Parameters	Points		
	1	2	3
AOX (kg/t paper)	<0.1	0.10 ≤ AOX < 0.3	0.30 ≤ AOX < 0.50
COD (kg/t paper)	<20.0	20.0 ≤ COD < 50.0	50.0 ≤ COD ≤ 65.0
S (kg/t paper)	<1.0	1.0 ≤ S ≤ 1.50	1.50 ≤ S2.50

Ecological debtday. How far into a year we can advance before natural resources fail to support consumption. In 1987, debtday was 19 Dec.; 1995, 21 Nov., 2006, 9 Oct. www.neweconomics.org.

Ecological efficiency. The ratio between the amount of energy flow at different points along a food chain.

Ecological footprint. The impact that a settlement has as a result of its fuel, food and waste disposal needs. These typically extend beyond the settlement itself and so delineate its ECOLOGICAL SPACE (UK 5.6 hectares per person, US 10 hectares per person). If there were fair shares each person would have 1.8 ha of bio capacity. www.footprint.network.org.

The Biffa (waste management company) Environmental Trust sponsored the following study ('Island State') with analysis performed by Best Foot Forward, and Imperial College of Science, Technology and Medicine, with support from the Isle of Wight Council.

An ecological footprint analysis of the Isle of Wight

Description

The project explored a methodology for measurement of natural resources usage on the Isle of Wight, and links this with environmental aspects of sustainability. First, data were collected and analysed to determine the Island's consumption of energy and materials. The throughput of these was examined to understand the volume and nature of the material and energy flows and waste arisings. Secondly, using these consumption data and other relevant data sets, an Ecological Footprint Analysis of the Isle of Wight was conducted to demonstrate the pressures that the Island's population places on the local and global environment, thereby providing a measure of ecological sustainability.

Mass flows

The Isle of Wight population 'consumed' 753 368 t of materials in 1998/99. This represents 5.8 t per capita. The largest single category of materials consumed was bulk stone, aggregates, etc (368 838 t or nearly 3 t per capita, of which about two-thirds was imported). The majority (around 3.5 t per capita) of these consumed materials were retained in the economy, primarily as buildings. Around 1 t per capita was disposed of as solid waste.

Key notes

If everyone lived like the population on the Isle of Wight, we would need nearly 2.5 planets. If the Isle of Wight was to be self-sufficient or bioregionally sustainable whilst maintaining current lifestyles and technologies, the Island would need to be about 2.25 times its actual size, or the population would need to reduce consumption by 56 per cent.

The *per capita* ecological footprint of the Isle of Wight for 1998/99 was 5.15 ha. Of this, 0.68 ha is attributable to the tourist population and 4.47 ha to Island residents.

This is greater than the global average ecological footprint for 1998/99 of around 2.28 ha per capita.

It is also greater than the global average 'earth share' of available bioproductive land for 1998/99 of around 1.87 ha per capita, which includes 12 per cent of available land allocated to biodiversity.

A total 169 497 t of waste is received by Island Waste Services, of which 64 per cent is commercial waste and 36 per cent domestic waste. This represents the vast majority but not all of the waste produced on the Island. A number of private operators also collect inert wastes. In total 138 814 t of waste goes to landfill; 72 411 t for disposal and 66 403 t of inert waste used mainly for landfill cover and restoration. The remainder is either recycled, composted or used for energy recovery.

Conclusion

The average Islander consumes almost 2.5 times the sustainable average 'earth share' and currently:

- Produces over 480 kg of domestic waste, just over half of which is sent to landfill (10 000 m^2);
- Uses more than 2000 kWh of electricity at home per year, most of which is supplied by fossil fuels (5000 m^2);

- Wastes food and pays little attention to where food is produced. $(10\,000\,m^2)$;
- Travels mainly by car and takes one holiday a year requiring travel by aeroplane. $(75\,000\,m^2)$;
- Has a low domestic usage of gas $(1000\,m^2)$;
- Is relatively frugal with the use of hot and cold water $(100\,m^2)$;

This results in the individual's contribution to the Island's footprint of 3.36 ha or 33 600 m^2. The economy and public services on the Island add another 1.11 ha or 11 100 m^2 to the Islander's *per capita* footprint.

Recommendations

To be ecological sustainable, an Islander would need to live within the 'average earth share' (roughly 1.9 ha in 1998). Assuming current technologies prevailed, they would need to:

- Produce little or no waste and reuse and recycle wherever possible $(3500\,m^2)$;
- Conserve energy and buy electricity from renewable sources $(200\,m^2)$;
- Eat locally grown, vegetarian food and compost food waste $(3200\,m^2)$;
- Travel mostly by foot, bicycle or public transport, and holiday closer to home $(2000\,m^2)$;
- Use heating sparingly and have excellent home insulation $(1000\,m^2)$;
- Be frugal with their use of hot and cold water $(100\,m^2)$.

This would give an individual's contribution to the Island's footprint of 1 ha or 10 000 m^2. The remaining 10 000 m^2 could be used by public services and for the benefit of the wider Island economy.

Materials consumption – overview

Materials	Tonnes
Bulk stone, aggregates, etc	252 163
Liquid fuel – petrol, etc	62 579
General goods	270 253
Agricultural products (local)	51 698
Aggregates (local)	116 675
Total (rounded)	753 368

This excellent report deserves to be read in depth.

The Living Planet Report 2006, from WWF International, confirms that we are using the planet's resources faster than they can be renewed – the latest data available (for 2003) indicate that humanity's Ecological Footprint, our impact upon the planet, has more than tripled since 1961. Our footprint now exceeds the world's ability to regenerate by about 25 per cent.

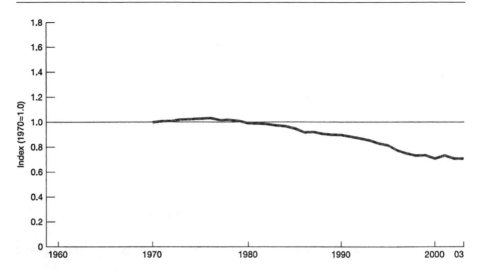

Figure 60   Living Planet Index.

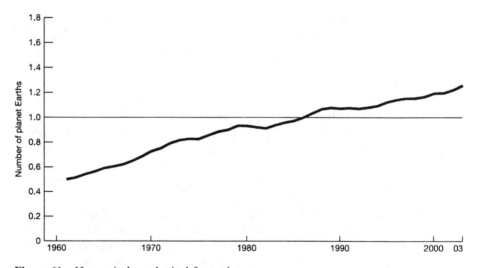

Figure 61   Humanity's ecological footprint.

The consequences of our accelerating pressure on the Earth's natural systems are both predictable and dire. The other index in this report, the Living Planet Index, shows a rapid and continuing loss of biodiversity – populations of vertebrate species have declined by about one third since 1970. This confirms previous trends.

We know where to start. The biggest contributor to our footprint is the way in which we generate and use energy. The Living Planet Report indicates that our reliance on fossil fuels to meet our energy needs continues to grow

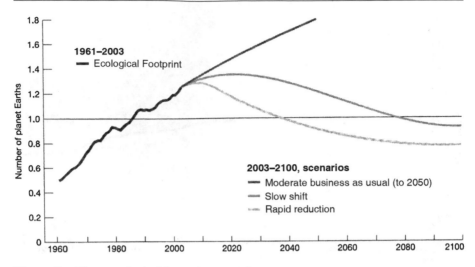

Figure 62   Three ecological footprint scenarios.

and that climate-changing emissions now make up 48 per cent – almost half – of our global footprint. But we must all do more. The message of the Living Planet Report 2006 is that we are living beyond our means, and that the choices each of us makes today will shape the possibilities for the generations which follow us.

This report describes the changing state of global biodiversity and the pressure on the biosphere arising from human consumption of natural resources. It is built around two indicators: the Living Planet Index, which reflects the health of the planet's ecosystems; and the Ecological Footprint, which shows the extent of human demand on these ecosystems. These measures are tracked over several decades to reveal past trends, then three scenarios explore what might lie ahead.

The Living Planet Index measures trends in the Earth's biological diversity. It tracks populations of 1313 vertebrate species- fish, amphibians, reptiles, birds, mammals – from all around the world. Separate indices are produced for terrestrial, marine and freshwater species, and the three trends are then averaged to create an aggregated index. Although vertebrates represent only a fraction of known species, it is assumed that trends in their populations are typical of biodiversity overall. By tracking wild species, the Living Planet Index is also monitoring the health of ecosystems. Between 1970 and 2003, the index fell by about 30 per cent. This global trend suggests that we are degrading natural ecosystems at a rate unprecedented in human history.

Biodiversity suffers when the biosphere's productivity cannot keep pace with human consumption and waste generation. The Ecological Footprint tracks this in terms of the area of biologically productive land and water needed to provide ecological resources and services – food, fibre, and timber, land on which to build, and land to absorb carbon dioxide (CO_2) released by burning fossil fuels. The Earth's bio-capacity is the amount of biologically productive area – cropland, pasture, forest and fisheries – that is available to meet humanity's needs.

A moderate business-as-usual scenario, based on United Nations projections showing slow, steady growth of economies and populations, suggests that by mid-century, humanity's demand on nature will be twice the biosphere's productive capacity. At this level of ecological deficit, exhaustion of ecological assets and large-scale ecosystem collapse become increasingly likely.

Two different paths leading to sustainability are also explored. One entails a slow shift from our current route, the other a more rapid transition to sustainability. The Ecological Footprint allows us to estimate the cumulative ecological deficit that will accrue under each of these scenarios: the larger this ecological debt, and the longer it persists, the greater the risk of damage to the planet.

Ecological demand and supply in selected countries, 2003

	Total Ecological Footprint (million 2003 gha)	Per capita Ecological Footprint (gha/person)	Biocapacity (gha/person)	Ecological reserve/deficit (–) (gha/person)
World	14073	2.2	1.8	–0.4
USA	2819	9.6	4.7	–4.8
China	2152	1.6	0.8	–0.9
India	802	0.8	0.4	–0.4
Russian Federation	631	4.4	6.9	2.5
Japan	556	4.4	0.7	–3.6
Brazil	383	2.1	9.9	7.8
Germany	375	4.5	1.7	–2.8
France	339	5.6	3.0	–2.6
UK	333	5.6	1.6	–4.0
Mexico	265	2.6	1.7	–0.9
Canada	240	7.6	14.5	6.9
Italy	239	4.2	1.0	–3.1

Source: Living Planet Report 2006, WWF International.

The human being's average footprint is calculated to be 2.2 hectares ($5\frac{1}{2}$ acres) per capita but only 1.8 hectares of each person's consumption can be regenerated by the planet each year.

Source: WWF Living Planet Report, Oct 2006.

Ecological indicators. Organisms whose presence in a particular area indicates the occurrence of a particular set of water, soil and climatic conditions. (◊ BIOTIC INDEX; GLADIOLI; LICHENS)

Ecological niche. Each organism has a special task in an ecosystem – known as a niche. No two species of plant or animal can occupy the same niche for long; competition ensues and one species eventually adapts to occupy a different niche, or dies out.

Ecological pyramids. Diagrams which show the overall flow of energy through an ECOSYSTEM. The producer (green plant) level forms the base and the successive trophic or feeding levels (herbivores, carnivores, etc.) occupy the remaining tiers. There are various types of pyramids:

1. *The pyramid of numbers*, which shows the numbers of organisms at each level of a food chain.
2. *The pyramid of biomass*, which is constructed using the total weight of organisms at each level of the pyramid of numbers. It shows the BIOMASS at a particular time, not over a period of time. The pyramid sometimes looks as if it is the wrong way up (see Figure 63), as in the case of phytoplankton and zooplankton in the English Channel. This is because the phytoplankton are eaten almost as soon as they are formed. (The rate of production of the phytoplankton is very much in excess of the growth rate of the zooplankton. The mass of zooplankton can therefore be in excess of the mass of phytoplankton, which are consumed almost immediately.)
3. *The pyramid of energy*, which presents the best overall picture of energy flow. It shows the rate at which food is produced, as well as the total amount. The pyramid is not affected by the size of the organisms, nor by how quickly energy flows through them. (◊ FOOD CHAIN, FOOD WEB)

Ecological space. The area from which a settlement draws the physical resources it uses and in which it disposes of its waste products. London household wastes are disposed of in landfills in Essex, Bedfordshire, Bucks., and Northants. Meanwhile, the Greater London Authority (GLA) 'bans' new incinerators as it wishes to further increase costly recycling.

'Transport for London' estimates (2006) that road vehicles moving London's waste emit ca 250 000 tpy CO_2. [Dumble P. CIWM Wastes Management p. 63, July 2006.]

Ecology. The study of the relationships between living organisms and between organisms and their environment, especially animal and plant communities, their energy flows and their interactions with their surroundings.

Eco-Management and Audit Scheme (EMAS). EMAS is a voluntary scheme for individual industrial sites, introduced in April 1995. The scheme is established by European law. In the UK, the Secretary of State for the Environment is responsible for its administration.

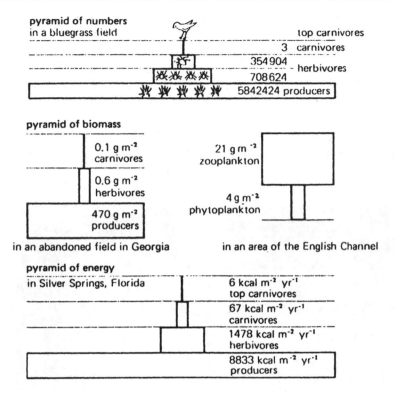

Figure 63 Ecological pyramids. A pyramid of numbers shows the totals of individual organisms found in an area at a particular time. A pyramid of biomass shows the total weight, in grams per square metre, or organisms in an area at a particular time. A pyramid of energy shows the amounts of energy, in kilocalories per square metre per year, available to other organisms in a year's time.

It is designed to provide recognition for those companies who have established a programme of positive action to protect the environment, and who seek continuously to improve their performance in this respect.

To register a site under EMAS a company should have a clearly defined strategy for environmental management, complete with quantified objectives. Initially, EMAS will apply only to Europe's industrial sites. However, in the UK, the scheme has been extended to include local authorities.

The following must be implemented at the relevant site:

1. An environmental policy, for the whole organization, which commits it to compliance with existing legislation and to reasonable, continuous improvement of environmental performance.
2. An environmental review, which covers all aspects of the industrial site or local authority service registered.

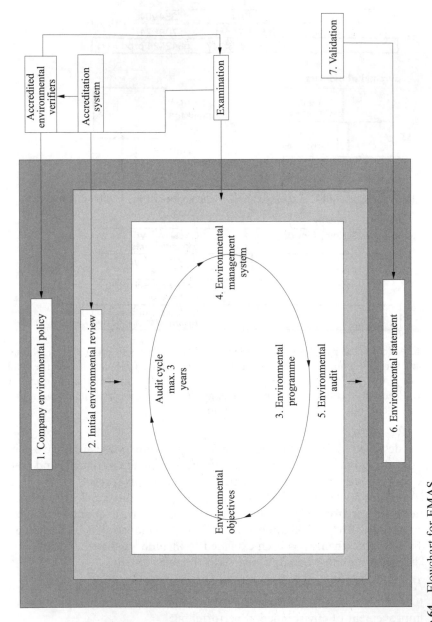

Figure 64 Flowchart for EMAS.
Source: Eco-management & Audit Scheme – An Introductory Guide, Department of the Environment, 1995.

3. An environmental programme, that sets quantified objectives.
4. An environmental management system, to give effect to the policy and programme.
5. An environmental audit cycle, to provide regular information on the progress of the programme.
6. An environmental statement, a concise and comprehensible statement of progress, prepared with the general public in mind.
7. Validation, by an independent verifier. (Check their credential & past clients adoption of their records)

($\lozenge$ ENVIRONMENTAL REPORTS)

Economic level of leakage (ELL). The level of leakage at which it would cost more to make further reductions in leakage than to produce the water from another source. Operating at the ELL means the total cost to the customer of supplying water is minimized and companies are operating efficiently.

Economics. The study of the ways in which people choose what to make of whatever resources they can get, and the ways in which they conduct and organize matters of exchange. The basic propositions of economics relate to matters of human psychology – such as 'values', offers to buy (DEMAND), offers to sell (SUPPLY), tendencies to consume or to delay consumption, and the effects of different inducements in changing these kinds of behaviour. Among those who have criticized the role that economics has come to play in our society are a number of eminent economists, including Galbraith, Mishan, Boulding and Schumacher. One can do no better than quote E. F. Schumacher in this context, *Small is Beautiful*, Blond & Briggs, 1973, Ch. 3:

> It is hardly an exaggeration to say that, with increasing affluence, economics has moved into the very centre of public concern, and economic performance, economic growth, economic expansion, and so forth have become the abiding interest, if not the obsession, of all modern societies. In the current vocabulary of condemnation there are few words as final and conclusive as the word 'uneconomic'. If an activity has been branded as uneconomic, its right to existence is not merely questioned but energetically denied . . .
>
> In fact, the prevailing creed, held with equal fervour by all political parties, is that the common good will necessarily be maximized if everybody, every industry and trade, whether nationalized or not, strives to earn an acceptable 'return' on the capital employed. Not even Adam Smith had a more implicit faith in the 'hidden hand' to ensure that 'what is good for General Motors is good for the United States'.
>
> However that may be, about the *fragmentary* nature of the judgements of economics there can be no doubt whatever. Even within the narrow compass of the economic calculus, these judgements are necessarily and *methodically* narrow. For one thing, they give vastly more weight to the short than to the long term, because in the long term, as Keynes put it with cheerful brutality, we are all dead. And then, second, they are based on a definition of cost which excludes all 'free goods', that is to say, the entire God-given environment, except for those parts of it that

220 Ecosystem

have been privately appropriated. This means, that an activity can be economic although it plays hell with the environment, and that a competing activity, if at some cost it protects and conserves the environment, will be uneconomic.

... it is inherent in the methodology of economics *to ignore man's dependence on the natural world.*

Sources: T. Congden and D. McWilliams, *Basic Economics: A Dictionary of Terms,* Arrow, 1976.
J. K. Galbraith, *The New Industrial State,* André Deutsch, 1972; Penguin Books, 1968.

Ecosystem. The plants, animals and microbes that live in a defined zone (it can range from a desert to an ocean) and the physical environment in which they live comprise *together* an ecosystem. The ecosystem embraces the FOOD CHAIN through which energy flows, together with the biological CYCLES necessary for the recycling of essential nutrients. Thus an ecosystem has the means of producing both ENERGY and materials for life going on continuously. Taken on a global basis, all the separate ecosystems are the life-sustaining processes on which our survival depends. Their integrity must be preserved, otherwise biological communities can die out, or essential services not be performed, such as the self-purification of a river or the control of greenfly by ladybirds. Most ecosystem are extremely complex: for example, a deciduous forest can support over 100 species of birds, as well as many wild flowers, grasses and shrubs. Human intervention by means of planting conifers reduces bird species, animal variety and the types of flowers, ferns and grasses. In general, ecosystems are extremely resilient and where a consumer has more than one source of food, if one becomes less readily available the consumer can still survive. However, if the consumer depends on or only has one source of food available, then its survival can be in jeopardy. (◊ FOOD CHAIN)

Figure 65 shows an aquatic ecosystem where the primary inputs are energy, by PHOTOSYNTHESIS fixed by the primary producers algae and green plants, *plus* the organic and inorganic materials carried by the river. The primary and micro-consumers eat dead organic matter; the intermediate consumers, worms and insect larvae, feed on the primary consumers and ALGAE. The herbivores eat the green plants. The carnivores (fish) eat the intermediate consumers. The scavengers eat the bottom debris and dead organic matter too, so that virtually nothing is wasted. Thus, the ecosystem is a complex interlinking arrangement with its own form of *equilibrium,* i.e. a balance is struck between the total production of living material and the rate of death and decay over a period of time. This is ecological equilibrium. The more complex the ecosystem, the greater the stability. (◊ DECOMPOSERS; PROTOZOA)

Human intervention often simplifies ecosystems dramatically, for example by mono-crop agriculture where some fields grow wheat for up to 30 years with the result that diseases tend to build up. These are often kept at bay by

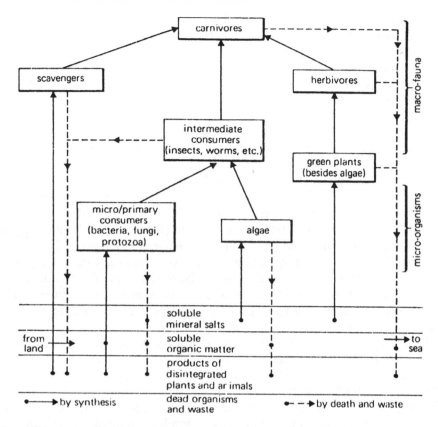

Figure 65 A simplified aquatic ecosystem for a river-bed community.

synthetic chemicals whereas the traditional rotation of crops would have ensured (usually) that disease did not build up. Other common examples of the dangers involved in modification of ecosystems are the conversion of prairie land to wheat, which destroys stable grassland ecosystems, and the logging of deciduous forests to leave barren land which can lead to soil erosion, or, if replanted with conifers, to a modification of forest ecosystems. Large-scale changes in the use of land also result in a change in the ALBEDO, often increasing it, which in turn affects the MEAN ANNUAL TEMPERATURE and thus, cumulatively, the earth's climate. The spraying of crops with INSEC-TICIDES can be a major violation of an already simplified ecosystem. The animals at the top of the food chain are smaller in number than those below. Predators which eat pests, e.g. carnivores, are more likely to be destroyed in a blanket-spraying programme compared with the pests, as there are many more pests than predators. Thus, once spraying is initiated and the equilibrium upset, it can become a self-perpetuating cycle, spraying then becoming a necessity as the predators have been destroyed. Pests seem to develop resistance to pesticides more quickly than their predators. (◊ DDT)

In accepting the undoubted short-term benefits of modifying ecosystems, we may in the long term lay ourselves open to the risk that the original ecosystems have been destroyed or gene banks of *naturally* resistant crops or animals lost forever.

Eden Project. A project near St Austell in Cornwall, which consists of giant conservatories called 'Biomes' where visitors can experience climates such as tropical rain forests, the Mediterranean, South Africa and California.

Effective chimney or stack height. The effective height of a chimney (H) is the sum of its actual physical height (h) and the rise of the emitted plume caused by its BUOYANCY and EFFLUX VELOCITY as shown in Figure 66. The effective chimney height is used in air pollution dispersion calculations for the calculation of maximum ground level concentrations (Figure 67). It should be noted that a negative plume rise can occur under some conditions.

Efficiency[1]. The ratio of energy output to energy input of a process using consistent energy units of kWh. For boiler plants, the ratio of useful output to total energy input. (*Note*: this can be approximated to fuel input but there are often substantial energy inputs from pumps, feed heaters, etc. as well.) (◊ THERMAL EFFICIENCY, CARNOT EFFICIENCY)

Efficiency[2]. For gas filters, dust separators, and droplet separators, the ratio of the quantity of particles retained by a separator to the quantity entering it. (◊ ELECTROSTATIC PRECIPITAIOR)

Effluent[1]. Any fluid discharged from a source into the environment.

Effluent[2]. Any liquid which flows out of a containing space, but more particularly sewage or trade waste, partially or completely treated, which flows out of a treatment plant. For example, treated sewage is the liquid finally dis-

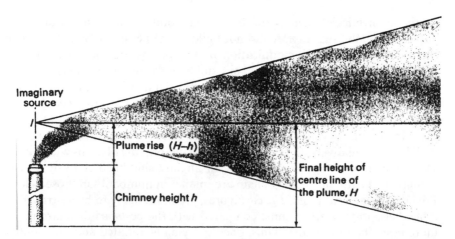

Figure 66 A diagram to show the imaginary source, *I*, of an effluent plume originating above a chimney exit.

charged from a sewage treatment works. Industrial effluents can require complex treatment processes, as described below.

Effluent, physico-chemical treatment. The effluents from food-processing industries are often extremely noxious and costly to treat; yet they contain fats and proteins (soluble and insoluble) whose recovery can yield valuable products. The separation of the fats and proteins can be done by flotation and flocculation followed by ION EXCHANGE, i.e. by physico-chemical means. One such plant is shown in Figure 68 for the treatment of poultry-processing wastes.

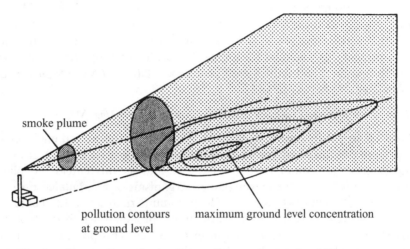

smoke plume

pollution contours at ground level   maximum ground level concentration

Figure 67 A pollutant dispersing under conditions of neutral stability.

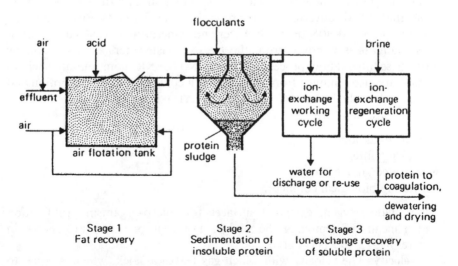

flocculants

air   acid   brine

effluent

ion-exchange working cycle   ion-exchange regeneration cycle

air

air flotation tank   protein sludge

water for discharge or re-use   protein to coagulation, dewatering and drying

Stage 1
Fat recovery

Stage 2
Sedimentation of insoluble protein

Stage 3
Ion-exchange recovery of soluble protein

Figure 68 Physico-chemical treatment of livestock processing wastes for fat and protein recovery.

The incoming effluent is acidified to pH 3 and a coagulating agent added, so that when air is bubbled through the first chamber the fat floats and is skimmed off. The protein-rich liquor is neutralized (pH 7) and flocculants added which make the insoluble PROTEIN 'floc' or come together. The protein then settles and is sent for stabilization. The soluble protein can only be removed by ION-EXCHANGE techniques and the problem is to obtain the correct material for the solution to be treated. (♢ pH)

CELLULOSE ion-exchange media can selectively remove the high-molecular-weight protein molecules and the final clear water from the plant can be re-used or discharged. The ion-exchange medium must be recharged by brine which removes the protein molecules from the medium in a protein-brine solution which can then be coagulated and dried. After treatment the recovered protein has potential for animal feedstuffs, thereby recovering a valuable by-product and reducing effluent BIOCHEMICAL OXYGEN DEMAND by as much as 90 per cent.

Source: Paul Butler, 'Processes in action', *Process Engineering*, May 1975, p. 65.

Effluent biological treatment. Treatment plants are designed and engineered to create an optimized environment in which micro-organisms can grow. The liquid growth medium is the wastewater and as this enters the plant it becomes mixed with the existing micro-organisms or recirculated ones as in the ACTIVATED SLUDGE process. The pollutants represent a food source for the micro-organisms and so these are rapidly broken down through BIODEGRADATION and thereby removed from the water. This provides a source of ENERGY and building blocks so that further microbial growth and replication can occur. Thus soluble ORGANIC compounds polluting the wastewater are biodegraded and converted into microbial cells, i.e. biomass. Biodegradation also produces CARBON DIOXIDE, water and mineral NUTRIENTS as by-products. Biomass is removed from the treated wastewater, as a SLUDGE by natural sedimentation processes. Many INORGANIC compounds, such as metals, are also removed from the water as a consequence of absorption and adsorption to the biomass. (♢ SEWAGE TREATMENT)

Four variations are:

- aerated lagoon
- trickling filter
- activated sludge
- rotating biological contactor.

All provide oxygen, extended surfaces for micro-organism proliferation and a means of removing the sludge. The four processes are shown in Figures 69–72. (♢ REED BEDS)

Efflux velocity. The velocity with which gas leaves a stack, which is equal to the volume of gas issuing from the stack mouth per second divided by the

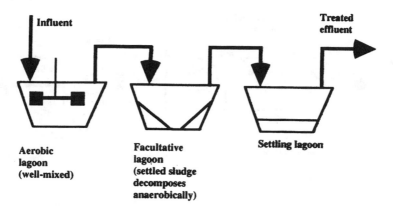

Figure 69 Aerated lagoon.

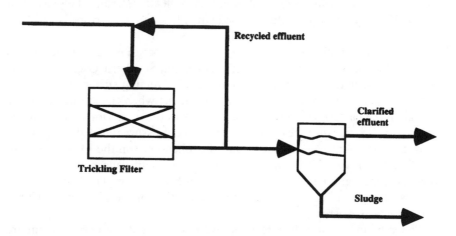

Figure 70 Trickling filter.

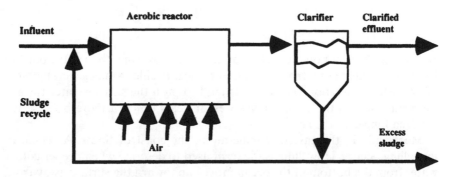

Figure 71 Activated sludge.

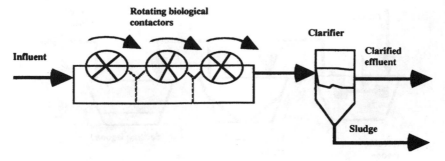

Figure 72   Rotating biological contactor.

cross-sectional area of the mouth. This is a major parameter in air pollution dispersion. If the efflux velocity if low, DOWNDRAUGHT can take place. If it is too high and a wet de-duster used, droplets can be stripped from the stack.

El Nino. El Nino, an abnormal warming of surface ocean waters in the eastern tropical Pacific, is one part of what's called the Southern Oscillation. The Southern Oscillation is the see-saw pattern of reversing surface air pressure between the eastern and western tropical Pacific; when the surface pressure is high in the eastern tropical Pacific it is low in the western tropical Pacific, and *vice-versa.* Because the ocean warming and pressure reversals are, for the most part, simultaneous, scientists call this phenomenon the El Nino/Southern Oscillation or ENSO for short. South American fisherman have given this phenomenon the name El Nino, which is Spanish for 'The Christ Child,' because it comes about the time of the celebration of the birth of the Christ Child, i.e. at Christmas.

To really understand the effects of an El Nino event, we need to compare the normal conditions of the Pacific region and then see what happens during El Nino.

Normal conditions (Non-El Nino)

In general, the water on the surface of the ocean is warmer than at the bottom because it is heated by the sun. In the tropical Pacific, winds generally blow in an easterly direction. These winds tend to push the surface water toward the west also. As the water moves west it heats up even more because it's exposed longer to the sun.

Meanwhile in the eastern Pacific along the coast of South America an upwelling occurs. Upwelling is the term used to describe when deeper colder water from the bottom of the ocean moves up toward the surface away from the shore. This nutrient-rich water is responsible for supporting the large fish population commonly found in this area. Indeed, the Peruvian fishing grounds are one of the five richest in the world.

Because the trade winds push surface water westward toward Indonesia, the sea level is roughly half a meter higher in the western Pacific than in the east. Thus you have warmer, deeper waters in the western Pacific and cooler, shallower waters in the east near the coast of South America. The different water temperatures of these areas affects the types of weather these two regions experience.

In the east the water cools the air above it, and the air becomes too dense to rise to produce clouds and rain. However, in the western Pacific the air is heated by the water below it, increasing the buoyancy of the lower atmosphere thus increasing the likelihood of rain. This is why heavy rain storms are typical near Indonesia while Peru is relatively dry.

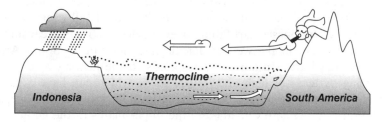

Figure 73a Normally, strong trade winds blow from the east along the equator, pushing warm water into the Pacific Ocean. The **thermocline** layer of water is the area of transition between the warmer surface waters and the colder water of the bottom.

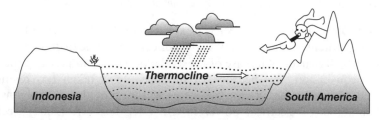

Figure 73b An El Nino condition results from weakened trade winds in the western Pacific near Indonesia, allowing piled-up warm water to flow toward South America.

El-Nino conditions

El Nino happens when weakening trade winds (which sometimes even reverse direction) allow the warmer water from the western Pacific to flow toward the east. This flattens out the sea level, builds up warm surface water off the coast of South America, and increases the temperature of the water in the eastern Pacific.

The deeper, warmer water in the east limits the amount of nutrient-rich deep water normally surfaced by the upwelling process. Since fish can no longer access this rich food source, many of them die off. These conditions are called 'El Nino', or 'the Christ Child', which is what Peruvian fisherman call the particularly bad fishing period around December. More importantly, the different water temperatures tend to change the weather of the region.

What happens to the ocean also affects the atmosphere. Tropical thunderstorms are fuelled by hot, humid air over the oceans. The hotter the air, the stronger and bigger the thunderstorms. As the Pacific's warmest water spreads eastward, the biggest thunderstorms move with it.

El Nino can have impacts on weather at various locations around the globe. Off the east coast of southern Africa, drought conditions often occur. In countries such as Zimbabwe, the effects of drought can be devastating.

The clouds and rainstorms associated with warm ocean waters also shift toward the east. Thus, rains which normally would fall over the tropical rain forests of Indonesia start falling over the deserts of Peru, causing forest fires and drought in the western Pacific and flooding in South America. Moreover the Earth's atmosphere responds to the heating of El Nino by producing patterns of high and low pressure which can have a profound impact on weather far away from the equatorial Pacific. For instance, higher temperatures in western Canada and the upper plains of the United States, colder temperatures in the southern United States. The east coast of southern Africa often experiences drought during El Nino.

Source: http://kids.earth.nasa.gov/archive/nino/intro.html. Accessed 3 January 2007.

Electricity. Omnibus term for electrical power delivered by means of a voltage (electrical potential) difference. It is 100 per cent available to do work whereas heat is not. By virtue of its ease of use as the power plant has done the initial energy conversion, this bedrock of modern life is taken too readily for granted. Currently, despite the hype, renewables provide relatively insignificant amounts of electrical power in the UK. The UK (2004) electricity sources are set out below (DTI data). Much derided coal may well make a come back. (◊ CLEAN COAL TECHNOLOGY)

The total UK (2004) electricity supply is 42 GW of which 19 per cent is nuclear i.e. 8 GW. Now, the technical bit. The UK (2004) energy supply (gross) is 333 GW (213 net) hence by sleight of hand it may be claimed that nuclear energy provides 3.76 per cent of the UK's energy needs. This is comparing apples (kW_{th}) with oranges (kW_e).

Source: *Times Column*, Commentary, by M A Sieghart, 23 June 2006 and subsequent correspondence 6 & 10 July 2006. Back to basics.

A systematic review of the literature on electricity generation unit cost estimates is given in the facing table.

Statistics for predominant technologies – costs per MWh

	Coal	Gas	Nuclear	Wind	Wind – offshore
Mean	£32.90	£31.20	£32.20	£39.30	£48.00
Median	£31.90	£30.50	£31.30	£35.90	£47.90
Inter-quartile range	£13.10	£9.50	£16.50	£24.20	£33.60
Standard deviation	£9.70	£8.90	£10.50	£16.60	£20.00

It is worth noting how close the measures of central tendency for each of the main technologies are to each other. The mean values for coal, gas and nuclear are within approximately 5 per cent of each other, with wind significantly higher, and offshore wind higher still.

Source: P. Heptonstall, UK Energy Research Centre, A Review of Electricity Unit Cost Estimates, Working Paper, December 2006, updated May 2007, UKERC/WP/TPA/2007/2006.

Electric vehicles. A battery electric vehicle (BEV) is an electrical vehicle that utilizes chemical energy stored in rechargable battery packs. Electric vehicles use electric motors instead of, or in addition to, internal combustion engines (ICEs). Vehicles using both electric motors and ICEs are called hybrid vehicles and are usually not considered pure BEVs. Hybrid vehicles with batteries that can be charged and used without their ICE are called plug-in hybrid electric vahicles and are pure BEVs while they are not burning fuel. BEVs are usually autombiles, light trucks, motorized bicycles, electric scooters, folrklift trucks and similar vehicles, because batteries are less appropriate for larger long-range applications.

BEVs were among the earliest automobiles, and are more energy efficient than all internal combustion vehicles. BEVs produce no exhaust fumes, and minimal pollution if charged from most forms of renewable energy. BEVs are quieter than internal combustion vehicles, and do not produce noxious fumes. Battery technology advancements have addressed many problems with high costs, limited travel distance between battery recharging, charging time, and battery lifespan, drawbacks traditionally attributed to the limited adoption of the BEV.

Battery cell cars are powered by simply plugging into any electrical mains outlet. This only costs 1p per mile, which obviously compares favourably with both petrol and diesel. Battery cell cars are currently the most commercially popular type of electric car.

Hybrid vehicles are a combination of petrol and electric. The electric element of the car is used as an assist to the petrol engine, and runs the vehicle to speeds up to 6 mph, after which the petrol takes over. This particular type of electric car is useful in city environments, where a lot of time is spent at lower speeds. It also eliminates emissions from slow moving traffic jams when the engine is kept running. In London, electrically powered vehicles are exempt from the London congestion charge. In most UK cities, low-speed electric milk floats (milk trucks) are used for the home delivery of fresh milk.

Electrodialysis. The separation of ionic components from solutions by means of ion-selective membranes under the influence of an electric field. Electrodialysis is used in DESALINATION of low-salinity waters for human consumption, kidney machines, and specialized effluent-treatment processes.

Electrolysis. The chemical splitting of an electrolyte – a solution that can carry an electric current – by the passage of an electric current through it. The solution is ionized into positively and negatively charged IONS which move towards the oppositely charged electrodes immersed in the solution. Once at the electrode, they give up their charge and can be collected. Thus, the electrolysis of water gives HYDROGEN and OXYGEN as separate components. The hydrogen can then be used in a FUEL CELL.

The electrolysis of brine – sodium chloride (NaCl) and water (H_2O) – gives free CHLORINE (Cl) plus hydrogen (H_2) plus a solution of caustic soda (NaOH). The electrolysis of brine is the main process for CHLORINE manufacture.

Electromagnetic radiation. A form of radiation that involves variations in electric and magnetic effects, and is emitted by all objects with a temperature above zero Kelvin. Electromagnetic radiation is divided into subranges according to the wavelength of the waves that comprise the radiation, and includes radio waves, microwaves, infrared radiation, visible light, ultraviolet (uv) radiation, X-rays and gamma-rays.

Electron. A negatively charged elementary particle which is present in the orbital structure of all ATOMS.

Electroremediation (ER). ER is based on the use of an electric current, between electrodes planted in a contaminated site, to remove contaminants from soil and groundwaters. The process can be used in combination with other techniques such as groundwater extraction and BIOREMEDIATION. ER can be used for the reclamation of both inorganic contaminants, such as HEAVY METALS, NITRATES and PHOSPHATES, and organic pollutants including petroleum hydrocarbons. It is particularly suited to CLAY or fine grain soils, where hydraulic resistance is high, thus making them unsuitable for pump and treat, or soil washing.

A combination of voltage and pH desorbs contaminating IONS attached to soil particles, which then migrate to electrode wells. Control of pH keeps the contamination in a solution in the well, which is pumped through a decontamination system that removes the contamination and returns clean water to the soil (see Figure 74).

Electrostatic precipitator. The gases from combustion processes may contain large burdens of PARTICULATE matter. Before discharge to atmosphere, they must be cleaned. An electrostatic precipitator is one method. The dirty gases are passed through an intense electric field and become electrically charged as shown in Figure 75. The charged particles are attracted to the collector

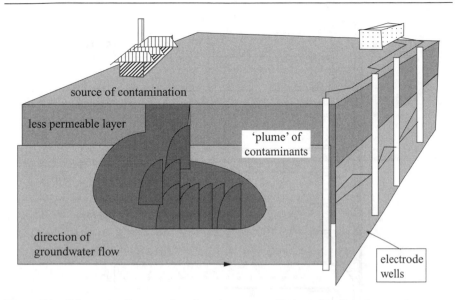

Figure 74   Schematic diagram showing electroremediation of polluted groundwater. *Source*: A. Turner, 'Electroremediation of Polluted Groundwater', *Envirotec*, Vol. 4, issue 3, June/July 1996, p. 14.

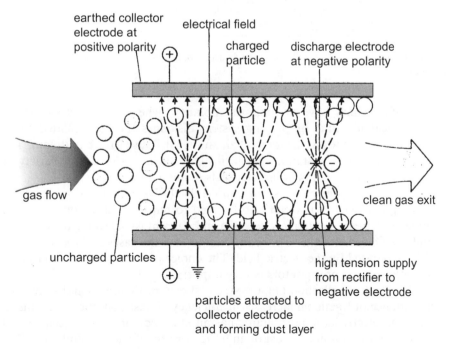

Figure 75   The principles of electrostatic precipitation.

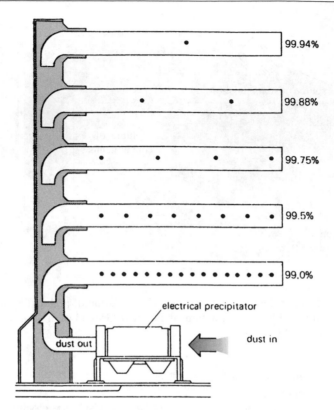

99.94%

99.88%

99.75%

99.5%

99.0%

electrical precipitator

dust out   dust in

Figure 76 Dust collection efficiency (each dot represents a unit of mass flow per unit of time).

electrodes which have an opposite polarity, where they accumulate. A mechanical handling or rapping system dislodges the accumulated DUST which by this time has agglomerated, and it falls out of the gas stream. The product from power stations is called FLY ASH and is used in a variety of land reclamation and building materials manufacture.

Electrostatic precipitators can remove 97–99.5 per cent and more of the initial grit burden. They can also be used to remove acid or oil mists and fumes and are thus extremely versatile and efficient gas-cleaning devices. For higher efficiencies, water can be injected to cool the gases and the solid particles separated by the electric field. The choice of construction materials in wet electrostatic precipitators is very important.

The effect of collection EFFICIENCY is shown in Figure 76 and shows the importance of fractional increases in efficiency in dust-collection equipment, e.g. if an electrostatic precipitator has an efficiency of 97.5 per cent it will emit twice the amount of dust of an 98.75 per cent efficient installation. Precipitators of 99.7 per cent efficiency are now used on new power station installations or have been replaced by BAG FILTERS.

Element. A simple substance which cannot be broken down into simpler substances by chemical methods. An element consists of ATOMS all with the same ATOMIC NUMBER.

An element may be represented by a CHEMICAL SYMBOL which is used in formulae and equations. The symbol is composed of the initial letter of the name, plus another letter where necessary. Usually the initial letter of the English name is used, but sometimes it is that of the Latin name. The symbol is used to represent an atom of the element as well as the element itself. See Appendix II. (The Periodic Table)

The symbols representing the elements may be combined to form chemical formulae which represent compounds: for example, sodium chloride (formula NaCl) consists of equal numbers of atoms of sodium (symbol Na, from the Latin name *Natrium*) and chlorine (symbol Cl). When a molecule of the compound contains different elements in unequal proportions, their ratios are indicated by subscripts: for example, water (formula H_2O) consists of HYDROGEN (symbol H) and OXYGEN (symbol O) in the ratio 2:1, two atoms of hydrogen joined to one atom of oxygen. These are examples of molecular formulae in which the individual and total number of the constituent atoms in the molecule are indicated. ($\lozenge$ ISOTOPE, PERIODIC TABLE)

ELISA. Enzyme-linked immunosorbent assay, a specially developed test used to detect mammalian protein in animal feed.

Elutriation. A method of separating particles using specific gravity differences between the particles when they are suspended in a fluid. In practice, the particles are usually allowed to settle against an upward-moving flow of fluid (e.g. water or air); heavier material settles to the bottom, while fine material remains suspended and is removed with the fluid, e.g. air classifiers used in REFUSE-DERIVED FUEL production use this principle to separate the light combustibles from the heavy non-combustibles.

ELV – End-of-Life Vehicle Directive. The European Union ELV Directive which requires Member States to reuse and recycle 85 per cent, and reuse and recover 95 per cent of vehicles by weight by 2015. Each manufacturer ('producer') is responsible for its own products and has to establish Authorized Treatment Facilities. Recovery may embrace gasification of appropriate shredder residues, for use as a fuel.

Emission. The amount of pollutant discharged per unit time. Or the amount of pollutant per unit volume of gas, or liquid emitted.

Emission factor. The mass of a given pollutant produced per unit mass of fuel consumed, or per unit of production (e.g. per million litres of petrol produced) or per item of production (e.g. per motor car off the assembly line).

Emission inventory. The compilation either by measurement or (more usually) by estimations, of a map of the distribution of emissions over a given area, showing the positions of the more important sources and the amounts they emit. The emission per unit area for each of the smaller sources is usually given as well.

Emission source. Any factory, furnace, chemical process, etc. that discharges pollutants into the air. A *point source* is plant whose total emission of a pollutant is sufficiently large to warrant separate consideration in an emission inventory. (An emission of the order of 100 tonnes per year or more may lead to a plant being considered as a point source.) In the USA any stack is usually considered a point source; stacks with emissions exceeding 100 tonnes per year are considered major point sources.

Emission standard. The amount of pollutants not to be exceeded in the discharge from a pollution source. Emission standards are commonly described in one or more of the following ways for liquid, solid or gaseous pollutants:

1. Mass of pollutants over a certain time: e.g. kilograms per hour, tonnes per day, or pounds per hour.
2. Mass of pollutants per unit mass of material processed: e.g. 20 kilograms per tonne (2 per cent).
3. Mass of pollutant per unit volume of discharged gas at specified conditions of temperature and pressure: e.g. milligrams per cubic metre.
4. The concentration specified as the volume of pollutants (if gaseous) per unit volume of discharged gas, again at specified conditions of temperature and pressure.

The emission standards for NOISE and IONIZING RADIATION are dealt with under their respective entries. (◊ THREE-MINUTE MEAN CONCENTRATION; CONCENTRATION)

Endosulfan. An ORGANOCHLORINE insecticide. It can be toxic to fish in concentrations as low as 0.00002 parts per million, i.e. 1 million litres water would require 0.02 g of Endosulfan, if evenly distributed, to render it toxic to fish. This amount (and more) can easily be left in discarded chemical containers. The potency of many agro-chemicals is such that empty containers should be returned to the supplier for environmentally acceptable disposal. (◊ RED LIST)

Endrin. ◊ CHLORINATED HYDROCARBONS; ALDRIN; RED LIST.

Energy. The ability to do work. Energy has many forms: mechanical, chemical, thermal (heat), nuclear, electrical, commonly measured in JOULES or kilowatt-hours (1 kWh = 3.6 MJ).

It is common practice to consider energy from thermal sources, such as the heat released from the combustion of fossil fuels or nuclear fission, to be primary energy as it can only be used for heating and must be converted by means of an engine such as a steam turbine before it can do work. The conversion of heat to electrical or mechanical energy is governed by the second law of thermodynamics. (◊ LAWS OF THERMODYNAMICS)

Approximate energy conversion ratios for heat to work for a thermal power station range from 30 per cent (poor) to 55 per cent (excellent). Thus a 33 per cent efficient power station requires an input of 3 kilowatt-hours thermal ($3\,\mathrm{kWh_{th}}$) for 1 kilowatt-hour electrical ($1\,\mathrm{kWh_e}$) output. Thus the form in which energy is supplied must be specified when ENERGY ANALYSIS is being performed. A process which uses $2\,\mathrm{kWh_{th}}$ per unit output is much more energy efficient than a competitive process which requires $1\,\mathrm{kWh_e}$ per unit output, as the competitor process in effect requires a thermal input at the power station of about $3.3\,\mathrm{kWh_{th}}$ to make $1\,\mathrm{kWh_e}$. (◊ ENERGY RESOURCES; WATT; JOULE EFFICIENCY)

The 2006 UK power generation spectrum is as follows:

Coal	37%
Gas	36%
Nuclear	18%
Renewables	4%
Oil/other	5%

Source: Department of Trade and Industry, 2007.

Energy – Potential Energy. This is commonly the energy available from, for example, a 'head' of water, or a mass suspended from a rope which is used to help lift a load at the bottom end. However, an advertisement (*The Times*, Business and Climate Supplement, 30 March 2007) by General Electric uses the term to mean the energy possibilities from wind as follows:

$$P = \frac{1}{2}\ [\rho A V]\ V^2 = \frac{1}{2}\ \rho A V^3$$

where

P = potential energy in Joules
ρ = air density in $\mathrm{kg\ m^{-3}}$
A = swept area of blade
V = wind velocity in $\mathrm{m\ sec^{-1}}$

This tells us that the possible maximum energy for wind power varies as V^3 – This cannot be equated to a constant head, as in hydroelectric power.

$$\text{e.g. } V = 5\,\mathrm{m\,sec^{-1}},\ P \propto 125 \text{ units}$$

$$V = 10\,\mathrm{m\,sec^{-1}},\ P \propto 1000 \text{ units}$$

i.e. a halving of wind speed, reduces the possible power output by a factor of 8. Ever wondered why wind turbines are so often idle?

Energy analysis. Every operation carried out on materials, in the mining of fuels or metals, refining or working of metals, transporting, manufacturing and construction of all kinds, involves the use of energy. Hence, any object, from a power station to a milk-bottle, from a barrel of oil to a newspaper, requires the expenditure of energy to bring it into existence – and also to dispose of it.

It is possible to examine the details of the processes by which something is made and to determine the amount of energy needed to produce it. This involves adding up the amounts of energy needed to make each part and that needed to assemble the product. For each part the energy required to form it and the energy required to provide the necessary materials are included, and allowances must be made for some fraction of the energy needed to make any machines involved in the various processes. This is called 'energy analysis by process analysis'.

Other methods, usually less accurate since they involve the use of non-physical variables such as financial costs, may be used for energy analysis.

Energy analysis constitutes a valuable supplement to economic analysis when attempts are made to estimate the effects of changes in the availability of energy upon a national economy. Energy analysis is based extensively upon physical reasoning; this enables it to evade many of the difficulties that confront economic analysis, and which derive from such matters as absolute and relative variations in price. Nevertheless, energy analysis can only be carried out in accordance with one or another of several possible sets of conventions – concerning, for example, how much of the energy expended should be attributed to each of the products of a process with several outputs.

Source: S. Nilsson, 'Energy analysis', *Ambio*, vol. 3, no. 6, 1974.

Differentiation between whether the energy flows are from finite or renewable resources is also necessary. For example, it is claimed that:

> in energy-balance terms, the energy saved through recycling many materials exceeds the energy which could be produced through energy recovery.
>
> Paper recycling leads to energy savings of 28–70% or 6–15 MJ/kg on the basis of the total energy consumed in the manufacture of paper in the UK in 1989, whereas the calorific value of dry newspaper is 16 MJ/kg which could generate up to 4 MJ of electricity or provide 12.8 MJ for process or district heating.

However, the basis of the calculation is on primary (FUEL) energy saved, hence the 4 MJ electrical energy is irrelevant. Furthermore, the recycling of

PAPER in the UK consumes finite fossil fuels whereas pulp/paper manufacture in Scandinavia uses renewable fuels (forest and plant residues). Hence, paper recycling in the UK cannot really be said to be saving energy at all, if the boundary is drawn around the UK. Not that this inconvenient fact is allowed to get in the way of recycling promotion. Integrated waste management is required not the snow-white aura imputed to recycling as opposed to energy recovery in today's state of the art plants. Both can work together and are not mutually exclusive.

(a) Differentiation between renewable and finite sources of energy.
(b) All calculations to be done on a primary (fuel) basis. Purists may add in a factor for the procurement of the finite energy (e.g. oil) from the ground and the subsequent processing.

As an example, the US container glass industry purchased 118.5×10^{15} J in 1985: 78 per cent was in the form of natural gas, 19 per cent in the form of electricity, and the remaining 3 per cent in the form of distillate oil. When the efficiency of electricity generation is taken into account, these figures imply primary energy consumption of 164×10^{15} J. Accounting for transportation and raw-material production energies gives a grand-total primary energy consumption of 192×10^{15} J for 1985. ($\lozenge$ BOTTLES; PAPER; RECYCLING)

Sources: The Local Authority Recycling Advisory Committee's (LARAC) Policy Statement on 'Recycling, Energy Recovery & the National Waste Management Strategy', September 1994.
E. Babcock *et al.*, *The US Glass Industry: An Energy Perspective*, Energetics, Inc., prepared for Pacific Northwest Laboratory, September 1988.

Energy conservation. $\lozenge$ ENERGY EFFICIENCY.
Energy demand. It is precisely those technologies which are regarded as giving us our productive EFFICIENCY, measured in economic terms, that are responsible for the alarming rate of growth of energy consumption in the western world. For example, modern agricultural methods require large fossil fuel subsidies (in the form of OIL, PESTICIDES, artificial FERTILIZERS, etc.). In the USA this has been estimated (Perelman, 1972) as equivalent to an input–output ratio of 5:1 in ENERGY terms. Furthermore, as populations continue to grow, demands for increased food output will require large increases in the use of nitrate fertilizers.

But present methods require about 2 kg of COAL energy equivalent to fix 1 kg of nitrogen. The energy demands of sea water DESALINATION plants, required for increasing water production in the many arid areas of the earth, are also very high but they can be coupled to power stations to use the *waste heat*.

Future prospects are of increasing energy demands as essential industrial materials become scarce. The table gives the total direct energy inputs for producing certain vital metals, neglecting transport costs for both products and raw materials. Bear in mind that the combustion of good quality coal yields about 29 megajoules per kilogram (MJ/kg):

Production process	MJ/kg
Copper from 1.0% sulphide ore in place (1940s)	54
Copper from 0.3% sulphide ore in place (1980s)	98
Aluminium from 50% bauxite in place (1970s)	204
Magnesium from seawater (anytime)	360
Titanium from ilmenite in place (1970s)	593

From J. C. Bravard, H. B. Flora II, and C. Portal, *Energy Expenditures Associated with the Production and Recycle of Metals*, ORNL-NSF-EP-24, Oak Ridge National Laboratory, November 1972.

As these ores become progressively more scarce and the market price rises, it will become economically attractive to work leaner sources, with a corresponding requirement for much higher energy inputs. Unfortunately, it appears likely that for most non-structural metals there is not a continuum of progressively poorer ores in growing quantities but rather an abrupt grade gap, i.e. there is good ore and then there is rubbish. Extracting metals from the 'rubbish', even if economically feasible because of scarcity, is almost certainly impossible in terms of total energy availability.

The UK energy demand is predicted to be as shown in Table below.

Historic and projected growth of energy demand by sector (in million tonnes oil equivalent)

	Residential sector	Transport (including aviation)	Industry	Service sector	Total
1990	40.8	58.0	38.7	19.2	156.7
2010	43.0	75.3	33.8	19.5	171.6
2020	45.7	83.5	36.8	21	187.0
2030	46.8	93.0	37.5	21.6	198.9
2050	49.7	112.2	38.9	23.0	223.8
Growth 2010–2050	+16%	+49%	+15%	+18%	+23%

Energy demand by fuel type is shown below.

Energy Demand, by fuel type (in million tonnes oil equivalent)

	Electricity	Natural gas and Coke Oven Gas (used in iron and steel sector)	Oil (includes aviation fuel supplied in the UK and used for international flights)	Solid fuel	Renewables	Total
1990	23.6	46.0	72.7	13.8	0.45	156.55
2010	29.6	54.0	85.6	1.9	0.43	171.53
2020	33.4	58.3	93.1	1.7	0.49	186.99
2030	36.0	60.9	99.3	2.0	0.66	198.86
2050	40.9	62.9	117.1	2.1	0.78	223.78

Source: UK Energy and CO_2 emissions Projections, July 2006. www.dti.gov.uk/files/file31861.
pdf.
Accessed 16 February 2007.

Annual demand for energy in China will climb to 1.94 bn tonnes of oil equivalent by 2015 if present policies of decentralization and progressive price lib-
eralization continue (according to a study by DRI/McGraw Hill, 1995). The
investment required to meet this demand could reach $1000 bn, of which
slightly more than half would go on electric power generation. Foreign
capital would account for some 20 per cent of the total, the report forecast.

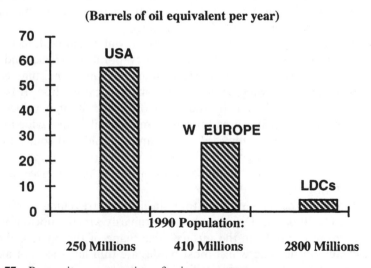

Figure 77 Per capita consumption of primary energy.
Source: J. B. Wyebrew, Shell UK Ltd., Paper presented at IMech E:Power Generation
and the Environment, 13–14 November 1990.

Source: 'China's Energy in Transition', DRI International Energy Consulting, 8–10 tue Villédo, 75001 Paris, France, as reported in *Financial Time* 5/4/95 and Department of Trade & Industry, Energy Paper 65, March 1995.

Related UK carbon dioxide emissions scenario

Energy-related CO_2 emissions can be obtained using estimates of the carbon content of individual fuels in the projections of energy demand, already referred to. The table below shows the CO_2 projections associated with different sectors.

Projected CO_2 emissions (in million tonnes Carbon)

	Residential sector	Transport sector	Industry	Services	Total CO_2 emissions
1990	40.3	40.0	56.4	23.8	160.5
2000	38.8	41.1	48.9	20.7	149.5
2010	36.7	42.4	45.8	19.5	144.4
2030	41.1	41.5	46.6	21.6	150.8
2050	47.3	40.5	50.3	23.8	161.9

Source: UK Energy and CO2 emissions Projections, July 2006. www.dti.gov.uk/files/file31861.pdf.
Accessed 16 February 2007.

Energy efficiency case study

Up to 20 per cent of the energy used by industry in the UK is expended on water removal or drying. There are two basic drying mechanisms: mechanical dewatering and evaporation by thermal means.

In terms of energy expended per unit mass of water removed, mechanical dewatering is always preferable to the use of evaporation. Hence, industrial drying processes should be designed for maximum mechanical moisture removal.

The air knife (or air wipe) is primarily a mechanical method. It is a precisely controlled jet of air delivered by a discharge at high velocity and relatively low pressure. It is used for moisture removal from a wide range of materials and also for the removal of unwanted liquid and solid films and other surface deposits. A good air knife can dry a product using one-tenth of the power, one-tenth of the time and in one-tenth of the space of a conventional drying system.

The 'blade' of the air knife is a thin sheet of air travelling at speeds of approximately 100 km/h. Drying action is primarily kinetic, not thermal, as the jet wipes moisture from the surface. The shape of the blade is dictated by the plenum orifice and whilst most blades are thin flat sheets of air, any shape can be created by suitable orifice design.

The action of the air knife in the simple case of a rigid flat sheet is shown schematically in Figure 78.

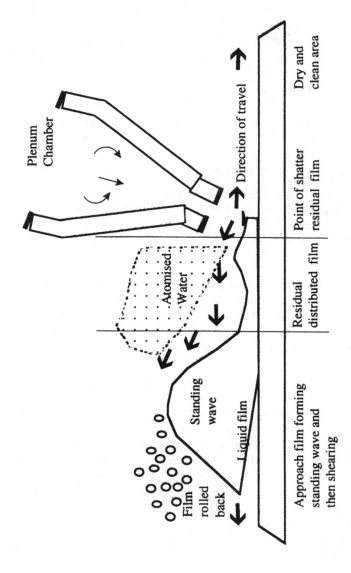

Figure 78 Schematic air knife action of the high velocity, close proximity jet.
Source: Introduction to Air Knife Technology, Midlands Electricity plc, Halesowen.

It is usual to fix the air knife in position and pass the material to be dried, through the blade. The horizontal component of the air stream rolls back the oncoming liquid. This creates a standing wave at the point where the bulk of the liquid is sheared off, leaving a thin residual layer. This residual layer is shattered by the momentum of the air stream to form minute liquid droplets that are then carried away by the air stream to leave a clean, dry surface. A small proportion of the total flow is set up as a high velocity air stream close to the surface to be dried and in the direction of the process material. This effect can be beneficial in providing a high-speed evaporative or warming process if required.

Primary energy consumption increased by 2.4 per cent in 2006, down from 3.2 per cent in 2005 and just above the 10-year average. China alone generated almost half of the world's energy growth over the last five years, of which 73 per cent was coal. Growth in energy demand has been fuelled by global economic growth, most notably in China.

Source: The BP Statistical Review of World Energy, 2007.
Source: A. B. Lovins, *World Energy Strategies*, Earth Resources Research for Friends of the Earth, 1973.
M. J. Perelman, 'Farming with petroleum', *Environment*, vol. 14, no. 8, October 1972, p. 8.

Energy, effect of conversion on climate. The influence of heat produced as a result of our energy conversion activities is already significant on a local scale and will soon become significant globally if present trends continue. Most large industrial areas already add 10 per cent to the sun's heat input, producing significant local variations in climate, and such areas are likely to grow and proliferate.

An extreme example of such 'heat islands' is Manhattan Island in New York City, with a human-made power density of over 700 watts per square metre, compared with 93 watts per square metre from the sun. (◊ TEMPERATURE INVERSION)

On a global scale, significant changes in climate may be triggered by relatively small variations in the heat balance in critical areas such as the floating Arctic pack ice. At present growth rates, one can foresee the possibility of such anthropological changes in the next century, resulting from the combined effects of increases in anthropological heat, and increased levels of CHLOROFLUOROCARBONS and METHANE which are both GREENHOUSE GASES.

It is unfortunate that most of the basic questions about climatic mechanisms are at present unresolved, and it is therefore almost impossible to make assertions on sound scientific bases regarding the specific climate effects of human activity. However, present rapid growth rates are viewed with considerable disquiet by many climatologists, and a prudent policy of energy conservation on a global scale, while perhaps impossible politically, is evidently urgently required.

Energy efficiency. The rational use of ENERGY RESOURCES as part of an overall energy policy is now a pressing problem, especially as a means to combating the GREENHOUSE EFFECT. A wide range of options is available, ranging from RECYCLING of non-fuel materials to the avoidance of competition between various sources such as coal, gas and oil. Some areas where improvements can be made or new techniques applied are shown in Figure 79. A study by the UK Department of Energy in 1989 conceded that the scope for energy savings could be as high as 60 per cent, although a more realistic scope for saving is 20 per cent. Of this, typically 10 per cent energy consumption may be reduced by 'good housekeeping' steps. These include minimizing waste by dealing with leaks, checking thermostat and time-control settings, checking combustion efficiency, cleaning lights and heat emitters, and switching off equipment not in use. All are essentially no-cost improvements, but depend on individual motivation. A further 10 per cent saving is often possible by investing in 'retrofit' systems or equipment which will often give a pay-back within two years. Insulation of building fabric, adding time controls and installing pipe and tank insulation are examples.

More energy-efficient factories means reduced energy costs for manufacturing and thus greater long-term product competitiveness in the marketplace. In the domestic sector, the immediate benefit is reduced fuel costs. For government and public utilities it means a reduction of ENERGY DEMAND.

An excellent example of domestic energy efficiency is provided by the Urban Villa project which consists of 16 low-energy apartments in Amstelveen, the Netherlands. The project has demonstrated the technical feasibility of minimizing the energy consumption of a home without sacrificing living comfort. A well-co-ordinated design has resulted in apartments with predicted annual energy costs about 45 per cent lower than those of conventional apartments of comparable size.

The external walls have 200 mm insulation, thermal bridging is kept at a minimum, 40 per cent of the south facade is glazed for passive solar heating with vents for summer ventilation, inlet air is heated by an exhaust air heat exchange and hot water is provided by solar collectors.

The performance, achieved by well-proven techniques, is impressive:

	Energy consumption (m^3 natural gas)		
	Urban villa: predicted per 100 m^2 apartment	Average 100 m^2 apartment	Savings m^2
Heating fuel	131	1180	1049
Domestic hot water	167	380	213
Total	298	1560	1262
Electricity	3159 kWh	3350 kWh	191 kWh

* 1 m^3 natural gas is equivalent to 31.6 MJ.
Source: Cadett IEA/DECD, Technical Brochure 64, 1998.

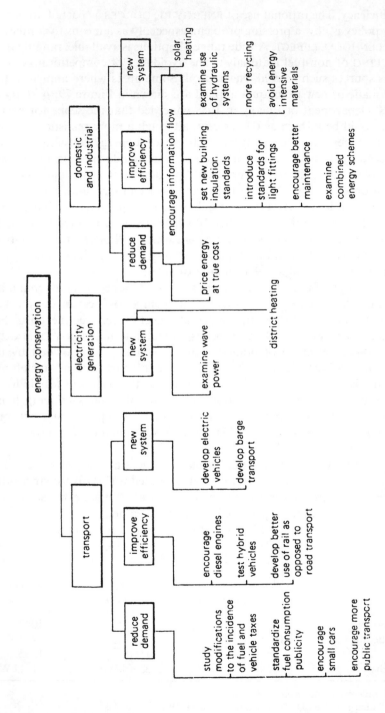

Figure 79 Proposals for energy conservation.

The demonstration project pay-back time is 33 years at today's Netherlands fuel costs, but this could be halved if commercially adopted. Also, it is quite likely that fuel costs will increase in real terms.

It is sometimes surprising what simple changes in operating techniques/instrumentation can achieve, e.g. in two-stage 'firing' of industrial boilers the flow (hot water output) and return (cooler input to boiler) temperature difference along with the heat transfer rates in the boiler firing stages, can be very closely monitored and a decision made by a 'load compensating sequence controller' as to whether it is more economical to fire the first stage only or both.

It is claimed that fuel savings of up to 35 per cent may be achieved by making the boilers more thermally-efficient, as well as prolonging boiler life.

Case Study data are provided below to illustrate the before and after consumption rates after weather correction at various locations run by the Down Lisburn Health & Social Services Trust.

Column 1: Consumption for the year 2000 (prior to M2G System control);
Column 2: Consumption weather-corrected for the year 2002 (all boilers controlled by M2G System);
Column 3: The percentage reduction in fuel;
Column 4: The payback in on investment based on December 2003 energy prices;
Column 5: The reduction in CO_2 tonnes emissions due to M2G System control.

Natural gas	Year 2000 MWh	Year 2002 MWh	% Saving	Payback in years	CO2 tonnes reduced
Laurelhill House	557.30	412.99	25.9	0.99	27.4
Lindsay House	182.5	145.59	20.2	3.86**	7.0
Lisburn ATC	613.50	446.10	27.3	0.85	31.8
Class D Oil	Year 2000 Litres	Year 2002 Litres	% Saving	Payback in years	CO_2 tonnes reduced
Finniston House	197 621	162 788	17.7	0.30	94.1
Hillsborough Health Centre	13005	8472	34.9	2.14	12.2

Source: www.sabien-tech.co.uk/case2.html. Accessed 17 January 2007.

Energy recovery. A form of resource recovery in which the combustible fraction of waste is converted to some form of usable energy, or the recovery of waste heat from industrial or domestic premises/processes for further use. (⟡ INCINERATION)

Energy resources. There are two forms of resources, income and capital, or renewable and non-renewable. Once a capital resource is spent, it cannot be recovered; oil, coal and uranium are capital resources. SOLAR ENERGY is an income source and comes in continually. The derivatives of solar energy such as grass, trees, etc., are forms of income but are finite in the rate at which they can be exploited. A finite or capital resource is one which has been laid down or formed over geological time. Many such resources are being exploited and quite probably will be depleted over the time-scale of several generations. World production of crude oil, rose from almost zero in 1880 to around 3000 million tonnes per year in 1988 (approx. 23 000 million barrels). This is an example of exponential growth (with a few hiccups when there was an oil crisis). When the amount of the proved available reserves in the ground is divided by the production rate, the RESERVES–PRODUCTION RATIO is obtained. Figure 80 shows the curve of the upper and lower estimates for world crude-oil production (based on the virtual exponential growth since 1880) for values of ultimate total oil production of 1350×10^9 and 2100×10^9 barrels respectively. Eventually the rate of production cannot be increased any more and rate of discovery is less than production and so the production will decline to zero – in other words, it is a finite cycle. Figure 80 shows that the middle 80 per cent of oil production on the lower estimate (1350×10^9) would occupy a mere 58 years; for the more optimistic estimate (2100×10^9) it would occupy 64 years. One can argue about the rate of decline as new

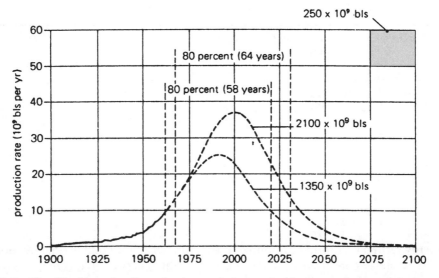

Figure 80   World crude-oil production cycle for total ultimate production estimates of 1350×10^9 bls and 2100×10^9 bls respectively.

resources are discovered or new technology allows further extraction from existing fields, but the ultimate end result will still be a substantial curtailment in oil reserves in the 21st century. (◊ EXPONENTIAL CURVE)

The long-term availability of usable sources of energy and our ability to use them wisely are the factors which will be predominant in deciding the length and nature of the human race's habitation of our planet. In considering the availability of energy sources one must remember that social, economic, political and geographical factors can be as important as the actual abundance or scarcity of a particular resource. As an example, it seems likely that Britain's North Sea oil reserves, instead of being properly husbanded, will, because of economic demands (a guaranteed minimum return on capital invested) be used up as rapidly as they can be extracted. *The Times*, 20 March 1989, in an article on North Sea oil output cutbacks stated 'Last year, Britain lost 61 million barrels of output from the North Sea and this year an estimated 91 million barrels will be lost. In Government revenues, £600000 in North Sea taxation will be lost and as much as £2 billion in export earnings.' Such a policy appears quite rational to the conventional wisdom in economic policy, which tends to regard money and goods as equivalent. *The Times* article also stated:

> Since March 1983, Opec has been trying to get Britain to reach some sort of agreement over cutting output.

> Opec has argued that while it has cut its output by half since the days of the second oil crisis to about 18 million barrels per day, Britain at one point was producing 2.7 million barrels per day.

Economic attitudes to energy sources also disguise the very nature of these resources, i.e. that they are a *degradable* resource which, to be used effectively, must take into account the continuous chains of conversion from a high-energy form to low-temperature heat (e.g. car exhausts). It is major UK economic considerations which prevent, for example, the electrical power generators from designing power stations to make the most thermodynamically efficient use of their waste heat. And, in energy terms, it hardly makes sense to burn gas to generate steam to produce electricity to boil a kettle of water or heat a house, when those operations could be carried out so much more energy efficiently by using gas directly. (◊ DISTRICT HEATING)

The way in which we use our energy supply is extremely patchy. The USA, with only 6 per cent of the world's population, uses over one-third of the world's energy. This gross inequality of distribution is one of the most disturbing features of the present situation. Further, it is in those countries with the largest GROSS NATIONAL PRODUCT that the rate of growth of energy consumption is greatest, due to increased material standards of living and the use of energy intensive technologies. It is these countries whose economies are at greatest risk when the realities of energy limitations become apparent. (◊ GAS; COAL; OIL; HYDROELECTRICITY; NUCLEAR ENERGY; SOLAR ENERGY; OIL SHALES; TAR SANDS)

Energy storage. The Holy Grail of renewable energy proponents. This would enable energy supply fluctuations to be ironed out and also the release of top-up power supplies at periods of high demand. This can be done (at substantial cost), for example, in a Pumped Storage scheme, whereby water released through hydroelectric systems in high demand periods, can be pumped to a high-level reservoir during periods of low power demand, using cheap base load plant electricity. Other energy storage possibilities include hydrogen production and storage, vast compressed air reservoirs or perhaps, 'enormous' rechargeable battery banks. The problem is that what may work on a small scale, may just not be practicable on a large scale.

Energy Walk Round. This is a walk around a building or site to assess energy use, with a view to making savings. A walk round lets you see what is happening on the ground, and enables you to identify wasteful energy use, thus identifying opportunities for savings. It also brings to light maintenance issues that need addressing. High energy-consuming processes are prioritised for particular attention.

Tips for energy saving include:

• Ensuring that equipment that needs to stay on during the day for occasional use have an energy-saving mode.
• Ensuring that equipment is not running when there is no demand for it
• Ensuring equipment is not being misused(e.g. using a compressed air hose to blow dust from machinery)
• Capturing waste heat expelled by equipment (such as compressors) and using it for space or process heating
• Switching off lights when there is sufficient daylight
• Closing windows and doors in heated or air-conditioned areas
• Switching off monitors and desk lighting when desks are empty for any length of time
• Checking that room thermostat settings are correct
• Checking that heaters and boilers have been serviced in the last 12 months
• Listening and touching motors and gearboxes (If they are noisy, rough-sounding or are uncomfortably hot, they may require maintenance or replacement)
• Ensuring that flames in boilers and burners are blue (Yellow or smoky flames indicate incomplete combustion)
• Checking that timers and programmers are working, and are at the correct settings
• Checking compressed air, steam, and refrigeration systems for leaks

- Ensuring that heaters and air-conditioning units are not operating in the same space at the same time
- Ensuring that filters and grills associated with heating, air-conditioning and ventilation systems are cleaned at intervals recommended by the manufacturer
- Ensuring that there are no obstructions in front of radiators, heaters, grills, fans or vents
- Checking for air leaks around windows, doors, skirting and eaves
- Ensuring that heating or air-conditioning units are switched off at the end of the day
- Replacing any damaged or damp insulation
- Keeping light fittings and glazing clean
- Replacing traditional tungsten light bulbs with compact fluorescent bulbs
- Ensuring external lights are switched off during the day
- Ensuring all unnecessary equipment is switched off overnight and at weekends.

The steps involved in an energy-saving plan, are as follows:

(a) Carry out a site walk round
(b) Involve staff in the walk round wherever possible
(c) Make a note of any housekeeping problem areas, and what actions are required
(d) Analyse findings and prioritise actions in terms of business benefit against cost and payback
(e) Report the walk round results to staff and management
(f) Produce a plan of action, with dates and named individuals responsible
(g) Implement quick wins and report successes
(h) Maintain momentum - keep people informed of progress and continue to raise awareness of energy use
(i) Follow-up outstanding actions and implement in line with the action plan
(j) Schedule the next walk round
(k) Feed the findings from further walk rounds into current plans and activities.

It is a good idea to conduct a walk round when there is no production or activity. This will highlight equipment that is running out of production hours. Any equipment left on should be absolutely necessary. It is also wise to walk round at different times of the year to find out if there is a seasonal pattern to the way energy is used and wasted.

Entrainment. The entrapment and transportation of a material by a flowing fluid e.g. dust particles entrained in a CYCLONE DUST SEPARATOR.

Entropy. A measure of disorder or a measure of unavailable ENERGY. Thermo-dynamically, there is constant entropy only in a reversible process (a theoretical construct used in the LAWS OF THERMODYNAMICS).

For all real world processes, there is irreversibility, e.g. heat transfer in practice requires a TEMPERATURE difference for the heat to be transferred. An electric kettle element is hotter than the surrounding water. Hence irreversibility is built into everyday processes and entropy as a result increases; that is why the heat in the water cannot be reconverted to the equivalent amount of electrical ENERGY used to heat it in the first place. Some of the energy is unavailable.

For the idealized reversible process, the change in entropy $(S_1 - S_2)$ is the ratio of the amount of HEAT (Q) transferred in going from higher temperature (condition 1) to lower temperature (condition 2), i.e. $(Q_1 - Q_2)$ to the thermodynamic TEMPERATURE at which the heat is supplied, i.e. the temperature of the source. Hence:

$$(S_1 - S_2) = \frac{(Q_1 - Q_2)}{T_{source}}$$

The above equation is commonly presented as $\Delta Q = T\Delta S$, where ΔT is a very small change at temperature T.

The concept of entropy is extremely useful in analysing energy conversion processes in which a FUEL is combusted to raise the temperature of a working fluid in order to obtain POWER.

The clear implication of entropy creation is that a given amount of energy can be used only once: there are no magic perpetual motion machines. All earthly processes are such that entropy increases. You don't break even! (◊ CARNOT EFFICIENCY)

Environment. Defined by the UK Environmental Protection Act 1990, Part I, Section 1, it is 'Consists of all, or any, of the following media, namely, the air, water and land; and the medium of air includes the air within buildings and the air within other natural or man-made structures above or below ground.'

Environment Act 1995. An Act to provide for the establishment of a body corporate to be known as the ENVIRONMENT AGENCY and a body corporate to be known as the SCOTTISH ENVIRONMENT PROTECTION AGENCY; to provide for the transfer of functions, property rights and liabilities to those bodies and for the conferring of other functions on them; to make provision with respect to contaminated land and abandoned mines; to make further provision in relation to National Parks; to make further provision for the control of pollution, the conservation of natural resources and the conservation or enhancement of the environment; to make provision for imposing obligations on certain persons in respect of certain products or materials; to make provision in relation to fisheries; to make provision for certain enactments to bind the Crown; to make provision with respect to the application of certain enactments in relation to the Isles of Scilly; and for connected purposes. [19th July 1995].

Source: Environment Act 1995, Chapter 25.

The Environment Agency combines the functions of the former National Rivers Authority, Her Majesty's Inspectorate of Pollution, the Waste Regulation Authorities and parts of the former Department of the Environment. The Agency's principal aim is: 'to protect and improve the environment and make a contribution towards the delivery of sustainable development through the integrated management of air, land and water'. Its duties include:

- Licensing of waste management facilities;
- registration of activities and facilities exempt from licensing (such as incineration);
- environmental monitoring of licensed and exempt facilities;
- regulation of waste handling and transport;
- 'cradle to grave' regulation of wastes that present a significant health risk (Special Wastes);
- regulation of waste imports and exports;
- enforcement against illegal waste management activities;
- response to emergencies and incidents;
- enforcement of Producer Responsibility Regulations;
- formulating Regulations;
- provision of statutory information;
- research and development;
- encouragement of best practice and waste minimisation;
- advice on planning consultations;
- provision of advice to government departments;
- provision of advice to local government;
- provision of advice to industry;
- production of strategic waste management assessments.

In Scotland, waste regulation and the functions of the river purification authorities and HMIPI were absorbed within the Scottish Environment Protection Agency (SEPA). It was intended to provide a comprehensive approach to the protection and management of the environment by combining the regulation of land, air and water.

- The Environment Agency has responsibility for:
- Regulating over 2000 industrial processes with the greatest pollution potential, using the best available techniques not entailing excessive cost to prevent or minimise pollution.

- Advising the Environment Secretary on the Government's National Air Strategy, and providing guidance to Local Authorities on their local Air Quality Management Plans.
- Regulating the disposal of radioactive waste at more than 8000 sites, including nuclear sites, and the keeping and use of radioactive material and the accumulation of radioactive waste at non-nuclear sites only.
- Regulating the treatment and disposal of controlled waste, involving 8000 waste management sites and some 70 000 carriers so as to prevent pollution or harm to human health.
- Implementing the Government's National Waste Management Strategy for England and Wales in its Waste Regulation work.
- Preserving and improving the quality of rivers, estuaries and coastal waters through its pollution control powers, including 100 000 water discharge consents and regulation of more than 6000 sewage works.
- Action to conserve and secure proper use of water resources, including 50 000 licensed water abstractions.
- Supervising all flood defence matters, involving over 43 000 km of defence works.
- Maintenance and improvement of salmon, trout, freshwater and eel fisheries, including issue of some 1 000 000 angling licenses.
- Conserving the water environment, including areas of outstanding natural beauty or environmental sensitivity extending to nearly 4 million hectares, and promoting its use for recreation.
- Maintaining and improving non-marine navigation, including licensing of some 40 000 boats.
- Regulating the management and remediation of contaminated land designated as special sites.
- Providing independent and authoritative views on a wide range of environmental issues which may involve analysis and comment beyond the Agency's specific regulatory remit.
- Liaison with international counterparts and Governments, particularly within the European Union, to help develop consistent environmental policies and action world wide.

Environmental Assessment Level. For many substances released to the ambient atmosphere there are no environmental quality standards. For the purposes of Integrated Pollution Control under the provisions of the Environmental Protection Act 1990, the concept of a regulatory limit known as an Environmental Assessment Level (EAL) has been introduced. An EAL is the concentration of a substance in a particular environmental medium that the regulator uses for comparison purposes. In the absence of a comprehensive list of EALs, interim values have been based on a variety of data sources including EPAQS, EU guidelines, WHO guidelines, other international guidelines (e.g. from the UN Economic Commission for Europe), HSE Occupational Exposure Limits (OEL values) and expert judgement.

The use of OEL values is addressed in regulatory guidance. These assessment levels are derived from the guidelines listed above after incorporating safety factors to account for exposure periods, susceptibility to pollution and so on. While these derived limits for environmental assessment levels are not based on robust toxicological data, they have been subject to public scrutiny, but still must be treated with caution and are subject to review.

The table below gives some examples of environmental assessment levels for releases to air.

Environmental assessment levels. (EALs)

	Long-term EAL		Short-term EAL	
	$\mu g/m^{-3}$	Source	$\mu g/m^{-3}$	Source
Ammonia	100	EPA	2400	EH40 (OES)
Arsine	2	EH40 (OES)	60	EH40 (OES)
Benzene	3.25	EPAQS	960	EH40 (MEL)
Cd and compounds	0.005	WHO	1.5	EH40 (MEL)
Ethylene oxide	20	EH40 (MEL)	600	EH40 (MEL)
1,1,1-Trichloroethane	3.8	EH40 (OES)	49 000	EH40 (OES)
Trichloroethylene	1070	EH40 (MEL)	1 000	WHO

Environmental audit. This is an account by manufacturers of the products produced, of their effects on the environment – energy use policies, materials use policies, waste output and their effects on the environment. Purchasing procedure monitoring to obtain environmentally sound goods and services, etc. may also be adopted as part of an environmental audit.

One offshoot is waste auditing in which waste producers keep records of the composition and volume of wastes produced and the disposal routes used for them. This is a 'cradle to grave' practice which is widely employed in the USA. (◊ POLLUTION PREVENTION PAYS; WASTES – SOLID AND HAZARDOUS. VII – MINIMIZATION)

Environmental burden (EB). A pioneering approach by ICI to rank the potential environmental impact of its different emissions. They believe this method helps to improve their environmental management and reporting. It follows three stages.

First, identify a set of recognized global environmental impact categories upon which ICI's various emissions to air and water may exert an effect:

- acidity
- global warming
- human health effects
- ozone depletion

- photochemical ozone (smog) creation
- aquatic oxygen demand
- ecotoxcity to aquatic life.

Second, assign a factor to each individual emission which reflects the potency of its possible impact. (These factors are obtained from recognized scientific literature.)

Third, apply a formula, based on the weight of each substance emitted multiplied by its potency factor, to calculate the EB of these emissions against each environmental category.

Because it assesses potential harm, EB cannot be used to establish the impact of wastes sent to land. This is because nothing should be landfilled unless it is safe to do so. ICI, therefore, continues to report waste deposits to land, divided into those considered hazardous and those considered non-hazardous.

Other key points about the EB approach are:

- individual chemicals can be assigned to more than one environmental impact category;
- each chemical has a specific potency factor for each category and these factors can differ;
- each category has its own characteristics and units of measurements;
- burdens for each category cannot be added together to give a total EB – it is not appropriate since they are as different as 'chalk and cheese';
- EB assumes that all individual operations comply with local regulations;
- EB does not address local issues such as noise and odour.

Example of EB calculation. Consider a hypothetical contribution to atmospheric acidification.

First, record the weight (W) of each single substance emission which has the potential to impact on atmospheric acidification. In this example, the total weight is 32 tonnes.

Second, ascribe a potency factor (PF) to each emission:

	W (tonne)	PF
Ammonia	20	1.88
Hydrogen chloride	3	0.88
Nitrogen chloride	4	0.70
Sulphur dioxide	5	1.00

In this category, sulphur dioxide is the reference substance, so the units of the calculated burden are tonnes SO_2 equivalent.

Third, apply the calculation $EB = W \times PF$ to each substance, and then add the numbers together to obtain the EB for atmospheric acidification.

	$W \times PF$
Ammonia	$20 \times 1.88 = 37.60$
Hydrogen chloride	$3 \times 0.88 = 2.64$
Nitrogen chloride	$4 \times 0.70 = 2.80$
Sulphur dioxide	$5 \times 1.00 = 5.00$

The $EB_{\text{Atmospheric Acidification}}$ is 48.04 units SO_2 equivalent.

The EB changes if the mix of substances emitted changes even though the overall tonnage remains the same. For example, if the emissions of ammonia and nitrogen dioxide are transposed (4 tonnes of ammonia and 20 tonnes of nitrogen dioxide), the total tonnage stays at 32 tonnes.

However, the EB number is smaller (29.16 compared to 48.04), reflecting a lower potential impact on atmospheric acidification by this group of emissions.

	$W \times PF$
Ammonia	$4 \times 1.88 = 7.52$
Hydrogen chloride	$3 \times 0.88 = 2.64$
Nitrogen chloride	$20 \times 0.70 = 14.00$
Sulphur dioxide	$5 \times 1.00 = 5.00$

The $EB_{\text{Atmospheric Acidification}}$ is 29.16 units SO_2 equivalent.

This methodology is invaluable for external reporting and internal environmental management and is to be commended.

Source: *Environmental Burden: the ICI Approach*, ICI Public Affairs, London.

Environmental damage – monetary valuation. Various methods have been proposed for placing monetary value on environmental damage.

Preventative expenditure. The amount paid to prevent or ameliorate unwanted effects; an example is expenditure on insulation and double-glazing to keep out noise.

Replacement/restoration cost. The amount public bodies or individuals spend to restore or replace a lost amenity or landscape.

Property valuation. Differences in the market value of similar properties which reflect differences in the local environment due to traffic or industrial activity.

Loss of earnings. Loss of productive output through injury or ill-health.

Changes in productivity. The monetary value of a reduction in renewable resources (crop or fishery) yield due to environmental damage.

Contingent valuation. The amount people say they would be willing to pay to avoid unwanted effects.

Maintenance expenditure. The additional cost of maintaining capital structures such as bridges and buildings or the additional costs imposed to control deterioration.

Most studies indicated a cost between 0.4 and 0.7 per cent of Gross Domestic Product (GDP) attributable to nitrogen oxides and volatile organic compounds from transport.

In view of the higher level of exposure to noise in Britain, and its higher level of population density, the cost of noise and vibration is likely to be higher than in many of the countries covered in these studies. A small allowance for the cost of vibration damage and disturbance also has to be added. A range of 0.25 to 1.0 per cent of GDP has therefore been taken as a likely range of costs for noise from transport.

Source: Royal Commission on Environmental Pollution, 18th Report, 1994.

Environmental Impact Assessment (EIA). Industrial processes involve the conversion of raw materials and the consumption of water/energy/air to provide the output goods/services.

As the conversion can never be total (LAWS OF THERMODYNAMICS), residues in the form of solid wastes and gaseous/liquid effluents as well as WASTE HEAT, NOISE and vibration will be produced. If the residues are not utilized, they are discharged into the BIOSPHERE and can become POLLUTANTS.

As well as affecting the physical environment, industry and industrialization can have impacts on society. These are generally complex and often cannot be determined at the initial stages because of interactions which do not follow any fixed guidelines or laws of nature.

The principal method of ensuring that environmental considerations are taken into account at the planning stage is to conduct an *environmental impact assessment* or EIA, which is then embodied in an *environmental impact statement* (EIS) which is now often required by law before a new project can proceed. Typically, an EIS will embody the following information:

(a) a description of the development proposed, comprising information about the site and the design and size or scale of the development;
(b) the data necessary to identify and assess the main effects which that development is likely to have on the environment;
(c) a description of the likely significant effects, direct and indirect, on the environment of the development, explained by reference to its possible

impact on human beings, flora, fauna, soil, water, air, climate, the land-scape, the interaction between any of the foregoing material assets, the cultural heritage, employment, transport, education resources, housing, etc.;

(d) where significant adverse effects are identified with respect to any of the foregoing, a description of the measures envisaged in order to avoid, reduce or remedy those effects;

(e) a summary in non-technical language of the information specified above, where public participation is involved.

This is not an insignificant task and three outline examples of EIA applications are given below for a tidal barrage, coal and nuclear power generation, and toxic waste incineration, respectively, to illustrate what needs to be done.

(i) *Tidal Barrage (for Severn Estuary)*
A proposed BARRAGE for the Severn Estuary may involve a 16 km dam fitted with 200 turbine generators. The maximum output would be 8640 MW. The proposed barrage would generate about 5.4% of current England and Wales demand of 350 TW hours per year, and cut 18 million tonnes of greenhouse gases per year. Clearly this can have a substantial environmental impact.

The first step is to thoroughly understand the behaviour of the existing waves, tides and currents as these influence the nature and amount of sediment deposited, and the water quality including OXYGEN content and SALINITY. These factors in turn influence estuarine plant and animal life. The natural and human being-centred environments are then studied along with their interlinkages, e.g. some of the problems posed by the barrage if constructed could cause changes in patterns of human activity which could lead to more developments which in turn could lead to higher pollutant levels and/or increased public health risks for those in contact with the water. The relationships between the major subject areas involved in the environmental impact of the proposed barrage are shown in Figure 81.

(ii) *Power generation*
The major environmental impacts (all figures approximate) of coal and nuclear power stations are tabulated on pages 258–259.

(iii) *Regional Waste Treatment Centre (RWTC) (Courtesy of Leigh Environmental Ltd, 1989)*
This proposed centre (a public enquiry turned down this proposal on the grounds of potential contamination of water resources; the technology was not queried) is designed to treat and recycle a wide variety of industrial residues by high temperature incineration and a variety of physical and chemical processes.

The plant briefly comprises:

(a) a thermal treatment unit for liquid and solid industrial residues
(b) a solvent recovery plant
(c) an acid neutralization plant.

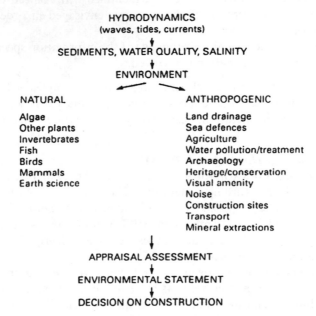

Figure 81   The relationships between the major subject areas involved in the environmental impact of the proposed barrage.
Source: *Quarterly Journal of Renewable Energy*, Issue 6, Jan. 1989, Dept. of Environment.

Impacts of coal and nuclear power stations.

Emissions (*limestone/gypsum flue gas desulphurization fitted*)	Coal (*1800 MW output*)	Nuclear (*1150 MW output*)
Gaseous	11 000 000 tpy* carbon dioxide	—
	16 000 tpy sulphur dioxide	—
	27 000 tpy nitrogen oxides	—
	1000 tpy dust	—
Solids	21 000 tpy sludge	530 m³ low level radioactive waste
	500 000 tpy gypsum	100 m³ intermediate level radioactive waste
	1 000 000 tonnes ash	3.5 m³ high level waste

Emissions (limestone/gypsum flue gas desulphurization fitted)	Coal (1800 MW output)	Nuclear (1150 MW output)
Inputs	5 000 000 tpy coal 250 000 tpy limestone 1 200 000 000 m³ cooling water	186 tpy uranium 1 000 000 000 m³ cooling water
Land	80 hectares coastal or riverside plus 90 during construction	20 hectares isolated coastal or river site

Note: Construction and later on fuel traffic movements are not included. For coal, this aspect is substantial and can be a major factor in the environmental impact as well as the visual intrusion of high (100+) stacks.
*tpy = tonnes per year.
Source: Council for the Protection of Rural England, 1989.

Thermal unit. This is a single stream, thermal unit capable of treating 20 000 tonnes per year of chemical solid, liquid and sludge residues. The plant will initially operate 35 weeks of the year on a three-shift system, 24 hours per day, 7 days per week. It comprises the following elements:

(a) material handling systems for solids, liquids and sludges
(b) rotary kiln thermal unit
(c) secondary combustion chamber
(d) gas cleaning systems
(e) residuals handling system
(f) instrumentation and control system.

Materials handling system. The materials handling systems receive incoming residues, which are stored and blended so that specific mixtures will burn under optimum conditions in the rotary kiln. Solid materials are fed into a hopper and enter the rotary kiln via an airlock and feeding chute. Sludge is atomized by steam in a lance and injected into the rotary kiln. Liquids are atomized by air in special burners and lances and injected into the rotary kiln and secondary combustion chamber.

Rotary kiln. The rotary kiln is a near-horizontal refractory lined drum which rotates about its axis at between 0.1 and 0.6 revolutions per minute. Residues and fuel are introduced at one end and exhaust gases leave the other end. Solvents and other high calorific residues are atomized in special burners and lances as the main source of combustion. Temperatures in the kiln up to 1250 °C (sufficient to destroy POLYCHLORINATED BIPHENYLS, for example) are maintained by the addition of fuel oil through auxiliary burners.

Secondary combustion chamber (SCC). The SCC is designed to create a large reaction volume for the secondary combustion of flue gases leaving the rotary kiln and also to make the complete oxidation of material in the SCC possible. The SCC has auxiliary burners operating with fuel oil to ensure that combustion temperatures are maintained. Past experience with this design has shown that there is about 5 seconds' residence time for flue gases. Under these conditions the destruction efficiency of polychlorinated biphenyls (PCBs) will meet the Environment Agency's requirement of not less than 99.9999 per cent.

Flue gas cleaning. The exhaust gas cleaning system consists of the following stages:

(a) saturation venturi to reduce the exhaust gas temperature;
(b) variable throat venturi to remove particulate matter;
(c) first scrubber to remove hydrofluoric, hydrobromic and hydrochloric acid;
(d) second scrubber to remove sulphur and phosphorus oxides and reduce the water burden on the outlet gases;
(e) mist eliminator.

After cleaning, the exhaust gases are diluted with preheated ambient air and discharged from the chimney by an induced draft fan.

Residuals handling system. The residual solid material, i.e. inert slag and ash, is removed from the end of the rotary kiln through a water quench and conveyed into a skip for disposal off-site in a licensed facility.

Instrumentation and control. A number of plant safety and reliability features are built into the design of the control system to ensure that emissions to the atmosphere are kept within the design limits. The control systems include automatic emission monitoring linked to the residue feed auxiliary burners and scrubber dosing systems. In the event of an abnormal situation occurring, the plant will be shut down automatically. An auxiliary generator covers loss of mains power and standby scrubber liquor pumps will maintain gas cleaning efficiency in the event of a pump failure.

Solvent recovery plant. Deliveries of solvents in drums and in bulk will arrive at the reception area where they will be transferred to reception tanks by vacuum system. Solvent will then be transferred to the feed tanks and then on the vacuum evaporator. After distillation the various fractions will be blended for sale. Residues are passed to the thermal unit for disposal. The solvent recovery plant is generally held under negative pressure by a vacuum pump, the exhaust from these pumps is scrubbed in an oil scrubber before being discharged to ATMOSPHERE. Emissions of non-methane HYDROCARBONS to the ATMOSPHERE will probably be of the order of 5–10 milligrams per second averaged over the day.

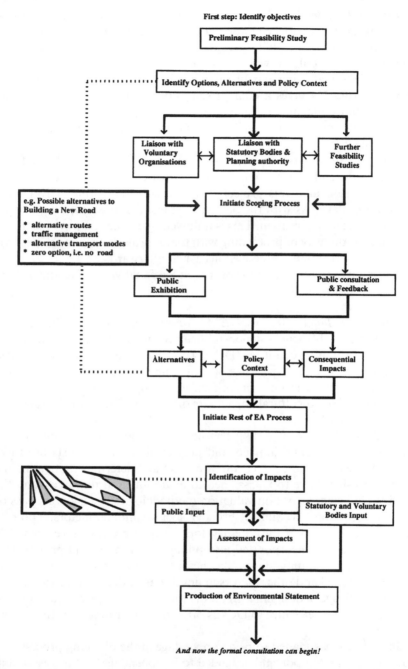

Figure 82 The scope for public involvement prior to producing the environmental statement.
Source: Council for the Protection of Rural England (CPRE) Environmental Statement, November 1990.

Acid neutralization plant. The acid neutralization plant will treat acids as well as the exhausted scrubber liquors, adjust their pH and remove the precipitated solids. The latter will be prepared for removal and disposal at a landfill site. The cleaned effluent will be discharged to sewer.

One of the major impacts of the plant is its air pollution potential. The table on page 264 gives a comparison between the expected ground level concentrations (glc) and air quality standards and existing air quality determined by monitoring and also illustrates the great care taken in the preparation of this EIS.

Similar assessments for land use, water protection, visual impact and effects on the community, noise, traffic increase and employment were also performed to enable this project to proceed to the formal UK planning application stage. As with the Severn Barrage, the most important impacts are identified and appraised. This is followed by public scrutiny. Clearly, this is a much better way of proceeding with major projects than in the past when waste was buried first, and questions asked afterwards.

The inputs and outputs to the proposed Regional Waste Treatment Center are shown in Figure 83.

Environmental Impact Appraisal (EIA) – developing countries. The techniques developed for EIA generally depend heavily on data which is a major difficulty for developing countries (DCs). Techniques employing such things as value judgements, the experience of local communities and public participation – which may be more appropriate for developing countries, where socioeconomic and socio-cultural impacts are often the greatest – are less developed (or non-existent).

Trade-offs have to be established between various sectoral impacts, and between environmental impacts and project modification costs in order to arrive at the environmental cost of a project and determine mitigating measures. So values have to be assigned to environmental resources, and there needs to be further work on valuation methodologies to make this possible.

EIA rarely takes account of broader socio-economic factors. Too often, assessments begin after the core development components have been identified. EIA must be incorporated into policy, programme and project design at the earliest planning stage. Reconciling competing physical, economic, ecological, social and other factors in development decision-making remains the key challenge in designing EIA tools towards development sustainability. This has a major relevance to DCs as often mining or logging is the principal activity.

Just as EIA has tended to be an appendage in the planning process, so its socio-economic component has tended to be appendaged to the rest of it and is frequently not reflected in it. But people are the very centre of the process of sustainable development. The Hague Symposium of 1991 established a

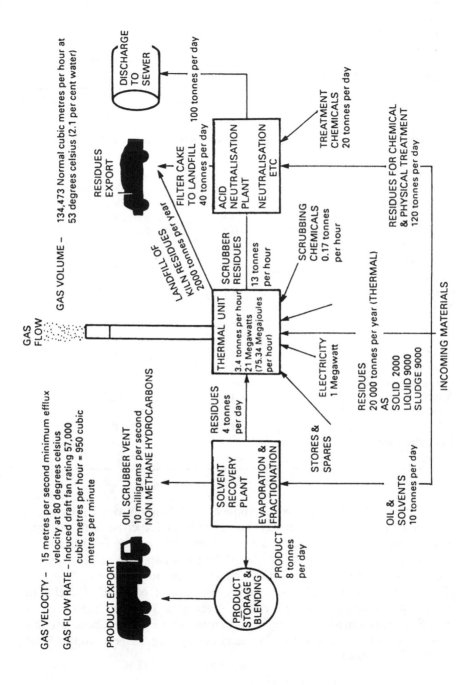

Figure 83 Inputs and outputs to proposed Doncaster regional waste treatment centre.

guiding principle: 'Every environmental measure must be tested against that yardstick: to what extent does it add to the human welfare of the majority of the world's population?'

Comparison between the predicted ground level concentrations and air quality standards and existing air quality.

Substances	Existing glc ($\mu g/m^3$)	Additional future glc ($\mu g/m^3$)	Air quality standards ($\mu g/m^3$)
Hydrogen chloride	—*	7.4	6
Hydrogen fluoride		0.4	62.5
Sulphur dioxide	40	7.4	100 (mean)
Smoke	17	3.6	100 (mean)
Nitrogen dioxide	68	37	200**
Lead	—	<0.3	0.5
Zinc	—	<0.3	125
Mercury	—	<0.3	0.3
Cadmium	—	<0.3	3
Copper	—	0.2	25
Chromium	—	0.3	1.25
Nickel	—	0.06	25
Tin	—	1.5	50
Cobalt	—	0.06	2.5
Silver	—	0.0096	0.25
Arsenic	—	<0.3	5
Hydrogen bromide	—	0.3	250
Antimony	—	<0.3	12.5
Phosphorus	—	<0.3	50
Silicon	—	<0.3	20
Deposited dust (mg/m/day)	20	0.6	200

* Data not available. ** 98 percentile limit.

There is no single best practice applicable to all countries, and an EIA must be tailored to national needs and capabilities. Expertise in its management is lacking in developing countries. Capacity building for sustainable development should be based on local requirements and socio-economic conditions in developing countries and countries in transition. It should begin with a multifaceted and dynamic assessment of needs. Developing country practitioners must be involved in actual activities so as to enhance their skills in conducting, managing and monitoring EIAs and be aware that they do not convey certitude as they may be based on suspect data or value judgements, or both.

Source: *Our Planet*, Vol. 7, No. 1, 1995.

Environmental Management Systems (EMS). A proactive approach to environmental management, implementing an Environmental Management System (EMS) enables companies to reduce their risks and liabilities, whilst enhancing their corporate image with the public, customers and investors. The elements required for an EMS are embodied in environmental quality standards BS 7750 or ISO 14001.

The main requirements are:

- definition and documentation of the organizational structure;
- drawing up an inventory of releases, wastes, energy; raw materials usage to be documented;
- inventory of legislative and regulatory requirements;
- environmental effects; assessment;
- setting objectives and targets;
- environmental management plans;
- management, documentation and records;
- environmental audits; audit plan plus reports and follow-ups;
- verification and testing;
- personnel factors of awareness, training and qualifications.

Environmental improvement targets are essential for spurring change and measuring achievement.

A good target has five elements:

- *quantifiable* – allowing measurements of progress, and clearly defined completion;
- *defined time-scale* – preferably no more than a year or two;
- *defined responsibilities* – achieving a target has to be someone's job;
- *integration with the main business planning cycle* – this ensures that environmental targets don't get trampled by commercial priorities;
- *publication* – a target which can be swept under the carpet is worse than useless. Publication brings commitment and focus.

Source: P. Scott and M. Wright, *Tomorrow*, Vol. V1, No. 5, 1996, p. 55.

EMS are usually ISO 14001 compliant. They embody the following:

Certification audits are conducted by auditors/assessors for a certification body prior to certification to a recognized EMS standard of an organization. Usually these audits are conducted at an advanced stage in the development and implementation of EMS. They include a detailed review of the internal audit programme but are not intended to duplicate the internal audit which has its own important purpose in checking that the EMS has been properly implemented and maintained.

EMS certification audits are traditionally conducted in two phases in accordance with guidance specified by UK accredited services. These phases may be combined; however, they have distinct purposes – '*Do as required*' and '*Do as you say*'.

Phase 1. An organization, or site for EC Eco-Management Audit Scheme (Regulation 1836/93) (EMAS) is expected to comply with specified requirements as set out in ISO 14001 or EMAS and with current statutory and other requirements.

Phase 2. On the assumption that the specified requirements mentioned above are complied with, the auditor(s) may subsequently conduct a Phase 2 audit to establish that the system conforms to planned arrangements and these have been properly implemented and maintained.

> And finally, a cautionary note: an EMS in itself does not guarantee that a company will obtain excellent environmental performance levels, nor sustainability. The latter depends on the objectives and targets a company sets for itself through the EMS.

Source: *Environmental Protection Bulletin*, Issue 056, September 1998, p. 16.

The overall certification scheme.

Process	Carried out by	Applicable standard to control the process
Operation of EMS	Organizations/sites	ISO 14001/EMAS Regulation
Assessment of EMS	EMS auditors/assessors working for certification bodies	ISO 14010, ISO 14011 and IS 14012
Certification	Certification bodies	ISO/IAF guidance on application of ISO/IEC guide 62 and EAC/G5
Accreditation of certification bodies	Recognized accreditation bodies	ISO/IEC guide 61
Assessment of accreditation bodies	Peers	Multilateral agreements between accreditation bodies

Environmental policy formulation. The first step for a company wishing to improve its environmental performance is to carry out an assessment of the effects of a company's operations and, possibly, also its products, on the environment.

This should take account of both short- and long-term effects and include assessment of the effects of future developments such as new production facilities.

The assessment should be comprehensive and include a review of energy usage and, where applicable, the production and disposal of waste.

The principal purposes of the review are to:

- provide the basis for an ENVIRONMENTAL POLICY STATEMENT;
- establish performance targets.

To be effective, such statements should set numerical targets.

Environmental policy statement. The former Railtrack's Environmental Policy Statement is an excellent example of what should be covered by policy statements:

> We will comply fully with and keep abreast of all legal obligations covering our operations past, present and future, requiring our employees and contractors to act in accordance with our environmental policy, for which we will provide appropriate training and communications.
>
> We will also communicate this policy to our customers and seek their help in implementing it. We will ensure that new projects, maintenance and renewals are managed professionally in a way which incorporates assessment of environmental impact and takes appropriate action to keep any adverse impacts to a minimum.
>
> We will seek to minimise emissions and reduce waste from our activities, concentrating on areas where there is most room for improvement in order to make most impact.
>
> We will aim to be sensitive in our management of natural and heritage features, taking into consideration the views of all those with an interest in our activities and working with them where appropriate.
>
> Implementation of Railtrack's environmental policy will be the overall responsibility of line managers; reports will be provided to Railtrack's Board of Directors on a quarterly basis. Annual reports of our progress against targets and objectives will be published and audited as appropriate. This policy will be made available to staff and the public.

Source: Railtrack Environmental Policy Statement, Railtrack Service Information Series, 1999.

The environmental report (voluntarily published by industry as a form of public accountability to all stakeholders) is normally independently verified and forward thinking companies concerned about the TRIPLE BOTTOM LINE now follow up their accredited ENVIRONMENTAL MANAGEMENT SYSTEM with a report designed for public consumption. Major companies see great advantage to this activity:

- Akzo Nobel mentioned an enhanced ability to track progress against specific targets;
- Electrolux said that reporting helps in implementing its environmental strategy;
- GSK noted that reporting brings greater awareness of broad environmental issues throughout the organization;
- J. Sainsbury valued the improved ability to clearly convey the corporate message.

Source: The CER Report, *Tomorrow*, September/October 1998, p. 59.

National Power publishes an environment performance review for its installations, e.g. that for West Burton (coal-fired power stations) lists the significant issues and response:

Significant issues and responses

Burning coal releases carbon dioxide which contributes to global warming.	The aim is to improve efficiency to obtain more electricity from the fuel. We continually monitor the way we operate to identify cost effective improvements; for example, in our own power use.
Sulphur and chlorine in the fuel are released as sulphur dioxide and hydrogen which contribute to acid rain.	We comply with the SO_2 emission limit set by the Environment Agency for the station. This is a part of the company-wide limit which will reduce National Power's total SO_2 emissions. We ensure that the sulphur content of the coal purchased is within authorized limits.
Oxides of nitrogen are formed in the combustion process. They contribute to acid rain and may affect air quality.	We ensure compliance with the limit set by the Environment Agency for the station. This is part of a company-wide limit which will reduce National Power's total NO_x emissions by 30% by 1998 from a 1980 baseline.
Particulate matter (coal ash) emitted with the flue gas could affect local air quality.	We operate our dust control equipment to ensure compliance with EA's requirements.
Liquid effluent from the plant is discharged to the river.	We regularly sample and analyse our effluents to ensure compliance with the Environment Agency's requirements.
Oil spills and losses from equipment can find their way to watercourses.	Careful management and control of our processes helps to minimise oil losses.
The combustion of coal leaves an ash residue.	We aim to maintain our tight control of the ash disposal facility, complying with our legal requirements. We will maintain our efforts to find new markets and uses for the ash, aiming to sell all the ash we make.
Waste.	We will maintain a close management control of waste, aiming to identify ways of reducing costs whilst maintaining environmental standards.

The power station performance in 1993 and 1995 is given in the table below.

	1993/94	*1994/95*
Total emissions (kilotonnes)		
Carbon dioxide	10 632	11 196
Sulphur dioxide	158.2	162.6
Oxides of nitrogen	27.7	28.9
Hydrogen chloride	16.5	17.7
Dust	3.4	3.4
Total fuel used and electricity generated		
Electricity generated (GWh)	12 205	12 996
Coal burned (kilotonnes)	4 754	4 963
Oil burned (kilotonnes)	53.1	95.5
Water abstraction		
River water abstracted	48.6	48.5
Rain water and borehole water	1.3	1.2
Water returned	30.3	29.7

If you do your sums, these show that net reductions of CO_2, NO_x and SO_x of 9, 0.1 and 0.4 g/kWh, respectively, were achieved between the years of 1993/94 and 1994/95.

Another example is provided by Aylesford Newsprint Ltd. (ANL) in their 1999 CER.

The use of NATURAL GAS as the predominant FUEL has led to significant reductions in both SULPHUR DIOXIDE and PARTICULATES as shown below.

Process emissions to land were significantly reduced in 1999. The Table gives the relevant breakdown. Water consumption has reduced to 13.9 m³/te (1998 15.2 m³/te). This shows what a highly motivated company can achieve.

ANL have very helpfully given an independently verified statement of the resource use and emissions from 1 tonne of newspaper in their plant (the data are not applicable to any other pulp/paper mill):

Resource use		*Emissions to air*	
Used paper	1.24 tonnes	NO_x	0.89 kg
Freshwater	16 m³	SO_2	0.12 kg
Energy	960 kWh	Particulates	0.006 kg
Emissions to water		CO_2 (fossil)	643 kg
COD	2.7 kg	CO_2 (biogenous)	130 kg
BOD	0.12 kg	*Emissions to land*	
Suspended solids	0.24 kg	Process waste	179 kg
Nitrogen	0.014 kg		
Phosphate	0.008 kg		

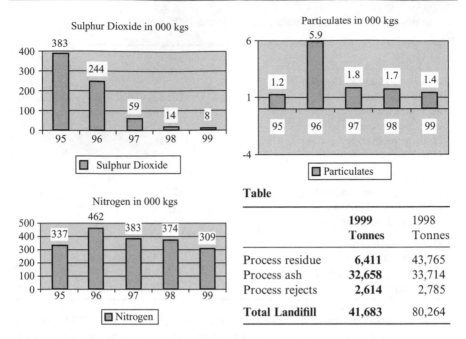

Figure 84 Annual emissions to air and to land, 1994–97.
Source: Aylesford Newsprint Environmental Report 1999.

Aylesford Newsprint Ltd. detail further improvements in their 2006 Safety, Health, Environmental and Fire Prevention Report. The Company saves about half a million tonnes of paper from going to landfill each year, and recycles 93 per cent of the water it uses in the paper manufacturing system. Water intake in 2006 was reduced by an average of 10 per cent. The payload on trucks has been increased by 11.5 per cent in 10 years, leading to 2171 fewer journeys. This cuts down on the associated emissions.

Source: Aylesford Newsprint, Safety, Health, Environmental and Fire Prevention Report 2006, http://www.aylesfordnewsprint.co.uk/ReportFiles/SHEF_2006.pdf. Accessed 15 October 2007

Environmental Protection Act (EPA) 1990. The law to make provision for the improved control of pollution arising from certain industrial and other processes; to re-enact the provisions of the Control of Pollution Act 1974 relating to waste on land with modifications regarding the function of the regulatory and other authorities concerned in the collection and disposal of waste and to make further provision in relation to such waste. A major feature is INTEGRATED POLLUTION CONTROL.

Environmental reports. A major feature of environmentally aware industry is the annual publication of their environmental state of health which contains information on waste minimization, recycled or reused energy savings, emissions reduction, water consumption reduction, etc.

A new breed of 'environmental verifier' has emerged as a consequence prompted by the EU's ECO-MANAGEMENT AND AUDIT SCHEME.

The verifier conducts the following operations:

- Confirming whether data are gathered in an appropriate way.
- Ensuring they are reproducible.
- Noting whether data are presented in a fair way; for example, are diagrams and other statistical methods of presentation telling the real story or are they deliberately misleading.
- Assessing what methods company laboratories use and ensuring instruments have traceable calibration records.
- Assessing what calculation methods are used and whether they are applied correctly at the different sites.
- Confirming that the calculation methods and factors are recognized by the company's industrial sector and hence are defensible.

It should be noted that verification is not necessarily a guarantee of soundness at all times, and in all matters, but a prudent measure for acceptability.

Source: D. E. Evans and A.-M. Warris, 'Verification of Corporate Environmental Reports', *Environmental Assessment*, Vol. 3, No. 1, March 1995.

Environmental taxes. Simply, taxes to benefit the environment, e.g. LANDFILL TAX which is a levy on each tonne of waste landfilled. In the UK this was created through constructive dialogue with industry. In this case, the Environmental Services Association (a trade body) recommended that environmental taxes must:

- meet the general tests of good taxation;
- be based on sound science–cost benefit and cost-effectiveness analysis, which includes second order effects (e.g. impact on competition, transaction costs, transitional arrangements);
- be revenue neutral – not adding to the UK tax burden;
- take account of the impact on UK competitiveness;
- fit with existing fiscal instruments;
- be constantly reviewed to assess effectiveness.

As a result of constructive dialogue, landfill tax clearly demonstrates that partnership can develop environmental tax systems which:

- are enforceable with minimum scope for avoidance;
- are certain in their effect;
- have low transaction costs;
- allow sufficient lead time to develop alternative facilities and their services;
- incorporate an option to appeal to an independent arbitrator;
- offer a direct environmental benefit through a tax credit scheme.

The UK landfill tax (in 2007) on a tonne of active waste (i.e. biologically active as opposed to inert), was £24.

The rate increase builds on the important role played to date by the tax with almost a third of companies considering waste minimization, reuse and recycling. Analysis to date suggests that a long-term programme of increases in landfill tax could significantly reduce the proportion of waste expected to be landfilled.

Examples of environmental taxes are: AGGREGATE TAX; battery tax; CARBON TAX; end of life vehicle tax; energy tax; fuel duty; incineration tax; LANDFILL TAX; packaging ordinance; packaging tax (non-returnable or deposit returns system tax); road fund tax; sulphur emissions tax; waste water charges.

Plenty of ammunition in the above armoury to enable 'environmental' improvement.

Source: *Environmental Services Association Bulletin*, 2, 1999.

Enzymes. Proteinaceous substances that catalyse microbiological reactions such as decay or fermentation. They are not used up in the process but speed it up greatly. They can promote a wide range of reactions, but a particular enzyme can usually only promote a reaction on a specific substrate.

Exposure to enzymes is common in the textile and brewing industries, but most acutely of all in bakeries, where powdery alpha-amylase is added to baking agents. Alpha-amylase is an enzyme that breaks down starch and intensifies the effect of the yeast, making bread spongier and easier to bake.

An innovative redesign of workplace ventilation has achieved a 140-fold reduction in concentrations, as shown in Figure 85. (◊ CATALYSIS; ENZYME TECHNOLOGY)

Enzyme-linked immunosorbent assay. ◊ ELISA.

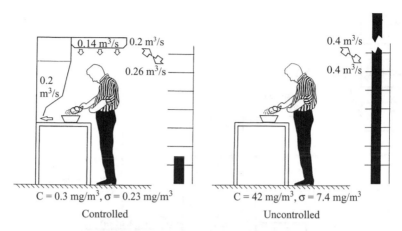

Figure 85 Redesign of workplace.
Sources: K. Heinonen, L. Kulmala and A. Säämänen, 'Local ventilation for powder handling – combination of local supply and exhaust air', *American Industrial Hygiene Association Journal*, No. 57, April 1996.
Work Health Safety, 1998, p. 27. (Reproduced by permission of AIHA)

Enzyme technology. Enzyme technology covers a wide commercial area which embraces drug manufacture, beer malting, stabilization of foods, and PROTEIN production. Thus the enzymes known as cellulases ('*-ase*' indicates an enzyme; '*-ose*' is the substrate or product, e.g. cellulose) are the class of enzymes that promote the decomposition of CELLULOSE, in this case to glucose. There are some 13 000 micro-organisms that live on cellulose and they accomplish this by secreting enzymes which break down the substrate for a food source. One organism (*Trichoderma viride*) can produce a cellulase-rich component capable of breaking down crystalline cellulose.

Using cellulose as an example, a commercial process for cellulose exploitation (CELLULOSE ECONOMY) requires that the micro-organism (in this case *Trichoderma viride*) is grown on a culture medium containing spruce pulp and nutrient salts. The culture is filtered leaving an enzyme solution which is then available to hydrolyse cellulose substrates, that is, convert the cellulose to sugars (HYDROLYSIS). So far tests have been conducted on milled newspapers; the cellulase solution and newspapers are placed in a reaction vessel under strictly controlled conditions (pH 4.8, 50 °C). The time taken to break down the cellulose is of the order 20–80 hours and a product of glucose syrup obtained. Any non-utilized cellulose is recycled along with the enzymes. The yield of glucose is roughly 50 per cent of the original cellulose. The glucose is then used for FERMENTATION products and/or the production of

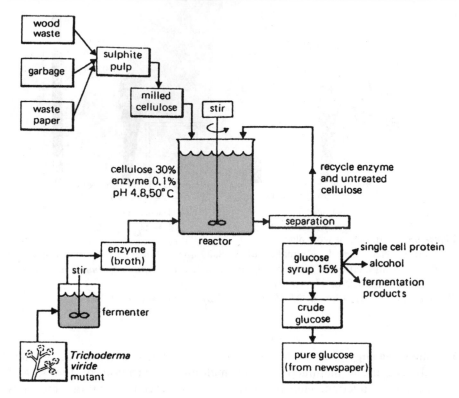

Figure 86   Natick process for enzymatic hydrolysis of cellulose to glucose.

SINGLE-CELL PROTEIN. Figure 86 shows a flow diagram for the process, based on work undertaken at the US Army Natick Laboratory.

Enzymes require strictly controlled conditions and are easily destroyed or inactivated by many substances. Feedstock purity and processing conditions must be closely monitored. Although not consumed in the reaction, enzymes eventually degrade and must therefore be continually synthesized. In the cellulase reaction, product inhibition can eventually occur and the reaction cease, with the result that all the cellulase cannot be used in the one go. Multi-charging must then be used if all the cellulose is to be converted to glucose.

This area of enzyme technology has great promise for the production of single-cell protein from low-grade starchy materials, thereby upgrading animal feedstocks, or the production of BIOETHANOL.

Epidemiology. The study of categories of persons and the patterns of diseases from which they suffer so as to determine the events or circumstances causing these diseases. If a cause is discovered, then those responsible for PUBLIC

HEALTH policy can take appropriate steps to prevent the disease in question.

A classic case of epidemiology was Snow's identification of contaminated water from the Broad Street pump in 1854 as being responsible for the spread of CHOLERA in London.

Source: J. J. Snow, *On the Mode of Communication of Cholera*, 2nd edn, Churchill, 1885.

Another example of the role of epidemiology is the evidence that smoke and SULPHUR DIOXIDE act together in causing and aggravating bronchitis. The Clean Air Acts brought a striking reduction in extra deaths due to these pollutants as shown in the data for London in the following table.

Year	Maximum daily concentration (micrograms per cubic metre)		Extra deaths in Greater London
	Smoke	*Sulphur dioxide*	
1952	76 000	3500	4000
1962	3 000	3500	750
1972	200	1200	—

From Royal Commission on Environmental Pollution, Fourth Report: *Pollution: Progress and Problems*, Cmd 5780, HMSO, December 1974.

The epidemiologist sifts the evidence and identifies patterns and causes. We therefore know with reasonable certainty that in the UK, the CLEAN AIR ACT has done the job it was intended to do.

Numerous epidemiological studies have shown that the incidence of morbidity increases in the observed sectors with pollutant concentration (see Figure 87).

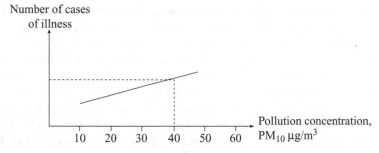

Figure 87 Effect of increasing concentration of PM_{10} (vehicle emission particulates below 10 μm diameter).
Source: *Monetarization of the external health costs attributable to transport*, Consolidation Report, GVR Report no. 272, Switzerland, 1998.

An example of this is set out in the table below.

Epidemiological evidence of the effects of air pollution on health.

Short-term effects of air pollution	Long-term effects of air pollution
Decrease in	*Decrease in*
Lung function	Lung function
Performance	Symptom-free intervals
Increase in	*Increase in*
Eye irritation	Bronchial reactivity
Bronchial reactivity	Chronic bronchitis
Headaches	Chronic respiratory symptoms
Respiratory symptoms (coughs, etc.)	Frequency of cancer
Pseudocroup	Total mortality
Exacerbation of chronic respiratory disease	
Asthma attacks	
Absences from school	
Absences from work	
Hospitalization	
Daily mortality	

Case study

A proportional mortality and proportional cancer mortality analysis of deaths in US workers who had contact with solvents and other chemicals, between 1969 and 2001, in IBM, was carried out in 2006. IBM manufactures electronic office equipment, mainframe and personal computers, computer parts, components, accessories and software products. Solvents and several other chemicals, such as arsenic, nickel, chromium, cadmium, lead, benzene and hydrochloric acid have been used in electronics manufacturing. In addition, workers are exposed to electromagnetic fields (especially ultraviolet and radio frequency). The 'clean rooms' in which semiconductors such as silicon wafers (or chips) are produced, are carefully designed to protect the fragile wafers from contaminants. Each layer of silicon molecules is exposed to highly-toxic chemicals to achieve maximum precision in etching and layering. Workers wear 'bunny suits' to prevent any hair or speck of dust from fatally damaging the chips, and the air is filtered to eliminate any particles. But the chemical by-products of the processing, which do not pose a hazard to the chips, are permitted to circulate freely in the clean room. Based upon available literature, workers exposed to the chemicals and processes in computer and semiconductor manufacturing would be expected to have elevated mortality from NON-HODGKINS LYMPHOMA, brain cancer, kidney cancer, lung cancer and breast cancer. These expectations were evaluated in the present study, by calculating the overall and sex-, age-, and time-specific standardized PROPORTIONAL MORTALITY RATIOS (PMRs) and standardized PROPORTIONAL CANCER MORTALITY RATIOS (PCMRs).The data was analysed using the OCCUPATIONAL COHORT MORTALITY ANALYSIS PROGRAM software package (OCMAP-PLUS Version 3.10) from the University of Pittsburgh. Standardized proportional mortality ratios were calculated from the ratio of the observed number of deaths due to a particular cause amongst the workers, to the expected number of deaths based on US population mortality data. The study base comprised 31 941 decedents (27 272 males and 4669 females), who had worked for at least five years for IBM, and whose death information was collected in the company's corporate mortality file (which was a list of over 30 000 IBM employee deaths, tracked by work location and cause, between 1969 and 2001). Proportional Mortality Ratios and Proportional Cancer Mortality Ratios and their 95 per cent CONFIDENCE INTERVALS (CIs) were computed for 66 causes of death in males and females. The PMRs for all cancers combined were elevated in males (PMR = 107; 95 per cent CI = 105–109) and females (PMR = 115; 95% CI = 110–119). Proportional cancer mortality ratios (PCMRs) for brain and central nervous system cancer were elevated (PCMR = 166; 95% CI = 129–213), kidney cancer (PCMR = 162; 95% CI = 124–212), melanoma of skin (PCMR = 179; 95% CI = 131–244) and pancreatic cancer (PCMR = 126; 95% CI = 101–157) were significantly elevated in male manufacturing workers. Kidney cancer (PCMR = 212; 95% CI = 116–387) and cancer of all lymphatic and haematopoietic tissue (PCMR = 162; 95% CI = 121–218) were significantly elevated in female manufacturing workers.

Source: Mortality among US employees of a large computer manufacturing company: 1969–2001, *Environmental Health: A Global Access Science Source 2006*, 5:30.

◊ NON-HODGKINS LYMPHOMA
◊ PMR, Proportional Mortality Ratio
◊ PCMR, Proportional Cancer Mortality Ratio
◊ OCMAP, Occupational Cohort Mortality Analysis Programme
◊ CI, Confidence Interval

Equivalent continuous A-weighted sound pressure level. The value of the A-weighted sound pressure level in decibels (◊ DECIBELS A-SCALE). Hypothetical continuous steady sound that within a specified time interval results in the same total ENERGY as the actual varying sound level. The value of this level is dominated by short bursts of high levels or peaks in the noise.

Erosion. The lowering of the land surface by weathering, corrosion and transportation, under the influence of gravity, wind and running water. It can also apply to the eating away of the coastline by the sea. Raindrops can hit exposed soil with an explosive effect, launching soil particles into the air. In most areas, raindrop splash and sheet erosion are the dominant forms of erosion. It is intensified on sloping land, where more than half of the soil contained in the splashes is carried downhill. (◊ DESERTIFICATION)

Error. The difference between an experimental result and the 'true' value. The uncertainty of an experimental result. If used, the meaning should be made very clear.

***Escherichia coli* (*E. coli*).** A species of bacteria found in the gut of humans and other mammals. It is found in abundance in the faeces of these creatures. It is capable of easy isolation, identification and numerical estimation, and is unable to grow in the aquatic environment. It is thus an ideal indicator organism to use for detection of pollution by faecal matter of human or animal origin. It can multiply at 44 °C. When shown to ferment lactose or mannitol at 44 °C with the production of ACID and gas within 24 hours, and produce indole from tryptophane, it is referred to as confirmed *E. coli*.

Estuarial storage. The storage of water for domestic water supplies in estuaries when suitable inland sites for reservoirs have been exhausted. In the UK, Morecambe Bay, the Solway Firth and the Wash have been suggested as potential sites, with a BARRAGE used to control or prevent the entry of sea water.

Ethanol. Ethyl alcohol (C_2H_5OH). It has a boiling point of 78.4 °C and a specific gravity of 0.789. Ethanol is mainly synthesized from ethene in the petro-chemical industry. It can also be made by the FERMENTATION of natural sugars, or those produced by HYDROLYSIS of refuse or cellulose waste. Its main use is as a SOLVENT or chemical raw material. It can also be used as a motor FUEL. Gross CALORIFIC VALUE 30 MJ/kg. ◊ BIOETHANOL.

Ethylene. C_2H_4 or H_2CH_2, a fundamental 'building block' for the petrochemical industry (i.e. obtained from OIL or NATURAL GAS), manufactured by 'catalytic cracking'.

It is used in the manufacture of POLYETHYLENE and polypropylene as well as 'downstream' products such as DIETHYLENE GLYCOL which is widely used as a solvent and in PLASTICIZER manufacture as well as ethylene glycol (antifreeze).

EU legislation. The European Commission has the sole right of initiation of proposals. This does not mean that all ideas originate from the Commission, but wherever they come from, they must be turned into proposals by them.

The European Parliament's role is relatively limited. Under the main environment article in the Treaty (130) it is consulted only once. It has no responsibility for initiating proposals and no influence on the composition of Commissioners who are ultimately responsible for proposals that emanate from the Commission.

EC environmental measures have wide implications. The following are some examples:

- new waste controls, including liability for environmental damage caused by waste;
- product safety controls;
- regulation of biotechnology;
- emission limits for small combustion plants;
- limits for discharges of dangerous substances into water;
- public access to environmental information;
- restrictions on the marketing and use of cadmium;
- environmental labelling of consumer goods.

European Chemical Industry Council. ◊ CEFIC.

European Environment Agency (EEA). Agency of the European Union devoted to establishing a network for the monitoring of the European environment, governed by a Management Board composed of representatives of the governments of member states, a European Commission representative and two scientists appointed by the European Parliament, assisted by a committee of scientists. It was established by EEC Regulation 1210/1990, as amended by EEC Regulation 933/1999, and became operational in 1994. It is headquartered in Copenhagen, Denmark.

European Pressurized Reactor (EPR). Basically a Pressurized Water Reactor with enhanced efficiency and increased safety features (viz. containment barrier should the core rupture, and aeroplane crash strengthening). (◊ NUCLEAR REACTOR DESIGNS). This is one of the favoured reactor designs for the UK which is (naturally) indecisive about what direction to take. The reality is that just as recycling and energy from waste are needed to initiate the use of landfill for waste disposal, NUCLEAR POWER, RENEWABLES, CARBON CAPTURE AND STORAGE (for fossil fuelled power station's CO_2 emissions) are all vital.

Recall 'peak oil' production is mooted for ca. 2012, and 'peak gas' 2025 – no time to lose.

Eutrophic. Term describing water that is rich in nutrients.

Eutrophication. The natural ageing of a lake or land-locked body of water which results in ORGANIC material being produced in abundance due to a ready supply of nutrients accumulated over the years. Eutrophication can be greatly increased by humans as a result of nitrates and phosphates from fertilizer run-off and sewage treatment processes. (◊ MODERN FARMING METHODS; SEWAGE TREATMENT, PHOSPHATES)

A eutrophic lake is highly productive in organic material and can result in algal blooms which are short lived and whose decay imposes a heavy oxygen demand on the water. Thus nutrients from SEWAGE and fertilizers can ruin a lake and cause the loss of a body of water which may be of use to humans. This has happened in parts of the Great Lakes system where it has been estimated that Lake Erie received 37 500 tonnes of NITROGEN from run-off and 45 000 tonnes from sewage in 1968 alone. This is supposed to have aged the lake 15 000 years quicker than if it were left to its own devices. (◊ ALGAE; DISSOLVED OXYGEN; OLIGOTROPHIC)

Evapotranspiration. The process by which water moves from the land surface into the atmosphere as vapour. It combines both direct evaporation from wet surfaces and transpiration from vegetation.

Excess air. The additional COMBUSTION air required to ensure complete combustion of a FUEL. This is either stated as a percentage of the theoretical air required or given as % OXYGEN in the flue gas stream (by mass %, dry flue gas conditions, 273 K and 101.3 kPa).

This ensures virtually complete combustion as measured in the exhaust gases by the low levels of CARBON MONOXIDE (whose presence indicates incomplete combustion).

Exponential curve. In many natural phenomena, in which, say, yeast organisms are reproducing, their rate of increase is not constant but rather the rate of increase itself is continually increasing because the organisms themselves double in number, in say 10 hours, and double again from the 10-hour number in 20 hours. Thus, at time zero, if the number of organisms per cubic metre is 1 million, then in 10 hours there will be 2 million, in 20 hours 4 million, in 30 hours 8 million and in 40 hours 16 million.

Such a rate of growth is called *exponential growth* and a curve such as Figure 88 is known as an exponential growth curve. It is plotted as:

$$y = e^{kt}$$

where y is the number measured at time t, and k is a constant of proportionality determined by experiment; e is the base number of what are known as natural LOGARITHMS and has the value 2.7183.

Using the world population as our model, let

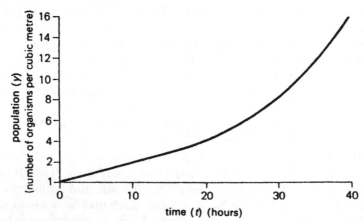

Figure 88   Exponential growth curve for the number of organisms doubling every ten hours.

$$n_0 = \text{population at time zero}$$
$$n_t = \text{population at time t}$$
$$k = \text{growth constant}$$
$$t = \text{time in years}$$

then

$$N_t = N_0 e^{kt}$$

i.e. the population at any time t is a function of the population at time $t = 0$ (that is, N_0) and the exponential constant k.

If we wish to know how long it will take for the world population to double given that the annual growth rate is 2 per cent (this gives us k), we proceed as follows:

$$N_t = N_0 e^{0.02t}$$

and for population doubling

$$\frac{N_t}{N_0} = 2$$

that is,

$$2 = e^{0.02t}$$

Using natural logarithm tables where the natural logarithm of e = 1, we find that value which gives us 2, so that:

$$\text{nat. log } 2 = 0.02t$$

From the tables, nat.log 2 = 0.6931, therefore

$$t = \frac{0.6931}{0.02} = 34.65 \, \text{years}$$

Thus the doubling time for a 2 per cent increase is 35 years. This gives rise to a handy rule of thumb that for percentage growth rates less than 10 per cent, i.e. doubling time (years) =70/(annual growth rate %). For example, with an annual growth rate of 5 per cent, the doubling time would be 14 years.

We have dealt with exponential growth, but we can have exponential decay as well, which is represented by the equation

$$y = e^{-kt}$$

and is just as important in natural systems. For example, the atoms of radioactive elements emit particles (radiation), and in doing so decay to atoms of a different mass. The rate of decay at any time is proportional to the total number of unchanged atoms at that time and is characterized by the HALF-LIFE which is the time taken for half the number of atoms of any radioactive element to decay from an initial condition. The concept of half-life is illustrated in Figure 89.

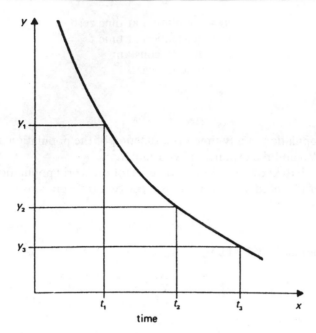

Figure 89 Exponential decay of a parameter y. The figure shows a graph of $y = e^{-kt}$ upon which three points, (t_1, y_1), (t_2, y_2) and (t_3, y_3), are picked out. The values of y_1, y_2 and y_3 are related by $y_1 = 2y_2$ and $y_2 = 2y_3$. The properties of the exponential curve are such that for this case $t_3 - t_2 = t_2 - t_1$. In other words, the time taken for y to be reduced of half its initial value is the same, whatever that initial value y is taken to be.

Exponential decay. ◊ EXPONENTIAL CURVE.

Exposure–dose effect relationships. The effect of exposure to a pollutant is a function of the pollutant, the type of target (e.g. animal or vegetable), and the concentration of pollutant and duration of exposure.

   Short-term exposures to high CONCENTRATIONS are not usually equivalent to one-hundredth of the concentration for 100 times as long. The low-concentration long-period exposure may well have minimal effect, whereas the former may have a serious effect. It is for this reason that MAXIMUM EXPOSURE LIMITS are set. The converse can also be true in some circumstances. (◊ TIME-WEIGHTED AVERAGE)

Exposure limits. Substances hazardous to health may cause adverse effects (e.g. irritation of the skin, eyes or respiratory tract, narcosis or cancer) through accumulation of the substances in the body or through the development of increased risk of disease with each contact. It is important to control both short-term and long-term exposure so as to avoid both types of effect. Two types of exposure limit are recognized. The long-term exposure limit is concerned with the total intake over long periods and is therefore appropriate for protecting against the effects of long-term exposure. The short-term exposure limit is aimed primarily at avoiding acute effects or at least reducing the risk of their occurrence. Specific short-term exposure limits are listed for those substances for which there is evidence of a risk of acute effects occurring as a result of brief exposures. For those substances for which no short-term exposure limit is listed, it is recommended that a figure of three times the long-term exposure limit averaged over a 10-minute period be used as a guideline for controlling exposure to short-term excursions.

Both the long-term and short-term exposure limits are expressed as air-borne concentrations averaged over a specified period of time. The period for the long-term limit is normally eight hours; when a different period is used, this is stated. The averaging period for the short-term exposure limit is normally 10 minutes; such a limit applying to any 10-minute period throughout the working shift. Ideally, monitoring should be continuous using fast response instruments. Ref. UK Health & Safety Executive 1991. (◊ OCCUPATIONAL EXPOSURE STANDARD)

External combustion engine. An engine where the heat source or FUEL COMBUS-TION is outside the engine as opposed to inside the engine as in the INTERNAL COMBUSTION ENGINE. External combustion allows a wide variety of fuels to be used efficiently to supply thermal ENERGY to the engine. However, this class is mainly dominated by STEAM turbines and steam engines which are hardly suitable for motor cars but very suitable for ships, power stations, etc. An exception is the Stirling engine, where air or gas is trapped in a dual piston cylinder. When the gas is heated (externally), the working piston moves, doing work. As the motion continues a displacer piston moves hot gas to the cool end of the cylinder where, on cooling, it is compressed by the working piston and transferred by the displacer back to the hot end. This method of using two pistons in the one cylinder causes complexities and expense, but the Stirling engine is quiet, virtually non-polluting, can use any fuel and has prospects of increased thermal EFFICIENCY compared to the internal combustion engine.

External costs. Costs not *fully* accounted for in conventional economic costing systems. Often called environmental costs, or external diseconomies such as LITTER on a beach. Figure 90 gives an example of the external costs of road transport.

Source: *Monetarization of the external health costs attributable to transport*, GVR Report no. 272, Switzerland, 1998.

Extraction of oil from shales. The gasification or extraction of oil from shales is a technique that may well increase as ENERGY reserves decline. The process is carried out in retorts in four separate stages (Figure 91).

1. The recycled gas stream is preheated by the spent shale.
2. The gas is ignited and the heat of the combustion causes the oils in the shale to be driven off in the retorting zone.

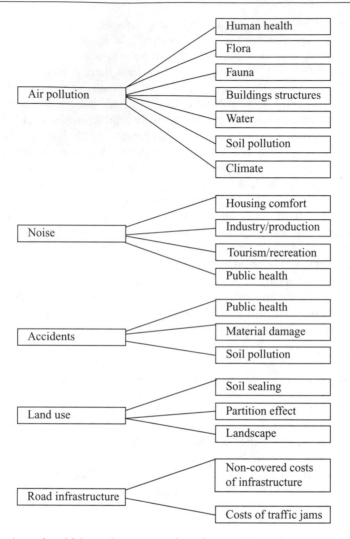

Figure 90   Areas in which road transport gives rise to external costs.

3. The hot gases and oils are used to preheat the incoming shale.
4. The product gases and oils are cleaned in oil separators and an electro-static precipitator is used to remove oil mist if necessary.

Part of the gas is recycled to run the retort and the remainder plus the oil is sold.

For questions concerning the applicability of the process on a scale suffi-cient to provide a significant amount of oil, see under 'oil shales'. Proposals

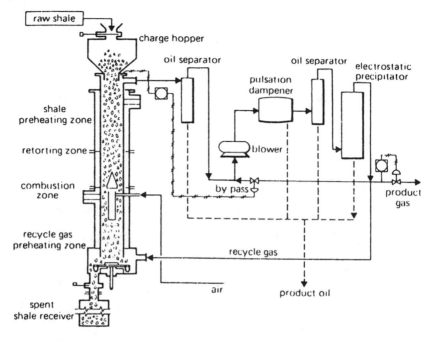

Figure 91 Gas production from oil shales.

are in hand (Alberta) for large scale underground heaters to effect *in-situ* oil extraction to avoid vast mining operations. (◊ OIL SHALES)

Exxon Valdez. An oil tanker, which struck Bligh Reef in Prince William Sound, Alaska, in March 1989 spilling an estimated $50\,000$–$150\,000\,m^3$ of crude oil. Thousands of sea creatures were killed immediately. This was the biggest environmental disaster in US history, soiling 1300 miles of coastline.

F

Factor 4. Factor 4 stands for a four-fold reduction of the environmental burden, or four-fold increase in ECOLOGICAL EFFICIENCY of current products and services in order to achieve SUSTAINABLE DEVELOPMENT.

Every unit of a product or service should use only one-quarter of its actual consumption of raw materials and energy and create only one-quarter of its original waste and emissions throughout its life cycle.

Factor 4 is said to be a minimal requirement to achieve more equity while remaining within the boundaries of the earth's carrying capacity, and needs to be achieved within a relatively short term (5–10 years).

Factor 10 stands for a ten-fold increase in ecological efficiencies. The OECD is striving to obtain this within 30 years.

Fallout (radioactive). (i) The descent to the earth's surface of the radioactive particles produced in a nuclear explosion or accidental release. (ii) The radioactive particles themselves. (Fallout is sometimes broadened to include washout, i.e. the removal of any particulate matter from the ATMOSPHERE by rainfall.)

Far East. The ultimate destination of much of the UK's household derived segregated waste paper, plastics and (growing) obsolete waste electronic and household electrical equipment. There is (so far) very little confidence that anything approaching UK standards of environmental protection or worker health and safety requirements are implemented. Indeed, it has been reported [Environmental News and Data Services 379 Aug 06, p.15] that of 190 international waste shipments inspected by the EA between Sept 2004 and May 2006, 152 (80%) were found to be illegal. A major concern is that poor working conditions can expose people to the hazardous substances found in old components. The UK waste industry hopes to have a 'recycling registration services scheme' up and running fully by 2007. This is bolting the door after the horse has fled! – Recall the above, when ministers trumpet how we are meeting our recycling targets ('6000 coal miners killed in China's mines in 2006' ITN 6:30 News Bulletin, 30/11/06).

Fatal Accident Rates (FAR). This is a measure of fatalities per 100 million hours spent on one activity.

The UK Health and Safety Executive classifies RISKS as intolerably high, as low as reasonably practicable (ALARP) and broadly tolerable. The FAR values for tolerable risks are:

Workers of all occupations 1/1000 per year (FAR 48)
Public 1/10 000 per year (FAR 1.15)
Nuclear industry 1/100 000 per year (FAR 0.1)

Highest fatal causes of death.

Cause of death	FAR
Disease 40–44 age group	17
Car (all ages)	15
Home	1.5
Coal mining	8
Oil and gas rigs	20
Skiing in France (1974–76)	130
Rock climbing (rock faces)	4000

Such a computation puts the nuclear industry's stringent safety measures in perspective. Similarly, the risk of cancer from DIOXINS from INCINERATION to a maximally exposed individual (i.e. maximum exposure 24 hours per day 365 days per year) is

$$3 \times 10^{-7} \text{ or } 3/10\,000\,000 \text{ per year}$$

i.e. a FAR of 0.00345 or negligibly small. However, perception is another matter. But, if rationality were to prevail, smoking would be totally banned and cars severely curtailed, yet the number of motor vehicles is predicted to double in the UK by ca. 2025.

Sources: L. Roberts, 'The Public Perception of Risk', Lecture at RSA, 26 April 1995.
C. C. Travis and B. P. Blaylock, 'Health Risks Associated with Air Emissions from MS Combustors', *American Chemical Society Conference*, Washington DC, 21–26 August 1994.

Feed-conversion ratio. The ratio of weight of feed to gain in weight (in kilograms) in fattening cattle or poultry, etc. Typical values lie between 3 and 10, depending on animal and feed analysis. The use of hormones to promote growth is one method of reducing the feed-conversion ratio. However, growth hormones are suspected CARCINOGENS. Even if this is not the case, it raises the question of just how far we can tamper with natural systems and living creatures for short-term gains but with unknown long-term consequences.

Feedstock energy. A term used in LIFE CYCLE ANALYSIS (LCA) which accounts for the 'combustion heat' of raw material inputs that are not used as an energy source in the system being analysed, e.g. newsprint has a feedstock energy of 15 MJ/kg which is not made 'available' in the recycling process, but is partly recoverable as thermal energy on COMBUSTION by INCINERATION.

The conversion of this recovered thermal energy can be for heating only, in which case >90 per cent is available, or for electricity production, in which case 25 per cent is available, or a mixture of both for COMBINED HEAT AND POWER, in which case ca. 85 per cent is available, if the DISTRICT HEATING scheme is optimized.

Hence LCAs have to be very careful to delineate their SYSTEM BOUNDARY and both energy units and their form (i.e. thermal or electrical).

Fermentation. The decomposition of ORGANIC substances by micro-organisms and/or enzymes. The process is usually accompanied by the evolution of heat and gas, and can be aerobic or anaerobic. Examples:

Alcoholic fermentation (anaerobic).
$$C_6H_{12}O_6 = 2C_2H_5OH + 2CO_2$$
sugars ethyl carbon
alcohol dioxide

Lactic fermentation (microaerobic or anaerobic) – a common process for the production, preservation and seasoning of food.
$$C_6H_{12}O_6 \rightarrow 2CH_3CH(OH)COOH$$
sugars lactic acid

Acetic fermentation (aerobic).
$$C_2H_5OH + O_2 = CH_3COOH + H_2O$$
ethyl alcohol acetic acid

Aerobic fermentation of glucose solutions (and other substrates, e.g. HYDRO-CARBONS) by yeasts also leads to yeast growth which can then be harvested as a source of SINGLE-CELL PROTEIN.

The crucial parameters in industrial aerobic fermentation are the oxygen requirement per tonne of substrate and the removal of the heat evolved during fermentation. If the oxygen requirement per tonne of substrates is taken as 1 for molasses, then *n*-paraffins need 2.5 and methane 5 times as much oxygen. Thus a cheap substrate may have high energy costs due to oxygen transfer requirements. ($\lozenge$ AEROBIC PROCESSES; ANAEROBIC PROCESS, BIOETHANOL)

Ferric sulphate. $Fe_2(SO_4)_3$, used in water purification. Severn Trent Water admitted leaking ferric sulphate into one of the best stretches of salmon river in Wales, killing all but 2 per cent of the stock, some 33 000 young salmon. They were fined £175 000. The polluter paid!

Source: *The Independent*, 6 August 1996.

Fertilizer. A chemical which promotes plant growth by enhancing the supply of essential NUTRIENTS such as nitrogen, phosphate and potassium. Fertilizers can be inorganic, such as ammonium sulphate ($(NH_4)_2SO_4$) or lime, or ORGANIC, such as SEWAGE sludge or manure. They can be added to the soil or, at low concentrations, sprayed on foliage as a foliage-feed.

Fertilizer run-off can be a major threat to watercourses and has already caused enrichment in the Great Lakes. ($\lozenge$ EUTROPHICATION)

Where natural fixation of NITROGEN is taking place, as in a clover crop, the addition of inorganic nitrogenous fertilizers can inhibit the natural fixation. ($\lozenge$ NITROGEN CYCLE)

Fixation of nitrogen by humans for fertilizers now equals the natural fixation rate, and the attendant increase in NITROGEN OXIDES may be a threat to the OZONE SHIELD. Such effects are cumulative, and farmers often find that, with time, they are using increasing amounts of fertilizer to achieve the same yield. In addition, nitrate leaching from fertilizers has been blamed for increasing nitrate levels in water supplies, so much so that the EC limit of $50 \, mg/m^3$ has been substantially breached in parts of the UK. A novel solution of paying farmers not to use nitrate fertilizers in so-called water protection zones has now been implemented. ($\lozenge$ METHAEMOGLOBINAEMIA)

Plant breeders are now attempting to develop plants which will fix nitrogen directly from the atmosphere, and so eliminate or decrease the need for nitrogenous fertilizers.

Although some agronomists believe that further increases in farm production are possible by increased use of fertilizers, some leading agricultural chemists are convinced that we may be close to the economic limit of fertilizer use and that, with the exception of grassland, the economic limit may already have been exceeded. Furthermore, there is reason to suspect that artificial fertilizers alter the soil ecology in a detrimental manner.

Fertilizer supplies. The basic raw material of nitrogen fertilizers (e.g. ammonium sulphate, ammonium nitrate) is ammonia (NH_3), manufactured from atmospheric NITROGEN and HYDROGEN obtained from HYDROCARBONS such as methane or naphtha (a petroleum derivative). The price of nitrogen fertilizers has recently risen in line with the increasing price of oil, and developing countries are having great difficulty in maintaining their essential supplies without external aid.

Fibrosis. A scarring of the lung tissues, caused by dust inhalation. Almost all dust diseases, such as pneumoconiosis (coal-dust disease) and silicosis (stone-dust disease caused by mining, quarrying, shot blasting and stone-dressing operations) are characterized by a scarring of the lungs.

Other dusts which have been implicated in lung disorders include talc, fireclay, mica, china clay, graphite and ASBESTOS. Flax and hemp dust give rise to a disabling lung disease called byssinosis.

Field moisture capacity (Field capacity). The equilibrium amount of water retained in the SOIL after excess water has drained away. In the UK the soil generally reaches the field moisture capacity in winter and early spring. Thereafter, evaporation and plant growth removes moisture and causes a SOIL MOISTURE DEFICIT.

Film badge. A method of detecting individual exposure to IONIZING RADIATION. Consists of a masked photographic film which becomes progressively more 'exposed' on increasing or prolonged exposure to radiation.

Filter medium. This is the material, such as sand or anthracite, used, for example, in water filtering.

Fines. In the coal industry, coal having a maximum particle size usually less than 1.5 mm and rarely above 3 mm. In general, that fraction of a material that has been broken down into particles too small for a given use. If not properly contained, fines can cause dust pollution or the possibility of dust explosions.

Fire-water. Water stored for use in fire-fighting operations.

First law of thermodynamics. ◊ LAWS OF THERMODYNAMICS.

Fischer-Tropsch Process. A catalysed chemical reaction at high temperature and pressure in which carbon monoxide and hydrogen are converted into liquid hydrocarbon of various forms. Typical catalysts used are based on iron and cobalt. The principal purpose of this process is to produce a synthetic petroleum substitute for use as synthetic oil or as synthetic fuel.

The mixture of carbon monoxide and hydrogen is called synthesis gas or syngas. The resulting hydrocarbon products are refined to produce the desired synthetic fuel. The carbon dioxide and carbon monoxide is generated by partial oxidation of coal- and wood-based fuels. The utility of the process is primarily in its role in producing fluid hydrocarbon or hydrogen from a solid feedstock, such as coal or solid carbon-containing wastes of various types. Non-oxidation pyrolysis of the solid material produces syngas which can be used directly as a fuel without being taken through Fischer-Tropsch transformations. If liquid petroleum-like fuel, lubricant, or wax is required, the Fischer-Tropsch process can be applied. The process was invented in petroleum-poor but coal-rich Germany in 1923, by Franz Fischer and Hans Tropsch to produce liquid fuel. Germany's annual synthetic fuel production reached more than 124 000 barrels per day from 25 plants (~6.5 million tons in 1944).

The process was used in South Africa to meet its energy needs during its isolation under Apartheid. This process has received renewed attention in the quest to produce low sulphur diesel fuel in order to minimiz the environmental impact from the use of the diesel engine.

Shell has pioneered this technology to produce GTL (gas-to-liquid) fuel which is virtually sulphur-free, and which has a high CETANE NUMBER. GTL makes an excellent low-emission fuel for DIESEL engines.

Source: Hibbert, L. 'Bright Sparks', *Professional Engineering,* 10 January 2007, pp. 25–25.

Fission. The spontaneous or induced splitting of heavy atoms (uranium, plutonium) into two roughly equal parts, thereby releasing large quantities of energy. (◊ NUCLEAR ENERGY; NUCLEAR REACTOR DESIGNS)

Fixation. ◊ NITROGEN CYCLE.

Fixed carbon. A measure of the primary productivity of an ecosystem based on the amount of carbon fixed by PHOTOSYNTHESIS per unit area. Also used in fuel analysis. (◊ CARBON CYCLE, PROXIMATE ANALYSIS)

Flare. The flame produced by the burning of surplus and residual gases at the top of a flame pipe at an oil refinery or other chemical industry factory or LANDFILL SITE. The gases cannot be released into the atmosphere owing to their unpleasant odour and to the explosion risk that would result. Flares can produce a great deal of smoke; this can be reduced by injecting steam or a water spray close to the point of ignition (this procedure, however, renders the flame noisy).

Flash point. The temperature at which a flammable liquid gives off sufficient vapour to catch fire when ignited.

Flocculation. The bringing together of divided material in effluents so that a floc is formed (a gelatinous mass) which allows the solids to separate from the effluent by settling. (◊ SEWAGE TREATMENT; POLYELECTROLYTE).

Floods Directive. Under this Directive EU members have to assess the flood risks they face, produce flood risk maps, and draw up plans to manage these risks. Member states are not allowed to pass problems from one area to another.

Flue gas desulphurization (FGD). FGD is now implemented on a large scale in order to comply with EU legislation. In 2004, the UK emitted 833 thousand tonnes of sulphur dioxide, of which 539 thousand tonnes came from large combustion plants (power stations, refineries and some industrial sources). The levels in 2004 were much lower than they were in 1980 (4838 and 3457 thousand tonnes, respectively). The reduction in emissions from Large Combustion Plants over the period 1980–2004 was 84.4 per cent. This has exceeded the target of a 60 per cent reduction needed in 2003, in line with the EC Large

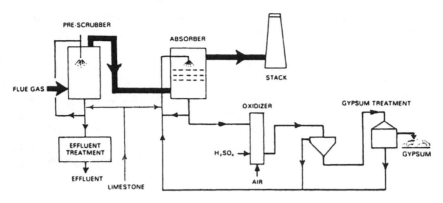

Figure 92   Schematic limestone/gypsum FGD plant.

Combustion Plant Directive. The FGD system at the 4 GW Drax Power Station, one of six UK coal-fired power stations fitted with FGD, is estimated to cost £30 million a year to run.

Sources: Defra (2006) The environment in your pocket 2006, Defra, London. ENDS Report 383, December 2006.

The wet limestone/gypsum FGD process is shown in Figure 92 in which the sulphur dioxide is absorbed by a limestone slurry spray to form calcium sulphite which is then oxidized to calcium sulphate where, together with water it forms gypsum ($CaSO_4 \cdot 2\ H_2O$). This FGD option is much to the fore for power stations as it is less costly than alternative methods where there is a ready supply of limestone available. It also has gypsum as a possible saleable end product for wallboard manufacture.

Source: W. D. Halstead, *Flue gas desulphurisation – problems and potential*, CEGB Research, No. 22, September 1988.

To get some idea of the dimensions of the problem, and FGD plant or plants that might be needed, consider Drax coal-fired power station with its output of almost 4000 MW. It consists of six boilers, each of 660 MW. Each boiler burns around 72 kg of coal per second, that is 432 kg of coal per second for the whole station. The coal contains, on average, 0.3 per cent chlorine and 2 per cent sulphur. In other words, the station produces 8.6 kg of sulphur per second, which is around 17 kg of sulphur dioxide per second, and about 1.3 kg of hydrochloric acid per second. Each boiler produces 1000 actual cubic metres (acm) of flue gas per second (the volume of a large four-bedroomed detached house and garage).

If we envisage each boiler having two FGD units, each FGD unit with its prescrubber has to handle 500 acm of flue gas per second carrying about

1.5 kg of sulphur dioxide and just over 100 g of hydrochloric acid, together with whatever fly ash makes it past the ELECTROSTATIC PRECIPITATORS at the exit to the boiler.

In terms of input for the whole plant: 17 kg of sulphur dioxide per second will require 27 kg of limestone per second on a stoichiometric basis, that is, 850 000 tonnes of limestone per year, which produces, in theory, around 1.5 million tonnes of gypsum cake per year. In practice, ca. 800 600 tonnes gypsum per year will be produced as the station will not be on 100 per cent load continually and not all the SO_2 will be removed. The actual limestone requirements will approximate 0.5 million tonnes per year, which is to be compared with the ca. 200 million tonnes of limestone quarried in the UK annually.

The possible environmental disbenefits of the gypsum FGD process are the need for high grade limestone (up to 1.5 million tonnes per year for the currently planned programme) and the need for tipping space for the calcium sulphate if there is not enough wallboard manufacturing capacity or if the manufacturers' specifications cannot be met. The alternative (regenerative FGD process) which produces H_2SO_4 as a by-product, suffers from possibly high running costs rendering the sulphuric acid produced perhaps too expensive in an already established market.

The continued growth of gas-fired power stations and flue gas desulphurization use should ensure a progressive reduction in sulphur dioxide emissions to one-fifth of their 1980 level by 2010. Coal burning in the UK will probably stabilize at 30 million tonnes per year, roughly half of it consumed in the Drax and Ratcliffe-on-Soar power stations, both of which have flue gas desulphurization fitted. The UK is well in advance in this area and when the gas runs out, *clean coal* gasification can take over.

Source: J. Redman, 'FGD. The Wet Limestone/Gypsum Process', *The Chemical Engineer*, Oct. 1988, 29–36 (and subsequent correspondence).

Flue gas recirculation. A means of effecting NITROGEN OXIDE reduction in exhaust gases by recirculating a proportion (usually 20 per cent) of the exhaust gas with the COMBUSTION air.

Flue gas scrubbing. The removal of acidic pollutants from the flue gases from installations such as incinerators. A single, two- or multi-stage Venturi scrubber using large quantities of cleaning fluid can be utilized. The gases are cooled and saturated with slaked lime or sodium hydroxide depending on the treatment required.

There are other methods available using AMMONIA as well as so-called regenerative methods.

Advantages:
– high degree of removal of acid gases.

Disadvantages:
– extensive equipment required
– flue gas reheating is necessary
– wastewater may require treatment
– highly visible wet plume (unless reheating is carried out)
– greater running costs than dry or semi-dry systems (which can have lower removal efficiencies).

The flue gases from the furnace are conveyed through the boilers to the flue gas cleaning system by means of induced-draught fans. Before entering the Venturi scrubber, the hot flue gases are usually circulated through a regenerative glass tube exchanger and thereby transfer some of their heat to the cold flue gases.

Inside the Venturi scrubber the hydrochloric acid (HCl), sulphur dioxide (SO_2) and hydrogen fluoride (HF) contents are reduced by their interaction with the dosing chemicals and in the process they are cooled down to saturation temperature and cleaned. Following this, the flue gases are reheated in the regenerative gas preheaters before being discharged through the chimney. Then necessary limestone powder is mixed in the mixing tank with filtrate from the gypsum drain, and pumped to the absorbers. Typical reactions in the scrubber are given below.

Acid gas removal with slaked lime ($Ca(OH)_2$):

$$Ca(OH)_2 + 2HCl \rightarrow CaCl_2 + 2H_2O$$
$$Ca(OH)_2 + 2HF \rightarrow CaF_2 + 2H_2O$$
$$Ca(OH)_2 + SO_2 + \tfrac{1}{2}O_2 \rightarrow CaSO_2 + H_2O$$

This may be more widely used in the future as emissions legislation is progressively tightened.

Alternatives to the wet method include the semi-dry process which uses a limestone water SLURRY. A typical semi-dry layout is given in Figure 93, for an EfW incinerator. Or the dry method may be used, in which the reaction agent (limestone) is either mixed with the fuel or refuse and introduced separately into the furnace.

Advantages:
– little apparatus needed
– no reheating necessary
– no wastewater

Disadvantages:
– degree of acid gas removal relatively low compared with semi-dry or wet methods
– high limestone demand
– the reaction product is mixed with ash

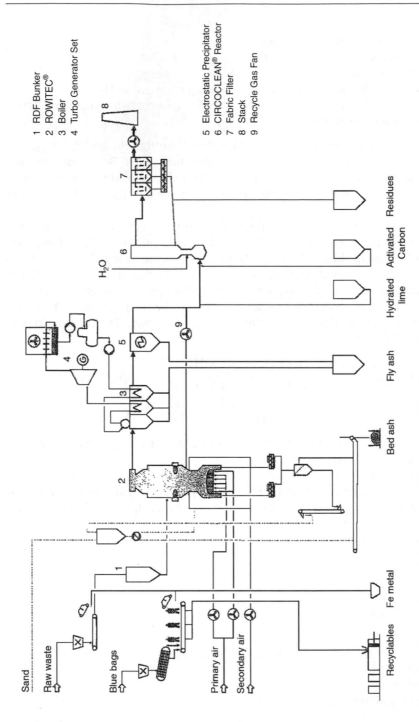

1 RDF Bunker
2 ROWITEC®
3 Boiler
4 Turbo Generator Set

5 Electrostatic Precipitator
6 CIRCOCLEAN® Reactor
7 Fabric Filter
8 Stack
9 Recycle Gas Fan

Figure 93 Layout for semi-dry process of flue gas scrubbing for a waste incinerator.
Source: Lentjes GmbH, Ratingen, Germany, www.lentjes.de.

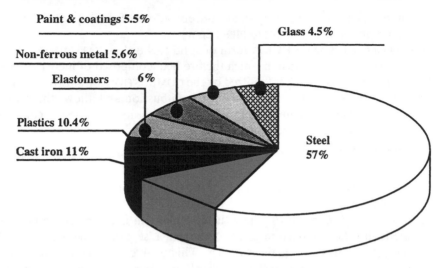

Paint & coatings 5.5%

Glass 4.5%

Non-ferrous metal 5.6%

Elastomers   6%

Plastics 10.4%

Cast iron 11%

Steel
57%

Figure 94   Breakdown of components used in car manufacture.
Source: Financial Times, 26 April 1995.

Fluff. The shredded rubber, plastics and carpets residues from car recycling. Currently, it is mainly used as a FUEL in INCINERATION plants (or landfilled in the UK).

The increasing use of PLASTICS in car manufacture means that RECYCLING efforts could be increased. However, it is important to realize that there are limited markets for low grade plastic materials and the use of waste to ENERGY can be both a cost and environmentally effective solution and much better than LANDFILL. Figure 94 gives a typical breakdown of the components used in car manufacture, about 16 per cent of which can result in fluff production. (◊ INCINERATION; GASIFICATION)

Fluidized bed biofilm reactor (FBR). A biofilm reactor where BACTERIA grow attached to solid carrier particles with a diameter of about 1 mm. They are very useful for slowly degrading substances, e.g. CHLOROPHENOL. This method prevents them from being washed out of the system.

The fluidization of the carriers provides a very large surface area for the bacteria to colonize in a small volume with resulting high bacterial densities.

Compact reactor volumes and large surface areas mean thin biofilms, decreasing diffusion resistance which is enhanced by turbulent flow regime. This allows a high BIODEGRADATION rate. Since the incoming water is highly diluted by recirculated water, the bacteria actually experience much lower chlorophenol concentrations than are present in the influent. This selects a population that can utilize very small concentrations of chlorophenols resulting in a high quality effluent. ($\lozenge$ PIPE FLOW)

Fluidized bed combustion (FBC). A form of solid fuel or low CALORIFIC VALUE waste COMBUSTION system in which the fire bed, composed of inert particles to which the fuel is added, is fluidized by the COMBUSTION air blown upwards through it. This produces highly efficient combustion and allows the use of low-grade FUELS and combustible wastes that are not suitable for conventional combustion plant designs.

Fluidized beds have evolved as:

1. *Bubbling fluidized beds* (BFB) where the combustion is in a conventional bubbling bed.
2. *Circulating fluidized beds* where the bed medium is entrained and circulated with the combustion gases. A CYCLONE separates the bed material and returns it to the main chamber. This enables excellent combustion efficiency at low excess air levels, as well as under-staged combustion and low furnace temperature conditions, qualities which ensure complete combustion and extremely low NO_x emissions. The vigorous fluidization and intensive mixing achieved in the furnace ensure highly uniform distribution of the fuel/air mixture across the combustion zone, minimizing the risk of a reducing atmosphere and the attendant risk of furnace wall corrosion. The large quantity of circulating particulates effectively equalizes the vertical temperature gradients, creating a uniform temperature throughout the furnace. In addition, the bed acts as a thermal 'flywheel' to level out the effect of variations in the fuel quality on the combustion process.
3. *Revolving fluidized bed* where the use of differential air pressures and special bed geometry enables the bed to 'revolve'. This ensures high efficiency combustion of difficult materials such as municipal solid waste. Figure 95 illustrates a revolving fluidized bed.

FBC is receiving considerable interest as part of solid waste management solutions. The Solid Waste Authority of Palm Beach, Florida, employed this option as it allows optimal RECYCLING followed by the FBC of the residual WASTE DERIVED FUEL. The (1991) mass balance and performance data are given in Figure 96.

Fluoridation. The addition of a fluoride SALT in trace quantities to public drinking water for improving the resistance to dental caries. The aim is to adjust the fluorine content to the optimum amount of 1 part per million. Many studies have shown that when a population is supplied with water containing such a concentration the incidence of dental decay is at a minimum in the population.

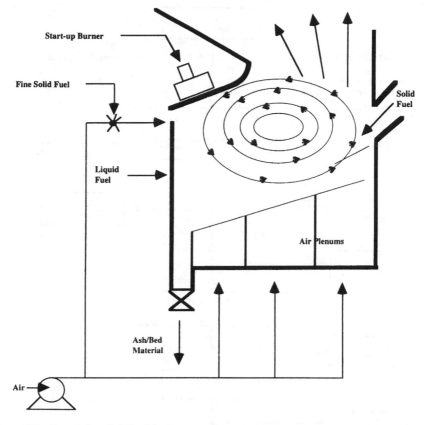

Figure 95   Revolving fluidized bed.

Fluorides. SALTS of hydrofluoric ACID. They are released into the environment by ceramic works, ALUMINIUM smelting and phosphate plants. The fluoride particles can be deposited on the herbage in the vicinity of process plants, and there have been many recorded cases of cattle suffering fluorosis which leads to loss of teeth and bone growths at joints, giving rise to lameness. Fluorides can also affect plants by entering the stomata and then moving to leaf margins where they accumulate. Extremely small concentrations in air, as low as 0.005 part per million, will blight maize, and 0.001 part per million lowers citrus productivity (US data). (◊ GLADIOLI)

The release of HYDROGEN FLUORIDE in concentrations in excess of 1 microgram per cubic metre $(1\,\mu g/m^3)$ can lead to herbage fluoride levels in excess of 30 to 35 parts per million, at which level dairy cattle are at risk.

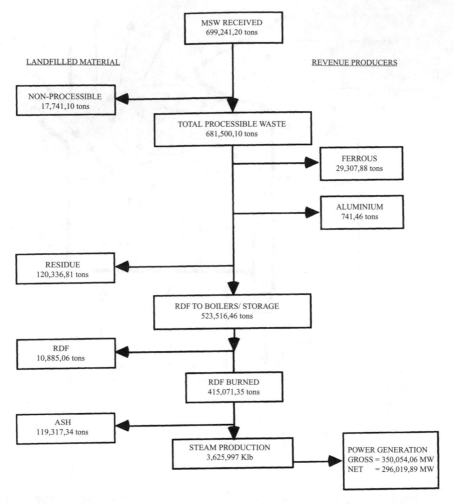

Figure 96 Palm Beach, Solid Waste Authority mass balance (US tons)/performance data (1991).
Source: The Solid Waste Authority of Palm Beach, Florida, 1991.

Fluorine (F). A greenish-yellow gas, and the most reactive element known. It is highly poisonous. Fluorine is used in the manufacture of HALOGENATED FLUOROCARBONS or CHLOROFLUOROCARBONS used in AEROSOL PROPELLANTS.

Fluorosis. ◊ FLUORIDES.

Flushing bioreactor. Fancy name for a LANDFILL design in which LEACHATE is recirculated uniformly throughout the loosely packed MUNICIPAL SOLID

WASTE mass so that BIODEGRADATION is enhanced. The concept is fraught with difficulties, e.g. seven to eight times the landfill volume needs to be recirculated as leachate over 30 years throughout all the filled void interstices. Landfill gas control may be hindered.

The House of Lords Select Committee Report on Sustainable Landfill (September 1998) stated:

> On the basis of the available evidence, we are doubtful whether the processes within a flushing bioreactor are capable of precise specification or their effect measurable with any certainty. The principle, however, of recirculation of leachate and otherwise maintaining optimum conditions for progressive decomposition, coupled with collection and use of landfill gas within existing mixed waste sites should, in our view, be supported.

The waste management industry is split on the merits of the flushing bioreactor concept. Meanwhile the debate has generated lots of heat but very little light. The technical merits and economics are still to be determined by a brave company which puts its own money on the line.

Flux[1]. The rate at which a particular quantity is transferred per unit area, e.g. $m^3/m^2/day$.

Flux[2]. Additive to smelting furnaces which combines with impurities to form SLAG.

Fly ash. The finely divided particles of ash readily entrained in the flue gases arising from the COMBUSTION of fossil FUELS (mainly COAL). The PARTICLES of ash may contain unburnt fuel. ($\Diamond$ ELECTROSTATIC PRECIPITATOR). Particulate matter is now the preferred term.

Flywheel. A form of energy storage whereby surplus energy from, say, locomotive deceleration, is stored in a flywheel rather than dissipated in the heat of braking. The flywheel is then used to accelerate the locomotive when required. This is a form of clean energy storage.

Fog. Microscopic water droplets, varying from 2 to 20 micrometres in diameter, suspended in air and reducing visibility. One form, ice fog, occurs in regions where the ambient temperature is less than −20°C. It is formed by the spontaneous nucleation and freezing of water vapour in combustion gases from power stations or vehicular exhausts in Arctic regions.

Both forms of fog substantially reduce visibility. Ice fog has been known to cut visibility down to less than 2 metres, as can an extremely dense 'pea-souper'.

In urban areas the density of fog is closely related to the amount of particulate material present in the air and upon which moisture can condense easily as droplets producing SMOG.

Food chain. A series of organisms through which ENERGY is transferred. Each link feeds on the one before it (except for the first one which is herbage – see

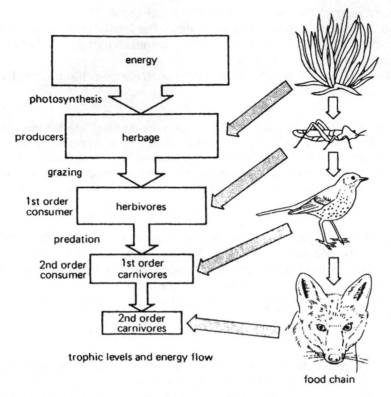

trophic levels and energy flow

food chain

Figure 97 Trophic levels and energy flow, and an example of an associated food chain.

Figure 97) and is eaten by the one following it. Herbage is said to belong to the class known as producers, which transform SOLAR ENERGY and CARBON DIOXIDE into sugars via PHOTOSYNTHESIS. The remaining links are consumers. Consumers are ordered first, second, and so on. Thus a three-stage food chain would be grass consumed by cattle consumed by humans, i.e. producer, first-order consumer, second-order consumer. A food chain is essentially an energy conversion scheme and as in any energy conversion device there are conversion efficiencies and transfer losses. Hence, the energy fixed by producers will always be greater than that fixed by first-order consumers, which in turn will always be greater than that fixed by second-order consumers, and so on (Figure 97).

Those organisms whose food is obtained from green plants and have the same number of links in their chain are said to occupy the same trophic or energy level. Thus a cow and a rabbit are both at the same trophic level. The shorter the chain, the more food there is available, e.g. 5 kilograms (5 kg) grain may at best produce 1 kg live-weight gain on cattle, and 10 kg meat may produce only 1 kg gain in human. The feed-conversion efficiency is usually very much less than this. If a link can be cut out of the chain, e.g. by humans consuming grain directly (a protein supplement may still be needed), then 5 kg grain may produce 1 kg in humans instead of 0.1 kg as formerly. ($\lozenge$ FOOD WEB, ECOLOGICAL PYRAMIDS, ECOSYSTEM)

Food miles. As with most attempts to reduce complex issues to one specific number, the use of food miles overlooks that, for example, roses grown in Kenya require no fossil fuels for heating purposes whereas those grown in Holland do. Hence, gross CO_2 emissions are greatly in favour of Kenyan rose cultivation. However, the ecological effects of intensive Kenyan rose growing are another matter. It has been claimed that Lake Naivasha in Kenya is in danger of drying out in 10–15 years due to rose cultivation, with knock-on effects on hundreds of thousands of workers in the region.

Another single parameter confounder is the use of CO_2 'saved' by recycling waste. This neglects costs and environmental impacts in developing countries and the health of both workers in the UK and the Far East where much of UK waste is sent for recycling. ($\lozenge$ MRF, RECYCLING)

Food web. A system of interlocking FOOD CHAINS. A food web comprises all the separate food chains in a community, including DECOMPOSERS whose role is vital in recycling essential materials and nutrients. (CARBON CYCLE; NITROGEN CYCLE)

Force. That which produces unit acceleration in unit mass. It is measured in Newtons, symbol N ($kg\,m\,s^{-2}$). In formulaic terms, $F = ma$, where F = force, m = mass, a = acceleration.

Hence, 1 kg falling through a clear drop under the influence of gravitational acceleration, $a = 9.81\,m\,s^{-2}$, exerts a force of 9.81 N. ($\lozenge$ HYDROPOWER)

Forest management. The Forest Stewardship Council, a Scandinavian consortium of forest owners and some environmental groups, has drawn up Forest Stewardship Council guidelines.

Forest management operations should:

1. respect the laws of the country it is operating in and international treaties and agreements to which the country adheres;
2. define and legally establish long-term tenure and rights to use land and forest resources;
3. respect the legal and customary rights of indigenous peoples to own, use and manage their lands and resources;
4. support the social and economic well-being of forest workers and local communities;
5. encourage the optimal use of forest products and services to ensure economic viability;

6. maintain the critical ecological functions of the forest and minimize adverse impacts on biological diversity, water resources, soils, non-timber resources and ecosystems and landscapes;
7. write, implement and keep up to date an appropriate management plan;
8. conduct regular monitoring that assesses the conditions of the forest, the yields of forest producers, the chain of custody and management operations;
9. not replace natural forests by tree plantations. Plantations should complement natural forests and reduce pressures on them.

Source: *Tomorrow*, Spring, 1995.

Fossil fuels. Term for coal, oil or natural gas, i.e. fuels derived from ORGANIC MATTER deposited over geological time-scales. (Uranium and vegetable materials are not fossil fuels.)

Free radical. A molecular fragment or an ION that has one or more unpaired electrons, rending it highly reactive. Free radicals are very short-lived in gaseous systems, but their high reactivity enables them to take part in chemical reactions that would not otherwise occur, e.g. certain free radicals play a significant role in the production of the constituents of PHOTOCHEMICAL SMOG.

Freezing. ◊ DESALINATION.

Freon. A proprietary brand of CHLOROFLUOROCARBONS; not a generic name.

Frequency. A measure of rate of vibration or oscillation given by the inverse of the period of simple harmonic motion. It represents the number of complete oscillations or cycles per second. The frequency of sound is measured in hertz (Hz).

Frontal depression. A travelling low-pressure weather disturbance, usually associated with widespread cloud and precipitation. It has warm, cold and very often, occluded, fronts.

Froth flotation. Process of using a froth of water and oil to effect the separation of finely divided minerals. Also proposed for separating ground glass for CULLET purposes from ash and other fines in DOMESTIC REFUSE.

Fuel. A source of thermal ENERGY. Fuel can be bacterial, fossil, vegetable or nuclear in origin. (◊ CELLULOSE ECONOMY; COAL; ENZYMES; ETHANOL; GAS, NATURAL; HYDROGEN; METHANE; METHANOL; NUCLEAR ENERGY; OIL; OIL SHALES; TAR SANDS)

The list below shows the gross CALORIFIC VALUE of various fuels.

Gas oil	45.5 GJ/t
Light fuel oil	42.5 GJ/t
Heavy fuel oil	41.8 GJ/t
Anthracite (rank 100)	30 GJ/t
Industrial boiler (rank 80)	25 GJ/t
Butane	49 GJ/t
Natural gas	38.60 MJ/m^3

Fuel, authorized. An authorized fuel, which under regulations made under the UK Clean Air Act 1956, can be burned in smoke control areas. The combustion of such fuels is not necessarily completely smokeless.

Fuel cells. One of a class of devices known as DIRECT ENERGY CONVERTERS, they are electrochemical devices which convert the chemical energy of a fuel, such as hydrogen, and an oxidant, such as oxygen, into electrical energy in the absence of combustion. There are several types of fuel cell but they are all based on a common design of an electrolyte sandwiched between two electrodes (an anode and a cathode).

The fuel cell directly converts chemical energy to electrical ENERGY without the intermediate step of random molecular energy (e.g. heat-raising of steam which needs a turbine and electrical generator before electricity is available). Thus, the fuel cell is not subject to the CARNOT EFFICIENCY restriction of the Second Law of Thermodynamics, which means that conversion efficiencies of 90 per cent and greater can, in theory, be obtained.

Most fuel cells use hydrogen as the fuel. In the hydrogen/oxygen fuel cell (see Figure 98), hydrogen is supplied to the anode, where it dissociates on the anode catalyst (typically platinum-based) to produce protons (H^+) and electrons (e^-) as depicted in the following equation:

$$H_2 \rightarrow 2H^+ + 2e^-$$

The electrons pass to the cathode via the external load, while the protons are conducted to the cathode through the electrolyte. The protons react with the oxygen supplied to the cathode and the electrons to produce water:

$$\tfrac{1}{2}O_2 + 2H^+ + 2e^- \rightarrow H_2O$$

Thus, overall, we have

$$H_2 + \tfrac{1}{2}O \rightarrow H_2O$$

The voltage from the reaction is 1–1.5 volts per cell, and any voltage can be obtained by connecting the fuel cells in series, as with ordinary batteries. ($\lozenge$ LAWS OF THERMODYNAMICS)

The cost of power from fuel cells is not cheap, and the hydrogen is usually produced by reforming from hydrocarbon fuels.

There are other types of fuel which are also used in fuel cells. For instance, methanol is used in the Direct Methanol Fuel Cell.

Current developments are aimed at lowering costs and increasing efficiency. One fuel cell (UTC Power) claims an average electrical efficiency of 40 per cent and if the heat from the cell (at 220 °C) is utilized, an overall efficiency of 90 per cent may be possible.

Source: www.utc.com.

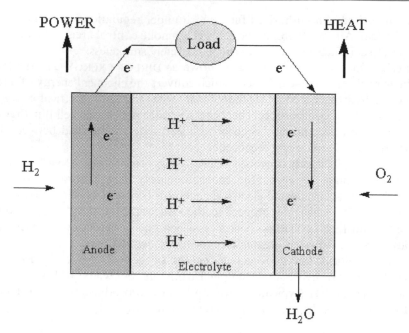

Figure 98   Diagram of a hydrogen/oxygen fuel cell.

Fuel cycle. The sequence of steps involved in supplying and using fuel for nuclear power generation. The main steps are mining and milling, extraction, purification, enrichment (where required), fuel fabrication, irradiation (burning) in the reactor, cooling, reprocessing, recycling, waste management and disposal. (▷ NUCLEAR POWER)

Fujita scale. The Fujita scale (F-Scale), or Fujita-Pearson scale, is a scale for rating tornado intensity, based on the damage tornadoes inflict on human-built structures and vegetation. The official Fujita scale category was determined by meterologists and engineers after examining damage, ground-swirl patterns, radar tracking, witness testimonies, media reports and damage imagery, as well as photogrammetry/videogrammetry if video was available.

The scale was applied retroactively to tornado reports from 1950 onward in the US, and occasionally to earlier infamous tornadoes.

Although the Fujita scale was a scientific improvement over estimates of tornado strength from earlier in the century, the wind speeds contained in the scale were, at best, educated guesses. Research, especially in the 1980s and 1990s, showed that tornado wind speeds were actually much lower than the F-scale indicated. Also, while the scale gave general descriptions for the type of damage each tornado could cause, it gave no leeway for strength of construction and other factors that might cause a building to receive higher damage at lower wind speeds.

On 1 February 2007, the Fujita scale was decommissioned in favour of the more accurate Enhanced Fujita (EF) Scale, which replaces it. The EF Scale improved on the F-scale on many counts – it accounts for different degrees of damage that occur with different types of structures, both anthropological and natural. It also provides much better estimates for wind speeds, and sets no upper limit on the wind speeds for the strongest level, EF5.

Fume. There is no generally accepted definition of the word, but it is usually taken to mean minute particles less than 1 micrometre in diameter suspended in air or flue gases or, in the UK Clean Air Act 1956 definition (Section 34) 'fume is any airborne solid matter smaller than dust'. Fumes are usually released as a result of certain metal working and chemical processes.

The term is often used to describe the vapours given off by a liquid, especially if it is offensive or toxic. (◊ FUMES, EFFECTS OF; VOLATILIZATION)

Fumes, effects of. When certain metals are heated, they tend to volatize and fumes (i.e. particles less than 1 micrometre in size) can spread easily. Those of manganese and zinc can produce an effect called metal fever.

Fumes released in plastics manufacture can also cause numbness, cramps and impotency. Teflon (PTFE) has been so documented, and recently vinyl chloride disease has been identified as being linked to the fumes from vinyl chloride monomer (VCM), as has ANGIOSARCOMA, a rare form of liver cancer. (◊ POLYVINYL CHLORIDE)

Fungi. Simple plants either unicellular or made up of cellular filaments; they contain no CHLOROPHYLL. They are agents of decay in all natural organic materials, food, timber, plant debris, etc.

Furans. In an environmental context this term refers to a group of 135 compounds known as chlorinated dibenzofurans. They resemble the chlorinated dibenzo-*p*-dioxins (DIOXIN) in their chemical structure, toxicity and behaviour in the environment. Chlorinated dibenzofurans are formed during the incomplete combustion of PCBs and are thought to be responsible for some of the adverse health effects of these compounds. Below is illustrated the structure of dibenzofuran: chlorine can be added to the carbon atoms marked X.

Furfural (C$_4$H$_3$ OCHO). Organic solvent or intermediary which can be obtained by acid HYDROLYSIS of PENTOSANS contained in waste agricultural products to pentoses and thence to furfural. Almost any plant material can be used and the production of furfural from these sources is receiving considerable attention as a means of utilizing 'income' resources. In the USA, over 2 million tonnes could be produced from maize cobs alone. Almost any petrochemical can be made from it including nylon intermediates. (The potential yield from sugar cane alone is 50 million tonnes per year.) (◊ SOLAR ENERGY)

Fusion. ◊ NUCLEAR ENERGY; NUCLEAR FISSION; NUCLEAR REACTOR DESIGNS.

G

Gaia. A Greek name for the goddess of the earth. It also refers to a scientific hypothesis formulated by James Lovelock whereby all living matter on the earth is believed to be a single living organism. In such a scheme, humanity is considered the nervous system of the living earth.

This concept has led to a heightened environmental concern by many lay persons and hence can only be encouraged. Meanwhile, engineers will continue their best endeavours to make do with less, harness resources ever more efficiently and in general provide practical solutions for society's needs.

Source: D. L. Brown (ed.), *A Brief Dictionary of New Age Terminology*.

Garbage. ◊ DOMESTIC REFUSE.

Gas, inert. A gas that does not react with materials with which it is in contact, e.g. NITROGEN, which is widely used in industry because of its inert properties.

Gas, liquefied petroleum (LPG). ◊ LIQUEFIED PETROLEUM GAS.

Gas, natural. ◊ NATURAL GAS.

Gaseous pollutants, control. ◊ THRESHOLD LIMITING VALUE; THREE-MINUTE MEAN CONCENTRATION; EMISSION STANDARD.

Gases, properties of. Gases are compressible and their density is proportional to pressure (all other things remaining constant). They expand in proportion to temperature and thus the greater the temperature, the lighter or less dense is the gas. A minimum gas exit temperature may be specified in air pollution

discharge consents, to increase plume buoyancy (plus avoiding ACID DEW-
POINT effects).

Gasification. The production of gaseous FUELS by reacting hot carbonaceous
materials with air, STEAM or OXYGEN. (◊ PYROLYSIS)

The process takes place at high temperature. The gasification product
is a mixture of combustible gases and tar compounds, together with particles
and water vapour. Depending on the gasification method, the proportion of
components varies, but common to all the processes is the fact that the gas
has to be purified before it can be used directly in a gas engine or a gas
turbine.

Gasification of wastes may make it possible to achieve high electrical efficiency. In smaller plant sizes 25 per cent electrical EFFICIENCY can be reached
by means of gas engines. As far as larger plants are concerned, combined
cycles (gas turbines in combination with heat recovery boilers and steam
turbines) could achieve efficiencies of 45 per cent or more, provided that the
gas purification is of a very high order.

There are three main parts in a gasification system: the gasifier, gas clean-up
and the power producer. Thus, end use dictates the amount of clean-up.

GASIFIERS can be divided into two main types, namely fixed bed and fluidized bed. In the fixed bed, large amounts of fuel are constantly present in the
gasifier. In the fluidized bed the residence time for the fuel in the gasifier is
short. The fixed bed technology is more suitable for small/medium size (1–
20 MW$_t$) and non-homogenous fuel, the fluidized bed for medium/large size
(20–200 MW$_t$) and well-defined and finely partitioned fuel. Typically, a 200
tonne per day plant operating on screened municipal solid waste will produce
8300 Nm^3 (Normal m^3) with a calorific value of 5 MJ/Nm^3. Full-scale plant
experience is still scant.

The untreated gas has a high tar and PARTICULATE content and has to be
purified before being used in a gas engine. If the gas is burned in a boiler or
chamber, a simple CYCLONE could be sufficient, whilst for use in a gas turbine
it is necessary to have a high quality gas without tar, particles and alkaline
metals. (◊ PYROLYSIS, FLUIDIZED BED COMBUSTION)

Gasification effectively 'separates' the conversion of a carbonaceous feedstock into a gas from the actual combustion or utilization of that gas.
Burning can only release heat whereas gasification is a starting point for a
range of possible products.

Gasification is a currently proven technology on homogenous feedstocks
such as PLASTICS, TYRES, and SEWAGE SLUDGE. It has the potential for
MUNICIPAL SOLID WASTE components such as paper and plastics.
(◊ INCINERATION)

Sources: The Chemical Engineer, 10 October 1996, p. 21. A. Donati, 'Different technologies for waste treatment', *Waste to Energy*, March 1995.

P00L10459

ST HELENS COLLEGE

UK 51909003 F

Ship To:
ST HELENS COLLEGE
TECHNOLOGY CAMPUS, WATERS
POCKET NOOK STREET
ST HELENS
MERSEYSIDE
WA9 1TT

Volume:
Edition: 4th ed
Year: 2008
Pagination: xxvii, 794 p.
Size: 23 cm

ISBN	Qty	Sales Order
9780470061954	1	F 9926852 1

Customer P/O No   **Cust P/O List**

532/0008   24.95 GBP

Title: Dictionary of environmental science and technology. Andrew Porteous

Format: P (Paperback)
Author: Porteous, Andrew
Publisher: John Wiley
Fund:
Location: Technology
Loan Type:
Coutts CN: 6491327

Order Specific Instructions

The table below gives syngas compositions and calorific values for various gasifiers and feedstocks.

Fuel type	Bone meal		Wood chips		MSW	
Property	10kWe scale	1 MWe scale	10kWe scale	1 MWe scale	10kWe scale	1 MWe scale
CO %	19.46	19.6	16.32	16.21	15.32	18.6
O_2 %	1.09	1.85	1.07	2.62	2.28	2.34
N_2 %	55.45	49.81	53.38	52.82	57.60	41.15
H_2 %	13.26	10.14	13.38	13.73	7.32	16.84
CH_4 % + HC %	1.37	5.53	2.64	2.26	2.60	5.54
CO_2 %	9.36	13.07	13.22	12.36	14.88	15.53
Calorific value (MJ/Nm^3)	4.70	5.97	4.82	4.83	4.25	6.70

Source: Akay, G., Dogru, M. and Calkan, O., The Chemical Engineer, Dec 06/Jan 07, pp 55–57.

The syngas was used in a modified commercially-available gas engine rated at 1048 kW$_e$ of natural gas but which produced 624 kW$_e$ using 1240 Nm^3/h product gas (CV = 5.2 MJ/m^3) at 40 °C. Emissions were in compliance with the EU Waste Incineration Directive.

Genetic engineering. A popular term for techniques which interchange DNA sections between individuals of the same or different species. (◊ RECOMBINANT DNA TECHNOLOGY)

Genetic load. ◊ IONIZING RADIATION.

Genetic pollution. The process by which genes from RECOMBINANT DNA organisms may escape and become incorporated into wild species. For instance, genes conferring resistance to herbicides may escape, in pollen, from a crop plant and be expressed in a related weed species, rendering the herbicide ineffective for controlling that weed.

Genetically modified organisms. The definition of a genetically modified organism (GMO) given in Article 2 of the Deliberate Release Directive (90/220/ EEC) is 'any biological entity capable of replication or of transferring genetic material, in which the genetic material has been altered in a way that does not occur naturally by mating and/or natural recombination'.

Source: EEC Council Directive on the deliberate release to the environment of genetically modified organisms, *Official Journal of the European Communities*, L 117, 8 May 1990, pp. 15–27.

Geographic Information System (GIS). A geographic information system (GIS) is a system for creating, storing, analysing and managing spatial data and associated attributes. In the strictest sense, it is a computer system capable of integrating, storing, editing, analysing, sharing, and displaying geographical information. In a more generic sense, GIS is a tool that allows users to create interactive queries (user created searches), analyse the spatial information, and edit data.

Geothermal power. The use of ENERGY from the earth's interior conducted to the surface in a few areas of the globe where igneous rocks are in a molten or partly molten state usually within 10 km of the earth's surface. To be useful the energy must be available in superheated water or STEAM form and it may then be used in a conventional power plant. Italy, USA, Iceland, USSR and New Zealand all have suitable geothermal energy fields. This method of power generation is a strictly local and usually small-scale affair although very useful to those areas where it occurs. As a fraction of global energy requirements it is very small.

The use of so-called 'hot rocks' up to 10 km under the earth's crust could also be considered a free energy source – but at what price to extract the thermal energy? Just as with other renewable sources of energy, economic feasibility is all important, and careful evaluation is vital as the money spent on prestigious projects may be more effectively applied in say, domestic energy saving through improved house insulation.

Gladioli. Flowers which are useful indicators of the presence and concentration of airborne FLUORIDES. The leaves mottle and turn yellow-brown in the presence of CONCENTRATIONS as low as 0.5 micrograms per cubic metre. Skilled interpretation is of course required. The variety 'Snow Princess' is most commonly used as the most sensitive, while other varieties are progressively more resistant to fluoride damage. (◊ BIOLOGICAL INDICATOR)

Glass. Common (window) glass is a mixture of 70 per cent silica (SiO_2), 14 per cent lime and magnesia ($CaO + MgO$), 12 per cent soda (Na_2O) and 1–2 per cent alumina, ferric oxide and trioxide. For glass bottle manufacture, CULLET may be added both as a means of reducing energy consumption per unit of product and to encourage the development of a market for recycled empty bottles collected via BOTTLE BANKS. (◊ RECYCLING)

European Glass Recycling Rates.

Country	Rate
Switzerland	91%
Sweden	84%
Netherlands	84%
Norway	81%
Germany	81%
Finland	69%
Austria	65%
Denmark	63%
France	55%
UK	24%

Source: www.foe.co.uk/resource/press_releases/20000314121853.html.
Accessed 27 February 2007

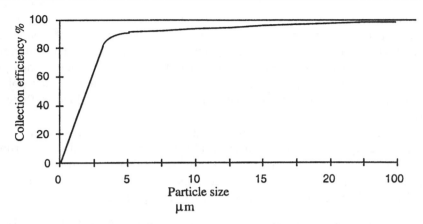

Figure 99   Grade efficiency curve versus particle size.

Glucose. Often called the key sugar as it is important to both plants and animals as an energy producer. It is a monosaccharide hexose ($C_6 H_{12} O_6$) and is present naturally in fruits. It also results from the HYDROLYSIS of other sugars and starches such as sucrose, lactose, maltose, cellulose and glycogen. An example of this in commercial use is the enzymic conversion of maltose and sucrose to glucose by yeast in the brewing process. (◊ SUGARS; ENZYME TECHNOLOGY)

Grade. Measure of the quality of an ore. The term is also used in expressing the size of particulate matter as a 'grade distribution'.

Grade-efficiency curve. Figure 99 shows a typical curve for a SPRAY TOWER used to remove particulates from a gas stream.

Granulated activated carbon (GAC). ◊ ACTIVATED CARBON.

Granulomas. Multiple accumulation of cells distributed in nodules in the lung. It is not particularly common. BERYLLIUM, tungsten carbide and zinc dusts have been responsible.

Gravel pack. A gravel screen around a perforated well casing. This increases the effective well diameter and filters out sediments. Gravel packs are also used in LANDFILL GAS extraction wells.

Gray (Gy). ◊ IONIZING RADIATION DOSE MEASUREMENT.

Green taxes. These should:

1. reflect costs of environmental damage in the prices of goods, services or activities;
2. create incentives for producers *and* consumers to move to environmentally friendly behaviour;
3. provide incentives for producers to use less energy, fewer raw materials (or different ones), develop new modes of transportation, housing and energy use, and change consumption trends;
4. raise revenue which *may* be used for environmental improvement.

(◊ ENVIRONMENTAL TAXES; FACTOR 4; AGGREGATE TAX)

Greenhouse effect. The mechanism whereby incoming solar radiation is trapped
by a glass sheet or the presence of CARBON DIOXIDE and other GREENHOUSE
GASES in the atmosphere. As these gases are transparent to solar radiation,
the short-wave incoming radiation is transmitted. However, they are opaque
to long-wave re-radiation from the earth's surface or from any other object
underneath, thus heat is trapped and the underlying surface is thereby
warmed.

This effect is put to good use in a greenhouse, but fears have been expressed
that the earth's surface temperature could rise due to the build-up of carbon
dioxide and other greenhouse gases in the atmosphere. The attendant solar
heat gain could disrupt CLIMATE patterns. The best estimates by the UN
Panel on Climate Change for temperature changes by 2030 are:

> Central North America: increased temperature between 2 °C to 4 °C in winter and
> 2 °C to 3 °C in summer. South-east Asia: warming between 1 °C and 2 °C through-
> out the year. Southern Europe: up 2 °C in winter and between 2 and 3 °C in
> summer. Australia: up 1 °C to 2 °C in summer and 2 °C up in winter. Sahel (area
> south of Sahara): up between 1 °C and 3 °C generally.

There are sufficient reserves of fossil FUELS which when burnt will increase
the atmospheric carbon dioxide concentration by up to 8 times. (◊ ATMO-
SPHERE; SOLAR HEATING)

For further information, please see www.royalsoc.ac.uk/climateguide

Greenhouse gases. Collective term for those 'gases' that have influence in the
GREENHOUSE EFFECT, i.e. CHLOROFLUOROCARBONS, CARBON DIOXIDE,
METHANE, NITROUS OXIDE, OZONE and WATER VAPOUR. This term links all
gases together, although some components are much more damaging than
others, e.g. a *molecule* of methane is ca. 60 times more powerful in 'greenhouse' terms than a molecule of carbon dioxide, and CFCs are estimated to
be up to 10 000 times more powerful than carbon dioxide. Yet public debate
tends to focus on the latter.

The best estimate of the relative contributions to the greenhouse effect of
these gases is that carbon dioxide is the greatest single contributor (50 per
cent), CFCs 15 per cent, methane 20 per cent, ozone 10 per cent, water
vapour and nitrous oxide 5 per cent.

Much attention has been focused on carbon dioxide and the claim that a
greatly expanded NUCLEAR ENERGY programme will safeguard the environment by containing the greenhouse effect. The reality is that coal-burning
emissions contribute around 15 per cent of the global greenhouse effect
(power generation 7 per cent). In the UK context, annual consumption of
coal is ca. 61.8 million tonnes out of a total world-wide production of 4.9
billion tonnes (2005 figures).

Carbon dioxide and nitrogen dioxide emissions from aircraft engines are
currently thought to be responsible for about 5 per cent of the greenhouse

effect, and air transport world-wide is expected to double in 15 years. However, the Intergovernmental Panel on Climate Change (IPCC) recommended that international air traffic be excluded from national targets for greenhouse gas emissions as it is too difficult to allocate emissions from international flights to individual countries.

The IPCC has issued an update on methane and nitrous oxides as shown in the table below. Note that time-scales are very important and figures can be fudged by using 100 year time-scales as some UK bodies are prone to do.

Greenhouse warming-up potential (GWP) relative to 1 kg carbon dioxide.

	Lifetime (years)	GWP (20 years)	GWP (100 years)	GWP (500 years)
Methane	14.5 ± 2.5	62	24.5	7.5 units
Nitrous oxide	120	290	320	180

Source: *Greenhouse Issues*, IEA Publication No. 17, March 1995.

The UK coal-fired power industry contributes less than 0.5 per cent to the total global greenhouse effect. The time-scale for the construction and full commissioning of a nuclear power station is ca. 10 years, so we had better not hold our breath for nuclear power to come to the rescue of the greenhouse effect!

Measures which can be taken fall into the short term and long term. Action on CFCs (and related HALONS) production and use would have a very substantial impact on both the greenhouse effect and ozone layer depletion. Production of CFCs has now ceased and only supplies from CFC recycling will be available. The end is in sight!

The UK is well on the way to stabilizing or even reducing its CO_2 emission due to gas-fired generation (40 per cent reduction CO_2 coal fired/kWh) and nuclear power generation.

The UK contributes 2% of global CO_2 and hence any CO_2 reduction will be marginal in global terms but essential if UKPLC is to have any meaningful environmental clout in international conferences.

Some CO_2 facts:

- The generation of 1 kilowatt hour of electricity from a coal-fired power station produces 1 kg of CO_2 (1000 kilos = 1 tonne) (a kilowatt hour = 1000 watts of electricity supplied for 1 hour or a 100 W bulb lit for 10 hours).
- The burning of 1 THERM of natural gas produces 6 kilos of CO_2 (1 therm will boil a pint of water on a gas burner 160 times, or keep a burner on full for 9 hours).
- The combustion of 1 litre of petrol produces around 2.5 kg of CO_2.

- For the same amount of useful energy, oil emits 38 to 43 per cent more CO_2 than natural gas, and coal emits 72 to 95 per cent more.

Data of this nature can form the basis of a carbon tax. On this basis coal would be taxed more than oil, which would be taxed more than gas. However, resource conservation considerations may dictate that some coal is used in order to conserve oil and natural gas.

The market on its own is unlikely to deliver more than a small fraction of the full potential for energy efficiency because of barriers such as:

- The lack of information on how to improve energy efficiency;
- The fact that frequently the people who pay the fuel bills (tenants) are not those responsible for capital improvements (landlords);
- The significant difference between the 2 to 3 year pay-back periods required before businesses or individuals will invest in energy efficiency, and the 20 year pay-back period on which the energy supply industries often plan. Because of this difference, a disproportionate amount of investment is going into expanding supply (to meet the inefficient demand) rather than into improving the efficiency of energy use.

It would appear that climate change is now inevitable and action to contain the growth of the greenhouse effect (and a parallel OZONE layer depletion) may 'mean disruption to the world's economic system'. Certainly, fossil fuel combustion and other industrial CO_2 releases are up by a factor of 3 since 1950. Globally, these total ca. 6 billion tpy as carbon and any substantial reduction may mean disruption.

Technology is catching up. The greenhouse gas reduction potential of automotive fuels has been assessed for two time frames: short term and long term. Some results of the evaluation are summarized below. This shows the amount of reduction expected from use of various fuels in the long term.

Fuels that will reach the indicated levels of reduction:
50–75%/km
   Gasoline
   Diesel
   Liquefied natural gas
   Methanol from cellulose
   Ethanol from cellulose
   Biodiesel
   Dimethyl ether

75–100%/km
   Ethanol from cellulose
   Biodiesel

The basis for comparison is 'well-to-wheel': CO_2 equivalent emissions compared to a gasoline passenger car with today's technology, based on improvements which can be foreseen today.

In the long term, gasoline and diesel cars could reduce their greenhouse gas emissions by 50 per cent because the fuel consumption of these cars can be reduced by 50 per cent. However, in the case of hydrogen, for instance, technology is important because hydrogen could only reduce CO_2 equivalent emissions by more than 75 per cent if it were produced with 'clean' electricity, i.e. non-fossil fuel generation (nuclear power?). Therefore there are always difficult choices to be made and it is unlikely today's consumers will give up their cars.

Projected global mean surface air temperature changes from 1990 to 2100 predict an increase of about 2.5 °C (range from 1.5 °C to 4 °C). Comparing this with the temperature changes we commonly experience, it does not seem a very large rise. But remember, it is a rise in the average over the globe. Between the middle of an ice age and the warm periods in between ice ages, the global average temperature changed by only about 5 or 6 °C. So, 2.5 °C over a century is large in the context of climate change; it would in fact represent a change of climate more rapid than has been experienced by the earth at any time during the last 10 000 years. Hence the real cause for concern.

(◊ CARBON DIOXIDE; CHLOROFLUOROCARBONS, HALONS)

Sources: D. Everest, *The Greenhouse Effect Issues for Policy Makers*, Royal Institute of International Affairs, London, 1988.
T. M. Wigley, `Relative Contributions of Different Trace Gases to the Greenhouse Effect', *Climate Monitor*, vol. 16 (1), 1987.
Solving the Greenhouse Dilemma – A Strategy for the UK, UK Association for the Conservation of Energy, London, 1989.
Greenhouse Issues, No. 35, March 1998, p. 3.
J. Houghton, 'Global Warming and Climate Changes – a scientific update', *IChemE Environmental Protection Bulletin*, No. 20, September 97, pp. 3–8.

Greenhouse Gas Reduction Target (UK). Twenty per cent by 2020 – The binding EU target for a 20 per cent cut in GREENHOUSE GASES by 2020 (actually CO_2 reduction). This is based on 1990 levels [A 'possible' 30 per cent by 2020 if US and other 'major players' sign up]. Twenty per cent renewable energy use by 2020 is also pledged. Recall that in 2005, the UK achieved only 1–6 per cent from renewables. Sixty per cent CO_2 reductions are also promised by 2050. Fudging of statistics may be anticipated. Also, long will the political protagonists still have long-haul holidays. Will the substantial excess CO_2 from the additional concrete and steel manufacture, access roads, duplicated transmission line provisions required for 'renewables' be ignored?

It has been stated (Secretary of State for the Environment, interview, Newsnight, 13 March 2007) that up to 50 per cent of the UK's CO_2

reductions could be obtained from Carbon Credit purchases (e.g. from India).

'*Quis custodiet custodes*' – Who watches the watchers?

Green Revolution, The. The most widely recommended means of increasing agricultural yields is through the increased use of FERTILIZERS and the introduction of new 'high yield' varieties of grain.

Fertilizers are easily produced (although the ENERGY required in their production is considerable) and have been used intensively for many years now. However, the environmental consequences of the intensive use of fertilizers and the effects of a really large-scale increase in their use are incalculable. In addition, the difficulties of implementing the proposed increase in fertilizer use on the scale required are immense. Ehlich has calculated that if India were to apply fertilizer at the *per capita level* employed by the Netherlands, India's fertilizer needs alone would amount to nearly half the present world output.

The second proposal for increasing yields is to develop new high-yield or high-PROTEIN strains of food crops. Such new strains have had considerable success over the past few years, particularly in Asia. They mature early and are relatively insensitive to the length of day, making the production of two or three crops a year possible. But these new strains usually require high fertilizer inputs in order to realize their full potential, and we have already pointed out the problems involved there. Vast amounts of capital are required for fertilizer production and distribution. Abundant water is also necessary, as are PESTICIDES and mechanical planting and harvesting machinery. It is just those countries that need these crops most who are desperately short of capital. Furthermore, we are unsure how resistant these new strains are to the attacks of insects and plant diseases.

Therefore, although high-yield agriculture is promising, it is unlikely that it will ever fulfil its promise. Lack of capital, expertise, and most of all, lack of time and the will to act quickly will probably mean that at best the Green Revolution will only allow us to keep pace with population growth for a couple of decades. (◊ GENETICALLY MODIFIED ORGANISMS)

Green sand. A naturally occurring mineral which can be used as filter bed material for iron removal from drinking water.

Grey List. ◊ List II substances.

Grit. A general term for coarse particulate matter. BS 3405 defines grit as solid particles retained on a # 200 mesh BS sieve, nominal aperture $76 \mu m$.

Gross national product (GNP). A measure of the total flow of goods and services produced by the economy over a particular period – usually a year. It is obtained by adding up, at market prices, the total national output of goods and services. 'Intermediate' products are not included since it is assumed that their value is implicitly included in the prices of the final goods. To this total figure (frequently termed the 'gross domestic product') is added any income accruing to residents arising from investment abroad, and from it is deducted any income earned *in* the domestic market by foreigners abroad. The final figure is called the 'gross national product' and it is generally supposed to be a measure of economic success.

However, as a realistic guide to a nation's economic well-being there is a lot wrong with the GNP. First, it includes the very considerable expenditure on arms and the military, which is totally non-productive. Second, it includes such things as pollution. When somebody pollutes the environment and somebody else cleans it up, the cleaning-up process is included in the GNP. Similarly, if you are an urban dweller whose health is affected by the pollution, then your hospital bills also contribute to the GNP. But the main defect with the GNP as an indicator of economic well-being is its preoccupation with indiscriminate production.

Ground effect. In sound propagation this refers to the result of interaction between direct and ground-reflected sound. Over grassland the result leads to extra attenuation of the noise level at frequencies between 250 Hz and 1000 Hz depending on range.

Ground-level concentration. The concentration of a pollutant in air to which a human being is normally exposed, i.e. between ground and a height of some 2 metres above it. It does not mean the concentration in a layer of air in direct contact with the ground, where the concentration may be low if absorption of the pollution by the ground is occurring. The upper limit of height is flexible if concentrations do not change very much with height from 2 m upwards. When the major source of air pollution is a single stack or a small group of stacks the measured ground level concentration at any point will vary very rapidly with time on account of small but rapid changes in wind speed and direction, and 'ground level concentration' is meaningless without specification of the time over which the concentration is averaged.

Groundwater. Water occurring within the SATURATION ZONE of an AQUIFER is the only part of all subsurface water which is properly referred to as groundwater (or phreatic water). Groundwater present within this zone can be considered to be occupying a large natural reservoir or series of reservoirs whose capacity is the total volume of the PORES or VOIDS in the rocks that it fills.

Groundwater may be of variable chemical quality ranging from wholesome potable waters to highly mineralized BRINES.

It is the duty of the Environment Agency (EA) in England and Wales to monitor and protect groundwater and conserve it for water resource usage. It is also the EA's duty to maintain and conserve surface waters, which in many cases depends upon the proper management of groundwater.

The EA can influence planning decisions which may damage groundwater. As the way the land is used and developed is one of the greatest and most consistent threats to the quality of groundwater, land-use planning policies can play a significant role in protecting groundwater. Therefore, the EA must keep in close contact with the local planning authorities.

As some activities (e.g. LANDFILL) present a particular risk of pollution, the closer an activity is to a well or borehole, the greater the risk of the pumped water being polluted.

Around each groundwater source, the EA has defined three source Protection Zones. These vary in their size, shape and relationships according to the particular situation at any one place. The type of soil, the geology, the rainfall and the amount of water pumped out of the ground are all taken into consideration. Policy statements dealing with the new developments within each of these three zones, in addition to other areas where groundwater is generally at risk, have been published; for example:

Policy B (*Physical Disturbance*). Surface mineral exploitation is the main form of physical disturbance. The extraction of minerals, including limestones, sandstones, clay, sand and gravel can mean dewatering activities pose a threat to both the quality and quantity of groundwater.

Policy C (*Waste Disposal*). Virtually all landfill sites are located at disused mineral extraction sites. The majority of these historical landfill sites have been located and designed on the 'dilute and disperse' principle, without the same regard for groundwater that is at present required.

Policy E (*Disposal of Sludges*). The application of sewage sludges to agricultural land requires careful control to avoid polluting groundwater by bacteria. (◊ NATIONAL RIVERS AUTHORITY; ENVIRONMENT AGENCY)

Source: *Policy and Practice for the Protection of Groundwater*, summary document, National Rivers Authority.

Groundwater Action Programme. At a European Commission ministerial seminar on groundwater, held at The Hague in 1991, participants agreed to establish a programme of actions at the national and Community level aimed at sustainable management and protection of groundwater resources. An action programme was drawn up and adopted by the European Commission on 10 July 1996. It includes:

- integrated planning and management between surface water and groundwater and between quantitative and qualitative management;
- abstraction of freshwater (establishment of a freshwater conservation policy, with encouragement for water saving and reuse);
- diffuse sources of pollution (establishment of a Community framework for the development of codes of good practice, sustainable use of pesticides, reduction of nitrates and measures to deal with sewage sludge);
- control of point sources of pollution from activities and facilities that may affect the quality of groundwater.

The main associated objectives are:

- mapping and characterization of groundwater systems;
- integration of groundwater protection and management into spatial planning, including establishment of zoning of vulnerable and important areas;
- preparation of inventories of point sources and polluted groundwater and soil;
- a comprehensive regulatory system, rules for freshwater abstraction and for activities and facilities that may lead to pollution of groundwater;
- possibilities for use of financial incentives or sanctions, including taxes and levies;
- introduction of measures to promote water saving, reuse and sustainable use of freshwater resources, and, where appropriate, reduction of consumption of water.

Source: *CBI Environmental Newsletter*, September 1996.

Groundwater Regulations. Regulations governing substances that must be controlled to protect groundwater.

LIST I SUBSTANCES include many PESTICIDES, sheep dip, SOLVENTS, HYDROCARBONS and toxic metals. These must be prevented from entering groundwater.

LIST II SUBSTANCES are the less dangerous chemicals (which could be harmful to groundwater if disposed of in 'large amounts'). Heavy metals, all remaining pesticides, and sewage effluent fall into this category. Entry of these substances into groundwater must be restricted to prevent pollution.

Regulation is by discharge consents or integrated pollution control authorization or by an authorization for disposing or discharging List I or List II substances or potentially polluting chemicals.

Prohibition notices can be served.

Source: *Wet News*, 3 February 1999.

Groundwater remediation technologies. Groundwater can sometimes be contaminated due to seepage of SOLVENTS, etc. into the AQUIFER. In such cases, the groundwater is pumped out and treated in an above-ground process, e.g. air stripping. If the pollutant is an ionic species, e.g. seawater intrusion into freshwater aquifers, REVERSE OSMOSIS can be utilized to remediate the groundwater. (◊ BIOTECHNOLOGY)

An excellent example of air stripping of VOLATILE ORGANIC COMPOUNDS (VOCs) at low concentrations and combined air stripping and activated carbon polishing for high VOC concentrations (up to $160\,000\,\mu g/l$) is given in an article by Fellingham *et al.* on remediation of groundwater contamination at Harwell. The contaminants included chloroform, carbon tetrachloride, 1,1,1-trichloroethane, trichloroethene, tetrachoroethene, benzene, toluene and xylene. These had routinely been disposed of at two sites from 1946 to 1977 in six unlined shallow pits at one site and 25 larger ones at the second. Enough said.

Source: *Environmental Protection Bulletin*, Issue 062, September 1999.

Groundwater vulnerability maps. A series of maps showing the UK's aquifers, classified according to the properties of the rocks and the overlying soils.

Guidance Notes. These are issued for relevant processes and are at two basic levels.

(a) Those covered by the Environment Agency;
(b) Those covered by Local Authorities (Secretary of State).

Process Guidance Notes include for example:

Merchant and In House Chemical Waste Incineration, IPR 5/1
Municipal Waste Incineration, IPR 5/3
Clinical Waste Incineration, IPR 5/4
Furnaces, IPR 1/1 (large boilers = 50 MW thermal or more).

Requirements to be met are specified, such as:

(i)   Releases into air \
(ii) Releases into water | quantities, concentrations, monitoring, record-keeping and reporting are all pre-scribed for specified releases. In addition, specific plant operating conditions must be met.
(iii) Releases into land /

The notes are issued with the objective that BATNEEC will be used and met, e.g. for MSW INCINERATION.

By 1 December 1996, existing municipal incineration plant with a capacity of at least 6 tonnes per hour must comply with the following combustion conditions: the gases resulting from the combustion of the waste must be raised, after the last injection of combustion air and even under the most unfavourable conditions, to a temperature of at least 850 °C for at least two seconds in the presence of at least 6% oxygen. However, in the event of major technical difficulties, the provisions concerning the two-second period must be implemented at the latest when the furnaces are replaced.

The guidance notes may be used by applicants and other interested parties as guidance on the criteria against which judgement will be made on:

(a) the acceptability of an application or variation
(b) the conditions to be included in an authorization or variation notice.

Source: Waste Disposal & Recycling, Chief Inspector's Guidance Notes, IPR 5/3.

Gypsum. Hydrated calcium sulphate $CaSO_4 \cdot 2H_2O$ used in the manufacture of plasterboard. Also a by-product of the wet limestone FLUE GAS DESULPHURIZATION process. For example, the proposed FGD plant for the 4 GW UK coal-fired Drax power station will consume 500 000 tonnes of limestone and produce around 860 000 tonnes of gypsum each year for the building industry.

H

Haber process. ◊ NITROGEN CYCLE.

Habitat. The environment in which an organism lives. The term is most strictly applied to the range of environments a particular species inhabits, but in practical use it is often defined by reference to a community rather than an individual species.

Haemolysis tests. Haemolysis is the breaking of blood corpuscles by the action of a poisonous substance. It may be used as a means of testing for the biological activity of a suspected substance by incubating the substance with red blood cells, and then measuring the amount of haemoglobin released by the breaking of cells.

Haemolysis tests on PVC powder suggest that PVC dust is as biologically active as blue asbestos or crocidolite. *If* this is the case, it has serious ramifications concerning the handling of PVC powders which are used in many plastics plants. (◊ ASBESTOS)

Half-life.
1. Time needed for half a quantity of ingested material to be eliminated from the body naturally. Also radiobiological half-life: time needed for ingested radioactive material to deliver half its radiation dose. This allows for both decay in activity and the time in the body.
2. The time taken for half the quantity of a substance to disappear from the environment, e.g. by biodegradation or by discharge from a biological system. For example, inorganic MERCURY has a half-life of six days in humans; organic mercury compounds have a half-life of 70 days. Thus,

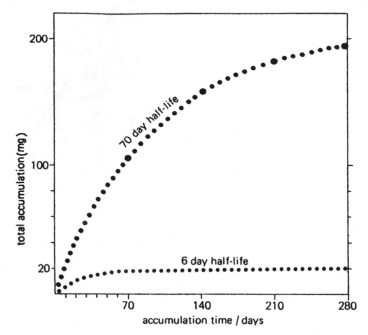

Figure 100   The effect of different periods of half-life on accumulation in the body. Organic mercury reaches ten times the level of inorganic mercury in nine months. At the 2 milligram per day ingestion level shown here symptoms of severe poisoning from organic mercury would in fact appear before the third month.

the ingestion of mercury in food will result in totally different body burdens depending on the form (see Figure 100).
3. Radioactive half-life is the time taken for half the number of atoms of a radioactive substance to decay to atoms of a different mass. Different RADIONUCLIDES have different half-lives: plutonium-239 has a half-life of 24 400 years; strontium-90 has a half-life of 28 years; xenon-138 has a half-life of 17 minutes. (◊ EXPONENTIAL CURVE)

Radiological half-life. In estimating the dose of radiation from radioactive matter that has been ingested, both the radioactivity of the matter and the residence time in the body are important and give rise to the concept of radiological half-life.

Various organs have different half-lives for the presence of the same compound or element. For example, an ingestion of organic mercury may have a half-life of 50 days in the liver, but 150 days in the brain. Thus, even though the brain will get a much smaller proportion of the input, the bulk going to the kidneys, at the end of approximately a year the brain concentration is higher.

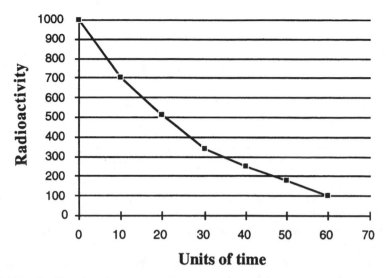

Figure 101 Radioactive decay.

This accounts for the effect of heavy metal poisoning on the central nervous system as at Minamata. ($\lozenge$ MINAMATA DISEASE)

Source: A. Tucker, *The Toxic Metals*, Pan/Ballantine; Earth Island, 1972.

Radioactivity decay. The example in Figure 101 shows a 'half-life' – that is, half the radiactivity fades away – of 20 'units of time'. Some radioactive matter has a very long half-life and the units would be commensurately large, e.g. for plutonium a time 'unit' would scale 1200 years on the curve. $\lozenge$ EXPONENTIAL CURVE.

Hall process. The extraction of ALUMINIUM by ELECTROLYSIS in a solution of molten ALUMINA (which is produced from the BAUXITE ore) and CRYOLITE. The aluminium falls to the bottom of the electrolysis cell. Needs cheap and abundant electrical power.

This accounts for the large ENERGY debt in aluminium manufacture as opposed to steel. On a tonne for tonne basis, aluminium is up to five times more energy intensive, depending on how the boundaries are drawn in the analysis.

Halogenated fluorocarbons. Group name for ethane- or METHANE-based compounds, in which some or all of the hydrogen in their structures is replaced by chlorine, bromine and/or fluorine. CHLOROFLUOROCARBONS (CFCs) are a major part of this family. They are used as refrigerants, aerosol propellants in fire-fighting and as foam expanders because of their low boiling point. They break down when subjected to high levels of ultraviolet radiation, to chlorine monoxide which reacts with OZONE and turns it into OXYGEN, leading to depletion of the ozone layer. Examples include the refrigerant Freon-12 (CCl_2F_2) which has a boiling point of 28 °C and TRICHLOROFLU-OROMETHANE (CCl_3F) used as an aerosol propellant and designated Freon-11.

Halogenation. Reaction with a member of the halogen group: namely, fluorine, chlorine, bromine, iodine.

Halons. A HALOGENATED FLUOROCARBON which contains bromine used especially in fire extinction. Ozone damage potential relative to CFCs is 3–10 times greater. (◊ HALOGENATED FLUOROCARBONS)

Hardness. A property of water usually manifested as 'needing more soap to get a lather' that is classed as either temporary or permanent. Temporary hardness can be removed by boiling and deposits a carbonate scale. Non-carbonate hardness cannot be removed by boiling and is classified as permanent.

   Hardness values are expressed as an equivalent amount of calcium carbonate ($CaCO_3$). Water with a hardness of less than 50 ppm is soft. Above 200 ppm, domestic supplies are usually blended to reduce the hardness value.

Harm. Defined by the UK Environmental Protection Act (1990) as meaning: 'harm to the health of living organisms or other interference with the ecological systems of which they form part and, in the case of man, includes offence caused to any of his senses or harm to his property'. 'Harmless' has a corresponding opposite meaning.

Hazard. A circumstance that poses a threat. This will need RISK assessment.

Hazard Analysis and Critical Control Points (HACCP). HACCP is a systematic preventative approach to foodborne illness that addresses physical, chemical and biological hazards as a means of prevention rather than finished product inspection. HACCP is used in the food industry to identify potential food safety hazards, so that key actions can be taken at Critical Control Points (CCPs) to reduce or eliminate the risk of the hazards being realized. The system is used at all stages of food production and preparation processes. Today HACCP is being applied to industries other than food, such as cosmetics and pharmaceuticals. This method, which in effect seeks to plan out unsafe practices, differs from traditional 'produce and test' quality assurance methods which are less successful and inappropriate for highly perishable foods.

Hazardous pollutants. This is a classification used by the United States Environmental Protection Agency. A hazardous pollutant is one to which even slight exposures may cause serious illness or death. MERCURY, ASBESTOS and BERYLLIUM have all been so classified.

Hazardous waste. ◊ WASTES.

Hazardous waste incineration. This is the INCINERATION under very strictly controlled conditions of WASTES deemed to be hazardous. They are highly dangerous to life, when present in the environment. They include spent solvents, PCBs, and pesticide residues.

The hazardous waste incineration sector performs a valuable service as these wastes are destroyed to values of 99.999 per cent destruction and removal efficiency. Typically a ROTARY KILN, and multiple stage SCRUBBING of the flue gases is employed, as shown in Figure 102, overleaf.

Hazardous Waste Regulations. The Hazardous Waste Regulations (England and Wales) 2005, came fully into force on 16 July 2005. These replaced the Special Waste Regulations 1996. They implement the Hazardous Waste Directive (HWD, Council Directive 91/689/EC). In 1994 a comprehensive list of all wastes, hazardous or otherwise was produced. The list is known as the European Waste Catalogue (EWC). Council Decision 94/904/EC then identified which wastes on the EWC were deemed to be hazardous, resulting in the Hazardous Waste List (HWL). Finally, the EWC 1994 and HWL were updated and combined and resulted in the European Waste Catalogue (EWC 2002). EWC 2002 catalogues all waste types according to generic industry, process or waste type. EWC 2002 differentiates between non-hazardous and hazardous wastes by identifying hazardous waste entries with an asterisk.

Definition of Hazardous Waste

The Hazardous Waste (England and Wales) Regulations 2005 define hazardous waste as:

- any waste listed as hazardous in the List of Waste (England) Regulations 2005;
- any specific batch of waste that the Secretary of State determines is exceptionally to be classified as hazardous;
- any specific batch of waste produced in Wales, Scotland or Northern Ireland that the Welsh Assembly Government, the Scottish Executive, or the Northern Ireland Department of the Environment determines is exceptionally to be classified as hazardous.

The Secretary of State can also make additional types of waste as hazardous by virtue of making regulations under Section 62A of the Environmental Protection Act 1990. The Environment Agency has published technical guidance – WM2, Hazardous Waste, Interpretation of the definition and classification of hazardous waste (second edition, version 2.1). This Technical Guidance on hazardous waste has a similar purpose to WM1 Special Wastes: A technical guidance note on their definition and classification. This

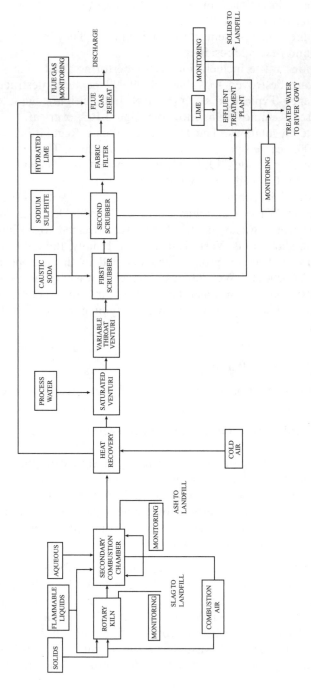

Figure 102 Hazardous waste incineration flow diagram for Cleanaway Ellesmere Port Plant (Courtesy of Cleanaway Limited).

document defines hazardous waste for regimes that refer to hazardous waste. WM1 will continue to be used to provide Guidance on the assessment of waste according to the criteria contained in the Special Waste Regulations as appropriate in England, Wales, Scotland and Northern Ireland. It is intended as a reference document for use by the waste management industry, producers and regulators of hazardous waste. This Technical guidance has been produced by the Environment Agency, SEPA and the Environment and Heritage Service. In this document, they are known collectively as 'the Agencies'.

Hazardous Properties

Hazardous Properties are defined in HWD Annex 111 as follows

H1	Explosive
H2	Oxidising
H3A	Highly Flammable
H3B	Flammable
H4	Irritant
H5	Harmful
H6	Toxic
H7	Carcinogenic
H8	Corrosive
H9	Infectious
H10	Toxic for reproduction
H11	Mutagenic
H12	Substances and preparations, which release toxic or very toxic gases in contact with water, air or an acid.
H13	Substances and preparations capable by any means, after disposal, of yielding another substance, e.g. a leachate, which possesses any of the characteristics listed above.
H14	Ecotoxic

The Hazardous Waste Regulations do not apply to domestic waste (waste from accommodations used purely for living purposes) via normal mixed domestic refuse collections. (◊ HOUSEHOLD HAZARDOUS WASTE)

Haze. Dust, salt or smoke particles often suspended in water droplets in the atmosphere. Haze can also be caused by substances called terpenes given off naturally by vegetation and forests.

Hazen number. A unit of measurement for colour in water. (Based on the colour produced by 1 mg platinum per litre in the presence of a cobalt-based compound.)

Head. The total height of water against which a pump has to work. Comprises the sum of the static head plus friction and velocity heads.

Health and Safety Commission (HSC). The Health and Safety Commission is a statutory body, established under the Health and Safety at Work etc. Act 1974, responsible for health and safety regulation in Great Britain.

Health and Safety Executive (HSE). The Health and Safety Executive is a statutory body, established under the Health and Safety at Work etc. Act 1974. It is an enforcing authority working in support of the HSC. Local authorities are also enforcing authorities under the Health and Safety at Work etc. Act 1974.

Health effects of air pollution. BENZENE and 1,3-butadiene have been identified as genotoxic carcinogens.

Exposure to low concentrations of CARBON MONOXIDE has been shown to be associated with adverse effects on the cardiovascular system.

Lead is a recognized neurotoxin and exposure may retard intellectual development in children.

For nitrogen dioxide, there is some evidence that raised levels may be associated with increases in respiratory hospital admissions and daily deaths, although separating the effects of nitrogen dioxide from those of co-pollutants, especially particles, is difficult. However, nitrogen dioxide contributes to the formulation of ozone and it is accepted that ozone has clear effects on health in terms of both morbidity (hospital admissions) and daily mortality. OZONE also damages materials, crops and other ecosystems. Abatement of nitrogen dioxide and SULPHUR DIOXIDE will have considerable beneficial impact on acidification.

Particles have been associated with increased respiratory hospital admissions and changes in daily mortality.

Sulphur dioxide causes bronchoconstriction and has been shown to be associated with increased respiratory hospital admissions and changes in daily mortality. There is evidence for adverse impacts upon other receptors, including reduction of crop yield, corrosion of building materials, and damage to forests, fisheries and natural ecosystems. Reductions in sulphur dioxide should have significant benefits for human health. Sulphur dioxide is a major contributor to ACID RAIN.

Hearing. The ear responds in varying degrees to sound of different frequencies. The range of frequencies it can encompass is roughly from 20 to 20 000 hertz (Hz).

For a given sound pressure level a high frequency sound appears to be stronger than a low frequency one. The loudness of sounds as judged by the ear depends on sound pressure level and frequency. Thus the measurement of sound is weighted in a way similar to the frequency response of the ear and the standard weighting used is called the A-scale. Meter readings on such A-weighting are called sound levels measured in decibels on the A-scale or dB(A). (◊ DECIBEL)

There is only one important source for which dB(A) are not normally used, and that is aircraft. The sound generated by aircraft engines has predominant components in particular frequency bands, and these can have a significant *subjective* effect for which A-weighted sound levels do not adequately account. Measurements are therefore made (in decibels) in each of a number of restricted frequency bands: from these a total level is calculated, giving due emphasis to the predominant components. The calculated total is called the 'perceived noise level', the units of which are PNdB. The precise numerical difference between PNdB and dB(A) varies depending on which frequency bands contain the predominant sound. The bands differ with different types of aircraft engine. However, values of PNdB are higher than values of dB(A) would be for the same sound, and as a rough guide a difference of 13 can be assumed. Certain types of new aircraft are now subject to noise certification before they come into service. For this purpose the perceived noise level is adjusted to take account of any pure tones (single frequency notes) in the noise, and of the length of time for which the higher noise levels are experienced. The result is termed the 'effective perceived noise level', the units of which are EPNdB. (◊ NOISE; SOUND ROAD TRAFFIC NOISE; AIRCRAFT NOISE; INDUSTRIAL NOISE MEASUREMENT; NOISE INDICES)

Hearing loss. A person's hearing is checked with an audiometer. The audiometer, in fact, compares the threshold of hearing of the subject under test with that of a normally hearing individual. Sounds of different frequency (usually 0.5, 1, 2, 3, 4 and 6 kHz) are produced and the subject of the test is asked to pinpoint the intensities at which the sounds are just audible. So if the audiometer records a hearing level of 30 dB at 2000 Hz, the threshold of the individual being tested differs by 30 dB from a statistical 0, which is held to represent normal hearing at that frequency.

Results of an audiometric test are recorded on an audiogram which indicates the hearing loss. Figure 103 shows audiograms indicating hearing losses typical of workers in the weaving trade after various durations of exposure to the noise of the looms. (Note the typical initial damage at about 4000 Hz and the way the damage extends over a wider frequency range as exposure continues.)

Heat. The classic definition is that the heat (J) contained by a body is the product of its mass (kg) multiplied by both its specific heat capacity ($J\,kg^{-1}\,K^{-1}$) and its thermodynamic temperature (K).

Latent heat is the thermal energy required to change the state (solid, liquid or gas) of a substance, e.g. water to steam at constant pressure. It is specified in $kJ\,kg^{-1}$.

Sensible heat is the product of the specific heat capacity, the mass of the 'body' (e.g. boiler water) and the change of temperature from 0 °C.

Specific heat is the quantity of heat (thermal energy, J) which is required to raise the temperature of unit mass (kg) of a substance by 1 kelvin (K).

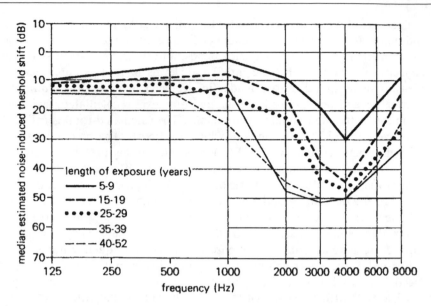

Figure 103 Noise-induced hearing loss as a function of years of exposure for weavers. Note that the initial loss is at 4 kHz.

The specific heat of water at 1 atmospheric pressure is approximately $4.2 \, kJ \, kg^{-1} \, K^{-1}$ (density $1 \, g \, cm^{-3}$).

Heat capacity. When the temperature of a body is increased by 1 K, the ENERGY (J) required is said to be the heat capacity. Specific heat capacity refers to unit mass, i.e. heat capacity per unit mass ($J \, kg^{-1} \, K^{-1}$).

Heat island. ⟡ TEMPERATURE INVERSION.

Heat : Power Ratio (HPR). For condensers, heat and power level assessments, the output heat: power ratio and the TEMPERATURE level at which heat is recovered are key parameters. HPRs can be achieved up to 10:1 for a steam turbine with supplementary firing.

Heat pump. A device for pumping heat from a low temperature source, e.g. the surrounding ground or a stream, and delivering it at a higher temperature; in other words, a refrigerator in reverse. Heat pumps require a mechanical or electrical energy input and therefore the performance coefficient or ratio of heat produced to electricity consumed is very important. As explained, under the Second Law of Thermodynamics, the conversion of heat to electricity in most power stations is about 3 heat units for one of electricity, so to obtain wide applicability heat pumps must have COEFFICIENTS OF PER-FORMANCE (COP) greater than 3, otherwise they would consume a greater heat equivalent in electricity than they were producing. Theoretically, heat pumps can have cop. of 8–10, but practically these are much less. (◊ LAWS OF THERMODYNAMICS)

Heavy metals. MERCURY, LEAD, CADMIUM, ZINC, CHROMIUM and PLUTONIUM are among the so-called heavy metals – those with a high atomic mass. (The term is rather loose and is taken by some to include arsenic, BERYLLIUM and SELENIUM, which are not really metals, in addition to those listed above.) Some of the above, if ingested in minute quantities, can be harmful. In the case of plutonium a few micrograms can be lethal. With the exception of plutonium, the others are in common use in industrial processes and therefore may be discharged to the environment to be mobilized by the action of air and water and concentrated by the action of living organisms and, as in the classic case of MINAMATA DISEASE, can cause deformity and death.

As some of these elements are neither heavy nor true metals, the term toxic metals is sometimes preferred.

Because of the tendency of heavy metals to accumulate in selective organs such as the brain or liver, average intake estimates are of little use. The dose to each individual organ must be considered separately as is the case with IONIZING RADIATION, because some parts of the body are more vulnerable than others. (◊ HALF-LIFE)

Ecosystems can be affected by their discharge, e.g. marine and freshwater diatoms can lose 50 per cent of their growth rate at concentrations as low as 1 part per thousand million (10^9) of mercuric compounds.

The damage that can be caused by heavy metals entering the food chain is illustrated by the November 1989 UK cattle feed incident in which ca. 1000 farms had their milk cattle quarantined because of cattle feed pellets which had become contaminated with lead, after contact with lead sulphate during transport, resulting in contaminated milk production.

Hepatitis. A disease that can be caught by drinking sewage-contaminated water, and which causes liver inflammation. (◊ PUBLIC HEALTH)

Her Majesty's Inspectorate of Pollution ((HMIP) now subsumed by the Environment Agency.

Herbicides. Also known as defoliants. Chemicals used for the elimination of unwanted plant or the total elimination of all plant growth. They are extensively used in agriculture, ranging from selective control of couch grass or wild oats, to ground clearance of overgrown areas. Other uses are in roadside maintenance, railway track clearance, etc. There are many chemical classes of herbicide. Of the phenoxyaliphatic acids, 2,4-D and 2,4,5-T are common, and cause changes in plant metabolism. A mixture of these two herbicides was used as Agent Orange, of which certain samples were found to be contaminated with DIOXIN. Other important herbicides include the substituted ureas (e.g. diuron) and the heterocyclic nitrogens (e.g. the triazines such as simazine), all of which interfere with photosynthesis. Other classes include carbamates, thiocarbamates, etc.

While the direct effect on animals may be minimal, herbicides have enormous ecological implications by destroying food sources, habitats, etc. Their use is yet another example of humans changing ECOSYSTEMS to suit themselves. The use of defoliants, as in Vietnam or in jungle clearance for agriculture, can permanently destroy tropical forests. Once the tree cover is removed, the soil is subjected to EROSION and precious nutrients are rapidly leached away. (◊ LEACHING)

Herbivore. Organism that primarily eats plant material (e.g. the water vole, *Arvicola terrestris*).

Herring catch. ◊ MAXIMUM SUSTAINABLE YIELD.

Hertz (Hz). SI unit of frequency, i.e. number of occurrences or cycles per second, as used in the frequency of radio waves, sounds, vibrations, etc. (◊ NOISE INDUCED HEARING LOSS)

Heterotrophic organism. An organism that requires organic material (food) from the environment, i.e. all animals, fungi, yeasts, and most bacteria are heterotrophic. For their food supply these organisms eventually rely on the activities of the AUTOTROPHIC ORGANISMS.

Hexa-chrome. ◊ CHROME WASTE.

High rate filter. A BIOLOGICAL FILTER which operates at a loading in excess of 3 cubic metres of effluent per cubic metre of bed per day.

High volume sampling. A means of measurement of airborne PARTICULATE matter in which large samples of air are sucked through appropriate filters and volumetric or flow metering equipment. This provides samples for subsequent analysis.

Hot rocks. Another form of renewable energy which claims that the thermal energy under the Earth's crust can be used. This was attempted in Falmouth UK in the 1980s when $2 \times 5\,km$ boreholes were sunk and water pumped in for heating and subsequent power generation. The temperature at the 5 km depth was ca. 100 °C and 'the real problem was that a substantial proportion of the water pumped down was lost – and did not return as hot water'.

Source: Francis Otway, Fellow I Mech E, Letter to Professional Engineering, 26 July 2006, Vol. 19, No. 14, p19.

(Follow up). There are claims that deeper drilling would have yielded 200 °C water.

Source: I MechE. Professional Engineering Letters, 20 September 2006.

Household Hazardous Waste. Good practice guidance states the minimum service provision for hazardous waste facilities at Household Waste Recycling Sites as follows:

'All Household Waste Recycling Sites (HWRSs) should have facilities for the more common types of Household Hazardous Waste, including hazardous waste electrical equipment (cathode ray tube televisions and/or computer monitors, fridges and fluorescent tubes), gas bottles, automotive batteries, engine oil and household batteries.'

'At least one HWRS in any Waste Disposal Area should provide facilities for asbestos and for household and garden chemicals.'

Source: The Haz Guide, National Household Hazardous Waste Forum, http://www.nhhwf.org.uk/.

Case study: West Sussex County Council

West Sussex County Council (WSCC) has a network of 11 Household Waste Recycling Sites (HWRSs) and five Transfer Stations, and provides facilities for collection of a wide range of hazardous household waste at all of these sites, including smaller sites in rural areas. All sites have a security system installed, comprising CCTV with a number plate recognition system and a trained security officer.

WSCC has contracted Viridor Waste Management to operate its municipal waste services for a period of 25 years. This contract, which started in April 2004, includes the delivery of waste reception services at HWRSs. Viridor, WSCC and the District and Borough Councils in the county have formed a waste management partnership and work together under the banner of West Sussex Reclaim. The seven Waste Collection Authorities in West Sussex are Adun District Council, Arun District Council, Chichester District Council, Crawley Borough Council, Horsham District Council, Mid-Sussex District Council and Worthing Borough Council.

There are currently 11 HWRSs in West Sussex, serving approximately 750 000 residents. The HWRSs vary from relatively large urban sites with total waste throughputs of more than 20 000 tonnes, to smaller, more rural sites. An extensive renovation programme of HWRSs in West Sussex is being undertaken and is currently nearing completion.

All sites in West Sussex have facilities for the following hazardous materials:

Asbestos (facilities available at four sites)
Automotive batteries
Household batteries
Mineral oil
Gas bottles
Fluorescent tubes and energy efficient light-bulbs
Household and garden chemicals
Fridge/ freezers
TVs/ Monitors

Hazardous Household Waste Arisings in West Sussex

Haz item	2005/ 06		2006/07	
	Weight (tonnes)	Percentage	Weight (tonnes)	Percentage
Asbestos	253.26	7.57	229.53	5.64
Hazardous chemicals	3.93	0.12	10	0.25
Car batteries	473.39	14.15	496.79	12.20
Household Batteries	5.09	0.15	14.22	0.35
TVs and Monitors	923.79	27.62	1539.66	37.81
Fridges/ freezers	1480.81	44.27	1582.91	38.87
Fluorescent tubes	6.17	0.18	5.07	0.12
Gas bottles	59.28	1.77	46.88	1.15
Mineral Oil	139.52	4.17	146.82	3.61
Total	3345.24	100.00	4071.88	100.00

	2005/06	2006/07
Total HWRS Arising (tonnes)	150 957	151 154
Total hazardous waste %	2.21	2.69
Total collected household waste in the County (tonnes)	415 508	414 765
Total hazardous waste %	0.80	0.98

Source: Annual BVPIs\Arisings\2006–2007\HWRS Arisings 2006–2007.
Please note figures for 2006/07 are based on 11 months' actual data, March data based on esti-mate/rolling figure.

Case study: The Life Cycle of a TV in West Sussex

Under the new hazardous waste regulations, from 16 July 2005, TVs and monitors became classed as hazardous waste. If such items are delivered to a Household Waste Recycling Site (HWRS) and presented separately then they should be segregated from other waste.

Reception/Collection Arrangements

First, the householder brings their TV/monitor to the HWRS. The site has the relevant licensing in place to accept hazardous waste and is registered with the Environment Agency as a hazardous waste producer (under the Hazardous Waste (England and Wales) Regulations 2005, Part 5 – Notifica-tion of Premises).

Secondly, the site operative directs the householder to the designated col-lection container for TVs and monitors.

Finally, the TVs/monitors are collected by a registered waste carrier (under the Environmental Protection Act 1990 – Section 34) and the load is consigned as a hazardous waste (under the Hazardous Waste (England and Wales) Regulations 2005, Part 6, Section 35).

Reprocessing

The TVs are transferred to an appropriately licensed Cathode Ray Tube reprocessor. The non-hazardous components of the TV/monitor are dismantled into its constituent components – shown in the table below: The hazardous component of the TV/monitor – the Cathode Ray Tube – is then left. This is then split to separate the two types of glass:
    Funnel (hazardous) glass (containing heavy metals such as phosphorus).
    Screen/Panel (non-hazardous) glass.

Major Component Parts of a TV

Output material	End product
Plastic – cases and base units	Plastic goods, e.g. garden furniture
Ferrous and non ferrous metal	Sheet metal used in manufacturing
Circuit boards	Metal and plastic
Screen and funnel glass	New CRT glass
Wood/rubber/paper/phosphors	Landfill – material cannot be recycled

The recycling rate for a 32 kg TV is approximately 95 per cent.

Approximately 85 000 TVs and monitors were collected in West Sussex during the period April 2006–March 2007. (◊ HAZARDOUS WASTE REGULATIONS)

Household Waste Disposal Charges. The following extract of a letter by David Campbell, Fellow CIWM, published in the CIWM Journal of December 2006, illustrates one aspect of this issue:

'If personal/household charges are to be levied, the charge should cover all waste generated and reflect the overall costs for collection and for complete downstream processing, composting, recycling, thermal treatment or landfill. I do not understand why charging for residual non-recyclable waste alone is required to encourage more and more recycling, nor how this would reduce waste in our 'throwaway society'. Where it is non-economic to recycle separated combustible waste the materials should be collected and combusted with the energy generated recovered and residues re-used where possible.

I have worked in several countries that include 'zero waste' policies within their national waste management legislation but all I saw was mountains of dumped wastes. So can someone please define clearly what is meant by 'a zero waste UK' because I suspect it does not mean what is implies, as might be interpreted by the vast majority of the general public, and we must not mislead them'.

Hub height. The height of a wind turbine tower from the ground to the centre-line of the turbine rotor.

Humic acid. Omnibus term for a wide range of organic compounds resulting from the decay of vegetable matter such as peat. They make water acidic, frothy, and impart colour.

Humus. Biologically stable dead organic material resulting from aerobic decomposition of sewage and other organic matter. The stable, dark-coloured organic material that accumulates as a by-product of decomposition of plant or animal residues added to soil. The term is often used synonymously with soil organic matter. (◊ SLUDGE, SEWAGE)

Hydrate. In chemical terms, SALTs which contain water of crystallization. For the oil and gas industry purposes, METHANE hydrates are formed when water and methane are in contact at high pressure and temperature $0–300\,°C$. The gas is effectively trapped in the water molecules. $1\,m^3$ of methane hydrate can provide ca $170\,m^3$ methane (at NTP) when dissociation takes place. The dissociation can be by depressurization, heating, salt injection, or possibly CO_2 injection. Care must be taken to ensure that Methane (with a greenhouse warming potential 62 times that of CO_2, over a 20-year time scale) is not released thereby nullifying any CO_2 storage benefits. One estimate is that there are 2830–8500 trillion m^3 hydrate deposits.

Source: B. Sampson, Professional Engineering, 13 June 2007, p36.

Hydrated lime. Alternative description to SLAKED LIME, for the chemical CALCIUM HYDROXIDE $Ca(OH)_2$.

Hydrated lime is used extensively in the chemical and water treatment industries for process chemistry, water pH control, effluent treatment, acid and gas scrubbing. etc. (◊ SCRUBBER)

Hydraulic conductivity (permeability). The ease with which water moves through an aquifer, i.e. the rate of flow of GROUND WATER through unit cross-sectional area of an aquifer under unit HYDRAULIC GRADIENT. It can be measured in feet per second, centimetres per second or inches per day. Commonly, hydraulic conductivity is measured in metres per day. Hydraulic conductivity is usually denoted as k and possesses a wide range in values from clays ($k = 10^{-5}$ to 10^{-7} metre per day) to gravels ($k = 1$ to 10^{-4} metre per day or more). Thus, a LANDFILL SITE lined with clay will mainly contain LEACHATE whereas one on gravel will allow virtually instant dispersal of any pollutants. The EC Landfill Directive posits a maximum hydraulic conductivity (or permeability coefficient) of 1×10^{-9} metres per second for new landfill sites.

Hydraulic gradient. Hydraulic gradient is the difference in hydraulic head divided by the distance along the fluid flow path. Ground water moves through an AQUIFER in the direction of the hydraulic gradient.

Hydro-power (small-scale). Hydro-power is one of the oldest methods of harnessing renewable energy. The first water-wheels were used for irrigation over 2000 years ago. Hydro-power systems convert potential energy stored in water held at height to kinetic energy to turn a turbine to produce electricity.

Micro-hydro plants below 100 kW are available. Useful power may be produced from even a small stream. The likely range is from a few hundred watts (possibly for use with batteries) for domestic schemes, to a minimum 25 kW for commercial schemes. Hydro power requires the source to be relatively close to where the power will be used, or to a suitable grid connection. Hydro systems can be connected to the main electricity grid or as a part of a stand-alone (off-grid) power system. In a grid-connected system, any electricity generated but not used can be sold to electricity companies. In an off-grid hydro system, electricity can be supplied directly to the devices powered or through a battery bank and inverter set up. A back-up power system may be needed to compensate for seasonal variations in water flow.

Provided the resource is there, community hydro projects can also be a viable proposition. Energy available in a body of water depends on the water's flow rate (per second) and the height (or head) that the water falls. The scheme's actual output will depend on how efficiently it converts the power of the water into electrical power (maximum efficiencies of over 90 per cent are possible but for small systems 50 per cent is more realistic). A small turbine on a hill stream with a flow of say, 15 litres per second, and a head of 15 m, will generate about 1 kW, enough to meet the basic needs of a house. Turbines can have visual impact and produce some noise, but these can be mitigated relatively easily. The main issue is to maintain the river's ecology by restricting the proportion of the total flow diverted through the turbine.

Hydrocarbon. An organic compound containing the elements CARBON and HYDROGEN. The carbon ATOMS may be arranged either in open-ended chains which may or may not be branched, or in closed rings. Some may also have minor or trace quantities of oxygen, nitrogen, sulphur and other elements. There are many groups of different hydrocarbon compounds but the three commonest in the FOSSIL FUELS are the paraffins, naphthenes and benzene-type compounds. Some are solid, others liquids or gases.

Hydrochlorofluorocarbons (HCFCs). Used as a replacement for CFCs in refrigeration, foam blowing and aerosols. OZONE DEPLETION POTENTIAL relative to CFCs is 0.02–0.1. ICI is concentrating on HFA-134a (CF_3CH_2F) as a replacement for CFC-12 (CF_2Cl_2), and on HCFC-123 (CF_3CHCl_2), as a replacement for CFC-11 ($CFCl_3$).

Hydroelectricity. ◊ WATER POWER.

Hydrogen (H_2). Colourless, odourless gas, normally existing as the molecule H_2. It is the lightest substance in existence, relative atomic mass 1. It is extremely flammable and when burned combines with oxygen to form water with heat liberated in the process.

It can be readily converted to liquid fuels, e.g. METHANOL, AMMONIA, hydrazine, and can of course be pumped readily by pipeline. These attributes have made many people propose the 'hydrogen fuel economy' as an alternative source of energy when conventional energy resources are no longer available. The hydrogen would be manufactured electrolytically or by hydrogenation of remaining coal supplies. Currently: 90% of today's hydrogen is manufactured by 'reforming' natural gas, CH_4, which in addition to yielding hydrogen, also produces CO_2 which is capable of sequestration. The hydrogen fuel economy is based on the premise that large quantities of NUCLEAR ENERGY will be available for the generation of the necessary power for the ELECTROLYSIS and the ready availability of hydrogen-absorbing metallic matrices (as stores) which release it when warmed up.

Hydrogen is increasingly viewed as a low pollution energy source via FUEL CELLS. However, gaseous H_2 occupies roughly four times the volume of the equivalent petrol tank, hence currently, storage under very high pressures (10, 000 psi) for use in motor vehicles and safety must be paramount. Still no free lunch (yet?).

Hydrogenation. The chemical combination of HYDROGEN with another substance, usually by the action of heat and pressure in the presence of a catalyst, e.g. the hydrogenation of coal combines its CARBON with hydrogen to produce HYDROCARBON.

Hydrogen chloride (HCl). A colourless gas with a choking odour, highly soluble in water to form hydrochloric acid. Both the gas and its solution are poisonous, corrosive, and phytotoxic (although much less injurious to plants than hydrogen fluoride). (◊ ACID)

Hydrogen fluoride (HF). A colourless FUMING gas or liquid (boiling point 19.5 °C). The aqueous solution is known as hydrofluoric acid. Both are highly corrosive and toxic. Many salts of hydrofluoric acid (fluorides) are also highly poisonous. (◊ FLUORINE; FLUORIDES)

Hydrogen ion concentration. ◊ pH.

Hydrogen peroxide (H_2O_2). Powerful disinfectant and bleaching agent which rapidly gives off OXYGEN. Measured in volume strength, i.e. 20 volume H_2O_2 will give off 20 times its volume of oxygen.

Hydrogen sulphide (H_2S). Dense colourless gas with a smell of rotten eggs. It is extremely toxic. It is produced under anaerobic decay conditions and can accumulate in sewers. This has accounted for several fatalities. It is also produced in industrial processes such as oil refining, chemical manufacture and wood pulp processing. Gas absorption is often employed to effect removal. (◊ TOXIC WASTES)

Hydrogeology. The study of subsurface water with the emphasis on direction of flow, interconnections and interactions with the surrounding strata. A hydrogeological survey is a must before a LANDFILL for wastes can be constructed.

Hydrographs. Graphical representations of either surface stream discharges or water level fluctuations in wells. For example, a stream hydrograph portrays the characteristics of the flow of a stream. Discharge is plotted vertically and time horizontally.

   Hydrographs of water levels in wells can be computed from measurements of water depth below ground surface. Long-term records may allow estimates of the ultimate yield of the aquifer and the rate of its replenishment.

Hydrological cycle. The means by which water is circulated in the biosphere. Evapo-transpiration from the land mass plus evaporation from the ocean is counterbalanced by cooling in the atmosphere and precipitation over both land and oceans (see Figure 104). The hydrological cycle requires that on a world-wide basis the evaporation and precipitation are equal. However, oceanic evaporation is greater than oceanic precipitation, thus an excess precipitation is given to the land. Eventually this land precipitation ends up in lakes and rivers and thus eventually returns to the sea, so completing the CYCLE. (◊ TRANSPIRATION)

   Humans intercept the land precipitation by means of RESERVOIRS or river and ground-water abstraction, and so obtain water supplies, but after use this abstracted water still ends up in the sea – its arrival there is merely

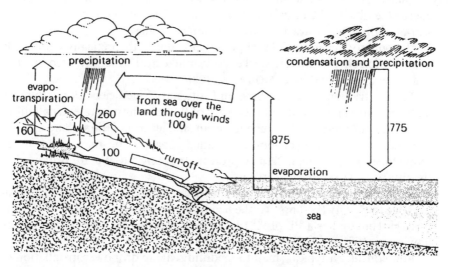

Figure 104   Hydrological cycle in cubic kilometres per day.

delayed. Water is also required for PHOTOSYNTHESIS but the fraction of water so used compared with that which is transpired by green plants is less than 1 per cent.

Hydrology. The science concerned with the occurrence and circulation of water in all its phases and modes and the relationship of these to humans.

Hydrolysis. The breaking down of a substance by interaction with water. The hydrolysis of carbohydrates, e.g. conversion of starches and sugars, is of major importance in the food and brewing industries. Carbohydrates may be hydrolysed biologically with the aid of ENZYME TECHNOLOGY or chemically at high temperatures with an acid or alkali present as a catalyst. The hydrolysis of CELLULOSE produces fermentable sugars which can be processed by fermentation to yield ethyl alcohol, butanol, acetic acid or PROTEIN. Hydrolysis has been proposed as a refuse disposal process and is a means of using renewable resources to obtain organic molecules such as ethanol, one of the building blocks of the chemical industry. It could also be used as a source of ethene, e.g. for PVC manufacture. Protein production is economically feasible. Laboratory work is proceeding in both the UK and the USA on the processing of all classes of cellulosic materials by both biological and chemical means.

Hydrolysis can also be the reaction between water and a compound (commonly a salt or mineral) in which the hydroxyl from the water combines with the cation from the compound undergoing hydrolysis to form a base; the hydrogen ion from the water combines with the anion from the compound to form an acid. In this way, metal cations within mineral structures are replaced by hydrogen ions in soil water.

Hydropower. A form of RENEWABLE ENERGY which relies on the potential energy (i.e. due to the head or potential difference) of impounded water in a storage dam which is used to power a water turbine.

The potential energy of the impounded water is theoretically completely available to convert into electrical energy, unlike heat which is not completely available. ($\Diamond$ LAWS OF THERMODYNAMICS)

Pipeline friction losses, turbine and alternator inefficiencies combine to make overall EFFICIENCIES of the order of 90 per cent for large installations (>5 MWe) but these can be much lower in smaller installations. The UK has a small (<3 per cent) hydropower installed capacity and there is not much scope for more, although 22 new projects totalling 9 MWe are proposed in the fifth NON-FOSSIL FUEL ORDER.

The potential energy (PE) in joules is the mass in kilograms × the acceleration due to gravity ($9.81 \, m \, s^{-2}$) × the height in metres, e.g. 1 kg of water falling through 1 metre has a PE of 9.81 J. If the water flows at $1 \, kg \, s^{-1}$, the energy flow rate is $9.81 \, J \, s^{-1}$ or 9.81 watts.

Now consider a typical small to medium scale hydropower installation of $5000 \, kg \, s^{-1}$ ($5 \, m^3 \, s^{-1}$) falling through 30 metres and assuming 100 per cent efficiency. The power output is $5000 \times 9.81 \times 30\,W = 1\,500\,000\,W$ or 1.5 MW.

As a comparison, UK National Power, Eggborough, power station has four steam turbine turbo alternator sets, each rated at 500 MWe output. Our 'medium' hydropower station would need to be replicated 1333 times.

Hydropower in the UK is very much a local opportunity which can supply isolated communities very effectively.

Hydrophilic. Literally 'water-loving'; describes the chemical groups that tend to make molecules soluble in water (e.g. $-OH$, $-NH_2$).

Hydrophobic. Literally 'water-hating'; describes the chemical groups that tend to make molecules insoluble in water (e.g. $-CH_3$, $-C_6H_5$).

Hydropulper. A wet pulping device for converting dry pulps and waste papers into a fibrous slurry by the addition of water. It is used in the recycling of PAPER and is the basis of a recycling process for DOMESTIC REFUSE. The pulp recovered from domestic refuse is of a very low quality and so far has little commercial value except for low-grade uses such as roofing felt where the pulp is subsequently tarred or coated to make it waterproof.

Hydrosphere[1]. The portion of the earth's crust covered by the oceans, seas and lakes. When the complete earth's crust is meant, it is often referred to as the hydro-lithosphere. (⟡ LITHOSPHERE)

Hydrosphere[2]. The body of (liquid) water on or near the surface of the earth. The oceans are the largest single component in the hydrosphere, cover over 70 per cent of the earth's surface and have an average depth of nearly 2.5 miles.

Hygiene Assessment System scores. The Hygiene Assessment System (HAS) was developed as a management rather than a consumer tool. HAS is a risk-based method of assessing hygiene standards at licensed abattoirs and cutting plants. The Official Veterinary Surgeon (OVS), a vet who works for the Meat Hygiene Service but who is based at the meat plant, looks at all aspects of production, weighted according to how important they are to hygienic production of fresh meat. An assessment produces an HAS score out of 100, but the score is not a percentage.

HAS scores are not required by law. However, meat plants have to be assessed under the hygiene regulations to make sure they are operating satisfactorily. HAS is one of the tools that is used for this task.

Like all assessment methods that require an individual to use his or her professional judgement, HAS contains an element of subjectivity. Measures have been taken to reduce its effects as far as possible. These include:

- extensive training of inspection staff;
- external audit by the Veterinary Public Health Unit of the Joint Food Safety and Standards Group in England and the State Veterinary Service in Scotland and Wales;

- a review system to allow plant operators to query the basis for a particular score;
- publication of scores wherever possible, as a 3-month rolling average;
- introduction of an advisory panel to standardize guidance and scoring.

However, it is impossible to remove the effects of subjectivity altogether. This is another reason why the scores should not be treated as a 'league table': a plant scored slightly higher by one OVS may be scored slightly lower by another. Scores ranged from 98 to 53 (November 98 to January 99, 1361 plants).

Source: *HAS Scores*, Meat Hygiene Enforcement Report, MAFF Issue 23, March 1999.

Hyperactivity (associated with behavioural disturbances). Overactiveness, especially in young children and youths. Many experts believe LEAD poisoning to be a major cause of this syndrome in many cases and recommend 'de-leading' as a cure. (◊ CHELATING AGENTS)

Ideal gas law. One mole of an ideal gas has a volume of 22.4143 litres at a temperature of 0 °C and a pressure of 1 atmosphere (STP). For example, the molecular mass of air is 29, so 29 g of air has a volume of 22.4143 litres and its density is 29/22.4143 = 1.29 g/l at 0 °C and 1 atmosphere.

For many gases this approach works well. The density can be corrected for temperature and pressure using the ideal gas law: $pV = nRT$.

Physical properties of gases and symbols.

Property	Symbol	Unit
Density (gas)	ρ	kg/m^3
Density (liquid)	ρ	kg/m^3
Heat capacity (gas)	C_p	J/kg K
Thermal conductivity (gas)	λ	W/m K
Thermal conductivity (liquid)	λ	W/m K
Dynamic viscosity (gas)	μ	Pa s
Dynamic viscosity (liquid)	μ	Pa s
Kinematic viscosity (gas)	ν	m^2/s
Kinematic viscosity (liquid)	ν	m^2/s
Vapour pressure	P_{vp}	bar
Heat of vaporization	H_{vp}	J/mol
Normal boiling point	T_b	K

IGCC. ◊ INTEGRATED GASIFICATION COMBINED CYCLE.

Imhoff cone. A graduated glass cone (1 litre capacity) used in the laboratory for the measurement of settleable solids in a sewage effluent.

Improvement notice. A Statutory Notice that can be issued when informal action has not achieved results, and there remains a risk to public health (e.g. in meat processing). It specifies improvements to plant, procedures or equipment that need to be made to remove a risk to public health.

Improvement Notices are one of a range of means which enforcing authorities use to achieve the broad aim of dealing with serious risks, securing compliance with health and safety law, and preventing harm. An Improvement Notice allows time for the recipient to comply.

Incineration. A waste volumetric reduction process that relies on COMBUSTION under suitable controlled conditions to reduce the volume and/or mass of material for disposal. Incineration falls into two main classes: MUNICIPAL SOLID WASTE (MSW) and HAZARDOUS WASTE. Each has differing design requirements.

The two most important aspects of MSW as a FUEL are that it has a low CALORIFIC VALUE, typically 30–40 per cent of that of an industrial bituminous coal, and a density, as fired, of about 200kg/m^3 or 20 per cent of that of coal.

In order to illustrate the factors involved in MSW incineration a material analysis of 'as received' MSW, by weight, is given in table (a).

For fuel purposes, the ULTIMATE ANALYSIS by weight is given in table (b) with the gross calorific value plus the breakdown of moisture, combustibles and inerts on a weight/weight basis.

It should be noted that the composition of MSW varies seasonally with regard to the amount of vegetable matter and the moisture content respectively (also related to the amount of vegetable matter). Hence, calorific values can fluctuate by as much as ±6 per cent of the mean and due allowance should be made for this in well-regulated energy recovery incinerators.

Combustion air requirements. Assuming an excess air supply of 80 per cent to ensure complete combustion and the prevention of reducing conditions (which promote corrosion), an air supply of 5.14 kg air/kg MSW is needed. If it is anticipated that industrial waste will also be burnt in the incinerator, it is necessary to also know the composition and throughput and the effect of this industrial waste on the waste composition.

(a) Material analysis of sample of MSW by weight.

Material	Weight %
Dust and cinder	9.0
Vegetable matter	24.0
Paper and cardboard	31.0
Metals	7.8
Textiles	4.9
Glass	8.0
Plastics	11.0
Unclassified	4.7

(b) Ultimate analysis and gross calorific value of MSW.

Material	% By weight
Carbon	24
Hydrogen	3.2
Oxygen	15.9
Nitrogen	0.7
Sulphur	0.1
Water	31.2
Chlorine	0.7
Ash and inerts	24.2
Net calorific value as fired 10600 MJ/t	
Moisture	31.2 w/w
Combustibles	44.6 w/w
Inerts	24.2 w/w

Sources: A. Porteous, 'Why energy from waste incineration is an essential component of environmentally-responsible waste management', *Waste Management*, 25 (2005), pp. 451–459.

The peculiarities of MSW combustion are such that a secondary or overfire air supply is necessary to burn the volatiles released in the primary combustion stage. The underfire air both dries the refuse and supplies the air necessary for the primary or bed-based combustion. For calculation purposes, the working assumption can be made that the primary:secondary air ratio is 3:1. The gas, moisture and solid flows to be calculated per tonne of MSW are shown in Figure 105 (numbers rounded off for ease). ($ CALORIFIC VALUE; ULTIMATE ANALYSIS; MSW; INCINERATION PERFORMANCE; INCINERATION GRATES)

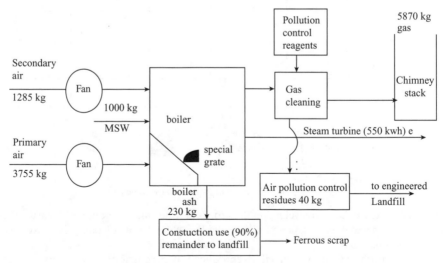

Figure 105 Schematic mass burn waste to energy plant.

HAZARDOUS WASTE INCINERATION requires features such as air locks through which the hazardous substances (e.g. PCBS) are loaded, supplementary fuel and perhaps a ROTARY KILN for thorough destruction of wastes. In addition stringent air pollution control measures are necessary, including gas scrubbing stages and ACTIVATED CARBON injection followed by filtration.

Incineration is now practised to the highest standards. Table (c) gives full performance and Waste Incineration Directive details.

Incineration of waste to energy can cope with any level of RECYCLING as set out in Figure 106. This cannot be emphasized enough.

Waste to energy is totally compatible with any level of recycling.

The disposal costs associated with incineration are typically greater than short haul landfill costs, but where there is a market for power and HEAT, as in Europe, costs approaching or bettering those of long haul landfill are possible, especially when landfill tax is taken into consideration. Table (d) gives outline cost data. Indeed, incineration may be the only MSW disposal option

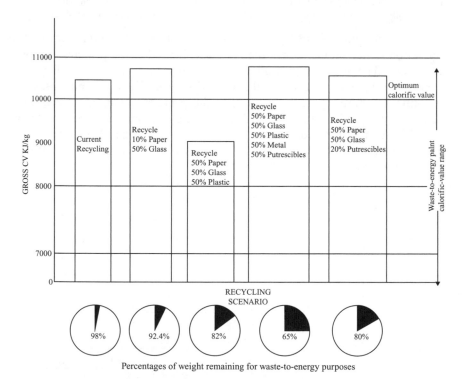

Percentages of weight remaining for waste-to-energy purposes

Figure 106 The effect of recycling on the gross calorific value of household waste (percentages are percentages of potential recyclables (e.g. as not all paper is recyclable: usually one-third is).

Source: Hampshire County Council. (Reproduced by permission)

(c) A comparison of UK (best practice) and European mean EfWI emissions and percentage improvement over UK 1991 performance.

Component	Emission to air in mg/Nm³ – dioxins in ng/Nm³ – dry gas 11% O₂			UK 1991 Mean emissions mg/Nm³	% Reduction $\frac{(4)-(1)}{(4)} \times 100$	Emission burden g/t 'Best' practice emissions g/te (based on (1) plus NO_x of 200 mg/Nm³ and HC1 of 10 mg/ Nm³ dioxins ng TEQ
	Measured (UK) (best practice)	European (mean)	Waste Incineration Directive			
	(1)	(2)	(3)	(4)	(5)	(6)
Particulates	0.9	2.2	10	500	99.8	4.95
HCl	20	1.6	10	689	97.1	55
HF	<0.1	0.03	1	N.A	—	0.55
SO_2	36	7.2	50	338	89	198
NO_x as NO_2	274	29	200 (plant >3 tph)	N.A	—	1100
CO	5	—	50	220	98	27.5
VOC	<5	—	10	NA	—	<27.5
Hg	<0.02	<0.001	0.05	0.26	99	0.11
Cd	<0.001	<0.001	0.05 (Cd and Ti)	0.6	99.8	0.0055
Σ7HM/Σ12 (heavy metal summation)	Σ7	Σ12	0.5	>11.0	99	<0.55
Dioxin I-TEQ ng/Nm³	0.006	0.16	0.1	>225	99.9	33 ng
NH_3	<0.1	<0.01	—	—	—	

Non-biogenic CO_2 132 kg

readily available due to the nature of a locality and the difficulty of having new landfill sites accepted by the public.

(d) MSW incineration costs: privately financed 220 000 tonnes per year installation. (2007 estd costs).

	Costs (£ million)	Disposal cost (£/tonne)
Capital cost (to full EC standards + site work)	135	
Operational costs	9.3	
Capital charges (12%, over 20 years' return on investment)	15.44	
Total cost **without** power generation	24.74	112.00
Income		
Electricity sales (550 kWh/t at 5 p/kWh)	6	
Net cost at 5 p/kWh$_e$		85.00
at 7 p/kWh$_e$ net cost	8.4	74.00

Comments: (i) Population 600 000. (ii) Design capacity 35 tonnes/hour. (iii) Up to 600 kWh/t may be available depending on MSW calorific value. (iv) Electricity tariff and return on investment crucial to disposal cost. (v) Landfill tax will shortly be £35/te. (vi) Recycling cost anything from £85 (paper) to £250 tonne for plastics.

Table (e) shows the percentage composition of exiting flue gases for a typical MSW incinerator. Note that 85% of the CO_2 is bioderived due to the organically derived reduction in waste. Table (f) gives the CO_2 reduction attainable with EfWI components with coal and gas power generation, after allowance is made for the CO_2 reduction from the bioderived portion of the MSW.

(e) Percentage composition of exiting flue gases.

Component	%
CO_2 contributes (85% bioderived)	12.3
Neutral H_2O contributes	8.37
Neutral O_2 contributes	10.3
Neutral N_2 contributes	68.4
NO_x	0.014
HCI (after clean-up)	0.001
Total (rounded)	100.00

Source: A. Porteous, 'Why energy from waste incineration is an essential component of environmentally-responsible waste management', *Waste Management*, 25 (2005), pp. 451–459.

(f) Typical CO_2 emissions from industrial boilers and power generation plant and EfWI CO_2 reduction comparison.

Coal-fired	410 g/kWh thermal (ca. 950 g CO_2/kWh electricity
Gas-fired	226 g/kWh thermal (ca. 525 g CO_2/kWh electricity
Combined cycle gas turbine (CCGT)	ca. 400 g CO_2/kWh electricity
MSW incineration net CO_2 (after bioderived CO_2 deduction)	ca. 264 g CO_2/kWh electricity
CO_2 saving achieved by EfWI electrical power generation is:	
Coal	(950–264) = 686 g CO_2/kWh electricity
Gas	(525–264) = 261 g CO_2/kWh electricity
CCGT	(400–264) = 136 g CO_2/kWh electricity [ca. 90% reduction in particulates compared with coal fired power generation is also achieved]

Source: A. Porteous, 'Why energy from waste incineration is an essential component of environmentally-responsible waste management', *Waste Management*, 25 (2005), pp. 451–459.

Incineration grates

The properties of crude unsorted municipal solid waste (MSW) are such that thorough controlled bed agitation is required. In addition, both undergrate and overfire air are necessary, as well as a reliable means of ash removal. One well-tried arrangement is given in the VKW roller grate which is installed worldwide.

Figure 107 shows the layout of a typical roller grate incinerator (along with allied boilers and gas cleaning plants in cross-section). It comprises a series of six hollow rollers which transport the MSW downwards through the furnace, whilst an angle of about 30° ensures sufficient agitation for complete combustion. Primary air is supplied to assist the combustion through holes in the grate. Each roller is about 1.5 m diameter and wide enough to carry the design throughput. The design of the grate segments permits expansion and contraction, and allows small particles to drop through whilst allowing efficient distribution of the primary underbed combustion air supply. The speed of rotation of each roller is separately controlled through a variable speed gearbox to ensure efficient combustion. The speed range varies from 5 revs/hour near the entry point to 0.5 rev/hour at the final roller, in order to match the changing waste densities. This system allows a high degree of combustion control as feed and combustion rates can be evenly balanced.

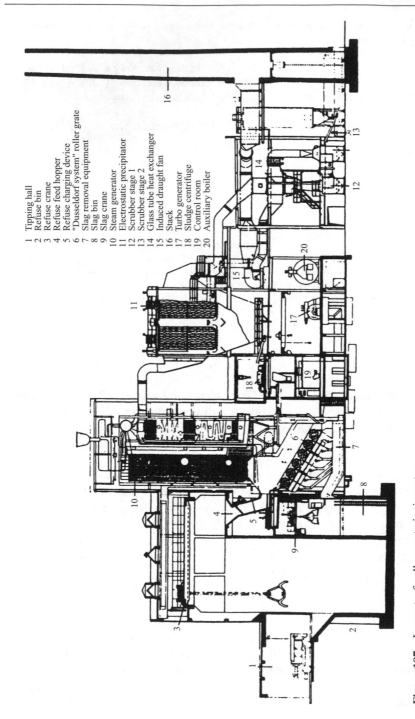

1 Tipping hall
2 Refuse bin
3 Refuse crane
4 Refuse feed hopper
5 Refuse charging device
6 "Dusseldorf system" roller grate
7 Slag removal equipment
8 Slag bin
9 Slag crane
10 Steam generator
11 Electrostatic precipitator
12 Scrubber stage 1
13 Scrubber stage 2
14 Glass tube heat exchanger
15 Induced draught fan
16 Stack
17 Turbo generator
18 Sludge centrifuge
19 Control room
20 Auxiliary boiler

Figure 107 Layout of roller grate incinerator.
Source: Motherwell Bridge Engineering, PO Box 4, Logans Road, Motherwell, ML1 3NP, Scotland.

Ash removal. Ash removal (see Figure 107) is usually effected by quenching the hot ash. This causes disintegration of the clinker and gives a granulated ash suitable for further processing and ferrous metals extraction. Incinerator ash is extremely abrasive and requires extremely robust handling systems.

Fluidized beds. Several types of fluidized bed grates have been developed for burning low calorific value fuel including lignite, wood waste, peat, pulverized MSW (with metals extracted) and others. The main operating principles are that particles of fuel which are usually required to be smaller than 25 mm are fed into a sand bed which is agitated either by means of primary air entering from beneath or by a combination of a mixing motion and primary air, such that the fuel is virtually floating in a bed of fluidized hot sand during combustion. This reduces corrosion and erosion problems and increases the efficiency of combustion. The sand bed is recycled for reuse and it is claimed that the sand can easily be cleaned after mixing with the fine particles of grit and dust resulting from the combustion of waste. Crushed limestone can also be added to the bed to reduce acid emissions. Rotary kilns (Figure 108) can also be used for difficult wastes, e.g. clinical or hazardous wastes.

Incineration performance

Waste reduction. Based on the MSW weight composition (see INCINERATION entry, Table (a)) incineration can achieve a 75 per cent weight reduction, and given that complete combustion has been achieved, an 85 per cent volume reduction (see Figure 109). However, the degree of BURN OUT must also be assessed and this involves measuring both the amount of fixed carbon and

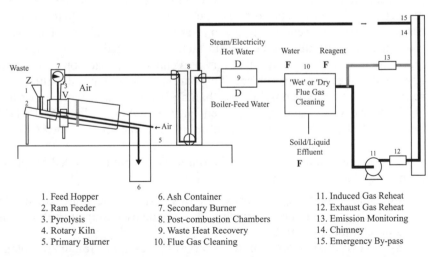

1. Feed Hopper
2. Ram Feeder
3. Pyrolysis
4. Rotary Kiln
5. Primary Burner
6. Ash Container
7. Secondary Burner
8. Post-combustion Chambers
9. Waste Heat Recovery
10. Flue Gas Cleaning
11. Induced Gas Reheat
12. Exhaust Gas Reheat
13. Emission Monitoring
14. Chimney
15. Emergency By-pass

Figure 108 Typical diagram of rotary kiln system (Courtesy: Motherwell Bridge Envirotec).

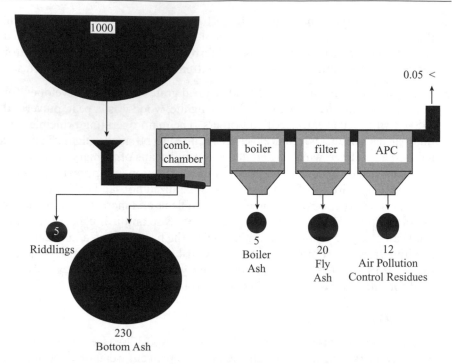

Figure 109   Streams of solid masses in a municipal solid waste incinerator with wet flue gas scrubbing (values in kg/Mg waste).

putrescible matter in the residue by chemical analysis and fermentation, respectively. The normal performance guarantees are no more than 3 per cent carbon and 0.03 per cent putrescible matter by weight in the ash. This means that the ash is virtually sterile and, in fact, is scavenged in some European plants for both ferrous metals recovery and clinker for road construction, or as a sand and gravel replacement. This considerably reduces the amount of residues for ultimate disposal to less than 5 per cent of the incoming waste.

The South East London Combined Heat and Power Plant (SELCHP), opened in 1994, is a typical modern installation. Table (g) gives the SELCHP emissions performance.

Waste incineration plants in England and Wales are controlled by public authorities from four main aspects:

(a) by the local planning authority (for this purpose, the county or metro-politan district council in England and the district council in Wales), as a development of land.

(b) by the ENVIRONMENT AGENCY (Part A processes) or for smaller plants (Part B processes) by the district council and other authorities, as a source of pollution.

(c) by the Health and Safety Executive (HSE) where there is a potential hazard to workers on the site and persons outside it.

(g) Details of the SELCHP plant.

SELCHP	420 000 tonnes per annum in 2 × 29 tonnes per hour refuse burning streams.
Storage capacity	4 days of full plant capacity – 5000 tonnes.
Number of tipping bays	11.
Steam output	144 tonnes of steam per hour at 395 °C and 46 bar with refuse net calorific value of 8500 kJ/kg.
Flue gas treatment	Each stream fitted with CNIM semi-dry lime scrubbers followed by high performance Bag House type filters, ejecting into a double flue 100 metre chimney.
Availability	Guaranteed 85% each refuse burning stream.
Operating staff	55 persons.
Site area	5.5 acres.
Design and construction cost	£85 million (1992 basis).
Combustion temperature	In excess of 850 °C with 2 seconds residence time for dioxin destruction.

(h) Comparison of current operational mean plant emissions with authorization limits for SELCHP, December 1996. (Note SELCHP is fully Waste Incineration Directive (WID) compliant. This shows the plant's rigorous standards well before WID implementation.)

Parameter	Concentration limit (mg/m^3)	Unit 1 (mg/Nm^3)	Unit 2 (mg/Nm^3)
Total particulate matter	20	1.4	1.4
Hydrogen chloride	30	13.7	9.3
Hydrogen fluoride	2	<0.01	<0.01
Sulphur dioxide	80	5.9	10.8
Nitrogen oxides[a]	450	385	358
Volatile organic compounds	20	<3	<3
Cadmium	0.1	0.001	0.001
Mercury	0.1	<0.008	0.003
Other metals total[b]	2.0	0.062	0.004
Carbon monoxide	80	9	9
Dioxins (ng/m^{-3}) (TEQ)	1.0	0.02	0.02

All concentrations are expressed at reference conditions of 273 K, 101.3 kPa, 11% O_2 dry gas.
[a] DENOX equipment now fitted.
[b] Other metals are: arsenic, chromium, copper, lead, manganese, nickel and tin taken together.
Source: G. Atkins, 'Incineration and Energy from Waste, The BPEO for MSW: The SELCHP experience', *Waste Planning*, No. 27, June 1998, pp. 25–27.

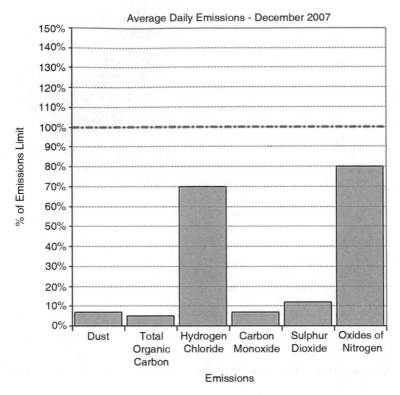

Figure 110   Average daily emissions from the Veolia Energy-from-Waste Plant, Sheffield, December 2007.
Source: www.veoliaenvironmentalservices.co.uk/sheffield/pages/emissions.asp Accessed 3 January 2008.

Figure 110 shows the emissions from the Energy-from-Waste Plant at Sheffield.

The Secretaries of State for the Environment and for Wales decide appeals, and their Departments provide advice and guidance to local authorities. In important respects, the legal provisions applying to incineration plants are now contained in EC legislation, rather than purely national legislation. If a plant is designed to produce electricity for public supply and has a net capacity of more than 50 MWe, its construction and operation require the consent of the President of the Board of Trade.

In Scotland the legislation is broadly similar. The planning authority may be the regional, district or islands council; the waste regulation authority is the district or islands council. The Scottish Environment Protection Agency (SEPA) controls emissions to air from large incineration plants; the district or islands council is responsible for smaller plants.

The Environment Agency regulates as Part A processes for waste disposal the incineration of:

(a) wastes arising from the manufacture of chemicals or plastics;
(b) specified waste chemicals (bromine, cadmium, chlorine, fluorine, iodine, lead, mercury, nitrogen, phosphorus, sulphur, zinc);
(c) any substances if the incinerator is rated at 1 tonne an hour or more;
(d) chemicals contaminating reusable metal containers.

Other processes involving wastes are currently regarded by the Environment Agency as having the primary purpose of providing energy. These include combustion processes with a net rated thermal input of greater than 3 MW such as waste-derived fuel, tyres, wood waste, straw or poultry litter.

The 1990 Act also made improvements in the legislation on local authority control of air pollution. District and metropolitan district councils regulate emissions to air from smaller incineration and combustion plants and from crematoria (Part B processes). However, incinerators not regulated by the Environment Agency and designed to dispose of 50 kg an hour or less of substances other than clinical waste, sewage sludge, sewage screenings and municipal waste, do not require an authorization from the local authority.

The fate of dioxins in modern incineration plants is given in Table (i). Emissions to air are greatly reduced. (concentrations of $<<0.1$ ng/Nm^3 are now de rigeur). A small increase in the bottom ash has been posited, but these are consigned to either engineered LANDFILL or construction use and are in effect removed from the environment. Modern energy from waste incineration plants may be considered as dioxin sinks, as they destroy more dioxins than they create (0.7 g in 100 000 tonnes input waste, less than <0.03 g to air per 100 000 tonnes of waste (2007).

(i) Fate of dioxins in energy-from-waste plant.

	Old plant (%)	Modern plant (%)
Waste in 50 μg/mg	100	100
Bottom ash	<10	<10
Boiler residue	32	<3
Gas cleaning	800	<20
Flue gas	100	<<1

Source: IAWG, December 1994.

Vogg has stated:

Thermal waste treatment as a pollutant sink is no longer an illusion but already reality in a number of cases. If one has the basic statements expressed in this paper interact in a reasonable manner and combines them intelligently under the aspect of reasonableness, a substantial amount of pollutants are undoubtedly eliminated from the ecosystem by thermal treatment. This applies both to inorganic and to organic noxious chemicals.

The future of a new generation of incineration facilities has already come.

Sources: H. Vogg, 'Thermal waste treatment reasons and objectives', *Waste Management* 95, Copenhagen, 16–17 March 1995.

Department of the Environment, Transport and the Regions, *Sustainable Energy from Waste*, October 1999.

Municipal waste incineration plant must comply with the following conditions: the gases resulting from the combustion of the waste must be raised, after the last injection of combustion air and even under the most unfavourable conditions, to a temperature of at least 850 °C in the presence of at least 6 per cent oxygen for a sufficient period of time to be determined by the Inspector. This, together with ACTIVATED CARBON injection, ensures very low levels of dioxin in the stack gases: levels of <0.1 ng/Nm^3 are now continually achieved. Pre-1995, UK MSW plants emitted in the order of 40 ng/Nm^3. This is a reduction of >99.75 per cent and represents a major dioxins clean-up achievement which puts incineration gas 'clean up' well ahead of all UK solid fuel combustion processes.

Incineration case study (1)

The Uppsala Waste-to-Energy plant is a good example of the Swedish philosophy regarding recovery from waste and minimizing the use of landfill:

- The municipal solid waste (MSW) is a resource. After hazardous things and economically valuable material have been sorted out in households, etc., the residual municipal solid waste (RMSW) is a biofuel which shall be used in waste to energy plants.
- Optimal design of the waste to energy plant minimizes the impact.
- Effective utilization of the energy in the waste by steam/hot water production maximizes the income for the plant and minimizes the use of alternative fossil fuels.
- The plant has a very high reliability mainly due to interested and well-educated personnel.
- The company is open and is accepted by the public, politicians and authorities.

Figure 111 shows the 1980 & 1987 respective district heating fuel mix for the Uppsala Waste-to-Energy plant. Note the striking 86% reduction in oil dependency, with waste combustion contributing a 42% reduction.

Incineration case study (2)

An excellent example of household waste incineration with the combustion of other fuels (Figure 112) is provided by the Mabjerg-Vaerket CHP station in the Danish Holstebro/Struer region.

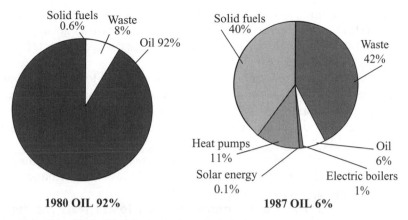

Figure 111   District heating fuel mix for Uppsala.

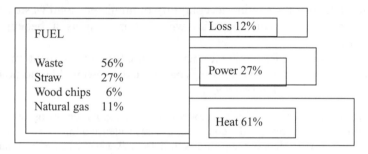

Figure 112   Fuel use and output from the Mabjerg-Vaerket CHP station.

Details of the plant are as follows:

Power output, net	28 MW
Efficiency (electricity)	27%
Electricity generation, annual	160 × 106 kWh
Heat output, net	67 MJ/s
Efficiency (heat)	61%
Heat production	1 439 000 GJ
Flow temperature	75–90 °C
Return temperature	35–55 °C
Heat requirement coverage	Approx. 90%
Steam temperature, turbine	520 °C
Steam pressure, turbine	65 bar
Station price, inc. building	DKK 600 million

Source: Vestkraft Information Leaflet on Mabjerg-Vaerket CHP power station.

The power station forms part of the district heating system, as the basic load unit. The heat it produces covers about 90 per cent of district heating requirements in the principalities of Holstebro and Struer where the total heating basis is ca. 1600 TJ per year.

The waste boilers are in continuous operation at full load throughout the year, except during periods of overhaul or operational stop. Annual continuous operation is estimated as 7500 hours per waste line.

Straw and wood chips are mainly used during the winter period (September–May).

An annual overhaul of each furnace/boiler is planned. These overhauls are carried out in different periods of comparatively low heat demand so that waste burning capacity is always available.

The station turbine is to be overhauled every fourth year. The duration of such overhauls is expected to be 3–4 weeks. During these periods heat will be produced by bypassing the turbine and using a separate heat exchanger.

This highlights the routine acceptability of municipal solid waste incineration in most European countries where it is seen as a benefit to the community.

The UK Medical Research Council Institute for Environment and Health conducted population health surveys based on data from old (and now obsolete) incinerators and concluded:

> Epidemiological studies in incinerator workers and populations living near incinerators have not demonstrated a consistent excess incidence of any specific disease.
>
> However, the ability of these studies to detect adverse effects has been limited by various factors including their relatively small sizes, difficulties in accounting for possible confounding exposures (e.g. smoking and socio-economic factors), health worker effects (the bias that occurs when comparing mortality and morbidity of people who are fit enough to be in employment with those of the general population).

The new generation of incinerator plants can only be that much safer.

Source: Medical Research Council for Environment and Health, Report R7 1998.

The Swiss Environmental Protection Agency (BUWAL) and the Department of Waste, Water, Energy and Air (AWEL) under the canton of Zürich conclude in a scientific study that the electricity produced from the incineration of waste only has a very modest environmental impact, even in comparison with electricity from other renewable energy sources.

Source: Warmer Bulletin, Issue 105, July 2006.

Last word: Proof that both high recycling rates and incineration can work together are given in the table overleaf for Sweden. Note only 9.1 per cent of the MSW was sent to landfill in 2004. Mission accomplished!

Waste management in Sweden (2003–2004)

Waste management option	2004(t)	2003(t)	2004 (kg/inh)	2003 (kg/inh)	2004 (%)	2003 (%)
Recycling	1 384 760	1 313 760	153.7	146.4	33.2	31.4
Biological treatment	433 830	402 780	48.1	44.9	10.4	9.6
Waste to energy	1 944 290	1 867 670	215.8	208.1	46.7	44.6
Landfill	380 000	575 000	42.2	64.0	9.1	13.8
Hazardous waste	25 700	26 660	2.9	3.0	0.6	0.6
Total	4 168 580	4 185 870	462.7	466.4	100	100

Source: Warmer Bulletin, Issue 106, September 2006.

Indexing. The adjusting of the terms of capital cost or long-term contracts to take account of inflation.

An alternative interpretation (for engineering uses) is the scaling of plant from known costs and capacity to suit other ranges, e.g. a 50 per cent increase in capacity may be accompanied by a 33 per cent price increase. All very well till the design limits are attained and then a new 'line' needs to be installed.

Indoor air quality. A growing body of scientific evidence indicates that the air within buildings can be more seriously polluted than ambient air in even the most industrialized cities. Moreover, people spend up to 90 per cent of their time indoors and those who may be exposed to indoor air pollutants for the longest periods are often those most susceptible to the effects of air pollution. They include the young, the elderly and the chronically ill, especially those suffering from respiratory or cardiovascular disease. So, for many, the risks to health due to exposure to air pollution may be greater indoors than outdoors.

Immediate effects include irritation of the eyes, nose and throat, headaches, dizziness and fatigue. Such immediate effects are usually short term and treatable. Symptoms of some diseases, including asthma, hypersensitivity pneumonitis (a group of respiratory diseases that cause inflammation of the lung (specifically granulomatous cells); most forms of hypersensitivity pneumonitis are caused by the inhalation of organic dusts, including moulds) and humidifier fever, may also show up soon after exposure to some indoor air pollutants. Other health effects, including some respiratory diseases, heart disease and cancer, may show up either years after exposure has occurred or only after long or repeated periods of exposure.

Indoor pollution sources that release gases or particles into the air are the primary cause of indoor air quality problems. These sources include combustion of oil, gas, kerosene, coal, wood and tobacco products, building materials and furnishings as diverse as deteriorating asbestos-containing insulation, wet or damp carpet and furniture made of chipboard and the like. Organic chemicals are widely used as ingredients in household products. Paints, varnishes, and wax all contain organic solvents, as do many cleaning, disinfecting, cosmetic, degreasing and hobby products. Fuels are made up of organic chemicals. All of these products can release organic compounds during and after use and, to some degree, in storage. Outdoor sources of RADON, PESTICIDES, and other air pollutants can also enter a building.

   Inadequate ventilation can increase indoor pollution by not drawing in enough outdoor air to dilute emissions from indoor sources and by not carrying indoor air pollutants out of the building. High temperature and humidity levels can also increase concentrations of certain pollutants.

Industrial ecology. The concept of a closed industrial system in which products (and it is claimed, energy) are returned for reprocessing and reuse. Guidelines are as follows.

1. *Efficient use of resources.* 'Doing more with less' is common sense for business – and good for the environment. Reducing the amount of material and ENERGY used to make or supply goods can directly cut the cost of doing business, through lower production and waste management costs, or avoiding the cost of potential environmental liabilities.

2. *Extending the life of products.* Planned obsolescence is itself obsolete. The goal should be to achieve longer product lifespans by, for example, designing upgradable and repairable products and producing more durable goods. Extending a product's life means fewer replacements, less waste and the use of fewer material and energy resources. While longer-life products are not new, environmental issues are casting a new light on them, causing companies to re-evaluate their strategies and implement new programmes to increase product life expectancies and recycle materials at the end of their useful life.

3. *Pollution prevention.* Pollution prevention, aimed at avoiding waste in the first place, can be achieved through operational, technical or behavioural change at any point in the product life-cycle. It may include product substitution, product reformulations or housekeeping improvements, and is applicable to service sectors, offices and consumers as much as to industrial processes.

4. *Recycling and reuse.* Reducing the consumption of virgin materials and diverting wastes from final disposal is the essence of industrial ecology. So, the RECYCLING and reuse of materials and goods to divert wastes back into productive uses is crucial. Recycling and reuse strategies apply to all business sectors, and to private households.

5. *Eco-industrial parks.* Kalundborg in Denmark provides a good illustration of the concept of industrial ecology. There, outputs from several industrial processes serve as inputs for others. Now, new industrial parks

in other parts of the world are looking at Kalundborg as a model of how to attract businesses that can 'feed' off one another. (◊ FACTOR FOUR)

Source: World Business Council Sustainable Development, *Tomorrow*, August 1996.

Industrial noise measurement. The procedure used to assess whether a factory or item of industrial plant is likely to present a noise nuisance, is BS 4142: 1997 (Method for rating industrial noise affecting mixed residential and industrial areas). It is based on the extent to which the noise level created by the source exceeds the noise level prevailing when it is not in operation.

Industrial waste. ◊ WASTES.

Infections, water-borne. These include bacterial infections such as CHOLERA, the enteric diseases, TYPHOID and para-typhoid, and Weil's disease; the viral infections poliomyelitis and hepatitis, protozoan infections such as amoebic dysentery, and SCHISTOSOMIASIS (bilharzia) caused by a trematode worm. All the agents responsible are water-borne; bilharzia is parasitic in some snails in irrigation ditches in developing countries and Weil's disease may be transmitted by rats in sewers. (◊ DAM PROJECTS)

PUBLIC HEALTH practice is intended to prevent, among other things, the spread of disease. Thus, drinking water and sewage are kept separate. However, in less sanitary countries, or even in a crisis in more developed countries, if life-support systems break down, water withdrawal from some sources must cease immediately.

Infiltration[1]. The downward flow of water through the soil surface into subsurface strata.

Infiltration[2]. The term is also used to denote the contamination of wells by salt water or the spoiling of an aquifer by pollutants, e.g. by LEACHATE from a LANDFILL SITE.

Infrared radiation. Radiation whose wavelength is in the infrared radiation spectrum. It can be used to conduct energy surveys using a special camera and so enable sources of waste heat to be pinpointed.

Infrasound. Low-frequency sound which usually emanates from low revving ship or heavy transport engines in frequencies below 100 Hz. A typical living-room will resonate around 12 Hz thus amplifying heavy traffic noise. One noise measurement in London recorded a value of 85 DECIBELS at 5 Hz coming from ships in the docks. Infrasound is not catered for on the dB(A) scale which is weighted to represent the response of the human ear and therefore takes little account of noise below 100 Hz.

The only way of cutting down infrasound is to stop it at source as insulation is rather ineffective in this frequency range.

Inhibitor. A substance, naturally occurring or added, whose presence in small amounts retards or prevents the occurrence of certain phenomena considered undesirable, e.g. gum formation in stored gasolines, colour change in lubricating oils, corrosion in turbines, scale formation in boilers, etc.

Innovation. An example of innovation is the world's first Solid Insulation Distribution Transformer (SIDT). For years, many electrical utilities have been distributing power via underground systems. However, equipment corrodes fairly quickly, and this deterioration causes a wide range of problems. One of the most serious consequences is undoubtedly the risk of leakage of the dielectric oil contained in conventional distribution transformers and other equipment. In addition to causing the eventual breakdown of the transformer, these leaks can lead to fires and violent explosions. Another significant impact is the risk of soil and groundwater contamination.

SIDT, developed jointly by ABB and Hydro-Quebec, Canada, enables electric utilities to operate their power distribution systems economically while meeting strict environmental standards. SIDTs are corrosion-free, silent, withstand severe weather, and contain non-flammable or volatile liquids. Since they eliminate the risk of soil or groundwater contamination, SIDTs are an ideal solution for underground systems. Units with a full-load rating of 167 kVA have been successfully tested in service in Hydro-Quebec's underground network since March 1996.

The SIDT features four fully integrated components:

- a solid insulation system;
- a composite material outer shell;
- a simple but highly efficient internal cooling system based on HEAT PIPES;
- magnetic core cushioning.

This is an excellent example of environmental innovation.

Source: ABB Review, No. 5, 1998, p. 21.

Inorganic matter. Matter which is mineral in origin and does not contain carbon compounds, except as carbonates, carbides, etc. (◊ ORGANIC MATTER)

Insecticides, synthetic. ◊ CHLORINATED HYDROCARBONS; ORGANOPHOSPHATES.

Insolation. The amount of direct solar radiation incident per unit horizontal area, measured in W/m^2.

Insulation. Materials provide thermal or acoustic insulation if they prevent transmission of HEAT or SOUND. Important to distinguish from ABSORPTION.

Intergovernmental panel on climate change. Recognizing the problem of potential global climate change, the World Meteorological Organization (WMO) and the United Nations Environment Programme (UNEP) established the Intergovernmental Panel on Climate Change (IPCC) in 1988. It is open to all members of the UN and WMO. The role of the IPCC is to assess on a comprehensive, objective, open and transparent basis the scientific, technical and socio-economic information relevant to understanding the scientific basis of risk of human-induced climate change, its potential impacts and options for adaptation and mitigation. The IPCC does not carry out research nor does it monitor climate-related data or other relevant parameters. It bases its assessment mainly on peer reviewed and published scientific/technical literature. ◊ KYOTO AGREEMENT.

Integrated Gasification Combined Cycle (IGCC). IGCC is expected to be the most promising technology for generating electricity from coal giving an increase of 10 per cent in thermal efficiency over conventional coal burning stations. Coal consumption will be some 30 per cent less than normal and CO_2 emissions will be reduced by nearly 25 per cent.

In an IGCC plant, coal is pulverized, dried and fed to the gasifier where STEAM and OXYGEN are added to oxidize it partially to HYDROGEN and CARBON MONOXIDE. HEAT from GASIFICATION produces steam used in a steam turbine driving generator. The gas is cooled to solidify the ash and generate more steam. Sulphur in the process is converted to sulphuric anhydride from which it is recovered as sulphur for sale. The gas and steam turbines constitute the combined cycle of electricity production. The turbine exhaust gases pass to a recovery boiler to produce more steam for the turbine. (◊ COMBINED CYCLE POWER STATION; GASIFICATION)

Integrated Pollution Control (IPC). Introduced in the UK Environmental Protection Act 1990 so that all major emissions are considered simultaneously and not in isolation, i.e. the reduction of pollution in one environmental medium can have effects on another. The Best Available Techniques Not Entailing Excessive Costs (BATNEEC) are required to minimize pollution of the environment as a whole. The concept is now enshrined in INTEGRATED POLLUTION PREVENTION AND CONTROL (IPCC) in order to achieve EU compliance with the IPPC Directive (96/61/EC).

The IPPC Directive aim is 'to achieve integrated prevention and control of pollution' from listed activities. The Directive applies to energy, metal, mineral and chemical industries waste management, pulp, paper and board production, dyeing of fibres, tanning slaughterhouses and disposal of animal carcasses, various food processes, and intensive poultry and pig rearing installations.

Integrated pollution control (IPC)

The prescribed processes to be controlled under IPC, the timetable for their introduction and the prescribed substances are detailed in the Environmental Protection (Prescribed Processes and Substances) Regulations 1991.

The Environment Agency is required to grant an authorization, subject to any conditions which the Act requires or empowers it to impose, or to refuse it. The Environment Agency must refuse to issue an authorization unless it is considered that the operator will be able to carry on the process in compliance with the conditions in the authorization.

In setting conditions, the Act imposes a duty on The Environment Agency to meet certain objectives. Conditions should ensure that:

- the best available techniques (technology, operation and management practices) not entailing excessive cost (BATNEEC) are used to prevent or, if that is not practicable, to minimise the release of prescribed substances which are released and any other substances which cause harm,
- releases do not cause, or contribute to, the breach of any direction given by the Secretary of State to implement European Community or international obligations relating to environmental protection, or any statutory environmental quality standards or objectives, or other statutory limits or requirements,
- when a process is likely to involve releases into more than one environmental medium (the case for most prescribed processes), the best practicable environmental option (BPEO) is achieved, i.e. the releases from the process are controlled through the use of BATNEEC to give the least overall affect on the environment as a whole.

An important element is the requirement to maintain a register of applications, authorizations issued and information received from operators as a result of conditions in authorizations. The register is available for public scrutiny at all reasonable times.

Waste disposal and recycling (which embraces incineration) is one of the main categories of process prescribed by the regulations. Specifically, they cover:

(a) waste chemicals and plastics (arising from their manufacture),
(b) specific chemicals and their compounds, and
(c) any other waste including animal remains, at a rate of 1 tonne or more per hour.

Comprehensive guidance has been prepared, the most useful being Chief Inspector's Guidance Notes (CIGNs) which provide inspectors and industry with the main emission standards for prescribed substances arising from each process. They specify the minimum standards to be attained by existing plant and what constitutes BATNEEC for new plant. They take into consideration the results of BAT research reviews commissioned by the Inspectorate.

They provide guidance on:

- the general provision of the Act, the Regulations, applications and authorizations,
- the process descriptions to which they apply,
- a suggested timetable for upgrading plant to current standards,
- definitions of substantial change,

- BATNEEC/BPEO process/abatement techniques, and
- release levels corresponding to those techniques.

The importance of the Guidance Notes cannot be underestimated. A developer proposing to build an incineration plant should ensure that the proposal meets, at least, current best practice. Consideration should also be given to known or likely future standards. (◊ GUIDANCE NOTES, ENVIRONMENT AGENCY)

Integrated Pollution Prevention and Control (IPPC). The European Union (EU) environmental legislation applies separately to different aspects of the environment – air, water, waste and so on. However, controls applied to solve only one problem can mean that the damage comes out in another way. So it is often more effective to deal with a variety of environmental impacts together rather than treating them all separately.

The IPPC Directive, based mainly on UK law, requires major industrial installations to be licensed in an integrated manner, controlling emissions to air and water and the management of waste to protect the environment as a whole. It also takes account of other environmental issues, such as noise, energy efficiency, consumption of raw materials and water, prevention of accidents, and restoration of the site to a satisfactory state upon closure. This approach encourages industrialists to think about the whole process and adopt 'cleaner technology' rather than just adding 'end-of-pipe' controls.

Integrated waste management. A strategy for the management of waste utilizing a range of environmentally sound systems and processes. Typically it would include the promotion of waste minimization, materials recycling, and resource recovery, with landfill as a last resort, i.e. the first three options would be optimized and landfill minimized.

Intelligent packaging. Most of the technical, scientific and legislative problems that active packaging materials may pose are shared also by intelligent packaging materials. They have been defined as:

"Concepts that monitor to give information about the quality of the packed food"

Examples are indicators of time-temperature storage history; carbon dioxide or oxygen status; spoilage status, and tamper-breakage information. Consequently, for intelligent packaging, the food is intended to influence the packaging.

Source: CSL Report FD 03/28 for the Food Standards Agency: Active Packaging – Current Trends & Potential for Migration, April 2004.

Interceptor. A trap which intercepts and separates oil, grease, or grit from sewage or surface water.

Intermediate processor. A company which purchases source-separated materials for further processing. (◊ RECYCLING)

Internal combustion engine. The internal combustion engine is characterized by combustion of the fuel inside an enclosure, usually a cylinder, which causes rapid gas expansion which in turn forces a piston to move and do work. The characteristics of an internal combustion engine are that there is a need for a homogeneous, easily-ignited fuel (petrol or diesel oil) which, on ignition, does work due to the expansion of the gases which are then released to the atmosphere. The cyclic nature of the operation plus poor fuel distribution in the cylinder means that the THERMAL EFFICIENCY is low, combustion is often incomplete and thus unburnt HYDROCARBONS, CARBON MONOXIDE and CARBON DIOXIDE are released to the atmosphere. NITROGEN OXIDES are also released at concentrations which depend on the temperature and duration of the combustion flame. Despite this, the internal combustion engine is in mass production and has market dominance. Attempts to replace it will meet with strong opposition due to the capital invested in its manufacture and aftercare. (◊ EXTERNAL COMBUSTION ENGINE; AUTOMOBILE EMISSIONS)

International Statistical Classification of Diseases (ICD). The International Statistical Classification of Diseases and Related Health Problems (commonly known by the abbreviation ICD) is a detailed description of known diseases and injuries. Every disease (or group of related diseases) is described with its diagnosis and given a unique code, up to six characters long.

The ICD has become the international standard diagnostic classification for all general epidemiological and many health management purposes. These include the analysis of the general health situation of population groups and monitoring of the incidence and prevalence of diseases and other health problems in relation to other variables such as the characteristics and circumstances of the individuals affected. It is used to classify diseases and other health problems recorded on many types of health and vital records, including death certificates and hospital records. In addition to enabling the storage and retrieval of diagnostic information for clinical and epidemiological purposes, these records also provide the basis for the compilation of national mortality and morbidity statistics by World Health Organization Member States.

ICD is used world-wide for morbidity and mortality statistics, reimbursement systems and automated decision support in medicine. The system is designed to promote international comparability in the collection, processing, classification, and presentation of these statistics.

International Committee on Radiation Protection (ICRP). Committee charged with setting permissible DOSE levels of IONIZING RADIATION to both the population at large and the workforce (who are 'permitted' greater doses). (◊ IONIZING RADIATION)

International Nuclear Event Scale (INES). Launched on 1 November 1990, this classifies nuclear events at seven levels. Basically, the lowest three levels will be for incidents of increasing seriousness and the upper four apply to accidents of increasing seriousness. Events which have no safety significance are classified as below scale or level 0 and may not be given much (if any) publicity.

Intrinsic energy value. Energy either directly consumed in the manufacture of a product or the energy needed to produce feedstock. See table below for comparison of the intrinsic energy and calorific values for various commodity plastics.

Material	PVC	PET	PP	PS	LDPE
Intrinsic energy value (MJ/kg)	53	84	73	80	69
Calorific value (MJ/kg)	18	22	41	38	43

Source: Association of Plastics Manufacturers.

Inversion layer. ◊ TEMPERATURE INVERSION.

Ion. An electrically charged atom or molecule produced by a loss or gain of electrons. Gases can be ionized by electrical discharge or ionizing radiation. Ions are also formed in solution.

Ion exchange. The removal of IONS from solution. It can be carried out by the use of a suitable ion-exchange medium or bed, which has the power to remove or capture the desired ions from solution. The choice of medium determines the type of ion removed (positive or negative). Once the medium is saturated it must be taken off-stream and recharged by passing an alkaline or acidic solution through the bed which replaces the captured ions with ions from the recharging solution. Ion exchange has many uses. For example, in water softening, calcium ions are replaced by sodium ions, and the water is thereby softened. The bed is then recharged with a brine solution which replaces the captured calcium ions and the bed is then ready for reuse. Normally one bed is 'on-stream' while the other is 'off-stream' for recharging.

Ion exchange is used in water softening, water purification, SOLUTION MINING, metallic effluent treatment, etc.

Ion-selective electrode. An electrode which has a high degree of selectivity for one ION over other ions which may be present in a sample. A potential difference is obtained that is proportional to the LOGARITHM of the activity of a particular ion. Ion selective electrodes are available for H^+, Na^+, K^+, NH_4^+, etc.

Ionizing radiation. Certain radiations in their passage through matter are capable of causing ionization, i.e. they can 'knock' electrons out of atoms or molecules or create ions either directly or indirectly. The radiations that do this directly are either fast-moving particles (for example, electrons, protons and α-particles) or electromagnetic rays such as X-rays or γ-rays. (γ-rays are similar to X-rays but have much shorter wavelengths and higher energies. The difference is more one of name than of physics.) There are numerous sources of naturally occurring ionizing radiation including cosmic rays that arrive continuously from outer space, but the sources of environmental significance are certain radioactive substances liberated into the environment by humans.

Ionizing radiation, dose measurement. Ionizing radiations may be divided into two main groups:

1. *Electromagnetic radiations* (X-rays and gamma rays), which belong to the same family of electromagnetic radiation as visible light and radio waves.
2. *Corpuscular radiations*, some of which – alpha particles, beta particles (electrons) and protons – are electrically charged, whereas others – neutrons – have no electric charge. This distinction between the two groups becomes blurred, however, when their mode of absorption in materials is considered.

Whilst the exact nature of the biological effects of these radiations is not fully understood, they are related to the ionization that the radiations are capable of producing in living tissue. Thus, the biological effects of all ionizing radiations are essentially similar. However, the distribution of damage throughout the body may be very different according to the type, energy and penetrating power of the radiation involved. α-particles from radioisotopes have ranges of only about 0.001–0.007 centimetres in soft tissue and less in bone. β-particles have ranges in soft tissues of the order of several millimetres, i.e. much greater than those of α-particles in such tissues. X-rays and γ-rays are not stopped by tissues.

The dose of radiation is the amount of energy absorbed per unit mass of material. The unit is the RAD (radiation absorbed dose (rad)), which is 0.01 joules per kilogram (10^{-2} J/kg). The SI unit of absorbed dose is the GRAY (Gy) and corresponds to an energy absorption of 1 J/kg of matter. So,

$$1\,\text{Gy} = 100\ \text{rad}$$

However, the rad or gray is a physical measure of energy absorbed and does not indicate the biological effects of irradiation. These vary according to the type and energy of the radiation involved.

The most significant factor in estimating the biological effectiveness of radiation is the LINEAR ENERGY TRANSFER, LET. Significant penetration into soft tissues (typically up to 2 m) is achieved by X- and γ-rays, which have lower LETs compared to β-particles, which can typically penetrate a few millimetres, or to α-particles which usually have very high LETs and can only penetrate a few hundredths of a millimetre. Thus α-particles are likely to produce a much greater amount of damage than β-particles or γ-rays but this damage is usually summarized by a quality factor, Q (before 1962 this was known as the relative biological effectiveness or RBE). This is defined as the ratio of the effectiveness of the radiation to the effectiveness of 200 kV

X-rays. For β-particles, X- and γ-rays, $Q = 1$, for α-particles $Q = 20$, and for neutrons, $Q = 10$. Thus for an absorbed dose R (in rad or Gy), the biologically effective dose D is given by

$$D = RQ$$

In addition to the absorbed dose and the quality of the radiation absorbed, biological changes can be influenced by a large number of other factors; some organs are more susceptible to damage than others, internally ingested radioactive materials may concentrate in certain organs and some organs may be regarded as more vital than others.

To allow for these different circumstances, the use of a dose equivalent is recommended. The dose equivalent is defined as the absorbed dose R (in rad or Gy) multiplied by the quality factor Q and any other modifying factor N; hence

$$\text{Dose equivalent} = RNQ$$

Dose equivalent is measured in rem (Rœntgen Equivalent Man) if absorbed dose is measured in rad. It is measured in SIEVERT (Sv) if the absorbed dose is measured in gray.

By way of illustration, for neutrons impinging on the eye $Q = 10$ and $N = 3$. Hence an absorbed dose of 0.01 Gy (1 rad) corresponds to a dose equivalent in this instance of 0.335 Sv (30 rem). The same absorbed dose falling on the skin has $N = 1$ and so has a dose equivalent of 0.1 Sv (10 rem) indicating that a dose to the eye is three times more damaging than the same dose to the skin. In both cases, the Q factor indicates that neutrons produce 10 times the effect of 200 kV X-rays. Neither the quality factor, Q, nor the modifying factor, N, in the above equation can be calculated; both are empirical constants. (◊ GRAY; LINEAR ENERGY TRANSFER; RAD; REM; SIEVERT).

The data below show the relationships between the SI units associated with radiation and those preceding 1978, but which are still in common use.

Quantity	New named unit and symbol	In other SI units
Exposure	—	$C\,kg^{-1}$
Absorbed dose	Gray (Gy)	$J\,kg^{-1}$
Dose equivalent	Sievert (Sv)	$J\,kg^{-1}$
Activity	Becquerel (Bq)	s^{-1}

Quantity	Old special unit and symbol	Conversion factor
Exposure	rsntgen (R)	$1\,C\,kg^{-1} \approx 3876\,R$
Absorbed dose	rad (rad)	$1\,Gy = 100\,rad$
Dose equivalent	rem (rem)	$1\,Sv = 100\,rem$
Activity	curie (Ci)	$1\,Bq \approx 2.7 \times 10^{-11}\,Ci$

Ionizing radiation, effects. Exposure to ionizing radiation can be harmful as the radiation can cause cancers in the living population and genetic changes that may produce heritable defects in future generations. Ionizing radiation causes mutations, i.e. random changes in the structure of DNA, the long molecule that contains the coded genetic information necessary for the development and functioning of the human being. As the mutations are random events, they are almost certainly harmful to some degree.

The outcome of radiation exposure may be such that the cell suffers so much damage that it dies, or that the cell continues to function but in a modified way. The second outcome can lead to uncontrolled growth of a colony of cells derived from the affected one. This is a cancer, and the upset is called somatic (bodily) mutation.

For reasons not fully understood, cancer resulting from irradiation appears only after a long delay. The incidence of cancer began to show up as abnormal among the survivors of the nuclear explosions at Hiroshima and Nagasaki after ca. 30 years. An abnormal incidence of leukaemia (the uncontrolled growth in numbers of white blood cells), which has a much shorter latency period, was in evidence long ago in the studies of the Atomic Bomb Casualty Commission.

Now if the mutation is in a germ cell (sperm or egg), the entire organism arising from the germ cell together with its progeny will be affected. Many mutations are sufficiently severe to prevent their carriers from living, or at least from reproducing and handing on the defect to later generations. Many others, though, have their effects masked if they are inherited from only one parent. Such mutations are recessive. Their effects appear only in about a quarter of the off-spring of parents both of whom carry the mutant gene.

There are many undesirable recessive genes and most of us carry a few. Consequently from time to time a child of normal healthy parents shows a terrible defect such as gross malformation. The hidden undesirable genes in a population are often called its genetic load. The tendency of ionizing radiation is to increase the genetic load.

Not all mutations have such dramatic effects. They may result merely in a loss of vigour, a susceptibility to disease, or a reduction in life-span. Such genes may spread widely in human populations that are sustained by medical attention.

Ionizing radiation also tends to produce a reduction in life-span. The effect is often regarded as premature ageing. The figures below, published in 1957, are *possible* examples of this premature ageing.

The radiologists referred to in the table were practising their profession over a period in which the dangers of X-rays were far less clearly appreciated than they are today. The age distribution of the control was also

Group	Average age at death
USA population	65.6 years
Physicians with no known contact with radiation	65.7 years
Radiologists	60.5 years

different from that of the radiologists which casts some doubt on drawing *absolute* conclusions from the data. Nowadays much more sensitive X-ray film is used which requires less intense radiation, and the relevant equipment is carefully shielded to protect its users from undesirable radiation.

It is usual to compare levels of anthropological radiation to that of the natural background radiation from cosmic rays and other natural sources. The inference is then drawn that as humans are currently contributing radiation amounting to around 0.1 per cent of the natural background, we therefore have a lot of scope for increasing nuclear installations before we need become concerned about the level of human-made radiation. However, because the natural background is irreducible, it does not necessarily follow that this sets an allowable scale for human-made radiation emissions as radionuclides can be concentrated in the food chain, for example.

For the UK, natural background radiation accounts for 87 per cent of the average annual dose to the UK population (2.5 million sieverts in all), artificial (medical X-rays 12 per cent, nuclear discharges 0.1 per cent, fallout 0.4 per cent, work and miscellaneous 0.6 per cent). There are large variations from the average; the actual dose received by any individual will vary depending upon altitude (because of an increased contribution from cosmic rays) and the type of rocks occurring in the locality: granite, for example, emits some five times more radiation than clays.

The effects of an increase of radiation are proportionately very small indeed, but applied to a total population can give rise to large numbers of cancer incidences and genetically defective births. For example, the genetic handicap risk (Maryland Academy of Sciences) for a radiation dose of 1.7 mSv (170 millirem) per year is estimated as 4.5×10^{-5}. The number of cases per annum is then (for the US population of 200 million) $4.5 \times 10^{-5} \times 200 \times 10^6$, and for a birth-rate of 4.6 million per year, the absolute numbers of genetically unsound births per 1000 births is therefore

$$\frac{4.5 \times 10^{-5} \times 200 \times 10^6}{4.6 \times 10^6} \times 1000$$

or 2 per 1000 births.

It is worth emphasizing that knowledge of the effects of ionizing radiation on mutation rates is slight and growing only slowly. Estimates for human populations are based on readily-observed cases, but mutations that give rise to slightly impaired individuals whose characteristics are nevertheless within the ordinary range of variation are an even more unpleasant prospect. The attrition of human capacities and the deterioration of health and vigour could proceed unnoticed for indefinitely long periods.

It is now generally agreed that, at least for the purposes of estimating hazards, any dose of ionizing radiation, no matter how small or how slowly delivered, must be regarded as prospectively harmful, able to induce genetic mutations and cancer, and generally to erode the life-span of its recipient.

The costs of increased exposure to radiation (as in any form of pollution) must be balanced against the social benefits to be derived from the activities leading to the increased exposure. The debate is whether these risks are justifiable and whether we have the right to saddle future generations with our long-life radioactive wastes. (◊ COST-BENEFIT ANALYSIS; NUCLEAR ENERGY; NUCLEAR REACTOR DESIGN; NUCLEAR REACTOR WASTES; IONIZING RADIATION, DOSE MEASUREMENT).

Source: National Radiological Protection Board, *Radiation Doses – Maps and Magnitudes*, Harwell, Oxon, undated.

Ionizing radiation, dose limits. All exposure to radiation should be kept 'As Low As Reasonably Practicable' (ALARP). The Ionising Radiations Regulations 1999 (IRR99) came into effect on 1 January 2000, and specify the following limits:

1) Radiation workers >18 years:
   The effective dose shall be 20 mSv in any calendar year
2) Members of the public:
   The effective dose shall be 1 mSv in any calendar year
3) Trainees aged under 18 years:
   The effective dose shall be 6 mSv in any calendar year

To prevent deterministic effects the limits are

1) Radiation worker >18 years:
   The limit on equivalent dose for the skin, hands, forearms, feet and ankles shall be 500 mSv/year
2) Radiation worker >18 years:
   The limit on equivalent dose for the lens of the eye shall be 150 mSv/year
3) Trainee aged <18 years:
   The limit on equivalent dose for the lens of the eye shall be 50 mSv/year
4) Trainee aged <18 years
   The limit on equivalent dose for skin, hands, forearms, feet and ankles shall be 150 mSv/year

5) Women of reproductive capacity
   The limit on equivalent dose for the abdomen shall be 13 mSv in any consecutive period of three months. Once a pregnancy has been confirmed and the employer notified, the equivalent dose to the foetus should not exceed 1 mSv during the remainder of the pregnancy.

On average, each person in the UK is exposed to about 2.6 mSv of radiation a year. Soils and rocks can contain uranium and thorium. The daughter nuclides from these elements emit gamma radiation which can irradiate our bodies. In granite areas of Cornwall, they can contribute an annual dose of up to 0.6 mSv per year.

Decay of uranium in soil can give rise to radon gas, which is responsible for nearly half of the public radiation dose from natural sources. The radon gives rise to short-lived alpha emitters which attach to dust particle which can be trapped in the alveoli in lungs, contributing a dose of 1.2 mSv per year, possibly leading to lung cancer. The average annual indoor dose of radon is 1 mSv. If this level reaches 10 mSv a year, it is advisable to install additional vents to reduce the concentration. It is estimated that 100 000 houses in the UK (mainly in Cornwall, the Highland Regions, Derbyshire, Devon and Northampton) have a level above 10 mSv a year.

Ionization. ◊ IONIZING RADIATION, EFFECTS.

Ionosphere. The region of the upper atmosphere (starting from 90 km above the earth) which includes the highly ionized Appleton and Kennelly–Heavyside layers, the existence of which enables intercontinental radio transmission to be achieved.

Irradiation, uses of. The major potential uses of irradiation in the environment are:

1. *Insect sterilization.* Irradiated males can be released and, as the mating in some species only takes place once, it is possible to reduce the population of pests without recourse to pesticides.

2. *Digestibility aids.* Many cellulosic wastes can be used by ruminants as an energy source especially if the cellulose structure is broken down. Irradiation is one means of doing this.

3. *Food preservation.* The storage life of iced fish on board ship can be doubled by a dose of 100 krad without any apparent adverse effects. This technique may enable trawlers to use distant fishing grounds, extend the duration of voyages and deliver a more hygienic product to the customer. Food preservation is an area in which major advances are expected as irradiation has not been shown to produce dangerous radioactivity in foods when a cobalt-60 radiation source is used. (This produces γ-rays of 1.17 meV (the electron-volt is the general unit of energy of moving particles) which are suitable for food processing.) However, there are serious doubts about its use in treating food in which decay has already com-

menced and the subsequent sale of the produce as sound. Should this method proliferate, the subsequent disposal of the spent cobalt sources, whose half-life is ca. 5 years, will produce yet more radioactive waste for secure long-term disposal.

4. *Radioactive tracers*. The pathways of pollutants can be traced by releasing radioactive monitors. This has been done in studies of air pollution and effluent dispersal. They are also of great use in following fluid flow and leak detection in industry, the ingestion/digestion of food in animals, monitoring engine wear and many other applications.

5. *Disinfestation of seeds*. The removal of pests from stored grains.

6. *Sterilization*. Large doses of radiation will produce a completely sterile product, e.g. pharmaceutical products and material for transplants.

Irrigation. The application of water to arable soils, so that plant growth may be initiated and maintained. Irrigation is used not only in many arid and semi-arid lands, but in the UK to rectify SOIL MOISTURE DEFICITS (SMDs). In the past, eastern England could have benefited from irrigation every three years in ten. The practice may be expected to grow in the UK due to the high SMDs now being experienced.

Irrigation with brackish water over long periods can eventually render soils unfit for crop growth, as plant evapo-transpiration acts as a distillation process and leaves the dissolved solids in the soil where they accumulate. The spread of Middle Eastern deserts has been partially attributed to early irrigation attempts which resulted in salinization of once fertile soils.

ISO 1400 Series. A voluntary framework (and tools) for applying ENVIRONMENTAL MANAGEMENT SYSTEMS (EMS) in all varieties of organizations.

There are three separate (but related) auditing standards:

ISO 14010 – general guidance
ISO 14011 – procedures for auditing emissions
ISO 14012 – qualifications for environmental auditors.

The ISO standards enable world-wide EMS specifications and frameworks to be established which is also complementary to the EU ECO-MANAGEMENT AND AUDIT SCHEME, which requires a verified environmental statement.

Source: B. Sadler, M. Baxter, *Guide to Environmental Management*, Environment Agency, June 1998, pp. 15–20.

Isobar. A closed curve on a weather map which connects places of the same value of pressure.

Isomer. Isomers are molecules with the same chemical formula but differ in constitution or structure by the different arrangement of atoms within the molecules.

Isotherms. Contours of constant temperature.

Isotope. Atoms of the same ELEMENT which have the same number of protons in their nuclei but differing numbers of neutrons. Hence, they have the same atomic number but different MASS NUMBERS. Isotopes are written as mass number (i.e. number of protons and neutrons) followed by the symbol, e.g. ^{235}U, which is the fissile isotope of uranium and is capable of sustaining chain reactions in a nuclear reactor.

Isotopes of the same element have identical chemical behaviour but differing physical behaviour, e.g. cobalt-60, used as a γ-ray source for food irradiation, is manufactured in a nuclear reactor by irradiating ordinary cobalt-59 with neutrons. ($\lozenge$ IRRADIATION, USES OF; NUCLEUS; RADIOACTIVE ISOTOPE)

Itai-itai disease. Literally, 'ouch-ouch' disease, caused by cadmium poisoning. Softening of the bones and kidney failure results. It was discovered in Japan in 1950, where cadmium was released into rivers by mining companies.

J

Jet, overfire. A jet of AIR or STEAM designed to promote TURBULENCE above
a fuel bed in a furnace and so reduce the formation of smoke.

Joule (J). International System of Units (SI) unit of WORK and ENERGY. It is
defined as the work done when a force of 1 NEWTON acts through 1 metre.
Commonly, the megajoule (MJ) is used because it is a more convenient size
for the measure of energy supplied. It is 1 million joules.

The measure of POWER is the WATT, which is 1 joule per second, or 1 J/s.

There are several forms of ENERGY, e.g. thermal, chemical, electrical, and
joule is applicable to all of them. The form of the energy is important,
however, because thermal energy and electrical energy are very different in
their ability to do work. For this reason, in ENERGY ANALYSIS, it is very
common to denote the form the energy is in, i.e. whether it is primary (fuel
energy) or electrical energy. The conversion factor for FUEL energy to electrical energy is commonly taken as 30 per cent but may be less where internal
COMBUSTION engines are in use. (◊ LAWS OF THERMODYNAMICS, FORCE)

K

Kappa number. A universal pulping measure which denotes the extent of LIGNIN removal in the 'cooking' of chemical pulp. It is measured in the number of millilitres of a 20 mmol/l solution of potassium permanganate consumed by 1 g of pulp (dry weight) under controlled test conditions. (◊ KRAFT PAPER; PULP).

Kelvin (K). The SI unit of temperature interval. The Kelvin or absolute scale of temperature starts from absolute zero temperature. 1 kelvin = 1 Celsius degree = 1.8 Fahrenheit degrees. 273.15 K = 0 °C = 32 °F.

Kerbside collection. The regular pick-up of recyclable materials and DOMESTIC REFUSE from residential kerbsides.

Kerbside recycling. Distinct from bring recycling, where the public take waste materials for collection by scheme operators, in kerbside recycling operators collect recyclable waste from individual homes. However, this form of waste recycling does require a level of participation from the public in that households are required at least to separate recyclable waste from non-recyclable waste.

Although some collections are of one material, many involve the collection of several categories of waste material. Households are usually required to separate waste into these categories, although further separation is often carried subsequent to collection at material recovery facilities (MRFs). Alternative terminology includes 'doorstep collection' and 'door-to-door collection'.

Kilogram mole. The mass in kg occupied by 1 mole of a substance, e.g. 1 kg mole CO_2 is 44 kg.

Kilowatt-hour (kWh). Unit of ENERGY equal to 1000 WATT-HOURS or 3.6 megajoules. The kilowatt-hour can be used as a measure of ENERGY input or output in energy analysis, in which case, kWh_{th} is used for energy in the form of FUEL (or thermal energy). If in the electrical form, the designation kWh_e is used which, depending on the efficiency of conversion, is equal to 3–4 kWh_{th}. ($\lozenge$ WATT; EFFICIENCY)

This concept can best be illustrated by considering the energy value of a tonne of high grade COAL. It can be stated as 8000 kWh_{th} or, if burnt in a high efficiency power station, it provides around 2600 kWh_e. This is the basis of ENERGY ANALYSIS which looks at the energy required for, say, the production of a tonne ALUMINIUM on primary (fuel) energy basis and thus enables the identification of those areas where energy saving is possible. In the case of ALUMINIUM, RECYCLING of used beverage containers can save 95 per cent of the energy required to manufacture the metal from the bauxite ore. This provides a national resource conservation argument for either recycling aluminium or substituting other less energy intensive materials (if available) if recycling is not possible.

Kinematic viscosity. A measure of how easily a fluid flows. In general, kinematic viscosity decreases with increasing temperature.

Kinetic energy. The energy of motion. An object of mass m and speed v will have kinetic energy $\frac{1}{2}mv^2$.

Kraft paper. Term applied to paper manufactured almost entirely by kraft (sulphate pulping process) pulp whereby a chemical pulp is manufactured by cooking the raw wood chips with sodium hydroxide and sodium sulphate. The 'sulphate' denotes the use of sodium sulphate as a make-up chemical which replaces any process chemical losses. This is a dominant Scandinavian pulping process which is often virtually self-sufficient in energy usage by virtue of the use of forest residues as a fuel, and for producing electricity. Kraft paper is a strong brown paper with long cellulose fibres, used for packaging and wrapping. It is easily recycled. ($\lozenge$ PAPER RECYCLING; PULP)

Kyoto Agreement: Framework Convention on Climate Change. A major international meeting of heads of state in Kyoto (1997) resulted in agreement that the main GREENHOUSE GASES and principally CARBON DIOXIDE would be reduced by 12.5 per cent from 1990 levels by 2010. The UK has pledged that carbon dioxide emissions will be cut by 20 per cent. The UK's baseline figures are set out in the table overleaf. UK emissions in 2010 are predicted to surpass its target of 12.5 per cent, and be 23.6% below its 1990 level.

There is a counter-argument that as the 12.5 per cent cut is universally shared, taking the UK to 20 per cent reduces competitiveness and the additional 7.5 per cent contribution from the UK is miniscule when climate

Kyoto Agreement: UK baseline figures.

Sector	Sector (MtC equivalent)	National (MtC equivalent)	Major sectoral emission sources
Carbon dioxide	63	168	Fuel combustion
Methane	7.6	25	Offshore oil and gas production, coal mining, gas distribution
Nitrous oxide	1	18	Refineries
Hydrofluorocarbons	0	4.2	
Perfluorocarbons	0	0.2	
Sulphur hexafluoride	0.03	0.2	Electrical switching gear
Total	72	216	

Base year is 1995 for HFCs, PFCs, and SF_6.

change is global. A CARBON TAX is often seen as an easy option but this is fraught with difficulties (e.g. favouring NUCLEAR POWER). And where is the point of application to be: at end use or power generator? If businesses are taxed the primary producers of chemicals, paper, cement, etc. could have disproportionate tax burdens compared with others and therefore they will attempt some form of exemption and leave a greater burden on others. This is a textbook example of simple nostrums being very difficult or inequitable in their implementation. (◊ GREENHOUSE GASES; GREENHOUSE EFFECT)

In March 2006, the UK dropped its target of 20% emissions by 2010. Other countries, e.g. Canada, have done the same to their targets. The US is beyond the pale, so far. The doubletalk in this area is staggering: patio heater sales abound, cheap flights proliferate, and Virgin have been offering half price hot air balloon flights ['Unforgettable' – Metro Christmas Reader offer, p. 40, 21 November 2006]. Happy Days. And let's not forget the 12 000 mile round trip planned for prawns (langoustines) from Scotland to be shipped to Thailand for hand shelling and then back to Scotland for breading and cooking for sale in supermarkets. You couldn't make it up. [Times, 16 November 2006 S. English.]. You could sign up to www.flightpledge.org.uk and voluntarily commit to reduce your flight frequency, or even not fly altogether.

Finally, the 2007 IPCC report is a worthy reminder of the scale of the problem. The 2007 Report by the Intergovernmental Panel on Climate Change, based on over 29 000 pieces of data measuring actual physical and biological changes, concludes that the observed increase in the global average temperature was 'very likely' due to anthropological greenhouse gas emissions. World greenhouse gas emissions grew by 70 per cent between 1970 and 2004. Emissions are expected to grow by between 25 and 90 per cent between

2000 and 2030. Developed countries accounted for 46 per cent of all emissions in 2000 but just a fifth of the world population. Up to about 75 per cent of the projected growth in emissions of carbon dioxide from energy use will come from developing nations. Limiting concentrations of greenhouse gases to 445 parts per million of the atmosphere, would reduce annual GDP growth rates by less than 0.12 per cent a year. (◊ IPPC)

L

LATS. ♭ Landfill allowance trading scheme whereby L.A.'s with a notional surplus landfill capacity can sell it on to those whose landfill quota has been exceeded.

LC$_{50}$. ♭ LETHAL CONCENTRATION.

LD$_{50}$. ♭ DOSE.

Leq. ♭ NOISE INDICES.

L$_{10}$. ♭ NOISE INDICES.

L$_{90}$. ♭ NOISE INDICES.

Laminar flow. A fluid flow pattern in which each layer of liquid or gas flows smoothly and is not broken up. When the flow becomes mixed up it is said to be turbulent. (♭ TURBULENCE)

Landfill Allowance Trading Scheme (LATS). Landfill Allowance Trading Schemes convey the right for a waste disposal authority to landfill a certain amount of biodegradable municipal waste in a specified scheme year, as part of the Government's efforts to meet the demands of the EU Landfill Directive. The LATS system sees progressively tighter restrictions on the amount of biodegradable municipal waste – defined as paper, food and garden waste – that disposal authorities can landfill. These allowances are tradable, so that high landfilling authorities can buy more allowances if they expect to landfill more than the allowances they hold. Similarly, authorities with low landfill rates can sell their surplus allowances. Councils which exceed their Allowance are liable for a fine of £150 for every tonne they landfill beyond the limit.

Landfill Directive (EU). A major piece of legislation which will result in substantial environmental improvements.

Background. The European Commission submitted a Proposal for a Directive on the Landfill of Waste to the Council on 22 July 1991. In 1995 a Common Position was reached by the Council. The European Parliament, however, rejected the Common Position on 22 May 1996. The Parliament objected to the high number of derogations on the grounds that it severely limited the effect of the Directive. The Parliament particularly objected to the exclusion of more than 50 per cent of the European Community territory by the derogation for areas with a population density of less than 35 persons per square kilometre. The Commission subsequently withdrew its Proposal. In June 1996, in the Resolution on the Waste Strategy, the Council invited the Commission to present a new Proposal on the Landfill of Waste without delay.

Following a significant delay, European Commissioners voted to transmit a revised proposal for a Landfill Directive to the Council of Ministers.

The EU agreed a common Position on this very important proposal (March 1998) which is outlined below.

Key proposals

(1) *Biodegradable waste.* The proportion of biodegradable municipal waste going to landfill must be reduced to 75 per cent by 2006, 50 per cent by 2009 and 35 per cent by 2016, based on 1995 levels. In recognition that some member states now send 90 per cent of their wastes to landfill, unless they choose to, the UK amongst others will not have to meet the first target until 2010 and the final target until 2020.

The Commission had proposed to reduce biodegradable municipal waste disposed to landfill to 75 per cent of 1993 levels as far as possible by 2002, with further reductions to 50 per cent by 2005 and 25 per cent by 2010. The European Parliament had recommended making the targets *obligatory* and extending the categories to cover biodegradable wastes from industry and public institutions.

(2) *Treatment.* A further requirement is that putrescible waste must be pretreated (this includes sorting) before landfilling, provided such pretreatment has the beneficial effect of reducing the quantity or hazard to human health or the environment. This sensible concession allows

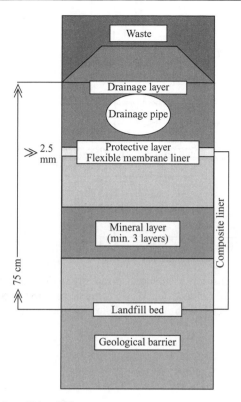

Figure 113   German class II landfill.

homogenous wastes (e.g. soil and wastes from effluent treatment) to be lanfilled directly.

(3) *Mixing waste*. Mixing of hazardous and non-hazardous waste has at long last been banned. But an exemption will be allowed for landfilling vitrified hazardous waste in separate cells in non-hazardous sites.

The adoption of much higher engineering standards for all new landfills will be a welcome outcome of the Directive and ENVIRONMENT AGENCY action. The standard of construction of a German class II landfill is illustrated in Figure 113.

Landfill gas (LFG). LFG is generated in landfill sites by the anaerobic decomposition of domestic refuse (municipal solid waste); see Figure 114 (◊ METHANE). It consists of a mixture of gases and is colourless with an offensive odour due to the traces of organosulphur compounds. A typical landfill gas composition for mature refuse (% by volume) is given below.

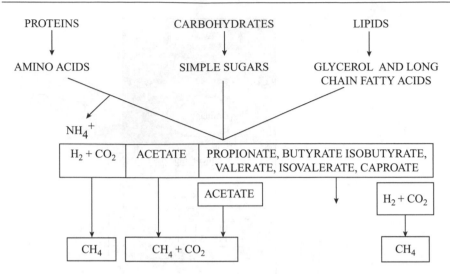

Figure 114 Decomposition of materials occurring in domestic waste.
Source: *The Control of Landfill Gas*, Waste Management Paper No. 27, HMSO. (Crown copyright is reproduced with the permission of the Controller of Her Majesty's Stationery Office.)

Component	Typical value (%)
Methane	63.8
Carbon dioxide	33.6
Oxygen	0.16
Nitrogen	2.4
Hydrogen	<0.05
Carbon monoxide	<0.001
Saturated hydrocarbons	0.005
Unsaturated hydrocarbons	0.0009
Halogenated compounds	0.00002
Organosulphur compounds	<0.00001
Alcohols	<0.00001
Others	0.00005

Aside from its unpleasantness, it is highly dangerous as methane is explosive in concentrations in air between 5 per cent, the Lower Explosive Limit (LEL), and the Upper Explosive Limit (UEL) of 15 per cent.

Waste Management Paper 27 'Control of Gas', Revised Draft 3rd Peer
Review Group Meeting, 01/08/88, gives the following recommendations for
action should LFG migrate to dwellings or buildings.

Gas level		Procedures
Per cent gas in air	*Per cent lower explosive limit*	
<0.5	<10	No action
0.5	10	Audible alarm
		Verification of concentration with a portable instrument
		Ventilation of buildings using windows and doors
		Turn off/extinguish potential ignition sources
1.0	20	Evacuate building, notify emergency services
5.0	100	Lower explosive limit (LEL)

The volume of LFG generated with time is given in Figure 115. Both the
time-scale and the amounts are variable, but 1 tonne of domestic refuse can
generate up to 450 m^3 (in addition, putrescible wastes that are co-disposed in
the landfill site can substantially increase the volume of LFG generated).
LFG will appear within 3 months to 1 year after waste deposition, peaking
at 5–10 years and tapering off over 20–40 years. However, many factors
interplay and it is for this reason that closed landfill sites must be monitored

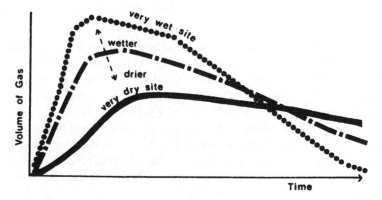

Figure 115A   Volume of landfill gas generated vs. time.

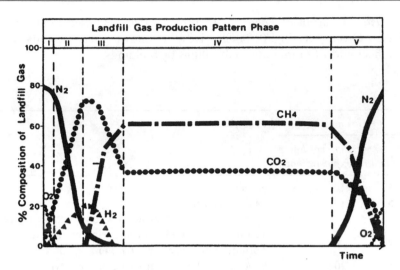

Figure 115B   Landfill gas composition vs. time.

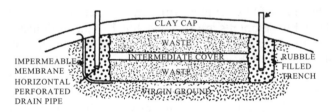

Figure 116A   Shallow site venting system.

for up to perhaps 30 years after site closure to ensure that there is no repetition of the infamous Loscoe UK LFG explosion which took place on 24 March 1986 destroying a bungalow and severely injuring the occupants. This occurred $3\frac{1}{2}$ years after the Loscoe landfill site was closed.

Landfill gas must be controlled at all operational landfill sites, whether actively or passively vented or both, especially in the case of deep sites.

Figure 116 shows schematic venting systems for shallow and deep sites respectively. It should be noted that barrier systems to prevent lateral gas migration are not enough and that for deep sites extraction will also be required.

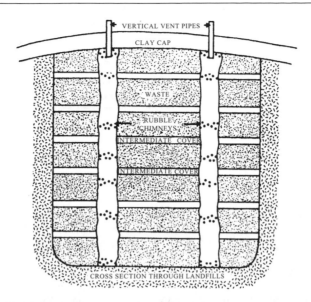

Figure 116B   Deep site venting system.

LFG utilization in the UK centres on using it for cement or brick kilns or on-site electricity generation in modified reciprocating engines or gas turbines. Good results have been achieved with all these uses. However, it must be said that only around one-third of the available refuse energy can be utilized by LFG extraction because of losses. INCINERATION with energy recovery is much more energy efficient than LFG collection as it uses all of the refuse for fuel purposes. Also, waste incineration prevents the formation of the LFG in the first place, thereby stopping the powerful greenhouse gas, methane, being formed. This has a greenhouse effect of up to 65 times that of carbon dioxide (on an equal mass basis and 20 year time-scale). UK Landfill sites have been variously estimated as emitting 1 to 2.5 mtpy methane, depending on the assumptions made in the 'collection efficiency' and the percentage of carbon which biodegrades in the landfill – an unsatisfactory state of affairs.

As many old landfill sites are now considered to be prime candidates for development, the options are either waste removal and redeposition elsewhere ca. £30/m³ + landfill tax or piling of the site plus provision of a ventilation space between the floor slab (sealed against gas) and the landfill cover.

The extent of the UK's LFG problems are only now coming to light. The first Annual Report (1989) of the UK HM Inspectorate of Pollution revealed that 1390 active or recently closed landfill sites posed a potential gas risk, as set out below.

	Total	Distance to housing or industry (metres)		Gas control installed
		0–250	250–500	
Active landfills:				
England	581	319	70	135
Wales	21	5	3	2
Total	602	324	73	137
Landfills closed since 1977:				
England	761	426	67	81
Wales	27	6	5	0
Total	788	432	72	81

Although somewhat historic now, the above data shows that the UK's former laissez faire policy has bequeathed a legacy of risk.

The seriousness (both to lives and financial consequences) with which methane gas migration must be viewed is exemplified by the UK Abbeystead, Lancashire, pumping station explosion in 1984. Sixteen people were killed and 28 injured. Compensation claims totalling up to £4 million (by those involved and North-West Water) have resulted against the firm of consulting engineers who designed and supervised the construction of the pumping station into which naturally occurring methane gas had migrated. An explosion was triggered when the pumps were switched on.

Sources: *Methane Emissions from UK Landfills*, AEA Technology, March 1999; A. Porteous and S. Burnley, Letter to *Waste Manager*, November 1999.

Landfill gas controls. Landfill gas (LFG) production is controlled by the following.

1. Waste composition. It is for this reason that the EU LANDFILL DIRECTIVE limits the amount of organic content in municipal solid waste which is landfilled.
2. Physical dimensions of site. ANAEROBIC processes dominate the site at depth >5 m.
3. Progressive restoration coupled with high input rates encourage more rapid development of the anaerobic process.
4. Daily or intermediate cover and the use of low permeability materials in cell wall construction may encourage perched WATER TABLES to develop and have effects on moisture movement, transmission of gases and

buffering of LEACHATES. Such effects will be important in terms of gas production, migration pathways and proposed methods of gas control.

5. The degree of saturation of the waste and its density depend on both the void space and the absorptive capacity of the waste. The greater the waste density in a landfill the higher the theoretical yield of landfill gas per unit volume of void space. However, water movement within the waste is necessary to permit the free movement of nutrients for bacteria to flourish. High waste densities reduce the permeability of the waste to gas, resulting in a build-up of gas pressure.

6. A moist environment encourages high rates of gas production. Waste moisture, rainfall, surface and groundwater infiltration and the products of waste breakdown all contribute. The recirculation of leachate practised on some sites maintains high moisture contents and provides a source of nutrients and bacteria which will accelerate gas production rates.

7. Methanogenesis proceeds optimally in a pH range of 6.5 to 8.5 and is only inhibited when the pH value is outside this range.

8. Shallow landfill site variations in production rates may among other factors reflect seasonal changes in ambient temperatures and residue gas production rates.

9. Gas production may be recommenced if changes occur at the site which reactivate microbial activity. This could occur if development occurs on the site or liquid levels within the wastes are allowed to rise by cessation of pumping of leachate.

10. The FLUSHING BIOREACTOR concept, if adopted, will require a considerable rethink on LFG controls.

Finally, flare emissions must now be controlled. The table below gives typical results available from well-designed flares.

Substance	TA-Luft (mg/Nm³)	Biogas HTE Flare Peak Level (mg/Nm³)	Biogas HTE Flare Mean Level (mg/Nm³)
Carbon monoxide (CO)	50	26	14
Nitrogen oxides (NO$_x$)	200	77	46
Sulphur dioxide (SO$_2$)	50	18	4
Hydrogen fluoride (HF)	1	1	0.7
Hydrogen chloride (HCl)	30	25	13
Unburnt hydrocarbons (as C)	10	ND	ND
Dioxins and furans (TEQ)	0.18	ND	ND

Source: Biogas Technology Ltd, personal communication, 25 June 1996; Environment Agency Draft on Landfill Gas Flare Emissions, 1999.

EA UK limits for emissions (for flares commissioned after 31 December 2003) based on normal operating conditions and load (Temperature: 0 °C (273 K); pressure: 101.3 kPa; oxygen: 3% (dry gas)) are shown below. (Note that no dioxin levels are currently enforced by the EA.)

Component	$(mg/N\ m^3)$
NO$_x$ (expressed as NO$_2$)	150
CO	50
Total Volatile Organic Compounds (VOCs)	10
Non-methane Volatile Organic Compounds (NMVOCs)	5

Landfill gas surface emissions monitoring. As discussed in relation to LANDFILL GAS, the US EPA now requires landfill gas surface emissions to be monitored. This has been done in California since April 1985. (◊ LANDFILL GAS MONITORING)

Californian landfills must be monitored in several categories: ambient air and weather station monitoring; instantaneous and integrated surface sampling; landfill gas migration perimeter probe monitoring; and landfill gas collection system sampling.

The air samplers are placed at predetermined locations. Ambient air samples are taken monthly over a 24-hour period; wind data are continuously recorded.

To prohibit photochemical reactions, ambient air bag samples are enclosed in light-sealed containers, usually cardboard boxes. Within 72 hours of collection, the Tedlar bag samples must be analysed for total organic compounds (TOC) and toxic air contaminants.

Composite surface sampling, also called integrated surface sampling, is required each month. The landfill surface is divided into monitoring/sampling grids, each approximately 50 000 square feet (4600 m^2).

Sampling a portion of the landfill area is required each month; the sample size depends on the total landfill area. California's limit for TOC concentrations is 50 ppm, measured as methane.

In addition, the Californian Air Quality Management District (AQMD) system includes probes installed outside the refuse deposit area and parallel to the landfill's perimeter; probe locations are more than 1000 feet (320 m) apart. At least one probe is installed at a pre-approved depth at each location.

Probe sampling is required monthly. Methane concentration limits of 5 per cent methane are set and enforced by the California Integrated Waste

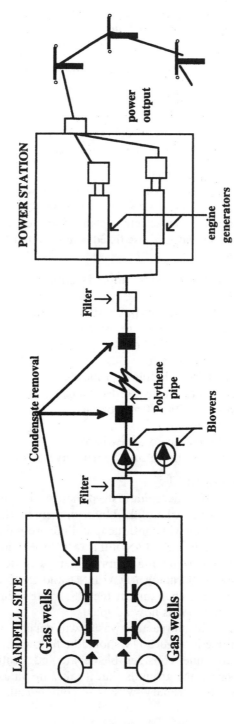

Figure 117 Schematic of a landfill gas plant.
Source: Waste Management International plc, News Release, January 1995.

Management Board. The depth of each probe is determined by the refuse depth (within 160 m of the probe), as follows:

First depth: (3 m) below surface.
Second depth: 25 per cent of refuse depth or (8 m) below surface, whichever is deeper.
Third depth: 50 per cent of refuse depth or (16 m) below surface, whichever dimension is deeper.
Fourth depth: 75 per cent of refuse depth or (24 m) below surface, whichever is deeper.

Landfill site. A disposal site as defined in the UK ENVIRONMENTAL ACT 1990 for the disposal of controlled wastes. Figure 118 shows an 'ideal' landfill operation with both LANDFILL GAS and LEACHATE collection systems installed. Once the site has been selected, planned and engineered it should comply with guidelines similar to those listed below so that the enviornmental impact is minimized:

1. All deposits of waste made in individual layers which are compacted on deposition.
2. Layers no more than 2.5 metres in depth.
3. Each layer to be covered with earth, or similar, at least 255 mm thick.
4. Waste to be covered within 24 hours.
5. No waste to be tipped in water.
6. Screens erected to collect windblown rubbish.
7. Precautions taken to prevent fire and vermin.
8. Organic waste covered with 600 mm of earth.
9. Each deposit to be kept tidy.
10. Adequate competent labour to be available.
11. Each layer allowed to settle before the next layer is started.

(◊ WASTES CLASSIFICATION)

UK landfill practices are undergoing a steep change with highly engineered landfills now the norm for new landfills. Unfortunately there is a continuing legacy of 'leaky sieves' which will continue to pollute groundwaters and emit landfill gas for many decades. The LANDFILL TAX (or levy in which the revenues are returned for environmental improvement) can do much to minimize waste arising by encouraging RECYCLING and INCINERATION (with energy recovery). Times are changing in the UK waste industry and a systematic approach (Figure 119) is now adopted.

The design approach for bioreactive wastes will depend on a number of factors. The decision tree in Figure 119 should be followed.

Landfill practices are under considerable debate and scrutiny. The use of cover has been queried on the grounds that it takes up valuable void space which could be used for primary wastes – commercial, industrial and

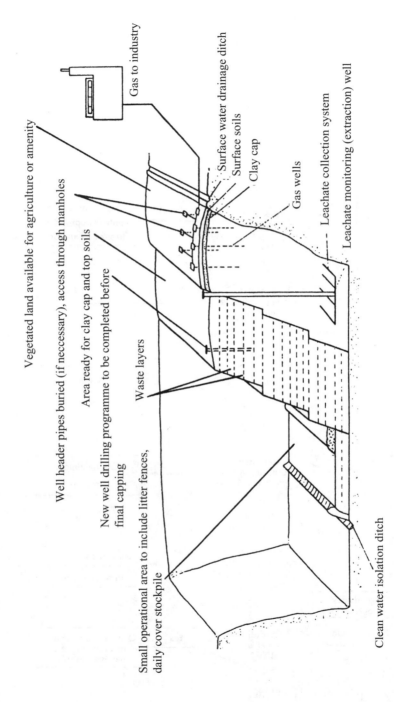

Figure 118 Cut away section of 'ideal' landfill operation with gas utilization.

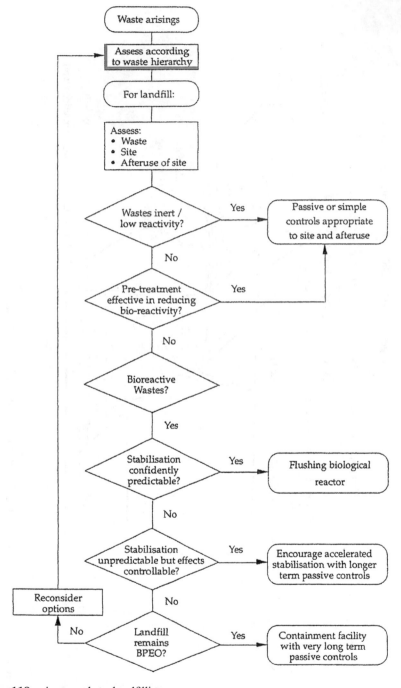

Figure 119 Approach to landfilling.
Source: Department of the Environment, 1997, Waste Management Paper 26B, Landfill
Design, Construction and Operational Practice, The Stationery Office, London.

household. If applied at the minimum depth prescribed by licences of 150 mm and placed for every 2 m depth of waste, some 7.5 per cent of the void space would be lost. Further to this, faces and flanks have to be covered resulting in even more void being lost. It is likely that the total void space lost will be well in excess of 25 per cent. This void space loss could be reduced if operators could move away from the 2 m lift of waste prescribed in the licence. With appropriate compaction techniques there is no good reason why the depth of the layer of waste could not be doubled or even trebled.

There is obviously a commercial disadvantage for operators if they are filling their valuable resource, the void, with low revenue wastes. There is also a distinct advantage to the community as a whole as the void would be filled faster than perhaps is desirable.

The void is a valuable resource and to fill it with inappropriate materials can only be counterproductive. Suitable voids, those capable of being engineered for containment purposes, are a fast diminishing resource. Therefore, it is essential that those capable of receiving a wide range of wastes are used to their best advantage.

Source: *Waste Management*, February 1996.

The House of Lords Inquiry into Sustainable Landfill, 1998, stated that: 'In our view, the directive [EU LANDFILL DIRECTIVE] does not explicitly advance the cause of sustainability.' Useful though the 'hierarchy' may be as a concept, it is perhaps better to see it less as a chain (or ladder) and more as a matrix of opportunities and options, together forming an integrated strategy in which the Best Practicable Environment Option (BPEO) will guide policies in specific circumstances – for example, reflecting the widely differing conditions of scattered rural communities and large conurbations. This has been well put by the Scottish Environment Protection Agency: 'At its very simplest, sustainable development can be construed as attempting to move away from a linear-based system of resource utilisation–process–waste generation to a more closed-loop approach maximising the efficient use of resources and minimising waste.' We should like to see this matrix approach better reflected in the Directive or at least in its national transposition.

Landfill will continue to play an essential part in the portfolio of waste disposal practices which, in the right combination and developed over the right time-scale, must feature in sustainable waste management policy. But it must be accepted that in the relatively small islands which make up the United Kingdom, both the population pressures in the centres of economic activity and the need to protect an irreplaceable natural heritage present a powerful challenge to landfill in the long run. Both landfill and land-raising are land-hungry: sites which are acceptable to their neighbours are increasingly difficult to find. There may be no immediate threat to the supply of traditional holes in the ground; and where they rest on substantial clay deposits they will continue to be the BPEO for a substantial proportion of total waste disposals. But it should be recognized that their continued availability depends, in part, on unsustainable practices in the construction industry. The construction industry cannot take it for granted that raw materials can be quarried and extracted from the environment on demand without regard to the alternative of using recycled construction wastes as aggregate and hardcore. *Waste reduction is not just an option; it is an imperative.*

Figures for 2005–06 were landfill 64%, recycling and composting 27%, energy from waste 8%. The UK is gradually moving away from reliance on landfill (2006) to a much more integrated approach as per ECO-CYCLE practices.

Case study: integrated waste management

A 25-year contract has been let (December 1997) involving the formation of Kirklees Waste Services Ltd (KWSL), a joint venture company, in conjunction with Kirklees Metropolitan Council. KWSL took over immediate responsibility for two sites and two transfer stations and will, over the next five years, provide the following:

- a new waste-to-energy plant on the site of the former incineration unit;
- a new materials recovery facility;
- a new transfer loading station;
- two new green waste composting facilities;
- conversion of three Civic Amenity sites to household waste recycling centres.

In so doing, KWSL anticipates capital investment of £40 m and, by 2002, the diversion of over 60 per cent of the waste arising away from landfill.

Features of the contract include the following.

(a) A fixed fee will be payable by the Council to the contractor regardless of which facility the waste passes through. Consequently, the Council can look forward to long-term stability with full knowledge of future charges.

(b) The authority has entered into a partnership with KWSL and will be fully involved in modifying arrangements in accordance with future conditions. Furthermore, the contract is flexible enough to cope with changing needs and demands brought about by new legislation and other variabilities in conditions.

(c) A combination of technologies has been used, making possible a significant shift away from the current dependence on landfill which is fully consistent with the principles of sustainable waste management.

Sources: C. Burford, 'Integrated Waste Management in the UK', Integrated Waste Management Session, IWM Annual Conference, June 1998. House of Lords European Communities Committee 17th Report, *Sustainable Landfill*, 1998.

A joint approach to raising recycling rates

In recent years the shift in the UK waste strategy away from landfill to a more sustainable system based on higher levels of recycling and recovery has highlighted the need for an even more integrated management approach.

Reclaim

In January 2000, work began on procuring a 25-year PFI contract to improve recycling within West Sussex. It has been branded Reclaim and was awarded to Viridor Waste Management in April 2004. The outcome of this work is an innovative solution that has recycling and composting at its heart.

In developing the approach, a Memorandum of Understanding (MOU) was agreed between the County Council and the seven district/borough councils to underpin the working relationships and the delivery of the contract.

Since April 2004, four Household Waste Recycling Sites have been totally rebuilt and the first brand new facility in West Sussex for more than a decade has been provided. A further three sites are undergoing major upgrades and two new Waste Transfer Stations and a Materials Recovery Facility (MRF) will complete the build programme. The new infrastructure has supported an increase in recycling from 22 to 34 per cent and will enable the local authorities to achieve their target of recycling 45 per cent by 2015.

End disposal has not been ignored and the local authorities are currently in the process of procuring innovative solutions to waste minimization and end treatment processing under the Materials Resource Management Contract.

Strategy

The local authorities within West Sussex have a strong history of producing joint waste management strategies. The first strategy 'A Way with Waste' opened the doors to encouraging the reduction of waste produced via increased composting which, together with recycling, started the change in reducing how much waste was being sent to landfill. In July 2006, the local authorities adopted a 30-year strategy known as the Materials Resource Management Strategy. This document was the subject of an award-winning programme of public consultation, and is a long-term strategy covering the period until 2035. It is one of the first strategies within the UK to undergo a full Strategic Environmental Assessment along with a MORI quantitative survey to determine the public's perception of waste management practices.

Conclusion

The strength of the partnership between each of the local authorities and the private sector means that there is the will to develop integrated solutions to further drive waste away from landfill and to increase waste reduction and recycling. The heavy emphasis on considering generated waste as a resource will ensure that the focus within West Sussex will always be on recycling and recovery rather than on disposal.

Source: P. Russell, Head of Waste Management Services, West Sussex Council

Landfill sustainability. The UK Chartered Institute of Wastes Management has developed the following objectives for a sustainable landfill:

- the contents of the landfill must be managed so that outputs are released to the environment in a controlled and acceptable way;
- the residues left in the site should not pose an unacceptable risk to the environment, and the need for aftercare and monitoring should not be passed on to the next generation;
- future use of groundwater and other resources should not be compromised.

In addition, it has emphasized that it is necessary to consider approaches both for existing sites and for new sites; it could be suggested that slow leakage to the environment can be better than a total containment if slow improvement and stabilization are achieved without any irreversible harm being caused; sustainability may need to be considered on a site-specific basis, in conjunction with a risk assessment for the site concerned. (⟡ FLUSHING BIOREACTOR)

Source: *The Role and Operation of the Flushing Bioreactor*, IWM Sustainable Landfill Working Group, April 1999.

Landfill tax. A levy per tonne or cubic metre of WASTE sent to LANDFILL. This is used, for example, in Sweden to encourage the use of RECYCLING and WASTE MINIMIZATION.
    Landfill tax has been in place in the UK since October 1996.
    Landfill tax kills three birds with one stone:

- by deterring landfilling of waste, we can help our environment,
- by encouraging recycling, we can save resources,

- by raising revenue from waste disposal, we will be able to make further cuts in employers' national insurance contributions and so create more jobs.

As an illustration, if it were to raise £3500 million, that could be used to finance a cut in the main rate of employer National Insurance Charges of 0.2 per cent or a reduction of more than 1 per cent in the lower rates of employer NICs.

The tax, collected by HM Revenue & Customs, is designed to use market forces to protect the environment by increasing the cost of disposing of waste in landfill sites. Waste disposal companies pass the additional costs on to waste producers who, in turn, are made aware of the true costs of their activities and so have an incentive to reduce waste and make better use of the waste they produce.

The tax provides an additional incentive for RECYCLING, which reinforces the Government's general policy of sustainable waste management. It reflects Ministers' views that taxation should play an important role in protecting the environment.

The tax is charged per tonne of waste and inert waste is subject to a reduced rate of tax.

A disposal of material is a disposal of waste for landfill tax if:

- The material is disposed of as waste – this means that the waste producer disposes of it intending to discard it or throw it away. Even if material could have been reused, it is still waste if it has been discarded.
- The disposal is by way of landfill – this means that the waste has been disposed of on land or under the surface of land.

Where soils used to engineer landfill sites are bought in specifically for site engineering (for example, to line or cap the site), they are not regarded as waste, and so will not be taxable.

The tax on a tonne of active waste (i.e. biologically active as opposed to inert) in 2007 was £24. It is subject to an escalator of £8 per tonne each year as an incentive for business to reduce, reuse & recycle wastes.

The rate increase builds on the important role played to date by the tax with almost a third of companies considering waste minimization, reuse and recycling. Analysis to date suggests that a long-term programme of increases in landfill tax could significantly reduce the proportion of waste expected to be landfilled.

Lapse rate, temperature. The rate of change of air temperature with increasing height: usually an average rate over a distance (commonly 100 metres). It is often stated as the ADIABATIC lapse rate of dry air, which is very close to that of moist, unsaturated air. Its value is about 1 °C/100 m. (There is also the saturation adiabatic lapse rate, the adiabatic lapse rate of air that is saturated with water vapour. This is less than that of the dry adiabatic lapse rate.)

Super-adiabatic lapse rate is a temperature lapse rate which is greater than the dry adiabatic lapse rate. This occurs over a land strongly heated by the sun, and is usually very favourable to the dispersion of air pollutants. However, it causes looping of elevated PLUMES and can lead to pollutants being brought down to ground level in high concentrations for short periods.

Latent heat. The energy associated with evaporation of water without a change in temperature.

Laws of thermodynamics. The laws of thermodynamics apply to all known phenomena.

The First Law of Thermodynamics. ("You don't get something for nothing")
This law is a statement of the conservation of ENERGY, i.e. energy can neither be created nor destroyed. Energy has many forms – electrical, chemical (the combustion of coal is release of chemical energy), thermal, nuclear, etc. The first law tells us that it is the total energy in all its various forms which is a constant. Physical processes only change the distribution of energy, never the sum. While, the first law only tells us energy is conserved, it does not tell us the direction in which processes proceed. This is the province of the Second Law of Thermodynamics.

The Second Law of Thermodynamics. ("You get very little for 5 pence")
This law specifies the direction in which physical processes proceed. One form of this is that heat transfer takes place spontaneously from a hot body to a cooler body. Note that the opposite process, spontaneous heat flow from a cold body to a hot body (the hot body becoming warmer; the cold body becoming colder) *does not violate the first law*. It does not occur because it violates the second law.

The second law is also frequently formulated in terms of order and disorder. It tells us, for instance, that concentrations will decrease, e.g. a sugar lump will dissolve in water, or in general that order becomes disorder. The general statement of the second law is that everything proceeds to a state of maximum disorder, and by implication is less useful in the event of the disorder occurring, e.g. a sugar lump is more useful when it is not dissolved in the water. A kettle of boiling water is more useful than the same amount of water mixed in a bath of cool water. All these examples are manifestations of the second law. It applies to all biological and technological processes and cannot be circumvented. It explains that the final outcome of energy consumption is the eventual production of heat at very low and near useless temperatures, such as the waste heat discharge from power stations or heat losses from the body.

The laws of thermodynamics are often used to determine the ideal efficiency for a given process and explain why the conversion of heat to work cannot be done on a one-for-one basis, i.e. why 1 kilowatt-hour of electrical energy output from a generator requires 3 or 4 kilowatt-hours of thermal energy input. ($\lozenge$ CARNOT EFFICIENCY)

Leachate. The seepage of liquid through a waste disposal site or spoil heap. Leachates from municipal waste landfill sites (early years of decomposition) can be characterized by high BODs (up to 20 000 mg/l), high BOD:COD ratios, several hundred mg/l ammonia and several hundred mg/l organic nitrogen, high concentrations of volatile fatty acids, acidic pH and an unpleasant smell.

The final 'methanogenic' or 'stabilized' phase is characterized by:

- BOD < 200 mg/l; COD several hundred mg/l;
- low BOD:COD ratio; NH_3 several hundred mg/l;
- very low concentration of fatty acids and neutral to alkaline pH.

If household wastes have been landfilled, the leachates will have a high oxygen demand which requires treatment either on site or at a sewage treatment works before being discharged into any receiving body of water.

Landfill site leachates have been recorded in strengths ranging from 100 to 54 000 mg/l total organic carbon. These values show that regular leachate monitoring is required both during operation and after closure of landfill sites in order to guard against water pollution incidents. This is particularly important after a period of heavy rainfall.

A historical analysis of the UK Pitsea waste disposal site leachate (1984) is given below (all in mg/l except pH).

pH	8.0–8.5
Total organic carbon (TOC)	200–650
Chemical oxygen demand (COD)	850–1350
Biochemical oxygen demand (BOD)	80–250
Ammonia nitrogen, NH_3–N	200–600
Organic nitrogen (as N)	5–20
Oxidized nitrogen (as N)	0.1–10
Alkalinity (as $CaCO_3$)	2000–2500
Phosphate (as P)	0.2
Total suspended solids (105 YC)	100–200
Volatile suspended solids (550 °C)	50–100
Fatty acids, C_1–C_6 (as C)	20

Source: K. Knox, Leachate production control treatment, Ch. 4 in *Hazardous Waste Management Handbook*, A. Porteous (Ed.). Butterworths, 1985.

Leachate values from the German (Dusseldorf) landfill before and after treatment, are tabulated below.

Parameter	Raw leachate (mg/l)	Plant effluent (mg/l)
pH	8.7	4.6
Conductivity	17000 mS/cm	33 mS/cm
COD	3900	<15
BOD	1400	2.4
AOX	2150	47
NH_3–N	1066	1.3
Chloride	2200	1.8
Magnesium	210	<0.1
Calcium	190	<0.5
Lead	0.1	0.02
Chromium	0.6	0.017
Copper	0.08	0.01
Zinc	0.4	0.4
Cadmium	0.007	<0.0002
Mercury	0.0003	<0.0002
Nickel	0.27	0.001

These extremely high discharge standards are achieved by the use of REVERSE OSMOSIS.

The standards now adopted in Germany are typified by that at Flotzgrün, where BASF operates an industrial waste landfill. In 1986 Bilfinger + Berger was awarded the contract for the construction of the sixth section of the landfill site. The contract was for the construction of the base of the site using a system developed by Bilfinger + Berger and BASF; it included the seepage shafts and the piped drainage system as shown in Figure 120.

The site base is formed by a double sealing 'quilt' comprising 2.2 mm thick plastic membranes of LUCOBIT (BASF's of civil engineering grade plastic material) spaced at a distance of 40 cm, with an intermediate layer of drainage gravel. This package is provided with pipes for leak detection, ventilation and injection if necessary (see Figure 121). The 'quilt' is divided into individual fields of approximately 50 × 50 m, which are linked to seepage shafts in groups of four. The shafts are made of individual sections which can slide into one another telescopically and have been designed for a final height of 55 m. The pipes within the 'quilt' and the overlying leachate collection pipes drain into these shafts. The leachate is fed through closed pipes into the collectors which, laid underneath the site base in protective tubes, interconnect

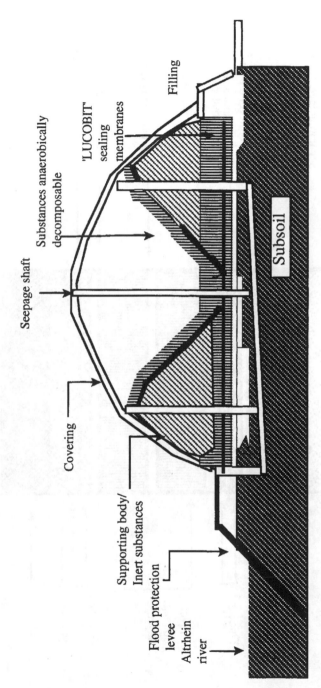

Figure 120 Double sealing quilt for landfill site.

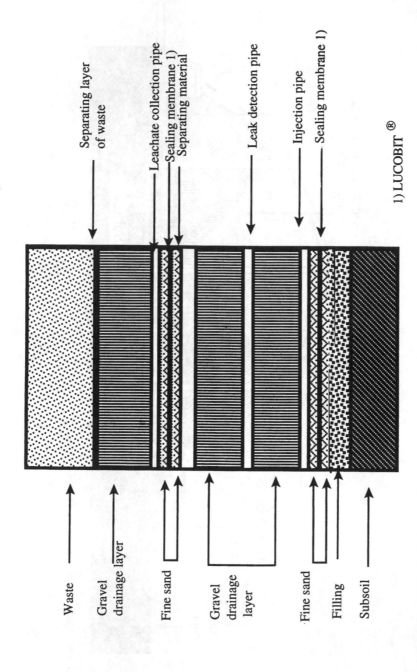

Figure 121 CONTREP base sealing system.

the individual shafts. The design of the protective tubes and the collectors is such that they can follow ground settlements without suffering damage. During operation of the site the leachate is pumped from a storage tank and transported to the BASF wastewater treatment in Frankenthal.

In order to protect the waste brought to the facility against floods, the new section, like the existing ones, is protected by a levee. When the waste has reached its final height it will be covered with another lucobit plastic membrane and landscaped soil.

Sources: *The Chemical Engineer*, Number 551, October 1993.
B. Croft, D. Campbell, 'Characterisation of 100 Landfill Sites', Paper presented at *Harwell Waste Management Symposium*, 1990.
Bilfinger + Berger promotional leaflet for CONTREP Base sealing system for BASF site at 'Flotzgrün'.
Waste Management International plc, News Release, January 1995. J. G. Dobrowlski, 'Controlling LFG Surface Emissions', *World Wastes*, March 1995.

Leaching. Commonly, the removal by water of any soluble constituent from the soil or from waste tipped on land. Generally, the gradual dissolution of a material from a solid containing it, e.g. metal from ore. Leaching of domestic refuse tips can give rise to a LEACHATE with a high BIOCHEMICAL OXYGEN DEMAND which is offensive and can spoil the amenity of neighbouring watercourses if untreated. Leaching often occurs with soil constituents such as nitrate fertilizers with the result that NITRATES end up in potable waters. Changes in farming practice such as not ploughing up grassland and programmed fertilizer application can lead to stabilized nitrate levels. However, in 1990, over 1.6 million Britons drank water that exceeded EC limits for NITRATES.

Leaching of calcium ions, both from the soil and from the leaves of plants which have already absorbed the calcium ions through their roots, is accelerated in areas where there is acidic rainfall due to industrial pollution. When this occurs the plant suffers as calcium is an essential material for cell structure and an enzymatic activator. (◊ ACID RAIN; ENZYMES)

Lead (Pb). A HEAVY METAL. A soft, blue-grey, easily worked metal, which has a multitude of uses, from lead acid accumulators to early water pipes, printer's metal and glazes. It is manufactured by roasting the ore (galena) in a furnace. World production is over 3.5 million tonnes annually.

As a pollutant, lead is a systemic agent affecting the brain. It has been suggested that it is associated with mental retardation and hyperactivity in infants living in soft-water areas where old lead piping is still in use, as the lead readily enters into solution. Anti-knock additives in petrol may also be implicated in infant retardation, and the implementation of lead-free motoring in Europe and the USA has ensured removal of this pollutant source. In addition, lead which is stored in the bones replacing calcium can be remobilized during periods of illness, cortisone therapy, and in old age. Research indicates that lead may blunt the body's defence mechanisms, i.e. the immune system. If this is so, then both the rate of infections by bacteria and viruses and the incidence of cancer can be enhanced by severe exposure. (◊ SYNERGISM)

An approved code of practice for the control of lead at work requires a worker to be suspended from work if a blood level of $80\,\mu$g/dl is found, or $40\,\mu$g/dl in the case of a woman. Even more stringent precautions apply in the US because of the recognized CHRONIC effects. Under EEC surveys of blood lead concentration carried out in the early 1980s, reference levels for triggering action to identify and reduce exposure were set at:

- no more than 2 per cent of any group to have blood lead > $34\,\mu$g/dl
- no more than 10 per cent of any group to have blood lead > $30\,\mu$g/dl
- no more than 50 per cent of any group to have blood lead > $20\,\mu$g/dl

Lead intake can come from drinking water, from food, and from inhaled particles from motor exhausts. The particles are less than $1\,\mu$m in size, and therefore can enter the lungs easily. A person breathing traffic-congested air may absorb close to the toxic level, excluding any intake from food and drink. Other sources are lead-based paints, and for some children, PICA results in them swallowing paint flakes or contaminated dust. (◊ CHELATING AGENTS)

Lean combustion (or lean burn). Technique for the reduction of AUTOMOBILE EMISSIONS which makes use of gasified petroleum as the engine fuel compared with the conventional method of fine droplets. This enables the fuel to be burnt in air to petrol ratios of around 20:1 as opposed to the conventional 15:1, thus greatly reducing emissions of NO_x and CO. ◊ UNBURNT HYDRO-CARBONS without resort to a CATALYTIC REACTOR.

Legionnaires' disease. Legionnaires' disease is a bacterial disease that may cause pneumonia. The majority of cases are reported as single (sporadic) cases but outbreaks can occur. The illness occurs more frequently in men than women. It usually affects middle-aged or elderly people and it more commonly affects smokers or people with other chest problems. Legionnaires' disease is uncommon in younger people and is very uncommon under the age of 20. It is called Legionnaires' disease because the first outbreak occurred in Philadelphia in 1976, among people attending a state convention of the American Legion. Subsequently, the bacterium causing the illness was identified and named *Legionella pneumophila*. It is widely distributed in the environment, and has been found in ponds, hot and cold water systems, and water in air conditioning cooling systems. Outbreaks often occur from purpose-built water systems where temperatures are warm enough to encourage growth of the bacteria, e.g. in cooling towers, evaporative condensers and whirlpool spas (tradename

Jacuzzi) and from water used for domestic purposes in buildings such as hotels.

To prevent the occurrence of Legionnaires' disease, companies which operate these systems must comply with regulations requiring them to manage, maintain and treat them properly. Amongst other things, this means that the water must be treated and the system cleaned regularly.

The disease is spread through the air from a water source. Person-to-person spread does not occur. Breathing in aerosols from a contaminated water system is the most likely route of transmission. The early symptoms include a 'flu-like' illness with muscle aches, tiredness, headaches, dry cough and fever. Sometimes diarrhoea occurs, and confusion may develop. These symptoms frequently lead on to pneumonia. Deaths occur in 10–15 per cent of otherwise healthy individuals, and may be higher in some groups of patients. The incubation period ranges from 2 to 10 days but is usually 5 to 6 days. In rare cases some people may develop symptoms as late as three weeks after exposure. Antibiotics against the infection are effective in treating the disease.

Leptospirosis. Leptospirosis (or Weil's Disease) is a disease that can be passed from animals to humans – such diseases are called zoonoses. It is more common in tropical areas of the world but is also found in temperate areas such as Europe, including the UK. Leptospirosis is caused by bacteria of the genus *Leptospira* (referred to as leptospires) which infect a variety of wild and domestic animals. The animals can then spread the leptospires in their urine. All mammals can probably carry some type of leptospire, and may therefore spread the disease among others of their own kind, and to other species, including humans. Common animal reservoirs (maintenance hosts) include rodents, cattle and pigs.

Human infection occurs through exposure to water or an environment contaminated by infected animal urine, and has been associated with a variety of occupations such as farming which can involve direct or indirect contact with infected urine or recreational pursuits. In the UK, such activities include canoeing, windsurfing, swimming in lakes and rivers, pot holing, and fishing.

Leptospirosis is primarily a disease of tropical and subtropical regions; it is uncommon in temperate climates. Leptospires are naturally aquatic organisms and are found in fresh water, damp soil, vegetation, and mud. Flooding after heavy rainfall may spread the organism because, as water saturates the soil, leptospires pass directly into surface waters. Infected animals carry the bacteria in their kidneys. They can excrete leptospires in their urine for some time, and spread infection to other animals or humans coming into direct or indirect contact with the urine. Often the infected animal does not become ill. For example neither rats, which carry the type known as *Leptospira icterohaemorrhagiae*, and cattle, which carry another strain (*L. hardjo*), appear ill. In general, herbivorous animals seem more likely to become, and remain, infected. Cattle urine is neutral or slightly alkaline (high pH), whereas the urine of carnivores tends to be acidic (low pH) – this acidity may damage the leptospires in the kidney, clearing any infection. Humans are considered to be a dead-end or accidental host of leptospires. Infection may be acquired by direct or indirect contact with infected urine, tissues, or secretions. Leptospires enter the body through cut or damaged skin, but may also pass across damaged or intact mucous membranes, and the eyes. They are also known to pass through intact skin. Person-to-person spread is very rare, if it occurs at all.

Leptospirosis can be used to describe infections in both man and animals caused by any pathogenic strain of leptospire. In humans it causes a wide range of symptoms, although some infected people appear healthy. All forms of leptospirosis start in a similar way. Leptospirosis is an acute biphasic illness. Some cases may be asymptomatic or may present in the first phase with the abrupt onset of a flu-like illness, with a severe headache, chills, muscle aches, and vomiting. This is known as the bacteraemic phase, when the leptospires spread through the blood to many tissues, including the brain. This phase may resolve without treatment. In some cases, an immune phase may follow with a return of fever, jaundice (yellow skin and eyes), red eyes, abdominal pain, diarrhoea, or a rash. In more severe cases, there may be failure of some organs, e.g. the kidneys, or meningitis. Generally, cases will recover fully within two to six weeks but some may take up to three months. After infection, immunity develops against the infecting strain, but this may not fully protect against infection with unrelated strains. Typically, symptoms develop seven to 14 days after infection, though rarely the incubation period can be as short as two to three days or as long as 30 days.

Leptospirosis is treated with antibiotics such as penicillin or doxycycline, which should be given early in the course of the disease. Intravenous antibiotics may be needed for people with more severe symptoms. There is no human vaccine available in the UK that is effective against leptospirosis. For people who may be at high risk for short periods, especially through their occupation, taking doxycycline (200 mg weekly) may be effective.

General advice includes taking whatever measures are feasible to reduce rodent populations, such as clearing rubbish and preventing rodent access into buildings. Immunising and treating infected animals is worthwhile. The risk of acquiring leptospirosis can be greatly reduced by not swimming or wading in water that might be contaminated with animal urine. Protective clothing should be worn by those exposed to contaminated water or soil because of their job or recreational activities. The use of gloves or protective footwear in potentially contaminated environments is recommended. Cuts or abrasions should be covered with waterproof dressings before possible exposure, and cuts or abrasions received during activities should be thoroughly cleaned. Showering promptly after immersion in surface waters is recommended.

Lethal concentration. The CONCENTRATION of a substance in air or water that can cause death. This is usually expressed as the concentration that is sufficient to kill 50 per cent of a sample within a certain time, and is abbreviated to LC_{50}.

The toxicity of materials is commonly investigated by the use of survival curves which plot the numbers of population of, say, fish surviving at various times with the concentration of the suspected toxic agent. For a short time almost all fish survive even at high poison concentrations. Very few survive for long periods, hence the LC_{50} concept is very useful as it spans the two extremes. Figure 122 shows survival curves for various concentrations of a toxic material. These curves are plotted as percentage survival against time. The time taken to kill half the fish is the intersection of the broken line and the concentration curves, e.g. a 75-week LC_{50} would be 20 grams per cubic metre.

Lethal dose. ◊ DOSE.

Leukaemia. A CANCER associated with changes in the lymphatic system. Acute lymphoblastic leukaemia is common in children and myeloid leukaemia is associated with workplace exposure to BENZENE.

Childhood leukaemia has been blamed variously on IONIZING RADIATION discharges and 'new town' viral infections when young families are gathered from diverse areas in close proximity to a new environment.

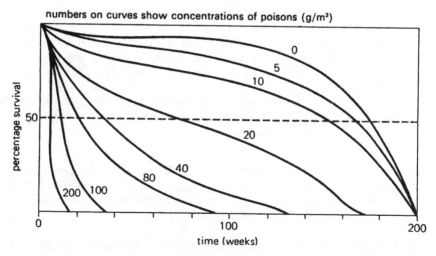

Figure 122 Percentage survival against time for toxicity tests on various concentrations. The points of intersection of the broken line with the curves are lethal concentrations for 50 per cent of the population (LC_{50}).

Statistically significant clusters have been found in new towns as well as near SELLAFIELD where nuclear fuel reprocessing takes place in the UK, but which itself has undergone a major population shift with large numbers of newcomers living in the area. It would be difficult to draw firm conclusions on the causes.

Work by Dr L. Kinlen of the Cancer Research Campaign's Epidemiology Unit has, on the basis of detailed examination of large rural construction sites, reinforced the hypothesis that leukaemia and non-Hodgkins' lymphoma are infectious diseases unconnected with radiation.

Dr Kinlen stated in the abstract of his joint paper:

Objective – to determine whether population mixing produced by large, non-nuclear construction projects in rural areas is associated with an increase in childhood leukaemia and non-Hodgkin's lymphoma.

Design – A study of the incidence of leukaemia and non-Hodgkin's lymphoma among children living near large construction projects in Britain since 1945, situated more than 20 km from a population centre, involving a workforce of more than 1000, and built over three or more calendar years. For period before 1962 mortality was studied.

Setting – Areas within 10 km of relevant sites, and the highland counties of Scotland with many hydro-electric schemes.

Subjects – Children aged under 16.

Results – A 37% excess of leukaemia and non-Hodgkin's lymphoma at 0–14 years of age was recorded during construction and the following calendar year. The excesses were greater at times when construction workers and operating staff over-lapped (72%), particularly in areas of relatively high social class. For several sites the excesses were similar to or greater than that near the nuclear site of Sellafield (67%), which is distinctive in its large workforce with many construction workers. Seascale, near Sellafield, with a nine-fold increase had an unusually high proportion of residents in social class I. The only study parish of comparable social class also showed a significant excess, with a confidence interval that included the Seascale excess.

Conclusion – The findings support the infection hypothesis and reinforce the view that the excess of childhood leukaemia and non-Hodgkin's lymphoma near Sellafield has a similar explanation.

Source: L. J. Kinlen, M. Dickson and C. A. Stiller, *Childhood leukaemia and non-Hodgkin's lymphoma near large rural construction sites, with a comparison with Sellafield nuclear site*, internal report for Cancer Research Campaign Epidemiology Unit, University of Oxford, Radcliffe Infirmary, Oxford, 1995.

Levies. Charges made against products to provide resources to support the recycling of the component materials. (◊ RECYCLING, FINANCIAL INCENTIVES FOR)

Liability. Most (UK) liability is fault based, i.e. requiring proven negligence of a liable party. As it is often very difficult to prove that any one party is liable, especially in CONTAMINATED LAND cases where sites have been subject to multiple contaminative uses or multiple ownership, the European Commission would like to complement any fault-based system with a 'strict liability system' where no fault need be proved.

This is akin to the US Superfund approach where clean-up can be mandated.

See below for an example of strict liability and the consequences:

R. v. *Yorkshire Water Services Limited* (July 1994)
Court of appeal

Yorkshire Water Services Limited ('YWS') pleaded guilty to offences under Section 107(1)(c) of the Water Act 1989, namely causing sewage effluent to enter controlled waters. YWS did not play any direct part in the discharges to the controlled waters as the day-to-day management had been delegated to its agent, Scarborough Borough Council. YWS accepted that it was strictly liable pursuant to Section 107(1)(c) of the Act, but appealed against the amount of the fines imposed by the Magistrates' Court, which were £50000 for the more serious incident and £25000 for the second incident.

The Court of Appeal accepted YWS's case and stated that the fines should have been substantially lower, particularly, it seems, as this was a case of strict liability on the part of a principal in respect of actions of his agent. The fines were reduced to £10000 and £5000 respectively.

Source: *Times Law Reports*, 19 July 1994.

Lichen. A compound plant formed by the symbiotic association of two organisms: a fungus and an alga (◊ SYMBIOSIS). They occur on a variety of surfaces, e.g. tree trunks, rocks, walls and the ground. Nutrition is derived from the dissolved solids in rain which are absorbed through the whole body surface. For this reason lichens are intolerant of poisonous substances and are therefore sensitive to air pollution, especially to sulphur oxides. Classification of surviving lichen flora has been used to monitor maximum air pollution levels reached over the country (◊ BIOLOGICAL INDICATOR) and also enabled levels of radioactivity from CHERNOBYL to be monitored in Poland as they absorbed radioactive CAESIUM-137; an increase of 165 times previous levels was found.

Lidar. A method for the detection of cloud patterns using the particle scatter radiation in a tuned laser beam. The method is also suitable for the detection of atmospheric pollutants (including hydrocarbons) and can deliver instantaneous measurements of traffic pollution.

Life cycle analysis (LCA). LCA is a method for evaluating 'the whole life of a product', that is all the stages involved, such as raw materials acquisition, manufacturing, distribution and retail, use and reuse and maintenance, recycling and waste management, in order to create less environmentally harmful products. LCA consists of three parts (Figure 123): inventory analysis (selecting items for evaluation and quantitative analysis) (Figure 124), impact analysis (evaluation of impacts on ecosystem), improvement analysis (evaluation of measures to reduce environmental loads).

Before carrying out an analysis, the boundaries of the operations that together produce the process or product have to be defined. This is important

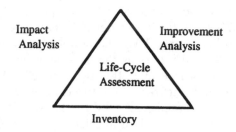

Figure 123 Sub-systems of LCA.

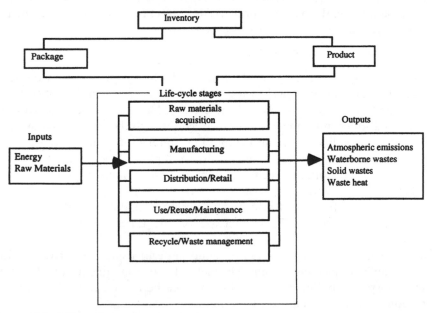

Figure 124 Life-cycle inventory procedures.

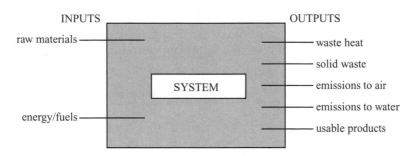

Figure 125

because if any part of the system contained within the boundaries is changed, all the other inputs and outputs will also change. This is shown schematically in Figure 125.

Life cycle analysis is only one of the names for this sort of analysis; it is also known as eco-balance, cradle-to-grave analysis, resource analysis, and environmental impact analysis.

The main purpose of an LCA is to identify where improvements can be made to reduce the environmental impact of a product or process in terms of energy and raw materials used and wastes produced. It can also be used to guide the development of new products.

It is important to distinguish between life cycle analysis and life cycle assessment. Analysis is the collection of the data – it produces an inventory; assessment goes one stage further and adds on an evaluation of the inventory.

An LCA does not define or explain actual environmental effect. For example, an LCA will tell us how many grams of limestone are used to make a bottle for mineral water and how much energy was used to extract it. But it does not tell us the environmental impact of this action, such as whether limestone is a scarce resource or whether its extraction causes pollution, blights housing or destroys scenic landscapes.

An LCA is especially applicable in the packaging field because, unlike washing machines, cars or many other consumer goods, packaging is a comparatively simple item made of few materials, and follows a fairly simple life cycle. (⟩ ENVIRONMENTAL IMPACT ASSESSMENT)

Life cycle inventory (LCI). The end result of a LIFE CYCLE ANALYSIS with aggregate emission such as NO_x, SO_2 or other impacts determined for a particular product and presented as an inventory of impacts. Great care is needed in appraising LCIs, e.g. in the cumulated LCI of a product, the figure of NO_x emissions is very high. This is a typical short-term, regional problem and the highest single contribution to the overall figure is required (e.g. emissions of a special waste incinerator). See Figure 124.

For the development of possible improvement options the decision maker wants to know:

1. Is the emission at the site we are talking about interpreted as an ecologically relevant problem?

   This would be visible in a low regional threshold, e.g. NO_x. Therefore, the location of the site has to be known in order to determine the valid threshold and the share of the whole critical air volume of this step attributable to the NO_x emissions, weighted with the emission threshold and the NOEC (No Observable Effect Concentration).

2. How good is the data quality of the overall figure? For example, is the scale of the NO_x emissions of the special waste incineration only high because the NO_x emissions of most other steps are not known and are set to 'zero' by default, rather than because of reality? This information helps to identify data gaps and is very useful to avoid sub-optimization due to bad data quality, or an expensive remediation of a minor ecological problem.

LCAs and the associated LCI compilations are fraught with value judgements. Care must be taken to avoid a 'flavour of the month' approach and certainly the 'hype' which accompanies said analyses in consumer products is to be ignored.

Source: H. Brunn, *LCA News*, Vol. 5, No. 2, March 1995.

Light water reactor. ◊ NUCLEAR REACTOR DESIGNS.

Lignin. The organic glue that holds the cell walls and cellulose fibres of plants together. It is a phenolic polymer which is resistant to biological attack. It forms 25–40 per cent of the wood of trees and must be dissolved chemically in PULP manufacture to obtain high grade pulp. (◊ DIGESTOR)

Its presence in agricultural wastes makes many of them unsuitable for feedstocks unless chemically treated. (◊ STRAW)

Limestone. One of the major classes of sedimentary rocks, consisting mainly of calcium carbonate, formed from the skeletal remains of fossil plants or animals, or by chemical precipitation.

Lindane. The active ISOMER of hexachlorocyclohexane is the γ-isomer, also known as lindane, gammexane, γ-BHC and γ-HCH. It is used as an insecticide. It is made by chlorinating benzene, which produces several isomers, but the γ-form is the only active ingredient of technical grades which include a mixture of isomers. It is these impurities which give lindane its undesirable, characteristic musty odour, which may taint food products. Lindane is a neurotoxicant with mammalian toxicity greater than that of DDT. The UK Advisory Committee on Pesticides, which regulates their use, is now considering new evidence on Lindane, which it cleared from accusations of a link to the human blood disorder, aplastic anaemia, in 1981. (↷ ENDOSULFAN; CHLORINATED HYDROCARBONS)

Linear Energy Transfer (LET). The linear rate of ENERGY transferred (i.e. dissipated) by the passage of particulate or electromagnetic radiation through an absorbing material. It is measured in units of energy deposited per unit length, e.g. eV (electron volt) per micron or joules per micron. (↷ IONIZING RADIATION)

Liquefied petroleum gas (LPG). A gaseous mixture of light HYDROCARBONS, whose maintained principal components are propane, propene, butanes and butenes, liquefied by increased pressure or lowered temperature.

List I substances. This is a list of 129 substances drawn up in the EC Dangerous Substances Directive (75/464/EEC) on pollution caused by certain dangerous substances discharged to the aquatic environment. The substances are considered to be sufficiently toxic, persistent or bioaccumulative (see BIOLOGICAL CONCENTRATION) that steps should be taken to eliminate POLLUTION by them. This is achieved by the setting of limit values through Daughter EU Directives. List I is also known as the Black List. From the 129 substances, the UK government has drawn up an initial list of 23 of the most dangerous substances and put into a RED List. Their discharges to water have to be minimized using BATNEEC principles.

List I contains the individual substances which belong to the families and groups of substances shown below, with the exception of those which are considered inappropriate to List I on the basis of a low risk of toxicity, persistence and bioaccumulation.

1. Organohalogen compounds and substances which may form such compounds in the aquatic environment.
2. Organophosphorus compounds.
3. Organotin compounds.
4. Substances which possess carcinogenic, mutagenic or teratogenic properties in or via the aquatic environment (including substances which have those properties which would otherwise be in List II – see below)
5. Mercury and its compounds.
6. Cadmium and its compounds.
7. Mineral oil and hydrocarbons.
8. Cyanides.

(↷ GROUNDWATER REGULATIONS)

Source: EC Groundwater Directive.

List II substances. Further to the entry for List I above, a substance is considered to be in List II if it could have a harmful effect on groundwater and it belongs to one of the following families or groups of substances:

(a) the following metalloids and metals and their compounds:

1.	Zinc	6. Selenium
2.	Copper	7. Arsenic
3.	Nickel	8. Antimony
4.	Chromium	9. Molybdenum
5.	Lead	10. Titanium
11.	Tin	16. Vanadium
12.	Barium	17. Cobalt
13.	Beryllium	18. Thallium
14.	Boron	19. Tellurium
15.	Uranium	20. Silver

(◊ GROUNDWATER REGULATIONS)

(b) biocides and their derivatives not appearing in List I
(c) substances which have a deleterious effect on the taste or odour of groundwater, and compounds liable to cause the formation of such substances in such water and to render it unfit for human consumption
(d) toxic or persistent organic compounds of silicon, and substances which may cause the formation of such compounds in water, excluding those which are biologically harmless or are rapidly converted in water into harmless substances
(e) inorganic compounds of phosphorus and elemental phosphorus
(f) fluorides
(g) ammonia and nitrites

List II is also referred to as the 'Grey List'.

Source: EC Groundwater Directive.

Lithosphere. The earth's crust, that is, the layers of soil and rock which comprise the earth's crust. When the complete earth's crust is meant, it is often referred to as the hydro-lithosphere. (◊ HYDROSPHERE)

Litter. The control of the unsightliness and associated public health hazards of litter, especially on beaches where sewage-related debris is often deposited along with plastics and paper, is of great concern. Below are data for UK beaches as an example.

The top 20 litter items recorded during Beachwatch 2005 accounted for almost 80 per cent of all litter items found. Most of the items in the top 20 appear year after year. A large number of the items are made of plastic.

Top 20 Litter Items found during Beachwatch 2005 surveys

Position 2004	Position 2005	Item	% of total litter	Items/km
1	1	Plastic Pieces >1 cm–50 cm	11.8	234.6
6	2	Plastic Pieces <1 cm	6.5	128.9
7	3	Crisp/Sweet/Lollywrappers	6.4	126.5
4	4	Plastic Caps/lids	6.1	119.9
3	5	Cotton Bud Sticks	5.6	111.5
N/A	6	Fishing net <50 cm	5.4	107.9
N/A	7	Rope	5.4	107.2
5	8	Polystyrene pieces	5.0	99.8
8	9	Plastic Drinks Bottles	3.7	73.1
9	10	Cigarette Stubs	3.7	72.3
Top 10 Items			59.7	1181.9
10	11	Glass Pieces	2.6	51.5
11	12	Fishing Line	2.3	45.6
16	13	Paper Pieces	2.3	45.3
20	14	Cloth Pieces/String	2.3	45.3
13	15	Plastic Bags (Including Supermarket)	2.3	45.1
12	16	Metal Drinks Cans	2.1	41.9
15	17	Plastic cutlery/Trays/Straws	1.9	37.3
17	18	Wood Pieces	1.4	28.2
18	19	Foam/Sponge	1.4	27.9
N/A	20	Fast food containers/cups	1.0	20.2
Top 20 Items			79.3	1570.2

Mysterious litter items

Occasionally, unusual finds are reported on beaches around the UK. Some of the more interesting or unusual litter items found during Beachwatch 2005 include:

False teeth
manhole cover
colostomy bag
5 road signs
carpet
2 ship's hatches
toupee

motor bike helmet
Christmas tree
4 hub caps
lumps of asbestos
door bell
3 shopping trolleys
kitchen sink

The Marine Conservation Society (MCS) has also published very useful guidelines (set out below), which are of great utility for all forms of litter control and are strongly recommended.

Reduce pollution from ships and boats

- Effective and rapid implementation of international and national legislation on the dumping of litter at sea under the International Convention for the Prevention of Pollution from Ships, 1973 (MARPOL 73/78).
- Port authorities must provide adequate and user-friendly port reception facilities involving communication with user groups and waste management bodies.
- Government, trade and member associations should inform commercial and recreational boat users and fishermen of legislation and correct disposal procedures.
- Shipboard and port waste management plants must be developed and implemented. Government should encourage non-signatory countries to ratify MARPOL.

Reduce the input of plastics

- Incentives should be increased for manufacturers and retailers to reduce the use of plastics and excessive packaging.
- Government must expand the network of plastics recycling schemes.
- Government should encourage research into degradable alternatives.
- Educate and enable the public to Reduce-Reuse-Recycle.

Reduce the input of sewage-related debris

- Water authorities must improve sewage treatment facilities, with secondary treatment as a minimum at all coastal sewage outfalls.
- UV disinfection should be used at all sewage works affecting bathing waters.
- Water authorities must rapidly achieve full compliance with the EC Bathing Water Directive (76/160/EC) 1976 and the EC Urban Waste Water Treatment Directive (97/27/EC) 1991.

- The public must be informed to 'Bag it and Bin it' rather than flush sanitary waste.
- Councils, hoteliers, etc. must provide sanitary disposal bins in all public toilets.
- Sanitary product manufacturers should reduce the use of plastics in sanitary products and label packaging with the environmentally friendly disposal method.

Individual responsibility

- Every individual must dispose of waste correctly wherever facilities are available.
- Relevant authorities must ensure that facilities for disposal are adequate.
- The impacts of marine debris should be included in awareness raising campaigns and education in schools on general waste reduction, disposal and recycling.

Local authority responsibility

- Integrate beaches into local authority waste management plans.
- Local authorities must provide adequate disposal facilities on beaches and in towns.
- Local authorities should increase and support volunteer beach cleaning operations.

Sources: S. Pollard (MCS), *Wastes Management Journal*, May 1996. Nationwide Beach-Clean & Survey Report, Reader's Digest Marine Conservation Society, Beachwatch '98.

Load factor. The average load on an electrical generating system throughout the year expressed as a percentage of the highest load that occurs at any time during the year. The estimation of total pollutant emission by power stations is dependent on the load factor.

$$\text{Load factor} = \frac{\text{Average load over a period}}{\text{Peak load for that period}}$$

Loading rate, organic. The ratio of food to micro-organisms (FM) in a biological treatment process for wastewater.

$$\text{FM} = \frac{\text{Mass of BOD added to the system each day (food)}}{\text{Mass of micro-organisms persent (biomass)}}$$

The FM value determines the required volume of the biological treatment unit.

Local Agenda 21. ◊ RIO EARTH SUMMIT.

Logarithms. When two quantities, x and y, are related by an EXPONENTIAL CURVE, that is, by the equation

$$y = e^x$$

x may be seen to be the power to which e (e = 2.718) must be raised to give y. Alternatively, x is called the *natural logarithm* of y, and is represented by the symbol '$\ln y$' or '$\log_e y$'.

There are many types of logarithms to bases other than e. The logarithms that are most frequently used are those to base 10, in other words, those which result from an equation, $y = 10^x$. These are called *common logarithms* and are represented by the symbol '$\log y$'. Thus:

$$10 = 10^1, \text{ i.e. } \log 10 = 1$$
$$100 = 10^2, \text{ i.e. } \log 100 = 2$$
$$1000 = 10^3, \text{ i.e. } \log 1000 = 3$$
$$10000 = 10^4, \text{ i.e. } \log 10000 = 4$$

Logarithms to base e and base 10 are chosen to suit the system under consideration. For example, the DECIBEL scale for SOUND measurement uses logarithms to base 10, as does pH (◊ pH). Logarithmic graphs are in common use for values that span ranges of 1000 or 10000 units of measurement. 'e' is used for exponential growth considerations, e.g. the time t required for a quantity to double from its original value at a constant growth rate r is given by $t = \frac{\ln 2}{r} = \frac{0.6931}{r}$. Logarithmic scales (e.g. RICHTER SCALE) are used where the data span several orders of magnitude, as opposed to a linear scale (e.g. TEMPERATURE where the range could be 0–100 °C). (◊ EXPONENTIAL CURVE)

London convention. A prohibition on the disposal of waste in the marine environment. This will require minor tweaking to allow CARBON CAPTURE & STORAGE in sub-seabed formations which have been assessed as suitable, e.g. depleted oil & gas formations.

Long sea outfall. Generally, a pipe over 500 m long discharging wastewater into the marine environment.

Long-term exposure limit (LTEL). ◊ TIME-WEIGHTED AVERAGE.

Low Carbon Economy. Phrase used to encapsulate 'green' energy supplies, CO_2 reduction, etc. The UK ex-Prime Minister Tony Blair claimed (*The Times*, 23 May 2007) that markets for low carbon technologies will be worth at least $500 billion by 2050. Quite probably, but where is the nurture and the care of the vital engineering skills needed for UK PLC to benefit from this anticipated boom?

$$M$$

Macronutrient. A chemical element necessary in large amounts (usually >50 mg kg^{-1} tissue) for the growth of plants. Includes C, H, O, N, P, K, Ca, Mg and S.

Magnetherm. This is one of the main methods for producing magnesium and is based on the reduction of magnesium oxide when it is heated in a partial vacuum with oxides of calcium, ALUMINIUM and silicon. The oxides form a molten SLAG which reacts with a metallic reducing agent, such as ferrosilicon, to generate magnesium, which vaporizes in the vacuum. The magnesium is captured by condensing the vapour in a crucible.

The yield of magnesium depends on the temperature and pressure of the furnace, and the composition of the slag.

An unforeseen benefit is that a modified version of the process can make use of certain kinds of hazardous waste as part of the feedstock, such as ASBESTOS and cement waste containing asbestos which, when blended correctly, contains the correct proportions of the various oxides.

Source: 'Magnesium smelter puts toxic waste to work', *New Scientist*, March 1995.

Magnetohydrodynamic generator (MHD). A device for DIRECT ENERGY CONVERSION which generates electricity by the passage of a gas at high temperature and pressure through a magnetic field. The gas is ionized and this is enhanced by seeding with caesium or potassium. The gas is then passed through a nozzle at right angles to a magnetic field. The magnetohydrodynamic generation is essentially a 'topping' cycle – that is, a means of extracting greater electrical energy from a given amount of fuel by burning the fuel at as high a temperature as possible (2500 °C or greater) followed by a conventional steam-turbine generation at a lower temperature.

Magnetohydrodynamic generation received great attention in the 1950s and 1960s but has now dropped from favour on grounds of capital cost and scale, as it can only be usefully employed in the range from 10 megawatts upwards.

Magnox.
1. A magnesium alloy which contains small amounts of aluminium and beryllium used for cladding natural uranium metal fuel used in Magnox reactors.
2. The name given to the type of gas-cooled graphite moderated reactor using Magnox-clad fuel, on which Britain's first NUCLEAR POWER programme was based.

Major Accident to the Environment. Defra has established threshold criteria defining a 'Major Accident to the Environment' (MATTE), based on Schedule 7 (part 1) of the Control of Major Accident Hazards Regulations 1999.

Malathion. ◊ ORGANOPHOSPHORUS COMPOUNDS.

Mammalian meat and bone meal. Mammalian protein obtained by rendering, or an equivalent process in the case of imported products, whole or part of a dead mammal.

Managed realignment. Sometimes known as 'managed retreat', it means allowing the sea to flood areas of land that were previously protected by walls or embankments. The flooded area soon becomes saltmarsh and mudflat, creating a natural buffer between the sea and inland areas.

Manning Equation. An empirical equation that relates speed of water flow in a channel to the hydraulic radius and the gradient of the water surface, through the Manning roughness coefficient.

Manning Roughness Coefficient. The proportionality constant of the Manning equation that relates the speed of water flow in a channel to the hydraulic radius and the gradient of the water surface.

Marginal cost. This is a measure of the cost to society of producing one more unit of output. If the marginal cost is less than the marginal utility (a measure of the increase in social benefit from consuming one more unit of output), it is worthwhile having the extra unit of output, because the additional benefit exceeds the extra cost.

Marginal analysis is the technique for investigating marginal cost and marginal utility. In a nutshell it may be paraphrased as if we change our economic behaviour in such and such a fashion, would we benefit or not – this type of analysis, of course, assumes that no cataclysmic changes would be made and that business as usual is the *modus operandi*.

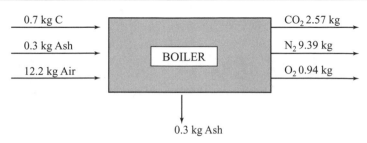

Figure 126

Marl. A soft rock mixture of clay or silt and calcite (calcium carbonate, $CaCo_3$) mud. Marls are plastic when wet and friable when dry.

Marsh gas. ♢ METHANE.

Mass balance. The balancing of the inputs and outputs of a process (e.g. boiler plant) which is operating in the steady state mode.

For example, a boiler plant burning coal of 70 per cent carbon and 30 per cent ash at 50 per cent EXCESS AIR has the inputs and outputs shown in Figure 126 on the basis of 1 kg coal input.

Another example is the mass balance for an MSW processing plant: the inputs total 100 t; the outputs (including ash) total 100 t, i.e. they balance (Figure 127).

Mass burning. ♢ INCINERATION.

Mass number. The number of protons and neutrons in an atom. (♢ ISOTOPE)

Material unaccounted for (MUF). This is of particular reference to PLUTONIUM stocks and is the difference between a book figure and a physical stock-take at certain points in time. Negative MUF does not mean any material is actually missing. MUF can give rise to apparent gains as well as losses because no measurements in any industrial process can be made with total accuracy.

All civil nuclear materials in the UK are subject to the inspection of international safeguard authorities to ensure that there is no loss or diversion, e.g. the cumulative plutonium MUF figure for Sellafield for the past 25 years was an apparent loss (not an actual loss) of 48 kg.

Source: R. Howsley, former Head of Security, British Nuclear Fuels, letter to *Daily Telegraph*, 28 December 1997.

Materials, energy consumption. The energy required in the manufacture of some common materials and manufactured products is shown in the table overleaf. The high energy input and its effects on costs are very apparent and it is obvious why 'alternative technology' is based on both low cost and low energy input materials.

Energy consumption in basic materials processing.*

Material	Energy for unit production ($kWh_{TH}\,tonne^{-1}$)	Machinery depreciation ($kWh_{TH}\,tonne^{-1}$)	Transportation ($kWh_{TH}\,tonne^{-1}$)	Total ($kWh_{TH}\,tonne^{-1}$)
Steel (rolled)	11 700	700	200	12 600
Aluminium (rolled)	66 000	1000	200	67 200
Copper (rolled or hard drawn)	20 000	800	200	21 000
Silicon, metal and high-grade steel alloys	58 000	1000	200	59 200
Zinc	13 800	700	200	14 700
Lead	12 000	700	200	12 900
Miscellaneous electrical produced metals	50 000	1000	200	51 200
Titanium (rolled)	140 000	1000	200	141 200
Cement	2000	50	50	2300
Sand and gravel	18	1	2 (short distance hauling)	21
Inorganic chemicals	2400	100	200	2700
Glass (plate finished)	6700	300	200	7200
Plastics	2400	300	200	2900
Paper	5900	300	200	6400
Lumber	1.47 per board foot	0.02 per board foot	0.02 per board foot	1.51 per board foot
Coal	40	2		42

* Accuracy ±20 per cent.
From 'Economical use of energy and materials', *Environment*, vol. **14**, no. 5, June 1972, p. 14.

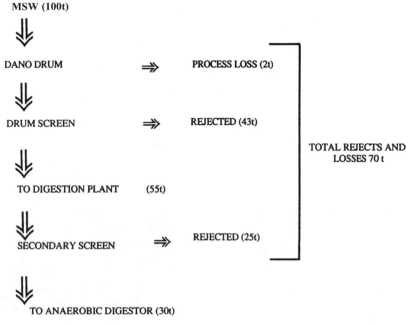

MSW (100t)

DANO DRUM ⟹ PROCESS LOSS (2t)

DRUM SCREEN ⟹ REJECTED (43t)

TOTAL REJECTS AND
LOSSES 70 t

TO DIGESTION PLANT (55t)

SECONDARY SCREEN ⟹ REJECTED (25t)

TO ANAEROBIC DIGESTOR (30t)

**i.e. Digester Accepts c. 30% of incoming material
(of which approx. 25% will be converted to Biogas).**

⟹COMBUSTION.

Figure 127 Outline mass balance for 100 tonne MSW input to a Dano drum MSW processor (courtesy of Motherwell Bridge Envirotec).

Materials reclamation/recovery/recycling facility (MRF). Site where the PLASTICS, GLASS, metal and PAPER components of solid waste are, either mechanically or manually, separated, baled and stored prior to reprocessing.

Figures 128 and 129 show two different layouts for the provision of an anaerobic digester feedstock and paper recovery, respectively.

Dust and other exposure measurements at MRFs are given in table (a) for crude MSW and in table (b) from kerbside recycled materials.

Case study

The new facility (Figures 128 and 129) at Cardiff County Council's Lamby Way depot is the first council-operated plant of its scale and kind in the UK. The speed of separation of recyclable waste at the new Material Recycling Facility (MRF) will be an impressive 10 times faster than the Council's previous facility.

The operation has the ability to handle 60000 tonnes per annum of co-mingled, mixed, dry recyclable materials comprising:

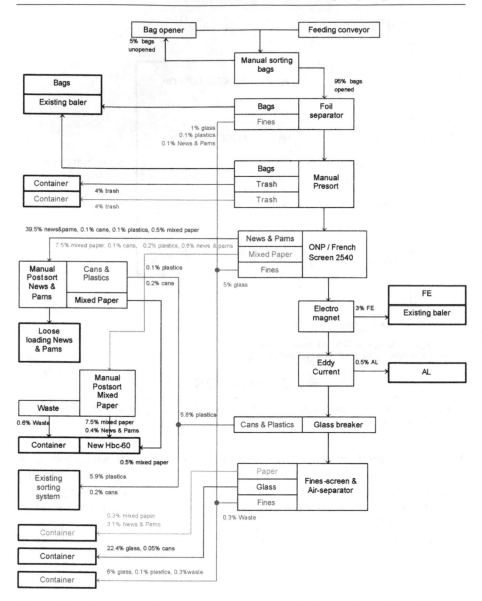

Figure 128a   Schematic of the Mass Balance of Cardiff County Council's MRF.
Source: Bollegraaf UK Ltd., 93 William Street, West Bromwich, West Midlands, B70 0BG, UK.

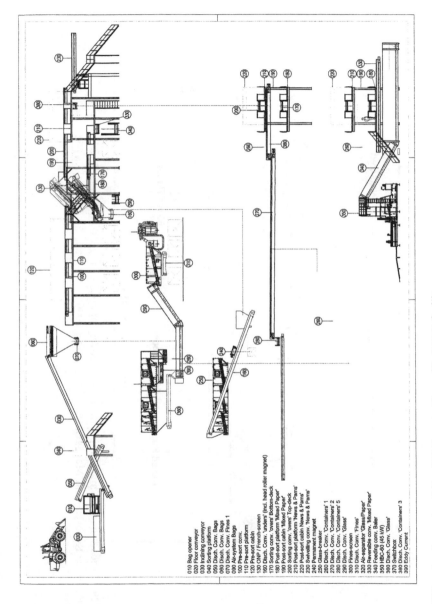

010 Bag opener
020 Floor conveyor
030 Inclining conveyor
040 Sorting platform
050 Disch. Conv. Bags
060 Disch. Conv. Bags
070 Disch. Conv. Fines 1
090 Air-system Bags
100 Pre-sort conv.
110 Pre-sort platform
120 Pre-sort cabin
130 ONP / French-screen
160 Disch. Conv. 'unders' (incl. head roller magnet)
170 Sorting conv. 'overs' Bottom-deck
180 Post-sort platform 'Mixed Paper'
190 Post-sort cabin 'Mixed Paper'
200 Sorting conv. 'overs' Top-deck
210 Post-sort platform 'News & Pams'
220 Post-sort cabin News & Pams'
230 Swivelling conv. 'News & Pams'
240 Permanent magnet
250 Glass-breaker
260 Disch. Conv. 'Containers' 1
270 Disch. Conv. 'Containers' 2
280 Disch. Conv. 'Containers' 5
290 Disch. Conv. 'Glass'
300 Fines-screen
310 Disch. Conv. 'Fines'
320 Air-separator 'Glass/Paper'
330 Reversible conv. 'Mixed Paper'
340 Feeding conv. Baler
350 HBC-60 (45 kW)
370 Switchbox
380 Disch. Conv. 'Glass'
390 Disch. Conv. 'Container' 3
390 Eddy Current

Figure 128b   A schematic of Cardiff County Council's MRF facility.
Source: Bollegraaf UK Ltd, 93 William Street, West Bromwich, West Midlands, B70 0BG, UK.

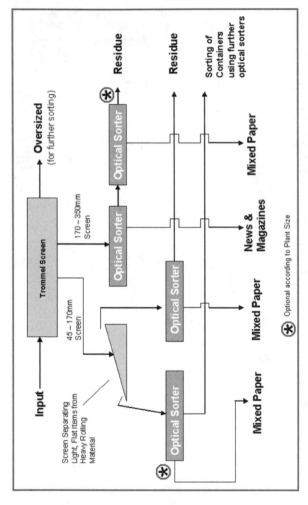

Figure 129 Process diagram highlighting the paper recovery options in a UK commingled MRF.
Source: Resource Recovery and Management, 22 September 2006.

Glass, bottles and jars (30 per cent)
Plastics, (PET clear and colour + HDPE + mixed plastic) bottles and containers (5 per cent)
Metal Can and Packaging, Fe and Al (5 per cent)
Newspapers and magazines (45 per cent)
Mixed paper including cardboard and card packaging (8 per cent)
The MRF system has been designed to operate at a throughput of 15 tonnes (with glass) per hour but has the ability to run consistently at 18 tonnes per hour.

Source: Courtesy of Bollegraaf UK Limited, 93 William Street, West Bromwich, West Midlands, B70 0BG, UK.

Cardiff County Council carried out a 'test sort' on 61 240 kg of material, on 21 July 2006.

The material was processed in approx. 5 hours.

Breakdown of material:

Loose paper	26 905 kg
Baled paper	7 980 kg
Glass	16 020 kg
Cardboard	600 kg
Fe cans	1 360 kg
Al cans	412 kg
LDPE	1 317 kg
HDPE	1 571 kg
PET clear	942 kg
PET colour	848 kg
Mixed plastics	837 kg

Total recyclables recovered = 58 792 kg (96 per cent)
Total residue to landfill = 2448 kg (4 per cent)

The health of workers in MRFs is of concern and detailed measurement of dust, endotoxin (a cell wall component in bacteria) and $(1 \rightarrow 3)$ β-D-glucan (a cell wall component of fungi and some bacteria present in organic dust) were made. The measurements were conducted on four classes of MRFs:

Box collection of source-segregated household waste materials;
Bag collection of source-segregated household waste materials;
Twin-bin collection of source-segregated household waste materials;
Mixed collection of materials and unsorted household wastes.

The results are given in Table (a) overleaf.

(a) Exposure Measurements in the MRFs in England and Wales

MRF	Total dust (mg/m³)					Endotoxin (ng/m³)					(1–3)-β–D–Glucan (ng/m³)				
	N	Min	Max	Mean	Median	N	Min	Max	Mean	Median	N	Min	Max	Mean	Median
a: Exposure measurements in the MRFs in England and Wales															
1	18	0	6.47	2.09	1.37	12	1.36	20.22	6.40	2.95	6	1.32	26.39	15.40	18.23
2	36	0.3	8.23	3.59	3.55	16	0.31	27.25	4.97	2.24	16	0.13	35.50	7.41	4.84
3	33	0	28.33	5.65	3.52	15	1.61	36.11	12.31	8.78	15	6.79	81.82	25.56	15.48
4	30	2.86	62.61	20.26	13.67	16	1.29	198.17	30.99	11.43	13	4.02	73.91	20.82	12.78
5	24	0	22.31	5.85	5.07	11	0.19	3.95	1.86	1.77	11	5.64	38.53	14.22	11.30
6	42	0	8.95	2.76	1.86	20	1.29	40.15	8.46	5.61	20	2.29	47.17	12.51	8.25
7	19	2.01	17.85	6.55	5.47	9	1.63	103.85	18.82	11.31	9	4.69	37.01	19.11	14.10
8	23	0	18.59	3.6	2.21	12	2.70	17.50	8.05	6.55	12	5.17	137.37	44.16	40.05
9	35	0	45.02	6.12	3.45	17	0.75	17.93	6.17	5.70	17	0	36.06	10.41	8.48
6 and 9	77	0	45.02	4.29	2.77	37	0.75	40.15	7.41	5.70	37	0	47.17	11.54	8.48
b: Exposure measurements in the four different types of MRFs in England and Wales															
Box	84	0	8.95	2.51	2.21	46	1.29	40.15	6.25	4.11	33	0.32	47.17	13.99	15.17
Bag	60	0	18.59	3.60	2.85	28	0.31	27.25	6.29	4.17	28	0.13	137.37	23.16	11.07
Twin-bin	48	2.01	62.61	15.16	8.21	25	1.29	198.17	26.61	11.31	22	4.02	73.91	20.12	13.26
Mixed	92	0	45.02	5.88	4.09	43	0.19	36.11	7.21	4.74	43	0	81.82	16.67	11.45

Note: MRF 6 and 9 were small and all measured values were within normal ranges.
Source: TL Gladding, J Thorn, and D Stott, 'Organic dust exposure and work-related effects among recycling workers', *Am J Ind Med*, 43, (2003): 584–591.

The relationship between length of time working in an MRF in England and Wales, and Prevalences (%) of reported symptoms is given in Table (b).

(b)

Symptoms	<6 months	6–18 months	>18 months	P-values (Linear test for trends)
Subjects (n)	38	53	68	
Breathlessness	5.3	1.9	11.8	0.124
Short of breath	2.6	3.8	10.3	0.092
Irritated nose/sneezing	42.1	60.4	54.4	0.322
Cough with phlegm	21.1	24.5	29.4	0.332
Hoarse/parched throat	15.8	18.9	32.4	0.039*
Itching/burning/watering eyes	7.9	15.1	27.9	0.009*
Stuffy nose	63.2	62.3	54.4	0.338
Itchy red skin	2.6	20.8	13.2	0.235
Flu symptoms	18.4	32.1	35.3	0.085
Unusually tired	5.3	17.0	17.6	0.109
Difficulty concentrating	2.6	7.5	14.7	0.036*
Stomach problems	2.6	11.3	16.2	0.038*
Diarrhoea	10.5	49.0	50.7	0.000*

* Indicates significant association between length of time working in an MRF and reported symptoms ($P < 0.05$).

The researchers concluded that the results suggest MRF workers exposed to higher levels of endotoxin and $(1 \rightarrow 3)$–β–D–glucan at their work sites exhibit various work-related symptoms, and that the longer a worker is in the MRF environment, the more likely he is to become affected by various respiratory and gastrointestinal symptoms. The results further suggest that differences in MRF operations may have an impact on exposures to workers. MRFs with high residue rates are mainly twin-bin MRFs, which also had a higher exposure.

A typical breakdown of kerbside recyclables by weight and by size

Total input breakdown	+320	−320 + 160 mm	−160 mm + 40 mm	−40 mm	Total wt %
Paper & Card	14.5	43.4	20.3	0.5	78.6
Plastic Film	0.4	1.2	2.0	0.0	3.6
Dense Plastic	0.2	1.4	9.0	0.1	10.7
Textiles	0.0	0.0	0.0	0.0	0.0
Misc. Combustibles	0.0	0.1	0.3	0.0	0.5
Non-Combustibles	0.0	0.0	0.0	0.0	0.0
Glass	0.0	0.0	0.3	0.0	0.3
Ferrous Metal	0.1	0.3	4.1	0.1	4.5
Non-Ferrous metal	0.0	0.0	0.9	0.0	0.9
Putrescibles	0.0	0.0	0.5	0.3	0.8
Hazardous household waste	0.0	0.0	0.0	0.0	0.0
Waste electrical and electronic equipment	0.1	0.0	0.0	0.0	0.1
Total wt %	15.3	46.4	37.3	1.0	100.0

Note: Above are exact sizes and do not take account of screen efficiency.
Source: *Resource Recovery and Management*, 22 September 2006, p13.

Maximum allowable concentration (MAC). In the USA, the upper limit for the concentration of noxious or toxic emissions in workplaces. Exposure to the MAC of a pollutant should not cause any distress to anyone except the occasional sensitive individual.

Maximum exposure limits (MEL). Maximum level of exposure to a harmful substance allowed for a worker (or member of the population at large).
  The criteria for setting limits are:

1. A substance not able to satisfy the criteria for an OCCUPATIONAL EXPOSURE STANDARD (OES) and which has or is liable to have a serious risk to humans including acute toxicity and/or potential to cause serious long-term health effects; *or*
2. Socio-economic factors, which indicate that, although the substance meets the criteria for an OES, a numerically higher value is necessary if certain uses are to be reasonably practicable. (◊ OCCUPATIONAL EXPOSURE STANDARD; TIME-WEIGHTED AVERAGE)

Maximum sustainable yield. The maximum yield that can be obtained from a given crop or species if it is to maintain equilibrium. The management of a forest or fishery or farm should avoid over-grazing, over-cropping or over-harvesting. Thus, fishery catches should be strictly controlled so that the fish population can have a sufficient breeding mass and thus give a sustained yield for future generations. This philosophy is particularly applicable to whaling, herring and salmon fishing, but is relevant to many other areas, such as forestry, and is basically the application of good stewardship of natural resources that cannot regenerate if exploited too far. The loss of WHALE species is an example of exploitation of a renewable source.

Counterparts abound everywhere: a river should not be so depleted of DISSOLVED OXYGEN by domestic or industrial pollution that it cannot recover. The land should not be 'mined' for short-term agricultural profit. THE COMMONS must be husbanded and good husbandry enforced.

Mean, arithmetic, geometric. Generally similar to the idea of an average, a central value about which individual values, greater or less, are distributed. The arithmetic mean, or the average, of a set of values is their sum divided by their number. The geometric mean of n values is the nth root of the product of all of the values.

Mean annual temperature (global). This is the mean temperature of the earth (averaged over a year) for individual latitudes leading to a composite value which is a measure of the earth's radiation balance. If more radiation is absorbed by the land mass and oceans this year than was absorbed last year without a compensating change in the outgoing radiation, the mean annual temperature will rise a fraction of a degree. If the opposite process occurs, it will drop. The mean annual temperature has risen 0.5°C in this century. An unstoppable rise of 1°C is forecast with a possible 4.5°C increase by 2030 if GREENHOUSE GAS emissions are not reduced, as the longer measures are delayed, the worse will be the effect. (◊ GREENHOUSE EFFECT)

Means, best practicable (BPM). According to the Alkali, etc. Works Regulation Act 1906, and subsequently the Health & Safety at Work, etc. Act 1974, a scheduled process must be provided with the *best practicable means* for preventing the escape of noxious or offensive gases and smoke, grit, and dust to the atmosphere and for rendering such gases, where necessarily discharged, harmless and inoffensive. The words 'best practicable means' are often used to describe the whole approach of British anti-pollution legislation towards industrial emissions: the cost of pollution abatement and its effect on the viability of industry are taken into account. Thus, while better pollution control can often be achieved, the 'best practicable means' philosophy can allow lower standards of control to keep a sector of industry viable, having regard to established practice, the area in which it is sited, etc. In effect this may mean that no firm is legally compelled to modify its activities or install pollution-abatement equipment until it is economically convenient for the firm to do so. In the meantime, the social costs of the consequent pollution continue to be borne by the community. The firm may well be urged to investigate means of suppressing pollution, but may not be forced to do so. The BPM concept has been replaced by that of the BEST PRACTICABLE ENVIRONMENTAL OPTION (BPEO) which may well be BPM in practice for many established industries. (◊ BATNEEC; EXPOSURE–DOSE EFFECT RELATIONSHIP; THRESHOLD LIMITING VALUE)

Meat Hygiene Service (MHS). The Meat Hygiene Service (MHS) is responsible for the protection of public health and animal health and welfare in Great Britain, through proportionate enforcement of legislation in approved fresh meat premises. It provides verification, audit and meat inspection services in approved slaughterhouses, cutting plants, farmed and wild game facilities, and co-located minced meat and meat products premises. The MHS has a statutory duty to provide these services on demand, 24 hours a day, 365 days a year, throughout England, Scotland and Wales. It provides export certification where required by importing countries or community rules. The MHS is an Executive Agency of the Food Standards Agency. It was first established as an Executive Agency of the Ministry of Agriculture, Fisheries and Food (now part of the Department for Environment, Food and Rural Affairs) on 1 April 1995, when it took over meat inspection duties from some 300 local authorities.

Source: *BSE Enforcement Bulletin*, no. 1.

Mechanical Biological Treatment. Mechanical biological treatment (MBT) is a combination of mechanical and biological processes employed to achieve recovery of resources from, and stabilization of domestic wastes. The resources, including metal, plastic and glass can then be recycled. It consists of any number of combinations of mechanical sorting and biological treatment processes for waste. MBT generally, but not exclusively, is used to process unsorted household waste.

The 'mechanical' element is usually an automated mechanical sorting stage. This either removes recyclable elements from a mixed waste stream (such as metals, plastics and glass) or processes them. It typically involves factory style conveyor, industrial magnet, Eddy current separators, trommel, shredder and other tailor-made systems. The mechanical element has a number of similarities to a materials recovery facility (MRF).

Some systems integrate a wet MRF to recover and wash the recyclable elements of the waste in a form that can be sent for recycling. MBT can alternatively process the waste to produce refuse-derived fuel (RDF). RDF can be used in cement kilns or power plants and is generally made up from plastics and biodegradable organic waste.

The 'biological' element refers to either anaerobic digestion or composting. Anaerobic digestion breaks down the biodegradable component of the waste to produce biogas and digestate. The biogas can be used to generate renewable energy. Some processes through MBT enable high rates of gas and green energy generation without the production of RDF. This is facilitated by processing the waste in water.

Biological can also refer to a composting stage. Here the organic component is treated with aerobic microorganisms. They break down the waste into carbon dioxide and compost. There is no green energy produced by systems employing only composting treatment for the biodegradable waste. By processing the biodegradable waste either by anaerobic digestion or by composting MBT technologies help to reduce the contribution of greenhouse gases to global warming.

Some of the advantages of MBT are:

- It is a proven technology in Europe;
- It reduces the volume of waste and thereby the landfill void space taken and thus the cost to the local authority of disposal;
- It reduces the biodegradability of the waste and thereby the disamenity of disposal by reducing the amount of gas, leachate, vermin, odour and on-site litter;
- It reduces the cost of managing the landfill and potentially the long-term liability of the site;
- It reduces the production of methane and thereby the impact on climate change;
- The RDF produced has a higher calorific value than untreated residual waste;
- Good quality metals can be recovered for recycling.

Note any secondary fuels may require transport to CEMENT KILNS for COMBUSTION and their number is finite.

As of 2007, no definitive, independently verified costs have been published in the professional journals.

Melt-down. A (so far theoretical) nuclear reactor accident where the coolant escapes and the fuel becomes sufficiently hot that it melts and falls onto the reactor base. The nearest occurrence was in the USA at THREE MILE ISLAND.

Membrane. A thin sheet of ion-exchange material used in ELECTRODIALYSIS to remove impurities from brackish water. In REVERSE OSMOSIS the membrane is semi-permeable, i.e. it has minute pores through which water can diffuse but through which the passage of salts is prevented, again effecting purification.

Membrane Filtration (Micro/Ultra filtration).
Membrane filtration covers a wide range of processes and can be used for various source water qualities, depending on the membrane process being used. Microfiltration, used for treatment of surface waters, can remove a wide range of particulate matter, including bacteria, protozoan cysts and oocysts, and particles that cause turbidity. Viruses, however, are so small that some tend to pass through the microfiltration membranes; therefore a smaller pore size is required, offered by ultrafiltration membranes. The membrane surface is kept clean though aeration and membrane back-pulsing. Diffused air is introduced from the bottom of the membrane module, where it is released and bubbles up the outside of the fibre, scouring particulate matter from the membrane surface as it rises. Periodically, the membranes are back-pulsed. This is accomplished by briefly reversing the flow of permeate through the membrane to remove any particles that may have obstructed the pores during membrane operation.

Typical Performance:

Turbidity <0.1 NTU
Bacteria >4 log removal
Giardia cysts >4 log removal
Cryptosporidium oocysts >4 log removal
Virus rejection >2.5 log to >4 log removal
Total suspended solids <1 mg/l
Total Organic Carbon 50–90 per cent removal, with appropriate biological design and/or chemical addition
Colour <5 Hazen Units

Mercaptans. A family of foul-smelling SULPHUR compounds produced by decaying ORGANIC matter, emitted at sewage works, food-processing plants, brick-making works and oil refineries. The most offensive, ethyl mercaptan (C_2H_5SH; boiling point 30 °C) can be smelt at concentrations as low as 1 part in 50 000 million in air.

Mercury (Hg). A HEAVY METAL that exists as a liquid at normal temperatures; atomic mass 200.59. It is extracted from an ore (cinnabar). Mercury is used as a catalyst in many industrial processes as well as in barometers, thermometers, etc. Over 10 000 tonnes are produced annually.

Provisional tolerable weekly intake for humans:

Substance	Intake per week expressed as	
	mg per person	mg/kg body-weight
Mercury		
total mercury	0.3	0.005
methyl mercury (expressed as mercury)	0.2	0.0033
Lead[1]	3	0.05
Cadmium	0.4–0.5	0.0067–0.0083

[1] These intake levels do not apply to infants.

Source: Food and Agriculture Organization, *Evaluation of Mercury, Lead, Cadmium and the Food Additives Amaranth, Diethylprocarbonate and Octy Gallate*, Rome, 1973.

As a pollutant it is a SYSTEMIC AGENT, affecting the brain, kidneys and bowels (◊ MINAMATA DISEASE). The organic forms, e.g. methyl mercury, are particularly toxic. World Health Organization food regulations state 0.05 parts per million as the highest allowable concentration of mercury in foodstuffs.

Inshore fish in the Mersey Estuary and Morecambe Bay have been found to contain up to 1.1 mg/kg. Fish from distant waters contain an average 0.06 mg/kg (Ministry of Agriculture, Fisheries and Food, *Survey of Mercury in Food*, HMSO, 1971). It is possible that people with very restricted diets of local fish could ingest more than the provisional *tolerable* weekly intake for humans shown in the data above for mercury, lead and cadmium. (The term 'tolerable' signifies permissibility rather than acceptability since the intake of a contaminant is unavoidably associated with the consumption of otherwise wholesome and nutritious foods: the term 'provisional' expresses the tentative nature of the evaluation (FAO, 1973). (◊ CRITICAL GROUP)

The provisional tolerable weekly intake that is used for metallic contaminants in food as an ACCEPTABLE DAILY INTAKE cannot be readily determined. (◊ TOLERABLE DAILY INTAKE)

There is a paucity of information on the effects of sub-lethal doses of heavy metals but, in view of the adverse properties of LEAD, there is obviously a strong case for strict control of all heavy-metal discharges and emissions.

Major emission sources are coal burning (where mercury is always present), the paper industry and chemical plants.

Mesopause. The boundary between the stratosphere and the mesosphere.

Mesosphere. The layer of the atmosphere above the stratosphere – meaning 'middle' sphere.

Metabolite. Either a simple product formed by the enzymatic decomposition processes of a living organism or a complex product also created enzymatically. (◊ ENZYME)

Metals. Solid materials generally possessing high melting points, which are good thermal and electrical conductors. Most pure metals are usually alloyed with others to improve their mechanical properties (e.g. steel: iron alloyed with carbon). The cost of metals reflects their relative abundance in the earth's crust (as ores) and the cost of extraction of the metal from the ore. ALUMINIUM, though very abundant in the form of alumina (Al_2O_3), is expensive owing to the large amounts of electricity needed to extract the metal from its ore (◊ MATERIALS, ENERGY CONSUMPTION). Iron is very abundant as Fe_2O_3, etc. and much less thermal energy is needed in manufacture of the raw metal using the blast furnace. Metals are widely recycled, particularly precious metals, gold, silver and platinum but also iron and steel alloys, owing to their abundant use (car bodies, tin cans, etc.). As with all materials RECYCLING problems are created with alloys. Some alloying elements are easily removed in the SLAG, particularly highly reactive metals like aluminium, but other, more noble, elements like copper and tin remain in the steel. They are known as *tramp elements*. The behaviour of alloy elements and impurities in steel making is shown below:

Elements almost completely taken up by slag	Silicon
	Aluminium
	Titanium
	Zirconium
	Boron
	Vanadium
Elements distributed between slag and metal	Manganese
	Phosphorus
	Sulphur
	Chromium
Elements remaining almost completely in solution in the steel	Copper
	Nickel
	Tin
	Molybdenum
	Cobalt
	Tungsten
	Arsenic
Elements eliminated from slag and metal	Zinc
	Cadmium

The degree to which tramp element impurities are harmful depends on the intended use of the steel. Reinforcing bars, for example, may contain impurities which would make the steel entirely unsuitable for car body sheet.

The economics of recycling metals depends on their purity, cost and degree of dissemination. If the main use of a given metal is only for one product, then recycling is usually straightforward since there is then an assured, continuous supply for the recycling industry. Examples include photographic film (for silver) and car batteries (for lead).

Metamorphic rock. Rock whose texture and/or mineralogy has been changed by the action of heat and/or pressure (usually both are involved). Metamorphic rocks can be derived from sedimentary rocks or igneous rocks (or preexisting metamorphic rocks.

Methaemoglobinaemia. A disorder, known as the 'blue baby disease', which affects the OXYGEN-carrying capacity of the blood. It is associated with drinking water containing NITROGEN in the form of nitrates. The amount of nitrate per litre of water at intakes on the River Thames and the River Lee has in the past (early 1970s) approached (and for the Lee exceeded) the 50 mg nitrate per litre (or 11.3 mg/l as nitrogen) limit considered to be safe. Above this limit, bottle-fed babies may develop methaemoglobinaemia; while above a concentration of 20 mg per litre, as nitrogen, the adult population may be at risk. So far, London's water supply has been kept well below the safe limit by diluting the Thames and Lee water with water from reservoirs.

Modern farming techniques involving heavy applications of nitrogen FERTILIZER, extensive cultivation of leguminous plants such as peas and beans, and sewage works effluent are all contributing to the high nitrate levels in rivers. The residues from agriculture are present in the ground in large quantities, and modern land drainage techniques mean that they will eventually find their way into the water supply. ($\lozenge$ LEACHATE)

Methane (marsh gas; CH$_4$). A gas which can easily be liquefied. It is chiefly used as a fuel. Methane occurs naturally in oil wells and as a result of bacteriological decomposition, e.g. the ANAEROBIC DIGESTION of sewage SLUDGE. It can also be synthesized. ($\lozenge$ LANDFILL GAS)

Methane reduction. Underground coal mines discharge large volumes of air containing dilute amounts of methane (0.1–1.0 v/v%). World-wide methane emissions in ventilation air amount to over 25 MT/y (equivalent to 525 Mt/y of CO_2 100 year time scale).

Natural Resources Canada (NRCan) and its industrial partners have shown that its catalytic flow-reversal reactor (CFRR, Figure 130) technology can eliminate methane as dilute as 0.1 v/v% in air, with no requirement for external heat. Indeed, for concentrations above 0.3 v/v% the methane heat of oxidation can be recovered with an efficiency between 40 per cent and 95 per cent (depending on the inlet methane concentration).

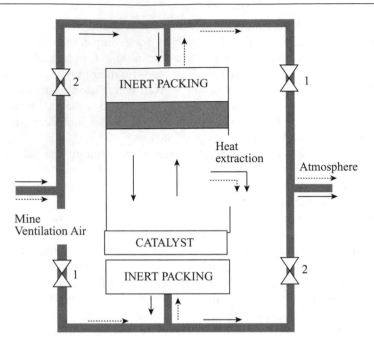

Figure 130   CFRR operating principle.

The CFRR operates by alternately using one of two CATALYST beds via two valves (1 and 2 in Figure 130). During the first half-cycle, ambient temperature mine air is heated by passing through the upper inert bed. It enters the upper catalyst bed at a temperature sufficient for exothermic methane oxidation to occur, which further heats the air. It then passes through the lower catalyst bed and finally heats the lower inert bed before exiting to atmosphere at near-ambient temperatures. During this first half-cycle, the upper inert bed, which was initially hot, cools down while the lower inert bed, which was initially cool, heats up.

After 5–30 minutes, the upper inert bed has cooled, and the flow is then reversed, to use the heat available in the lower bed. At the end of the second half-cycle, downward flow is re-established.

A large part of the methane's heat of reaction can be removed from the two catalyst beds, either with an internal heat exchanger or through hot air extraction (depending on site-specific considerations). This extracted heat is of high quality (at temperatures of 300–800 °C) and can be applied to site-specific ENERGY requirements or converted to electric power for the grid.

The economic viability of any technology to treat mine ventilation air depends on two components:

1. the intrinsic (capital and operating) costs of the methane oxidation technology; and
2. external variables (value of recovered heat, value of the environmental benefits attained).

The low capital and operating costs of the CFRR (including the relatively low catalyst replacement costs) have been shown to result in a 20–30 per cent intrinsic cost advantage over competing technologies.

The overall economic viability must, however, be established on a site-by-site basis.

Source: *Greenhouse Issues*, November 1998.

Methanogenesis. Process in the carbon cycle that occurs under extremely reduced conditions where bacteria use CO_2 or low-molecular-mass organic compounds as oxidizing agents, producing methane (CH_4).

Methanogenic bacteria. Bacteria that carry out methanogenesis.

Methanol (methyl alcohol or wood alcohol; CH_3OH). A chemical that can be produced by destructive distillation of coal or wood. Methanol is usually synthesized from CARBON MONOXIDE and HYDROGEN or from METHANE. It is used as an intermediary by the chemical industry, and also as a SOLVENT and denaturant.

One important use is as a substrate for the production of PROTEIN; a class of bacteria has been isolated that can utilize it. It is also being actively promulgated as a motor fuel to conserve oil reserves, either on its own or mixed with petrol in proportions between 15 and 85 per cent. It is claimed that higher mileages result, and that 50 per cent less carbon monoxide is produced from such fuel mixtures. However, its CALORIFIC VALUE (gross) is 21 MJ/kg compared to PETROLEUM'S 45 MJ/kg.

Methanol may also be classed as a renewable fuel if manufactured by combining hydrogen (from renewable sources) with recycled CO_2 (taken, for example, from the flue gases of power plants) to produce methanol. The methanol would act as an energy carrier for hydrogen. Methanol has twice the energy density per unit volume of liquid hydrogen. It is a liquid and can easily be stored and transported. Also, because the CO_2 has been recycled, there are no net CO_2 emissions caused by this cycle. It all sounds too good.

Methanolysis. ◊ PLASTICS RECYCLING.

Methanotrophic. Literally, 'methane-consuming'; bacteria that oxidize methane for energy and as a source of carbon.

Methyl tertiary butyl ether (MTBE). An unleaded fuel additive considered non-toxic but it can taint drinking water at concentrations of around 3–6 mg/l.

MTBE has already been banned in Alaska because gas station staff have been affected by headaches and nausea.

Metrolink. A light rail system operated by Greater Manchester Metro Ltd, using light rail vehicles. There are 26 vehicles in the fleet, each costing £1 million and carrying 86 passengers seated and 120 standing – 206 in total. The vehicles travel up to 50 mph out of the city and 30 mph through the city.

This has significantly cut inner city car use.

Microbes. A microbe or microorganism is an organism that is microscopic (too small to be visible to the human eye). The study of microorganisms is called microbiology. Microorganisms can be bacteria, fungi, etc, but not viruses and prions because they are generally classified as non-living. Micro-organisms are often described as single-cell or unicellular organisms; however, some unicellular Protists are visible to the human eye, and some multicellular species are microscopic.

Microorganisms live almost everywhere on Earth where there is water, including hot springs on the ocean floor and deep inside rocks within the Earth's crust. Microorganisms are critical to nutrient recycling in ecosystems as they act as decomposers. As some microorganisms can also fix nitrogen, they are an important part of the nitrogen cycle. However, pathogenic microbes can invade other organisms and cause diseases.

Microorganisms are used in brewing, baking and other food-making processes, such as in the production of soya sauce, yoghurt, and cheese. Microorganisms are also used to produce antibiotics through fermentation. Insulin has also been produced using microbes. They are very effective in breaking down organic materials, and are used in wastewater treatment and in composting processes. Soil which has been contaminated with oil (as in the environmental disaster in Kuwait, following the First Gulf War) can be cleaned up using oil-degrading bacteria.

Microbial Fuel Cell. This is a device to produce electricity by bacterial oxidation of organic matter in wastewater. When the bacteria consume the organic matter, electrons are produced and these induce a current. Simultaneously the wastewater is cleaned up. This system has been tried at small scale and efforts are underway to scale it up.

Source: Water and Waste Treatment, March 2007.

Microgeneration. The generation of low-carbon heat and power by individuals, small businesses and communities to meet their own needs. The size limit is typically 50 kW of energy or 45 kW of heat. A standard home needs a few kW of heat or power, so this means for businesses and small-scale multiple housing units such as a terrace or small block of flats, microgeneration is worth considering for economic reasons, as well as for the benefits to the environment.

Micron. One-millionth of a metre, hence the more correct term 'micrometre'. It is commonly used for particle sizing. Symbol μm in SI units.

Minamata disease. Minamata is a town on the west coast of Kyushu Island (Japan) where an extreme case of HEAVY METAL poisoning from methyl MERCURY ingested in the staple fish diet of the inhabitants caused severe disablement and death: 43 deaths and 68 major disablements were recorded between 1953 and 1956. (Environmentalists have claimed that almost 800 people have been killed in all by the discharges.)

The symptoms include numbness in fingers and lips and difficulty in speech and hearing. There is a marked inability to control limbs, followed by seizures. Children and old people are particularly vulnerable, as is the case with all heavy-metals ingestion.

The source of mercury in the bay was eventually traced to a PVC plant which used mercuric sulphate (an inorganic chemical) as a CATALYST and which discharged effluent containing both inorganic and organic mercury. The inorganic form was subsequently converted by marine life to the methyl form.

Japan's Supreme Court ruled in March 1989 that two former executives of a chemical firm were responsible for the mercury pollution in the 1950s. Each received two years' imprisonment.

The Minamata Disease Centre says there are 18 128 victims still waiting to be formally acknowledged as Minamata patients.

Minamata is now a classic case of industrial pollution and subsequent evasion of responsibility.

Mineral. A naturally occurring material with a local constant chemical composition. Usually a solid crystalline substance but can embrace SILT, CLAY and SEDIMENTS.

Mist. Microscopic liquid droplets suspended in air. Their diameter is less than 2 micrometres. ($\lozenge$ FOG)

Mixing layer. The lower part of the atmosphere within which air movement is much affected by the proximity of the earth's surface and where pollutants are dispersed and mixed. For convective conditions the mixing layer may extend up to several kilometres above the surface but for stable conditions its height may be only some 200 m.

Mogden formula. A formula agreed between the water industry and the Confederation of British Industries that links trade effluent charges to the costs imposed on customers, i.e. by paying according to the volume and strength of trade effluent discharged.

Moisture content. The quantity of water present in any substance, e.g. waste or SLUDGE. Expressed on a wet basis (as received) or dry basis (which is based on dry weight of sample).

Mole[1]. (US). The number of atoms or molecules equal to the 'Avogadro constant' (per mole, designated N_A).

Mole². (UK). The basic SI unit of substance which contains as many elementary units as there are atoms in 0.012 kg of carbon-12, e.g. 1 mole of a compound has a mass equal to its molecular weight in grams.

- 1 mole HCl has a mass of 36.46 g
- 1 mole H_2O has a mass of 18 g
- 1 mole of electrons has a mass of $mass_{electron} \times N_A$

Steam reforming

Steam reforming, hydrogen reforming or catalytic oxidation, is a method of producing hydrogen from hydrocarbons. On an industrial scale, it is the dominant method for producing hydrogen. Small-scale steam reforming units are currently subject to scientific research, as way to provide hydrogen to fuel cells. Steam reforming of natural gas, sometimes referred to as steam methane reforming (SMR), is the most common method of producing commercial bulk hydrogen, as well as the hydrogen used in the industrial synthesis of ammonia. It is also the least expensive method. At high temperatures (700–1100 °C) and in the presence of a metal-based catalyst, steam reacts with methane to yield carbon monoxide and hydrogen.

$$CH_4 + H_2O \rightarrow 3H_2O + CO$$

In this equation, 1 mole of methane reacts with 1 mole of water, to produce 3 moles of hydrogen gas and 1 mole of carbon monoxide.

Mole fraction. The ratio of the number of MOLES of a component of a mixture to the total number of moles in the mixture.

Molecular pollution trap. A purpose designed configuration of molecules which has the ability to remove phosphates/nitrates/chlorides from water.

One example uses a ring of four BENZENE molecules with two urea groups attached to the upper rim of the ring and four ethyl ester groups to the lower rim. The OXYGEN ATOM in one of the urea groups forms HYDROGEN bonds with the other urea group.

When a positively-charged ion comes into contact with the ring, it is caught by a slight negative charge. This pulls apart the hydrogen bonds linking the urea molecules and a negatively charged ion is then pulled in between the slightly positively charged urea molecules, and is held in place by electrostatic forces.

Source: *New Scientist*, 6 July 1996.

Molecule. The smallest part of an element or compound capable of existing independently which has all the chemical properties of the element or compound.

Monitoring. In environmental health, the repetitive and continued observation, measurement, and evaluation of health and/or environmental or technical data to follow changes over a period of time. Measurements of pollutant levels are made in relation to a set standard or to assess the efficiency of regulatory and/or control measures.

Monomer. A compound whose simple molecules can be joined together (POLYMERIZATION) to form a giant POLYMER molecule; e.g. VINYL CHLORIDE is polymerized to form the plastic POLYVINYL CHLORIDE.

Figure 131 shows a flow diagram of polymer manufacture from hydrocarbon feedstocks (left-hand side) via basic petrochemicals, intermediates and monomers to processors and users (right-hand side).

Monthly average. An average value of the concentration of a component (e.g. sulphur dioxide) on a monthly (30 day) basis.

Montreal Protocol. The Montreal Protocol on Substances That Deplete the Ozone Layer is an international treaty designed to protect the ozone layer by phasing out the production of a number of substances believed to be

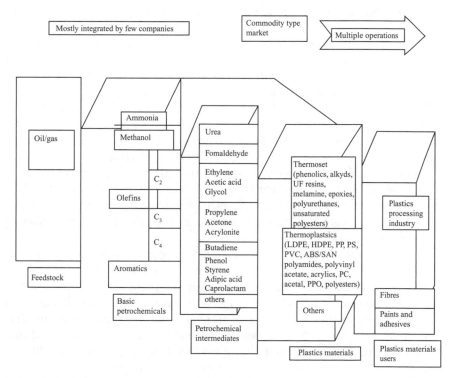

Figure 131 Polymer manufacture.
Source: *Introduction to Polymers*, Unit 1, The Open University.

responsible for ozone depletion. The treaty was opened for signature on 16 September 1987, and entered into force on 1 January 1989. Since then, it has undergone five revisions: in 1990 London, 1992 Copenhagen, 1995 Vienna, 1997 Montreal and 1999 Beijing. Due to its widespread adoption and implementation it has been hailed as an example of exceptional international cooperation.

Mosses. Multicellular, CHLOROPHYLL-bearing plants. Mosses are not complex but have a highly developed reproductive system. They occur in damp conditions, such as in woods and also on walls. The stems contain a central core of elongated cells which act as ion-exchange membranes. The intricate branching system with many leaves provides large surface areas on which particles can be trapped. This structure means that mosses can be used to concentrate airborne pollutants. Thus the presence (or absence) of certain species of mosses (and LICHENS) indicates the absence (or presence) of airborne pollutants such as SULPHUR DIOXIDE, FLUORIDES, and copper, LEAD and ZINC FUMES, etc. Where the mosses are absent, mossbags – i.e. bags of dry dead moss suspended in nylon nets – can be used to assess the presence. As the accumulation rate of the metals is a function of the concentration of the pollutant(s), in the air, the average concentration over the period of exposure can be found by analysis of the moss bags.

Most probable number (MPN). This is a method for estimating bacterial numbers in a sample (of water, food, soil, etc.) by multiple tube fermentation. Several tubes containing an appropriate NUTRIENT medium for growth of the particular bacterial group are inoculated with measured quantities of different dilutions of the test sample. The nutrient medium chosen is designed to indicate the presence of the bacteria after an incubation period at a specified temperature. Presence of the bacteria is usually indicated by a colour change in the medium or the production of gas. After the prescribed incubation time, examination of the tubes will reveal 'positive' and 'negative' ones indicating the presence or absence in each tube of the bacterial group in question. The number of 'positive' tubes for each of the sample concentrations originally selected can be converted into an estimate of the most probable number of bacteria in the original sample.

Mottling (of teeth). Tooth mottling, or 'mottled enamel', is a white or brown stain in the surface of the tooth sometimes associated with pitting of the surface. This should not be confused with stains on the surface. Mottled enamel is present when the tooth erupts and is caused by some defect during the formation of the tooth, whereas stains on the tooth surface occur after the tooth has come through, and are usually caused by food or drink. Fluoride can also cause tooth mottling. In this case it is called 'dental fluorosis'.

Mud. Wet loose mixture of particles less than $60\,\mu m$ in diameter. (⟡ CLAY; SILT)

Multiple flues. In large power stations or process plants, where there are separate boilers, the best practice for good plume dilution and dispersion is to take all gas discharge streams to one chimney with multiple flues so that emission velocity and temperature is maintained over a wide working range and interference from a multitude of chimneys, each with its own small plume, is avoided. The use of multiple flues means that one flue discharge does not affect the combustion draught conditions of an adjacent flue. (◊ PLUME RISE)

Municipal solid waste (MSW). The common name (US) given to the combined residential and commercial waste material generated in a given municipal area. (◊ DOMESTIC REFUSE; WASTES; GARBAGE)

Mutagen. Something (e.g. a chemical or radiation) that changes the genetic material (chromosomes) that is transferred to the daughter cells when cell division occurs. The result is that the new cells have changed characteristics.

The mutagen can act on a *germ* cell, e.g. sperm of man or that of any other sexually reproducing organism (rats, mice, fruit flies), and some of the offspring will carry the mutant genes in all their cells. A mutagen may also affect *somatic* cells, in which case the effects are to the person concerned and not to offspring. These effects depend on the type of cell affected. One effect could be to make cells grow and multiply faster than they can be removed by the blood; if the white cells are affected, the result is leukaemia. Another effect could be to start up cell divisions in cells that do not normally divide; if the division products displace or invade normal tissues, the result is a cancer. Both types of mutation are discussed further. (◊ IONIZING RADIATION)

Mutagens may also be carcinogenic or teratogenic, and therefore any substance that is found to be mutagenic must be tested for its carcinogenic and teratogenic properties as well. Although mutagens may not be carcinogens or teratogens, they must be considered suspect until cleared. *This conservative viewpoint is essential if human health is to be protected.* There is often a major time-lag between contact with a mutagenic agent and the onset of cancer. For example, VINYL CHLORIDE was only found to be carcinogenic after long use. It is now becoming apparent that many substances in common use are suspected as being carcinogenic, e.g. hair dyes. Because of the latency period before cancer manifests itself, these suspected substances may continue in use for many years. The principal area for research activity and concern is the use of the growing class of chemicals that affect nucleic acids, the basic chromosome component. (◊ CARCINOGEN; TERATOGEN)

The World Health Organization Scientific Group on the Evaluation and Testing of Drugs for Mutagenicity stated that the following should receive special attention:

1. compounds that are chemically, pharmacologically and biochemically related to known or suspected mutagens;
2. compounds that exhibit certain toxic effects in animals, such as depression of bone marrow; inhibition of spermatogenesis or oogenesis; inhibition of mitosis, teratogenic effects, carcinogenic effects, causation of sterility or semi-sterility in reproduction studies; stimulation or inhibition of growth or synthetic activity of a specific cell or organ; inhibition of immune response; and
3. compounds that are likely to be continuously absorbed into the body and retained by it for long periods.

The testing should be done with mammals, e.g. rodents, so that an indication of the potential effects on humans may be determined.

A mutagenic index is derived from the tests which reflect the percentage of dominant *lethal* mutations in the experimental group.

The finding of mutations in one species does not mean that it is mutagenic in all species, but a positive finding should certainly be considered as an indication of potential mutagenic activity in humans.

Mutagenic index. ◊ MUTAGEN.

Mutant genes. ◊ IONIZING RADIATION.

Mutation. The random alterations of reproductive cells of organisms which result in changes in the other cells of the organism. If a particular mutation protects the organism in a hostile environment, then the mutant has a good chance of survival.

N

Nanogram. One thousandth of one millionth of a gram (10^{-9} g).

National Radiological Protection Board (NRPB). A UK statutory body created under the Radiological Protection Act 1970 to provide information, advice and monitoring on all aspects of IONIZING RADIATION.

On 1 April 2005, it merged with the Health Protection Agency, forming its new Radiation Protection Division: It carries out health protection work related to ionising and non-ionising radiation. It also undertakes research, provided laboratory and technical services, runs training courses and provides expert information and advice.

National Survey of Air Pollution. A UK nationwide survey of SMOKE and SULPHUR DIOXIDE set up in 1960. The National Survey as such ceased in 1982 and is now called the UK Smoke and Sulphur Dioxide Monitoring Network. This smaller network includes a basic urban network monitoring for compliance with EU air quality limits, as well as some rural sites and *ad hoc* surveys.

Natural background radiation. ⟡ IONIZING RADIATION.

Natural capital. An omnibus term for vital natural resources/habitats such as the CARBON CYCLE, NITROGEN CYCLES, Water resources, and tropical forests. These are to be conserved through greater energy & water efficiency, renewable energy, and reductions in demand backed by appropriate enforcement measures.

Natural gas. Natural hydrocarbon gas(es) associated with oil production, principally METHANE (CH_4) and some ethane (C_2H_6).

It is a FUEL with a calorific value of approximately $30\,MJ/m^3N$ which burns cleanly with minimal SULPHUR DIOXIDE emissions. Typical emission values are $0.7\,mg\,SO_2$ per MJ compared with up to $910\,mg\,SO_2$ per MJ for coal-fired boilers.

This clean fuel aspect has given natural gas a major commercial advantage in the UK for power generation purposes using COMBINED CYCLE power plant with typical CARBON DIOXIDE emissions of 30 grams carbon per MJ (electricity) compared with 65–70 grams carbon per MJ (electricity) for coal-fired plant.

Natural gas reserves are finite. The UK's are predicted to last to ca. 2025 and supplies will have to be sought from Norway and Russia. We have, in effect, a short-term fix and the goal of SUSTAINABILITY will have to be pursued vigorously.

Natural turnover. The annual throughput of material 'processed' by a natural system in equilibrium. The natural turnover can be used to scale anthropogenic inputs of the same material in order to assess any risks that may occur if the anthropogenic input swamps the natural system and destroys its stability. ($\lozenge$ CARBON CYCLE)

Necrosis. Death of a cell while attached to a living body, e.g. SO_2 damage to leaves.

Nephritis. Inflammation of the kidneys which can be caused by drinking water with LEAD in solution, e.g. rainwater collected from lead-painted roofs.

Net present value. The economic value of a project, at today's prices, calculated by netting off its discounted cash flow from revenues and costs over its full life.

Neutral atmosphere. An atmosphere in which the ADIABATIC lapse rate matches the environmental lapse rate. ($\lozenge$ LAPSE RATE)

Neutralization (of a solution). The addition of ACID to an ALKALI or alkali to an acid to achieve neutrality, i.e. pH OF 7. An example would be the neutralization of sulphur dioxide gas from power station by absorption in alkaline calcium hydroxide solution. The product is calicum sulphate (gypsum) ($\lozenge$ GREEN TAXES)

Neutron. An elementary particle, which has no electric charge. It is a component of all atomic nuclei except HYDROGEN. On their own, neutrons are radioactive and decay with a half life of 12 minutes by beta-radiation emission to a PROTON.

Newton (N). Unit of force, i.e. the force required to accelerate a mass of 1 kilogram at 1 metre per second. Conversion factor $1\,N = 0.2248$ pounds force.

Nickel (Ni). A silvery-white magnetic metal which resists corrosion; atomic mass 58.91. It is obtained from various ores and used for nickel plating, in alloys, and as a catalyst. Metallic nickel and its soluble and insoluble salts are potent skin sensitizers. Also, occupational asthma due to respiratory sensitization has been recorded in the electroplating, metal polishing, catalyst reprocessing and stainless steel welding industries. A number of studies have shown clear evidence of lung and sino-nasal cancers in the nickel-refining industry. (◊ CONTROL LIMITS, OCCUPATIONAL)

Nitrates (NO₃). SALTS of nitric ACID which are formed naturally in the soil by micro-organisms from protein and nitrites, and in this form are available as a plant nutrient. Nitrates are also produced industrially and used as fertilizers. However, once spread on the land, not all may be used by crops and so may be leached down into aquifers or run into rivers. (◊ NITROGEN CYCLE; EUTROPHICATION, LEACHING).

The EU limit for nitrate in drinking water is 50 mg per litre (or 50 ppm). So-called Blue Baby Syndrome arose from water which contained in excess of 200 mg/litre nitrate. This level is never approached in the UK today.

Nitric oxide. ◊ NITROGEN OXIDES.

Nitrification. Conversion of nitrogenous matter into nitrates by bacteria, especially in soil. (◊ NITROGEN CYCLE) Also, the conversion of ammonia into nitrate in effluent treatment.

Nitrogen. The atmosphere contains 78 per cent by volume of nitrogen (N_2) and 21 per cent of oxygen (O_2). Its atomic weight is 14. Nitrogen is used in liquid form as a coolant.

Total nitrogen (TN) concentration is commonly used as an indicator of water quality and as a control parameter in wastewater treatment processes, where TN mostly consists of ammonia, nitrate and organic forms of nitrogen.

Nitrogen cycle. Nitrogen is an active and essential component for plant growth and proteins, yet it is chemically very inactive and before it can be incorporated by the vast majority of the biomass, it must be fixed. The fixation of nitrogen is its incorporation in a combined form such as ammonia (NH_3) or nitrate (NO_3^-), whereby it can be used by plants or animals. This fixation can be carried out industrially or naturally by the action of bacteria.

The industrial fixation of nitrogen now matches the bacterial fixation rate. This is an example of humans matching a natural cycle, so both routes must be discussed.

Industrial fixation (Haber process)

The fixation of nitrogen industrially is carried out at 500 °C and high pressure (200 atmospheres) and involves (a) the re-forming of methane to provide hydrogen, (b) the introduction of atmospheric nitrogen and oxygen. The oxygen reacts to form carbon monoxide and thereafter a catalytic reaction combines nitrogen and hydrogen to form ammonia, which can then be readily converted to nitric acid and then ammonium nitrate, a fertilizer of wide applicability. Current global industrial nitrogen fixation rates are around 80 million tonnes per year and are increasing.

Biological fixation

Nature accomplishes nitrogen fixation by means of nitrogen-fixing bacteria. The components of the natural nitrogen cycle are shown in Figure 132. They form a very intricate chain of interlocking activities and are essential to the maintenance of the atmospheric composition.

We start with plants such as peas, beans and clover (legume family) which have nitrogen-fixing bacteria living symbiotically in nodules on their roots and can fix as much as 0.24 to 0.36 tonnes nitrogen per hectare. The plant supplies the bacteria with food and energy, and the bacteria fix atmospheric nitrogen in the form of soluble compounds, some of which are excreted into the soil while the rest supply the plant with essential nitrates. The fixed nitrogen is incorporated in plant protein and if eaten becomes reincorporated as new proteins in the animal. Eventually this protein returns to the soil when the animal or plant dies (there is also a contribution from defecation) and is decomposed by bacterial action into its component amino acids. In aerobic soil conditions, many bacteria oxidize these amino acids to carbon dioxide,

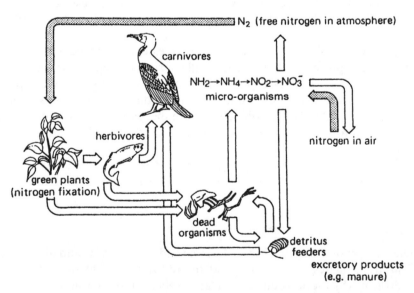

Figure 132   The nitrogen cycle.

water or ammonia. Two additional micro-organisms – the nitrifying bacteria – convert the ammonia to nitrite (NO_2^-) and thence to nitrate (NO_3^-). Once in the nitrate form, it is again available for plant food. Thus, the cycle nitrate to protein, protein to ammonia, ammonia to nitrite, nitrite to nitrate can be repeated. Similarly animal excreta are decomposed to ammonia and converted to nitrate by the nitrifying bacteria.

Superimposed on the cycle are the denitrifying bacteria which decompose (at a smaller rate than fixation) the nitrites or nitrates to molecular nitrogen (N_2). Without these denitrifying bacteria, most of the nitrogen would be locked up in the soil, as run-off to the oceans, or in sediments.

As humans fix 40 million tonnes of nitrogen per year with estimates showing that biological fixation is of the same magnitude, and as the denitrifying bacterial activities are in balance with the biological fixation, it must be assumed that a proportion of the human-made nitrates eventually end up as run-off in lakes, rivers or seas where EUTROPHICATION can result. The benefits of nitrate application are obvious in increased crop yields; however, the ecological implication of interfering with natural cycles has not been explored in sufficient depth. Industrial nitrogen fixation is expected to increase to cope with ever-increasing demands for food from finite land resources. In fact, humans can defeat nature in some cases, as the application of nitrogenous fertilizers stops the action of the nitrogen-fixing bacteria.

Breeding developments are in hand to induce these bacteria to live in symbiotic relation with other food crops such as cereals.

The nitrogen cycle is very closely linked to those of CARBON and OXYGEN, as some of the bacteria involved oxidize organic matter in the soil. There is, thus, a simultaneous cycling of carbon, oxygen and nitrogen.

In aquatic environments, nitrogen fixation can be accomplished by blue-green ALGAE which make some waters highly productive regions, rich in fish life because of the availability of food. Where nitrogen is the limiting NUTRIENT in a body of water, excess nitrates from agriculture or SEWAGE can stimulate the growth of ALGAE resulting in ALGAL BLOOMS in periods of hot weather.

Nitrogen dioxide. ◊ NITROGEN OXIDES.

Nitrogen fixation. The conversion of atmospheric nitrogen into biologically active nitrogen compounds by bacteria that contain a specialized enzyme capable of cleaving the triple bond in the molecules of nitrogen gas.

Nitrogen-fixing bacteria. ◊ NITROGEN CYCLE.

Nitrogen oxides. These are collectively referred to as NO_x. Three are of importance:

Nitrous oxide (N_2O) is a colourless gas used as an anaesthetic: background concentration is 0.25 parts per million. It is mainly formed by soil bacteria in decomposing nitrogenous material.

Nitric oxide (NO) is colourless, formed during the high temperature combustion of fuels which allows the nitrogen in the air to combine directly with oxygen. This takes place at temperatures above 1600 °C, but if the fuel also contains nitrogen, nitric oxide will be produced at temperatures above 1300 °C. NO is oxidized by ozone to NO_2:

$$NO + O_3 \rightarrow NO_2 + O_2.$$

Nitrogen dioxide (NO_2) is a reddish-brown, highly toxic gas with a pungent odour. It is one of the seven known nitrogen oxides which participate in PHOTOCHEMICAL SMOGS and primarily affect the respiratory system. Nitrogen dioxide is extremely poisonous and can pass into the lungs and form nitrous acid (HNO_2) and nitric acid (HNO_3), both of which attack the mucous lining. Nitrogen dioxide is also thought to act as a plant-growth retardant at normal atmospheric concentrations.

A UN protocol was signed in October 1988 by 24 countries to limit NO_x emissions by 1994 to 1987 levels.

The highest concentrations of nitrogen dioxide (1991) in London were recorded at the urban background monitoring sites at Bridge Place and Earls Court. The WHO guideline of 150 μg/m^3 averaged over 24 hours was exceeded, and at Bridge Place the guideline of 400 μg/m^3 averaged over an hour was also exceeded. In this case there was no breach of EC Directive because, although hourly means exceeding 200 μg/m^3 were recorded at the urban background sites at which compliance is measured, the requirement in the Directive is for 98 per cent compliance over a year.

The table below shows the NO_x emissions for the UK, in 2004.

Road Transport	595 000 tonnes
Energy Industries	445 000 tonnes
Other	581 000 tonnes

In London, ca. 80 per cent of NO_x emissions are transport related and as WHO guidelines can be breached, this means that point NO_x sources have to be strictly controlled. That said, traffic management (is your car really necessary?) will increasingly be called upon. It is unfair to burden industry totally whilst private motorists and heavy goods vehicles contribute most of the urban NO_x.

There are three methods of controlling power station NO_x emissions: burner control, injection of chemicals into the boiler, and selective catalytic reduction (SCR) where the flue gas is passed over a catalyst. The cheapest method is burner control which it is claimed can give 50 per cent reduction;

with new plant, 70 per cent may be possible. Retrofits will not achieve the same levels. Boiler injection and SCR systems are claimed to give up to 90 per cent reduction.

As coal-fired boilers produce 70–80 per cent of their NO_x from the oxidation of nitrogen in the fuel, burner design is paramount. The aim is to obtain combustion of the volatiles which form most of the NO_x in the 'cool' region of the flame with the char portion of the fuel burned at the 'hot' end of the flame as discussed by Redman (1989). (◊ AUTOMOBILE EMISSIONS; CATALYTIC CONVERTER)

Sources: J. Redman, 'Control of NO_x emission from large combustion plant', *The Chemical Engineer*, March 1989.

Nitrogen oxides – reduction. To protect global air quality, world-wide regulations are addressing the emission of oxides of nitrogen (NO_x) and carbon monoxide (CO) from industrial gas turbines. The Rolls Royce Dry Low Emissions System makes a major contribution to the environment through reductions of NO_x and CO emissions at source without the need for additional inputs such as steam or water and without the consumption of additional valuable natural resources and without the generation of alternative undesirable waste products.

The dry low emissions combustion arrangement selected is a unique series stage system comprising nine radially poistioned reverse flow tubular combustors. Pre-mixing within the primary stage is achieved by two radial swirlers in series, each swirler passageway having a number of points of fuel injection. A new fuel control system has been developed to distribute the overall fuel demand between the combustion zones to achieve the required temperatures, and to facilitate changeover between the various modes of operation. The results are up to a 50 per cent improvement over previous conventional arrangements.

Another example of engineers providing solutions to environmental problems!

Source: Engineering Council Environmental Awards 1998, Rolls Royce submission.

Noctilucent cloud. Vast clouds of dust at heights of ca. 75 km above earth which reflect light in the night sky.

Noise. SOUND that is socially or medically undesirable, i.e. any sound that intrudes, disturbs or annoys. Very high levels of sound can cause hearing damage. Noise terminology is complex as the physical properties of sound and the mental processes by which it is heard and reacted to must be considered. (◊ HEARING; ROAD TRAFFIC NOISE; AIRCRAFT NOISE; INDUSTRIAL NOISE MEASUREMENT; NOISE INDICES)

Typical Noise Levels

Location	Sound Pressure Level (dB)
Whisper	20
Quiet Bedroom	25
A Public Library	35
Conversational speech	60
An Office	65
Average street traffic	85
A road drill 7 m away	90
A noisy factory	100
A helicopter	110
A submarine engine room; a Rock Concert	120
25 m from a jet aircraft taking off	140

Noise at Work Regulations. The Control of Noise at Work Regulations (2005) define three action levels. 80 dB(A) or more first action level and 85 dB(A) or more second action level are both daily personal noise exposures; in addition there is a peak action level of 135 dB(C) (where 20μPa is used as the reference sound pressure).

Employers need to take some basic action at 80 dB(A), principally noise assessments, records of assessments, provision of information for employees and availability of personal protection. More will be required at 85 dB(A); there will be a duty to reduce exposure to noise by means other than personal protection so far as reasonably practicable, compulsory use of protectors and marking of ear protection zones. All measures required at 85 dB(A) will also be required where peak sound pressure exceeds 135 dB(C).

Employees will be required to use personal protection above 85 dB(A) and other protective measures and report any defects they discover.

Machinery suppliers will be required to provide adequate information concerning the noise levels likely to be generated. (◊ DECIBEL)

Noise certification. A procedure by which new aircraft must meet certain noise standards on take-off and landing before they are allowed to operate in the UK. They are based on engine power and unladen weight.

Noise control. There are three identifiable areas for noise control: source, path, and receiver. A source generates the noise. The path is the source–receiver separation and is characterized by the nature of the intervening space. The receiver is the listener. Noise control can be attempted at any one or all three of these areas.

1. *Control at source.* This is the most widely used form of noise control, e.g. motor vehicles are fitted with silencers and are required to emit less than a maximum permitted noise level (emission standard). For heavy vehicles in particular, control at source can mean tackling any or all of the components shown in the table below.

2. *Noise control between source and receiver.* In many instances, the source cannot be significantly altered, e.g. traffic noise, and therefore sound insulation such as absorbent reveal, double glazing, and fan ventilation instead of open windows, and air bricks have to be used. Noise barriers can also be erected. These should be as close to the source as possible and have a length of at least ten times the shortest distance between source and observer so that the sound waves do not merely bend round the ends of the barrier and reach the observer. They will bend over the top as well but this can be guarded by use of T-shaped barriers.

Source	Possible control measures for heavy vehicles	
Engine casing	Split in two to balance and 'break' transmission vibration	
Engine inlet	Engine inlet modification	
Engine exhaust	Redesign silencer	
Transmission and drive train	Anti-vibration	Total emitted airborne noise
	Redesign gear train	
Fan	Use variable pitch blades	
Tyre/roadway	Redesign suspension	
	Improve road surface	
Aerodynamics	Redesign body shell contours	

The effectiveness of sound barriers depends on frequency and density. The higher the frequency and the denser the material, the greater the reduction of SOUND. The standard for facade exposure to traffic noise in the UK is 68 dB(A) on the L_{10} INDEX. (◊ NOISE INDICES)

3. *Receiver control.* This means that ear protectors are required and these should be chosen for the particular sound pressure level that is to be reduced to a safe level. Ear protectors are mainly used in an industrial environment and are particularly important if NOISE-INDUCED HEARING LOSS is to be avoided.

Noise control boundary. The boundary around premises included in a noise abatement zone on which registered levels are fixed.

Noise dose. The total sound energy received at the ears of an exposed individual over the period of interest (time) measured in Pa^2H. Sometimes measured as sound exposure level (dB). Sound exposure level is equal to period Leq multiplied by the time of exposure.

Noise footprint. An area on the ground under a flight-path during take-off or landing that is bounded by the 90 PNdB or EPNdB noise contour.

Noise indices. Measures of the disturbing qualities of noise-loudness, variations in time, whines, bangs, etc. and the average subjective response. Individual responses to noise vary; noise indices are average measures. They are used in planning residential developments, motorways, etc. L_{10} is the level of noise in dB(A) exceeded for just 10 per cent of the time. For the measurement and prediction of noise from traffic the average of L_{10} values for each hour between 6.00 a.m. and 12.00 midnight on a normal weekday has been adopted by the Government as giving satisfactory correlation with dissatisfaction. This is known as the L_{10} (18 hour); L_{eq} is a measure of the equivalent continuous noise level from a site in energy terms over a specified period; the L_{90} level is exceeded for 90 per cent of the period of interest and is specified as a measure of background level in BS 4142, when measured in dB(A) with the industrial source of interest not operating.

Noise-induced hearing loss. Damage to the ear not caused by ageing or accident or disease. This is an insidious affliction. Many industries are inherently noisy – foundries, boiler shops, shipyards, press rooms. A worker first entering such an environment spends some time getting acclimatized and then may eventually accept the noise levels to which he or she is subjected. This casual acceptance encourages in both workers and management an attitude that noise is simply part of the job.

The first effects of exposure to excessive noise are ringing in the ears and a dulling of the hearing – temporary threshold shift, i.e. a 'louder' level of voice is required to conduct a conversation. These effects are often temporary, but if there is continued exposure or infrequent gaps between exposures, the ears do not recover and there is permanent damage. The inner ear can be damaged and in extreme cases the eardrum can be ruptured.

Industrial deafness is usually associated with inner-ear damage which may go unnoticed for many years until it is chronic. The usual result is that frequencies above 4000 Hz such as a high-pitched whistle or the full range of music cannot be detected. As speech is usually in the 500–2000 Hz range, it is only when an affected individual cannot follow speech readily that the damage is detected. To assess noise-induced hearing loss an audiometer is used which compares the subject's hearing threshold with that of a normal hearing individual. Thresholds are usually measured at 0.25, 0.5, 1, 2, 3, 4, 5, 6 and 8 kHz. Thus at, say, 2 kHz (speech frequency), if the subject is asked to identify the intensity at which he just hears 2 kHz and if this is 30 decibels, then the hearing level is said to be 30 dB at 2 kHz. Now normal hearing intensity at 2 kHz is 0 dB, so the subject's threshold differs by 30 dB. An audiogram can be obtained which plots noise-induced threshold shift against frequency as shown in Figure 103. (◊ HERTZ; NOISE-INDUCED HEARING LOSS)

The degree of handicap is given in table (a). Ordinary conversation is not the only criterion: the social dimension of hearing loss is extremely important and this is shown in table (b). Clearly excessive noise is also a damaging pollutant.

(a) Classification of handicap due to hearing loss.

Class	Degree of handicap	Average hearing level (dB)	Ability to understand ordinary speech
A	Not significant	Less than 25	No significant difficulty with faint speech
B	Slight	25 to less than 40	Difficulty only with faint speech
C	Mild	40 to less than 55	Frequent difficulty with normal speech
D	Marked	55 to less than 70	Frequent difficulty with loud speech
E	Severe	70 to less than 90	Shouted or amplified speech only understood
F	Extreme	90	Usually even amplified speech not understood

From W. Burns, *Noise and Man*, John Murray, 1968; revised edn, 1972.

(b) The social effects of hearing loss experienced by weavers with long exposure to noise. A population of weavers who had been working the trade for a number of years is compared with a control population who did not have a history of severe noise exposure.

Social effects	Weavers	Control
Difficulty in understanding family/friends	77%	15%
Difficulty in understanding strangers	80%	10%
Difficulty in use of telephone	64%	5%
Difficulty at public meetings, church, etc.	72%	6%
Own estimate of hearing below normal	81%	6%

From W. Taylor *et al.*, *Proceedings of a Conference on Occupational Noise in Medicine*, National Physical Laboratory, 1970.

Noise level. Rating level in BS 4142 (1995).

Noise measurement. ◊ DECIBEL; SOUND.

Noise measurement (industrial). The procedure used to assess whether a factory or item of industrial plant is likely to present a noise nuisance is BS 4142:1997 (Method for rating industrial noise affecting mixed residential and industrial areas). It is based on the extent to which the noise level created by the source exceeds the noise level prevailing when it is not in operation.

Non-Fossil Fuel Obligation. The NFFO was a policy initiated in 1990. It required electricity companies to purchase specific amounts of electricity produced by renewable means, at a premium price for a fixed period. It paid for the additional costs of nuclear power and renewables. Initially most of this went towards nuclear, but the proportion going to renewables, as opposed to nuclear, increased. The policy resulted in contracts being awarded to renewables generators in order to 'secure' a total of 1500 MW dnc of renewable by 2000. (For the meaning of dnc see

In all over 3600 MW dnc (Declared Net Capacity; a correction factor applied to the rated capacity of some intermittent generation sources to enable like-for-like comparisons with typical thermal plant capacity) of renewable capacity was awarded contracts, but much of it was never commissioned. Key features of this policy were the 'bankable' contracts which facilitated project financing, and the ability to stimulate renewables' deployment across a range of technologies at different stages of commercial development.

Non-Hodgkins lymphoma. Non-Hodgkins lymphoma is a cancer arising from lymphocytes, a type of white blood cell.

Non-renewable resources. Resources such as minerals or fossil fuels which are replaced on a geological time-scale.

Normal distribution. A very common SYMMETRICAL DISTRIBUTION found in statistics which can be used as a working approximation for many prediction purposes. It is easy to use with quite low error when deviations from non-normality are not too severe. It is characterized by two parameters: the MEAN and STANDARD DEVIATION. Both are obtained by sampling.

The graph of the distribution is given by

$$y = \frac{1}{\sigma\sqrt{2\pi}}\exp\left[-\frac{1}{2}\left(\frac{x-\mu}{\sigma}\right)^2\right]$$

where x can run from $-\infty$ to $+\infty$: μ and σ are the mean and standard deviation, respectively, of the distribution. The spread of the curve about μ depends on σ. The curve is symmetrical about the point $x = \mu$. (◊ VARIANCE)

Normal solution. A solution which contains the chemical equivalent weight in grams of the dissolved substance (solute) in a litre of solution.

Normal temperature and pressure (NTP). A temperature of 293.15 K (0 °C) and a pressure of 101 325 Pa; used as a reference standard for gas volumetric measurements. ⟡ (760 mmHg) STANDARD TEMPERATURE & PRESSURE.

North Sea Gas. (⟡ NATURAL GAS)

Nuclear energy. There are potentially two principal processes for obtaining ENERGY from nuclear sources – fission and fusion. (Fusion has not yet been engineered as a continuous process.)

Fission uses the energy released in a controlled nuclear reaction whereby an isotope (uranium-235) is split by the capture of neutrons, energy (plus more neutrons) is released, and mass consumed in the process. One gram of uranium-235 consumed in a fission reaction will release 81 900 million (8.19 × 10^{10}) JOULES or the equivalent of the heat of combustion of 2.7 tonnes of coal or 13.7 barrels of crude oil – hence a nuclear power station would consume about 3 kilograms of ^{235}U per day. While seemingly small, this is a significant amount of uranium in relation to the total supplies. The prospects for nuclear power would not be 'infinite' were it not for the various other types of reaction (such as the breeder reactor) and the possibility of an alternative long-term nuclear energy supply – fusion.

Fusion relies on the energy release when a heavier element is formed by the fusion of lighter ones. The sun's energy is a fusion process resulting from the formation of helium by the fusion of atoms of one or more of the hydrogen isotopes, i.e. hydrogen (H), deuterium (D) and tritium (T). As in fission, there is a loss of mass and it is this mass difference that appears as energy. Thus, the fusion reaction revolves round the hydrogen isotopes and, indeed, the energy released in an uncontrolled explosive manner by the fusion of D and T is the basis of the H-bomb. The potential of fusion power is illustrated by considering 1 cubic metre of water which ordinarily contains 1 D atom for each 6500 H atoms and has a potential fusion energy (D-D fusion) of 8 160 000 million (8.16 × 10^{12}) joules or the heat of combustion of 270 tonnes of coal or 1370 barrels of crude oil.

1 cubic kilometre = 1000 million cubic metres, so that 1 cubic kilometre of water contains the fusion (D-D) potential of 270 000 million tonnes of coal or 1 370 000 million barrels of crude oil, which approximates to the lower estimate of the world reserves of crude oil. (⟡ ENERGY RESOURCES)

The total volume of the ocean is about 1500 million cubic kilometres, and if 1 per cent of this were used, the potential energy release would be sufficient for 5 000 000 times that of the world's initial coal and oil supplies (see Hubbert, 1969). The reasons for the quest for controllable fusion power are now evident.

For the provision of nuclear energy as currently practised, i.e. fission, ⟡ NUCLEAR REACTOR DESIGNS. (⟡ IONIZING RADIATION; NUCLEAR REACTOR WASTES)

Source: M. K. Hubbert, 'Energy resources', *Resources and Man*, W. H. Freeman, 1969, Ch. 8.

Nuclear fission. In spite of the intense and very costly development programmes of the last three decades, the much heralded future of power from nuclear fission is still not in sight. Great Britain still generates ca 9% of electric power from nuclear fission (the extraction of heat either directly from fissile isotopes or indirectly from fertile isotopes). (◊ NUCLEAR REACTOR DESIGNS (*Fast Breeder*))

The problems of assessing the relative merits and demerits of an extensive nuclear fission technology are of an ethical nature and cannot simply be weighted by cost-benefit analysis. There is a tendency to over-emphasize the short-term benefits and undervalue the long-term problems. In attempting to assess the risks associated with fissile technology, in which 'no acts of God can be permitted', we cannot call on experience, since the magnitude of the problems confronting us are unique in human history. CHERNOBYL and the THREE MILE ISLAND pressurized water reactor incident have both demonstrated this. But technology and understanding of the risks have moved on. Long term secure storage of the very small fraction of highly radioactive wastes is now seen as a potential solution to this one outstanding problem. This has been adopted in Finland in very stable geological strata.

Nuclear power. Unlike any other human product, nuclear wastes will survive for periods that are more appropriate to geology than history. The HALF-LIFE of plutonium-239 is 24 400 years. For every kilogram of plutonium created today, there will still be 500 g left in 24 400 years, 250 g in 48 800 years, 125 g in 97 600 years, and so on. In view of the extreme toxicity of such a material, we cannot ignore serious doubts about our right to saddle our descendants with it – particularly as we have, as yet, devised no practical means of disposing of it. We have entered into a Faustian bargain whereby we are given an unlimited energy source in return for a pledge of eternal vigilance (Edsall, 1974).

Nuclear power is back on the agenda, as evidenced by the following quote from James Lovelock, the creator of the Gaia hypothesis: 'Nuclear power is the only green solution. When, in the 18th century, only one billion people lived on earth, their impact was small enough for it not to matter what energy source they used. But with six billion, and growing, few options remain; we cannot continue drawing energy from fossil fuels and there is no chance that the renewables, wind, tide, and water power, can provide enough energy and in time. Every year that we continue burning carbon makes it worse for our descendants and for civilisation'.

Source: Jowitt, P.W., 'Engineering Civilisation from the Shadows', 6th Brunel International Lecture, Institution of Civil Engineers, London, 2006.

We have no means of knowing whether the safety problems we are setting ourselves are capable of solution. The nuclear industry points with ample justification to the very strict controls and high levels of safety within the industry. But such statements avoid rather than answer the central question of whether the problems of plant safety and of containment and disposal that we are setting ourselves are inherently unsolvable. Unfortunately this dispute is not essentially a technical one and is not therefore resolvable on its technical merits.

People have to operate nuclear power-plants, no matter how much automation we introduce. People are forgetful; often they are irresponsible; and quite a few of them suffer from deep-seated irrational tendencies to hostility and violence . . . I believe that the confident advocates of the safety of nuclear power-plants base their confidence too narrowly on the safety that is possible to achieve under the most favourable circumstances, over a limited period of time, with a corps of highly trained and dedicated personnel. If we take a larger view of human nature and history, I believe that we can never expect such conditions to persist over centuries, much less over millennia (Edsall, 1974).

We can only hope that the safety of the public . . . will never be made dependent upon almost superhuman engineering and operational qualities (Cottrell, 1974).

Time has, however, moved on, and despite the CHERNOBYL and THREE MILE ISLAND incidents, UK nuclear power stations have an excellent record. The late
J. Collier (M.D. Nuclear Electric plc) has stated:

Taken overall, by good fortune as well as by design, nuclear power now seems poised to satisfy, on merit, the broad spread of policy aims set for the British energy sector:

1. It provides competition and consumer choice – with the full costs already included in the price.
2. It expects to face the disciplines of the capital markets – and meet the challenge.
3. The highest priority is placed on health and safety and the industry has a record second to none.
4. It already makes a major contribution to limiting the environmental impact of the energy sector – and offers the most dependable and cost effective route to energy supply without environmental risk. Energy conservation and energy efficiency are complementary measures not alternatives.

With its origins in science and nurtured in defence, the author has a vision that historians may one day look back on nuclear power as one of the more tangible examples of a 'peace dividend' from the twentieth century. The case for nuclear power is now very strong and the outcome of the Government's review is eagerly awaited. This really is 'the end of the beginning' and the start of the future clean energy for the twenty-first century.

UK Nuclear Electric has also brought attention to bear on the security of gas supplies and the emissions from conventional power generation systems (Nuclear Electric and The Environment, January 1995).

Gas demand is forecast to double by 2020. The Government forecasts that by 2020 nearly 60% of electricity will be generated using gas. The consequence will be a period of greatly accelerated increase in demand.

To sustain the anticipated increases in demand, gas will have to be imported from Norway and from the European gas transmission systems. But gas demand is increasing rapidly across the whole of Europe. This means that if the UK electricity supply is dominated by CCGT (combined cycle gas turbines) we could ultimately become reliant on gas from further afield – Northern Norway, Russia or the Middle East. This would expose the UK to supply and price risks as a result both of politics (at the well-head and *en route*) and of the economics of long distance pipelines.

Budget apportionment for Sizewell B Power Station.

Project	Value (£ million at April 1987 prices)
Nuclear steam supply system	566
Civil construction	406
Turbines and other mechanical plant	254
Control and instrumentation	129
Electrical plant	110
First fuel charge	65
Construction and commissioning	43
Software including engineering design fees	449
Setting up and launch costs	8
Total	2030

Source: *Prospects for Nuclear Power in the UK*, Conclusions of the Government's Nuclear Power Review, Department of Trade & Industry and the Scottish Office, May 1995.

Gas is an attractive fuel for power generation in the short term. It is relatively clean, and can be burnt in relatively inexpensive CCGT power stations. Those buying gas need not consider the state of the market beyond the 15 years over which their projects are typically financed.

Nuclear Electric and other companies in the nuclear power industry, however, are used to taking a longer view – 40 years or more. Decisions to burn gas liberally, taken on the basis of relatively short-term pay-back, could adversely affect the UK's energy supply and balance of payments in the medium and long-term. Also a valuable primary fuel and chemical feed stock

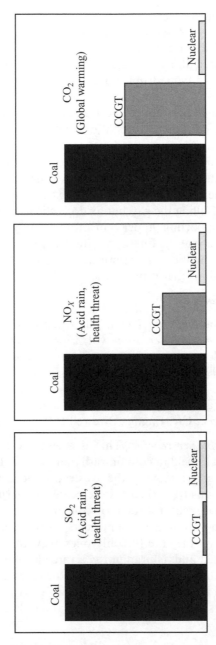

Figure 133 Relative atmospheric emissions from coal-fired (without FGD) and CCGT electricity generation. Nuclear energy has negligible atmospheric emissions.
Source: Nuclear Electric.

is being rapidly consumed when other sources of electricity with comparable long run economics are available.

Our judgement is that nuclear power will become increasingly attractive with time. The Company, therefore, welcomes the Government's commitment to retaining the nuclear option for the UK.

The 1995 review quoted above has had to be revised (2007) with the realisation that Peak Gas and quadrupled gas prices alone, force a review. HMG's carbon reduction commitment is also a driver. Add in, that in 2007 nuclear power supplied 18 per cent of the UK's electricity requirement and that all but one of the current power stations will be operational in 2023. Crisis measures resulting from the studious ignoral of the coming energy crunch (long predicted by experts) will see up to 30 (35 000 MW) new nuclear power plants constructed on the sites of the existing ones, with the first predicated for 2017. The need for secure generating capacity could see 'cheap' gas fired power stations filling the gap but as domestic gas production dwindles, Britain could be importing 90 per cent of its needs by 2020.

The important point about the Energy White Paper published in 2007 may be that for all its good intentions, it cannot deliver either the energy security or the emissions reductions that it promises.

Source: *Financial Times*, 24 May 2007.

A. Cottrell, Letter to the *Financial Times*, 7 January 1974.

J. T. Edsall, *Environmental Conservation*, vol. 1, no. 1, 1974, p. 32.

J. Collier (M.D. Nuclear Electric plc) extract from *Journal of Power & Energy*, vol. 207, 1993: and Nuclear Electric Briefing Note, 6 October 1994.

D. R. Davies, 'Planning and acquiring the UK's first PWR power station', *Proc. Instn. Civ. Engrs.*, Civ. Engng., Sizewell B Power Station, 1995.

Nuclear reactor designs. *Burner reactor*: This is essentially a pressure vessel contained in a biological shield to contain radiation release. It has the means for inserting the fuel and removing the waste products; moderators for slowing down the neutrons from their fast velocities to thermal velocities so that fission can take place; control rods for controlling the overall speed of the nuclear reaction; and canning, i.e. containers for the fuel. The drawback with burner reactors (and they are the only ones available commercially) is that the fuel is consumed and as uranium is a rare element and ^{235}U is less than 1 per cent of natural uranium, the burner reactors are considered a stop-gap measure until breeder reactors are available. Reactor designs vary in the fuels, moderators, coolants, canning, etc., and only the main designs are discussed.

1. Fuel can be natural uranium oxide or ^{235}U-enriched uranium oxide. The enriched fuel has a much higher energy release per unit mass.

2. Moderators can be light (ordinary) or heavy water or graphite. (Heavy water is water containing a substantial portion of deuterium oxide, D_2O.)

3. Coolants come from light or heavy water, carbon dioxide, helium and sodium. (Lithium has been proposed for a possible fusion reactor.) Carbon dioxide is readily manufactured. Helium is obtained from natural gas but could eventually be in short supply. Sodium is used as a coolant in fast breeder reactors and is obtained by electrolysis of molten salt.

4. Cans – fuel containers in metallic and non-metallic varieties: Magnox (magnesium oxide alloy), stainless steel, zirconium, silicon carbide/ graphite.

5. The materials of construction are steel, cement, copper, etc., and are the normal materials of conventional structures.

A summary of the common fission reactor types is given in Figures 135–141.

Source: G. R. Bainbridge and A. A. Farmer, 'Nuclear reactors for the future', *Symposium: Energy Resources or Misuse?*, University of Newcastle upon Tyne, Institute of Chemical Engineering, September 1974.

Fast breeder reactor: The fast breeder reactor uses liquid sodium as a coolant, but because of the inherent risks (however small) of a violent sodium–water reaction, special design techniques are employed to contain this reaction, if any, *outside* the reactor. Thus a primary liquid sodium cooling loop is

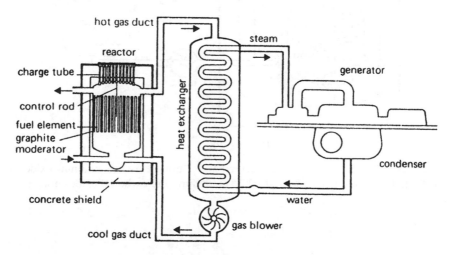

Figure 134 Magnox (magnesium oxide) reactor. The first British commercial reactor using metallic natural uranium metal fuel clad in magnesium alloy, Magnox, with graphite moderator and carbon dioxide coolant.

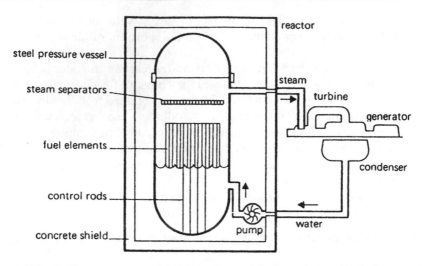

Figure 135 Boiling water reactor (BWR) using enriched uranium oxide fuel pellets clad in Zircaloy, with light water moderator and coolant.

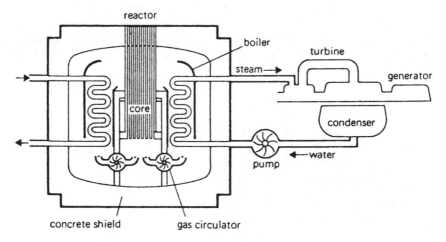

Figure 136 Advanced gas-cooled reactor (AGR) using enriched uranium oxide fuel pellets clad in stainless steel, with graphite moderator and carbon dioxide coolant.

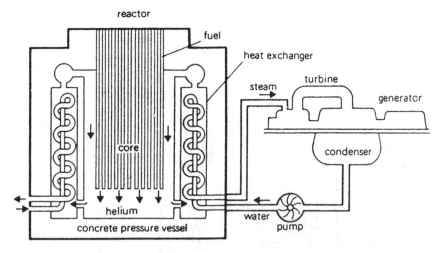

Figure 137   High temperature reactor (HTR) using enriched uranium oxide or carbide fuel granules clad in carbon-silicon carbide, with graphite moderator and helium coolant.

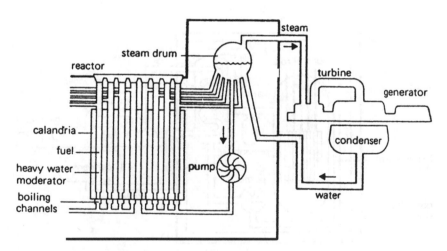

Figure 138   Steam generating heavy water reactor (SGHWR) using enriched uranium oxide fuel pellets clad in Zircaloy, with heavy water moderator and light water coolant.

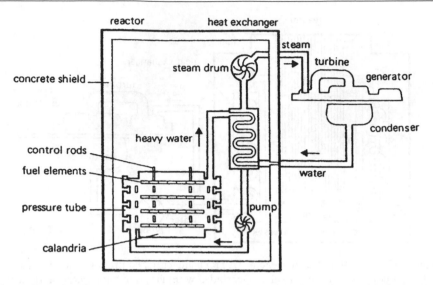

Figure 139 Candu reactor. Canadian reactor using natural uranium oxide fuel pellets clad in Zircaloy, with heavy water moderator and coolant.

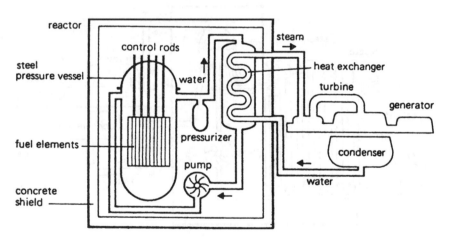

Figure 140 Pressurized water reactor (PWR) using enriched uranium oxide fuel pellets clad in Zircaloy, with light water moderator and coolant.

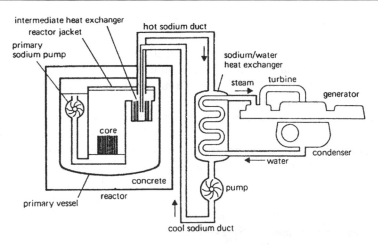

Figure 141 Fast breeder reactor with primary and secondary sodium circuits.

employed which transfers heat to a secondary liquid sodium loop, which transfers heat in turn to a high pressure water–steam circuit to separate steam from the turbines. The arrangement is shown in Figure 141.

A breeder reactor is one in which additional fissionable fuels are produced from the neutrons provided by an initial charge of uranium-235. The raw material for these new fuels is non-fissionable uranium-238 which comprises over 99 per cent of natural uranium, and thorium-232, which is essentially 100 per cent natural thorium. They are called fertile materials. The neutrons are not slowed down to thermal velocities as in burner reactors, hence the generic name *fast* breeder reactors.

The conversion of fertile materials to fissionable materials is called breeding and takes place for ^{238}U as follows:

(number of protons and neutrons)

$$\frac{238}{92}U + n \rightarrow \frac{239}{92}U \rightarrow \frac{239}{93}Np \rightarrow \frac{239}{94}Pu$$

(number of protons)

i.e. uranium-238 absorbs a neutron and becomes uranium-239. This changes spontaneously through two short-lived radioactive transformations to neptunium-239 and thence to plutonium-239.

Similarly thorium-232 and uranium-233 are fissionable in an identical fashion to uranium-235. The thermal energy released per gram of both materials is virtually the same as that of uranium-235. In a breeder reactor more fissionable material is produced than is consumed and in theory it is possible to convert all available fertile material, given that there is enough uranium-235 to start off with.

The breeding gain per charge currently achieved is about 1.2, and design improvements can get this up to 1.4. The doubling time is very important and currently about 20 years are required to double the initial amount of fuel. This could be shortened, but safety criteria may dictate otherwise. Prototype fast breeder reactors are now operating in France. The UK programme has been shut as there was little likelihood of the commercialization of the design within a 30–40 year time span.

Nuclear reactor economics. ◊ SIZEWELL.

Nuclear reactor wastes. Radioactive wastes are produced as a result of the controlled release of energy from fissile fuel and to a much lesser extent by neutron bombardment of otherwise neutral materials in the reactor, including reactor metals, coolant fluids, air, carbon dioxide or other gases. The mass of radioactive fission products produced in a reactor is very nearly equal to the mass of fuel consumed. Once a certain fraction of the fuel is spent, the elements are removed from the reactors and reprocessed; that is, the unspent fuel is separated from the fission products and reused.

Most radioactive waste emanates from the fuel-processing plant in liquid or slurry form and is stored in steel or concrete tanks for a preliminary cooling period prior to ultimate disposal. There are three classifications of radioactive waste slurries or liquids:

1. *High level* – radioactivity greater than 1 curie per gallon (4.5 litres). (2004 production $1890 \, m^3$).
2. *Intermediate level* – radioactivity in the range 1×10^{-6} curie to 1 curie per gallon. (2004 production $82\,460 \, m^3$).
3. *Low level* – radioactivity less than 1×10^{-6} curie per gallon. (2004 production $20\,850 \, m^3$).

The problems associated with nuclear wastes are highlighted in the extreme case of PLUTONIUM.

Up to the mid-1970s in the UK about 30 tonnes of plutonium had been produced by the nuclear power programme. The planned expansion of fast breeder reactors would mean that this amount would increase to several hundred tonnes by the end of the century. Clearly these quantities represent a very substantial hazard and must be effectively isolated from the biological environment for many thousands of years.

Conventionally, radioactive waste is stored until natural decay processes reduce the activity to acceptable levels. For plutonium, the storage periods are so long that artificial storage is only a means of buying time until more satisfactory solutions, e.g. VITRIFICATION, are found. Other proposals for high-level waste disposal include injecting them down deep boreholes into impermeable strata, or deep underground storage.

Storage, however, cannot be justified as a policy *ad infinitum*. High-level wastes (which have been conveniently ignored in the current debate) would require surveillance over about 250 000 years. At some time, disposal, or at the very least abandonment of surveillance, will become inevitable. Although attractive at first sight, a policy of storage is short-sighted and in the longer term intrinsically unsustainable.

A boundary must be drawn somewhere. It is not feasible, in the search for perfection, to reject the disposal of radioactive wastes for ever and a day. Society cannot be expected to supervise in perpetuity the wastes it has produced.

We must at some time relinquish control of them in the knowledge that there will be some uncertainty about the adequacy of the technology of disposal. There is no way of escaping the fate of imperfection: the point is to work to minimise uncertainties, not to ignore them.

Source: Berkhaut, 'The folly of perfection', *The Independent*, 1 January 1989.

The guiding principles for any long-term programme of nuclear waste disposal were laid down by the US National Academy of Sciences in 1967 as:

1. All radioactive wastes should be isolated from the biological environment during their periods of harmfulness.
2. No waste-disposal practice, even if regarded as safe at an initially low level of waste production, should be initiated unless it would still be safe when the rate of waste production becomes orders of magnitude larger.
3. No compromise of safety in the interests of economy of waste disposal should be tolerated.

This problem of what to do with long-lived radioactive waste poses in an actual form some of the dilemmas we discussed. Apart from radiation, nuclear reactors do not emit serious air pollution as do conventional power stations. They offer a long term prospect of cheap electricity, whereas coal and oil are in limited supply. Many people feel that to hold back on nuclear power because of the unsolved radioactive waste problems would be unduly cautious; technology in twenty years time, they argue, will surely have found a way round the difficulty.

The evident danger is that humans may have put all their eggs in the nuclear basket before discovering that a solution cannot be found. There would then be powerful political pressures to ignore the radiation hazards and continue using the reactors which had been built. It would be only prudent to slow down the nuclear power programme until we have solved the waste disposal problem, or until we have developed fusion power or some other type of reactor which does not leave dangerous wastes behind it.

Many responsible people would go further. They feel that no more nuclear reactors should be built until we know how to control their wastes. Since planned demand for electricity cannot be satisfied without nuclear power, they consider we must develop societies which are less extravagant in their use of electricity and other forms of energy. Moreover, they see the need for this change of direction as immediate and urgent.

Currently high-level waste disposal is handled strictly in accordance with the three principles listed above. However, both the first and the second principles are infringed by many present (UK and USA) practices involving the disposal of low-level liquid wastes into the sea and the release of gaseous wastes, after removal of most longer lived isotopes, through tall chimneys. This has resulted in highly radioactive plutonium particles turning up on UK beaches especially Dounreay, Scotland, where as of 2007 beach clean up was still being pursued. This was an experimental breeder reactor facility. Most low-level wastes such as contaminated clothing, concrete, pipes, etc. are containerized and placed in a licensed landfill at the Sellafield UK nuclear reprocessing plant. Long-term plans involve the construction of an undersea repository of 2.5 million m^3 volume which will take care of these wastes till 2050.

In the UK, United Kingdom Nirex Ltd is responsible for providing and operating a repository for the disposal, deep underground, of intermediate-level radioactive wastes and of low-level wastes requiring deep disposal. This is in accordance with government policy for wastes arising in the UK. Similar disposal methods are favoured by other countries producing substantial quantities of long-lived radioactive waste.

Following an extensive site selection exercise, which culminated in preliminary geological investigations from 1989 at two sites, an area near the Sellafield Nuclear Reprocessing Plant, West Cumbria was chosen in 1991 as the focus for further investigations.

An estimated 60 per cent of the waste volume destined for the repository arises at British Nuclear Fuel's Sellafield works.

Before Nirex can build and operate a repository it will have to show that a repository at the proposed site would satisfy the strict safety guidelines laid down by the Government.

Results show that the Borrowdale Volcanics Groups of rocks, the top surface of which is 400–600 m beneath the site near Sellafield, continues to hold good promise as an eventual location for the repository.

A Rock Lab is planned (despite planning rejection) whose aim is to help work out if and how radioactivity could get to the surface. There are tiny amounts of water in most rocks – it is possible this may move some of the radioactivity back to the surface over thousands of years. Gas will be released and this could also be a route back to the surface over many centuries. Therefore it is vital that Nirex has as much information as possible on the rocks at a potential site. One way of building up a picture is with deep boreholes.

Radioactivity fades with time. So, one could assume that intermediate-level waste, surrounded by cement, then placed in a steel drum, would be fairly safe. And if the drums are placed in hard rocks deep under the ground and more cement is poured around them, this is likely to be safe for thousands of years.

Source: www.nirex.co.uk. Accessed 23 May 2007.

(◊ IONIZING RADIATION, EFFECTS; IONIZING RADIATION, MAXIMUM PERMISSIBLE DOSE)

Source: *Pollution: Nuisance or Nemesis*, a report on the Control of Pollution, HMSO, February 1972.

Nucleus[1]. The positively charged core of an ATOM made up of NEUTRONS and PROTONS. Accounts for almost the whole mass of the atom.

Nucleus[2]. Focus for growth; for example, a dust particle acts as a nucleus of a fog droplet or ice particle formation.

Nuclide. Term for the ISOTOPE of an atom. If it is radioactive, it is known as a radionuclide. (◊ RADIOACTIVE ISOTOPE)

Nuisance. In law, that which annoys or hurts, e.g. excessive noise levels, emission of noxious odours, polluting discharges. A large body of case law exists. Section 80 of the Environment Protection Act 1990 states:

Where a local authority is satisfied that a statutory nuisance exists, or is likely to recur . . . the local authority shall serve . . . ('an abatement notice') imposing all or any of the following requirements (a) requiring the abatement of the nuisance or prohibiting or restricting its occurrence or recurrence; (b) requiring the execution of such works, and the taking of such other steps as may be necessary for any of those purposes, and the notice shall specify the time or times within which the requirements of the notice are to be complied with.

A powerful remedy exists for those affected by a nuisance viz. 'A claimant is *prima facie* entitled to an injunction to protect his or her legal right against a person who has committed a wrongful act such as a continuing nuisance'. R. Macrory 'Appeals Court reinforces role of injunctions in nuisance cases'. ENDS 382 Nov 2006 p. 59 (◊ DUTY OF CARE; *RYLANDS* v. *FLETCHER*)

Nuisance threshold. The lowest concentration of an air pollutant that can be considered objectionable.

Nutrient removal from wastewater. It has been recognized that the plant nutrients N and P in wastewaters need to be removed before discharge in order to prevent EUTROPHICATION in watercourses. The most economic method is by using biological systems. The N present as ammonium (NH_4) is oxidized to NITRATE (NO_3) and later denitrified to N_2 gas using the bacterial species *Nitrosomonas* and *Nitrobacter*. Phosphate removal is achieved normally through bacterial growth in biological wastewater treatment systems but the rate can be enhanced by applying a sequence of ANAEROBIC and AEROBIC conditions which induce polyphosphate storage in the microbes.

Nutrients. The raw materials necessary for life which are consumed during the metabolic process of nutrition. Their type and consumption vary according to the particular plant or animal species. The main categories are proteins, carbohydrates, fats, inorganic salts (e.g. nitrates, phosphates), minerals (e.g. calcium, iron), and water.

Nymph. The larval stage of aquatic insects such as may flies, stone flies and dragon flies. They are useful biological indicators of the purity of a watercourse. ($\diamond$ BIOTIC INDEX; BIOLOGICAL INDICATOR)

O

Occupational exposure standard (OES). The concentration of an airborne substance, averaged over a reference period, for which, according to current knowledge, there is no evidence that it is likely to be injurious to employees if they are exposed by inhalation, day after day, and which is specified in a list approved by the Health and Safety Committee (HSC). For a substance which has been assigned an OES, exposure by inhalation should be reduced to that standard. However, if exposure by inhalation exceeds the OES, then control will still be deemed to be adequate provided that the employer has identified why the OES has been exceeded and is taking appropriate steps to comply with the OES as soon as is reasonably practicable. In such a case the employer's objective must be to reduce exposure to the OES, but the final achievement of this objective may take some time. Factors which need to be considered in determining the urgency of the necessary action include the extent and cost of the required measures in relation to the nature and degree of exposure involved (ref. UK Health and Safety Executive, 1991).

The criteria for setting limits are:

1. The ability to identify, with reasonable certainty, a CONCENTRATION averaged over a reference period, at which there is no indication that the substance is likely to be injurious to employees if they are exposed by inhalation day after day to that concentration.
2. The OES can reasonably be complied with.
3. Exposures to concentrations greater than the OES, for the period of time it might reasonably be expected to take to identify and remedy the cause

of excessive exposure, are unlikely to produce serious short- or long-term effects on health.

Ocean dumping. Dumping of SEWAGE SLUDGE at sea is now banned under the EU Urban Waste Water Directive and land 'disposal' (recycling) is seen as one way forward despite potential HEAVY METAL accumulation.

However, today's sludge incineration plants have very low environmental impacts due to stringent emissions control. A UK tribal culture on land 'disposal' means that a polarization has currently taken place on the 'best' route when integrated management is actually called for.

Occupational Cohort Mortality Analysis Program (OCMAP). This computer program is a means of analysis of incidence or mortality rates and standardized measures in relation to multiple and diverse work history and exposure measures. (◊ PYROHYSIS)

Octane number. A measure of the knock-resistance of a FUEL for spark-ignition engines. Determined in test engines by comparison with reference fuels. (◊ CETANE NUMBER)

Odorant. A distinctive, sometimes unpleasant, odour added to odourless materials to give warning of their presence, e.g. to natural gas.

Odour threshold. The concentration of an odour-bearing gas at which half a panel of 'sniffers' can detect the smell. The odour is usually diluted in a dynamic system and presented to groups of volunteers at various dilutions. (◊ ODOURS)

Odours. The smell(s) produced by, usually, very small CONCENTRATIONS of ORGANIC vapours can produce violent aversion reactions in anyone exposed to them. The reactions range from nausea to insomnia. Many of these vapours come from operations such as animal by-products processing, farming, maggot breeding, brick-making and metallurgical processing. What is surprising is that the odour threshold in many cases occurs at very low concentrations of the substance in air and almost invariably these concentrations are well below the THRESHOLD LIMITING VALUE for the substance. For a 'normal' person, there is at least some comfort in knowing that mass poisoning is not taking place. (◊ MERCAPTANS)

Nevertheless, odour suppression at source can and should be practised on both public health and amenity grounds. The common methods are:

1. Absorption in a suitable liquid which may oxidize the offending vapour or neutralize it in the process. (◊ ABSORPTION)
2. Adsorption: ACTIVATED CARBON will adsorb organic molecules in preference to water vapour and can be tailored to provide optimum adsorption for the particular contaminant. (◊ ADSORPTION)
3. INCINERATION, i.e. oxidation at high temperature, will destroy most malodorous, gaseous and organic wastes but is usually very expensive.
4. After-burning of non-inflammable weak mixtures of organic vapours which heats the effluent air stream to temperatures greater than 750 °C. (◊ AFTER-BURNER)

5. Ozonation: the use of OZONE's very powerful oxidizing abilities to deal effectively with some odours such as that from hydrogen sulphide.
6. Masking by spraying a pleasant perfume. – Commonly used on LANDFILL SITES.

   (◊ ODOUR THRESHOLD)

Oestrogen mimicking pollutants. These are synthetic chemicals that have an oestrogen-like effect on aquatic creatures. Fish exposed to these chemicals produce a female egg protein called vitellogenin and develop abnormal sex organs containing egg cells. Amongst the chemicals thought to be responsible are polychlorinated organic compounds (such as dioxins), organochlorine pesticides, organotins, alkyl phenols, alkyl phenol ethoxylates and synthetic steroids.

Offensive trades. Under UK legislation offensive trades are subject to regulation by local authorities under the statutory provisions of the Public Health Acts. The processes involve animal products or residues (skin, bones, fat, blood, etc.) and are liable to give rise to an odour nuisance.

Offshore oil extraction, risks of. The extraction of offshore oil has five danger spots for large or catastrophic oil pollution:

1. Spill from the seabed or well-head.
2. Pipeline fracture from the seabed to the oil platform.
3. Pipeline fracture or leak from the platform to the shore or a mooring buoy.
4. Spillage at a mooring buoy when tankers couple up and uncouple.
5. Collision with well-heads or platforms.

The results can be a large OIL SLICK which has substantial effects on fisheries, sea birds and coastal amenities.

Oil. Our major source of fossil energy resources at the present moment. Roughly two-thirds of the world's known ultimately recoverable reserves are in the Arabian Persian Gulf area, and the total effective lifetime of world reserves will probably be 70 to 80 years using currently available technology. With the exception of additional resources in Russia and SE Asia it would appear that there is no geological justification for assuming that there will be further discoveries that will significantly alter the world picture, but the use of technological innovations in extracting more oil from the reservoirs may significantly extend existing oil field lifetimes. The proven UK North Sea reserves represent only ca. 3 per cent of ultimately recoverable world reserves (see Figure 142). However, one forecast by the UK Offshore Operators Association predicts that up to 1.4 million barrels per day could be available from the North Sea in 2013 compared with 2 million barrels per day in 1989. The UK would, by then, be a net importer of oil but around 60–70 per cent of demand could still come from the North Sea. Existing North Sea discoveries total 20.8 billion barrels and an expected further 5.8 billion barrels will be found. The Association is quoted by the *Financial Times* as stating that 'actual performance will depend heavily on the final environment, including

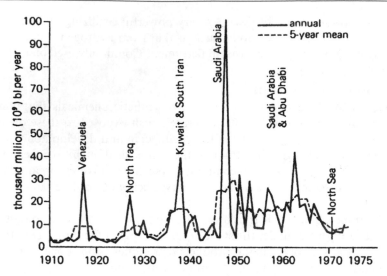

Figure 142   World oil discoveries (excuding Communist bloc). Note the importance of North Sea discoveries compared with Saudi Arabia.

the degree to which smaller and more complex field development is encouraged by tax arrangements . . . '. ($\lozenge$ RESERVES-PRODUCTION RATIO)

Laherrere analysed global reserves and stated:

> We conclude that the ultimate range of oil producible below $25/b is 1,500–1,750–2,300 Gb (Gigabarrels) with the mid number as the mean (expected value). It furthermore looks as if production is set to decline naturally at about 2.7 per cent a year from around 2000. If successful efforts are made to delay the onset of the decline, it simply means that the subsequent slope becomes a cliff. Heavy oil, tar sand and enhanced recovery will become important after 2000, and will mitigate but not reverse the decline, such that by 2050 production will have fallen to what it was in the 1960s.
>
> It is too axiomatic to state that oil production has a huge impact on the political and economic development of the world, including especially transport and agriculture, for which there are no easy substitutes. The weakness of the world's reserve and production data, much held confidential by governments and state companies, is a serious obstacle to the analysis of this important subject which is needed if the world is to have time to adjust and if its leaders are to secure the political consensus for the measures they will be forced to adopt.

This prophetic statement (1995) has yet to be disproved and serves as a much needed warning to our political masters. The RISK-AVERSE approach to new UK NUCLEAR POWER stations must need be reexamined and planned construction programmes initiated to provide a strategic capacity on power supply continuity, resource conservation and environmental grounds. The UK nuclear power electricity generating plants have an outstanding safety record.

Source: J. Laherrere, 'World oil reserves – which number to believe?', *OPEC Bulletin*, 9–13, February 1995.

Tahmassebi has analysed future oil demand as tabulated below – it is clear that a crunch is coming in energy supplies.

	Demand (*million barrels per day*)		
	Low-growth scenario	*Base-case scenario*	*High-growth scenario*
1994 estimated	67.8	67.8	68.1
1995 projected	68.6	69.0	69.3
2000 projected	73.2	75.3	77.3
2005 projected	75.1	80.3	84.6
2010 projected	78.9	86.1	93.9

Source: C. H. Tahmassebi, 'The changing structure of world oil markets and OPEC's financial needs', *OPEC Bulletin*, March 1995.

Oil is not only a FUEL, it is a fundamental resource for the chemical industry. Figure 143 illustrates how, from one basic raw material, a wide variety of end-products are made by means of different chemical processes.

Source: *'Crosslink' Bayer Group Journal*, Issue 7, 1995, p. 20 (Reproduced by permission of Bayer PIC).

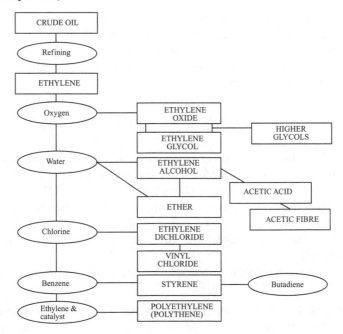

Figure 143 Uses of oil in the chemical industry.

Oil platforms, offshore installations removal of. Over the next 10 years ca. 50 North Sea offshore installations will come to the end of their useful lives and will be taken out of service.

The remainder will be decommissioned by 2080 when the oil runs out. The Brent Spar platform removal (1995) was a cause *célèbre* in that environmentalist miscalculation of the platform's residual contents gave rise to unnecessary public concern. We have been here before!

Times have moved on from considering disposal at sea as the only viable option, as exemplified by the Viking A Platform (which produced gas) dismantling and recycling programme which will ensure an exemplary 99.8 per cent recycling or reuse of the platform materials of construction (see tables, (a)–(c)).

International Maritime Organization (IMO) standards require installations in less than 75 m of water (and weighing less than 4000 tonnes) to be completely removed.

All post-1 January 1998 installations will also have to be removed entirely if they are in less than 100 m of water and be designed to be removable if placed in deeper water.

All other installations will be partially removed or left in place so long as there is at least 55 m of clear water above any submerged remains and regular maintenance to any remains above sea level in order to prevent structural failure.

The cut-off portions are likely to be dumped on the fringes of the North Sea, in depths of 2000 m where they can disintegrate slowly.

The dispersive powers of the ocean should ensure that minimal environmental damage occurs despite high decibel claims to the contrary from well-intentioned pressure groups.

Source: Conoco Information Sheet, Revision 0, 2 November 1995.

The scale of the current tasks are exemplified by the decommissioning of redundant facilities at the Ekofisk Field in the North Sea. This involves the removal of 11 steel platforms to shore for recycling by 2013. It also covers the original concrete storage tank and its protective concrete barrier, which is 106 m high, has a diameter of 140 m, and weighs 1.2 t. The four decommissioning options were alternate use, full removal, partial removal, and abandonment.

In-place abandonment won Norwegian government approval in 2002.

Source: Crook, J. 2006, Rebuilding Ekofisk for the 21st century, *The Chemical Engineer*, Dec 06/Jan 07, pp 40–41.

(a) Sources of materials on Viking A.

Topsides steelwork	Bulks:	All steelwork associated with topside structure, and including ladders, catwalks and walkways.
Mechanical systems	Plant:	All rotating/reciprocating equipment.
	Pipes:	All process vessels and piping associated with this equipment.
	Instrument:	All the instruments in these equipment packages.
	Electrical:	All the electric cabling in these equipment packages.
Piping systems	Equipment:	Valves and special items.
	Bulks:	Piping gaskets, paint on the pipes, not vessels.
	Supports:	Supports and hangers.
Architectural		Buildings for weather protection and accommodation.
	Fittings:	Beds, toilets, seats and partitions in accommodation blocks.
	Equipment:	Catering equipment and laundry and hot water systems in accommodation blocks.
	Bulks:	All structural steelworks, claddings and insulation in accommodation blocks.
Instrumentation	Equipment:	Instruments, DCS computers and stainless steel impulse lines.
	Bulks:	Other non-stainless steel impulse lines, supports, cables and valves.
HVAC system		Heating, ventilation and air conditioning.
	Equipment:	Fans, dampers, heater banks, chillers.
	Bulks:	Ducting and associated insulation.
Electrical	Equipment:	Motors, switch gear, junction boxes.
	Bulks:	Cabling, cable trays, lights, switches and glands, but excluding diesel generator sets.
Fire and safety	Equipment:	Lifeboats, life vests, fire extinguishers, smoke and gas detectors, fire hydrants and fire pumps.
	Bulks:	Piping, cabling and valves associated with fire and safety equipment.
Telecommunications	Equipment:	Telephone, radios (including panels) speakers, dishes, etc.
	Bulks:	All telecommunication cables, cable trays and telecommunications towers.

(b) Disposal of materials.

Material	Location	Use, purpose of origin	Approx. anticipated tonnage	Disposal
Structural cladding/steel	Topside subframe and basic structure/Wells of modules	Jackets and support frames for modules and skeleton of topsides structures/cladding and partitions	8454	Recycle
Concrete, grout	In partitions and as grout	Act as fireproofing in some areas, and used as grout in conductors and piles	100	Recycle
Zinc	Anodes, paints and claddings	Major component of some anodes, in paints and as cladding on some items	50	Recycle
Aluminium	Anodes, some external pipes	Major components of anodes, cladding	10–20	Recycle
Stainless steel	Pipework process equipment	High spec. parts of process system and instrument piping	1–5	Recycle
Copper	In electric cabling	Part of electrical system and some specialized parts of process pipework	1–5	Recycle
Residual process oil	Process pipework	Residual amount remaining after final flushing of pipework and present in compressor lubricating oil systems	<1	Recycle
Lead	Batteries and anodes	Component of some wet batteries and small component in some anodes	<1	Recycle
Halons	Fire fighting system	May be small quantities of cylinders and extinguishers on topsides	<1	Recycle
Plastics/rubber	Accommodation units, pipe liner, cabling cover/in hoses and cabling	Wide range of different types used/hoses and insulation, etc.	1–5	Landfill/ recycle
Asbestos/mineral wools	In panelling especially and in wide range of other locations	As insulation, sound-proofing and fire-proofing	1–5	Special landfill
Wood	In accommodation units	Furniture and fittings	<1	Landfill
LCA scale	Parts of process system	May be present in some parts of process system of pipes and tanks, deposited from solution	<1	Special landfill
PCB	In some electrical equipment	May be found in some electrical equipment	<1	Incineration

(c) Recycled products and disposal for reuse.

- Tubulars
- Plates and beams
- Pressure vessels
- Pumps and rotating equipment
- Power generation equipment
- Life boats and marine items
- Valves and well-head equipment
- Living accommodation units.

Oil shales. Large oil-bearing shale deposits in the Green River formation of Colorado. Wyoming and Utah have been estimated to contain 1800 thousand million barrels of oil – more than four times the crude oil discovered to date in the USA. However, only 6 per cent of the deposits are accessible, yielding more than 30 gallons of oil per tonne of rock.

Pilot projects have demonstrated the feasibility of recovering shale oil (kerogen) and converting it to crude oil in liquid form. Commercial extraction by surface retorting is now possible. ($\Diamond$ EXTRACTION OF OIL FROM SHALES)

Mining of the shale for commercial purposes must be carried out on a huge scale, however – in the region of 500 million tonnes per year. There is the problem of where to tip the spent shale, which is considerably greater in volume than the original rock. There is a danger of LEACHATE from the spent shale polluting watercourses.

One possible limitation on oil-shale production is thought to be the lack of sufficient water in the Colorado watershed as there are already large agricultural demands on the available water. The recovery of oil from oil shales on an industrial scale requires vast quantities of water for cooling and the effects on the immediate locality, whose economic structure depends largely on irrigation, are likely to be considerable.

Oil slick. A floating layer of oil on the seas, rivers, canals or lakes, usually as a result of accidental spillage from tanks, pipelines, etc. Left untreated, slicks can cause major local ecological disasters, especially to sea birds, and, if allowed to drift ashore, will foul beaches.

Common methods of eliminating slicks are:

1. To contain the oil slick by floating booms, and then transfer it into tanks by suction.
2. To adsorb the oil on nylon 'fur' and then squeeze it out of the fur for reuse or disposal.
3. To use detergents to break up the slick. However, this method has been attacked on the grounds that it does not remove the oil and that the

detergents contain toxic ingredients harmful to birds and marine life, particularly shellfish.

The catastrophic nature of oil spills was highlighted when the *Exxon Valdez* oil tanker discharged 50 000–150 000 m³ of crude oil after running aground off the South Alaskan coast in Prince William Sound on 24 March 1989. A 3000 square mile slick which contaminated at least 1300 miles of shoreline in an ecologically sensitive area was the result. Unpreparedness for such a disaster and the inability of physical cleaning methods meant that a massive detergent spraying programme was used. The clean-up operation took more than six months with incalculable effects on wildlife from both the oil pollution and the detergents. Rare species such as the humpbacked and killer whales were threatened from ingestion of oily water and eating polluted food.

A similar catastrophe took place off the Welsh coast when the *Sea Empress* spilled 72 000 tonnes of oil and damaged 200 km of Welsh coastline (February 1996).

The subsequent clean-up cost £100 m. The Milford Haven Port Authority was fined £4 m plus £1.8 m costs. This was reduced on appeal to £3.25 m (March 2000) – prevention can often be better than cure.

These incidents have highlighted that oil (or other large-scale natural resource extraction) operations can exact a substantial environmental toll. In the case of the *Exxon Valdez*, the oversight of an oil tanker captain caused the disaster. The human factor can never be entirely eliminated and the greater the potential environmental insult, the greater the precautions and back-up systems required. (◊ CHERNOBYL; VALDEZ)

In its seventh environmental conviction since 1990, Shell UK Ltd was fined £20 000 for polluting the Manchester Ship Canal with 140 tonnes of oil – the equivalent of 10 500 household buckets.

The company admitted the offence at Chester Magistrates' Court. It was also ordered to pay £4540 in costs.

The court heard that 188 tonnes of refined oil (known as olefin) was lost during a pumping operation at the Stanlow Manufacturing Complex in Ellesmere Port. About 140 tonnes entered the canal, some of it spreading five miles. The clean-up operation lasted 10 days.

In February 1990, Shell UK was fined £1 000 000 by the Crown Court for polluting the River Mersey with oil from the Tranmere Oil Terminal.

Source: Environment Agency *Environmental Action*, Issue 17, December 1998.

Oligotrophic. An aquatic environment which has low concentrations of nutrients present and therefore has low plant and animal life productivity. The opposite of eutrophic. A lake that is oligotrophic, being poor in producing organic matter, is a characteristic of 'young' lakes and reservoirs. (◊ EUTROPHICATION)

Open-cast mining. This is the process of surface mining in which large quantities of mineral-bearing rock are scooped out to produce, in effect, a very large hole, with terraced sides. The largest, in Bingham Canyon, Utah, is $1\frac{1}{2} \times 1\frac{2}{3}$ miles $\times \frac{1}{2}$ mile deep.

The two basic types of open-cast mining are quarrying, in which the rock is cut into large blocks and usually transported away from the site in this form, and metal-mining, where the large quantities of rock are ground to a powder on site and the metal-bearing minerals are removed.

In this way large quantities of poor ores may be processed in order to obtain relatively small amounts of the valuable minerals. This method also leaves a very large amount of useless ground-up rock.

The scale of some of these operations is illustrated by the Anaconda Company's mine at Twin Butte near Tuscon, Arizona, where 236 million tonnes of overburden and rock were removed to get to the low-grade copper ore 600–800 feet below ground. The ore itself has a copper content of 0.5 per cent, i.e. 200 tonnes ore are required to obtain 1 tonne copper, assuming 100 per cent recovery of copper. This leaves 199 tonnes of spoil for disposal. Lower and lower grades of ore cannot be worked as the energy required would escalate out of proportion to the material recovered. For example, if (say) 0.1 per cent ore were mined, this would require a minimum of 1000 tonnes of ore for 1 tonne of copper and produce 999 tonnes of spoil assuming total recovery of copper which is an impossibility. Clearly, there are limits.

Operator and Pollution Risk Appraisal (OPRA). A methodology for formal RISK assessment for processes subject to INTEGRATED POLLUTION CONTROL (IPC). OPRA has two components: Operator Performance Appraisal (OPA) which provides an appraisal of the probability of an incident, and Pollution Hazard Appraisal (PHA) which appraises the consequences of the incident. This enables the risk to be evaluated. ($\lozenge$ OPERATOR PERFORMANCE APPRAISAL (OPA), POLLUTION HAZARD APPRAISAL (PHA))

Operator Performance Appraisal (OPA). An OPA is carried out by Environment Agency for those processes subject to Integrated Pollution Control against the following indicators:

- Compliance with limits and adequacy of records.
- Knowledge of authorization requirements and implementation.
- Plant maintenance and operation.
- Management and training.
- Procedures and instructions.
- Frequency of incidents and justified complaints.
- Auditable environmental/management systems.

Each is graded on a scale from 1 to 5 (1 worst, 5 best) and then weighted to determine a final score.

It is part of the OPERATOR AND POLLUTION RISK APPRAISAL process.

The following are examples of scale values 1 and 5:

Scale 1 – plant maintenance in operation, no preventive maintenance programme in place, the company relies only on breakdown maintenance, and haphazard procedures implemented by operators.

Scale 5 – a suitable maintenance programme is in place based on industry standards and/or manufacturer's recommendations. This is fully implemented and plant operations are clearly defined and followed.

Ore. Any noticeable concentration of a metalliferous mineral (whether or not it is of economic value), e.g. the metal content of the various ores differs widely; currently a workable iron ore contains 20–30 per cent iron, whereas an ore containing only 0.5 per cent copper may be worked profitably. (◊ OPEN CAST MINING)

Ore dressing. The processing of raw material, won by mining, into a marketable form; the crushing, concentration and separation of the ORE mineral(s) from the waste residue. This may involve hand-picking, gravity concentration, magnetic separation, dense-media separation or chemical separation, e.g. separation of gold by the use of CYANIDE (and its attendant pollution risks).

Ore grade. The percentage of mineral in an ORE. This has a major effect on the amount of energy needed to win the pure metal.

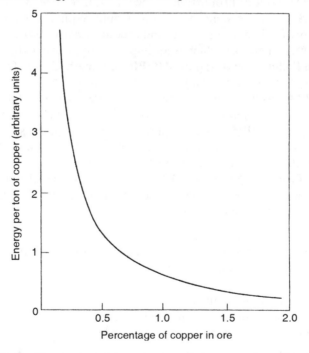

By the expenditure of increasingly large amounts of energy, per ton of copper produced, increasingly low-grade ores may be mined (based on P. Chapman, *Fuel's Paradise*, Penguin, 1975). Reprinted with permission of Edward Arnold (Publishers) Ltd.

Organic chemistry. The study of CARBON compounds. Does not include metal carbonates, oxides and sulphides and carbon.

Organic matter. Material containing CARBON combined with HYDROGEN often with other elements (OXYGEN, nitrogen), e.g. plastics, vegetable matter.

Organochlorines. A major class of chemicals emanating from the organic chemicals industry. It includes INSECTICIDES, AEROSOL PROPELLANTS, PCBs (POLYCHLORINATED BIPHENYLS), PVC, DDT and ENDOSULFAN. Organochlorines are characterized by persistence, mobility and high biological activity. They have very long HALF-LIVES: that for DDT is ten years or more, for DDE many decades.

All organochlorines have a high capacity to injure living systems and allied with their other attributes may possibly constitute the greatest threat to our life-supporting ECOSYSTEMS and associated biological cycles.

In many insecticide applications the organochlorines are being replaced by ORGANOPHOSPHORUS COMPOUNDS and the new synthetic PYRETHROIDS are likely to be produced at rates which might rise to 20 000 tonnes annually world-wide. (◊ PESTICIDES; CHLORINATED HYDROCARBONS)

Organophosphorus compounds. A group of pesticides which embraces such names as azodrin, malathion, parathion, diazinon, trithion and phosdrin. All are chemicals related to nerve gas developed during the Second World War, and block the central nervous system by inactivating the enzyme responsible for breaking down a nerve 'transmitter' chemical called acetylcholine. The result is hyperactivity resulting in death.

Organophosphorus compounds are generally very much more toxic to insects than mammals and also have a very much shorter HALF-LIFE than the ORGANOCHLORINES. For these reasons they are labelled as 'safe' insecticides, although they may be CARCINOGENIC. The long-term ecological effects of these chemicals is not known. They have only been in use for one generation which is insignificant on the evolutionary time-scale.

In common with most similar products, these chemicals buy time which must be put to use for the stabilization of populations and resource consumption and planned ECOSYSTEM management. (◊ PESTICIDES)

Orimulsion. A bituminous FUEL produced by the Venezuelan state oil company. It comprises 70 per cent bitumen and 30 per cent water with a claimed ENERGY value 9 per cent higher than power station grade COAL and can be burned in modified oil-fired boilers. Sulphur dioxide emissions when burning Orimulsion are about 2300 parts per million volume, compared with 1800 parts per million from heavy fuel oil, and typically about 1700 when using coal mined in Britain.

FLUE GAS DESULPHURIZATION will be required and was planned to be fitted out at the 2000 MW Pembroke oil-fired power station which was to burn between 4 m and 5 m tonnes of the fuel a year from 1998. This would remove 94 per cent of the SULPHUR DIOXIDE emissions.

The suppliers of this fuel were to build a facility in Germany to recover VANADIUM, magnesium and NICKEL from the power station ash which contains 19 per cent vanadium, 13 per cent magnesium and 2.3 per cent nickel. Claimed recovery rates were 99.8 per cent, 92 per cent and 91 per cent, respectively. Plans to use Orimulsion at Pembroke Power station were cancelled due to protests by environmentalists.

Osmosis. Process by which water moves from a solution of low solute concentration to a solution of higher solute concentration via a semipermeable membrane.

Overband magnet. An electromagnet which is positioned crosswise above a conveyor carrying solid waste or incinerated residue in order to recover ferrous metals.

Oxidation pond. A basin used for retention of wastewater in which biological oxidation of organic material is effected by natural or artificially accelerated transfer of OXYGEN to the water from air.

Oxide fuel. Nuclear fuels manufactured from the oxides of the fissile material. They can withstand much higher temperatures and are much less chemically reactive than metals.

Oxidizing agent. Chemical which either gives up oxygen in chemical reactions or supplies an equivalent element such as CHLORINE to combine with a REDUCING AGENT. Oxidation is also used for the removal of hydrogen from a substance. Atmospheric oxidizing agents include OZONE and NITROGEN DIOXIDE.

Oxygen (O). A colourless, odourless gas essential for all aerobic forms of life and for combustion. It forms 21 per cent by volume of the atmosphere. Chemical symbol O, formula O_2 (molecular weight 32). Unstable form OZONE, formula O_3.

Oxygen cycle. Oxygen is a major component of all living matter and is vital in the free state for the higher animals which require it in their metabolism. Its presence on earth is almost certainly due to the process of PHOTOSYNTHESIS in plants which is the assimilation of CARBON DIOXIDE and water for the production of carbohydrates and free oxygen. This free oxygen both supports and comes from life.

The emergence of free oxygen has its origin around 3000 million years ago when simple AUTOTROPHIC ORGANISMS evolved which were able to split water and release oxygen. (There is geological evidence of oxidized sediments around 1500 million years old in the form of ferric compounds, the oxidized form of ferrous rocks.)

With the emergence of free molecular oxygen (O_2), the sun's energy split the oxygen molecules into atomic oxygen (O), which is highly reactive and forms ozone (O_3) which has built up into the OZONE SHIELD. Thus the earth's atmosphere began to evolve and stabilize. However, oxygen produced by photosynthesis is used up by respiration either by the consumers in the FOOD

CHAIN or the decomposers. So, if the oxygen is recycled in this manner, how did atmospheric concentration build up? (◊ CARBON CYCLE) The answer lies in the carbonate and carbonaceous sediments; calcium carbonate (limestone) is an example of the former and coal an example of the latter. The sediments formed by the deposits of animal and plant bodies removed carbon from the carbon cycle and tied it up for geological time. For every atom of carbon laid down to sediment, two atoms of oxygen are left free. Thus, the bank of free oxygen that we have in the atmosphere was made possible by the formation of carbonaceous sediments. As the sediments were being deposited, the atmosphere evolved, the ozone shield grew, and we now have a stable atmosphere. The biosphere and the atmosphere evolved simultaneously due to the carbon and oxygen cycles operating together.

If photosynthesis were to stop tomorrow the reservoir of oxygen would be sufficient to sustain higher life for millions of years without significant depletion.

The carbonaceous sediments amount to many thousand million (US billion) tonnes, of which all the coal and oil likely to be used by humans is much less than 1 per cent. Therefore, the combustion of coal and oil, while leading to an increase in carbon dioxide content, will not significantly affect the oxygen content, although it may through the GREENHOUSE EFFECT alter the earth's radiation balance and climate.

The indivisibility of the biosphere processes is clearly illustrated in the oxygen cycle. We should take care not to abuse that which we do not understand and are unable to control.

Oxygen deficit. ◊ DISSOLVED OXYGEN.

Oxygen demand. The oxygen demand of liquid waste effluent is of crucial importance and two measures are commonly used:

1. BIOCHEMICAL OXYGEN DEMAND (BOD) is a measurement of the oxygen required by the microbes to reduce the wastes to simple compounds, as in SEWAGE TREATMENT. The BOD test in the UK is a standard test of the amount of OXYGEN required by a sample of effluent over 5 days at $20\,°C$ (or 7 days for 'difficult' effluents) and is stated as the parts per million of OXYGEN (milligrams per litre) taken up by the sample of effluent incubated in the dark. This measurement is denoted BOD_5 for 5-day tests and BOD_7 for 7-day tests.

2. CHEMICAL OXYGEN DEMAND (COD) measures the number of parts per million of oxygen taken up by a sample from a solution of boiling potassium dichromate in two hours. The BOD and COD tests differentiate between materials that can be oxidized biologically and those that cannot, and indicate what types of treatment will be required.

It should be pointed out that the above criteria are not in themselves adequate indicators of pollution. There are extremely toxic solutions of CYANIDE,

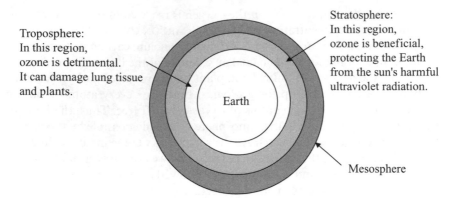

Figure 144   Ozone in the earth's atmosphere.

for example, that have acceptable BOD and COD values. ($\lozenge$ DISSOLVED OXYGEN)

Oxygen sag (curve). The decline and subsequent rise in percentage saturation of dissolved oxygen in a river downstream of a discharge of effluent containing biodegradable material. A graph of the percentage dissolved oxygen versus distance shows a characteristic dip or sag, and if re-aeration or recovery takes place, the dissolved oxygen content rises again (see Figure 54). The extent of the recovery is a function of the river length, biotic content, BOD loading of the effluent, and initial dissolved oxygen of the receiving stream. ($\lozenge$ DISSOLVED OXYGEN)

Ozone (O_3). Ozone contains three atoms of oxygen (O), whereas the atmospheric oxygen molecules contain two atoms (O_2). Ozone is formed naturally in the upper atmosphere (15–39 kilometres) by the action of the sun's ultraviolet rays (and also by lightning) (see Figure 144). This splits O_2 into single oxygen atoms which combine with other O_2 molecules to form O_3. The 'ozone layer' actually consists of a very few molecules of ozone per million molecules of air. The ultraviolet rays both destroy ozone and help create it. Absorption by stratospheric ozone protects living organisms from exposure to the harmful shorter ultraviolet wavelengths in solar radiation. Tropospheric ozone plays a key role in maintaining the oxidizing power of the lower atmosphere, but enhanced concentrations at ground level are viewed as a form of air pollution.

Ozone is removed by a series of chain reactions involving trace quantities of molecular and RADICAL species. The removal cycles have the general form:

$$O_3 + X \rightarrow OX + O_2$$
$$XO + O \rightarrow X + O_2$$

where X is OH, NO, Cl or Br. The regeneration of the original reactant, X, accounts for the remarkable efficiency of minute quantities of trace gases in destroying ozone, and the ozone concentration at the parts per million level is moderated by trace gases at the parts per billion level (that is, parts in 10^9 by volume).

Ozone concentrations in the stratosphere are naturally moderated in this manner such that 75 per cent of ozone is destroyed by the NO/NO$_2$ removal cycle (X = NO). At present about 2 per cent of ozone is destroyed by the Cl/ClO cycle (X = Cl) throughout the atmosphere, caused mostly by chlorine atoms which have been photochemically released from molecules such as chlorofluorocarbons (CFCs) like CFC11 (CFCl$_3$) and CFC12 (CF$_2$Cl$_2$).

$$CFCl_3 \rightarrow CFCl_2 + Cl$$
$$CFCl_2 \rightarrow CFCl + Cl$$

Molecules of chlorine will catalyse the destruction of large numbers of ozone molecules:

$$[Cl] + O_3 = [ClO] + O_2$$
$$[ClO] + [O] \rightarrow Cl_2 + O_2$$

The chlorine atoms are extremely reactive and it has been estimated that each chlorine atom will destroy about 10^5 molecules of ozone before being removed from the atmosphere by reaction with methane to form hydrochloric acid, which ultimately is precipitated in rain:

$$Cl + CH_4 \rightarrow HCl + CH_3$$

In contrast, inorganic forms of chlorine released at the earth's surface are removed by rain before they can enter the atmosphere.

Ozone is used as an oxidizing agent, in water treatment, for example. It is also produced by photochemical reactions involving hydrocarbons from car exhausts and NITROGEN OXIDES, when it becomes a dangerous irritant to eyes, throat and lungs. It can be formed in PHOTOCHEMICAL SMOG.

Individuals vary considerably in their response to ozone. Those who are sensitive experience temporary breathing difficulties if they take vigorous outdoor exercise when ozone concentrations are at or above about 160 μg/m^3; this level is often exceeded during hot summers, especially in southern England. In terms of lung function, people who suffer from asthma or other respiratory disorders are not more likely to be sensitive to ozone than other members of the population, although laboratory studies have shown that ozone may produce an enhanced inflammatory response in the airways of asthmatics.

During July 1994, periods of hot weather gave rise to prolonged, elevated levels of ozone across Europe. On a number of days the levels recorded at national monitoring sites exceeded the lower boundary, and often the upper boundary, of the WHO health-based guideline, which is 150–300 $\mu g/m^3$ averaged over an hour. In many cases, they entered the 'poor' air quality band (180–358 $\mu g/m^3$). The peak hourly levels recorded on those days are shown below.

Date	Place recorded	Levels of ozone ($\mu g/m^3$)
Friday, 1 July	Lullington Heath (Sussex coast)	190
Saturday, 2 July	London (Bridge Place)	190
Monday, 11 July	Lullington Heath	204
Tuesday, 12 July	Sibton (Suffolk)	238
Friday, 22 July	Lullington Heath	206
Saturday, 23 July	Harwell	204
Sunday, 24 July	Lullington Heath	218

RCEP 18th Report using data supplied by the National Environmental Technology Centre, Harwell; peak levels were converted from the ppb values supplied using a factor of 1 ppb = 2 $\mu g/m^3$.

Source: S. Penkett, 'The changing atmosphere', *Chemical Engineer*, August 1989.

There is evidence that background levels of ozone have doubled over the past 100 years. The provisional health objective in the UK is that the daily maximum 8 hour running mean should not exceed a concentration of 100 $\mu g/m^3$ on more than 10 days a year at any site. This objective was met at 56% of sites in 2005.

Ozone depletion potential (ODP). Measure of the potential for depletion of the ozone layer, e.g. most CHLOROFLUOROCARBONS have an ODP of 1, whereas HALONS can range from 3 to 10. The ODP value can vary as the bromine-based halons react synergistically (♭ SYNERGISM) with chlorine and therefore any increase in free chlorine levels in the ozone layer will increase the ODP of heavy halons released. (♭ CHLOROFLUOROCARBONS)

Ozone shield. A layer of OZONE surrounding the earth formed by ultra-violet radiation which splits molecular oxygen (O_2) to two atoms of oxygen which are highly reactive and form ozone (O_3). The ozone shield acts as a barrier to the radiation and protects the BIOSPHERE. The maximum concentration is found between 15 and 30 kilometres from the earth's surface. If the shield is reduced this may increase the incidence of radiation-induced skin cancer. Recent attention has also focused on nitric oxide (NO) which also attacks the layer. (♭ OXYGEN CYCLE; NITROGEN OXIDES)

P

Packaging

Introduction

Packaging comprises 18 per cent of household waste and is a soft target for environmentalists. Often ignored, it is essential in getting goods to consumers in optimum condition. It performs vital functions within the supply chain. Packaging protects more than ten times its weight of goods. This means that relative to the function it fulfils, the environmental impact of packaging is small. The dominant areas of environmental impact in the supply chain are producing goods and using them.

The way we live, work and play is changing. This means our consumption needs and habits are changing too. This leads to innovation in the products we buy and their packaging. A number of key lifestyle trends are dramatically affecting consumption and therefore the packaging/product supply chain. Most important of these trends are that more of us are living alone, people are living longer, and disposable incomes are rising. This means that while companies strive to do more to reduce their environmental impact, consumers are placing greater demands on packaging and the products it contains. Figure 145 shows the factors working for and against reducing the amount of packaging on the market.

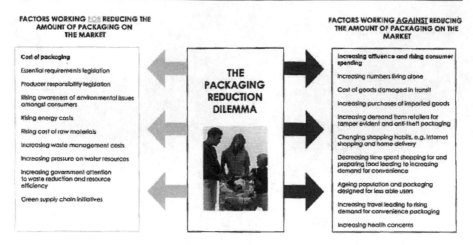

Figure 145   The packaging reduction dilemma.

Thus, packaging reduction is not amenable to Blue Peter-style solutions where the wave of a Presenter's hand can transform a toilet roll core into a highly-desirable household decoration.

Context is important, too. Since 1999, the amount of packaging used in the UK has increased by 4 per cent, much more slowly than the GDP. Sixty per cent of all UK packaging was recovered in 2006, including 1 Mt of used household packaging. Note what cannot be practically recycled can be combusted for energy recovery where 1 tonne of post-recycling household waste can provide 550 kWhe. This is energy which would otherwise have gone to waste. There are no one-shot solutions; recycling alone cannot cost-effectively cope with contaminated packaging, multi-layered materials, or plastic composites, all of which provide vital protection for foodstuffs, medicines etc. There are limits to everything in life (◊ LAWS OF THERMODYNAMICS)

Source: INCPEN, Towards Sustainable Distribution, Nov 2006.

Packaging, definitions.
DSD Commercial Recycling Objectives. Objectives of the German DSD Commercial Packaging Decree:

- A radical reduction in volume of packaging by avoiding use and stressing the value of reclamation.
- Manufacturers and retailers will be obliged to take responsibility for the packaging used as they are the originators.

- The local communities will be relieved from this part of waste disposal responsibility.
- Recycling will be given absolute priority over thermal recovery.

Group packaging or secondary packaging. Any packaging conceived so as to constitute at the point of purchase a grouping of a certain number of sales units whether the latter is sold as such to the final user or consumer or whether it serves only as a means to replenish the shelves at the point of sale; it can be removed from the product without affecting its characteristics.

Economic operators. In relation to packaging this means suppliers of packaging materials, packaging producers and converters, fillers and users, importers, traders and distributors, authorities and statutory organizations affected by the processing of packaging.

Non-returnable packaging. Any packaging for which no specific provisions for its return from the consumer or final user have been established.

One-way packaging. Any packaging not being used more than once for the same purpose.

Packaging. All products made of any materials of any nature to be used for the containment, protection, handling, delivery and presentation of goods, from raw materials to processed goods, from the producer to the user or the consumer. Non-returnable items used for the same purposes shall be considered to constitute packaging.

Packaging waste. Any packaging or packaging material covered by the definition of waste in Council Directive 75/442/EEC.

Packaging waste management. The management of waste as defined in Directive 75/442/EC.

Prevention. The reduction of the quantity and/or the harmfulness of materials used, packaging and packaging waste at production processes level and at the marketing, distribution, utilization and elimination stages, in particular by developing 'clean' products and technology.

Recovery. Any of the applicable operations provided for in Annex II.B to Directive 75/442/EEC (mainly as fuel).

Recycling. The recovery of the waste materials for the original purpose or for other purposes excluding energy recovery; recycling also means composting, regeneration and biomethanization.

Waste disposal. The collection, sorting, transport and treatment of packaging waste as well as its storage and tipping above or under ground, and the transformation operations necessary for its reuse, recovery or recycling.

Returnable packaging. Any packaging whose return from the consumer or final user is assured by specific means (separate collection, deposits, etc.), independently of its final destination, in order to be reused, recovered, or subjected to specific waste management operations.

Reusable packaging. Any packaging which has been conceived and designed to accomplish within its life cycle a minimum number of trips or rotations in order to be refilled or reused for the same purpose for which it was conceived; with or without the support of auxiliary products present on the market enabling the packaging to be refilled, such packaging will become packaging waste when no longer subject to reuse.

Sales packaging or primary packaging. Any packaging conceived so as to constitute a sales unit to the final user or consumer at the point of purchase.

Transport packaging or tertiary packaging. Any packaging conceived so as to facilitate handling and transport of a number of sales units or grouped packagings in order to prevent physical handling and transport damage.

Used packaging. The packaging itself left over once it has been emptied or the product has been unpacked.

Valpak. The Producer Responsibility Group's name for the industry organization which will implement packaging recycling. Valpak will pay the difference between the market price of the collected materials and the costs incurred in their collection. The aim is to provide reasonably stable prices and market confidence. (◊ VALPAK)

Waste. Any substance or object which the holder disposes of or is required to dispose of pursuant to the provisions of national law in force. (◊ PRODUCER RESPONSIBILITY OBLIGATIONS)

Packaging-derived fuel (PDF). PDF is produced from waste which has been 'source separated' in the household. It has a CALORIFIC VALUE of 80 per cent that of coal; while contributing just 20 per cent to the weight of municipal solid waste, packaging contains 40 per cent of the ENERGY.

Such fuel is usually combusted as a sole fuel in dedicated plants. However, it is also possible to co-combust up to 30 per cent of it with coal in conventional facilities.

Source: European Energy from Waste Coalition.

Evidence submitted by the British Plastics Federation (based on work by the UK Warren Spring Laboratory) to the House of Commons Environment Committee on RECYCLING of PLASTICS showed that plastics could account for 10.26 per cent by weight of household waste.

Plastics as a proportion of household waste.

Plastics source	Percentage of household waste	
Plastic film		
Refuse sacks	1.1	
Other plastics film	4.18	
Total	5.3	
Dense plastics		
Clean beverage bottles	0.63	
Coloured beverage bottles	0.12	Recyclable portion
Other plastics bottles	0.12	
Food packaging	1.9	
Other dense plastics	2.14	
Total	4.92	

This is a point well made in the study published by the Packaging and Industrial Films Association, Flexible Packaging Association, Oriented Polypropylene Film Manufacturers Association, *The Management of Waste Plastic Packaging Films (1993)*. This demonstrated that for plastic film waste arising at the domestic level the most sound method of disposal was as a high calorific component of the refuse fraction used as FUELS at INCINERATOR. Plants to merely set a material recovery target whilst ignoring the degree of contamination and, therefore, the major problems (in both cost and quality) that can be incurred in re-processing would be to impose unnecessary burdens on industry and the environment. This is especially so when an environmentally sound and financially viable option, energy recovery, exists to effect VALORIZATION as well. This point was accepted by the House of Commons Environment Committee in its Second Report on recycling when it stated 'We agree that energy recovery is appropriate for parts of the household waste plastics stream.' It is clear then that individual material appraisals need to be made. and where appropriate, the residual fuel value recovered and not consigned to LANDFILLS.

Sources: Packaging and Industrial Films Association, Oriented Polypropylene Film Manufacturers Association and the Flexible Packaging Association, *Management of Waste and Plastic Packaging Films*, January 1993.

UK packaging business recovery and recycling targets 2006–2010 (%)

Material	Paper	Glass	Aluminium	Steel	Plastic	Wood	Overall recovery	Minimum recycling*
2006	66.5	65	29	56	23	19.5	66	92
2007	67	69.5	31	57.5	24	20	67	92
2008	67.5	73.5	32.5	58.5	24.5	20.5	68	92
2009	68	74	33	59	25	21	69	92
2010	68.5	74.5	33.5	59.5	25.5	21.5	70	92

*The minimum percentage of recovery to be achieved through recycling.

Source: Defra.

Packaging and Packaging Waste Directive. The Directive on Packaging and Packaging Waste proposed by the European Commission on 15 July 1992 has been adopted (December 1994). The Directive aims to harmonize national measures concerning the management of packaging and packaging waste. This is the first step in a long-term process which will increase convergence gradually.

The relevant features are:

Scope. the directive covers all packaging placed on the market in the Community and all packaging waste, regardless of the materials used.

Targets. specific articles are included on preventative measures and reuse systems, and sets of quantitative targets for recovery and recycling of packaging waste.

Producer Responsibility Obligations. The Producer Responsibility Obligations require certain businesses to recover and recycle specified amounts of waste each year.

These obligations for packaging waste only apply to businesses that:

1. handle more than 50 tonnes of obligated packaging or packaging materials in a year, and
2. have an annual turnover of more than £2 million (based on the last financial year).

If the business belongs to a group of companies, these requirements apply to the total amount of packaging handled by the group and its total annual turnover.

Note: the obligations apply to the total amount of *packaging handled*, not the amount of packaging waste produced.

If the business meets the two criteria described above, it must demonstrate that a certain amount of packaging waste has been recovered or recycled each year.

In *England*, *Scotland* and *Wales*, the Producer Responsibility Obligations (Packaging Waste) Regulations 2005 apply. These regulations include provisions for people that lease property, franchisers and other licensors, which were previously exempt.

In *Northern Ireland* the Producer Responsibility Obligations (Packaging Waste) Regulations (Northern Ireland) 2006 apply.

If the business joins a registered compliance scheme, it does not have to meet the obligations itself. The scheme takes on the business' recovery and recycling obligations. This includes obtaining evidence of recovery and recycling, by providing Packaging Waste Recovery Notes (PRNs) or Packaging Waste Export Recovery Notes (PERNs), and reporting on compliance to the regulator.

The business has to supply the compliance scheme with relevant data (i.e. how many tonnes of packaging it handles) and pay a fee. This fee normally includes a reduced registration fee, which the scheme pays to the regulator on behalf of the business.

The Environmental Regulators have produced a list of registered compliance schemes in the UK.

It is up to the Member States to take necessary measures to establish specific return, collection and recovery systems in order to reach the objectives of the Directive. In compliance with the principle of subsidiarity, Member States are free to develop their own waste management schemes which have to be in conformity with the Treaty. Harmonized national databases have to be established to ensure a monitoring mechanism for the implementation of the objectives set out in the Directive.

The Directive lays out an important number of areas for standardization, regarding the essential requirements on the composition of reusable and recoverable, including recyclable, packaging. The Directive calls for all parties involved – consumers, industry and authorities – to co-operate in the spirit of shared responsibility. To this end the Member States shall ensure that users of packaging obtain the necessary information. (◊ PRODUCER RESPONSIBILITY OBLIGATIONS)

The recovery and recycling targets apply to glass, metals, plastics, paper and fibreboard (and wood from 2000). The UK must provide the following information to the European Commission for each material:

- tonnage of packaging produced in the UK;
- packaging exports;
- packaging imports;

- data about recovery and recycling of packaging waste;
- data about reusable packaging.

On 15 December 1995, British business leaders agreed a system of shared responsibility and so four specific stages of the packaging chain will become obligated to recover and recycle the following proportions of the packaging they handle:

- raw material handling, 6 per cent
- converting, 11 per cent
- packing/filling, 36 per cent
- selling (e.g. retailing), 47 per cent

A business carrying out more than one activity will have an obligation for each activity performed and an obligation will be placed on wholesalers from 2000.

Figure 146 gives a decision diagram on how to determine whether an item is packaging or not (expect modifications in due course).

Interpretation tests
Notes on steps A to F in the flow chart

Step A – Identify the sales unit
Interpretation begins with the 'sales unit'; this is the product and its packaging. The purpose of Step A is to identify it. For example, a silver spoon, preserves, jars, caps, labels, plastic sleeve, and price label all comprise the 'sales unit' of a presentation pack of preserves with serving spoon.

Step B – Remove the product
The product which is to be used or consumed. In the example given in Step A, this is the preserves and the spoon.

Some items are regarded by the Agencies as products in their own right although the boxes, bags, etc they come in are packaging; examples include tea bags, pencils, fire extinguishers, 35 mm film cassettes and toner cartridges.

Step C – Durable packaging for durable products
Durable products that cannot be used up (or consumed) may require durable packaging for long-term storage. An item that provides such long-term storage for a durable product is not regarded as obligated packaging by the Agencies.

The Article 21 Committee considers that a durable item is one that a majority of consumers use for longer than five years, and is intended for repeated use and not for disposal after the first use.

Preserves are consumable so no part of the 'sales unit' for them can be for long-term storage. Power tools on the other hand are durable products, so

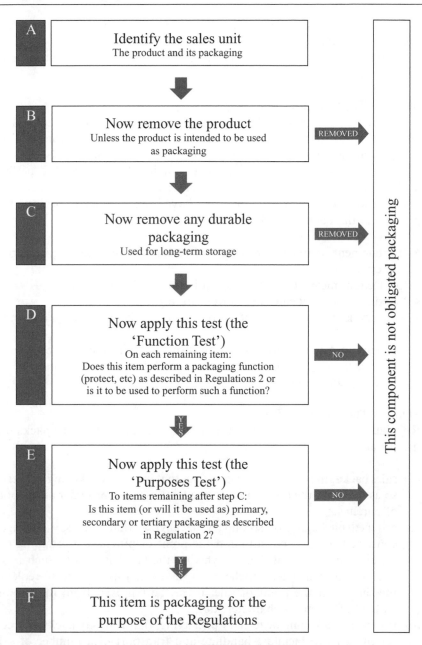

Figure 146 Determining whether an item is packaging.
Source: Producer Responsibility Obligations (Packaging Waste) Regulations 1997, SEPA, Environment Agency.

it is reasonable to assume their carrying cases as long-term storage. This also applies to durable carrying cases or moulded containers designed to last the lifetime of durable items such as spectacles, laptop computers, electric razors, cutlery, and cameras.

Step D – The 'Function Test'
Regulation 2 states that 'packaging' means 'all products made of any materials of any nature to be used for the **containment**, **protection**, **handling**, **delivery** and **presentation** of goods, from raw materials to processed goods. . . .'. The Agencies consider that a given item of packaging does not need to perform **all** of these functions, one function is sufficient.

Definitions of these specific functions have been adapted according to the purpose of the Packaging Directive:

- 'containment' is the act or process of restraining or enclosing e.g. drums and cans;
- 'protection' means the defence from harm, e.g. bubble wrap;
- 'handling' means facilitating movement, e.g. carpet cores;
- 'delivery' is the conveyance of the product(s) to the final user or consumer; and
- 'presentation' means to exhibit or display the product(s), which may include attracting attention to them e.g. a label, or a brightly coloured box containing an Easter egg.

Step E – The 'Purposes Test'
Regulation 2 sets out a second test. For part of a 'sales unit' to be packaging, it must also be:

(a) sales packaging or primary packaging, that is to say packaging conceived so as to constitute a sales unit to the final user or consumer at the point of purchase;
(b) grouped packaging or secondary packaging, that is to say packaging conceived so as to constitute at the point of purchase a grouping of a certain number of sales units, whether the latter is sold as such to the final user or consumer, or whether it serves only as a means to replenish the shelves at the point of sale; it can be removed from the product without affecting its characteristics;
(c) transport packaging or tertiary packaging, that is to say packaging conceived so as to facilitate handling and transport of a number of sales units or grouped packs in order to prevent physical handling and transport damage; for the purposes of these Regulations transport packaging does not include road, rail, ship and air containers. (Note: such containers are described in an Agencies' Explanatory Note.)

An important word found in the descriptions of primary, secondary and tertiary packaging is 'conceived'. For an item to be primary, secondary or tertiary packaging it does not have to have been conceived as such at the manufacturing stage, but at any stage. Once it passes the Function Test and the Purpose Test, it is deemed to have been packaging throughout the chain. This highlights the need for information to be passed between activities in the packaging chain.

Step F
All items which have reached this point in the flow diagram are considered to be packaging for the purposes of the regulations.

Packaging Recovery Notes (PRNs). PRNs are based on the EU PACKAGING AND PACKAGING WASTE DIRECTIVE and demonstrate compliance with same. They allow a company which has exceeded the amount it is supposed to recycle or recover to sell on the surplus to a company which has failed to meet its recycling obligations or finds it cheaper to purchase PRNs instead.

Packaging Waste Regulations (1997). ◊ PRODUCER RESPONSIBILITY OBLIGATIONS.

PAN. ◊ PEROXYACETYLNITRATE.

Paper. A matrix of CELLULOSE fibres usually free of non-cellulosic materials. UK (2005) consumption was of paper and board was 12.5 million tonnes (board is anything greater than $200 \, g/m^2$), of which 39% were produced nationally. The amount of recycled fibres was ca. 50 per cent mainly from specialist suppliers who reclaim the better grades of paper for repulping. The main classes of paper are:

Newsprint: made from mechanically ground wood PULP (which contains short cellulose fibres and some LIGNIN) and recycled newspapers. Up to 40 per cent of newsprint is recycled and this factor is expected to increase, provided market stabilization for reclaimed newspapers is established. Currently most recycled newsprint comes from printing plants, newsagents or is imported. Each tonne of recycled newspaper saves around 15 trees. However, the pulp-producing countries claim that trees are planted faster than they are consumed. This neglects the fact that recycling paper can save both water and energy, cf. virgin pulp manufacture. Although regulations differ, most US states and Canadian provinces are committed to phasing in a recycled content in newsprint of 40–60 per cent, availability permitting. (◊ DE-INKING; RECYCLING)

Printing and writing paper: usually bleached, with a fine texture and containing special fillers for ink absorption. Recycling is usually easily accomplished.

Speciality: this embraces papers having a high strength when wet, papers treated with resins, paper towels, papers for photographic emulsion, etc. Recycling is often difficult as resin types vary; therefore separation at source is essential.

With justification, recycling paper can save water and energy, compared to pulp manufacture. But as pulp manufacturers use renewable energy (forest and plant residues) and recycling consumes finite energy in the UK, it should be viewed as a cheap feedstock option for the paper industry, which can make economic sense for the paper manufacturers.

However, if the wastepaper collection incurs a net charge to the public, it is legitimate to ask why it should be subsidized when there are other pressing needs for public monies. (◊ ENERGY ANALYSIS)

Paper recycling: energy considerations

Considerable [interested party] studies have portrayed recycling as preferable to incineration. The way the studies are framed, a false dichotomy is usually set up (viz. recycling or incineration, instead of recycling and incineration). For example, not all paper is suitable for recycling, and, due to a continuing surplus of UK waste paper due to increased recycling, ca. 30–40 per cent may be exported to China and the Far East.

One recent desk study has concluded that recycling (paper) is better for the environment than burning it. Well, not necessarily if it is exported to the Far East where health and safety and environmental protection are not a high priority.

A closer examination of the energy savings of UK paper recycling *vis a vis* energy recovery/steam (at the boiler stop valve for heat/power, or both, preferably) has been presented to the House of Commons Environment Sub-Committee. The energy sums are summarized below.

An Energy Comparison of Waste Paper as a Fuel with Energy Saved by Recycling (A Porteous' evaluation).

One tonne of wastepaper combusted as fuel will produce 9.8 GJ of thermal energy as steam at the boiler stop valve.

One tonne newspaper produced using 'Best Available Technology' (BNMA 1995) would 'save' UK manufacturers 8.5 GJ of fossil fuel derived energy.

However, to produce one tonne of new fibres, 1.2 tonnes of old newsprint are required and therefore energy saving from newsprint combustion are even greater at 11.76 GJ if the comparison is done on a one tonne of 'new' (recycled) output newsprint requiring 1.2 tonnes of old newsprint as input.

If the reckoning is carried back to oil fuel equivalent i.e., conversion losses in refining and transport, etc are incorporated, the net energy saving from burning one tonne paper approximates to 13 GJ, which clearly outweighs industry recycling energy saving estimates.

Clearly, paper recycling is driven by market forces and is seen as a cheap source of fibres by the paper industry. Hence Local Authorities should not be 'out of pocket' in conducting paper recycling activities. There is also additional pollution from separately collecting the newspaper.

Kramer's (former EC DG for Environment) comments on cost benefit analyses are also apt:

'We gave this study to one of the most reputable consultancies in the community. It came out with a result that we considered most unsatisfactory. To be quite frank, when economic operators come to the Commission, they tell us that with cost benefit analyses, life cycle analyses and similar instruments, they can prove everything that they want and the contrary'.

In other words, Recycling is political and should be accepted as such.

A thorough study by the former highly-regarded, Warren Spring Laboratory, concluded that 'In certain circumstances there is a case for incineration with energy recovery of waste paper, particularly when in mixed contaminated form'.

Finally, the International Energy Agency, in an authoritative energy input study for the pulp and paper industry (5) demonstrated that for Scandinavian pulp mills, which use principally forest residues for fuel plus, mainly, hydroelectricity in many instances, nett fuel (oil) requirements per tonne of market KRAFT pulp is in the region 2–3 GS/t plus 320 kWhe (from various sources). It is not unusual for Scandinavian pulp mills to sell surplus energy. This means that burning Scandinavian sourced waste paper can be a nett renewable energy input to the UK. Paper recycling yes, within the UK, and energy recovery as well. Landfill diversion targets can be met by a combination of waste management practices which embrace waste minimisation, recycling, and energy recovery. Recycling dogma should be recognized for what it is. Decisions need to be made on rational economic and resource conservation criteria.

1) WRAP, The Environmental Benefits of Recycling – an Internal Review of Life-Cycle Comparisons, May 2006.
2) Porteous, A. Submission to the House of Commons Environment Sub-Committee, 'The Operators of the Landfill Tax', The Stationery Office, January 1999.
3) Kramer, L., Evidence in House of Lords Sustainable Landfill Report, Select Committee on the European Communities, 17th Report, 17 March 1998, p119, para 270.
4) Ogilvy, S.M., A Review of the Environmental Import of Recycling. DTI Report LR 911 (MR), July 1992.
5) National Swedish Board for Technical Development.

Source: IEA: Energy input analysis in the pulp and paper industry, Information 410–84, March 1984.

Paraffin. UK name for high grade kerosene (a medium light distillate produced in oil refining) used as a space heating FUEL. Also, the trivial name for 'alkanes' or saturated aliphatic hydrocarbons such as methane, ethane, propane and butane.

Parasites. Organisms that live attached to or in living organisms. They gain food and often shelter but the host gains nothing and usually suffers as a result.

Parathion. ◊ ORGANOPHOSPHORUS COMPOUNDS.

Particle, fundamental or elementary. This term, used in nuclear physics, refers to any particle of matter that is not composed of simpler units. The electron, proton and neutron, the main components of any atom, were the first to be discovered and researched. However, since then, others have been identified, such as the positron, the neutrino, mesons and hyperons.

Electrons, protons, positrons and neutrinos are stable; the remainder, mesons, hyperons and neutrons, decay spontaneously into fundamental particles of lower mass, accompanied by the liberation of energy. Neutrons will only decay spontaneously when isolated from the atomic nucleus. Mesons and hyperons, of which there are many different types with different properties, have mean half-lives of less than 0.000 003 seconds.

Particles and health. The Committee on the Medical Effects of Air Pollutants (COMEAP) has considered the possible effects of outdoor airborne non-biological particles on health. A wide range of sources includes primary emissions from motor vehicles, industrial sources or coal fires, and secondary aerosols derived from gaseous emissions, including sulphur dioxide and oxides of nitrogen, from industrial and vehicular sources. In the absence of strong evidence on the relative effects of different particles within the respirable range, current policy is based on PM_{10} (particles which are less than 10 micrometers in size) measurements. The Committee found clear evidence of associations between concentrations of particles similar to those encountered currently in the UK, and changes in a number of indicators of damage to health. These range from changes in lung function or exacerbation of asthma through increased symptoms and days of restricted activity, to hospital admissions and increases in deaths among the elderly or chronically sick. There was no evidence that healthy individuals are likely to experience acute effects on health because of exposure to concentrations of particles found in ambient air in the UK. (◊ EPIDEMIOLOGY)

Source: Committee on the Medical Effects of Air Pollutants (COMEAP), 1994 Report and Advisory Group on the Medical Aspects of Air Pollution Episodes: Activities Report 1994, HMSO, London, 1995.

Particulate pollutants, control. ◊ BAG FILTER; ELECTROSTATIC PRECIPITATOR; CYCLONE DUST SEPARATOR; THRESHOLD LIMITING VALUE; THREE-MINUTE MEAN CONCENTRATION.

Particulates. Fine solids or liquid droplets suspended in the air. The solids often provide extended surfaces due to their irregularities and therefore other pollutants can be carried along; for example, smoke particles and sulphur dioxide have greater effects on health when in combination than when emitted separately. It is postulated that the smoke particles, especially the PM_{10}, are carried deeper into the respiratory tract with the sulphur dioxide 'attached' (or adsorbed) and thus the medical effects are compounded. ($\lozenge$ SYNERGISM)

The term particulates as used in air pollution includes all the separate terms: GRIT, DUST, FUME, AEROSOL, SMOKE, etc.

Pascal. The SI unit of pressure (Pa); sea-level pressure is about 100 000 Pa, or 1000 hPa.

Pathogen. A living organism (usually a micro-organism) that causes disease.

Patulin. Because something is natural it must be good for us – this is a common mistake. The discovery, in 1992 by MAFF, that some apple juices contained relatively high levels of the naturally occurring contaminant patulin received wide publicity. Patulin is a mycotoxin produced by a number of moulds which grow naturally in and on fruits. It will always be present to some extent, since these foods are not produced and stored under sterile conditions.

MAFF has been monitoring apple juice for patulin since 1980. Levels were generally found to be low, until elevated levels were found in 1992 in some samples of apple juice. As a consequence, expert committees recommended that patulin levels be reduced to the lowest level technologically achievable and set an advisory level of 50 ppb.

Subsequent surveys between 1993 and 1996, to assess and monitor the effectiveness of subsequent industry actions, showed that the apple juice industry's voluntary measures to tackle the situation and reduce contamination had been most effective in all but a few cases. Two types of juice are routinely monitored.

Directly produced juices. High patulin levels were particularly associated with some directly produced juices. These juices are produced directly from apples rather than from apple juice concentrate. To meet demand throughout the year, apples have to be stored between harvests and pressed as required. The 1995 and 1996 patulin surveillance exercises were targeted at directly produced juices and showed that between 1992 and 1996, the proportion of directly produced juices exceeding the advisory limit of 50 ppb fell from 25 per cent to just 2 per cent (see Figure 147).

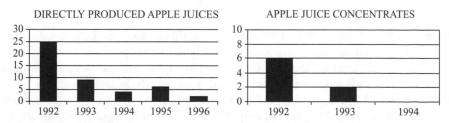

Figure 147 Percentage of samples containing patuling over advisory levels. (Reproduced by permission)

Apple juice concentrates. The majority of retail apple juice is produced from apples pressed at the time of harvest and turned into concentrate. The stable concentrate is then stored and diluted back to meet market demands and is commonly available as Tetrapacks™ on supermarket shelves. Surveys carried out in 1993 and 1994 indicated that the levels of patulin in juice made from apple juice concentrates were consistently well within the advisory level (see Figure 147).

Apple-based alcoholic beverages, such as cider, appear to be relatively free of patulin and it appears that the fermentation process used in their manufacture largely destroys any patulin.

Payback[1]. (Financing). The capital cost of a Project divided by the annual income or savings derived from it. While a Project which has a payback period of less than 5 years might receive consideration in low inflationary times, for many industries less than 3 years is considered desirable.

A good example is provided by the domestic use of wind turbines. $1kW_e$ output: Cost (after HMG subsidy) ca £1200, overall availability 30 per cent (allowing for fallow periods). Annual savings at 7.5p /kWh = £197. Payback = 6.1 years (domestic costing). This illustrates the effect of power rates and subsidies. Unsubsidised, and with a power rate of 3p/kwh (typical coal-fired power station revenue) payback could be in excess of 15 years. Industry would not invest at this level. A caveat is required for domestic wind power. Substantial fluctuations and downtime may be anticipated. Also, there can be wind interference from neighbouring buildings. (◊ CAVEAT EMPTOR)

Payback[2]. (Environmental). The amount of carbon dioxide expended/created by a renewable energy installation divided by the carbon dioxide saved annually, e.g. an estimate of the CO_2 created by installing a 2 MW wind turbine on peat moorlands 1 m deep divided by the CO_2 savings gives a low scenario payback of 8.2 years and a high scenario payback of 16 years. The CO_2 is principally created by the degradation of the excavated peat found on wind farm sites. Other environmental factors are considered in WIND ENERGY.

Source: Douglas, E. 'Gone with the wind', *New Scientist*, 8 July 2006, pp 37–39.

PCBs. ◊ POLYCHLORINATED BIPHENYLS.

Pebble bed nuclear reactor. Essentially a modified high-temperature reactor which uses circa 500 000 (or more) spherical, graphite-coated uranium fuel pellets, with helium coolant. Can be made into modules and is claimed to be inherently safe to operate. (◊ NUCLEAR REACTOR DESIGNS)

Pelleted fuel. A form of WASTE DERIVED FUEL in which the shredded waste (often DOMESTIC REFUSE) is compressed into solid fuel pellets by means of a press wheel. Figure 148 shows the flow sheet for a WDF pellet plant, processing 18 t/h input of domestic waste. The fuel properties of UK pellets

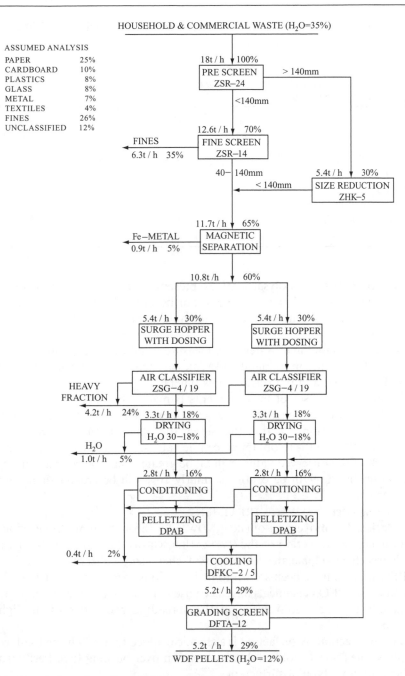

HOUSEHOLD & COMMERCIAL WASTE (H_2O=35%)

ASSUMED ANALYSIS

PAPER	25%
CARDBOARD	10%
PLASTICS	8%
GLASS	8%
METAL	7%
TEXTILES	4%
FINES	26%
UNCLASSIFIED	12%

18t / h 100%

PRE SCREEN
ZSR–24

> 140mm

<140mm

12.6t / h 70%

FINES
6.3t / h 35%

FINE SCREEN
ZSR–14

40– 140mm
< 140mm

5.4t / h 30%

SIZE REDUCTION
ZHK–5

11.7t / h 65%

Fe–METAL
0.9t / h 5%

MAGNETIC
SEPARATION

10.8t /h 60%

5.4t / h 30%

SURGE HOPPER
WITH DOSING

5.4t / h 30%

SURGE HOPPER
WITH DOSING

HEAVY
FRACTION

AIR CLASSIFIER
ZSG–4 / 19

AIR CLASSIFIER
ZSG–4 / 19

4.2t / h 24%

3.3t / h 18%

DRYING
H_2O 30–18%

3.3t / h 18%

DRYING
H_2O 30–18%

H_2O
1.0t / h 5%

2.8t / h 16%

CONDITIONING

2.8t / h 16%

CONDITIONING

PELLETIZING
DPAB

PELLETIZING
DPAB

0.4t / h 2%

COOLING
DFKC–2 / 5

5.2t / h 29%

GRADING SCREEN
DFTA–12

5.2t / h 29%
WDF PELLETS (H_2O=12%)

Figure 148 A flow sheet for a WDF pellet plant.

made from domestic refuse, shredded waste derived fuel and (UK) coal are given in the table below.

Comparison of the properties of a waste derived fuel (WDF) and coal, Warren Spring Laboratory (WSL), 1977.

Fuel	Calorific value as received (MJ/kg)	Moisture	Ash (%)	Volatile matter (%)
WSL pellets ex Doncaster refuse	15.7 (mean)	16.8 (mean)	14.6 (mean)	64 (mean)
Shredded WDF produced in USA	11.2 (mean)	26 (mean)	27 (mean)	
Typical UK household coal	30	6	6	34

Pentosans $(C_5H_8O_4)_x$. Polysaccharides present with cellulose in plant tissue including straw and sawdust. They can be utilized by HYDROLYSIS which converts them to pentoses (general formula $C_5 H_{10} O_5$) for the production of FURFURAL.

Per capita consumption of water. The measure of average water use per person. Water Companies in England and Wales are required to report estimates for both measured and unmeasured customers.

Percentile. When a set of measurements is made, the difference between the smallest (lowest) value and the largest (highest) is the range in which there are N observations. Within the range, the observations may be grouped, e.g. into 100 groups to obtain percentile values. As an illustration of use: the top 1% of the observed values will all be greater than the concentration represented by the 99th percentile; similarly 10% will be greater than the 90th percentile.

Percolating filter. ◊ SEWAGE TREATMENT.

Percolation. The movement of GROUNDWATER through an AQUIFER under the influence and direction of the HYDRAULIC GRADIENT. (◊ INFILTRATION)

Perfluorooctane sulphonates. A group of chemicals, collectively identified as PFOS, which have been shown to be hazardous (persistent, bioaccumulative and toxic). PFOS chemicals have been used in a diverse range of applications, including as an additive to aid the spreading properties of fire-fighting foam.

Perfluorosurfactant. A surfactant is a chemical added to fire-fighting foam which allows the foam to form a thin sealing film over burning fuel. Perfluorosurfactants are a type of surfactant.

Periodic table. All elements can be classified in terms of the periodic table, which arranges them in order of ATOMIC NUMBER (number of PROTONS in the

NUCLEUS) from left to right. Most naturally occurring elements up to URANIUM exist as mixtures of ISOTOPES (◊ RADIOACTIVE ISOTOPE), so their ATOMIC MASS is a non-integral number. Synthetically produced elements beyond uranium are highly radioactive and toxic (◊ PLUTONIUM). The vertical columns in the periodic table are known as Groups and are often named for their striking similarity in physical and chemical properties:

Group I: Li, Na, K, Rb, Cs – alkali metals
Group II: Be, Mg, Ca, Sr, Ba – alkaline earth metals
Group VI: O, S, Se – chalcogenides
Group VII: F, Cl, Br, I – halogens
Group 0: He, Ne, Ar, Kr, Xe – noble gases

The elements can also be classified in terms of their properties. Most of the elements are metallic (◊ METALS) and lie to the left-hand side of the table. The unreactive noble gases lie at the extreme right-hand side of the table, immediately adjacent to the non-metals. These elements are generally poor conductors of heat and electricity and usually occur as compounds. An intermediate group of elements (B, Si, Ge, As, Sb, Te) show intermediate

1 H																	2 He
3 Li	4 Be											5 B	6 C	7 N	8 O	9 F	10 Ne
11 Na	12 Mg	◄———— ·TRANSITION ELEMENTS————►										13 Al	14 Si	15 P	16 S	17 Cl	18 Ar
19 K	20 Ca	21 Sc	22 Ti	23 V	24 Cr	25 Mn	26 Fe	27 Co	28 Ni	29 Cu	30 Zn	31 Ga	32 Ge	33 As	34 Se	35 Br	36 Kr
37 Rb	38 Sr	39 Y	40 Zr	41 Nb	42 Mo	43 Tc	44 Ru	45 Rh	46 Pd	47 Ag	48 Cd	49 In	50 Sn	51 Sb	52 Te	53 I	54 Xe
55 Cs	56 Ba	57* La	72 Hf	73 Ta	74 W	75 Re	76 Os	77 Ir	78 Pt	79 Au	80 Hg	81 Tl	82 Pb	83 Bi	84 Po	85 At	86 Rn
87 Fr	88 Ra	89† Ac															

*LANTHANONS	58 Ce	59 Pr	60 Nd	61 Pm	62 Sm	63 Eu	64 Gd	65 Tb	66 Dy	67 Ho	68 Er	69 Tm	70 Yb	71 Lu

†ACTINONS	90 Th	91 Pa	92 U	93 Np	94 Pu	95 Am	96 Cm	97 Bk	98 Cf	99 Es	100 Fm	101 Md	102 No	103 Lr

├———► ARTIFICIAL ELEMENTS

physical properties and are hence termed semi-metals or semiconductors. (◊ APPENDIX II)

Permanent hardness. Water hardness that cannot be removed by boiling. Caused by the presence of carbonates and sulphates of calcium and magnesium. However, 'permanent' hardness can be removed by appropriate chemical treatment. (◊ HARDNESS; ZEOLITES)

Permeability. ◊ HYDRAULIC CONDUCTIVITY.

Permeation tube. Used for calibrating instruments and analytical methods. It consists of a sealed polymer tube containing a liquefied sample of the gas to be measured. At a fixed temperature, this diffuses through the walls of the tube at a constant rate, so that by allowing a stream of air to flow past the tube at a known rate, mixtures of this gas with air at very low, but accurately known, concentrations can be prepared.

Peroxyacetylnitrate (PAN). One of a number of complex compounds present in PHOTOCHEMICAL SMOG. It causes irritation to eyes and is toxic to plants.

Pervaporation. A membrane separation process in which the feedstock on the active side of the MEMBRANE is a liquid, and the permeate on the downside of the membrane is a vapour (see Figure 149).

There are three stages:

1. Selective sorption into the membrane on the feed-side.
2. Selective diffusion through the membrane.
3. Desorption into a vapour (◊ ABSORPTION; ADSORPTION; ZEOLITES).

Cost advantages of using pervaporation are as follows.

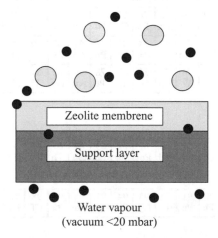

Figure 149   The operating principle of a pervaporation membrane.
Source: *The Chemical Engineer*, 16 May 1996, p. 13. (Reproduced by permission)

- ENERGY requirements – and hence running cost – are substantially lower than those of distillation, since only the aqueous component of the feedstock is vaporized.
- The energy advantages are greatest in cases where distillation would require a high reflux (recirculation) ratio.
- Operation at relatively low TEMPERATURES allows low-grade heat or waste heat to provide the energy. Pervaporation units have lower capital cost and are easier to install than distillation units.
- Pervaporation systems are especially cost effective in small-to-medium scale applications due to their modular nature.
- Distillation and pervaporation can combine in a hybrid process, minimizing both capital and running costs.

Source: A. Younges, 'British Gas and water treatment', IChemE *Avert*, Issue No. 13, Winter 1998/99, pp. 4–6.

Pest control. The term for the control of pests (e.g. tsetse flies, mosquitoes, cotton boll weevils) which affect public health, or attack resources of use to humans.

The techniques used other than PESTICIDES are:

1. Sterilization, i.e. use of irradiated males which controls or stops reproduction of the species. (◊ IRRADIATION)
2. Administration of juvenile hormones so that the species cannot metamorphose and therefore cannot reproduce.
3. Exposure to sex attractants: pheromones, specific chemicals produced by the female attract the male into an insect trap and hence the male can be destroyed.
4. Plant breeding to develop insect-resistant plants.
5. Modified planting practices: (a) a favourite food is planted nearby, therefore drawing the insect to it; (b) crop rotation.
6. Encouragement of predatory insects or animals, e.g. the use of small fish which prey on mosquito larvae.
7. Infection by viruses, bacteria or parasites, e.g. a strain of nematode (minute thread worm), which is a parasite of mosquitoes, is being introduced for mosquito control.
8. Deprivation of breeding ground, e.g. drainage of swamps to deny mosquitoes their breeding grounds, controlled tipping of DOMESTIC REFUSE to deny breeding grounds to flies.

Pesticide. A product or substance used in the control of pests such as vermin, mosquitoes, moulds, and weeds, all of which may affect public health or attack useful resources. (The term subsumes insecticides with which it is often used interchangeably.) There are three main classes:

1. Chlorinated hydrocarbons (e.g. DDT) which are long-lived and capable of being concentrated biologically.
2. ORGANOPHOSPHATES which are short-lived and degrade to 'harmless' end-products.
3. Artificial pyrethrins originally based on natural sources from pyrethrum flower heads but now being synthesized in very large amounts.

Widely used pesticides are:

Chlorinated hydrocarbons (persistent in the environment). DDT and its metabolites, e.g. DDE. Used widely to control malarial mosquitoes and houseflies, but strains have evolved with resistance. Acute oral toxicity to mammals low, but it becomes concentrated in fatty tissues, especially around and in vital organs.

Aldrin and dieldrin: formerly widely used as seed dressing, but use suspended because of death of birds and other wildlife. Their use is restricted in the UK because of effects on the FOOD CHAIN.

HCH and lindane (its gamma isomer): used for seed treatment. At one time widely used in gardening products.

Organophosphorus insecticides (short-lived in the environment). These are chemically related to nerve gases and some, such as parathion, can be dangerous to use. Others, such as malathion, are much less toxic and are in widespread use in agriculture, gardening and public health. The carbamate grouping of insecticides, which includes aldicarb and carbaryl, is similar in its effects to organophosphorus compounds.

Pyrethrins (◊ PYRETHROIDS): Pyrethrin I, allethrin and bioallethrin are natural products or synthetic chemicals closely related to natural products. Often a mixture, making build-up of resistance in houseflies, for example, more difficult. They are relatively safe.

The chlorinated hydrocarbon class concentrates in the fatty tissues and vital organs of birds. In time of stress, body fat is mobilized and death can result.

It is postulated that pesticides have synergistic effects (◊ SYNERGISM) in combination with air pollution and that dietary deficiency can markedly increase any hazard, particularly with the ORGANOCHLORINE and ORGANO-PHOSPHATE groups.

Another class, not now often used, is the inorganics, which are preparations of zinc, copper, arsenic or mercury. All are extremely toxic to human and animal life, but *may* be of use should resistance to organics develop.

Pesticides have undoubtedly contributed greatly to human health and increased food yields and will continue to do so – but unless caution is exercised, we will be in a race to develop new and/or more powerful insecticides as resistant strains of pest develop. There are dangers of over-zealous insecticide use. ($\lozenge$ ECOSYSTEM; CARCINOGEN; TERATOGENS; BIOLOGICAL CONCENTRATION)

PET. $\lozenge$ POLYETHYLENE TEREPHTHALATE.

Petrochemical. An intermediate chemical derived from petroleum, hydrocarbon liquids or natural gas, e.g. ethylene, benzene, propylene, toluene, xylene.

Petroleum. A range of distillate and residual liquid FUELS derived from OIL. The principal fuels in order of increasing boiling point are gasoline, kerosine and gas-oil, which includes diesel fuel. Residual oil is the refinery remainder after distillation and may be blended with gas-oil as a fuel in a large industrial furnace and for large slow-moving reciprocating engines. The average CALORIFIC VALUE (gross) is 45 MJ/kg.

PFA (Pulverised fuel ash). Inorganic silicaceous, powdery material which is the post-combustion residue from the UK's 18 coal-fired power stations. Used for building blocks, grouts and land cavity fill material. Currently subject to some waste regulation requirements because of its (very low) carbon content and its claimed potential to leach into groundwater.

pH. A measure of the alkaline or ACID strength of a substance. The pH value of any solution in water is expressed on a logarithmic scale to the base 10. It is defined and calculated as the LOGARITHM of the reciprocal of the hydrogen-ion concentration of a solution and may be expressed in symbolic form as:

$$pH = \log_{10}\left[\frac{1}{H^+}\right]$$

where H^+ is the concentration of hydrogen ions. ($\lozenge$ LOGARITHMS)

What this means in practice is that the pH scale ranges from 0 to 14 with the mid-point 7 indicating neutrality. If acid is added to water, the H^+ value increases and the pH decreases. Thus a pH value less than 7 is acidic. If greater than 7, it is alkaline (the opposite of acidity). Each unit increase in pH value expresses a change of state of 10 times the preceding state (because of the logarithmic scale). Thus, pH 5 is 10 times more acidic than pH 6; and pH 9 is 10 times more alkaline than pH 8.

pH measurements can be made by observing colour changes in special indicator chemicals or with indicator-impregnated paper (litmus paper), and also by pH electrodes. The pH value of effluents, toxic liquid wastes, etc. is a crucial parameter in effluent treatment. Acidic solutions, for example, would require neutralization with an alkali prior to disposal so that the pH is 7, and treatment can be undertaken without the added complications of acidity (or alkalinity). ($\lozenge$ ALKALI)

The pH of soils is an important factor in soil management – acidic soils require neutralization with limestone (calcium carbonate) which replaces two hydrogen ions in the soil with one calcium ion. However, as the calcium is leached, it must be replaced or the acidity returns. Other techniques, such as the use of CHELATING AGENTS, may also be used in soil management to remove or sequester unwanted ions. ($\lozenge$ LEACHING)

Phenols. A group of aromatic organic compounds which are highly toxic to living organisms. They can poison sewage treatment systems and taint water at very low concentrations.

Phosdrin. $\lozenge$ ORGANOPHOSPHORUS COMPOUNDS.

Phosphates (as pollutants). Essential inorganic substances for normal plant metabolism and applied as fertilizer to rectify a deficiency of phosphorus in the soil. Pollution can occur by too much being applied (or too little taken up by plants) and the surplus is leached into rivers, where it may contribute to EUTROPHICATION, and into groundwater.

Phosphorus (P). An element that plays an essential role in the growth and development of both plants and animals. In plants it is used in the energy exchange between adenosine diphosphate and adenosine triphosphate that provides energy, derived as a by-product of photosynthesis, at sites removed from the green parts in which photosynthesis occurs. In animals and plants, phosphorus is an essential component of DNA. It is impossible to conceive therefore of any substitute for phosphorus as a plant nutrient, and a phosphorus deficiency will retard growth and seriously affect yields.

In FERTILIZERS, phosphates may be associated with excessive quantities of fluorine, a cumulative plant poison.

Phosphorus cycle. The phosphorus cycle is sedimentary as are those of the elements calcium, iron, potassium, manganese, sodium and sulphur. In essence, the cycle operates as follows: compounds such as calcium phosphate, sodium phosphate, etc. are leached from the rocks into the soil and water, from where they are taken up by plant roots. The plants are subsequently consumed by herbivores, which in turn are eaten by carnivores. On the death of either herbivores or carnivores, decomposition takes place and the compounds are returned to the soil through the action of water – hence the sedimentary cycle.

In the sea there is a similar cycle in which phosphate compounds from sediments pass through the aquatic food chain, some fish being eaten by sea birds, whose droppings (guano) are rich in these compounds, and recycling eventually takes place.

Photochemical smog. Photochemical smog appears to be initiated by nitrogen dioxide. Absorbing the visible or ultra-violet energy of sunlight, it forms nitric oxide to free atoms of oxygen (O), which then combine with molecular oxygen (O_2) to form OZONE (O_3). In the presence of hydrocarbons (other than methane) and certain other organic compounds, a variety of chemical reactions takes place. Some 80 separate reactions have been identified or postulated, but two facts have been identified:

Stage 1: Smog concentration is linked to both the amount of sunlight and HYDROCARBONS.

Stage 2: The intensity of the smog is dependent on the initial concentration of NITROGEN OXIDES.

Many different substances are formed in sequence, including formaldehyde, acrolein, PAN, etc. The low-volatility organic compounds formed, condense to the characteristic haze of minute droplets which is called photochemical smog. The organics irritate the eye, and, together with ozone, can cause severe damage to leafy plants such as tobacco and endive. Photochemical smog tends to be most intense in the early afternoon, when sunlight intensity is greatest. In this respect it differs from traditional SMOG, which is most intense in the early morning and is dispersed by solar radiation.

This type of smog was first recognized in the Los Angeles area and its persistence has resulted in the passing of legislation to drastically curb automobile emissions. Control can be effected by controlling either hydrocarbon or industrial nitrogen oxide emissions. Photochemical activity of the type involved in smog formation also occurs in the UK, as indicated by measurement of OZONE, the usual indicator pollutant of photochemical reactions. (◊ AUTOMOBILE EMISSIONS; SMOG; NITROGEN OXIDES; VOLATILE ORGANIC COMPOUNDS)

Photodegradation. The process whereby ultraviolet radiation in sunlight attacks a chemical bond or link in a polymer or chemical structure, e.g. plastic.

Photon. A particle of light or other electromagnetic radiation. Each photon carries an amount of energy (E) that is determined by the frequency (f) of the radiation, as $E = hf$, where h (= 6.63×10^{-34} Js) is the Planck constant.

Photosynthesis. The means by which CHLOROPHYLL enables radiant energy to be used to accomplish the chemical conversion of elements in the atmosphere into organic matter. Chlorophyll is contained in organisms such as green and purple bacteria, blue-green algae (in fresh water), phytoplankton (at sea) and green plants (on land). The organisms live in those areas that receive sunlight such as the top few centimetres of soil, rivers and lakes. In the seas, sunlight can penetrate over 100 metres and this is the province of phytoplankton. On land, the green plants account for most of the photosynthesis.

Photosynthesis can be summarized as:

$$nCO_2 + 2nH_2A + \text{energy} \rightarrow (CH_2O)_n + nA_2 + nH_2O$$

or carbon dioxide plus hydrogen donor (H_2A) plus energy gives organic compounds (carbohydrates) plus a free, i.e. gaseous, compound plus water. Photosynthesis, as carried out by green plants and phytoplankton, uses carbon dioxide plus water for the hydrogen donor and the equation for such a reaction is:

$$nCO_2 + 2nH_2O + energy \xrightarrow[\text{green plant}]{} (CH_2O)_n + nO_2 + nH_2O$$

In this way the oxygen content of the atmosphere is maintained, carbon dioxide is fixed, and a carbohydrate source, $(CH_2O)_n$, is available for incorporation in cell structures, or as a source of energy directly or indirectly for all plants and animals. (◊ CARBON CYCLE; OXYGEN CYCLE)

The 'purple' and 'green' bacteria can use hydrogen sulphide (H_2S) as the hydrogen donor with sulphur as a by-product.

Photosynthetic efficiency. The percentage of total energy falling on the earth that is fixed by plants. It is approximately 6 per cent.

Phthalates. Organic compounds used as PLASTICIZERS in PVC manufacture. As they are not polymerized, they can migrate or volatilize and enter food from packaging, or drinking water from PVC pipes. They are persistent and bioaccumulative, and are accumulated in the fatty tissues of oily fish such as herring and mackerel. Concern has been expressed that the mechanism of accumulation is akin to that of DDT with attention being focused on dioctyladipate (DOA, a plasticizer used in food wrap films). The European Chemical Industry Federation (CEFIC) states that the acute toxicity of plasticizers is extremely low and puts this into perspective, by publishing data (see table on the facing page) showing a range of substances and their poison class. Plastic film producers are now using even less DOA.

Physico-chemical effluent treatment. The treatment of effluents by non-biological means, e.g. PRECIPITATION and settling.

Phytoplankton. Free floating minute plants in sea, lake and river surface waters where sufficient sunlight is available for PHOTOSYNTHESIS. Phytoplankton are said to be responsible for up to one-quarter of Europe's ACID RAIN emissions due to their production of dimethyl sulphide which converts to SULPHUR DIOXIDE and hence to acid rain. This process occurs in spring and summer. (◊ ZOOPLANKTON)

Pica. The compulsive habit of some children to eat non-food matter such as paint and soil. While not of widespread importance, this has caused serious problems when the material has contained substances such as LEAD from the pigment in old paintwork.

Pickling. The removal of scale from iron and steel usually by means of immersion in a hot hydrochloric or sulphuric ACID bath. WASTES include spent pickling liquor, sludges and rinse water. (◊ POLLUTION PREVENTION PAYS)

Acute toxicity of various substances.

Category*	Examples of toxic substances	Lethal dose (LD_{50}) in mg/kg body weight (for oral application)
Very toxic (less than 25 mg/kg body weight)	Clostridium botulinum toxin	0.00000003
	Hydrocyanic acid	0.7–1.0
	Arsenic (arsenic oxide)	1.4–4.3
Toxic (25–200 mg/kg body weight)	Sodium nitrite	57–86
	Barbiturates	47–143
Harmful (200–2000 mg/kg bodyweight)	Oxalic acid	375
	Carbon tetrachloride	457–686
Not classified as harmful (more than 2000 mg/kg body weight)	Ethanol	3300
	Common salt	7150–14300
	Plasticizers (e.g. dioctylphthalate (DOP))	More than 30 000

* Based on the EEC classification levels for acute toxicity – CEFIC Publication *Plasticizers*, January 1990.

Pilot plant. A small treatment plant which is built to obtain basic design reliability and cost data before the full-scale plant is designed.

Pipe flow. The factors which affect the flow of a fluid in a pipe are:

- Velocity of the fluid (mean velocity).
- Viscosity of the fluid.
- Density of the fluid.
- Diameter of the pipe.
- Friction where the fluid is in contact with the pipe.

If the effects of viscosity and pipe friction are ignored, a fluid would travel through a pipe in a uniform velocity across the diameter of the pipe. The velocity profile would be as in Figure 150. In practice, viscosity affects the flow rate of the fluid and works together with the pipe friction to further decrease the flow rate of the fluid near the pipe wall (see Figure 151).

The most important of the factors affecting fluid flow in pipes can be pulled together in one dimensionless quantity to express the characteristics of flow. This is known as the REYNOLDS NUMBER.

$$\text{The pipe Reynolds number} = R_e = \frac{\rho V D}{\mu}$$

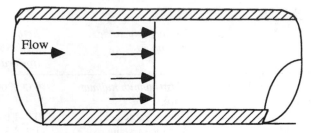

Figure 150   'Ideal' fluid velocity profile.

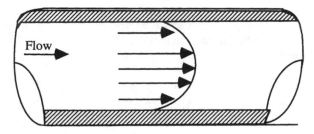

Figure 151   Real fluid velocity profile.

where ρ = density (kg/m³)
V = mean velocity in the pipe (m/s)
D = internal pipe diameter (m)
μ = dynamic viscosity (kg/ms).

It should be noted that the Reynolds number (R_e) is dimensionless. In simple terms:

$$R_e = \frac{\text{dynamic force}}{\text{viscous force}}$$

So, at very low velocities, the dynamic force will be low and, therefore, the Reynolds number will be small. Similarly, a high viscosity fluid will result in a low Reynolds number. This will result in viscous forces holding back the fluid flow at the pipe walls with the highest fluid velocity at the centre of the pipe. This is similar to the first figure except that the velocity profile is a parabola. It is known as LAMINAR FLOW and normally occurs at Reynolds numbers below 2000 (see Figure 152).

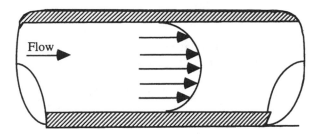

Figure 152 Laminar flow.

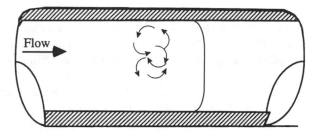

Figure 153 Turbulent flow velocity profile.

When Reynolds numbers are above 2000, i.e. high velocities and/or low viscosities, the flow breaks up and TURBULENT FLOW occurs, with a much flatter velocity profile (see Figure 153).

In the process world, unless very viscous fluids are being piped, turbulent flow is the norm. This is certainly the case for STEAM and compressed air where Reynolds numbers well in excess of 2000 are encountered and intimate mixing between particles of the fluid can be assumed. For mud slides and lava flows, laminar flow is usual. (◊ VISCOSITY)

Plasma. In engineering, a high-temperature gas stream (>30 000 °C) heated by an electric arc. Used in cutting of metals and decomposition of toxic wastes where expense is secondary (e.g. nerve gas stocks). However, recent developments have opened the door to RDF destruction as the RDF can be gasified and the gas and ash sent to a plasma-arc furnace. The end results are saleable power and a highly-stable vitrified ash. The claimed advantages of plasma-arc waste management technology are that:

- it is a high-intensity, clean, processing solution with minimal environmental impact
- it is a robust proven technology that is simple to operate and maintain
- it delivers high destruction and reduction efficiencies (DREs)

- it allows for the control of power input independently of process chemistry
- it produces a stable vitrified slag with a default non-hazardous designation, but with waste acceptance criteria performance to inert landfill status.

Source: Deegan, D.E., Chapman, C.D., Ismail, S.A., Wise, M.L.H., and Ly, H., 'The thermal treatment of hazardous waste materials using plasma arc technology', in Proceedings of the Institution of Mechanical Engineers Seminar on 'Delivering Waste Solutions – balancing Targets, Incentives and Infrastructure', IMechE, London, 2006.

Plasticizer. A high boiling point liquid (e.g. tricresyl phosphate, b.p. 420 °C) used in plastics or paints manufacture to confer flexibility. Widely used in PVC containers which need to be flexible. Plasticizers may inhibit high-grade RECYCLING of plastics unless a barrier construction is used whereby the recyclable plastic is effectively sandwiched inside a pure layer of virgin material.

Plastic. Any substance that is capable of plastic flow or deformation under certain conditions or at some stage of its manufacture and thus can be moulded into shape by heat and/or pressure. The common definition of plastic relates to those products of the chemical industry called 'polymers' which fall into two groups, thermoplastics and thermosetting materials.

Thermoplastics retain their potential plasticity after manufacture and can be re-formed by heating. The main types in this group are polyethylene, polypropylene, polystyrene and polyvinyl chloride (PVC) which embrace the whole range of domestic use. This group can be recycled to produce similar artefacts or be incorporated in lower grade ones, depending on the degree of separation obtained and product purity.

The thermosetting group includes resins such as the Bakelite or epoxy varieties, which are made into light switches, etc., and cannot be reused as there is permanent and irreversible change in the chemistry on setting.

It takes about seven times as much energy to make a cast iron pipe as it does to make the same pipe in PVC. Plastics consume less energy, first in their production and processing, then in their transport, and again in their durability. The energy saved by plastics is reflected in the price people pay for products made from plastics.

Plastics save energy because of their intrinsic properties. These include:

- thermal insulation
- electrical insulation
- high performance to weight ratios
- durability and chemical resistance

($\diamondsuit$ PLASTICS PACKAGING, RECYCLING)

Plastics, degradation. The very durability of plastics which is an excellent property for many purposes, makes them virtually indestructible when discarded, and as RECYCLING is difficult on both cost and quality grounds, means have been sought to make them degrade when their useful life is over. There are two postulated routes – photodegradation using ultraviolet radiation (solar radiation), and biodegradation.

- *Photodegradation* makes use of the fact that window glass *removes* ultraviolet radiation. Therefore plastic goods kept indoors are not exposed to ultraviolet rays, but when discarded on tips, etc. they would be exposed. Thus, the incorporation into the polymer of ultraviolet-sensitive groups would cause degradation on rejection.
- *Biodegration* is currently a long shot, but possible, as scientists have come up with a truly biodegradable plastic (one that disappears completely leaving no residue). It is made from a natural polymer called poly-3-hydroxybutyrate (PHB) which is found in the cells of certain bacteria. The bacteria can be grown on glucose substrates and the PHB extracted. This biodegradable polymer should have an important future. (There are still a number of practical problems to overcome, such as its brittleness.)

Plastics, disposal. The last phase in the life of a product is its disposal when it is no longer of any use. Disposable objects generally end up, with very few exceptions, in DOMESTIC REFUSE. Plastic objects, are disposed of by landfilling, by INCINERATION with or without energy recycling (energy recovery), or by recycling the raw material. The problem is that while plastics occupy 7 per cent by weight, they occupy 25 per cent by volume in municipal waste and recycling options are under urgent consideration. ($\lozenge$ LANDFILL)

Plastics packaging. Plastics will become the world's most popular packaging material, according to a study of the world's $800bn a year packaging industry.

It should be noted that much is contaminated and therefore unsuitable for materials recycling but excellent as a FUEL in INCINERATION with energy recovery.

The conclusion from the table overleaf is that to merely set a material recovery target whilst ignoring the degree of contamination and, therefore, the major problems (in both cost and quality) that can be incurred in reprocessing would be to impose unnecessary burdens on industry and the environment. This is especially so when an environmentally sound and financially viable option, energy recovery, exists to effect 'valorization' as well.

This point was accepted by the House of Commons Environment Committee in its Second Report on Recycling when it stated 'We agree that energy recovery is appropriate for parts of the household waste plastics stream.' It is clear then that individual material appraisals need to be made.

Assessed condition of used flexible plastics packaging in the UK.

	Total (×1000 tonnes)	'Clean' (×1000 tonnes)	Product contaminated (×1000 tonnes)	Ink contaminated (×1000 tonnes)	Mixed materials (×1000 tonnes)
Domestic use of polyethylene					
Refuse sacks	66		50	16	
Food contact packs					
Frozen/chilled food	29	10		19	
Moist products (i)	114	5	33	35	41
Dry products (ii)	157	40		70	47
Bread bags	12			12	
Non-food contact packs					
Carrier bags	80		10	70	
Counter bags	50	37		13	
Dry cleaning	11	10		1	
Mail envelopes	14			4	10
Subtotal	533	102	93	240	98
Industrial use of polyethylene					
Shrink/stretch film	129	78	38	13	
Sacks	31	2		23	6
Liners	25		25		
Bubble pack	7	3			4
Others	28	1	11	12	4
Subtotal	220	84	74	48	14
Total polyethylene	753	186	167	288	112
Polypropylene	61	8		23	30
PVC	15		15		
Polyester	3				3
Cellulose film	12	2		6	4

The above point is echoed in *A Way with Waste* (the waste management strategy for England and Wales (Part II), June 1999, DETR). In which it is stated 'the calorific value of most plastics is high and this energy can be recovered when it is impractical to recycle plastics from the waste stream'.

The Irish Government has instituted a plastic carrier bag tax to reduce LITTER.

Sources: Packaging and Industrial Films Association, Oriented Polypropylene Film Manufacturers Association and the Flexible Packaging Association, *Management of Waste and Plastic Packaging Films*, January 1993. The Consortium of the Packaging Chain (COPAC), *Action Plan to Address UK Integrated Solid Waste Management*, 26 October 1992. German Federal Bulletin, 'Survey of Sales Packaging Consumption 1991', 27 August 1992. House of Commons Environment Committee, Recycling Report, July 1994.

Plastics, recycling. Approximately 33 per cent of the 38.9 million tonnes per year plastics production (2004) in Europe has a very short life – the rest are virtually in captive use. The bulk of this 33 per cent is thermoplastics which can in theory be recycled. (◊ RECYCLING)

Direct recycling to obtain virgin plastics is difficult due to contamination, mixtures of grades, and the type of original product. To overcome this difficulty, selective collections are organized, as well as sorting by hand (PVC bottles, for example) at sorting centres. Thus, discarded plastic materials collected in this way are of a more uniform consistency and can be recycled more easily.

There are also extrusion and injection processes capable of utilizing mixtures of plastics as raw materials for making simple objects such as posts, mats, etc. The melting temperature for working with these mixtures must remain below 220 °C, at which temperature PVC decomposes. This is not a problem for the other mass-produced polymers, except for polyethylene terephthalate (PET), the working temperature of which is of the order of 260 °C.

An indirect recycling route is to take virtually as-received plastics waste and to grind it, mix in fillers and put it through a high-energy extruder and use the resultant material for 'low-quality' purposes such as fence posts, pallets and roof tiles, where finish is relatively unimportant but the plastics property of durability and resistance to decay is. This method has substantial promise and is expected to have a rapid growth in the USA where recycling is becoming a way of life. However, unless waste plastics can be collected at source, separation from DOMESTIC REFUSE is not economic as it comprises ca. 10 per cent by weight (but 25 per cent by volume). Thus the normal domestic refuse disposal processes will prevail, but with the proviso that PYROLYSIS or INCINERATION would be the most suitable energy recovery process for plastics because of their hydrocarbon content and high volatility. However, Germany has introduced mandatory deposits on plastic beverage containers which has resulted in PET bottles being withdrawn from the market or recycled. Restrictions have also been placed on many non-refillable drinks containers. An alternative approach is being tried in France where special public receptacles are provided for the deposit of PET bottled-water containers. These are ground up and reused for high-grade plastics manufacture and are eminently suitable for non-food grade container manufacture (see Figure 154).

An alternative approach is to break down the plastics waste by PYROLYSIS in order to recover a combustible gas as a fuel.

Other recycling methods, which go under the omnibus term feedstock recycling, embrace depolymerization, methanolysis, catalytic hydrogenation and thermal cracking.

The basic reaction for depolymerization of PET is the breaking of the ester linkage in the polymer chain. An example of this is the Eastman Chemical

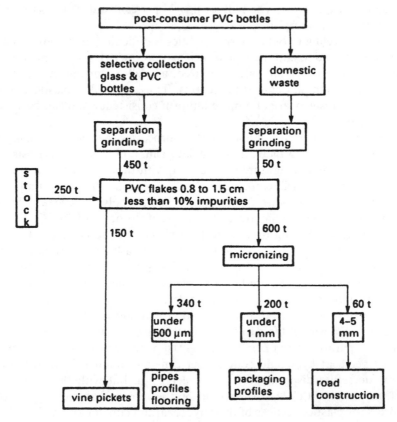

Figure 154   Uses for post-consumer PVC bottles in France.

methanolysis technology, where superheated methanol vapour is used as the degradative agent which apparently allows the process to be used with impure feedstock whilst maintaining the purity of the dimethyl terephthalate (DMT) and ETHYLENE GLYCOL (EG) produced. Feedstock impurities are removed as a SLUDGE residue.

The versatility of depolymerization chemical RECYCLING lies in the fact that operators can choose whichever method they require according to the products which they desire from the depolymerization processes.

The principle of catalytic hydrogenation is to reduce the large molecular weight plastic materials to smaller molecular weight HYDROCARBON oils for subsequent refining. Catalytic hydrogenation tests carried out in Kohleöl-Anlage, Bottrop, Germany, in the spring of 1992 have demonstrated the feasibility of chemical conversion of used plastics waste into a DIESEL-quality feedstock. All types of plastics can be treated and it is claimed that PVC presents no processing problems – the CHLORINE contained in the PVC is converted to sodium chloride by using a dry alkali injection system to remove acid gases formed during the process. The oils obtained by Kohleöl-Anlage, Bottrop, were of a quality suitable for conversion at the Ruhr Öl refinery in Gelseenkirchen into raw materials for the manufacture of plastics, thereby making it possible to close the processing cycle.

Source: *Opportunities and Barriers to Plastics Recycling*, AEA Technology, 1998.

There are substantial yield losses in conventional plastics recycling as seen in Figure 155.
(◊ POLYVINYL CHLORIDE, SPI MARKING SYSTEMS)

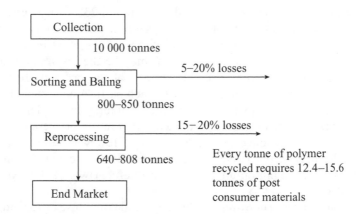

Figure 155   Yield losses in conventional plastics recycling.
Source: RECOUP brochure on Plastics Containers Markets, 1996, p. 9.

Plate count. The number of bacterial colonies which grow on a plate of nutrient agar from a diluted water sample incubated for a specified period at a certain temperature.

Plume. The stream of gases issuing from a stack which retains its identity and is not completely dispersed in the surrounding air. Near the stack the plume is often visible owing to water droplets, dust, or smoke that it contains, but it often persists downwind long after it has become invisible to the eye.

Plume rise. The rise of plume from stacks is a function of atmospheric conditions and the plume discharge (efflux) velocity, temperature and density. In still atmospheric conditions the plume will rise vertically; in strong WIND conditions it can be carried away horizontally or drawn downwards in the low pressure area behind the chimney. Further complications arise if the chimney is in a built-up area or in countryside of varying contours as the wind patterns can then cause downdraughts much more frequently.

To avoid downdraught, the exit velocity of the plume should be at least one-and-a-half times the wind velocity. Thus, where noxious fumes are concerned, a meteorological survey is required to determine the annual wind velocity profile so that the plume will not be drawn down for, say, at least 95 per cent of the year. The stack gas TEMPERATURE is also very important as this determines the gas buoyancy. The higher the temperature, the greater the buoyancy. The gas eventually cools to ambient and then the plume disperses downwind in an ever-widening cone.

One form of plume DISPERSION that can cause severe pollution under certain circumstances is a fanning plume, i.e. one that fans out horizontally into a thin layer under stable atmospheric conditions that restrict its spread vertically.

Plutonium (Pu). Plutonium-239 is an artificial radioisotope with a HALF-LIFE of 24 400 years. It is made by bombarding uranium-238 with neutrons. It is reactive and emits ALPHA RADIATION. It is a bone-seeking poison similar to radium but several times as toxic, and is one of the most hazardous substances known. Its most hazardous form may be the respirable plutonium dioxide particles produced by combustion – of the order of 10 thousand million particles per gram of metal – with each particle suspected of carrying a substantial risk (perhaps 1 per cent or more) of lung cancer. The plutonium in breeder reactor fuel is already oxidized into a refractory ceramic which, it is claimed, cannot produce respirable particles. However, the sodium coolant in a breeder may be reactive enough in an accident to reduce the plutonium dioxide fuel back to plutonium metal. (◊ NUCLEAR POWER; PLUTONIUM, THE HOT-SPOT CONTROVERSY)

Plutonium, the hot-spot controversy. The magnitude of a dose of ionizing radiation is defined as the energy absorbed from the radiation per unit mass of tissue. If a specified amount of a radioactive material undergoes decay in an organ, the energy released can be calculated from a knowledge of the physics of the isotope in question. The fraction of this energy absorbed within the organ can be estimated from a knowledge of the penetration capacity of the relevant radiation: for example, the energy of α-particles is entirely absorbed within less than a millimetre from the source, whereas X-rays are much more penetrating and only a small fraction of their energy will be absorbed within a metre or so of their source. Such considerations provide an estimate of the total energy absorbed within the organ. The dose to the organ is then estimated by dividing that amount of energy by the mass of the organ. In general, the risk of cancer is considered to increase in proportion to the dose of radiation received by any tissue.

The hot-spot controversy is about the potential for lung-cancer induction by very small particles of plutonium that may be inhaled and lodge in the lung. Plutonium emits α-particles and it is contended by critics of present recommendations of 'acceptable' doses that the conventional form of calculation may lead to gross underestimates of the dangers of such particles. They point out that all the energy of the α-particles emitted by the plutonium is absorbed in a small sphere of tissue surrounding the 'hot-spot'. This sphere is no more than a millimetre or so in diameter and the dose within is therefore thousands of times higher than is calculated by averaging the energy over the entire mass of the lung. Cells immediately adjacent to the 'hot particle' will simply be killed, which is of little importance, but doses of *all* lower levels down to zero will be delivered at increasing distances from the particle. Thus, any particular dose critical in causing cancer must occur somewhere in the vicinity of the hot spot.

PM$_{10}$. Particles with a diameter less than $10\,\mu$m which can be inhaled beyond the larynx. Those with a diameter less than $2.5\,\mu$m are called 'respirable particles' and can penetrate to the inner lungs. DIESEL vehicle (truck, bus and taxi) emissions are responsible for most PM$_{10}$s, with 87 per cent of UK black smoke emissions attributable to them, in London. Such emissions may be carcinogenic.

PNdB (perceived noise decibels). A frequency-weighted noise unit used for AIRCRAFT NOISE measurement. ($\lozenge$ NOISE; DECIBEL)

Point source. $\lozenge$ EMISSION SOURCE.

Poison, nuclear. Material in a reactor that absorbs neutrons strongly, thus reducing the reactivity. Some of the fission products (e.g. xenon) are nuclear poisons.

Pollutant. A substance or effect which adversely alters the environment by changing the growth rate of species, or interfering with the food chain, is toxic, or interferes with health, comfort, amenities, or property values of people. Generally pollutants are introduced into the environment in significant amounts in the form of sewage, waste, accidental discharge, or as a by-product of a manufacturing process or other human activity. A polluting substance can be a solid, semi-solid, liquid, gas or sub-molecular particle. A polluting effect is normally some kind of waste energy such as heat, noise or vibration.

Pollutants may be classified by various criteria:

1. *Natural or synthetic*: sulphur dioxide is an example of a natural pollutant; the class of CHLORINATED HYDROCARBONS is an example of a synthetic pollutant. Natural substances can be assimilated into biological cycles; they may often undergo BIODEGRADATION. Some synthetic substances such as DDT are not biodegradable; they are often toxic and accumulate in biological systems.
2. *Effect*: a pollutant may affect humans, a complete ecosystem, a single individual member or component of an ecosystem, an organ within the individual member, or a biochemical or cellular sub-system (e.g. crop growth rates).
3. *Properties*: e.g. toxicity, persistence, mobility, biological properties.
4. *Controllability*: the ease with which a pollutant can be removed from air or water is a very important factor. For example, most grit can be readily removed from flue gases, whereas sulphur dioxide cannot, without a great deal of expense.

In addition, the environmental attributes of the system into which a pollutant is to be discharged must be taken into account. If a watercourse is to be used for treated sewage discharge, the BIOCHEMICAL OXYGEN DEMAND imposed must be such that it does not swamp the DISSOLVED OXYGEN of the stream.

The maintenance of biological processes is fundamental to our continued existence and health, and they must not be grossly overloaded, or resources can be irretrievably lost. The gross pollution of the Great Lakes is one example of how humans have severely diminished a major natural resource.

(◊ PRIMARY POLLUTANT; SECONDARY POLLUTANT)

Polluter-pays principle. After the Second World War, when problems of pollution began to attract public attention and provoke protests, it was clear that the installation of plant and processes for pollution abatement would require financing from some source or other. Firms were evidently willing to install such plant if funds were provided from public sources, but such a policy did not have much appeal to taxpayers. Accordingly, much was made of the slogan 'the polluter must pay', interpreted as meaning that public money would not be used to subsidize pollution abatement for private industry. In practice, capital grants, subsidies, payments for reduced crop yields through using fewer fertilizers and tax allowances are all part of incentives to encourage industry and farmers to control pollution – in addition to increasing prices to consumers.

Pollution. Defined by the UK Environmental Protection Act 1990 as follows:

Pollution of the environment due to the release (into any environmental medium) from any process of substances which are capable of causing harm to man or any other living organisms supported by the environment.

Pollution control. The term for administrative mechanisms for control *and* the various technical processes and devices available for reducing emissions of waste streams.

In the UK, the administrative control is effected by legislation (e.g. The Alkali Act 1906, Clean Air Act 1956 and 1968, Control of Pollution Act 1974, Health and Safety at Work Act 1974, Environmental Protection Act 1990 and Environmental Act 1995) and its enforcement and implementation through statutory bodies such as the Environment Agency and the Health and Safety Executive.

The actual control of pollution is through process selection and plant construction. For example, in stack gas emission control, is it better to filter out particulates and scrub and neutralize the gases, or specify chimney height and efflux velocity and rely on the atmosphere for dilution and dispersion? These are typical considerations for only one problem – DILUTE AND DISPERSE versus CONCENTRATE AND CONTAIN. (◊ BEST PRACTICABLE ENVIRONMENTAL OPTION)

Pollution conversion. In the elimination of one or more sources of pollution, it is important that new ones are not created. If solid waste is incinerated, air pollution may occur instead, which may be more serious than the original problem. Similarly, disposal of sewage SLUDGE too liberally on agricultural land could result in the build-up of toxic metals. Washing or scrubbing of exhaust gases can lead to a water-pollution problem. The disposal of urban solid waste can pollute groundwater and produce METHANE. (◊ INTEGRATED POLLUTION CONTROL)

Pollution Hazard Appraisal (PHA). For each process regulated by The ENVIRONMENT AGENCY, in addition to an Operator Risk Appraisal, a PHA must also be carried out against the following characteristics:

- Hazardous substances
- Techniques for prevention and minimization
- Techniques for abatement
- Scale of process
- Location
- Frequency of operation
- Offensive substances in the process.

Each is assigned a factor on a scale from 1 to 5 to reflect its relative importance, where 1 is low hazard potential and 5 is high hazard potential.

Each result is then multiplied by a weighting factor which reflects the Environment Agency's view of the relative importance of each factor. From this a Pollution Hazard Value is derived for the process as a whole.

Examples of PHA

Location: remote, e.g. coastal and/or rural site, PHA rating 1.
Location: close proximity to areas of high population and/or where physical factors would exacerbate effects of any releases, PHA rating 5.

Source: Draft Report, *Operator and Pollution Appraisal*, (OPRA), consultative document, April 1995.

Pollution Index. Used in chimney height calculations to determine the limiting pollutant.

$$PI(m^3s^{-1}) = \frac{1000 \times \text{pollutant emission (g s}^{-1})}{(\text{guideline concentration} - \text{background concentration})(\text{mgm}^{-3})}$$

Pollution indicators, natural. ◊ BIOTIC INDEX; MOSSES; GLADIOLI; LICHENS.

Pollution of the Environment (Environmental Protection Act 1990). Pollution of the environment due to the RELEASE (into any environmental medium, from any process) of substances which are capable of causing harm to humans or any other living organisms supported by the environment.

Pollution permit. A system of permitting companies with excess pollution discharge 'allowances' (consents) to trade them with those who need to discharge more than they are allowed to. The net pollutant total is constant but the distributions vary according to market forces. Pollution permits are already in use in the US.

The pollution regulator sets a maximum level for emissions or discharges and companies can then buy and sell permits among themselves.

A sulphur dioxide scheme could be easiest as the UK has a national 'bubble' of sulphur dioxide which can be allocated among all power stations and oil refineries as long as the net allowable total is not exceeded.

Pollution Prevention Pays. Two examples are given below. The first from a compendium of *3P Success Stories* compiled by the Environment Engineering and Pollution Control Department, 3M, St Paul, Minnesota. The second on pickling waste minimization is from the Royal Commission on Environmental Pollution, *11th Report* (example submitted by the University of Aston). It is also worthwhile recording the 3M (UK) plc Corporate Environment Policy Statement (November 1989) as an example of enlightened corporate self and public interest: under its worldwide Environmental Policy, 3M will continue to recognize and exercise its responsibility to:

Waste stopper: pumice on copper, 3M Company, St Paul, Minnesota.

Problem	3M's electronic product plant in Columbia, Mo., makes flexible electronic circuits from copper sheeting. Before sheeting can be used in the production process, it has to be cleaned.
	Formerly, the metal was sprayed with ammonium persulphate, phosphoric acid and sulphuric acid. This created a hazardous waste that required special handling and disposal.
Solution	Cleaning by chemical spraying was replaced by a specially designed new machine with rotating brushes that scrubbed the copper with pumice.
	The fine abrasive pumice material leaves a sludge that is not hazardous and can be disposed of in a conventional sanitary landfill.
Payoff	40 000 pounds a year of hazardous waste liquid prevented.
	$15 000 first year savings in raw materials and in disposal and labour costs.
	In the third year of use, the new cleaning machine had saved enough to recover the $59 000 it cost. Because of increased production each year, costs saved and volumes of pollution prevented continue to rise.

Descaling of hot rolled steel RCEP 11th report.

The old technology	Acid pickling
Wastes produced	Depleted hydrochloric and sulphuric acid
	Acidified rinse water
Reasons for change	To achieve greater control over waste disposal
	To reduce rising waste disposal costs

Options for on-site waste reduction

Neutralizing acid liquors	Requires additional chemicals
	Residual sludges need disposal
Recovery of acid for reuse	Possible especially with modern ion exchange systems to recover both acids
Shot blasting	Physical rather than chemical process
	Leaves smaller volumes of inert wastes
	Allows savings of up to 50 per cent of original descaling costs

Options adopted

Shot blasting is now the preferred method for descaling drawn steel. Acid recovery still used in plants producing steel bars in a variety of shapes and sizes, for which shot blasting is not suitable.

- Solve its own environmental pollution and conservation problems
- Prevent pollution at source wherever and whenever possible
- Develop products that will have a minimal effect on the environment
- Conserve natural resources through the use of reclamation and other appropriate methods
- Assure that its facilities and products meet and sustain the regulations of all federal, state and local environmental agencies
- Assist, wherever possible, governmental agencies and other official organizations engaged in environmental activities.

Polonium. A radioactive isotope (Polonium −210) which is an ALPHA radiation emitter, with a HALF-LIFE of 138 days. Manufactured in nuclear reactors by NEUTRON bombardment of bismuth 209. Responsible for the death of former Russian spy Alexander Litvinenko on 23 November 2006, in London.

The Guardian of 29 November 2006 printed the following:

'It was stated in error in our front page report, 'The radioactive spy', November 25, that alpha radiation has to be inhaled, swallowed, or enter an open wound before causing harm. In fact, the material that emits the radiation, in this case Polonium 210, would have to be ingested or otherwise introduced into the body to take effect. Alpha radiation cannot be inhaled'.

Polychlorinated biphenyls (PCBs). Chlorinated hydrocarbons formerly used as plasticizers and in transformer-cooling oils to enhance flame retardance and insulating properties. They are highly persistent bioaccumulative pollutants found worldwide. Their use was banned in 1979 by law, but many old electrical components such as transformers and capacitors are around which means that PCBs will require specialist disposal by INCINERATION for many years to come. Many Third World countries also have substantial quantities of PCB-filled electrical equipment in use which is coming to the end of its useful life with no facilities for their environmentally effective disposal.

Great concern has been expressed over PCBs being disposed of in landfill sites. Where this has been done, 'eternal' vigilance would appear to be required unless the PCB contaminated fill is extracted and thermally treated. In accordance with the Landfill Regulations 2002 (as amended in June 2004), waste containing PCBs has to be tested before landfilling is permitted. Landfilling of PCB-containing waste is allowed only if the leaching limit value for PCB in the waste is $<1\,\mathrm{mg\,kg^{-1}}$.

The levels of PCBs in paper and board packaging in the UK are generally lower than those found in previous overseas surveys. Levels of up to 0.33 mg/kg were found, compared to up to 2 mg/kg in France and 1.2 mg/kg in Germany.

The Council of Europe's Committee of Experts on Materials and Articles Coming into Contact with Food, is continuing to consider paper and board used to package food. As part of this work the Committee has been discussing a limit set in Germany of 2 mg total PCBs/per kg paper and board. The Joint Food Safety and Standards Committee survey showed that levels in this country are well below that limit.

Source: *MAFF Food Safety Bulletin*, April 1999.

Historically PCBs were present in seawater such as the Clyde Estuary, where in 1969, concentrations of less than 0.01 micrograms per litre were found. Concentration in mussels was between 10 and 200 micrograms per litre, which gives a concentration factor of between 1000 and 20 000. They have been blamed for the death of sea birds in times of stress. PCBs have been found in the tissues of Arctic seals (1989) and maternal milk, and are showing a remarkable longevity in the environment due to their stability. Not only that, but concentrations are rising which perhaps pose serious threats for communities where fish is a major part of the diet. Also seals, walruses, etc. which are at the end of a FOOD CHAIN are at risk. PCBs and other persistent chemicals were thought to have impaired the immune system of the common seals in the North Sea, 70 per cent of whom fell prey to a distemper virus in 1988. This is a classic example of an anthropological compound being introduced without assessment of its environmental impact.

Under EU Council Directive 96/59/EC Disposal of Polychlorinated Biphenyls and Polychlorinated Terphenyls (PCBs and PCTs), Member States are required to take the necessary measures to ensure that used PCBs are disposed of and PCBs and equipment containing PCBs are decontaminated or disposed of as soon as possible. Member States must compile inventories of equipment with PCB volumes above a specified threshold and send summaries of such inventories to the Commission.

For equipment and any PCBs contained in it which are subject to an inventory, decontamination and/or disposal must be effected at the latest by the end of the year 2010.

PCB disposal undertakings are to keep registers of the quantity, origin, nature and PCB content of all used PCBs delivered. This information is to be communicated to the competent authorities and the registers are to be kept open to consultation by local authorities and the public.

Until such time as they are decontaminated, taken out of service or disposed of, the maintenance of transformers containing PCBs may continue only if the objective is to ensure that the PCBs they contain comply with technical standards or specifications regarding dielectric quality, and provided that they are in good working order and that they do not leak.

The new directive requires Member States to take the necessary measures to ensure that all undertakings engaged in the decontamination or disposal of PCBs, or equipment containing PCBs, obtain permits in accordance with the requirements of the Waste Framework Directive 91/156/EC. (◊ HAZARD-OUS WASTE INCINERATION)

Source: *Wastes Management*, December 1996; Council Directive 96/59/EC.

Polyelectrolyte. Long-chain organic compounds used to cause FLOCCULATION of dispersed non-settling matter in water which can then be removed by sedimentation.

Polyethylene (polyethene). A THERMOPLASTIC polymer of ethylene (ethene) (C_2H_4). It has good flexibility and is used very widely in packaging. (◊ PLAS-TICS, RECYCLING; PVC)

Polyethylene terephthalate (PET). Polyethylene terephthalate (PET, PETE) is a thermoplastic polymer resin of the polyester family that is used in synthetic fibre; beverage, food and other liquid containers; thermoforming applications; and engineering resins often in combination with glass fibre. It is one of the most important raw materials used in synthetic textiles. The majority of the world's PET production is for synthetic fibres (in excess of 60 per cent) with bottle production accounting for around 30 per cent of global demand. In discussing textile applications, PET is generally referred to as simply 'polyester' while 'PET' is used most often to refer to packaging applications.

PET can be semi-rigid to rigid, depending on its thickness, and is very lightweight. It makes a good gas and fair moisture barrier, as well as a good barrier to alcohol (requires additional 'Barrier' treatment) and solvent. It is strong and tough. It is naturally colourless and transparent.

While all thermoplastics are technically recyclable, PET bottle recycling is more practical than many other plastic applications. The primary reason is that plastic carbonated soft drink bottles and bottled water are almost exclusively PET which makes them more easily identifiable in a recycle stream. PET has a resin identification code of 1. PET, as with many plastics, is also an excellent candidate for thermal recycling (incineration) as it is composed of carbon, hydrogen and oxygen with only trace amounts of catalyst elements (no sulphur) and has the energy content of soft coal.

Polymer. A chemical compound made by the repeated joining of MONOMER molecules. (◊ POLYMERIZATION)

Polymerization. The joining together of MONOMER molecules by 'addition polymerization' in which case the POLYMER is a simple multiple of the monomer molecule, or by 'condensation polymerization', where the resulting polymer does not have the same empirical formula as the basic monomer constituent. The term is also used to cover the process of copolymerization,

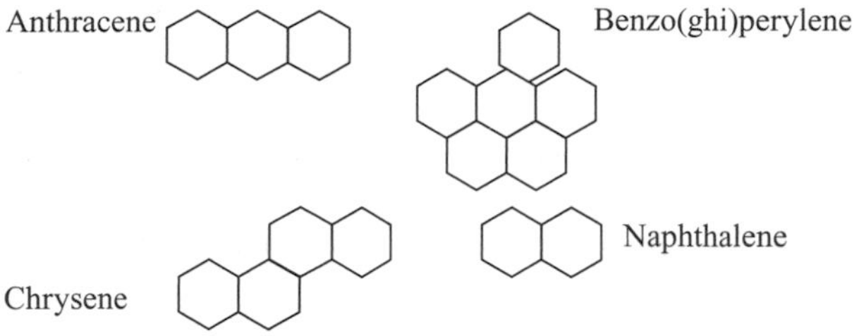

Figure 156 Examples of polynuclear aromatic hydrocarbons.

in which the polymer is built up from two or more different kinds of monomer molecules. Many plastics and textile fibres are made from natural or synthetic polymeric substances.

Polynuclear aromatic hydrocarbons. Hydrocarbons with multiple-ring structures are collectively referred to as polynuclear aromatic hydrocarbons, commonly abbreviated as PNAs or PAHs (see Figure 156).

This class of compounds is thought to be carcinogenic or mutagenic which has led to legislative restrictions on their release into the environment. They are mainly formed due to incomplete combustion of organic material such as fossil fuels.

Polyvinyl chloride (PVC). PVC is a thermoplastic polymer, that is, it can be softened for shaping by raising its temperature and then can be hardened by cooling without any chemical change taking place.

The monomer formula is $CH_2{=}CHCl$ which is chloroethene though it is commonly called vinyl chloride monomer (VCM). This monomer is usually produced in a three-stage process. Stage 1 is the production of ethylene dichloride by reacting ethylene dichloride (EDC) (1.2 dichloroethane) which has the formula CH_2Cl. The EDC is then catalytically cracked to produce VCM, hydrogen chloride and some chlorinated hydrocarbon by-products. The hydrogen chloride is usually reacted with more ethylene and oxygen in the oxychlorination process to produce more VCM. The oxychlorination process also gives rise to chlorinated by-products. The VCM is distilled, cooled and liquefied under pressure. It is transported and used as a liquefied gas. The raw materials for VCM manufacture are often transported as EDC rather than as ethylene and chlorine.

Though ethylene and VCM differ in only one atom, the polymers produced from them are very different. Unlike polyethylene (PE), PVC is never processed alone because its decomposition temperature is lower than its softening temperature. It is possible to process 'pure' PE for use, for example, in very low dielectric loss cable insulation. Most PE, like most other plastics including PVC, are processed in a physical blend with other additives or ingredients. The chlorine in the PVC molecule makes it much more polar than polyethylene. That polarity makes PVC compatible with a wider range of types and quantities of additive than any other commodity polymer.

While the necessity of processing PVC with other ingredients may seem a weakness, the wide range of properties these additives make possible, is the key strength of PVC. By a choice of additives, PVC formulation can be hard or soft, brittle or tough, flammable or flame resistant, matt or glossy, conducting or insulating, opaque or transparent, and they are available in a very wide range of colours. A PVC formulation is simply a physical mixture and no new chemicals are formed in the mixture. The other ingredients include:

- heat/light stabilizers – organic or inorganic metal compounds;
- impact modifiers – ethylene vinyl acetate (EVA), chlorinated polyethylene (CPE) or acrylic polymers;
- fillers – calcium carbonate;
- lubricants – metal compounds, waxes, oxidized polyethylene;
- plasticizers – phthalates, phosphates, adipates;
- pigments – organic or inorganic;
- fire retardants – antimony oxide, aluminium trihydrate.

The formulation ingredients are chosen to produce the physical and processing properties required in a particular application and shaping process. The ingredients are also chosen to meet national or international standards for safety critical applications such as toys, and food contact or medical devices.

While PVC is widely used in short-life applications, some 50–60 per cent of applications are long-life, for example in window frames, pipes and other construction applications. End-of-life PVC components may readily be recycled into second life applications though the economics of recycling are presently doubtful. The medium-life application of PVC in computer enclosures is already being profitably recycled and long-life applications will be readily recyclable when larger quantities become available.

Raw materials

The production of 1 tonne of polyethylene requires the fractionation of 18.7 tonnes of crude oil. Because 57 per cent of the basic PVC molecule is CHLO-RINE derived from common salt (sodium chloride) only 8 tonnes of oil are needed for each tonne of PVC polymer. Some PVC formulations include oil-based additives, such as the PHTHALATES, so their dependence on non-renewable fossil fuels is greater, but never at the level needed for polyethylene or polypropylene.

Since salt is neither a source of ENERGY nor a scarce raw material, PVC has a resource advantage over other polymers. Chlorine is produced from salt by the electrolysis of a solution of salt in water. This produces not only chlorine but two other very valuable raw materials, caustic soda and hydrogen. Caustic soda is essential to the production of numerous materials such as paper, soap, aluminium and viscose rayon. HYDROGEN is used as a raw material, for example in margarine manufacture, or as a fuel. Chlorine gas is toxic but once it is part of the PVC molecule it becomes an inert component. The PVC industry uses some 30 per cent of the chlorine made in western Europe and the rest finds very important applications throughout the rest of the chemical industry.

The energy cost of several polymers, including the energy content of the plastics themselves, have been calculated in a LIFE CYCLE ANALYSIS (LCA) project organized by the Association of Plastics Manufacturers in Europe. It should be noted that these are values for the polymers alone. For a ready-to-mould or extrude compound these values must be added in proportion to those of the other ingredients. Few of the polymer additive suppliers have produced LCA data.

Polymer	Energy cost (MJ/kg)
PVC	53
PE	69
PP	73
PS	80
PET	84

Health aspects

Various materials used in the production of PVC and PVC formulations have been under the spotlight of health concerns in the past 20 years or so.

Vinyl chloride monomer. Until the mid-1970s was known only as a low acute toxicity, anaesthetic gas which on long exposure at high levels could cause Reynards disease (restriction of finger arteries) and acreolysis of finger bones. Research into these phenomena led to an understanding that long-term worker exposure could cause angiosarcoma of the liver. About 175 cases have been diagnosed since the 1970s as being associated with PVC manufacture in the western world. Radical changes in working practices, agreed by industry, the unions and government have reduced worker exposure to well below the threshold of hazard. Residual VCM levels in PVC have never been high enough to represent a significant hazard to the public. Present levels are accepted by, for example, the European Pharmacopoeia, as representing an insignificant hazard in the most sensitive uses such as food-contact goods and medical devices.

Ethylene dichloride. This too is a carcinogen but it is made and processed in chemical plant designed to minimize emissions and so is acceptable to the regulatory authorities.

PVC dust. The health effects of PVC polymer DUST remain a matter for controversy. It has been suggested that very fine, respirable dust might cause the miners' disease, pneumoconiosis. This is not accepted by all the regulatory authorities but all agree that exposure to dust should be limited. The Health and Safety Executive in the UK has set a Workplace Exposure Limit (WEL) for PVC dust. The 8-hour Time Weighted Average limit for total inhalable PVC dust is $10 \, mg \, m^{-3}$, and for respirable dust, it is $4 \, mg \, m^{-3}$.

Heavy metal stabilizers. Governments in Europe are keen to reduce the quantity of HEAVY METAL compounds used and released to the environment. The regulations concerning sensitive applications of plastics forbid their use but there are long-life applications of PVC where they are still the additive of choice.

CADMIUM compounds are used as heat and light stabilizers in PVC exposed to the weather. The industry has significantly reduced the use of cadmium stabilizers by moving to lead or LEAD/barium/cadmium blends. During 1993 the EC enacted a Cadmium Directive (91/338/EEC) which limited the use of cadmium stabilizers to a shortlist of long-life applications including window-frame material. There is a limit of 0.01% for the cadmium content of finished products or components manufactured from specified plastics and liquid paints. Cadmium pigments are only allowed in high melting point plastics such as PP and PET. Some countries, such as Norway and Sweden, have banned cadmium completely.

Lead compounds have been more widely used than cadmium compounds and to date they are only nationally regulated. They still find wide use in PVC pipes, windows and conducts. There have been worries about their use in potable water pipes.

Organo-tin compounds have been widely used in PVC pipe and packaging compounds. No hazard in use is known but concern about the effects of related organo-tin anti-fouling paint on shellfish has brought questions about their use as stabilizers.

Plasticizers. Here the major group of compounds used are the PHTHALATES, the most used being di(2 ethyl hexyl)phthalate. Over the past 20 years the phthalates have been accused of being CARCINOGENS. They are carcinogenic for rodents at very heavy doses but no cancer effects have been proved for humans.

Phthalates have been grouped with a wide range of other chemicals as endocrine disrupters, also called oestrogen mimics.

End-of-life disposal

As indicated earlier, PVC formulation can readily be recycled, though such processes are not always economic. Alternatively PVC can be a minor component in feedstock recycling processes where polymer molecules are broken down into oil feedstock hydrocarbons. Alternatively PVC waste can be burnt, with heat and power recovery, in modern municipal solid waste (MSW) INCINERATORS equipped to extract hydrogen chloride from stack gases for second life application in steel cleansing. MSW incineration is not a significant contributor to acid rain, most of which (98 per cent) is due to emissions from power stations or motor transport. Only 0.3 per cent of acid rain can be attributed to MSW incineration and of that only 0.15 per cent can be linked to hydrogen chloride from incinerated PVC.

Figure 157 shows flows of energy, raw materials, products and by-products during the PVC lifecycle.

Ponding. The filling up of the VOID spaces in a biological filter, leading to pools of liquid which will not drain away.

Pool price. The normal price paid by regional electricity companies for marginal supplies of electricity. Computed on a half-hourly basis. Prices can range from 1.5 to 11 pence/kWh, depending on the cost of bringing in additional generating capacity.

The system marginal price (SMP), the price bid by the least economic plant expected to be called into use to meet demand, rises at times of peak demand.

By far the biggest element in the pool price is the capacity payment. This is calculated from the probability of loss of supply and the notional value that customers would put on that loss of supply.

Capacity payments are intended to cover the costs the generators face in making available generation plant which is seldom needed but which is essential to maintain an adequate level of security of supply.

Population. Number of individuals of a certain species that live in a particular area at a particular time.

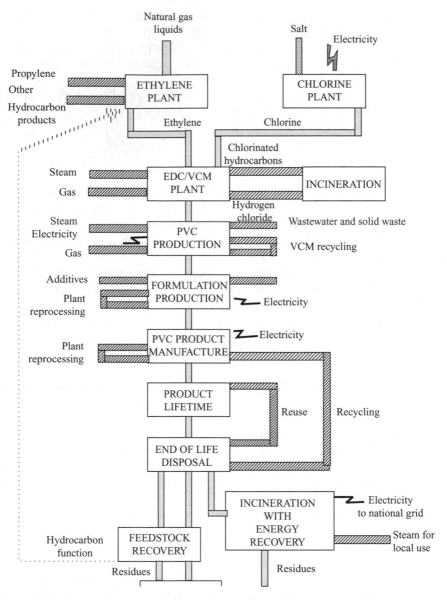

Figure 157   Energy flows, products and by-products during the PVC lifecycle.
Source: *PVC and the Environment 96*, Norsk Hydro, Petrochemical Division, 1996,
p. 18. (Reproduced by permission of Norsk Hydro ASA)

The tables below give the human population growth statistics and selected world fertility rates.

World human population growth.

Date	Human population
1 AD	170 000 000
1000	265 000 000
1500	415 000 000
1800	900 000 000
1900	1 625 000 000
2000	6 250 000 000
2025	predicted 8 466 000 000

Population equivalent. A term used in SEWAGE TREATMENT. This is derived from assuming each person contributes 0.06 kg BOD_5 per day. For instance, an effluent with a BOD_5 of 2000 mg/l from a dairy, with a volumetric flowrate of 500 m³/d, will have a population equivalent of:

$$2000 \frac{\text{mg BOD}}{1} \times 500 \frac{\text{m}^3}{\text{d}} \times 10^3 \frac{1}{\text{m}^3} \times \frac{1}{10^6} \frac{\text{kg}}{\text{mg}} \times \frac{1}{0.06} \frac{\text{d}}{\text{kg BOD}}$$
$$= 16666.7 \text{ or ca. } 16700$$

Total Fertility Rate. The single most important factor in determining future population is the total fertility rate (TFR). The TFR is defined as the average number of babies born to women during their reproductive years. A TFR of 2.1 is considered the replacement rate; once a TFR of a population reaches 2.1 the population will remain stable assuming no immigration or emigration takes place. When the TFR is greater than 2.1 a population will increase and when it is less than 2.1 a population, will eventually decrease, although due to the age structure of a population, it will take years before a low TFR is translated into lower population. The TFR is a synthetic rate, not something that is actually counted. It is not based on the fertility of any real group of women, since this would involve waiting until they had completed childbearing. Nor is it based on counting up the total number of children actually born over the lifetime, but instead is based on the age-specific fertility rates of women in their 'child-bearing years,' which in conventional international statistical usage is ages 15–49.

The TFR is therefore a measure of the fertility of an *imaginary* woman who passes through her reproductive life and is subject to *all* the age-specific fertility rates for ages 15–49 that were recorded for a given population in a given year. The TFR represents the average number of children a woman *would* have were she to fast-forward through all her childbearing years, assuming the age-specific fertility rates for a given year. This rate, then, is the number of children a woman would have if she was subject to prevailing fertility rates at all ages, and survives throughout all her childbearing years.

Country	Total Fertility Rate in 2000	Total Fertility Rate in 2006
Niger	7.16	7.46
Angola	6.52	6.35
Haiti	4.50	4.94
Bhutan	5.13	4.74
Iraq	4.87	4.18
Tajikistan	4.35	4.00
Honduras	4.26	3.59
Cambodia	4.82	3.37
USA	2.06	2.09
France	1.75	2.01
Sri Lanka	1.98	1.84
Australia	1.79	1.81
Mainland China	1.82	1.73
UK	1.74	1.66
Canada	1.64	1.61
Japan	1.41	1.40
Germany	1.38	1.39
Russia	1.25	1.28
Singapore	1.16	1.06

Pores. Small VOID spaces in rocks or aggregates. The volume of the pore space is an important parameter in estimating the volume of water, or oil held in storage in relation to the total volume of the aquifer or oil field respectively.

Poverty. The UK Government does not have a national definition of poverty; however, the European Council of Ministers defines poverty as the condition of 'persons whose resources (material, cultural and social) are so limited as to exclude them from the minimum acceptable way of life in the Member State in which they live'.

Power. The rate of doing WORK, measured in Js^{-1} or watts.

Power Factor. An electrical engineering measurement defined as the ratio of the total power (watts) supplied in an electrical installation to the product of volt x amps (watts) for the circuit. Unless the power factor is unity, a simple volts x amps calculation will not suffice. Considerable effort is expended to keep power factors close to unity to mitigate excess electrical loads and wastage.

Precautionary principle. Whereby preventative measures are taken, considering the costs and benefits of action and inaction, when there is a scientific basis to believe that release to the environment of substances, waste or energy is likely to cause harm to human health or the environment i.e. absolute proof of damage is not required. This principle is not to be construed as a charter for shutting down industry. Rigour is essential, as is peer reviewed experimentation and conclusions from same, before action is taken.

'The precautionary principle is too often used to justify inchoate fears of the unknown, and should be treated with the greatest caution by scientists.' (*Financial Times*, Editorial '*Fear of Phoning*', 12 May 2000.)

Precipitation:

1. General term for the release of water from the atmosphere. It can be in the form of rain, snow, hail, dew or hoar frost. In all cases, condensation is initiated by nuclei of water droplets or ice crystals forming in clouds of moist air cooled below the DEW-POINT. The nuclei then grow and coalesce in the clouds and then, provided the droplets are large enough to overcome any rising air currents, precipitation may take place.
2. The formation of an insoluble substance by chemical reaction which occurs in solution; a method used in treating some liquid hazardous wastes.
3. The process of removing suspended matter in water, sewage and industrial effluents by the addition of suitable coagulants such as aluminium sulphate, $Al_2 (SO_4)_3$, or ferric chloride, $FeCl_3$.

Precision. The closeness of agreement between the results obtained by applying an experimental procedure several times under prescribed conditions.

Predict & Provide. Typically, the use of future demand predictions as the basis for ensuring a steady supply of primary aggregates for the construction industry. This militates against the use of recycled materials from, say, construction & demolition wastes & against innovative building methods.

Prescribed processes. Defined by the UK Environmental Protection Act (1990) as: 'a process for the carrying on of which after a prescribed date an authorisation is required'.

Regulations may:

(a) prescribe separately, for each environmental medium, the substances the release of which into that medium is to be subject to control; and
(b) provide that a description of a substance is only prescribed, for any environmental medium, so far as it is released into that medium in such amounts over such periods, in such concentrations or in such other circumstances as may be specified in the regulations.

In relation to a substance of which a description is prescribed for releases into the air, the regulations may designate the substance as one for central control or one for local control.

Examples of prescribed substances include SULPHUR OXIDES, NITROGEN OXIDES, MERCURY, ALDRIN, ENDRIN, SOLVENTS, DIOXINS, and PESTICIDES.

Presbycusis. Aged-related hearing loss, which begins gradually after the age of 60. Over time, the detection of high-frequency sounds becomes more difficult.

Pressure. The force exerted per unit area. Atmospheric pressure is the amount of pressure exerted by the atmosphere above absolute zero pressure and which changes with elevation above sea level (atmospheric pressure decreases as altitude increases). Most pressure gauges indicate a pressure that is referenced to atmospheric pressure, i.e. atmospheric pressure = 0 bar gauge (more commonly shown as bar g) at sea level.

Pressure is measured in the SI system in newtons per square metre. *Note*: the BAR (1 bar = $10^5 N/m^2$) and millibar (1 mbar = $10^2 N/m^2$) are in very common use, the former for high pressures and the latter for variations in atmospheric pressure.

To obtain absolute pressure, the atmospheric pressure must be added to the gauge pressure (i.e. that which is measured on pressure gauges).

At sea level, the mean atmospheric absolute pressure = 1.013 bar. So at 0 bar g, the absolute pressure = 0 bar g + 1.013 = 1.013 bar abs. 2 bar g = 2 + 1.013 = 3.013 bar abs., etc.

Pressure, partial. The pressure exerted by one of the gaseous components of a mixture of gases on the assumption that it alone exists in the same volume as the mixture.

Pressure vessel. In NUCLEAR REACTOR DESIGNS, this is the reactor containment vessel built to withstand high internal pressure, constructed from steel or concrete.

Pressurized water reactor (PWR). ◊ NUCLEAR REACTOR DESIGNS; SIZEWELL.

Primary air. The air supplied to a FUEL in its early stages of combustion, i.e. at or on the grate.

Primary efficiency. ◊ PHOTOSYNTHETIC EFFICIENCY.

Primary pollutant. A pollutant emitted directly into the environment such as SO_2, CO. (◊ SECONDARY POLLUTANT)

Primary production. The biomass, or the energy it represents, produced by vegetation within a defined area over a defined time-scale. Gross primary production (*GPP*) is the total energy fixed by photosynthesis. Net primary production (*NPP*) is *GPP* minus the portion used by plants in respiration (*R*).

Primary sludge. The sludge consisting of settled solids from the primary stage of a SEWAGE TREATMENT plant. (◊ SLUDGE, SEWAGE)

Primary treatment. Treatment of wastewater by physical processes, generally involving settlement to remove gross solids and reduce suspended solids.

Priority list. ◊ RED LIST.

Probable reserves. Underdeveloped oil or gas reserves which are deemed to be recoverable from known formations but not yet proven due to lack of data. (◊ PROVEN RESERVES)

Process. Any activity carried on in Great Britain, whether on premises or by means of mobile plant, which is capable of causing POLLUTION of the environment. (◊ PRESCRIBED PROCESS)

Source: EPA 1990.

Production system. Term used in LIFE CYCLE ANALYSIS. A collection of materially and energetically connected unit processes, which perform one or more defined functions.

2-Propenol. A colourless liquid ALDEHYDE with a choking odour which occurs in PHOTOCHEMICAL SMOG.

Proportional Cancer Mortality Ratio (PCMR). This is the ratio of the observed number of deaths due to cancer among a sample cohort, to the expected number of deaths due to cancer based on the age-, sex- and time-specific mortalities of the control population.

Proportional Mortality Ratio (PMR). This is the ratio of the observed number of deaths due to a particular cause among a sample cohort, to the expected number of deaths based on the age-, sex-, time- and cause-specific mortalities of the control population.

Proteins. A group of nitrogenous organic compounds of high molecular weight (up to 10 000 000) which are essential components of all living matter. ENZYMES are an important class of proteins. (◊ AMINO ACID)

Proton. A stable elementary particle with an electrical charge equal and opposite to that of an ELECTRON.

Protozoa. Small unicellular animals having a well-defined nucleus (in contrast to bacteria). They are found in a variety of habitats, e.g. fresh and salt water and soil, and occupy an important position in these ECOSYSTEMS where they normally consume dead organic matter and wastes, although they can also be parasitic. (◊ ACTIVATED SLUDGE)

Proven reserves. The estimated quantities of oil or gas reserves (or other natural resource) which geological and engineering data demonstrate with reasonable certainty to be recoverable in future years from known reservoirs under existing technological and economic conditions. (◊ PROBABLE RESERVES)

Provisional tolerable weekly intake. ◊ MERCURY.

Proximate analysis. A simple method of analysis of solid FUELS for measuring the percentages of free moisture, volatile matter, ash and FIXED CARBON.

The last named is obtained by difference from 100 per cent. Proximate analysis is used to check quality of bulk deliveries and for daily works control. (◊ ULTIMATE ANALYSIS)

Proximity principle. The requirement to treat wastes close to where they arise, e.g. within the boundary of the plant or community in which they are generated. The problem is not to be exported to someone else's back yard. This poses major problems for cities if they do not wish to export their wastes and implies reduction, recycling and the adoption of waste to energy practices. ◊ ECOLOGICAL FOOTPRINT

Psychrometry. The measurement of atmospheric humidity. Usually done by the use of two thermometers: (a) dry bulb; (b) wet bulb. Readings are converted to a humidity value by means of charts or tables.

Public health. Public health embraces the health of both the individual and the community. In its widest sense it means the mental and physical health of the people, which in turn ensures the well-being of future generations – a resource to be husbanded just as much as the resources of land, air and water. All environmental factors are involved: food (type, quantity, wholesomeness), condition of work, home and play, pollutants, noise, etc.

Originally public health was primarily concerned with the prevention of diseases, e.g. CHOLERA or other sewage-borne diseases that used to abound in the UK and still do in many countries. We are still dependent on these early concepts and practices to prevent the return of such diseases, but the wider spheres outlined above have great importance once basic health is established.

Public health embraces an awareness of new hazards, RISKS or pollutants introduced by technology in the name of progress. The hazards or risks must be controlled so that the benefit is maximized to all. We also learn how pollutants or substances (e.g. VINYL CHLORIDE) are more dangerous than first realized. Thus, the field is dynamic and only through greater awareness of the risks of new substances or developments can progress be made. The aim of many of the entries in this book is to bring this awareness to the fore, so that we may use our new-found chemicals, processes and practices wisely for the individual, the community and future generations.

The basic definition of public health used here has been given by C. E. A. Winslow, *The Cost of Sickness and the Price of Health*, World Health Organization, Monograph Series No. 7, 1951.

Pulmonary irritants. A group of air pollutants which affect the mucous lining of the respiratory tract which comprises the nasopharynx, the tracheobronchial area and the lung tissue or alveoli.

The nasopharynx defence mechanisms include the hairs at the nasal entrance, which filter out large particles, and the mucous glands, which wash out many of those particles that escape the hairs. The bronchial tree contracts to prevent dust entry (e.g. coal dust) and hence reduces the amount of particulate matter. Coughing also ejects mucus in which the particulates are trapped. While bronchial spasms lessen the amount of dust, they also reduce the amount of air received. The alveoli consist of minute air sacs (around 300 million) filled with capillaries through which oxygen enters the blood and carbon dioxide is removed.

The main respiratory diseases are:

1. Emphysema, where the alveoli lose their oxygen/carbon dioxide exchange properties, due to smoking and chronic exposure to air pollutants.
2. Pulmonary fibrosis. Scarring of the lung tissue.
3. Pulmonary oedema. A drowning of the lung tissue in fluid caused by exposure to highly concentrated amounts of irritant or corrosive pollutants.

Severe pulmonary irritants include NITROGEN OXIDES, SULPHUR OXIDES and CHLORINE which can cause severe bronchial problems.

Source: G. L. Waldbott, *Health Effects of Environmental Pollutants*, 2nd edn, C. V. Mosby, St Louis, 1978.

Pulp. The raw material for PAPER making. Pulp is usually obtained from trees, esparto grass and other long-fibred, CELLULOSE-containing materials. There are two main classes of pulp:

1. Chemical pulp, obtained from wood by chemical means, i.e. by sulphite, sulphate or soda processes which dissolve the LIGNIN and release the long cellulose fibres.
2. Groundwood or mechanical pulp, made by grinding the wood so that the fibres are separated. Groundwood is mainly used for newsprint because of its low quality, due to short fibre length.

The effluent from pulpmills has a high BIOCHEMICAL OXYGEN DEMAND and SUSPENDED SOLIDS and requires treatment before discharge. One recycling and pollution abatement method in certain pulpmill effluents is the manufacture of SINGLE-CELL PROTEIN by growing and harvesting yeasts on the dilute sugar solutions in the effluent.

Elemental chlorine free (ECF) refers to a pulp bleaching process which does not use chlorine gas. This means that chlorine dioxide can be used. ECF bleaching is not chlorine gas free. During chlorine dioxide's reaction with lignin in the bleaching process, hypochlorous acid is formed:

$$ClO_2 (+lignin) \rightarrow HClO (+other)$$

Hypochlorous acid is in a pH dependent equilibrium with chlorine gas:

$$HClO + H^+ + Cl^- \rightleftharpoons Cl_2 + H_2O$$

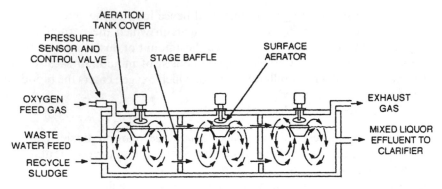

Figure 158 The UNOX™ pure oxygen activated sludge process in schematic form. Reproduced by permission of Wimpey Construction Ltd.

Chlorine gas reacts with organic substances (lignin, etc.) and forms chlorinated organic compounds.

Totally chlorine free (TCF) refers to a pulp bleaching process which does not use chlorine gas or chlorine compounds.

Source: *Response*, SÖDRA Cell Magazine, Autumn 1996, p. 6.

(◊ KRAFT PAPER, KAPPA NURBER, PAPER RECYCZE)

Pure oxygen activated sludge process. The use of pure OXYGEN instead of air in the ACTIVATED SLUDGE process to treat wastewaters. Treatment is usually carried out in covered, completely mixed tanks positioned in series (see Figure 158). In each stage mechanical aerators ensure that the oxygen is dispersed adequately within the tank contents. The reasons for using pure oxygen in place of air is to take advantage of the higher concentration driving forces, primarily within the liquid phase, to give higher oxygen transfer rates per unit reactor volume. This allows for higher ORGANIC loadings (0.4–1.0 kg BOD_5/kg MLVSS/day) compared to conventional air activated sludge systems (0.2–0.6 kg BOD_5/kg MLVSS/day). Thus using pure oxygen a significant increase in plant capacity can be achieved without a corresponding increase in reactor volume. This feature is attractive where severe space limitations are present.

Putrefaction. Uncontrolled ANAEROBIC decomposition of organic wastes, e.g. in refuse or in sewage.

Putrescible wastes. Wastes that can undergo PUTREFACTION such as food wastes, sewage SLUDGE, slaughterhouse residues.

PVC. ◊ POLYVINYL CHLORIDE.

Pyrethroids. These are the active insecticidal constituents of pyrethrum flowers (*Chrysanthemum cinerariaefolium* and *Chrysanthemum coccineium*). The

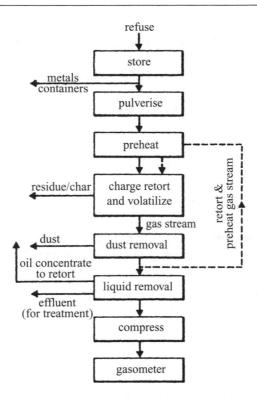

Figure 159   Flow diagram for pyrolysis of refuse for gas production.

insecticidal properties are due to five differing compounds: pyrethrins I and II, cinerins I and II, and jasmolin II. They are present in the achenes (single-seeded fruits) of the flowers in concentrations of 0.7 to 3 per cent, and may be extracted by organic SOLVENTS. As the cost of this extraction from the flowers is high, synthetic derivatives called allethrins are in common use.

Pyrethroids are highly unstable to the action of air, moisture and alkalis. The residues deteriorate rapidly after application. The characteristic insecticidal action is a very rapid knock-down of insect life followed by a substantial recovery due to detoxification enzymes present in the organisms.

Pyrethroids have low mammalian toxicity because of rapid detoxification by enzymes in the body.

Pyrolysis. The heating of wastes containing organic matter such as DOMESTIC REFUSE in a closed retort in the absence of air (see Figure 159). The subsequent volatilization produces combustible gases, a low-calorific value combustible char, a mixture of oils and liquid effluent.

The gas has a CALORIFIC VALUE half that of natural gas and requires modified appliances for its COMBUSTION. The OIL may not be present in sufficient quantities to justify a refinery stream in the UK and is therefore sent back for GASIFICATION in the pyrolysis reactor. The char can be upgraded to a fuel equivalent to low-grade COAL. The liquid effluent is treated to prevent water pollution. The process requires around 50 per cent of the fuel produced – the remainder is available for sale.

Pyrolysis is an embryonic waste disposal process which is receiving much attention in the USA as a means of conserving ENERGY resources and recycling organic wastes. It is already in use for energy recovery from TYRES and waste PLASTICS.

A rediscovered use of the technology is to heat SEWAGE SLUDGE to 400–500 °C in a pyrolysis reactor. The condensed gases yield a biofuel oil for heat/power use. Other potential feedstocks are chicken litter, cow manure, etc. An 80% efficiency is claimed by recycling hot gases to aid feedstock heating.

Source: 'Human waste used to create green fuel.' www.guardian.co.uk/environment 29/11/06.

Q

Q$_{10}$. The amount by which the growth or activity of an organism or an enzymic reaction increases per 10°C rise in temperature.

Quality assurance. The guarantee that the quality of a product or service is actually what is claimed on the basis of the quality control applied in creating the product or providing the service. Quality assurance is there to protect against lapses in QUALITY CONTROL.

Quality control. The maintenance of the quality of a product or service above a minimum standard based on known and accepted criteria.

Quality factor (Q). This ranges from 1 to 20 and is used to weight the biological effects that IONIZING RADIATION has on human tissue.

- for X-rays, gamma rays (γ), and beta rays (β), Q = 1
- for NEUTRONS Q = 10
- for alpha (α) particles Q = 20

(◊ IONIZING RADIATION; DOSE MEASUREMENT)

Quarrying. Means of mechanically extracting COAL, AGGREGATES, CLAY, etc. for commercial purposes. Potential environmental and cultural impacts include:

- transport, especially heavy goods vehicle (HGV) traffic;
- noise nuisance;
- dust nuisance;
- blasting vibration and air over-pressure as nuisances;
- diversion of surface water flows and pollution of groundwater;

- visual intrusion;
- loss of heritage;
- loss or damage to wildlife habitat or ecosystem function;
- loss of amenity;
- further nuisance from worked-out hole, e.g. water sports, LANDFILL.

(↻ AGGREGATE TAX; CONSTRUCTION AND DEMOLITION WASTE; PREDICT & PROVIDE)

Source: DETR, *Environmental Costs and Benefits of the Supply of Aggregates*, Executive Summary, April 98.

Quartz. Crystalline form of silica (SiO_2). Usually white coloured but can be colourless. Is a major constituent of granite, schist and sandstones. GLASS making requires quartz sand as a raw material.

Quicklime. Alternative description, burnt lime; chemical description, calcium oxide, CaO.

Quicklime is one of the most versatile chemical products available and is indispensable to many industries: steelmaking, construction products, industrial chemicals, oil, pharmaceuticals and many environmental treatment processes.

R

Rad. A measure of absorbed radiation dose = 0.01 joules per kg, now replaced by the gray (Gy) = 100 rads. (◊ IONIZING RADIATION (DOSE MEASUREMENT))

Radial flow tank. A circular tank with a central inlet used in sewage works and water treatment processes. The outlet consists of a weir around the circumference. The floor usually slopes to the centre where settled sludge can be drawn off.

Radiation. Figure 162 shows the sources of radiation to someone living in the UK. This is now 2.5 millisieverts (mSv) per year. The actual dose will depend on life-style and geographical location.

Air crew, and some frequent flyers, are at the top of the occupational exposure league with 4.6 mSv a year, compared with workers in nuclear energy plants who receive 3.6 mSv.

Two hundred hours of flight on a subsonic aircraft would produce an annual dose of 1 mSv, but the same duration on a supersonic aircraft would give 2 mSv, because it flies at a much higher altitude. (◊ IONIZING RADIATION)

Source: *Financial Times*, 12 April 1999.

Radiation sickness. The severe effects caused by a large dose of ionizing radiation to the whole body, e.g. from a nuclear reactor accident or nuclear weapon fallout. (Known as a non-stochastic effect by the nuclear professionals.) (◊ STOCHASTIC EFFECTS)

Radiative Forcing. Term used to describe the enhanced greenhouse effect of greenhouse gases emitted at high attitude by aircraft. In practice this means

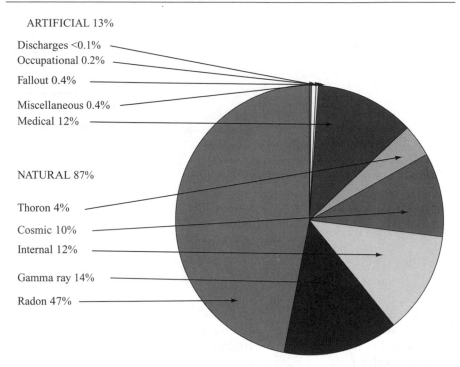

ARTIFICIAL 13%

Discharges <0.1%
Occupational 0.2%

Fallout 0.4%

Miscellaneous 0.4%
Medical 12%

NATURAL 87%

Thoron 4%

Cosmic 10%

Internal 12%

Gamma ray 14%

Radon 47%

Figure 162   Composition of the total radiation exposure to the UK population.
Source: Environment in Trust, *Radioactive Waste Management – A Safe Solution*,
Department of the Environment.

a factor of 3–4 times greater for CO_2 (Note: NO_x is also emitted). Hence when
HMG says that aircraft generate 32 mtpy carbon dioxide, the true number is
96 mtpy. In addition only flights out of the UK are considered. Hence if
inward flights were also considered, the best guesstimate could be as high as
180 mtpy – a not insubstantial contribution of the UK's claimed 670 mtpy
carbon dioxide emissions. What are we to believe? The UK's methane greenhouse gas effect can be also open to a variety of interpretations, depending
on whether a 20 or 100 year timescale is used. The 100-year timescale minimizes the effect. (Guess which one is frequently used, often with no indication
of the timescale?).

Radical (free radical). A reactive atom or portion of a stable molecule which
contains an odd number of bonding electrons. Characterized by very short
atmospheric lifetimes in the lower atmosphere but longer lifetimes in the
upper atmosphere. Typical atmospheric free radicals are HO, HO_2, Cl, NO_3,
NH_4. They are responsible for most atmospheric chemical reactions.

Radioactive decay. ◊ RADIONUCLIDE.

Radioactive half-life. ◊ HALF-LIFE.

Radioactive isotope. The nucleus of an atom is made up of NEUTRONS and PROTONS. The number of protons in the nucleus is called the ATOMIC NUMBER and characterizes the chemical element. The total number of neutrons (n) and protons (p) determines the mass of the nucleus. If two atoms have nuclei with the same number of protons but differing numbers of neutrons, they are said to be atoms of different isotopes of the same chemical element, and are shown with their respective mass number ($n + p$), e.g. ^{235}U and ^{238}U are isotopes of uranium. The isotopes of an element are identical in chemical properties and in all physical properties except those dependent on atomic mass.

Nuclear changes may result from neutron bombardment which transmutes an atom of one element to either an atom of another element or an atom of a different ISOTOPE of the same element.

For the use of the isotopes of uranium, ◊ HALF-LIFE; IONIZING RADIATION; NUCLEAR ENERGY.

Radionuclide. An atom that has an unstable nucleus which spontaneously disintegrates and emits ALPHA and BETA PARTICLES or gamma radiation, or both. Radium, uranium and strontium are examples of radionuclides, although radioactive isotopes of all common elements – carbon, potassium and hydrogen, for example – occur naturally. The process of spontaneous transmutation is called radioactive decay. (◊ HALF-LIFE; IONIZING RADIATION; RADIOACTIVE ISOTOPE)

Radon. A radioactive gaseous element emitted naturally from rocks and minerals where radioactive elements are present. It is released in non-coal mines, e.g. tin, iron, fluorspar, uranium. Radon is an ALPHA PARTICLE emitter as are its daughter or decay products, and has been indicted as a cause of excessive occurrences of lung cancer in uranium miners. New exposure standards have been adopted in the USA, Sweden and the UK based on a standardized dose measure called a 'working level month' or WLM. The new standards stipulate that a miner should not be exposed to more than 4 WLM per year.

Concern has been expressed at radon levels in some UK housing, usually adjacent to granite rocks or old tin mining regions. Levels of radon up to 300 BECQUERELS per cubic metre have been found. These can be reduced substantially by simply installing ventilation systems beneath the house. An action level of 200 Becquerels per cubic metre would warrant action to reduce the concentration.

Sir Richard Doll, who established the link between smoking and cancer 30 years ago, said:

> There are about 37,000 deaths from lung cancer in Britain and radon is responsible for around 1,800 of them. In the absence of smoking, there would still be 4,000 lung cancer deaths a year, of which 200 would be due to radon.

Sir Richard said someone who smoked 15 cigarettes a day had up to a 7 per cent risk of dying from the disease. If the radon factor was added to that, the risk would be about 9 per cent.

Source: *The Times*, 9 May 1998.

Rain. Precipitation of water from clouds. In the UK this can fluctuate greatly. For example, at the Welsh hydro-power plant at Rheidol, more than 178 mm (seven inches) of rain fell on the mountains above Nant y Moch reservoir between 10.00 am on the 26th and 10.00 pm on the 28th of December 1994 – 10 per cent of the average annual rainfall in just two-and-a-half days.

It was more than a third of the total rainfall for the month, in what was already the third wettest December since the station opened in 1963.

In just seven hours on the 27th, 3 000 000 cubic metres (675 million gallons) of water flowed into Nant y Moch. The reservoir has a capacity of 24 million cubic metres.

Downstream, water cascaded down the Rheidol at more than 40 cubic metres a second. Normal winter flow is three to four cubic metres per second.

The incidence of such heavy fluctuations in precipitation means that land-filling of wastes in high rainfall areas is an extremely difficult task which can lead to severe ground and surface water pollution. (◊ LANDFILL, HYDRO-LOGICAL CYCLE)

Source: *The GEN*, February 1995, p. 7.

Rating level. The measured noise level, adjusted for the nature of the noise (whether or not it is impulsive or irregular or contains definitive continuous note) is called Rating Level in BS 4142 (1990). This replaces Corrected Noise Level in the Standard. (◊ INDUSTRIAL NOISE MEASUREMENT)

Real-time measurement. A measurement that is made simultaneously with the event that is measured. In pollution studies it is to be contrasted with inte-grating, or time-averaging, methods in which the average concentration of a pollutant over a given period of time is determined. Real-time measurements are of particular importance when the pollutants involved are hazardous to health.

Reburn. Reburn is the injection of a secondary FUEL stream into a COMBUSTION system after the initial combustion is virtually complete. It is a potentially useful retrofit technology for the control of NO_x in industrial furnaces. ORIMULSION (a bitumen-in-water emulsion) is considered to be a prospective reburn fuel both economically and technically. A series of combustion tests were carried out in PowerGen's 1 MW Combustion Test Facility, to assess the potential of Orimulsion reburn in a coal-fired system fitted with a low-NO_x burner. With Orimulsion used to supply approximately 20 per cent of the thermal input to the system, NO_x emissions could be reduced by approximately 40 per cent while simultaneously improving combustion efficiency. Heat-release patterns within the combustion chamber were not significantly altered during reburn testing. On the basis of these tests, Orimulsion is concluded to be a potentially attractive reburn fuel.

Source: R. M. A. Irons and A. R. Jones, 'The effectiveness of Orimulsion as a reburn fuel for reduction of NO_x in a pilot-scale pulverised-fuel flame', *Journal of the Institute of Energy*, Vol. 69, September 1996, p. 163.

Recessive genes. ◊ IONIZING RADIATION EFFECTS.

Recharge. The replenishment of AQUIFERS by natural or artificial means. Natural recharge includes precipitation, seepage from streams and lakes or by underground leakage. Artificial recharge usually involves the construction of recharge basins on suitable porous strata with input from water purified from near-at-hand rivers or canals.

Recombinant DNA technology. A range of modern techniques in molelcular biology which enable sections of DNA to be transferred between individuals of the same or different species. This enables beneficial characteristics to be introduced into organisms which may be crop plants, sources of chemicals or microbes to be used for destroying toxic wastes. In this way tomatoes which ripen on demand, bacteria which produce human insulin, crops resistant to herbicides, and cocktails of organisms for decontaminating polluted land, have been produced. Concern has been expressed about the escape of such 'synthetic' organisms into the environment, particularly when bacteria resistant to antibiotics are concerned. (◊ GENETIC POLLUTION)

Record keeping. For pollution control purposes, statutory log of all monitoring and inspection required by regulatory bodies.

Recovery (of wastes). The recycling, composting and acquisition of energy from wastes.

Recovery and reuse case study

Cleaning in Place (CIP) of process equipment is an everyday part of dairy manufacturing that consumes significant quantities of water, chemicals and energy, as well as being a major contributor to process downtime. Typical CIP uses a sequence of caustic and acid-based cleaning agents and, depending on the process, sanitizing agents. The effluent from a plant CIP has high chemical oxygen demand (COD) from released fouling deposits as well as salts from the cleaning agents. While the COD is reduced by a primary and secondary effluent treatment, inorganic salts pass through into the treated water. The Dairy Process Engineering Centre (DPEC) based in Werribee, Victoria, Australia, used the CIP of large milk evaporators from a manufacturing site in Victoria, Australia as a case study to demonstrate the potential to recover, polish and reuse highly-fouled cleaning solutions. Samples taken during CIP helped profile the fouling removal and cleaning chemical concentrations, and allowed the selection of the best place of CIP polishing and reusing CIP. Effectively managing CIP reuse systems and treating recovered CIP solutions can substantially reduce chemicals consumption as well as salt content of the treated effluent.

Recycling. Reusing a material not necessarily in its original form. The natural recycling of the substances required for life are the keystone of our existence on earth. In the CARBON CYCLE the reuse of the same material is obtained with the aid of solar energy.

We cannot bypass the conservation laws of mass or ENERGY and in a world of increasing material scarcity it is important to make the best use of all resources. The intelligent adoption of recycling techniques allied with good design practice which allows for materials to be reclaimed after their useful life is over, can do much to conserve raw materials and energy, minimize pollution and save money. It would be too simplistic to expect people voluntarily to cut back on their standard of living so that major energy and materials savings can be made, but WASTE MINIMIZATION followed by recycling does offer the potential for considerable savings without major sacrifices on the part of the consumer.

Recycling falls into three classes.

Reuse: this is typified by the returnable bottle which makes several trips from bottler to consumer and back again where it is cleaned and refilled. Reuse may be allocated the highest availability in the recycling spectrum in that least energy and process complexity is normally expended in getting the material or article back into use. Typical reuse items are compressed gas cylinders, and the 44 gallon drum (a worldwide standard item).

Direct recycling: using the returnable BOTTLE as our example, once it is unfit for reuse it may be cleaned and broken down to CULLET at the glassworks and used to make more bottles. Direct recycling depends on the quality of the recycled material and its cost, which should not exceed that of the raw material. Currently most direct recycling occurs at the factory where the product is made, e.g. misshapen or broken bottles formed during glass manufacture are in fact fed back to the melting chamber. Industry calls this recycling and it is not to be confused with material reclaimed from waste or point of use. Thus paper with a 20 per cent recycled content may in fact be paper where surplus pulp fibres, mill offcuts and spoiled rolls have been internally re-routed back through the pulping process. Direct recycling has an intermediate availability in that both energy expenditure and process complexity may be required in getting the material back into use.

Indirect recycling: this practice often makes no pretence at reclaiming the material for use as such but rather gets a second bite at the cherry. Continuing with our glass bottle, it is quite probable that it will eventually end up in domestic refuse where it can be extracted by screening and separation in conjunction with other bottles. These bottles will probably be of different colours and varying degrees of cleanliness and are unsuitable for cullet use unless costly optical sorting is used. The bottles may, however, be ground up and used for a highly skid-resistant and durable road-surfacing material. Similarly, waste PLASTIC containers which *en masse* are unsuitable for direct conversion to new containers may be ground up and used for plastic fence-posts, pallets and chipboards, where appearance and structure are not primary considerations. Other forms of indirect recycling embrace the conversion of refuse to combustible gases, or the use of heat from the combustion of refuse for district heating by means of INCINERATION with heat recovery. Indirect recycling is the lowest form of recycling. Normally once processed in this phase, the material is no longer available for use except for landfill or incineration. The downgrading in use of several typical products is shown in Figure 163.

The use of LIFE CYCLE ANALYSIS can help to decide whether recycling saves resources or if it is better to incinerate with energy recovery as is the case with many PLASTIC materials in household waste.

The UK Environmental Services Association (ESA) has stated the following on recycling and recovery:

> Uncertain and variable prices for recovered materials and products are currently frustrating further progress towards major investment in new infrastructure and equipment.
>
> To invest in new recycling and recovery plants which involve commercial risk and significant expense, the waste management industry will require medium to long term secure contracts and more predictable market conditions.
>
> Demand for products containing recycled material must increase to create the expanded market which, in turn, will require investment in new recycling and recovery facilities.
>
> As rates of recovery increase, the task of achieving acceptable quality standards is likely to become progressively more difficult due to the contamination of a significant proportion of potential recyclable materials.
>
> To help Government and industry plan ways of meeting the demanding recycling and recovery targets, there is an urgent need for accurate, comprehensive data on the production of the various industrial and domestic waste streams.
>
> For certain waste streams which are particularly difficult to collect, sort and recycle, energy recovery from incineration may remain the best practicable environmental option in the short to medium term.

Source: *Environmental Services Association Bulletin*, 7, 1998.

Another commentator has written:

> The amount of metals recycled proves that as long as any commodity has an inherent value, there will always be someone prepared to separate it out for resale. If plastics, glass and other recyclables had sufficient value, the cleansing services would not have to worry about recycling them.

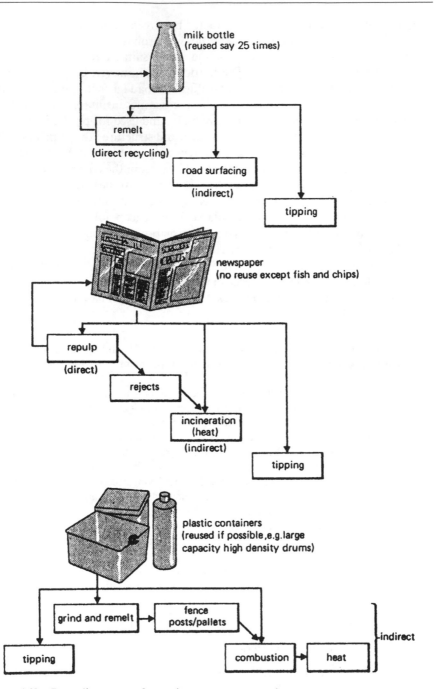

Figure 163   Recycling routes for various consumer products.

The realities of the 'market forces' logic also need to be recognised. I see little point in increasing the amount of material to be processed if the installed capacity for processing it is far below the potential quantities of material to be generated.

Source: J. Crawford, 'Recycling Reality', *Surveyor Magazine*, 29 October 1998, p. 11.

LANDFILL tipping is not a form of recycling, although many claims may be made for it to be thought of in this way; it is the sink for discarded materials, just as our surroundings form the last resting place for degraded energy. (◊ ENERGY; LAWS OF THERMODYNAMICS)

We need recycle or compost 40 per cent of household waste by 2010 and 50 per cent by 2015, according to the Waste Management Strategy for England and Wales (Defra 2007).

Recycling. High achievers: Top performers in Defra's local authority league tables for 2005/06

Local authority	Recycling rate (%)	Composting rate (%)	Total
Chiltern	32.1	4.3	36.4
Broadland	30.96	12.48	43.4
South Norfolk	29.57	0.81	30.4
Mole Valley	29.43	3.06	32.5

Note that recycling and composting are conflated, although much of the compost is not sold *per se*.

District and Borough Councils

Rank	Local authority	Recycling rate (%)	Composting rate (%)	Total	2005/06 target
1	North Kesteven	28.72	22.8	51.5	18
2	Rushcliffe	24.6	25.3	49.9	18
3	South Cambridgeshire	18.1	31.3	49.4	24

Recycling, design for. An excellent set of guidelines has been drawn up by RECOUP on PLASTIC container design for recycling. They are worthy of general application.

Think before you start

1. Have environmental concerns at the front of your mind while a new container is being conceived.
2. Consider the container's method of manufacture, materials used, its impact on the post-consumer waste stream and ability to be reused or recycled.
3. Make the pack design as simple as possible.

4. Try to ensure containers are manufactured from a single POLYMER.
5. Similarly, wherever possible, use the same material for the container, cap and non-integral features, such as pouring spout. Mixed materials can make segregation and recycling difficult.
6. Minimize the total ENERGY required for the production, distribution and choice of raw materials.
7. Where a material is used in a new form, ensure it is clearly identified to aid sorting.
8. Keep PACKAGING waste to a minimum.
9. Make sure that you do not over-specify the technical and mechanical properties of a container as this can lead to over-use of materials and energy.
10. Minimize the colour content in a container at every opportunity.
11. Examine the possibility of using post-consumer reclaimed material in the container, thus creating a bottle which is recycled as well as recyclable.
12. Remember also that recycling is not the only option: try therefore to encourage reuse of the container wherever this is practicable.
13. Caps, closures and decoration: there are many choices available, but remember some will have more environmental disadvantages than others.
14. Avoid mixing materials, for example using metal caps on plastic bottles. It is best not to use polystyrene and thermoset caps, or PVC liners in caps. Where possible, avoid the use of foil lidding bonded to the bottle neck.
15. Similarly, give some thought to the labels: try not to use PVC labels on PET bottles or PVC sleeving on HDPE.

Source: RECOUP, Metro Centre, Peterborough, PE8 5LN, UK.

Recycling, financial incentives for. A variety of taxes and levies could be imposed as incentives for recycling materials, and hence conserving resources, or to cover the costs of disposal of used materials.

Direct subsidies to reclaimers. Grants and price-support systems would ensure that recycled materials had the financial edge over original materials.

Disposal tax. A tax levied on a product according to the cost of disposal, e.g. a levy on a new car to allow for its eventual disposal. Suitable allowances (i.e. deductions or avoidance of tax) can be included for recyclable products. Another version is a LANDFILL or MINIMIZATION surcharge on every tonne of material sent for disposal (implemented by Sweden in 1993).

General pollution tax. A version of the POLLUTER-PAYS PRINCIPLE. If one assumes that recycling processes give rise to less pollution, this would favour recycling.

Packaging tax. This to some extent overlaps with the disposal tax. Products would be taxed on the packaging value added or the amount of packaging used over a certain limit adjudged sufficient for the purpose. This could work in conjunction with container deposits to encourage the return of used containers for recycling.

Virgin materials tax. A tax applied to limit the use of particular materials by raising the price artificially. This would reflect their lifetime at current rates of use, e.g. for oil based on the RESERVES–PRODUCTION RATIO.

Virgin materials levy. Levies imposed on primary materials by producer countries to restrict consumption. Britain is likely to be on the receiving end of such a measure. (◊ COLLECTION TAX)

Recycling levies. The table below shows Incpen's calculation of the fees payable by packers and fillers in four countries for recovering value from used packaging. One reason for the disparity is that the criteria on which levies are based and the method of calculation are not the same in all cases. Nonetheless, the figures are a good indication and represent a fairly realistic picture of the situation at present. French and Belgian levies may well rise in the future as their systems are not yet operating at full capacity. However, there is evidence to suggest that their fees will still end up considerably lower than those in Germany and Austria.

| | *Levy (£/tonne)* | | | |
	Belgium	*France*	*Austria*	*Germany*
Paper	9	37	161	170
Plastic	224	30	922	1255
Glass	4	2	45	64
Steel	30	16	265	240
Aluminium	50	33	394	638

Germany has very high recycling targets and places substantial hurdles in the path of waste to energy. France, on the other hand, is much more pragmatic, preferring VALORIZATION instead, which admits a spectrum of value recovery methods.

Source: *Incpen Newsletter*, Autumn 1994.

Figure 164 shows the effects of various levels of recycling on the FUEL value or CALORIFIC VALUE (CV) of MUNICIPAL SOLID WASTE.

It can be seen that the CV is noticeably unaffected by recycling and would still remain within the WASTE TO ENERGY (WTE) plant operating range. (◊ BOTTLES, PAPER)

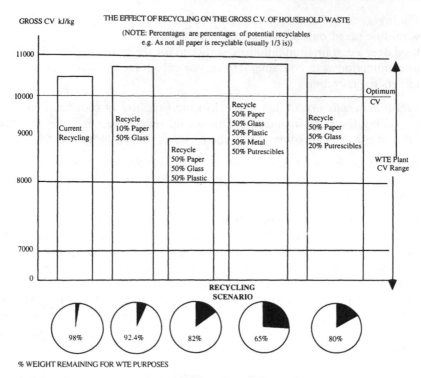

Figure 164   Recycling effect on calorific value.
Source: Hampshire Country Council. (Reproduced by permission)

It may reasonably be asked if the recycling of some of the principal components of MSW, especially mixed plastics and paper, actually saves energy or resources when ENERGY ANALYSIS is performed. An optimum needs to be struck which permits the recovery of the energy content of MSW too. (◊ VALORIZATION; INCINERATION)

Recycling, product specification, legislation for. This idea is based on the inherent wastefulness of the one-trip bottle and the throw-away-after-use syndrome. Legislation should be able to outlaw (in most cases) this wastefulness, and insist on a reusable or recyclable design. For example, the State of Minnesota has legislated for packaging controls which require all new packages and any changes in the packaging of existing products to be examined to see if they constitute a solid waste disposal problem or are inconsistent with the State's environmental policies.

Such legislation must be applied fairly; the glass, one-trip bottle should not be made the only banned form of container. Aluminium and plastics are much more energy intensive and thus they and other packaging materials must also be considered.

Product specification at the design stage can also achieve many of the desired ends of recycling legislation. The following criteria are suggested as being of use:

1. Increase the product's lifetime where possible, eliminating planned obsolescence. This should not only include increased durability but ease of repair and maintenance.
2. Design products for reuse or for multiple use where suitable. This would mean in some cases the redesigning of bottles to standard sizes and shapes with no non-standard moulding or closure methods so that they can be used by any bottler.
3. Design for ease of reclamation and recycling. Where a product presents unnecessary problems in reclamation, it should be replaced by a more easily reclaimed product. This will depend on many factors, not least the local conditions for reclamation. However, a number of products can be identified as difficult to reclaim and can be dealt with.
4. Design for disposal. If disposal is a problem, the product should be replaced by a more easily disposed product.
5. Use the least energy-intensive material that will do the job. Also, take into account the relative scarcity of materials and the lifetime of the product if it is to be reused.
6. Consideration should be given to the polluting effects of a material's manufacture.
7. Fully inform the public of the product's suitability for recycling and the raw materials and energy consumed in its production. Also, the environmental impact of its disposal.

Recycling case study – paper. In paper recycling, the environmental benefits of recycled fibres are all too often counteracted by the negative effects of the largely untreated residual products emitted from the recycling plants.

Stora Papyrus Dalum in Denmark is the first fine paper manufacturer to resolve this environmental dilemma. All the raw material (waste paper) is utilized and reused in different ways. No residues leave the mill to end up on the landfill or pollute the water or air.

The raw material for both Cyclus and Cyclusprint (both 100 per cent recycled fines) is waste paper from printing works and offices, mostly in Denmark, northern Germany, southern Sweden and Norway (the Oslo area).

The waste paper is converted into PULP at the Stora Papyrus DE-INKING plant at Naestved, south of Copenhagen. The waste paper is de-inked and the filler and coating are separated from the paper fibres.

The stock (mixture of fibre and water) is then pressed into thick sheets and packed in pulp bales, ready for delivery by rail to the papermill in Odense.

The actual de-inking plant itself is similar to the plants in most other mills making pulp from waste paper. What makes Naestved unique is the fact that the residual products from de-inking are 100 per cent recycled.

From 100 000 tonnes per year (tpy) waste paper, 75 000 tpy fibres are recovered, leaving 15 000 tpy clay and chalk which is sent to a nearby cement factory, and 10 000 tonnes of sludge, plastics, etc. The sludge is used as a land conditioner, and any plastics and combustibles are sent to the local district heating plant.

The company has put a lot of effort into making the whole process virtually closed, and has halved the water consumption in three years. Water which is released is pumped to Odense Council Sewage Works for final purification. The Council Technical Department has compared the quality of the water upstream of the mill with the water after purification. Their tests show that the water released is cleaner than the water taken in.

Source: *Sweden Today*, January 1995.

The UK paper recycling scene is shown in Figure 165. It can be seen that waste paper consumption as percentage of usage has increased from 50 per cent (1984) to 67 per cent (2006).

Some imports will always be necessary into the UK and it is to the great credit of the British paper industry that it achieves such a high level of recycling already. But there are limits hence export of waste paper to the FAR EAST.

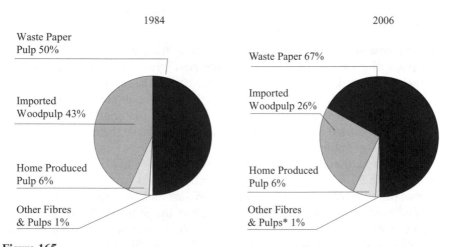

Figure 165
Source: Personal Communication, Confederation of Paper Industries, 2007
*Seasonal fibres such as straw, or rags and off-cuts from textiles, etc.

Recycling case study – US source separation.

Example of a US system achieving in excess of 50% diversion of waste from disposal.

	Per cent in MSW	Overall net recovery Percentage*	Recovered
Baseline recyclables			
Tin cans	3.0	70	2.1
Aluminium	0.5	70	0.4
Glass	8.0	70	5.6
News	12.0	70	8.4
Total	**23.5**	**70**	**16.5**
Yard waste**	15.0	70	10.5
Other wastes**			
Corrugated	10.0	70	7.0
Mixed Paper	20.0	70	14.0
Food Waste	5.0	70	3.5
Wood	2.0	70	1.4
Total	**37.0**	**70**	**25.9**
Totals	**75.5**	**70**	**52.9**

* The net recovery of the collection system and of the processing system.
** For example, diverted from landfill disposal via conversion to a compost product.

US experience on their domestic waste arisings shows that the processing of source-separated residential recyclables and of source-separated yard wastes typically results in the diversion of 20 to 30 per cent of generated solid wastes from landfill disposal and the recovery of valuable resources. If source-separated commercial waste recycling and processing is added to the overall programme and properly administered, an additional 5 to 10 per cent typically is achievable. The total diversion combining the above programmes is in the range of 25 to 40 per cent (from a higher waste production level than the UK).

A cautionary note: The markets must be available for the recycled materials and a net environmental gain should be clearly demonstrable.

Source: *Worldwide Waste Management*, Vol. 8, Issue 5, October/November 1998, p. 12.

Red List. A list of 23 dangerous substances, designated by the UK, whose discharges to water should be minimized under the BATNEEC principle and whose DISCHARGE CONSENTS are required to ensure that strict environmental quality standards are met and maintained in the receiving waters.

In addition, the *Third North Sea Conference*, March 1990 agreed a priority list of 39 substances (shown below) whose discharge to rivers and estuaries should be reduced by 50 per cent by 1995 (over 1985 levels). The list includes the Red List substances.

Red list substances	Additional substances on priority list
Mercury and its compounds	Copper
Cadmium and its compounds	Zinc
Gamma-hexachlorocyclohexane	Lead
DDT	Arsenic
Pentachlorophenol	Chromium
Hexachlorobenzene	Nickel
Hexachlorobutadiene	Chloroform
Aldrin	Carbon tetrachloride
Dieldrin	Azinphos-ethyl
Endrin	Fenthion
Polychlorinated biphenyls	Parathion
Dichlorvos	Parathion-methyl
1,2-Dichlorethane	Trichloroethylene
Trichlorobenzene	Tetrachloroethylene
Atrazine	Trichloroethane
Simazine	Dioxins
Tributyltin compounds	
Triphenyltin compounds	
Trifluralin	
Fenitrothion	
Azinphos-methyl	
Malathion	
Endosulfan	

($\natural$ list I substances; list II substances.)

Red tides. An accumulation of a toxic micro-organism in shellfish. Responsible for periodic deaths of seals/manatees, etc. in warm climates.

Reducing agent. Chemical which combines with oxygen, i.e. removes it from a substance. Term also used for the addition of hydrogen to a substance. The opposite of an OXIDIZING AGENT.

Reed beds. These are beds of soil or gravel media in a sealed pit containing reeds (usually the species *Phragmites australis*) for treatment of effluent (see Figure 166). They were first used for the treatment of sewage in Germany in the 1960s. The effluent to be treated is distributed through pipes and nozzles onto the bed. The rhizomes (underground stems) of the reeds grow vertically and horizontally, opening up the bed to provide a passage for effluent. OXYGEN is passed via the leaves and stems of the reeds, through the hollow rhizomes, and out through the roots to the rhizosphere (the area surrounding the rhizomes). Here large populations of aerobic bacteria effect the biodegradation of the organic matter in the effluent. Anaerobic bacteria in anoxic areas of the soil also contribute to breakdown of pollutants.

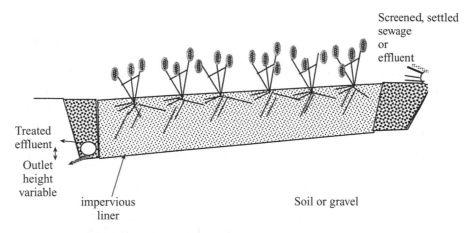

Figure 166   Cross-section through a reed bed.

In reed bed systems for sewage treatment, screened, settled sewage is passed through horizontal flow beds. BOD removal is in the range of 80–90 per cent.

Owing to the great diversity of microbial species in soil, flexibility is possible with reed beds; for instance, the fungal species *Actinomyces, Streptomyces* and *Basidiomyces*, which are capable of biodegrading many synthetic chemicals such as the common pesticides and chlorinated hydrocarbons, are found in soil but not normally in effluent treatment plants. Fully mature reed beds have been used to remove traces of phenol, methanol, acetone and amines from industrial effluents.

Unlike mechanical treatment plants, reed beds are not noisy or unsightly. They do, however, require a greater land area for equivalent treatment levels.

Reformulated gasoline. US term for petrol which has had oxygenation compound(s) such as ETHANOL added. This promotes cleaner combustion with claimed reductions in the 20–30 per cent range for CARBON MONOXIDE, BENZENE and UNBURNT HYDROCARBONS.

There is still some dispute in the US about the value of the oxygenates in the 'clean-burn' process. These are organic combustible liquids which contain oxygen atoms – not a natural component of petrol – such as ethyl alcohol or ethanol, and methyl tertiary-butyl ether. In reformulated petrol, these oxygenates replace aromatics, which are used to provide the OCTANE levels needed for smooth engine running. Benzene is among the most widely used aromatics.

Although formulas vary between refiners, the new products, with their lower concentrations of toxic compounds, zero content of heavy metals and reduced sulphur component, have been enthusiastically adopted.

Source: C. Parkes, 'US breathes more easily', *Financial Times*, 17 April 1997.

Refrigerants. A fluid medium (gas or liquid) for removing heat rapidly in refrigerating equipment.

Refrigerant choices

CFCs (mainly CFC 11 and CFC 12). These substances are highly damaging to the ozone layer, and also contribute to global warming. Further use of these materials will rely on existing stocks and recycling; inevitably, prices are rising fast. HCFCs (mainly HCFC 22 and HCFC 123) are substantially less damaging to the ozone layer, and were not controlled under the original protocol.

Ammonia. This was the workhorse of the refrigeration industry until about 30 years ago, when CFCs were introduced as the safer option. Many locations have continued to use ammonia (without serious incident). It is cheap and efficient, does not contribute to ozone depletion or global warming, but needs careful handling because it is toxic and mildly flammable. Often the best option, as it is a natural substance and is cheap and effective.

Other refrigerants. These mainly consist of the HFCs, a new class of materials which contain no chlorine and therefore do not damage the ozone layer. If released into the atmosphere they do, however, contribute to global warming. HFC 134a is by far the most widely used of the HFCs, and is generally considered to be non-toxic and non-flammable. There are also many HFC-blend refrigerants tailored for specific jobs, mainly for small to medium size commercial chillers, rather than large process cooling units. (◊ CFCs)

Source: Courtald's Environment Matters, March 1995.

Refrigeration – an environmentally friendly case study. The air-cycle system is an air-conditioning system using air as a refrigerant, and therefore it is extremely environmentally friendly when compared to vapour cycle systems.

The use of air as a refrigerant is based on the principle that a rapid reduction of the pressure of air also reduces its temperature so that the resulting cold air can then be used for cooling purposes. Conversely, when air is compressed its temperature increases and it can then be used as a heating medium. Figure 167 shows a basic open air-cycle system, commonly found on today's commercial aircraft. Pre-conditioned air from the bleed system enters a centrifugal compressor where its pressure is raised by applying input energy. The heat of compression is then removed by passing the compressed air through an ambient heat exchanger, cooled by outside air. The air is now cooler but still at pressure. By expanding it through a turbine the air generates mechanical energy in the radial turbine shaft and this results in a temperature drop in the expanded air. The expanded air leaving the turbine is significantly colder than when it enters, and this produces the effects of refrigeration.

The radial turbine and centrifugal compressor wheels are mounted on opposite ends of a common shaft and make up the cold air unit. The energy extracted by the turbine is used to drive the compressor to boost the initial

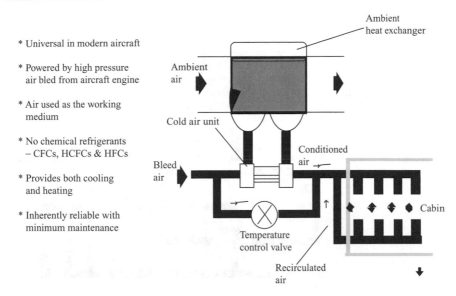

* Universal in modern aircraft

* Powered by high pressure
  air bled from aircraft engine

* Air used as the working
  medium

* No chemical refrigerants
  – CFCs, HCFCs & HFCs

* Provides both cooling
  and heating

* Inherently reliable with
  minimum maintenance

Figure 167   Typical aircraft air-cycle system.

pressure available and hence improve the performance of the system. A bypass valve provides temperature control.

In order to apply air-cycle technology to air conditioning of a train, a source of compressed air is required. This is provided by an electrically driven centrifugal compressor which replaces the gas turbine bleed air and forms the first stage of the compression cycle.

Normalair-Garrett have developed the electrically driven air-cycle system for passenger trains, thus eliminating all refrigerants (see Figure 168).

The air-cycle concept can be used as an open or closed loop cycle. In an open loop cycle, the refrigerant air is used to heat or cool a space directly. In a closed loop system, a heat exchanger transfers the heating or cooling effect to the carriage air. Figure 168 shows a closed loop system. This offers the advantages of variable capacity control within the closed loop without affecting the air flow to the compartment; no direct path for turbo-machinery noise to the passenger compartment; and the prevention of contamination or water entering the turbo-machinery.

The closed loop system shown consists of a motor compressor from which air is fed to the compressor side of the cold air unit. The air is then fed through the ambient heat exchanger from which it passes to the turbine side of the cold air unit. Following the expansion and hence cooling, it passes through the charge side of the load heat exchanger and then returns to the input of the motor compressor.

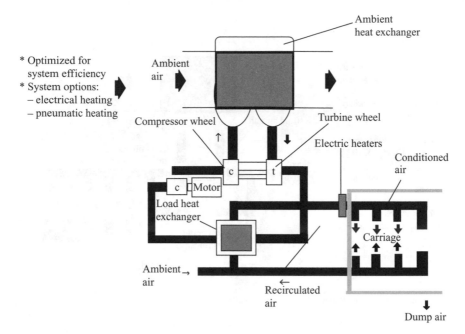

Figure 168   Air cycle configuration for trains.
Source: Normalair Garrett.

Cooling capacity control of the system is achieved by simply varying the speed of the motor compressor and hence the mass flow around the closed loop system.

Source: Normalair-Garrett information handout.

Refuse. ◊ DOMESTIC REFUSE.

Refuse derived fuels. ◊ WASTE DERIVED FUELS.

Reinluft process. A process for removing (and recovering) oxides of nitrogen and sulphur from stack gases. The gases are mixed with oxygen and converted to a higher oxidation state (a CATALYST may be used to increase the efficiency of conversion) and passed upward, in countercurrent fashion, over a bed of carbon in a multi-chambered adsorber. The oxides are adsorbed at different temperatures. Finally, they are swept out of the carbon bed at high temperature in a stream of nitrogen or carbon dioxide and reduced. The process may be carried out in the presence of water vapour, in which case the corresponding acids, rather than the gaseous oxides, are recovered.

Relative biological effectiveness (RBE). ◊ IONIZING RADIATION (DOSE MEASUREMENT).

Relative humidity. The ratio of the actual amount of moisture in the air to the amount needed for saturation at the same temperature.

Release. According to the UK Environmental Protection Act 1990, a substance is 'released' into any environmental medium whenever it is released directly into that medium, whether it is released into it within or outside Great Britain. 'Release' includes:

(a) in relation to air, any emission of the substance into air;
(b) in relation to water, any entry (including any discharge) of the substance into water;
(c) in relation to land, any deposit, keeping or disposal of the substance in or on land.

For this purpose 'water' and 'land' shall be construed in accordance with the subsections below.

(a) Any release into (i) the sea or the surface of the seabed, (ii) any river, watercourse, lake, loch or pond (whether natural or artificial or above or below ground) or reservoir or the surface of the riverbed or of other land supporting such waters, or (iii) groundwaters, is a release into water.
(b) Any release into (i) land covered by water falling outside (a) above or the water covering such land, or (ii) land beneath the surface of the seabed or of other land supporting waters falling within (a) (ii) above, is a release into land.
(c) Any release into a sewer (within the meaning of the Public Health Act 1936 or, in relation to Scotland, of the Sewerage (Scotland) Act 1968) shall be treated as a release into water. But a sewer and its contents shall be disregarded in determining whether there is pollution of the environment at any time.

'Groundwaters' means any waters contained in underground strata, or in (a) a well, borehole or similar work sunk into underground strata, including any adit or passage constructed in connection with the well, borehole or work for facilitating the collection of water in the well, borehole or work, or (b) any excavation into underground strata where the level of water in the excavation depends wholly or mainly on water entering it from the strata.

'Substance' shall be treated as including electricity or heat. (◊ PRESCRIBED PROCESSES)

Reliability prediction. It is on the human side that many spectacular failures have occurred, e.g. operator error in the THREE MILE ISLAND nuclear power plant incident, or the notorious unsuitable seal in the Apollo 13 mission.

Caveat emptor is needed when reliability is quoted.

The following is based on an article by N. Pascoe, Technical Director of Reliability Technology Limited.

The reliability of an item of age t [$t \geq 0$], denoted by $R(t)$, is defined in reliability texts as the probability that the item is still operating satisfactorily at that age.

Reliability prediction is often statistically based on the following assumptions:

- the failure rate of a system is the sum of the failure rate of its parts
- all failures occur independently
- all failures have a constant rate of occurrence
- every component failure causes a system failure
- all system failures are caused by component failures.

The measured failure intensity of a component is seldom due to a single repeatable process. It is most frequently attributable to many physical, chemical and human processes and interactions. For example, one or more of the following may cause failure of a transistor:

- bulk crystal defects
- diffusion defects
- faulty metallization
- faulty wire bond
- corrosion
- misapplication of test
- handling damage.

So there can be no single mathematical model for failure rate or time to failure.

Source: N. Pascoe, 'The uncertainty of reliability prediction', *Environmental Engineering*, December 1998, p. 22.

Rem (Roentgen Equivalent Man). A measure of the effective radiation dose absorbed by human tissue. It is the product of the dose in RADS and the QUALITY FACTOR. Now replaced by the Sievert (Sv) = 100 rem. (◊ IONIZING RADIATION (DOSE MEASUREMENT))

Renewable energy. Replenishable power source. (◊ SOLAR ENERGY; TIDAL POWER; WAVE POWER) See Figure 170 for renewable energy use in the UK.
A means of attaining SUSTAINABLE DEVELOPMENT.

The Renewables Obligation. The new Renewables Obligation and associated Renewables (Scotland) Obligation came into force in April 2002 as part of the Utilities Act (2000). It requires power suppliers to derive from renewables a specified proportion of the electricity they supply to their customers. The current (2007–08) level is 7.9%, rising to 15.4% by 2015–16. The cost to consumers will be limited by a price cap and the obligation is guaranteed in law until 2027.

Eligible renewable generators receive Renewables Obligation Certificates (ROCs) for each MWh of electricity generated. These certificates can then be sold to suppliers, in order to fulfil their obligation. Suppliers can either present enough certificates to cover the required percentage of their output,

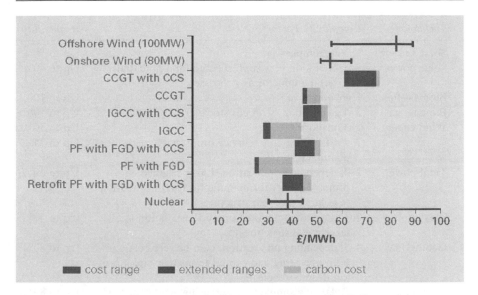

Figure 169 Comparative costs of electricity generating technologies. Base case costs with carbon price (€25/tCO₂) and high gas price (53 p/therm).
Source: World Coal Institute, Fact Focus #35, E. Coal Newsletter, July 2006, Vol. 58.

or they can pay a 'buyout' price of price of £30/MWh for any shortfall. All proceeds from buyout payments are recycled to suppliers in proportion to the number of ROCs they present.

Renewable Obligation Certificate (ROC). There is a legal requirement for electricity supply companies to buy a set percentage of their electricity from renewable energy sources. This percentage is set to rise annually, to a level of 15.4 per cent in 2015. The supply companies need not generate renewable electricity themselves, but can buy it from companies that generate electricity from renewable sources. The renewable energy company then obtains what is known as a Renewables Obligation Certificate (ROC) for each Megawatt hour (MWh) of electricity generated from renewable sources. Supply companies which buy electricity to supply consumers, then also have to buy the certificates which prove that they have used electricity from renewable energy sources, which is put toward meeting their obligation targets. The electricity goes straight into the National Grid from the generator.

Renewal Energy Technologies (RETs). RETs embrace a wide spectrum of heat/ power production options. The common factor is that each lays claim to not relying on fossi/nuclear fuels and to not producing excess carbon dioxide whilst in operation. Hence fuelwood (biomass) may be claimed to be carbon neutral as the CO₂ from combustion is offset by new forest growth in sustainable forests which absorb CO₂ in the growth process. The diversity of RETs is briefly summarized in the table below.

Technology	Complexity Factor	Availability
Solar heating	1; Very simple	Up to 50%
Solar power	3–5; Relies on very sophisticated (and potentially toxic) manufacturing processes.	Up to 40%
Biomass fuel	1 – woodstove	Up to 100%
Biomass gas	3–4 gasification (with gas storage)	Up to 100%
Wind energy	2–3 onshore	Up to 30%
	3–4 offshore. Large reliance on primary energy for manufacture.	Up to 35%
Tidal power	2–4, depending on location. Large reliance on primary energy for manufacture and installation (e.g. large concrete structures).	Up to 60%
Wave power	2–5, depending on location and installation (e.g. large metal structures).	Up to 50%
Geothermal	4–5, depending on location. Can be very costly in both operating and capital terms (e.g. sinking x km deep boreholes).	Up to 80%
Coal mine and Landfill Methane	2; Relatively simple but decreasing yield over 5–15 years.	Up to 80%
Waste incineration	2–3; Recycling lobbyists may dispute this entry but household waste can generate $550\,kWh_e$/tonne waste combusted, and as part of the waste is bioderived the technology effects a net CO_2 reduction/kWh_e compared with gas or coal-fired generation.	Up to 85%
Hydropower	1–2; Very limited in the UK. Normally simple technology but concrete dams are energy intensive.	Up to 100%
Anaerobic digestion	2; Local solution for suitable wastes, e.g. farmyard slurries, and food wastes. Methane gas may be stored.	Up to 100%
Perpetual motion	Infinity. There is no free lunch – ignore ($\Diamond$ LAWS OF THERMODYNAMICS)	
Combined heat and power	2–4; This raises the thermal efficiency of conventional power stations from about 35% to 80% plus. CHP conserves fossil fuels and hence deserves to be subsidized. ($\Diamond$ ENERGY CONSERVATION)	Up to 95%
Energy conservation	1–5; Energy conservation is self-explanatory and needs to be practiced in the greatest depth consistent with defensible paybacks. A great boost would be to give an allowance per kWh_e saved, akin to the subsidy levels accorded wind energy.	Up to 100%

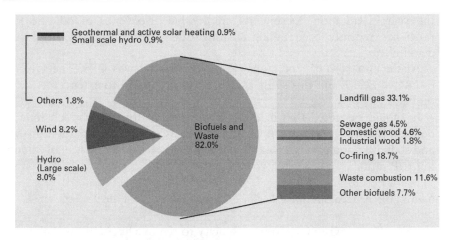

Figure 170   Renewable energy sources, 2006.
Source: http://www.berr.gov.uk/files/file39881.pdf.

Figure 169 above shows the costs of technologies under high gas prices (53 p/therm) with a carbon price of 25 Euros/t CO_2. The average price of gas in the UK in the first six months of 2006 was 52.3 p/therm and the current price of carbon is between 16–17 Euros/t CO_2.

The World Coal Institute's conclusion (July, 2006) is that under this scenario – and today's reality – clean coal technologies are all cheaper than the comparative gas technologies.

Repeatability. The closeness of agreement between successive results obtained with the same method, test material and under the same conditions (e.g. same laboratory, same apparatus).

Reprocessing. The reclamation of in-house arisings and those purchased from within the relevant industry for the subsequent manufacture of finished products. Also spent nuclear fuel 'reuse'.

Reproducibility. The closeness of agreement between individual results obtained with the same method on identical test material but under different conditions (e.g. different laboratories or apparatus).

Reserves. The amount of substance, e.g. OIL, that still remains available to be exploited at a cost below a certain level. They are normally referred to as recoverable reserves, i.e. the total quantities of oil that are likely to be brought to the surface for commercial use within a certain time-span and level of technology.

Reserves–Production ratio (R/P ratio). The ratio between the proved recoverable reserves remaining in the ground and the annual production rate, i.e. the number of years of production remaining at the rate of production in the year in question. Despite the impressive discoveries of oil, the R/P ratio has recently been falling steadily due to a continually increasing demand.

At the end of 2005, the global R/P ratio for oil was 40.6. The UK's is estimated at 6.1, but this shifts as new ways are found to extend the life of oil fields.

It should be noted that a low R/P ratio does not allow much time to develop alternative energy sources and life-styles, or contain economic growth and stabilize populations. ENERGY CONSERVATION is one sure way of extending R/P ratios.

Source: BP–Amoco *Statistical Review of World Energy*, June 1999.

Reservoirs. The provision of storage capacity to remove fluctuations in the flow of a material or component. A controlled reservoir has a means of discharging flow under control. An uncontrolled reservoir only has a spillway which allows discharge when capacity is attained.

Residence time. In air pollution studies, the length of time during which a given molecule of an air pollutant remains in the atmosphere (it may, at the end of that time, be replaced by another molecule, so the residence time is not the duration of air pollution). In chemical engineering, the term is applied to the length of time that a given material remains in a vessel through which it is flowing.

Residual fuel oil. The residue from the distillation of petroleum which is not economical to process further. Can be blended with 'heavy' distillates as fuel for low-speed reciprocating engines or used blended as a boiler fuel.

Resource recovery. Recovery of materials, fuel or energy from waste. (◊ PACKAGING)

Resources. Anything that is of use to humans. (◊ RESOURCES, RENEWABLE; RESOURCES, NON-RENEWABLE)

Resources, non-renewable. Substances which have been built up or evolved in a geological time-span and cannot be replaced except over a similar time-scale. Examples are copper, tin, COAL and OIL. It is often (erroneously) stated that when, for example, high-grade copper ore runs out, low-grade copper ore will become economically workable. However, this view neglects the facts of energy resources depletion and increasing pollution with lower grade burdens. Even if there were unlimited supplies of energy, the limitations imposed by the LAWS OF THERMODYNAMICS and climatic stability mean that there are limits to how much energy humans may use in working low-grade ore deposits. Furthermore, ore sources do not necessarily become more plentiful with lower grades. (◊ ARITHMETIC–GEOMETRIC RATIO)

RECYCLING is one method of conserving finite resources. Some resources such as land, water and air have definite limits on the amount of exploitation they can sustain. Water is a renewable resource, but it is finite in its rate of supply as dictated by the HYDROLOGICAL CYCLE.

Resources, renewable. Resources that derive from solar energy such as fish, trees, wind, rain. Plant-life such as timber or grass should be managed for MAXIMUM SUSTAINABLE YIELD. If the yield exceeds this rate, the system gives ever-diminishing returns. In a fishery pushed past its limit, the catch is maintained by collecting greater and greater numbers of younger fish until there is extinction. Resource management is now a necessity due to the pressures of population growth and affluence. (◊ WHALES)

Respiratory sensitizers. Respiratory sensitizers are substances which, when breathed, can trigger an irreversible allergic reaction in the respiratory system.

Sensitization is substance-specific and an employee becoming allergic to a particular substance will initially show the symptoms only when he or she breathes that substance. Once exposure to the trigger substance has stopped, the symptoms will cease but the individual remains sensitized. Currently an estimated 2000 new cases per year are occurring as a direct result of exposure to respiratory sensitizers (this is apart from the rise in the number of people suffering from asthma due to environmental factors).

Occupational asthma is a direct result of exposure to respiratory sensitizers in the workplace. But it is not the only breathing problem people face.

Other work-related respiratory complaints include rhinitis and conjunctivitis. Rhinitis is an inflammation of the mucous membrane that lines the nose. The symptoms of allergic rhinitis include nasal congestion, runny nose, sneezing and itching of the nose. Conjunctivitis is the inflammation of the conjunctival membrane which covers the outer area of the eyeball and inside of the eyelids. Victims suffer a gritty and burning sensation in the affected eye which feels worse on blinking and usually leads to inflammation accompanied by a sticky discharge. Again, the condition may become chronic if the cause is an allergic reaction to a sensitizer in the workplace.

Commons sensitizers include:
- isocyanates used in vehicle spraying and foam manufacturing;
- flour/grain/hay when handled or during processing such as milling, malting and at bakeries;
- soldering flux produced during electronics assembly;
- laboratory animals in research establishments;
- wood dust in sawmills and wood machining and finishing operations;
- epoxy resins and glues used in many industry sectors.

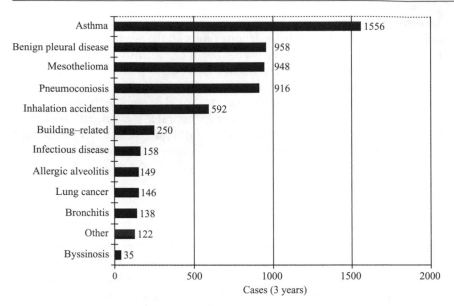

Figure 171   Cases of occupational lung disease reported to the UK SWORD project (1989–1991).

Figure 171 shows the number of cases of occupational lung disease reported to the UK SWORD (Surveillance of Work-related and Occupational Respiratory Disease) Project, for 1989 to 1991.

Source: 'Good Health is Good Business', *Health and Safety Executive Magazine*, 1996.

Reverse osmosis. A means of separating solutes such as sugar dissolved in water or desalting by means of a suitable semi-permeable MEMBRANE and the application of pressure to the solution. (◊ DESALINATION)

Reverse vending. Reverse vending schemes buy back recoverable materials from the public as a mechanism to generate higher recovery levels. The credit given to those providing materials can take a number of forms: a cash payment; a voucher redeemable for cash; a coupon providing a discount on a particular brand or in a particular retail outlet; or a chance to win a prize through participation in the scheme.

The consumer deposits containers into a reverse vending machine (RVM) through an entry point in the front of the machine. The containers are scanned to establish their type and colour. The scanning systems used depend on the model of RVM. There are two methods, which are sometimes used in conjunction with each other. (i) Material sensors are used to distinguish PET and PVC; colour sensors may be fitted to perform a colour sort on translucent plastics. Some multi-material RVMs offer magnetic sorting for cans. (ii) A laser scans the bar-code of the container; the resultant information is then cross-referenced with a product database. This database may either be held in the machine on a plug-in ROM card, or else the machine can be linked via a modem to a central control point. Here, the bar-code information is stored on a main database for analysis.

Source: RECOUP Information Leaflet, 1996.

Revolving screen. Separation device, used in gravel extraction, rock-crushing and domestic refuse RECYCLING, consisting of a perforated drum which allows materials of varying sizes to be graded and removed. (◊ AIR CLASSIFIER)

Reynolds number. A dimensionless parameter (i.e. a pure number which has no units) which is one of several such numbers used in the study of fluid flow. It is the ratio of inertial forces to viscous forces and is defined as RE = $\dfrac{\rho v D}{\mu}$, where ρ is the density of a fluid, μ its dynamic viscosity and v its velocity, and where D is a linear dimension that depends on the problem under study (for flow in a pipe, for example, it is the diameter of the pipe). The value of the Reynolds number helps to determine whether flow will be laminar or turbulent, i.e. 'streamlined' or 'mixed up'. The determination of a flow regime is of crucial importance in sewer and water pipeline design. (◊ LAMINAR FLOW; TURBULENCE)

Richter scale. A logarithmic scale developed in 1935 that measures how much energy is released by ground movement from the centre of a seismic shock. A 6.0 earthquake is 10 times more powerful than a 5.0 earthquake and a 7.0 earthquake is 100 times more powerful and so on. The October 1989 California earthquake measured 6.9 on the Richter scale. The greatest ever recorded is 8.9. (◊ LOGARITHMS)

Right to discharge. An alternative to the principle that the polluter must pay is the proposal that any firm wishing to discharge a pollutant to the environment would have to pay for the right to discharge such pollutants. The interpretation is essentially a change in emphasis, from curative to preventive legislation. (◊ POLLUTER-PAYS PRINCIPLE; BEST PRACTICABLE MEANS)

Ringelmann charts. A series of four charts of graduated shades from white to black for assessing the darkness of a plume of smoke by visual comparison. The standard chart (BS 2742) is a set of numbered grids differing from one another in the width and spacing of black lines printed on a white background.

Ringelmann **1** is equivalent to 20 per cent black, **2** is 40 per cent black, **3** is 60 per cent black, **4** is 80 per cent black. The scale may be extended to shade **0** (white) and shade **5** (black). The charts must be used at a certain specified distance so that the grid forms a grey scale. Micro and miniature smoke charts are also available and are more convenient to use.

In law 'dark smoke' is smoke which is as dark or darker than Ringelmann **2**, and 'black smoke' is as dark or darker than Ringelmann **4**.

Rio Earth Summit. A major, high level UNCED event held in Rio de Janeiro, Brazil, in 1992 to demonstrate the developed countries' concern for the global environment and how they could assist the less developed countries not to make the same mistakes as the west, and achieve sustainability.

The high-sounding declarations did help focus attention on GLOBAL WARMING. Over 160 nations signed the UN Framework Convention on Climate change which has become a useful lever in kick-starting the process of stabilizing GREENHOUSE GAS emissions.

The Rio Earth Summit also produced Agenda 21, a 500-page, 40-chapter tome. The overall impression is that it is 'UN Speak' but this admonitory document is at least a start on SUSTAINABLE DEVELOPMENT. It covers in varying degrees of resolution:

- social and economic development (from combating poverty to integration of environment and development in the political process);
- conservation and management of resources (from atmospheric protection, and ecosystem conservation, to the management of SEWAGE and RADIO-ACTIVE WASTES);
- major demographic and political group participation in decision-making, and e.g. women, children, NGOs, local authority initiatives, scientists, farmers.

The UK local authorities have seized on this aspect to encourage the formation of 'LA21 Committees', which act as think-tanks for them. Unfortunately, they may be seen as 'green up' advocates rather than framework planners (after all 'green up' doesn't cost too much).

Finally, the LA21 document ends with the means of its implementation. This ranges from financial resources, technology transfer, public education, and national and international mechanisms for co-operation.

Agenda 21 is a triumph of hope over experience and should perhaps be viewed as firing the starting gun for concerted action on conserving resources for future generations. It is capable of a multitude of interpretations and unfortunately self-serving trumpeting on actions taken in its name cannot be avoided, particularly if it can help raise taxes under a 'green cover' (pick your own example).

Risk. The chance of a particular adverse effect occurring in a given period of time, e.g. the chance of causing illness or death per year. Low risk can only be defined by comparison with every day life, e.g. heart disease (UK) accounts for approximately 50 per cent of male and female deaths; 50 children per day (UK) are admitted to hospital with cigarette smoke-related medical conditions. In 2005/06, 212 workers were killed, and 146236 injuried, due to accidents in the UK.

Risk assessment

When risk is assessed it is important to take on board that linear extrapolation of effects from high to lower doses is often not valid. In a third or more of instances in which a maximum tolerable dose elicited extra tumours in rodents, one-half that dose did not. Many scientists have pointed out that huge doses of non-genotoxic substances are accompanied by toxicity, cell death, and cell replacements. This creates conditions favourable for growth of tumours. At doses in which cellular death does not occur, tumours would not be produced by non-genotoxic substances. The majority of chemicals are not genotoxic, nor does metabolism of them give rise to genotoxic intermediates. Thus linear extrapolation is not applicable to the majority of chemicals. However, the extreme care taken is shown by the example of dioxins opposite.

Total daily intakes are of the order 1/100th of the ADI or less (incineration plants may contribute a negligible 0.006 per cent of dioxin ADI for a maximally exposed individual).

It has been said that:

> The current mode of extrapolating high-dose to low-dose effects is erroneous for both chemicals and radiation. Safe levels of exposure exist. The public has been needlessly frightened and deceived, and hundreds of billions of dollars wasted. A hard-hearted, rapid examination of phenomena occurring at low exposures should have a high priority.

Risk-aversion. A decision making straightjacket in which there are minimal rewards for taking successful risks and penalties for failure. The result is very conservative, cover-your-back, officialdom, e.g. the refusal of many planning authorities to allow waste incineration plants to be built on specious health risk grounds. Even worse, this mindset has led to schoolchildren being banned from playing 'conkers' with horse chestnuts on a length of string! Preserve us from jobsworths.

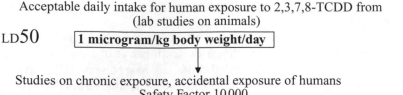

Acceptable daily intake for human exposure to 2,3,7,8-TCDD from
(lab studies on animals)

LD50 | **1 microgram/kg body weight/day** |

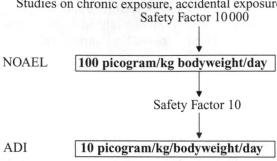

Studies on chronic exposure, accidental exposure of humans
Safety Factor 10 000

NOAEL | **100 picogram/kg bodyweight/day** |

Safety Factor 10

ADI | **10 picogram/kg/bodyweight/day** |

LD50   Lethal Dose (on Animals)
NOAEL   NO Observed Adverse Effect Level
ADI   Acceptable Daily Intake for Lifetime Exposure or TDI : Tolerable
       Daily Intake

i.e. A factor of $1/100,000$th of the LD_{50} is used in practice.

Source: P. H. Abelson, 'Risk assessments of low-level exposures', *Science*, vol. 2, 9
September 1994.

Figure 172 shows assessment of health, safety and environmental risks.

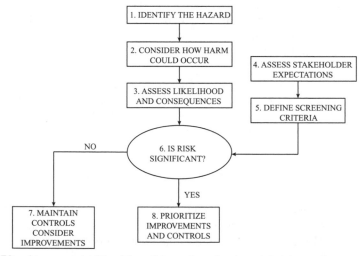

Figure 172   Assessment of health, safety and environmental risks.
Source: *Responsible Care Management Systems for Health, Safety and Environment*, The
Chemicals Industries Association Responsible Care series No. 3, July 1996, p. 26.
(Reproduced by permission)

River Ecosystem Classification. The Surface Waters (River Ecosystem) (Classification) Regulations 1994 (SI 1994 No. 1057), prescribe a system for classifying the quality of rivers and canals, to provide the basis for setting statutory water quality objectives (WQOs) under section 83 of the Water Resources Act 1991 in respect of individual stretches of water.

The River Ecosystem Classification comprises five hierarchical classes, in order of decreasing quality: RE1, RE2, RE3, RE4 and RE5. The criteria which samples of water are required to satisfy are set out, for ease of reference, in the table opposite.

River Quality Objectives (RQO). A series of measures which ensure that river quality is checked directly against all the quality standards needed to support the following uses.

- Abstraction for public water supply.
- Salmonid fishery.
- Cyprinid fishery.
- Amenity and conservation.
- Abstraction for industrial water supply.
- Spray irrigation of field crops.
- Livestock watering.

The determinands most often involved in the decision-making process are DISSOLVED OXYGEN, BIOCHEMICAL OXYGEN DEMAND, and AMMONIA. The impact of other substances, for example metals and pesticides, is also assessed against the standards set down in the River Quality Objectives. These substances are also prominent in several of the directives issued by the European Community. (◊ WATER QUALITY STANDARDS; ◊ RIVER ECOSYSTEM CLASSIFICATION)

River regulation. Upland reservoirs in the upper reaches of the river are used to regulate the flow and especially to increase it at times of low run-off. Abstraction is carried out in the lower reaches close to the main areas of demand – the yield can be much greater than if water were taken from the reservoir directly. River regulation is a common means of augmenting water supplies.

Occasionally AQUIFERS are used as extra sources of water to augment the flow of a river in dry weather. The conjunctive use of aquifers and rivers has resulted in increased yields of water from the Thames. (◊ WATER SUPPLY)

RIVPACS. The River Invertebrate Prediction and Classification Scheme. This is a computerized model which allows predictions to be made of the type of invertebrate community that would be expected in a river according to a range of natural features, assuming the river is not affected by pollution. The natural features include distance from the source of the river, channel width and depth, gradient of the river, altitude, velocity of flow, alkalinity, temperature, and the nature of the river bed.

Criteria for River Ecosystem Classification.

Class	Dissolved oxygen (% saturation) 10 percentile	BOD (ATU) (mg/l) 90 percentile	Total ammonia (mg N/l) 90 percentile	Un-ionized ammonia (mg N/l) 95 percentile	pH lower limit as 5 percentile; upper limit as 95 percentile	Hardness (mg/l $CaCO_3$)	Dissolved copper (μg/l) 95 percentile	Total zinc (μg/l) 95 percentile
RE1	80	2.5	0.25	0.021	6.0–9.0	≤10	5	30
						>10 and ≤50	22	200
						>50 and ≤100	40	300
						>100	112	500
RE2	70	4.0	0.6	0.021	6.0–9.0	≤10	5	30
						>10 and ≤50	22	200
						>50 and ≤100	40	300
						>100	112	500
RE3	60	6.0	1.3	0.021	6.0–9.0	≤10	5	300
						>10 and ≤50	22	700
						>50 and ≤100	40	1000
						>100	112	2000
RE4	50	8.0	2.5	—	6.0–9.0	≤10	5	300
						>10 and ≤50	22	700
						>50 and ≤100	40	1000
						>100	112	2000
RE5	20	15.0	9.0	—	—	—	—	—

RME (rape methyl ester). ♦ BIODIESEL.

Road traffic noise. The disturbing features of traffic noise are its general level and its variability with time. The latter refers to short-term variations due to the passage of individual vehicles, and to longer period variations at different times of day due to the general changes in traffic flow. It has been found that the dissatisfaction expressed by occupants of dwellings varies in accordance with the peak noise levels. Hence an index known as the '10 per cent level' (L_{A10}) is used. L_{A10} is the level of noise in dB(A) exceeded for just 10 per cent of the time. For the measurement and prediction of noise from traffic, the average of L_{A10} values for each hour between 6 a.m. and 12 midnight on a normal weekday has been recommended in the UK as giving satisfactory correlation with dissatisfaction. This is known as L_{A10} (18 hours), and permits accurate predictions for design purposes. Some values of L_{A10} (18 hours) and typical conditions in which they are experienced are listed below.

$dB(A)$	Situation
80	At 60 feet from the edge of a busy motorway carrying many heavy vehicles, average traffic speed 60 mph, intervening ground grassed.
70	At 60 feet from the edge of a busy main road through a residential area, average traffic speed 30 mph, intervening ground paved.
60	On a residential road parallel to a busy main road and screened by houses from the main road traffic.

(♦ AIRCRAFT NOISE; DECIBEL; HEARING; INDUSTRIAL NOISE MEASUREMENT; NOISE; NOISE INDICES)

Roentgen. A unit of exposure to radiation based on the capacity to cause ionization. It is equal to 2.58×10^4 coulomb per kg in air. Generally an exposure of 1 roentgen will result in an absorbed dose in tissue of about 1 rad. (♦ REM; IONIZING RADIATION (DOSE MEASUREMENT))

Rotary kiln. A slowly revolving drum, lined with refractory material and fired by gas or oil, used in the processing of ores and cement manufacture. Rotary kilns can also be used to incinerate sewage sludge and industrial wastes to obtain a high degree of BURN OUT. A recent recycling innovation is to use domestic refuse as part of the fuel supply for cement manufacture in rotary kilns as the needs of refuse combustion and cement manufacture complement each other.

ENEAL incineration process guaranteed pollutant emissions.

Emission	mg/m^3
Dust	<30
NO_x	<35
SO_2	<100
CO	<30
HCN	<0.05
HCl	<2
Total aromatic hydrocarbons	<30
Heavy metals	<50
TCDD + PCDD + TCDF + PCDF	<0.006
PCB	<0.1
O_2	10%

* Concentrations to standard temperature and pressure conditions.

One innovative use of rotary kilns adopted is the 'ENEAL' incineration process where whole scrap tyres (up to 1.2 m diameter) are used as the feed. The overall efficiency (based on the energy value of the scrap tyres) is claimed to be 90 per cent with air pollutant emission levels as set out in the table above. (◊ EFFICIENCY; INCINERATION)

Rotating biological contactor (RBC). A series of slowly rotating parallel disks on a common shaft whose lower halves are immersed in liquid effluent which requires biological treatment. The biological film on the disks is aerated as the disks rotate alternately in the effluent reservoir and out, in contact with the air.

Round Table on Sustainable Development. The UK Round Table on Sustainable Development was established in January 1995 to provide a forum for discussion on major issues of sustainable development. Its main purpose was to identify ways of achieving development in a sustainable manner. In May 1999 the Government published its sustainable development White Paper *A Better Quality of Life*. As well as setting out a fresh strategic framework for policy in this area, it announced the Government's intention to create a new and more powerful Sustainable Development Commission, into which the Round Table and the British Government Panel on Sustainable Development, were to be subsumed. The UK Round Table published its Fifth and final Annual Report in July 2000.

Royal Commission on Environmental Pollution (RCEP). The Royal Commission on Environmental Pollution is an independent standing body established in 1970 to advise the Queen, government, Parliament and the public on environmental issues. The Commission's terms of reference as set out in its warrant are to advise on matters, both national and international, concerning the pollution of the environment; on the adequacy of research in this field; and the future possibilities of danger to the environment. Within this remit the Commission has freedom to consider and advise on any matter it chooses; the government may also request consideration of particular topics. The Commission has interpreted 'pollution' broadly as covering any introduction by humans into the environment of substances or energy liable to cause hazards to human health, harm to living resources and ecological systems, damage to structures or amenity, or interference with legitimate uses of the environment. It now approaches issues within the framework of sustainable development.

The primary role of the Commission is to contribute to policy development in the longer term by providing an authoritative factual basis for policy-making and debate, and setting new policy agendas and priorities. This requires consideration of the economic, ethical and social aspects of an issue as well as the scientific and technological aspects. In reaching its conclusions, the Commission seeks to make a balanced assessment, taking account of the wider implications for society of any measures proposed.

What does the Commission do?

The Commission sees its role as reviewing and anticipating trends and developments in environmental policies, identifying fields where insufficient attention is being given to problems, and recommending action that should be taken.

The Commission's advice is mainly in the form of reports, which are the outcome of major studies. It also makes short statements, generally as news releases, on matters it considers of special importance or which arise out of studies. The First Report of the Commission made it clear that: 'We do not have the competence or the resources to act as environmental ombudsman, dealing with appeals against local or central government decisions about specific cases of alleged damage to the environment where there are already channels through which such appeals may be made; what we are able to do is to give advice on the general principles which should guide Parliament and public opinion.'

Although funded by the Department for Environment, Food and Rural Affairs, the Royal Commission is independent of government Departments. The Commission maintains links with government Departments, Parliamentary committees, pollution control agencies, research organizations, industry and environmental groups. It has an annual budget of around £900 000 (2006 figure).

Commission Members

The Members of the Royal Commission on Environmental Pollution are drawn from a variety of backgrounds in academia, industry and public life. They are appointed by the Queen on the advice of the Prime Minister. Contributing a wide range of expertise and experience in science, medicine, engineering, law, economics and business, Members serve part-time and as individuals, not as representatives of organizations or professions. The term of appointment is three years but Members may be reappointed. They are required to declare any interests which may conflict with their role as Commission Members.

The Secretariat

A full-time secretariat – the Secretary to the Commission, two Assistant Secretaries (one a scientist) and eight support staff (two of them scientists) – supports the Chairman and Members by arranging, preparing papers for, and recording meetings; by handling the Commission's finances, administration and correspondence; and by drafting and producing the Commission's reports.

Commission meetings

The Commission normally meets for $1\frac{1}{2}$–2 days a month. Additionally, smaller groups of members may meet to take forward particular aspects of studies. From January 1998 onwards, the minutes of Commission meetings have been made publicly available. The Commission's advice is mainly in the form of reports which are the outcome of studies. The current study (2007) is on the urban environment. The next study the Commission will be undertaking is on novel materials. The Commission is currently consulting on the scope of this study. The previous study on UK fisheries was completed in December 2004. The Commission's 24th report on chemicals was launched in June 2003. Prior to that, the planning report was launched in March 2002. In June 2000, the Commission published its report on energy (*Energy – The Changing Environment*). The Commission have also produced a summary booklet of the 21st report, the full report of which was launched in 1998.

The Commission has also undertaken special studies on specific topics and a report on crop spraying and the health of residents and bystanders was published on 22nd September 2005. The previous special study was on the use of biomass as an energy source and was completed in May 2004. The first special study was on aviation and was published on 29 November 2002.

Source: www.rcep.org.uk. Accessed 15 January 2007.

The Reports produced by the RCEP are listed below:

First Report	February 1971	Key Tasks for Attacking Pollution
Second Report	March 1972	Three Issues in Industrial Pollution
Third Report	September 1972	Pollution in some British Estuaries and Coastal Waters
Fourth Report	December 1974	Pollution Control: Progress and Problems
Fifth Report	January 1976	Air Pollution Control: an Integrated Approach
Sixth Report	September 1976	Nuclear Power and the Environment
Seventh Report	September 1979	Agriculture and Pollution
Eighth Report	October 1981	Oil Pollution of the Sea
Ninth Report	April 1983	Lead in the Environment
Tenth Report	February 1984	Tackling Pollution – Experience and Prospects
Eleventh Report	December 1985	Managing Waste: The Duty of Care
Twelfth Report	February 1988	Best Practicable Environmental Option
Thirteenth Report	July 1989	The Release of Genetically Engineered Organisms to the Environment
Fourteenth Report	June 1991	GENHAZ – A System for the Critical Appraisal of Proposals to Release Genetically Modified Organisms into the Environment
Fifteenth Report	September 1991	Emissions from Heavy Duty Diesel Vehicles
Sixteenth Report	June 1992	Freshwater Quality
Seventeenth Report	May 1993	Incineration of Waste
Eighteenth Report	October 1994	Transport and the Environment
Nineteenth Report	February 1996	Sustainable Use of Soil
Twentieth Report	September 1997	Transport and the Environment – Developments since 1994
Twenty-first Report	October 1998	Setting Environmental Standards
Twenty-second Report	June 2000	Energy – The Changing Climate
Twenty-third Report	March 2002	Environmental Planning
Special Report	November 2002	The Environmental Effects of Civil Aircraft in Flight
Twenty-fourth Report	June 2003	Chemicals in Products
Special Report	May 2004	The Use of Biomass for Heat and Power Production
Twenty-fifth Report	December 2004	Turning the Tide: Addressing the Impact of Fisheries on the Marine Environment
Special Report	September 2005	Crop Spraying and the Health of Residents and Bystanders
Twenty-sixth Report	March 2007	The Urban Environment

Information about the current work of the Royal Commission can be obtained from http://www.rcep.org.uk

The 18th Report (1994) is on transport and the environment and contains key recommendations for a sustainable transport policy. This implies major emissions reduction from road transport sources.

Royal Commission Standard. The treated sewage effluent standards proposed by the UK Royal Commission on Sewage Disposal (1898–1915) were no more than 30 mg/l SUSPENDED SOLIDS and 20 mg/l BOD, i.e. a 30/20 effluent. The Commission expected that the effluent would be diluted with eight volumes of clean river water of BOD 2 mg/l and hence no undue pollution of the receiving stream would result. Other discharge restrictions can be placed on HEAVY METALS, CYANIDES, PHENOLS, AMMONIA, etc. Where a limit for ammonia is specified, this is usually set at 10 mg/l. Hence, the Royal Commission Standard is often quoted as 30/20/10.

Effluent standards should be determined on the basis of the particular receiving river, but 30/20 is often the norm which is applied.

Run-off. The volume of water derived from snow or rain falling on a surface and which does not permeate into the soil.

Rylands v. *Fletcher* **(1968).** Classic law case which established that the liability for the consequences of non-natural or special operations on land resides with the owner of the land, i.e. if the use to which the land is put carries an increased danger to others, then the owner is liable for the consequences. In *Rylands* v. *Fletcher*, an escape of water from a reservoir flooded an adjacent mine. The ruling made stated

> In particular the rule states that anyone who brings or collects and keeps on his land anything likely to do mischief if it escapes must keep it at his peril and if he does not do so is *prima-facie* strictly liable for all that damage which is the natural consequence of its escape, and the defendant's use of the land must be non-natural.

This ruling could have consequences for all who pollute knowing that they do so or who could have foreseen that their operations could do so.

Cambridge Water Company v. Eastern Counties Leather

The decision in the case of *Cambridge Water Company* v. *Eastern Counties Leather* (water pollution from solvent spillage) is an important one in relation to potential liability in common law for environmental damage. The use of premises as a tannery was held to be a 'non-natural user of land' within the rule in *Rylands* v. *Fletcher*. As such, it was generally thought that liability would be 'strict', i.e. that it would not be necessary to prove fault, but only a causal link and that the damage suffered was not too remote. In the event the House of Lords treated the claim under the general heading of nuisance and held that it was necessary also to show that the damage suffered was forseeable in order to establish liability. Since at the time of the spillages of solvent in the Eastern Counties Leather premises it was not forseeable that the level of contamination in Cambridge Water's borehole would be in breach of some standards yet to be formulated, Eastern Counties Leather was held not to be liable at common law. However, it should be noted that Eastern Counties Leather subsequently found themselves in trouble with the National Rivers Authority acting under their statutory powers contained in the Water Resources Act 1991 and agreed to carry out remediation works.

Source: David Cuckson, Head of Environmental Law Group, London. (◊ DUTY OF CARE; LANDFILL GAS; NUISANCE)

S

Safe yield[1]. In water supply, the long-term rate at which water can be extracted from an AQUIFER without a continuing progressive decline in its water level or other adverse effects.

Safe yield[2]. In fisheries, the allowable catch which will not endanger the breeding stock. (◊ MAXIMUM SUSTAINABLE YIELD)

Safety and Environmental Risk Management (SERM). The SERM rating methodology looks at the risks from environmental issues in businesses, assesses other management issues, and comes up with a factor which is the residual RISK. This is converted into a score (which city analysts can understand), i. e. a credit rating for the environment. Hence, a company which routinely deals with inherently hazardous situations, but has a well-developed system for managing them, may gain a higher rating than a company which appears to be 'harmless' but has not developed adequate management systems.

   The concept of credit ratings to the field of safety and environmental risk management has been launched by a risk rating body, the SERM Rating Agency.

Source: *Industrial Environmental Management*, March 1998, p. 28.

Saffir–Simpson scale. A measure of the intensity of hurricanes that is based on the nature of the damage they produce. It ranges from category 1 (weak hurricane) to the very rare category 5 (devastating hurricane).

Salinity. Total amount of dissolved material expressed in terms of kilograms of material per million kilograms of feedwater, i.e. parts per million (ppm) of total dissolved solids.

Typical sea water has a salinity of 35000 ppm of which 30000 ppm is common salt (NaCl). Accepted potable water standards are a maximum of 250 ppm as salt and a total dissolved solids content of 500 ppm; these figures should preferably be lower. Bearing this in mind, if a desalting process can sustain only a 90 per cent separation of total dissolved solids from a feed of 35000 ppm, then the product water will have a dissolved solids content of 3500 ppm. Water of this salinity is obviously not potable. This restriction applies to both electrodialysis and reverse osmosis but not to distillation, as product purity can range from 1 to 100 ppm from a feed of 35000 ppm. (◊ DESALINATION)

The salinity of sample waters is given below:

Source	Salinity – total dissolved solids (ppm)
Potable well water	300–500
Approximate limit for irrigation	1000
Brackish well water	1500–6000
Water from the Baltic Sea	2000–3000
Water from the Arabian Gulf	44000
Typical sea water	35000

Salmonella. A strain of bacteria which can cause life-threatening dehydration in the very young or elderly by virtue of an electrolyte imbalance. In the UK news (23–25 June 2006) by virtue of the discovery of a rare strain of *Salmonella montevideo* (responsible for at least 58 people being severely ill) in Cadbury's chocolate bars. An absolute link has not been proven. However, this illustrates the views from the respective ends of the telescope. Cadbury's 'The level we found was so incredibly low that we decided not to inform the Food Standards Agency'. (Sunday Times 25/06/06) and the health professionals 'The acceptable level of salmonella in food is zero'.

Source: Prof. Hugh Pennington, BBC 2, 23 June 2006.

A million chocolate bars were destroyed thanks to a leaking wastewater pipe (◊ *E. COLI*).

Salmonid. Types of fish related to salmon (e.g. trout).

Salt. A compound which results from the replacement of the hydrogen ATOMS of an ACID, usually by metal atoms, to form, for example, chlorides (NaCl), sulphates ($CaSO_4$), and carbonates ($CaCO_3$).

Sample. A part of a population selected with the object of estimating some characteristic of the whole population. Can be random or spot. A *random sample* is a sample selected in such a way that all possible samples of the same size have the same chance of being chosen. A *spot sample* ('grab sample') is a sample of air, effluent, etc. collected over a short period of time and usually taken to a central laboratory for analysis.

Sampler, personal. A device attached to a person that samples air in the immediate vicinity so that personal exposure to pollutants may be determined.

Sampling, isokinetic. The taking of a sample of flowing gas (particularly gas flowing through a duct) in such a way that the sample does not undergo any change in either velocity or direction at the inlet of the probe. Used in taking samples of stack gases for measurement of their dust concentration.

Sampling, particulate. PARTICULATE sampling is dominated by two well-established methods. The reference test method (BS 3405) is to undertake isokinetic sampling and to determine the particulate burden in the gas by weighing filters. The most common continuous method is to shine a light beam across the duct or chimney and to measure the reduction in light intensity (opacity measurement). An alternative technique is to insert a probe into the gas stream and to monitor the particles by the charge transfer that occurs between particles and probe (the triboelectric effect); see Figure 173. (◊ SAMPLING, ISOKINETIC) Figure 174 gives the range of suitability for the different technologies.

Many gases, such as CARBON DIOXIDE, CARBON MONOXIDE and OXYGEN, present no particular problems for either sampling or analysis. Infra-red and other optical or electrochemical techniques of measurement have been widely used for many years. However, there are other gases which for such reasons as having a rather high DEW POINT (which causes condensation), being highly reactive or because they strongly absorb onto the surfaces of the sampling system, present great difficulties. HYDROGEN FLUORIDE (a pollutant of particular concern in the ceramic and brick industries) is such a gas which because of its high chemical reactivity, is especially difficult to sample and measure. An alternative route has been developed based on the pollutant removal from the gas stream into a liquid as early as possible in the sampling system, and subsequently analysed using the well-established technique of selective ion electrode. Figure 175 gives the flow diagram of the ETIS project fluoride analyser.

Sanitary landfill. ◊ CONTROLLED TIPPING.

Saturation (air). At a given temperature and (total) pressure, air is said to be saturated when the water contained in it is at the water vapour pressure which would occur when the water vapour is in contact with a free water surface at that temperature and total pressure.

Saturation (in soil). Soils that are charged with water to the fullest extent possible.

Saturation (solution). A liquid which can dissolve no more soluble solid (or occasionally a gas) in its bulk. Raising the TEMPERATURE permits a greater CONCENTRATION of solids to be dissolved but reduces that of gases.

Saturation zone. That portion of an AQUIFER which exists below the WATER TABLE, i.e. groundwater fills all the pore spaces and voids.

Scaling. The peeling off of oxides/corrosion products in the form of scales thus allowing a fresh metallic or mineral face for further attack.

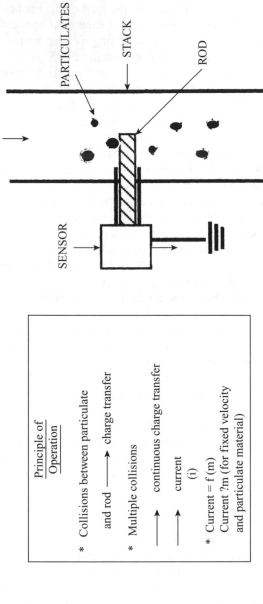

Figure 173   Particle impingement, principle of operation.

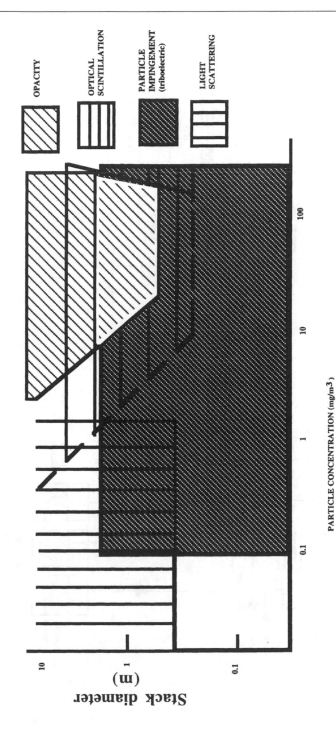

Figure 174   Suitability of dust emission technologies for different applications.

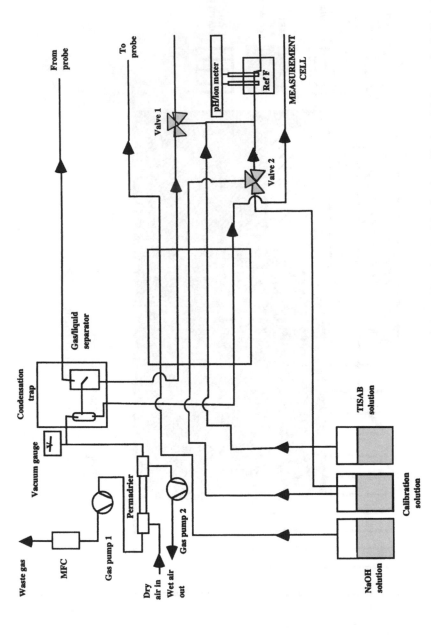

Figure 175 Flow diagram of fluoride analyser. *Source:* ETIS Seminar, 'Continuous Monitoring of Industrial Atmospheric Emissions', Kirton, Notts., 29–30 September 1993, from contributions respectively authorized by C. J. Regan, J. Fifer and W. Averdieck.

Scheduled Process. Processes listed under the UK Alkali etc. Works Regulation Act 1906 (termed 'works' in the Act) along with a list of 'noxious and offensive gases'. Later additions were consolidated in the Alkali etc. Works Order 1966. These were subscribed in the Control of Industrial Air Pollution (Registration of Works) Regulations 1989. (Renamed PRESCRIBED PROCESS)

A historical listing of scheduled processes (1986) is given in the following table.

Scheduled processes.

Scheduled Process	Substance controlled
Alkali works	HCl
Aluminium	Total fluoride
Amines	Total amine
	Trimethylamine
Ammonia	NH_3
Arsenic	As_2O_3
Beryllium	Be
Bisulphite	Total acidity
Cadmium	Cd
Cement	Particulate matter
	H_2S
Ceramics	Particulate matter
Chemical fertilizer	Total acidity
	HCl
Chemical incineration	HCl
	H_2S
	Particulate matter
Chlorine	Hg
	Chlorine
	HCl
Copper	Particulate matter
	Fume
	Cu
	Zn
	Pb
Diisocyanate	Diisocyanate
Electricity	Particulate matter
Gas and coke	Particulate matter
Hydrochloric acid	HCl
Hydrofluoric acid	Total gaseous F
	Total acidity
	Particulate matter
Iron and steel	

Scheduled Process	Substance controlled
Lead	Pb
	Particulate matter
Lime	Particulate matter
Metal recovery	HCl
	Particulate matter
Mineral	Particulate matter
Nitric acid	Total acidity
	NO_x
Petroleum	S
	H_2S
	Particulate matter
Phosphorus	P_2O_5
Sulphide	H_2S
Sulphuric acid	Sulphur
Tar	Tar
	H_2S
	Particulate matter
Vinyl chloride	Vinyl chloride

Discharge Authorizations are now used for all scheduled processes. They contain the following items:

1. The name and address of the owner of the works.
2. The location of the works and the name of the local authority.
3. The category of the works and the process description.
4. Details of the type and capacity of the abatement measures and chimneys.
5. Sources, types and volumes and presumptive limits of process emissions.
6. Details of continuous, manual or environmental monitoring.

With the implementation of the UK Environmental Protection Act 1990 INTEGRATED POLLUTION CONTROL is applied to PRESCRIBED PROCESSES by the Environment Agency, and by local authorities to less complex processes. The likely numbers are 5000 complex processes (EA) and 27 000 others (LAs).

Schistosomiasis (bilharzia). The disease caused by the worms *Schistosoma haematobium* and *Schistosoma mansoni*, which use human beings as a primary host and snails as secondary hosts. Irrigation ditches make ideal transmission networks for the dispersion of these snails. (◊ DAM PROJECTS)

Scintillation counter. A radiation detector in which the radiations cause individual flashes of light in a solid (or liquid) 'scintillator' material. Their intensity is related to the energy of the radiation. The flashes are amplified and measured electrically, and displayed or recorded digitally as individual 'counts'.

Scottish Environment Protection Agency. ◊ ENVIRONMENT AGENCY.

Scrap. Discarded material from manufacturing, and processing or remnants after an article's useful life has run out. (◊ RECYCLING; DOMESTIC REFUSE)

Scrapyard contamination. Since 1 April 1995, many UK Metals Recovery Sites (scrapyards and vehicle dismantlers) have been licensed under the Environmental Protection Act 1990 (EPA). Under this scheme, an operator wishing to return a Waste Management Licence can do so if it is 'unlikely that the condition of the land will cause pollution to the environment or harm to human health'.

The pollution potential of scrapyards and their associated activities is poorly understood and often underestimated. This is further complicated by the fact that many scrapyards occupy land subjected to other contaminating uses. The level of contamination on sites can be highlighted by one scrapyard in Leicestershire which has undergone an extensive investigation pending redevelopment (Site X). This investigation involved sampling 86 boreholes between 1.5 m and 9 m below ground level.

The levels of contamination found were compared to guidelines issued by the Interdepartmental Committee on the Redevelopment of Contaminated Land (ICRCL) (see table below). The results show that contamination is both significant and widespread. For example, copper and lead levels detected at the site were extreme.

The site illustrates that when contamination is looked for at long-established scrapyard sites, it can be found, though not always easily. The authors commented that: 'whether all or the most important pollutants were found is hard to know, but there is no reason to believe that the Leicestershire site is in any way unusual'.

This is an interesting facet of one form of RECYCLING.

Element concentrations (ppm) found within Site X compared to ICRCL values.

	Site X	ICRCL values	
		Contaminated land	Heavily contaminated land
Copper	500–6000	200–500	500–2500
Lead	400–4000	1000–20000	2000–10000
Nickel	50–800	50–200	200–1000
Phenols	40–440	4–50	50–250

Source: D. Hague and J. Knowles, 'Scrapyard contamination and site investigation', *Waste Management*, November 1995, p. 55.

Screen[1]. A device for size separation, e.g. screening of DOMESTIC REFUSE to +200 mm and −200 mm particle size. Or crushed road stone for quarry products.

Screen[2]. An array of fixed or moving bars used to remove large solids from sewage before treatment.

Screenings. The product from SCREENS. In sewage treatment, screenings range from 0.01 to 0.03 m^3/d per 1000 population.

Scrubber. Device for flue gas cleaning, such as spray towers, packed scrubbers and jet scrubbers. The gas is passed through wetted packing or a spray to ensure intimate contact with the scrubbing water. This removes particles down to 1 μm in diameter. The gas flow is usually countercurrent to the water flow. Scrubbers for particulate control produce SLUDGE that requires dewatering and disposal, as well as contaminated effluent which may be recirculated or treated before discharge. Scrubbers may also serve to control gaseous pollutants, in which case an alkaline solution is used. The table below gives details of the emissions from the Energy-from-Waste Plant in Portsmouth.

Average Daily Emissions from the Veolia Energy-from-Waste Plant, Portsmouth, in December 2007 (all values in mg/Nm^3).

Volatile organic compounds	<1
Active dust	<1
Ammonia	1.8
Carbon monoxide	5.3
HCl	8.2
SO$_2$	12.5
NO$_x$	182.4

Source: www.veoliaenvironmentalservices.co.uk/hampshire/pages/pdfs/portsmouth_2007_dec.pdf. Accessed 3 January 2008.

Scum. A layer of fats, oils, soaps, etc. which can collect on the surface of polluted lakes, rivers and the SEDIMENTATION tanks of sewage treatment plants.

Sea water. ◊ SALINITY.

Second law of thermodynamics. ◊ LAWS OF THERMODYNAMICS.

Secondary air. The air introduced above (and beyond) the bed of burning fuel or waste to promote the complete combustion of volatile materials which are released after the first stage of combustion.

Secondary liquid fuel (SLF). A blend of organic chemicals no longer suitable for recovery/original use. Substantial commercial interests are now pushing for its use in CEMENT KILNS which are substantial consumers of FUEL, e.g. a kiln input requirement of 127 MW is normal. Hence, a substantial replacement fuel market exists if the price is right and environmental standards are met, comparable to those of HAZARDOUS WASTE INCINERATION. Typical UK proposals are for up to 25 per cent of the heat input to be met by SLF, one brand of which is marketed as CEMFUEL™.

Secondary materials. Materials which have fulfilled their primary function and which cannot be used further in their present form. Also, materials which occur as by-products from the manufacturing or conversion of primary products.

Secondary pollutant. Formed from a PRIMARY POLLUTANT as a result of chemical changes such as photochemical and other reactions. Examples are OZONE and NO_2. ($\Diamond$ PRIMARY POLLUTANT; NITROGEN OXIDES)

Secondary treatment. Treatment of wastewater by a process generally involving biological treatment with a secondary settlement stage. Capable of producing a substantial reduction in BOD and suspended solids.

Sediment. The deposit of silt and accumulated organic and/or inorganic materials at the bottom of rivers, lakes, seas, etc. Sediments act as sinks for PESTICIDES and can contain concentrations as much as 800 times that of water in the case of DIELDRIN. They are therefore a source of secondary contamination, as well as allowing the bottom feeders (e.g. invertebrates) to accumulate the pollutants at much greater concentrations than water analysis would indicate, thereby contaminating the FOOD CHAIN. ($\Diamond$ MINAMATA DISEASE)

Sedimentation. The settling out of 'SETTLEABLE' SOLIDS from sewage in sedimentation tanks which allow very slow rates of flow, so that settling takes place.

Selective catalytic reduction (SCR). This is a process for reducing the NO_x content in flue gas by heating the scrubbed flue gas with a natural gas burner and then passing the hot gas with ammonia over a catalyst. The NO_x is reduced to N_2 gas and water. Using cleaned gas enhances the reaction, and ensures a longer catalyst life. Catalysts such as copper oxide (CuO) or titanium dioxide (TiO_2) are used. The latter is preferred as it is sulphur-resistant. Figure 176 shows how the system is positioned. The equation for the reaction is:

$$4NO + 4NH_3 + O_2 \Rightarrow 4N_2 + 6H_2O$$

Selective non-catalytic reduction (SNCR). In selective non-catalytic reduction, the NO_x is reduced to N_2 gas and water by injecting ammonia into the furnace (at 800–1000 °C) in the presence of oxygen (already there due to excess air being supplied). The reaction is slower than in SCR (see above), as no catalyst is used.

Selectivity[1]. The characteristic of radiation, toxic chemicals or heavy metals (if ingested) to affect certain organs of the body to a much greater degree than the whole body dose would indicate. ($\Diamond$ MINAMATA DISEASE; HALF-LIFE)

Selectivity[2]. The characteristic of plants and INSECTS to adapt to their environment by means of genetic selection of the most favourable strains.

Selenium (Se). Often referred to as a toxic metal, selenium is not, in fact, a metal but has certain metallic properties. It is a member of the sulphur group. It is produced as a by-product of the refining of copper, nickel, gold and silver ores. It is used extensively in the electronics industry, and also in paints and

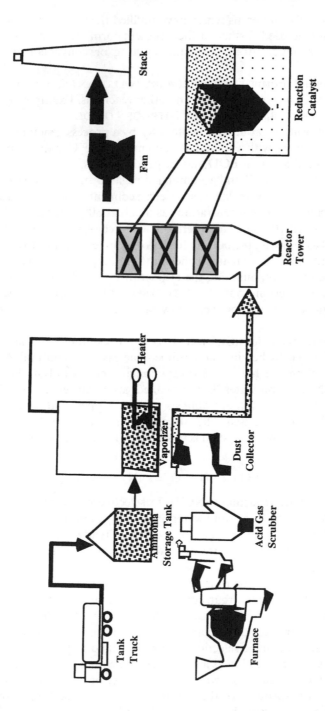

Figure 176 Selective catalytic NO$_x$ reduction system.

rubber compounds. It is a micro-nutrient at levels of 0.02 to 1 part per million. Maximum allowable concentration in drinking water is $10\,\mu g l^{-1}$. It can act as a systemic poison. The LD_{50} for one selenium compound is as low as 4 micrograms per kilogram body weight. Known cases of poisoning are rare. ($\lozenge$ LETHAL DOSE)

Settleable solids. Suspended solids which will settle out of an effluent (e.g. SEWAGE) in two hours when stationary.

Settling chamber. A chamber inserted between a furnace and its stack in which coarse particulate matter settles out of the gas stream. Also used in SEWAGE TREATMENT for the separation of grit and other solids before further treatment.

Seveso. $\lozenge$ DIOXIN.

Sewage. The liquid wastes from a community. Domestic sewage is from housing. Industrial sewage is normally from mixed industrial and residential areas. ($\lozenge$ SEWAGE TREATMENT)

Sewage treatment. The reduction of the ORGANIC loading that raw SEWAGE would impose on discharge to streams and watercourses. It is carried out by oxidation of the sewage, i.e. contact with air, which oxidizes most of the wastes to allow discharge. Several steps are required:

1. Preliminary screening to remove large suspended solids, metal and rags.
2. Grit removal.
3. Sedimentation to allow as much suspended organic solids as possible to settle.
4. Biological oxidation in either of two main types of plant:
   (a) Biological or percolating filter, i.e. a packed bed of clinker, stones, or plastic media, 2 metres deep, through which the sewage trickles and is biodegraded by microorganisms; or
   (b) ACTIVATED SLUDGE, in which the sewage is aerated in tanks by agitators or diffusers which maintain the level of dissolved oxygen as high as possible. Some sludge is recycled to seed the raw sewage and allow treatment to proceed faster.

   In both cases the aerobic bacteria are able to grow and the end result is sludge and an effluent which should be very low in biochemical oxygen demand (20 milligrams per litre or less) unless the plant is overloaded. Discharge of the effluent, provided there is at least an eight-fold dilution in the river, should not normally cause any problems following Step 5 below (and if necessary, Step 6 in cases where there are strict discharge requirements).
5. Sedimentation in sedimentation tanks where the discharge from biological oxidation has most of its residual suspended solids removed. The effluent flows upwards through the tank at very low velocities (1 to 2 metres per hour) so allowing the suspended solids to be removed as sludge, which is

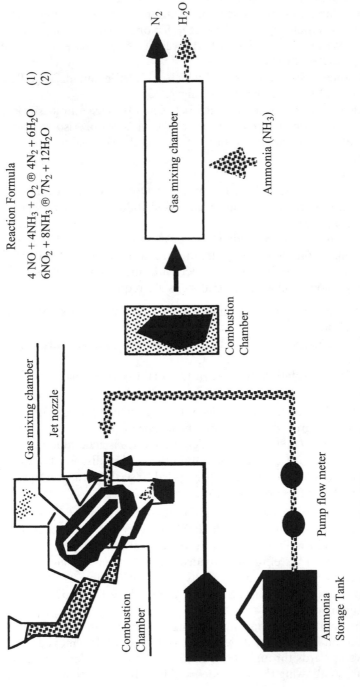

Reaction Formula

$4\,NO + 4NH_3 + O_2 \,®\, 4N_2 + 6H_2O$ (1)

$6NO_2 + 8NH_3 \,®\, 7N_2 + 12H_2O$ (2)

N_2

H_2O

Ammonia (NH_3)

Gas mixing chamber

Combustion Chamber

Gas mixing chamber

Jet nozzle

Combustion Chamber

Pump flow meter

Ammonia Storage Tank

Figure 177   Selective non-catalytic NO_x reduction system.

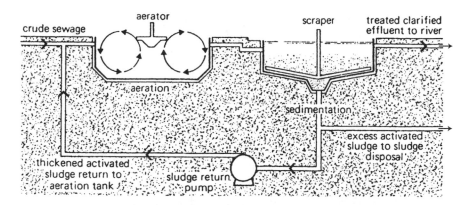

Figure 178 Activated sludge effluent treatment process.

then disposed of. For the activated sludge process, a portion is recycled as shown in Figure 178.

6. Tertiary treatment. If the treated effluent is of inadequate quality (e.g. too high in suspended solids) it can be micro-strained in special filters. Tertiary treatment also consists of other filters, lagoons and the use of REVERSE OSMOSIS. Tertiary treatment will grow in use as higher DIS-CHARGE STANDARDS are applied to reduce or contain the pollution of inland rivers from sewage treatment plant effluents sand in particular achieve a reduction in nitrate levels.

(◊ AEROBIC PROCESSES; BIOCHEMICAL OXYGEN DEMAND; ROYAL COMMIS-SION STANDARD; SLUDGE, SEWAGE)

Sewage treatment, deepshaft process. This technique was developed by ICI and is claimed to cut the costs of SEWAGE treatment by up to 50 per cent. It relies on a deep shaft (ap 30–150 m deep; 0.7–6.0 m in diameter) into which sewage and biodegradable effluents are admitted as shown in Figure 179. The mixture is circulated by air injection, and the mixing processes together with the increased pressure, enable the aerobic bacteria to reduce the biochemical oxygen demand of the wastes at rates of up to 10 times that of conventional sewage treatment plants. A baffle arrangement keeps untreated and treated effluent separate. The method still relies on aerobic bacteria but with an increased intensity of operation, thus allowing plants to be greatly reduced in area.

This plant was an offshoot of aerobic fermentation research. (◊ BIODEG-RADATION; BIOREACTOR)

Sewerage. A network of pipes and associated appliances for the collection and transport of domestic and industrial SEWAGE.

Shadow capacity. The standby reserve (fossil fuel) power station requirement needed to back up wind energy plants when they are standing idle. Currently

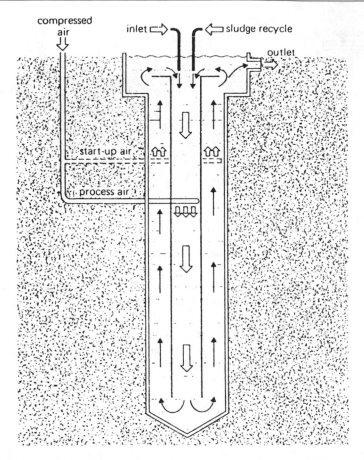

Figure 179   Deepshaft effluent treatment process (as developed by ICI).

wind energy in the UK is theoretically capable of supplying $1\frac{1}{2}$ per cent of the UK's electrical power demands. However, if the 2010, 10 per cent renewables target is to be reached, substantial reserve plant will be required. Recall that land-based wind energy, so far has produced electricity for 10–25 per cent of the year. Further recall that during the Christmas period of 2006, there was no usable wind for 7–10 days when power demands were at their peak. (◊ CAVEAT EMPTOR)

Shadow flicker. Under certain combinations of geographical position and time of day, the sun may pass behind the blades of a wind turbine and cast a shadow. When the blades rotate the shadow flickers on and off. The effect only occurs inside buildings where the flicker appears through a window opening. The seasonal duration of this effect can be calculated from the geometry of the machine and the latitude of the site. (◊ WIND ENERGY)

Shear. A machine used to reduce scrap metal materials to a small size for handling, transportation or smelting purposes. ($\diamond$ SHREDDER)

Shift reaction. Commonly refers to the reaction of CO with H_2O at high temperature (with the aid of a catalyst) to produce H_2 and CO_2. Seen as a means of capturing CO_2 for Carbon Capture and Storage (CCS).

Short rotation coppicing (SRC). SRC mainly consists of high-yielding varieties of willow which are planted at high density and harvested regularly: every three years on high quality arable or grass land, every five years on the poorer soils of reclaimed industrial sites.

The chipped product (up to 12 t/ha) may then be gasified for electrical POWER and/or DISTRICT HEATING. This is RENEWABLE ENERGY in action but (so far) it needs to be subsidised *vis-à-vis* conventional power generation. ($\diamond$ GASIFICATION)

Short-term exposure limit (STEL). $\diamond$ TIME-WEIGHTED AVERAGE.

Shredder. A machine used to break up materials into smaller pieces by tearing and/or impact.

Sick building syndrome. 'Sick building' syndrome is used to describe a situation in which a significant number (more than 20 per cent) of building occupants report illness perceived as being building-related. The complaints are characterized by a range of symptoms including, but not limited to, eye, nose, and throat irritation, dryness of mucous membranes and skin, nosebleeds, skin rash, mental fatigue, headache, cough, hoarseness, wheezing, nausea and dizziness. The introduction of new building materials, decreased ventilation and decreased air leakage have all contributed to the problem.

Sievert (Sv). $\diamond$ IONIZING RADIATION.

Silage. Green leaf cattle food which has had molasses added to promote fermentation and preservation. The liquors from silage production can have BODs up to 300 times that of domestic sewage effluent. They are highly polluting and can be a seasonal cause of fish deaths in small streams.

Silicosis. $\diamond$ FIBROSIS.

Silt. Normally a wet mixture of particles between 4 and 60 μm diameter, often found in the base of sewers and streams. Intermediate between CLAY and MUD.

Single-cell protein (SCP). Protein manufactured by microbial means from organic substances such as OIL, NATURAL GAS, CELLULOSE or SEWAGE. Microbial conversion offers the means of converting wastes into useful protein for both animals and human, because of the rapid rate of metabolism of micro-organisms and their subsequent growth and reproduction. Under suitable conditions many microbes, especially YEASTS, may be maintained in an exponential growth situation in continuous culture with a doubling time of 15–20 minutes. ($\diamond$ EXPONENTIAL CURVE)

The production technology uses aerobic FERMENTATION with copious supplies of air to obtain maximum growth rate. A single strain of organism is chosen because this allows the optimum production of protein from the substrate, e.g. the yeast, *Candida utilis* is used for protein production from molasses, and a similar strain is used on oil. The disadvantage of single strains is that sterility must be maintained.

Case study on Quorn

Quorn is the leading brand of mycoprotein food product (mycoprotein is the generic term for protein-rich foodstuffs made from processed fungus). Quorn is sold as a meat analogue or imitation meat. Produced as both a cooking ingredient and a range of microwave meals, it is marketed at the health-conscious as well as to vegetarians.

A shortage of protein-rich foods by the 1980s was predicted during the 1950s. In response to this, many research programmes were undertaken to utilize single-cell biomass as an animal feed. Contrary to the trend, Lord Rank instructed the RHM Research Centre to investigate converting starch (the waste product of cereal manufacturing undertaken by RHM) into a protein-rich food for human consumption.

Following an extensive screening process, the filamentous fungus *Fusarium venenatum* was isolated as the best candidate. In 1980, RHM was given permission to sell mycoprotein for human consumption after a ten-year evaluation programme – probably making Quorn one of the most tested foods in existence. The initial retail product was produced in 1985 by Marlow Foods (named after RHM's headquarters in Marlow, Buckinghamshire – a joint venture between RHM and Imperial Chemical Industries (ICI) who provided a fermenter left vacant from their abandoned single-cell feed programme. Patents for growing and processing the fungus, and other intellectual properties in the brand, were invested in Marlow by the two partners. Although the food sold well in the initial test market of the RHM staff canteen, the large supermarket chains were unconvinced until David Sainsbury, Baron Sainsbury of Turville, owner of the supermarket Sainsbury, agreed to stock the novel food. Quorn entered widespread distribution in the UK in 1994, and was introduced to other parts of Europe.

Although the mycoprotein was originally conceived as a protein-rich food supplement for the predicted global famine, the food shortage never materialized. In 1989 a survey revealed almost half of the UK population was reducing their intake of red meats and a fifth of young people were vegetarians. As a result, Marlow Foods decided to sell Quorn as a new healthy meat analogue which was free of animal fats and cholesterol. Marlow sells Quorn brand mycoprotein in ready-to-cook forms (as cubes and a form resembling minced meat), and later introduced a range of chilled vegetarian meals based on Quorn. Its range includes pizza, lasagna, cottage pie, and formed Quorn products resembling sliced meat, hotdogs, and burgers.

Quorn is made from the soil mould *Fusarium venenatum* strain PTA-2684. The fungus is grown in continually oxygenated media in large, sterile fermentation tanks. During the growth phase glucose is added as a food for the fungus, as are various vitamins and minerals (to improve the food value of the resulting product). The resulting mycoprotein is then extracted and heat-treated to remove excess levels of ribonucleic acid (RNA). Previous attempts at producing such fermented protein foodstuffs were thwarted by excessive levels of DNA or RNA; without the heat treatment, purine, found in nucleic acids, is metabolized producing uric acid, which can lead to gout.

The product is then dried and mixed with chicken egg albumen, which acts as a binder. It is then textured, giving it some of the grained character of meat, and pressed either into a mince, forms resembling chicken breasts, meatballs, turkey roasts, or into chunks (resembling diced chicken breast). In these forms Quorn has a varying colour and a mild flavour resembling the imitated meat product, and is suitable for use as a replacement for meat in many dishes, such as stews and casserole. The final Quorn product is high in vegetable protein and dietary fibre, and is low in saturated fat and salt. The amount of dietary iron it contains is lower than that of most meats.

Source: http://en.wikipedia.org/wiki/Quorn. Accessed 15 January 2007.

Sink[1]. The source of cooling water or air for power stations or other fuel-conversion devices where reject thermal energy is discharged.

Sink[2]. In air pollution, the receiving area for material removed from the atmosphere, e.g. in PHOTOSYNTHESIS, plants are sinks for carbon dioxide.

Site of Special Scientific Interest (SSSI). A Site of Special Scientific Interest or SSSI is a conservation designation denoting a protected area in the UK. SSSIs are the basic 'building block' of nature conservation legislation and most other legal nature/geological conservation designations are based upon them, including National Nature Reserves, Ramsar Sites, Special Protected Areas, and Special Areas of Conservation.

The process of designating a site as a Special Scientific Interest is called notification; it involves a number of steps, including consultation with the site's owner. If a site passes through this process and becomes a SSSI it is said to have been 'notified'. Sites which are notified due to their biological interest are commonly known as Biological SSSIs, and those which are notified for their geological interest Geological SSSIs. A minority of sites are notified for both their biological and geological interest.

Site licence[1]. Licence issued by the Nuclear Installations Inspectorate covering safety of design, construction, operation and maintenance of facilities at a nuclear site.

Site licence[2]. A waste disposal site licence issued by a Waste Disposal Authority.
(◊ WASTES)

Site Waste Management Plan (SWMP). From 2000, all construction projects
over £250 000 in England and Wales have required a Site Waste Management
Plan (SWMP). This is to help minimize the estimated 13 million tonnes of
completely unused building materials discarded as waste. The SWMP predicts the amount and type of waste that will be produced on a construction
site – and how it could be eliminated, reduced, reused, recycled or disposed
of. The plan is then updated during all stages of the construction process to
record how the waste is managed and to confirm its disposal at a legitimate
site. The aim is to improve resource efficiency by maximising the use of
materials and recovering waste.

Source: Envirowise Newsletter, March 2007.

Sizewell Nuclear Power Station. Site of UK's only PRESSURIZED WATER
REACTOR (PWR) Sizewell 'B'. A public inquiry hinged on the economic case
that was made for its construction. The proponents claimed that power
would be produced at comparable or better costs than those of a coal-fired
power station. This argument was exposed as fatally flawed after building
permission was obtained. In particular, the following factors were either not
examined properly or glossed over: the much lower cost of coal in the future;
under-estimation of the capital costs; alleged massive under-estimates of
decommissioning costs; the prospects for much increased imports of electricity from Scotland; the ready supply of British sector North Sea gas for fuelling high-efficiency and low cost COMBINED HEAT AND POWER stations.

A fitting commentary on the Sizewell 'B' public inquiry is provided by
Professor P. Odell's letter to the *Financial Times* (21 November 1989): 'The
inspector and his economic assessor simply failed to get to grips with the
prospects for nuclear power economics, and produced a report with wrong
conclusions.'

He further stated that on the *Financial Times'* own figure of expected full
generating costs (8p–10p per kWh), even a successfully completed (and completed on time) Sizewell station will cost taxpayers and/or electricity consumers some £400 million a year for every year of its life. Odell claimed it would
be cheaper to abandon construction.

An escalating cost analysis for the proposed Hinkley Point C PWR station
has been provided by Lord Marshall (in retrospect) after abandonment of
the UK's PWR construction programme. The tables opposite show the
'public sector price' and the 'private sector price' for Hinkley Point C
electricity.

The fluctuating fortunes of UK nuclear power are charted in the last table
(note the exclusion of interest during construction and the assumption of 40
year amortization).

'Public sector price' for Hinkley Point C electricity.

Price component	pence/kWh
Price quoted at public inquiry	3.09
Inflation	+0.33
Additional building costs	+0.10
Capitalized company overheads	+0.11
Operating cost increases	+0.68
Interest during construction to be recovered	−1.09
Total	3.22

'Private sector price' for Hinkley Point C electricity.

Price component	pence/kWh
Public sector price	3.22
Adjustment from 8% to 10% rate of return	+0.71
Uncertainties:	
70% availability (instead of 75%)	
Fuel reprocessing and decommissioning uncertainties	+0.54
New basis for calculating profit	+1.03
Increase for 20 year contract (instead of lifetime contract)	+0.75
Total	6.25

Source: 'What went wrong in the UK – Lord Marshall's view', *Nuclear Engineering International*, January 1990.

Lifetime generating costs of Sizewell C twin PWR Station (1993 money values).

Net output	1310 × 2 MWe
Capital cost (excluding interest during construction)	£3520 million
Unit costs	
Capital costs including interest during construction	1.9 pence/kWh
Fuel cost	0.4 pence/kWh
Operations and maintenance costs	0.6 pence/kWh
Decommission cost	0.01 pence/kWh
Total unit cost	2.9 pence/kWh

Principal assumptions: the capital cost of £3520 million excludes £110 million initial fuel charge; 85 per cent lifetime factor; 40 year amortization period/station lifetime; 8 per cent real rate of return on capital; 60 and 54 month construction times for the first and second reactors from first structural concrete to first fuel load; 2 per cent real rate of return on funded provisions.
Source: *The Prospects for Nuclear Power in the UK*, Department of Trade and Industry, HMSO, 1995.

So, it all depends on:

(a) what subsidies are available
(b) how optimistic or pessimistic the analysts are
(c) what is left out (e.g. full decommissioning costs).

Despite all the controversy, Sizewell has run successfully for many years and its contemporaries are a vital component of UK power supplies.

Sources: A. Cottrell, Letter to the *Financial Times*, 7 January 1974. J. T. Edsall, *Environmental Conservation*, vol. 1, no. 1, 1974, p. 32. J. Collier (M.D. Nuclear Electric plc) extract from *Journal of Power and Energy*, vol. 207, 1993: and *Nuclear Electric Briefing Note*, 6 October 1994, Davies, D. R., Bsc, CEng FICE, 'Planning and acquiring the UK's first PWR power station', *Proc. Instn., Civ. Engrs., Civ. Engng.*, Sizewell B Power Station, 1995.

Slag. The non-gaseous waste material formed in a metallurgical furnace; it is largely non-metallic but usually contains some metal. Slag contains as much of the undesirable constituents of the ore as possible and is withdrawn from the furnace in the molten form. Alkaline slags retain much of the sulphur in the fuel and the ore, and the gases issuing from the furnace, may be virtually sulphur free.

Slag is also formed in high-temperature incineration processes where refuse is converted to a molten slag which is removed from the bottom of a special cupola. Steelworks slag is often processed and recycled because of its high ferrous content. Other uses are as road foundations and in sintering plants for the manufacture of lightweight aggregate building blocks.

Slaked lime. A common name for CALCIUM OXIDE to which water has been added. It is used in INCINERATION plants and FLUE GAS DESULPHURIZATION to remove acid gases.

For example, the 400 000 tonnes per year South East London Combined Heat and Power Plant (SELCHP) uses 8000 tonnes per year of calcium oxide which is slaked on site and used in the gas SCRUBBERS to remove SO_x and HCl.

Slick. A floating patch of oil or SCUM on the surface of a body of water.

Sludge. An accumulation of fine solids which have been separated in SEWAGE treatment plants, oil refineries, tankers, etc.

Sludges pose difficult transport and treatment problems as they are typically 5 per cent solids and 95 per cent liquid, and slimy.

Sewage sludge should be treated (digested) to render it inoffensive. (◊ SLUDGE, SEWAGE)

Source: *Wet News*, 21 October 1998.

Sludge drying. Digested sludge cake at Colchester sewage works is conveyed directly from the dewatering plant to the inlet hopper of the dryer. Here, it is extruded into 'spaghetti' and distributed onto a slow moving belt and conveyed through the dryer. Drying gases heated from the burner, are recirculated around the belt at between 80 and 160 °C. Exhaust gases have their heat recovered and are condensed before being returned to the burner chamber for re-heating. Waste gases are scrubbed of any odorous compounds and discharged to atmosphere. The dried product is pelletized and stored for agricultural recycling.

Sludge, sewage. The term for the residue after aerobic treatment of SEWAGE has been completed. It is a slimy, offensive material, with 95 per cent or greater water content.

More and more efficient wastewater treatment plants are being built worldwide. As a result, sewage sludge volumes are on the increase (UK production 1.4 mtpy dry solids (2004)). Its disposal is a major problem and the three main methods are employed:

1. conversion into METHANE gas by digestion, and subsequent use of the residue as a soil dressing;
2. dewatering followed by incineration;
3. spreading on land.

Digestion is an ecologically sound method and is practised in large sewage works where the methane evolved is used to run the plant and heat the digesters (surplus gas may be available for sale). The process renders the sludge free from pathogenic organisms and offensive odours, and reduces by 30 per cent the mass of solid matter to be disposed of. Subsequent solids disposal – by land spreading or dumping – is very much simplified. Incineration can be costly and involve the use of fuel, as opposed to the generation of methane by the digestion process. It is often adopted where the sludge is produced in large volumes, as in inland cities, where metal content may be high. Land spreading is sometimes used in areas where the sludge is relatively free from HEAVY METAL contamination. This may restrict its use on the same land to once in 30 years to prevent accumulation of toxic heavy metals in excessive quantities. The average heavy metal concentration in UK sewage sludge is:

Metal	Concentration (mg/kg dry solids)
Zinc	1820
Copper	613
Nickel	188
Cadmium	29
Lead	550
Chromium	744

These values accord with EC permissible levels, but of course individual analyses are needed before application to land. OCEAN DUMPING was used

by large coastal cities: the sludge was dumped in deep trenches outside the
tidal flow to be diluted by the sea, but this has now been phased out owing
to EC pressure. Incineration in autothermic (self-sustaining) incinerators is
being canvassed as the most likely solution. FLUIDIZED BED COMBUSTION is
also an up and coming method with units designed to burn ca. 15 000 tonnes
of solids per year, extracted from 240 000 tonnes of wet sludge, which is
dewatered to 30 per cent moisture content. This permits autothermic combustion once the combustion process is started. The combustion temperature
is 850 °C which destroys any toxic compounds present. Alkaline scrubbing
of flue gases is employed. (◊ ANAEROBIC DIGESTION)

Sewage sludge incineration case study

The performance achieved by today's state-of-the-art sewage sludge incinerators is given in tables (a)–(c) for Dordrecht (Netherlands) plant.

(a) Limits of emissions.

Emission	Limits	Feb. 96	Jun. 96	Sep. 96	Dec. 96
Particulate	5 mg/m^3	<1	<1	<1	<1
CO	50	14	11	12	14
NO$_x$	70	113	121	102	194
C$_x$H$_y$	10	<2	<2	<1	2.2
SO$_x$	40	<1	<1	<1	1.4
HCl	10	<1	<0.2	0.3	0.5
Hg	50 µg/m^3	9	6	6	5
Cd	50	<2	<2	<3	<2
Heavy metals	1000	<25	<21	<19	<32
Dioxins/furans	0.1 µgTEQ/m^3	0.008	Not measured	<0.001	Not measured

* Denox plant in operation since 1997 resulting in values below 50 mg/Nm^3.

(b) Total emissions (kg/a) compared with release limits set by the authorities.

Emissions	1996 level	Authority release limits	(%)
CO	3 238	22 000	15
Particulate	108	2 200	5
HCl	359	4 400	8
SO$_2$	11 379	17 700	8
C$_x$H$_y$	600	4 400	14
NO$_x$	>46 570	31 000	>150
NO$_x$ 1997	223 000		<75
HG	8	440	2
Hg	2.8	22	13

Emissions	1996 level	Authority release limits	(%)
Cd	1.2	21	5
Heavy metals	10	440	2
Dioxins/furans (in mgTEQ/a)	2.01	44	5

(c) Total disposal costs based on annual throughput of 45 000 tDS. (1998)

Fixed costs	(%)
Investment costs	55
Staff and administration costs	10
Maintenance including staff for maintenance	10
Taxes and duties	4
Third party measurements and supervision costs	1
Insurance	1
Total fixed costs	81
Energy costs	6
Utility consumption	5
Disposal of residues	8
Total variable costs	19

Total disposal costs £130–140 per ton of DS of £25–28 per ton of sludge cake.

Sources: Courtesy of LURGI MG Engineering, from their seminar 'Thermal Disposal of Sludges, Hazardous and Special Wastes', 24 and 25 March 1999. B. Butterworth, *Wet News*, 21 October 1998.

Slurry. Fine PARTICLES suspended in water so as to form a paste with the properties of a viscous fluid, e.g. wet coal fines from a wet coal preparation plant. These can constitute a low grade FUEL.

Smog. Collective name for a FOG containing anthropogenically-derived air pollutants usually trapped near the ground by a TEMPERATURE INVERSION. Constituents can be smoke, sulphur dioxide, unburnt hydrocarbons and nitrogen oxides. Smog should not be confused with PHOTOCHEMICAL SMOG.

Smoke. Gas-borne solids resulting usually from incomplete combustion or chemical reaction. The particles are usually less than 2 micrometres in diameter. Other solid particles in smoke are compounds of silica, FLUORIDE, ALUMINIUM, LEAD, ACIDS, BASES, and ORGANIC compounds such as PHENOLS.

Smoke density is measured by photometric methods or estimated visually. (◊ RINGELMANN CHART)

Smoke is synergistic (◊ SYNERGISM) with SULPHUR OXIDES and can have significant adverse health effects at concentrations as low as 200 micrograms per cubic metre.

Smoke control area. In the United Kingdom, an area designated under the Clean Air Acts as one in which smoke emissions from domestic chimneys and industrial stacks are substantially reduced (only authorized FUELS may be burned and approved furnaces used).

Smoke shade. The measurement of the darkness of a plume of SMOKE by means of a shaded card such as the RINGELMANN CHART in which the observer finds the shade which provides the closest match to the plume.

Smoke stain. This is a patch of dark material accumulating on a white filter paper after a measured volume of air has been drawn through it. The darkness or change in reflectivity of light from the white paper can be used as a measure of the amount of dark particulate material in the air.

Soil. Soil is the upper layer of the earth's crust, existing at the boundary between rocks (lithosphere) and vegetation (biosphere). The topmost layer of decomposed rock and organic matter usually contains air, moisture and nutrients, and can therefore support life. Soil types include sand, clay, loam (a sand–clay mixture) and peat (which contains a large proportion of decaying plant matter). In tropical zones it may be lateritic, i.e. clay formed by the weathering of igneous rocks, which is notoriously difficult to cultivate on and once cultivated upon can set into a rock-like mass making further cultivation impossible.

It is a highly complex and variable material. About 50–56 per cent by weight is mineral matter derived from underlying rock, consisting of a variety of inorganic compounds of different forms and composition; 25–35 per cent is water; 15–25 per cent soil gases and 5–10 per cent organic material; however, these proportions can vary enormously. The organic component comprises material derived from living and dead organisms. Each gram of soil can contain 100 million bacteria, 500 000 fungi, 100 000 algae and 50 000 protozoa.

Soils are predominantly CATION exchangers, but some soils have also developed the capacity to adsorb ANIONS such as sulphate. The adsorption sites are usually oxides of iron and aluminium, predominantly in the lower horizons of the soil. Sulphate adsorption is very important in determining the response of soils to ACID RAIN.

Soil moisture deficit (SMD). The difference between the moisture remaining in the soil at any time and the FIELD MOISTURE CAPACITY. The SMD is important in agriculture and determines whether irrigation is required or not. In the UK the mean annual SMD ranges from less than 12 millimetres (west coast) to more than 150 millimetres (East Anglia) where irrigation is now extensively practised.

Soil types. The size of the soil particles ranges from large, sand (2.0–0.2 mm diameter) through silt (0.2–0.002 mm) to clay (<0.002 mm).

Soil is highly variable and will change over relatively large surface areas or locally (within a few kilometres) or vertically (giving a soil profile).

A soil classification system is used to group different soil types. There are 28 soil types world-wide, and eight within the UK. Soils in the UK are classified by being examined for stratification (irregular organic content), waterlogging, hardness, and amount of organic material.

Solar constant. The amount of solar radiation incident, per unit area and time, on a surface which is perpendicular to the radiation and is situated at the outer limit of the atmosphere (the earth being at its mean distance from the sun). Its value is approximately $1400\,\mathrm{J\,s^{-1}\,m^{-1}}$.

Solar energy. The prospect of a low maintenance cost, low environmental impact, 'free' energy source is obviously enormously attractive, and a considerable amount of research into practical methods of converting the sun's rays directly into usable energy has been mounted in the last few years. Such techniques would have the further advantage of helping to redress the energy imbalance between the developed and underdeveloped countries, since a majority of the latter are situated in tropical zones.

Engineering studies of large centralized solar electrical power systems suggest that capital costs per unit of energy output are within striking distance of traditional sources of energy, assuming the usual improvements of technique that would inevitably be associated with a serious commitment to such a system. Such schemes, however, would be in many ways a misuse of this ultimate, and in many ways ideal, resource, the principal advantage of which is that it is freely available in sufficient intensity between latitudes $\pm30°$. There would appear to be little point in designing vast centralized systems of energy conversion which then have to overcome distribution problems, when there is no technological reason why small individual units, capable of providing single dwellings with heating, lighting and cooking facilities could not be developed. Rudimentary forms of such devices are already in use in many countries and would almost certainly become competitive with conventional energy systems if properly developed on a commercial basis.

Such diffuse solar technology would obviously be ideal in developing countries where capital is not available for the development of large centralized power systems. Furthermore, such devices need not be limited to tropical areas. Modern 'selective black' surfaces, which are highly absorbent through the visible spectrum but poor radiators in the far infra-red, can attain very high working temperatures even on a cloudy day and bring solar energy use to the northern hemisphere, too.

Figure 180 shows a solar combined cycle power plant where heat transfer from concentrated solar radiation raises the motive air temperature to 600 °C (or more). Standby /start-up gas provision is also made for when solar radiation is inadequate. This ensures a high CAPACITY FACTOR and eliminates SHADOW CAPACITY fossil fuel power plant requirements which are necessary with WIND POWER.

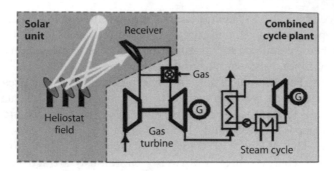

Figure 180 Solar combined cycle plant. *Source*: The Chemical Engineer, October 2006, p. 34, Inst. of Chemical Engineers, Rugby, UK.

Solar heating. A common means of putting the sun's energy to use is to cover a black water-filled metal panel with glass or plastic. The water is then heated by the GREENHOUSE EFFECT. Similar techniques apply to solar distillation where the sea or brackish water to be distilled is not enclosed and can therefore be evaporated. The glass sheet is cooler than the vapour evolved and therefore condensation takes place and the result is pure distilled water. Solar devices are essentially area intensive, because of the low solar radiation density. Production rates of 3–6 litres/m^2 are possible depending on location.

Other methods of solar heating include the use of black-coated pool bases, thus allowing the body of water to warm up very cheaply. Reflectors, 'concentrating' lenses, etc. have been tried mainly to generate high temperatures in special situations. (◊ SOLAR ENERGY)

Solar PV. Solar PV (photovoltaic) uses energy from the sun to create electricity to run appliances and lighting. PV requires only daylight – not direct sunlight – to generate electricity. Photovoltaic systems use cells to convert solar radiation into electricity. The PV cell consists of one or two layers of a semi-conducting material, usually silicon. When light shines on the cell, it creates an electric field across the layers, causing electricity to flow. The greater the intensity of the light, the greater the flow of electricity.

There are three main types of solar cells:

Monocrystalline – made from thin slices cut from a single crystal of silicon. This has a typical efficiency of 15 per cent.
Polycrystalline – made from thin slices cut from a block of silicon crystals. This has a typical efficiency of around 12 per cent.
Thin Film – made from a very thin layer of semiconductor atoms deposited on a glass or metal base. This has a typical efficiency of 7 per cent.

Individual PV cells are connected together to form a module. Modules are then linked and sized to meet a particular load. PV arrays come in a variety of shapes and colours, ranging from grey 'solar tiles' that look like roof tiles, to panels and transparent cells that can be used on conservatories and glass to provide shading as well as generate electricity. Typical domestic PV systems generate 1.5–2.0 kWp, and cover 10–15 m^2 of roof area.

Solar PV and your home

PV systems can be used for a building with a roof or wall that faces within 90° of south, as long as no other buildings or large trees overshadow it. If the roof surface is in shadow for parts of the day, the output of the system decreases. Solar panels are not light and the roof must be strong enough to take their weight, especially if the panel is placed on top of existing tiles. Grid connected systems require very little maintenance, generally limited to ensuring that the panels are kept relatively clean and that shade from trees has not become a problem. The wiring and components of the system should, however, be checked regularly by a qualified technician.

Stand-alone systems, i.e. those not connected to the grid, need maintenance on other system components, such as batteries.

Solar Pyramid. This draws in cool air at the bottom which is heated by SOLAR ENERGY and hence can be used to drive a turbine for electrical power generation. The SECOND LAW of THERMODYNAMICS cannot be circumvented, hence expect low EFFICIENCIES.

Solid waste. Any REFUSE, certain SLUDGES and other discarded materials, including solid and semi-solid materials resulting from industrial, commercial, mining, agricultural operations and domestic activities.

Solar water heating. Solar water heating systems use heat from the sun to work alongside conventional water heaters. The technology is well developed with a large choice of equipment to suit many applications. Solar water heating can provide almost all the hot water during the summer months and about 50 per cent year round.

It also reduces the impact on the environment – the average domestic system reduces carbon dioxide emissions by around 400 kg per year, depending on the fuel replaced. Solar water heating can be used in the home or for larger applications, such as swimming pools. For domestic hot water, there are three main components: solar panels, a heat transfer system, and a hot water cylinder. Solar panels – or collectors – are fitted to the roof. They collect heat from the sun's radiation. The heat transfer system uses the collected heat to heat water. A hot water cylinder stores the hot water that is heated during the day and supplies it for use later.

Solution mining. As the reserves of high-grade ores diminish, the prospects of recovering metals from low-grade ores and spoil heaps is receiving great attention. The means are varied but solution mining, whereby the derived metal is leached *in situ* and then recovered from the leachate, is one of the most promising. (◊ LEACHING)

This technique enables extremely low-grade ores to be mined. For example, copper ores with 0.5 per cent copper would require, by conventional means, extraction of the ore, crushing and milling, and at least 200 tonnes of rock would be removed per tonne of copper recovered. Solution mining would leach the copper from the rock *in situ* using a dilute sulphuric acid solution, after it had been fractured, say, by explosion, or crushed into large lumps (not milled which is extremely expensive) and piled into mounds. The copper ion solution can then be ponded, where it is placed in contact with scrap iron and ION EXCHANGE takes place. The reaction is as follows:

$$Fe + Cu^{++} \rightarrow Fe^{++} + Cu$$

iron	copper	ferrous	copper
	ions	ions	deposit
(in solution)		(in solution)	

The iron goes into solution and the copper is deposited on the bed of the pond. The pond is drained and the copper deposit removed. By this means, low-value scrap iron is substituted for high-value copper. Now, the process can stop there *but* a more useful technique is to regenerate the leach solution through the action of the bacteria *Thiobacillus ferroxidans*, which can oxidize the ferrous sulphate to ferric (a form of iron) and release sulphuric acid in the process. The leach liquor is thus regenerated. This method is used in the extraction of uranium from low-grade ores in spoil heaps, old mine workings, etc.

The bacterial re-oxidation process is the key to the development of this work and the leaching techniques are now being applied not only to low-grade copper but also nickel ores, and the possibility of recovering aluminium from non-bauxite sources is also under consideration. It also has obvious environmental benefits where mining is done *in situ*. This technique may grow as high-grade ore reserves decline. (◊ ARITHMETIC–GEOMETRIC RATIO)

Solvent extraction. A chemical separation method (used in nuclear fuel processing, etc.). Two immiscible liquids, one being a mixture of dissolved substances and the other a good solvent for the material to be extracted, are agitated together. The material required passes into the solvent from which it can be recovered when the two liquids are allowed to separate out. The method is the basis for the purification of uranium and the extraction of plutonium.

Solvents. Omnibus term for liquids which are used to dissolve other substances. The table overleaf gives a list of those in common use.

For water pollution purposes, the ORGANOCHLORINE solvents such as trichloroethylene, which is widely used for industrial cleaning purposes, are of concern as their presence in trace quantities can render groundwater non-potable on grounds of taste. Several boreholes in the UK have been closed because of penetration of industrial solvents into groundwater from either careless or improper disposal, or leakages from industrial wastes. Solvent recovery is a widely practised form of RECYCLING where spent solvents are distilled and reused. However, the cheaper solvents are often incinerated.

One of the WASTE MINIMIZATION targets in the chemical industry is the use of much reduced solvent quantities.

Ford's Halewood site introduced a computerized solvent management system in the paint shop, reducing their VOLATILE ORGANIC COMPOUNDS (VOC) emissions by almost 40 per cent and saving around £60000 a year. The savings were made through reduced material wastage, reduced disposal costs, reduced effluent treatment costs and reduced compliance monitoring costs. The payback period for their investment was approx. 6 months.

Source: www.envirowise.gov.uk/page.aspx?0=119216. Accessed 15 January 2007.

Somatic mutation. A mutation arising in a non-reproductive cell. ($\lozenge$ IONIZING RADIATION)

Soot. Finely divided carbon particles which adhere together. Soot is often left in flues when fossil fuels are incompletely burned.

Sootblowing. The use of jets of steam or compressed air to remove SOOT deposits from boilers, thereby increasing the heat transfer rate.

Sorption. Process of ADSORPTION or ABSORPTION of a substance on or in another substance. (Sorption covers both processes.)

Sound. Periodic wave-like fluctuations of air pressure. The amount by which the pressure changes is known as the sound pressure (which is the 'mean' of pressure peaks generated by the waves), and the rate at which the fluctuations occur is the frequency. High-frequency sound is characterized by screeches and whistles; low frequency sound by rumbles or booms.

The sound pressure level is a measure of the sound pressure differences using the logarithmic decibel (dB) scale. The nature of the decibel scale is illustrated in Figure 181. It will be seen that any ten-fold increase of sound pressure on a linear scale corresponds to a rise of sound pressure level of 20 on the decibel scale. Thus taking the pressure of a just-audible sound and ascribing to this a value of 0 dB, a sound of ten times that pressure has a level of 20 dB. A sound of a little more than three times the pressure of the just-audible one has a level of 10 dB. The sound pressure at which a sensation of pain begins in the ear is about one million times greater than that of the quietest sound that can be heard, and this has a level of 120 dB. Decibels therefore give a manageable way of measuring sound pressure, as well as having a close conformity to the ear's scale of response. The expression of a sound pressure level is always relative to the reference level of (in this instance) the quietest sound and is further discussed in the DECIBEL entry. ($\lozenge$ DECIBEL; HEARING; NOISE; ROAD TRAFFIC NOISE; AIRCRAFT NOISE; INDUSTRIAL NOISE MEASUREMENT; NOISE INDICES)

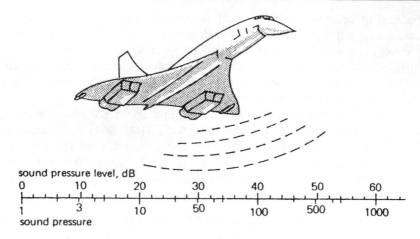

Figure 181   The decibel scale for sound pressures.

Source. The place, places or areas from where a pollutant is released into the atmosphere or water, or where noise is generated. A source can be classified as a *point source*, i.e. a large individual generator of pollution, an *area source*, or a *line source*, e.g. vehicle emissions and noise.

Special waste. ◊ WASTES.

Species[1]. In chemistry, a given kind of ATOM, MOLECULE or RADICAL which has a characteristic chemical structure and composition.

Species[2]. In botany or zoology, a group of closely-related individuals showing constant differences from allied groups.

Specific capacity. The specific capacity of a well is its yield per unit of drawdown, usually expressed as cubic metres per hour (i.e. difference in level between the pumping water level and the natural water level).

Specific gravity. The ratio of an object's or substance's weight to that of an equal volume of water. The property is useful in, for example, sink float separation of minerals or wastes.

Specific volume. The volume of unit mass, e.g. the specific volume of STEAM (saturated) at 0.1 BAR (pressure) is $14.56\,m^3/kg$; at 1 bar it is $1.673\,m^3/kg$.

Specified bovine material (SBM). Broadly speaking, the head (including brain but excluding tongue), spinal cord, tonsils and spleen from cattle over 6 months old and the thymus and intestines from cattle of any age.

SBM (No. 3) Order 1996 (UK), which came into force on 26 July 1996, defines and introduces various controls on the handling of SBM. Parallel arrangements exist in Northern Ireland under the SBM (No. 2) Order (Northern Ireland) 1996.

SPI (Society of Plastics Industry) marking systems. A system of labelling plastic bottles for ease of recycling, as set out below:

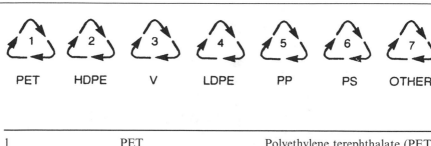

PET   HDPE   V   LDPE   PP   PS   OTHER

1	PET	Polyethylene terephthalate (PET)
2	HDPE	High density polyethylene
3	V	Polyvinyl/Vinyl chloride (PVC)
4	LDPE	Low density polyethylene
5	PP	Polypropylene
6	PS	Polystyrene
7	Other	Others

Spoil bank. A deposit of colliery or other mine or quarry waste. In addition to stone, spoil banks may contain sufficient fine and low-grade coal to enable spontaneous combustion to occur. If this happens, a very serious source of air pollution can occur. Acid water run-off is also a frequent companion which creates severe local stream pollution.

Spongiform encephalopathy. A progressive disease which causes microscopic holes in the brains of affected animals. The animals become uncoordinated, nervous and eventually die. BSE is bovine spongiform encephalopathy.

Spores. A reproductive structure that is adapted for surviving for extended periods of time in unfavourable conditions. They form part of the life cycle of many plants, & microoganisms. Spores are usually killed by heating at 121°C for 3 minutes.

Spray tower. Air pollution control device for PARTICULATE removal from DUST laden gases or for scrubbing of acid gases using an alkaline solution in which the gases are cleaned by contacting with spray (Figure 182). The effluent discharge is subsequently treated and the cleaning water recirculated. (◊ SCRUBBER)

Springs. A discharge of GROUNDWATER appearing at the ground surface as continuously flowing water. Springs appear where the WATER TABLE intersects the surface.

Stack gases. The gases discharged up a chimney stack for dispersion into the ATMOSPHERE. May also be called FLUE GASES or EXHAUST GASES (for motor vehicles).

Standard. Broadly, something used as a basis of comparison, often a unit of reference.

Standard Atmosphere. Unit of pressure exerted by a 760 mm high column of mercury at 0 °C and 101 325 pascal.

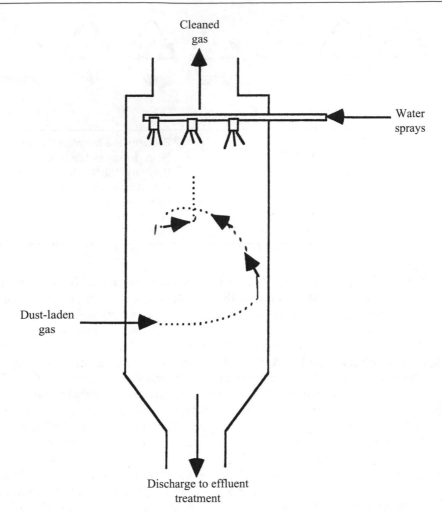

Figure 182   Spray tower scrubber.

Standard deviation. In statistics, the **standard deviation** of a set of values is a measure of the spread of the values. It is usually denoted with the letter σ. It is defined as the square root of the variance.

   Variance is the *average* of the squared differences between data points and the mean. Variance is tabulated in units squared. Standard deviation, being the square root of that quantity, therefore measures the spread of data about the mean, measured in the same units as the data.

   Said more formally, the standard deviation is the root mean square (RMS) deviation of values from their arithmetic mean.

The standard deviation is the most common measure of statistical dispersion, measuring how widely spread the values in a data set are. If many data points are close to the mean, then the standard deviation is small; if many data points are far from the mean, then the standard deviation is large. If all the data values are equal, then the standard deviation is zero.

The formula for standard deviation is:

$$\sigma = \sqrt{\frac{1}{N}\left(\sum_{i}^{N}(x_i - \bar{x})^2\right)}$$

where N is the number of samples taken.

The standard deviation of a discrete uniform random variable X can be calculated as follows:

1. For each value x_i calculate the difference between x_i and the average value.
2. Calculate the squares of these differences.
3. Find the average of the squared differences. This quantity is the variance σ^2.
4. Take the square root of the variance.

A simple example

Suppose we wish to find the standard deviation of the set of the number 4 and 8.

Step 1: find the arithmetic mean (or average) of 4 and 8

$$(4 + 8) / 2 = 6$$

Step 2: find difference between each number and the mean

$$4 - 6 = -2$$

$$8 - 6 = 2$$

Step 3: square each of the differences

$$(-2)^2 = 4$$

$$2^2 = 4$$

Step 4: sum the obtained squares

$$4 + 4 = 8$$

Step 5: divide the sum by the count of numbers (here we have two numbers)

$$8 / 2 = 4$$

Step 6: take the non-negative square root of the quotient.

$$\sqrt{4} = 2$$

So, the standard deviation is 2.

Source: http://en.wikipedia.org/wiki/Standard_deviation.
Accessed 3 January 2008.

Standard temperature and pressure (STP). As the density of gases depends on temperature and pressure, it is customary to define the pressure and temperature against which the volume of gases are measured. The normal reference point is standard temperature and pressure of 273.15 K, 760 milli-metres of mercury (101 325 pascal, the SI unit of pressure), respectively. All gas volumes are referred to these standard conditions. ($\Diamond$ GASES, PROPERTIES OF; NORMAL TEMPERATURE AND PRESSURE)

State Veterinary Services (SVS). A MAFF veterinary body dealing with all animal health and welfare matters. Covers Great Britain and plays a major role in advising on and implementing animal health policy.

Statistics. Numerical data, allegedly factual but sadly, *'caveat emptor'* applies all too often. It all depends (see CO_2 reduction – UK targets and read behind the headline numbers, also aviation CO_2 emissions, or the greenhouse weight-ing factor for methane from landfills used by HMG).

Two examples will suffice. The extract below from the *Sunday Times* of 15 April 2007 'Goldfinger Brown's £2 billion sell-of by HollyWatt & Robert Winnett.

'The Treasury intends to sell 125 tons of gold, 3 per cent of the total reserves, during 1999–2000 with the Bank of England conducting five auc-tions on the Treasury's behalf. Auctions will be held every other month starting in July.' – P.Hewitt, 7 May 1999.

Hewitt's figure of 3 per cent referred to 'total reserves' which, apart from gold, included tens of billions that the government borrows on the interna-tional currency markets, rather than the gold reserves actually owned out-right by Britain.

The answer while 'correct' was wholly misleading. Another example is the claim that recycling UK waste is environmentally beneficial. This disregards the fact that up to 40 per cent is sent to the Far East with unquantifiable environmental impacts. The boundary is usually drawn around the UK, which completely ignores all the health and safety problems and environmental impacts in the Far East recipient recycling operations.

Statutory Consultees. These are the Health and Safety Executive and, where water is concerned, the Environment Agency (or Scottish Environment Protection Agency). The following may be consulted where appropriate:

(a) The Minister at Defra.
(b) The Secretary of State for Wales.
(c) The sewerage undertaker for any process which may involve discharge to a sewer.
(d) Natural England, for processes located in England which may affect a site of special scientific interest; the Countryside Council for Wales for similar processes located in Wales; and the Scottish equivalent.
(e) The Harbour Authority for processes which may involve the release of a prescribed substance into a harbour under its control. For Scotland, the National Consultees are also used.

Statutory limit. A specific upper limit, which by law cannot be exceeded.

Steam. Water in the vapour state. It can be in the wet condition (i.e. containing water in the vapour), dry (i.e. free from water, usually called dry saturated where it is both dry and at the same temperature as the boiler water) or superheated (i.e. at a higher temperature than the boiler water). All three steam conditions are encountered in steam engineering and denote differing energy contents per unit mass, as well as densities.

Steam tables are a convenient method of showing the various related properties of saturated steam, for example see the table opposite. Steam tables show the properties of what is usually known as 'dry saturated steam'. This is steam which has been completely evaporated, so that it contains no droplets of liquid water.

In practice, steam often carries tiny droplets of water with it and cannot be described as dry saturated steam. Nevertheless, it is important that the steam used for process or heating is as dry as possible and correct STEAM TRAPPING and separation can improve steam quality.

Steam quality is described by its 'dryness fraction' – the proportion of completely dry steam present in the steam being considered. It is usually expressed as a decimal value less than 1, i.e. 0.95 represents 95 per cent dry steam. The volume of 1 kg of steam at any given pressure is termed its specific volume and the volume occupied by a unit mass of steam decreases as its pressure rises. The small droplets of water in wet steam have mass but occupy negligible volume.

Related properties of saturated steam.

| Gauge pressure (bar) | Absolute pressure (bar) | Temperature (°C) | Specific enthalpy | | | Specific volume steam (m³/kg) |
			Water (kJ/kg)	Evaporation (kJ/kg)	Steam (kJ/kg)	
3	4.013	143.75	605.3	2133.4	2738.7	0.461
5	6.013	158.92	670.9	2086.0	2756.9	0.315
10	11.013	184.13	781.6	2000.1	2781.7	0.177
15	16.013	210.45	859.0	1935.0	2794.0	0.124

Steam tables. ◊ STEAM.

Steam trap. A device for removing moisture from STEAM in order to supply dry process steam. Their correct operation is vital in the high EFFICIENCY running of various SYSTEMS as the following example illustrates.

Example

A process plant has 200 steam traps of which 10 per cent fail annually.

Thus, the average number of failed traps per annum = 20.

The plant operates with steam at 6 bar g and runs 12 hours per day, 6 days per week for 50 weeks per year = 3600 hours per year.

From the chart and the table shown, energy loss via 5 mm sharp edged orifice (for traps in use) = 50 kg/h.

Allowing an actual energy loss of 50 per cent of sharp-edged orifice = 25 kg/h per failed steam trap.

So, 20 × 25 × 3600 = 1 800 000 kg per year, i.e. 1800 tonnes of steam are wasted each year. Good maintenance pays off!

Source: Spirax Sacro Technology, Cheltenham.

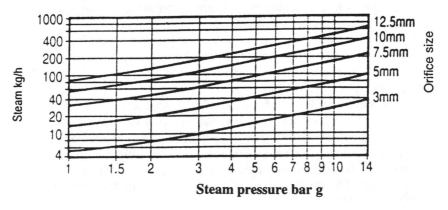

Figure 183 Typical wastage through stream trap leaks.

Trap size (mm)	Orifice size (mm)
15	3
20	5
25	7.5
40	10
50	12.5

Stern Review. An independent Review, 'The Economics of Climate Change' commissioned by the UK Chancellor of the Exchequer, reporting to both the Chancellor and to the Prime Minister, as a contribution to assessing the evidence and building understanding of the economics of climate change.

Released on 31 October 2006, the Review states that

(a) Global warming is a reality
(b) The UK and global economic well-being depends on rapidly moving to a global low carbon economy.
(c) Carbon pricing, carbon emissions trading, energy efficiency, renewables, and cessation of deforestation, are called for in abundance.

Various scenarios are also posited on a scale of 1–5 °C global temperature rise.

1 °C water supplies for 50 million people at risk.
2 °C ≈ 20 to 30 per cent less water in Africa and the Mediterranean.
3 °C mega drought in Southern Europe -4×10^9 people at risk.
4 °C ≈ 50 per cent less water in South Africa and the Mediterranean.
5 °C Glaciers vanish. Severe knock-on effects on food, environment, health are postulated.

A cynical mind set is required as no political party will ban 'gas guzzler' 4×4s and/or seriously curtail cheap flights. Many will argue that their respective CO_2 contributions are 'small' but like recycling such measures are bellwethers of environmental commitment. Although recycling in instances can be environmentally detrimental when, for example, UK waste ends up in China for so-called recycling. This is a country which pollutes on an enormous scale and pays scant attention to health and safety (e.g. ca. 4000 coal miners killed in 2005).

Perhaps a claim in *The Independent* (2 November 2006) which posits that the number of flights from the UK will treble by 2030 with a predicted increase in carbon emissions from 8.8 M tonnes (2000) to 18.8 M tonnes (2030) is a better indication of our and HMG's CO_2 reduction intentions.

Stochastic effects. As applied to radiation exposure, this refers to effects where the probability of an event (e.g. cancer) taking place is proportional to the dose received, i.e. no threshold is assumed. Non-stochastic effects occur only after a threshold has been exceeded, i.e. a 'large' dose has been received. (◊ IONIZING RADIATION)

Stoichiometric. The exact or fixed proportions of elements in a chemical compound, or of reactants to produce a compound. Thus in CARBON DIOXIDE the stoichiometric ratio of carbon atoms to OXYGEN is $1:2$. Stoichiometric amounts satisfy a balanced chemical reaction with no excess of reactants or products.

Stoker. A mechanical feeding mechanism to control the rate of solid fuel admission onto a boiler grate. Usually, for coal, a ram feed onto a continuous moving chain (chain grate) or ram feed. For DOMESTIC REFUSE, can be rotating cylindrical grates or various designs of rocking bars to ensure both feed control and agitation of the fire-bed.

Stokes' law. A mathematical expression for the drag of a small sphere falling through an infinite fluid: $D = 6\pi\mu ru$, where μ is the VISCOSITY of the fluid, r is the radius of the sphere, and u the velocity of the sphere. It is valid only for restricted conditions (laminar flow and low REYNOLDS NUMBER). Stokes' law is widely used in the study of the settling of PARTICULATE matter both out of the atmosphere and in water treatment plants.

Stomata. The small apertures of pores in a leaf epidermis (skin) or young stem of a plant, and also of externally secreting gland in animals. In plants they allow the passage of gases and vapours into and out of leaves.

Storage reservoir. A reservoir, normally constructed by damming a valley in an upland catchment area, for the provision of water for public supply.

Storm overflow. A device used in SEWERAGE systems to prevent overloading. Flows in excess of a predetermined quantity are discharged untreated (with the possible exception of screening) direct to a nearby receiving watercourse. Severe cases of local water pollution have been caused by this.

Stratification. The separation of a lake or sea into distinct layers of strata. These layers are characterized by 'warm' water on top and 'cold' on the bottom with an intermediate transitional band or 'thermocline' which is stagnant and seals off the bottom layer.

Figure 184 shows the stratification for Grasmere in the Lake District. Note that the amount of DISSOLVED OXYGEN has dropped to zero in the bottom layer and thus this ANAEROBIC zone will not support aquatic aerobic life. Stratification can also occur in estuaries when freshwater floats over salt water.

Stratosphere. The 'upper' portion of the earth's atmosphere above the TROPOSPHERE extending to a height of about 80 kilometres. The temperature increases with height in this region.

Straw. Plant stems remaining after cereal crops have been harvested. In the UK the varieties are wheat, barley and oats; each one has its own characteristics. Barley and oat straw can be chopped, milled and fed to ruminants (sheep

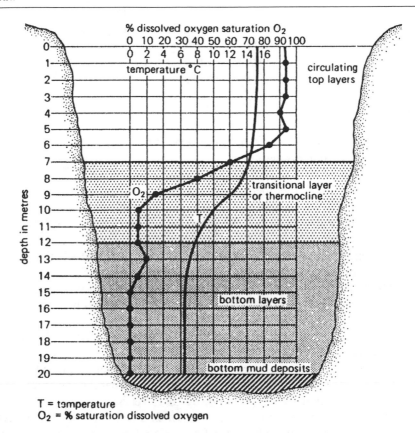

Figure 184 A stratified based on measurement of Grasmere, 2 September 1974 Temperature and dissolved oxygen are measured at metre intervals.

and cattle) as a carbohydrate source which requires PROTEIN supplementation for a balanced diet. As straw is produced in abundance in the UK (annual production ca. 9 million tonnes), many uses have been proposed, including papermaking, enzymatic decomposition to produce glucose, and as a high protein feedstock. Chemical delignification by boiling with caustic soda to produce an easily-digestible feedstock is one possible use being explored. (◊ LIGNIN)

As a fuel, straw has a CALORIFIC VALUE (GROSS) of 14.7 MJ/kg which compares with 28 MJ/kg for coal. However, it may be a cheap fuel, but it does require transporting and possibly briquetting or pelleting before combustion. It is normally burnt in a cyclone furnace with an associated waste heat boiler. The overall EFFICIENCY of this arrangement is ca. 75 per cent which is comparable with many coal-fired boilers of similar outputs (4 MW or less).

A microbiological method for breaking down straw for feedstuffs is to use selected strains of FUNGI. This technique can be carried out simply on the farm, so avoiding high transport and processing costs.

The process involves the following steps:

Straw
↓

Inoculation ← add cellulosic fungus + a scavenger organism to help break down mixed carbohydrates
↓

Incubation ← control temperature/moisture
↓

straw/protein feed

This technique can provide an upgraded ruminant feedstock that should require no protein supplement. (◊ CELLULOSE)

Strontium-90. Radioactive isotope of strontium produced in nuclear reactions as a fission product. It is chemically similar to calcium and has a half-life of 28 years. Like calcium it tends to be concentrated in bones. Thus, if milk contaminated with strontium-90 is drunk, the strontium-90 will be concentrated in the bones.

Structure plan. A written statement, approved by the Secretary of State for the Environment, of the County Planning Authority's general policies and main proposals for change over a period of up to 15 years.

Sublimation. Sublimation is the process by which a solid is converted directly to the vapour phase by application of heat. Condensation of this vapour to a solid, without an intermediate liquid phase, is known as desublimation.

Unlike DISTILLATION, which is applied to substances in the liquid phase, sublimation can be used to purify substances or mixtures which tend to decompose or polymerize at temperatures about their melting points.

For many gases this approach works well.

Substantial change. The meaning of substantial change is given in section 10/7 of the Environmental Protection Act 1990. Guidance of what constitutes a substantial change is also available from the Environment Agency. As a general principle, any change or increase in the release of any prescribed substance to a relevant medium should be considered as substantial. For IPC processes, any increase in rate of overall throughput beyond 5 per cent of the design capacity or authorization conditions, or changes in the process operating parameters, e.g. temperature or feedstock, or changes in process equipment are likely to be considered substantial changes.

Substitute liquid fuel (SLF). Also known as SECONDARY LIQUID FUEL.

SLFs and their use in CEMENT kilns are causing serious concern to the established HAZARDOUS WASTE disposal industry.

Subsurface water. That part of rainfall which is not evaporated or which does not flow away as surface RUN-OFF, penetrates into the ground and thereby becomes subsurface water.

Sugars. Carbohydrates, i.e. compounds of carbon, hydrogen and oxygen, which are usually crystalline and dissolve in water to give a sweet-tasting solution. They may be classified by molecular structure, e.g. as mono-, di-, tri- or polysaccharides. ($\lozenge$ CELLULOSE)

Monosaccharides $\left\{ \begin{array}{l} C_6H_{12}O_6 \text{ hexoses (6 carbons), e.g. glucose, fructose} \\ C_5H_{10}O_5, \text{ pentoses (5 carbons), e.g. xylose} \end{array} \right.$

Disaccharides – $C_{12}H_{22}O_{11}$, e.g. sucrose, maltose, lactose, cellobiose

Trisaccharides – $C_{18}H_{32}O_{16}$, e.g. raffinose

Sulphate. Salts of sulphuric acid containing the SO_4^{2-} group. SULPHUR DIOXIDE emitted from the burning of sulphur in fuels is oxidized slowly in the atmosphere to sulphur trioxide (SO_3) which forms sulphuric ACID with moisture and sulphates with basic materials such as AMMONIA or metals and their oxides. Sulphates so formed are PARTICULATES. Sea spray is a substantial natural source of airborne sulphate particles. The chemical reactions for acid formation (as well as acid attack on limestone) are given in equations [1], [2] and [3]:

$$SO_2 \quad + \frac{1}{2}O_2 \quad \rightarrow SO_3$$

(Sulphur (Oxygen) (Sulphur   [1]
dioxide)   trioxide)

$$SO_2 \quad + H_2O \quad \rightarrow H_2SO_4$$
(Sulphur (Water)   (Sulphuric acid)   [2]
trioxide)

$$H_2SO_4 \quad + CaCO_3 \quad \rightarrow CaSO_4 \quad + CO_2 \quad + H_2O$$
(Sulphuric (Limestone) (Calcium (Carbon (Water)   [3]
acid)   sulphate) dioxide)

Hence, buildings can be subjected to corrosive acid attack by the COMBUSTION OF SULPHUR-CONTAINING FUELS. ($\lozenge$ ACID RAIN)

Sulphate pulp. Chemical pulp manufactured by COOKING with a solution containing sodium hydroxide and sodium hydrogen sulphide. ($\lozenge$ COOKING)

Cooking involves treatment of the fibrous raw material at a minimum temperature of 100 °C with water and the addition of appropriate chemicals.

Sulphur. A non-metallic element, atomic number 16, relative atomic mass 32.6, symbol S. A constituent of all living matter. Occurs in all fossil FUELS and is emitted (in the form of sulphur oxides) on combustion. Attention has been given to the possibility of removing the sulphur from such fuels.

Sulphur cycle. The three principal forms of sulphur present in the lower atmosphere (sulphur dioxide, hydrogen sulphide and sulphates) have been formed and emitted by both natural and industrial processes. Other mechanisms return these compounds to the earth's surface. For example, atmospheric sulphur dioxide is oxidized to the trioxide, which combines with water and is washed out onto the earth's surface as sulphuric acid or sulphates; bacterial action converts the sulphates into hydrogen sulphide, which is then oxidized to sulphur dioxide. This cycle maintains the global atmospheric concentration of sulphur dioxide at a roughly constant level.

Sulphur dioxide. ◊ SULPHUR OXIDES.

Sulphur oxides. Sulphur dioxide (SO_2) and sulphur trioxide (SO_3). Of the two, sulphur dioxide predominates and in the presence of particular CATALYSTS, conversion to sulphur trioxide can take place.

Sulphur oxides occur naturally from sources such as volcanoes, sulphur springs, decaying organic matter and PHYTOPLANKTON activity. Global production by humans is about 100 million tonnes per year with over 90 per cent produced in the northern hemisphere. These sources are principally COMBUSTION of FUELS which contain sulphur, brickworks and spontaneous combustion in coal mine spoil heaps. The anthropological sulphur dioxide contribution is usually concentrated in industrial and domestic areas and can severely affect health during SMOG conditions. It is synergistic (◊ SYNERGISM) in combination with smoke. Together they affect the respiratory tract, and about 1 per cent of the population encounter bronchial spasms at concentrations of between 300 and 500 micrograms per cubic metre ($\mu g/m^3$). Above $57\,000\,\mu g/m^3$, waterlogging of the lungs takes place and eventually respiratory paralysis. During the 1952 London smog, the extra deaths of 4000 people in one week and another 8000 in the following three months were attributed to the combination of sulphur dioxide and smoke (see Figure 185). (◊ CLEAN AIR ACT)

The effects of sulphur dioxide and smoke, respectively, on humans begin at concentrations of $300–500\,\mu g/m^3$ for sulphur dioxide and $250\,\mu g/m^3$ for smoke. But the effect of sulphur dioxide on LICHENS begins at $40\,\mu g/m^3$. Thus the damage to lichens does not correlate with the health effects on humans. But where should the permissible level of exposure be set at?

It is clear that society puts up with different levels of effect, according to what is being affected – in this case plants or people – and it is apparent that sulphur dioxide levels are set for people. This may or may not be a wise judgement. The effects of lower sulphur dioxide levels may be increased crop yields, but the costs would be the costs of desulphurizing fuel before combus-

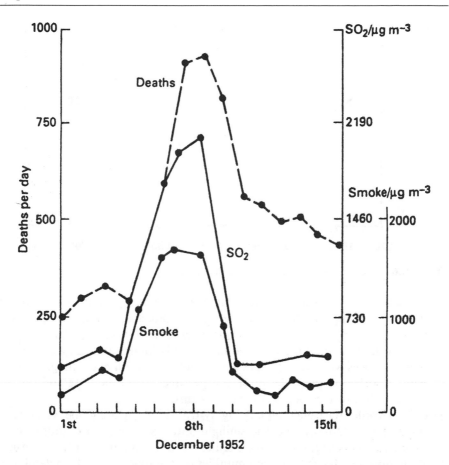

Figure 185 Health effects of SO₂ and smoke in December 1952 during the London smog.

tion, or the flue gases after combustion and before emission. COST-BENEFIT ANALYSIS could possibly be helpful in this area. This example illustrates how much is yet to be done to determine what is an acceptable level of any particular pollutant.

Removal of sulphur oxides from the atmosphere is usually accomplished in the form of ACID RAIN with serious effects on property through corrosion and in some cases on aquatic ecosystems due to the alteration of pH.

The emissions trend and the new target are shown in Figure 186.

Sulphur trioxide. ↝ SULPHUR OXIDES.

Sulphuric acid. A dense oily liquid, colourless when pure; formula $H_2 SO_4$. It is highly corrosive and poisonous. Sulphuric acid is the most widely used of all industrial chemicals. As a pollutant it occurs in the atmosphere in the form

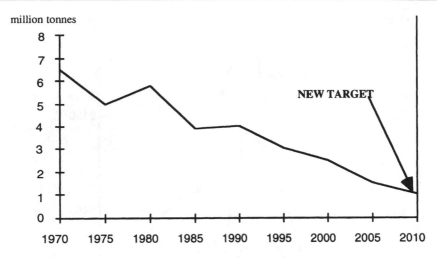

Figure 186 UK emissions of sulphur dioxide trends.

of an AEROSOL, called sulphuric ACID MIST, produced by the oxidation of atmospheric sulphur dioxide, as well as by direct emissions from stacks. These fine droplets are more difficult to remove from the air than gaseous sulphur dioxide, their life in the atmosphere is longer, and they can travel great distances with the wind. They can reach the alveoli in the lungs without being absorbed in the wider bronchial passages, or in the nose and throat; they can therefore be potentially very harmful.

Sunshine recorder (or heliograph). A highly-transparent glass globe which contains the means to focus incident sunlight onto a sensitized card, thus recording a trace. The length of the trace scales the duration, but not the intensity, of the sunshine. Indeed, two communities on the East Coast of England, viz. Hunstanton and Cromer, 35 miles apart, regularly have some 2–3 hours difference; in reported sunshine hours.

Example:

	Hunstanton	Cromer
15/06/06	2.4 hrs	4.7 hrs
19/6/06	6.6 hrs	9.7 hrs

Source: *Times Weather* (data, 24 hrs to 6 pm in the day before.

Clearly an element of interpretation is required, or Hunstanton and Cromer need to standardize their rulers! Freak results can occur, e.g. on 22/6/06 Hunstanton reported 4 hours sunshine and Cromer a magnificent 12.3 hours despite 0.11 inches rainfall.

Supercritical water oxidation. Supercritical water oxidation (SCWO) is a means of dealing with concentrated oxidizable waste streams containing toxic or otherwise intractable ORGANIC compounds, especially those containing CHLORINE and NITROGEN. It can also cope with HEAVY METAL pollutants.

The process operates above the CRITICAL POINT for water, and so involves high pressures and temperatures (>221.2 bar, >647 K).

The discharges from the process are a gas stream (containing CARBON DIOXIDE, nitrogen and excess OXYGEN) and a sanitized water stream containing some dissolved SALTS and suspended heavy metal oxides.

Some typical destruction efficiencies are shown in the table below:

Chemical	Destruction (%)
p-Chlorophenol	99.99
2,4-Dichlorophenol	99.7
2,4,6-Trichlorophenol	99.995
Pentachlorophenol	99.99
Pyridine	99.2
2,4-Dinitrotolene	99.0
Trichloroethylene	99.3
Ethylene glycol	99.9

Figure 187 is a schematic diagram of the SCWO process.

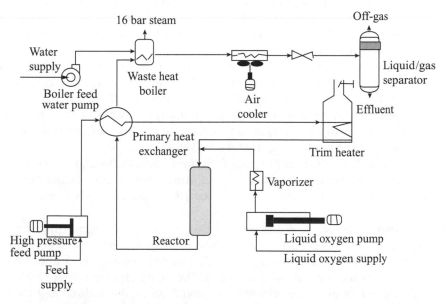

Figure 187   SCWO process schematic.
Source: Environmental Protection Bulletin, No. 44, Setember 1996. (Reported by permission of the Institution of Chemical Engineers).

A test plant has run smoothly on sewage SLUDGE in the US (Modell Environmental Corporation) heated with oxygen in a long, three-stage reactor tube. The supercritical water breaks the raw material down into simple oxides. After heating and cooling, the by-products are clean water, clean carbon dioxide gas and, an odour-free brown powder. As the system is completely sealed, there are no other emissions.

Source: S. Haig, 'Alchemy of sludge', *Financial Times*, 22 February 1995.

Supercritical power generation is a new means of achieving high efficiency. The Nordjyllands 3 Coal-fired Power Station in Denmark achieves an efficiency of 47% – technology has answers! (E Coal, Vol 64, February 2008, p7).

Superheating. A term used in STEAM engineering to show that it has had additional HEAT imparted at constant PRESSURE, i.e. it has a greater temperature than that of the boiler water (or saturation temperature). This enables the steam to be used to deliver more ENERGY to a steam turbine and thus permit more POWER generation per unit mass of steam. Greater EFFICIENCIES are made possible by the use of the higher temperatures. The steam may also be expanded in the turbine over a much greater temperature range before water droplets form in the steam and damage the turbine blades by erosion.

Supersonic flight. If the speed of an aeroplane is greater than the speed of sound in air then this is termed supersonic flight, and the plane is said to be travelling at greater than Mach 1 (the speed of sound in air). When this is the case, the plane creates a so-called shock wave, and a noise or 'boom' accompanies the flight path. The boom intensity is a function of the speed, altitude and plane design. (◊ CONCORDE)

Supply curve. A supply curve shows, for a range of prices, the quantities of a commodity that will be offered for sale. Such a curve may be drawn for a single seller or for a number of sellers in a market. (◊ DEMAND)

Surface inversion. A surface or ground inversion is a TEMPERATURE inversion based at the earth's surface, i.e. an increase of air temperature with height beginning at ground level. (◊ TEMPERATURE INVERSION)

Surface water. Run-off from rain that falls onto roofs, paths and driveways.

Suspended solids. The dry weight of solids captured by filtering a known volume of untreated SEWAGE, other effluents, or river water. Usually expressed in mg/l.

Sustainability. *Sustainable development*: formally it is 'development which meets the needs of the present generation without compromising the ability of future generations to meet their own needs'. This should make us think about how we currently use our natural resources at the future expense of our children.

The question of equity between the generations, industrialized and industrializing countries, and sectors in society is now firmly on inter-governmental agendas because, in reality, the pace and type of development for some is at the expense of others.

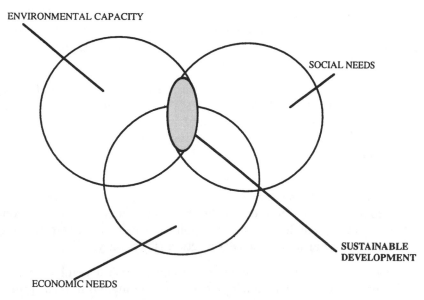

Figure 188 Measuring sustainable development.

There is now a growing recognition that social, environmental and economic needs must be fully integrated if sustainability is to be achieved (see Figure 188).

Measuring sustainable development: statistical indicators on a range of themes can be chosen which together give as good an indication as possible of progress. The themes are shown in the table below:

Environmental capacity	
• Resource use and waste	water, materials, land, energy, sources
• Pollution	of land, water, air, food, etc.
• Biodiversity	
Quality of life	
• Basic needs	food, water, shelter, energy
• Information/training/education	
• Leisure and culture	
• Freedom – political	participation
– personal	freedom from fear, crime or persecution
• Access/transport	to goods, services and people
• Income	adequate and fair
• Work	opportunity to, including voluntary
• Health	physical and mental
• Beauty/aesthetics	of places, spaces, objects

Source: Avon County Council Newsletter, 'The Environment in Avon', Issue 6, Spring 1995.

Sustainable development indicators: the role of key sustainable development indicators is fraught with problems as it is impossible to cover all aspects and a balance needs to be struck between 'resonance and completeness'. The UK Round Table have made the following recommendations.

Recommendation 1. Indicators of sustainable development should if possible have targets attached. Any target set for an indicator of sustainable development should have been arrived at by a transparent and logical process. Those with responsibility for achieving these targets should, if possible, be identified.

Recommendation 2. Those indicators in the 'state' category ('state' indicators are those which measure the state of the environment, the economy or social equity) should, wherever possible, have 'alert-zones' and/or 'red-zones' attached, respectively highlighting the point at which current trends should cause concern or at which immediate action is necessary.

Recommendation 3. The Government should publish a regular report on indicators of sustainable development, highlighting both their progress against targets and those indicators which are entering 'alert-zones' or 'red-zones'. Where appropriate, the report should include an analysis of reasons for the failure to meet targets, and engage all of those responsible for policy and implementation in identifying and carrying out necessary actions.

Recommendation 4. All of those developing indicators of sustainable development should regard an indicator as a tool. As such, its value and appropriateness depend on consideration of its intended use.

Recommendation 5. In seeking to engage the public, the Government should develop the coverage of its 1996 package of indicators of sustainable development in order to take some account of issues of high or increasing public concern, including those identified by its own Public Attitudes to the Environment surveys.

Recommendation 6. The Government should further develop its 1996 package of indicators of sustainable development to reflect the international impacts of domestic practices.

Recommendation 7. The Government is encouraged further to develop the methodologies within its Public Attitudes surveys which allow respondents to indicate their own concerns in addition to choosing issues from a preselected list.

Recommendation 8. The Government should further develop its 1996 package of indicators of sustainable development to fully incorporate economic and social as well as environmental indicators. These should include quantitative measures of social issues, including health, education, poverty, unemployment and crime, as well as measures of 'quality of life'.

Recommendation 9. The identification of a restricted set of 'key indicators of sustainable development' is supported. These indicators should cover each of the following areas of crucial importance:

- consumption of non-renewable resources
- pollution of air, water and land
- social issues
- biodiversity
- landscape and cultural resources.

The reasoning underlying the choice of specific indications, the setting of related targets and any subsequent changes to these issues should be fully explained.

Source: UK Round Table of Sustainable Development, Third Annual Report, March 1998.

The UK Round Table of Sustainable Development in a further 'blast' said on agriculture that it:

- urges the Government to promote environmental 'goods' based on best practice and positive management;
- recommends a duty of care, in respect of wildlife, landscape and natural features, upon the owners and managers of all undeveloped land in rural areas;
- calls for greater co-ordination between the strategic land use planning system and funding streams, such as EU structural funding, that affect rural land;
- recommends that the Government's Social Exclusion Unit take into account the distinctive features that give rise to deprivation in rural communities;
- calls for structural and cultural changes within Government so that departments charged with delivering agricultural and rural policy do so in a more integrated and co-operative way.

The UK Round Table on Sustainable Development has produced a number of Reports. One affirmed the key role that Small and Medium-Sized Enterprises can play in promoting sustainable development. The Round Table also contributed to a study on planning (Planning for Sustainable Development in the 21st Century) by the Royal Commission on Environmental Pollution, where it emphasized the value of the current planning process as a vehicle for delivering sustainable development, and suggested ways in which its weaknesses could be addressed.

The Report 'Not Too Difficult! – Economic instruments to promote sustainable development' argues that there is a strong case for further developing the use of economic instruments to deliver sustainable development. Concerns about social equity, international competitiveness and environmental efficiency can be addressed by careful design of the instrument.

The Report 'Indicators of Sustainable Development' identifies the need for urgent action on climate change, road traffic and other topics.

'Delivering Sustainable Development in the English Regions' commends the role of regional Round Tables of Sustainable Development and encourages the Regional Development Agencies to implement sustainable development fully in their action plans, programmes and projects.

Source: UK Round Table on Sustainable Development, Press Information 4/98, 9 July 1998.

Sustainability encouragement through resource pricing

In the short term the use of ENERGY and other RESOURCES responds only slightly to increases in price. But this should not be misinterpreted as suggesting that this is an inefficient instrument. Over the long term, price increases can be expected to have a very great effect, even on the consumption of car fuels, which is often said to be particularly unresponsive. Figure 189 shows a surprisingly strong relationship between per capita FUEL consumption and fuel prices that were consistently kept at different levels regardless of world market prices: the higher the price, the less energy was consumed.

Does not apply to conspicuous consumers, e.g. 160 mph power boat enthusiasts' boat engines (2 × 8.2 litre V12) which consume 600 litres of fuel per race. [Mail on Sunday, Night & Day Magazine, 3/12/06.]

Sustainability examples

Sustainable LANDFILL: a sustainable landfill will

- be located in an area that is not hydrogeologically sensitive, i.e. not close to water resources;
- be located in strata that do not facilitate rapid relocation of materials from the site;
- contain only those materials that will degrade or will remain immobile;
- exclude mobile persistent toxic species;
- be managed to promote degradation processes;
- be managed to collect and utilize LANDFILL GAS;

Sustainable landfill should result in the reintroduction in a controlled manner to the environment of the raw materials of society that have been exploited

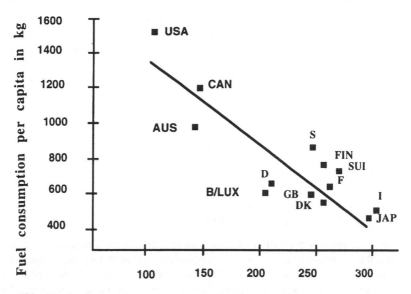

Figure 189   Fuel prices and consumption per capita.
Source: E. U. von Weizäcker, 'Let prices tell the ecological truth', *Our Planet*, (UNEP), Vol. 7, No. 1, 1995. (Reproduced by permission of the United Nations Environment Programme) Fuel prices (index USA 1998 = 100).

and ceased to have a useful function. The sustainable development policy of the UK government recognizes the inevitable need for landfill facilities and that it is the BEST PRACTICABLE ENVIRONMENTAL OPTION (BPEO) for certain waste materials, and is the only alternative for wastes which cannot be treated.

While in theory, sustainable landfill is possible, whether it is practically achievable is not certain. Even if the ultimate sustainable landfill proves unachievable, its pursuit will improve landfill practices. This in itself is a valid justification for the pursuit of sustainability.

M. J. Carter, M. J. Carter Associates, 'Towards sustainable landfill', *Scottish Envirotec*, February 1995, p. 22.

Sustainability and CHP

The UK throws away more thermal energy from power stations than it obtains in the form of NATURAL GAS from the North Sea. This energy flow (100 GW) provides more than enough energy to heat all the buildings in England, Scotland and Wales.

Source: D. Green, 'Sustainable energy policy demands efficiency', *Energy in Buildings and Industry*, March 1995.

One point is of pivotal significance. No long-term strategy of poverty alleviation can succeed in the face of environmental forces that promote persistent erosion of the natural resources upon which we all depend. And no environmental protection programme can make headway without removing the day-to-day pressures of poverty that leave people little choice but to discount the future so deeply that they fail to protect the resource base necessary for their own survival and their children's well-being.

To meet the challenges of 'Attacking Poverty, Building Solidarity, Creating Jobs', the global community must devise new and innovative strategies that attack the vicious cycle of poverty and environmental degradation. This will be just the first step.

Source: Elizabeth Dowdeswell, United Nations Under-Secretary General and Executive Director, UNEP, commenting on the World Summit for Social Development, 6–12 March 1995, Copenhagen, *Our Planet*, Vol. 7, No. 2.

There are no magic solutions. In the long run there is no alternative to restoring equilibrium between resources and population and the environment, working for a relatively steady state society. The first step to wisdom is to recognise the problem. The next is to do all possible to prepare for it. Governments must work in the first instance to manage it within their countries. They will need help from the international community, especially from those who have contributed most to the creation of the problem. Action to accommodate refugees across frontiers would be immensely more difficult.

Source: Sir Crispin Tickell, former United Kingdom Ambassador to the United Nations, *Our Planet*, Vol. 7, No. 3, 1995.

(◊ COMMON INHERITANCE)

The strongest ecological effect of ecological tax reform comes from making it more profitable to lay off kilowatt hours than to lay off people.

Source: E. U. von Weizäcker, 'Let prices tell the ecological truth', *Our Planet*, (UNEP), Vol. 7, No. 1, 1995.

Sustainable forestry

Criteria for sustainable forestry:

1. forest resources
2. health and viability of forests
3. productive functions of forests
4. biodiversity
5. conservation of forests
6. social and economic forests

Sustainable forestry maintains and augments these criteria. Quantitative indicators have been developed for each criterion. By monitoring changes in the indicators, the sustainability of the forest economy can be evaluated. Characteristics of forest diversity:

- Abundance of animal, plant and fungal species.
- Variety of habitat.
- Rich gene pool within a species.
- Variety of landscape.

In a typical forest landscape diversity is expressed by the amount of dead wood, the number of species, the incidence of endangered species and the number of key biotypes (a key biotype is an environment inhabited by a large number of specialized species, or which is essential to the species living in it).

Source: *Forest Forum*, Finnish Forest Industries, 1996, pp. 11, 15.

Waste sustainability indicators

The following have been reviewed (*The Waste Manager*, June/July 1999) based on work by Doreen Fedrigo of the UK Green Alliance, an independent charity.

1. Climate change
Total electricity consumption. Relatively easy to measure, electricity sources. Renewable sources may become available through the national grid. Measuring electricity alone will not give an accurate account of environmental impact, as this would be lower from renewable sources.

Total energy consumption. Measures all types of fuel for all uses. But still fails to discriminate between fuel sources environmentally.

CO_2 emissions. Sensitive to changes in fuel sources and to consumption reductions resulting from energy efficiency. It can also take into account emissions from sources other than energy use, such as from landfill gas. The only possible drawback may be difficulty in determining a widely accepted method for converting energy consumption into CO_2 equivalent.

Global warming potential. This indicator has the potential to include factors other than energy to reflect the combined environmental impact. The danger of using it as a single indicator to combine carbon dioxide, methane and other emissions is that it obscures progress, or lack of it, being made in reducing these emissions separately.

Tonnes of methane collected. Progress in reducing methane emissions from landfills could be considered in a climate change indicator that measured emission changes of the six greenhouse gases. However, it would be difficult to gauge progress on methane relative to the other gases. So there is a case for using separate methane indicators. It is technically difficult to estimate accurately the total methane a landfill emits, as it depends on the type of waste deposited and how fast it degrades.

Proportion of methane utilized. Electricity generation from landfill gas avoids methane emission into the atmosphere and reduces the reliance on fossil fuels.

Landfills where power is generated from gas. Could be expressed as a proportion of the total number of sites or, more accurately, only of those sites where gas reaches levels useful for power generation.

Energy produced as CO_2 avoidance
Methane emission savings expressed as CO_2 equivalent
Power from landfill gas in total megawatts. Expressed as a proportion of potential power from useful gas, a refinement on the previous indicator.

2. Transport
Total miles travelled by company vehicles.
Total fuel use by fleet vehicles.
Volume of waste moved by rail or water.

3. Water
Millions of cubic litres used.

4. Land use and wildlife
Actual planned end uses of restored land. Expressed by land area for particular uses, in turn expressed as a percentage of land used by a company.

Number of trees planted yearly and area covered. Could be expressed as proportion of land area worked.

Hedgerow planting. May help in returning the landscape to its form before industrial activity began.

Changes in species. If planting and management of a closed landfill could contribute to a local biodiversity plan, this could be monitored and reported as an indicator of progress.

5. Waste minimization
Contribution to education. This would measure donations to educational initiatives.

Contribution of a new product design. Companies could consider funding design.

Contribution through consultancy services. One approach to selling waste minimization is to propose to cut a client company's waste bills with a minimization and management package. Some companies offer a waste audit service. Measuring environmental impact would entail knowing how much waste had been avoided through a company's advice. To measure accurately a waste company's contribution requires that it is the only agent giving advice.

Tonnes of waste recycled. This might mean measuring the output of a materials recovery facility (MRF), if this is where sorting is carried out. It could also be a measure of materials collected which are already sorted and sent straight to a purchaser. (◊ WASTE)

6. Environmental management systems (EMS)
Number of sites registered to a recognized EMS.

7. Regulatory compliance
Number of prosecutions. Breaches of site licences and enforcement notices issued.

Environment Agency Operator and Pollution Rush Appraisal (OPRA) rating. Combines aspects of environmental performance and a company's ability to manage risk.

8. Neighbourliness
Number of complaints. Complaints could be followed up, with complainants asked to give a satisfaction rating. The average of these ratings could provide an indicator. Another way to get an objective assessment would be to use a polling organization to periodically interview neighbours.

Communications. Means of communicating with neighbours should be recorded alongside the number of complaints indicators but should not form part of it.

Funding of local community projects. Channelled through environmental bodies, this could be an important part of good relations with neighbours. A denominator could perhaps be units of turnover to offer valid comparison.

Suggested sustainability indicators for the oil industry are:

- proportion of gas to OIL held and sold on an 'ENERGY equivalency' basis;
- estimated CARBON content of the oil and gas portfolio;
- renewables content of the energy portfolio on an energy equivalency basis. (◊ RESERVES-PRODUCTION RATIO)

Source: J. Elkington, S. Fennel and H. Stibbard, 'Oil explorers', *Tomorrow*, September/October 99, pp. 58–61.

Symba. A process, developed by the Swedish Sugar Company, which uses starch as a substrate for the eventual production of SINGLE-CELL PROTEIN using the *Candida utilis* or *C. torula* yeast strain. For the purpose, *Candida utilis* is propagated with an organism, *Endomycopsin*, which produces an ENZYME (amylase) which can split starch into SUGARS. The organisms grow in symbiosis, the amylase activity of *Endomycopsin* converting the starch to sugars which the *Candida utilis* – the faster grower of the two – uses as a fermentation substrate, i.e.

$$\text{Starch} \xrightarrow{\;Endomycopsin\;} \text{Glucose} \xrightarrow{\;Candida\ utilis\;} \begin{array}{l} \text{Yeast cell substance} \\ \text{(single-cell protein)} \end{array}$$

The Symba process has been used in effluent treatment from potato processing plants and has reduced the BIOCHEMICAL OXYGEN DEMAND by 85 per cent in 10 hours. It reduces pollution and generates a useful product at the same time.

The process, can be used for single-cell protein production from waste carbohydrates (cellulose, starch, 'simple' sugars), and from the organic substances in chemical industry waste streams. The waste fibres and bark from pulp and board production are also fruitful sources for production of single-cell protein. (◊ SINGLE-CELL PROTEIN)

Symbiosis. A compatible association between dissimilar organisms to their mutual advantage. The classic case is the association of nitrogen-fixing bacteria with plants of the clover family. The bacteria occupy nodules on the roots of the plants, and fix the nitrogen from the air into nitrates for the plants, which in their turn supply the bacteria with carbohydrates as an energy source.

Symbiosis can be put to commercial use as in the SYMBA yeast process. (◊ SINGLE-CELL PROTEIN)

Symmetrical distribution. A set of values or observations which are distributed evenly about the MEAN. Often met as a bell-shaped curve or NORMAL DISTRIBUTION.

Synergism. A state in which the combined effect of two or more substances is greater than the sum of the separate effects, e.g. smoke and sulphur dioxides. It is the opposite of ANTAGONISM. (◊ PARTICULATES; SULPHUR OXIDES)

Synthesis gas. A gas formed from the catalytic combination, at high temperature (800 °C), of simple compounds such as CO_2 and CH_4 so that the resulting product has uniform and enhanced COMBUSTION properties which may mean in the case of waste GASIFICATION that the gas may be used as a FUEL in gas engines which need a more refined fuel than that of a boiler plant raising STEAM for POWER and/or DISTRICT HEATING.

System boundary. Term used in LIFE CYCLE ANALYSIS. The interface between a PRODUCT SYSTEM and the environment and other product systems.

Systemic agent. An agent that affects the body as a whole and not a particular part or organ. (◊ MERCURY)

T

Tailings. The residual fine-grained waste rejected after mining and processing of ore, usually after washing.

Take back. The taking back of consumer durables for RECYCLING. This is expected to grow.

Source: *Tomorrow*, September/October 1998.

Tallow. Rendered fat from the slaughter of livestock, can be used as a power station fuel or now as a feedstock for low sulphur diesel. US company Conoco Philips anticipates producing 62 Ml/y.

Source: tce, May 2007.

Tar sands. The Alberta tar sands (oil sands) constitute the largest known reserve of petroleum in the world (approximately 900 000 million barrels of 'in place' heavy oil). It will cost considerably more to extract this OIL than from conventional oil fields. Efforts are being made to improve existing processing technology.

Essentially, the extraction process is one of open-cast mining, in which the 'overburden' of vegetation and earth must first be removed in order to expose the bitumen-soaked sand deposits. These deposits are extremely difficult to handle, being sticky and corrosive in summer and rock-hard in winter. The capital investment required for the large-scale extraction of such material is huge, and present techniques are probably only capable of recovering about 10 per cent of the synthetic crude which is potentially available in such deposits. (◊ OIL SHALES)

TCDD. ◊ DIOXIN, TOLERABLE DAILY INTAKE.

Temperature. A property of a body or substance and a measure of how 'hot' it is, or how much thermal energy it contains. Temperature is measured on several scales, e.g. the centigrade or Celsius and Fahrenheit scales are both measured from a reference point – the freezing point of water – which is taken as 0 °C or 32 °F. The boiling point of water is taken as 100 °C or 212 °F respectively. For thermodynamic devices, it is usual to work in terms of absolute or thermodynamic temperature where the reference point is absolute zero, which is the lowest possible temperature attainable. For absolute temperature measurement the thermodynamic or KELVIN (K) scale which uses centigrade divisions is used.

Temperature is a fundamental measurement in most pollution work. The temperature of a stack gas plume, for example, determines its buoyancy and how far the effluent PLUME will rise before attaining the temperature of its surroundings. This in turn determines how much it will be diluted before traces of the pollutant reach ground level.

Temperature inversion. The TEMPERATURE of the air normally decreases at increasing heights above the ground. A variety of meteorological conditions can occur to reverse this trend and cause a layer of warmer air to overlie a cooler layer. The cooler air cannot then rise because it is heavier and so any air pollutants emitted below the inversion layer are trapped. Inversion of the normal temperature profile can begin at any height above the ground but the lowest are more noticeable in their effect as they trap the smoke and fumes from domestic chimneys. When they occur between 150 and 900 metres above the ground, they can trap the discharges from all chimneys (except those of the largest power stations).

Inversions commonly (but not always) occur in valleys or basins due to radiation cooling of the ground at night. This cools the air near the ground, and being cooler, it can drain into the low-lying areas as a katabatic flow.

Figure 190 shows the temperature profile with height for typical non-inversion and inversion conditions. In extreme circumstances, a halt may be required on industrial processes until the inversion lifts. In practice this temperature profile can assume a wide variety of forms.

A temperature inversion can also be associated with an area of high pressure or anticyclone. In this, air drifts downwards very slowly, warming as it does so.

Near the surface, however, the warm ground or sea sets off more vigorous upward motion.

These two air movements meet about one kilometre above the surface and a temperature inversion forms at their boundary.

A related phenomenon associated with industrialized cities is the *heat island*. This is most noticeable at night and early mornings and is caused by

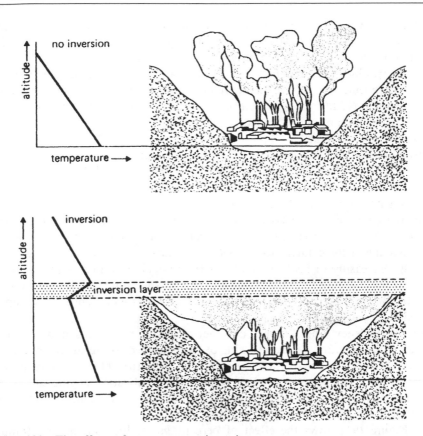

Figure 190   The effects of a temperature inversion.

hot air layers forming at building or chimney height which are warmer than both ground conditions and the air above the layer. This heat island can be 5–7°C warmer than ground conditions and, or course, can trap any pollutants emitted within it. The heat islands are dome-shaped and can disperse towards midday when the temperature increases, but in turn they may be replaced by a higher level inversion.

Temporary hardness. ◊ HARDNESS; ZEOLITES.

Teratogen. An agent that causes birth defects. A strong teratogen is DIOXIN found in herbicides. The LD_{50} for guinea pigs fed with dioxin is as low as 600 parts per million. Fish in Vietnamese waters (1970–71) had concentrations on average of 540 parts per million as a result of the use of herbicides as defoliants (1962–70). The use of defoliants in that country has been suggested as being responsible for rises in stillbirths and birth defects.

Terminal velocity. The maximum velocity at which a PARTICLE settles out in air or water.

Tertiary recovery. Recovery of oil or gas from a reservoir over and above that which can be obtained by primary and secondary recovery. It requires methods such as heating the reservoir to reduce the oil VISCOSITY.

Tertiary treatment. Effluent already treated to a standard achievable by biological treatment, may require further treatment in order to obtain higher quality objectives. Additional treatment may range from nutrient removal and the further reduction of suspended solids through to UV treatment to destroy pathogenic bacteria.

Tetraethyl lead (TEL). Tetraethyl lead and tetramethyl lead (TML) are the principal additives to petrol to raise the octane rating and thus reduce the tendency to 'knock' in spark-ignition internal combustion engines. Now being phased out in the EU and USA due both to suspected links with increased blood lead levels associated with particulate lead emissions and to the use of CATALYTIC REACTORS which require lead-free petrol to avoid poisoning of the catalyst. (◊ OCTANE NUMBER)

Therm. A measure of heat (thermal ENERGY) which was used as the common basis for charging for heat supplied by gas or steam in the UK. It is 100 000 (10^5) British Thermal Units or 1.055×10^8 joules. It will eventually be superseded by megajoules (MJ). (◊ BTU)

Thermal efficiency. The efficiency of a thermal ENERGY conversion device. It is defined as the ratio of energy value of fuel supplied to useful energy output from the device. The thermal efficiency of a modern coal-fired power station boiler is ca. 90 per cent; that of an internal combustion engine ca. 14 per cent; that of a coal-fired power station 38 per cent. (◊ CARNOT EFFICIENCY)

Figure 191 shows the effect of percentage loading on thermal (input–output) efficiency for two boiler designs ❶ and ❷. Clearly, the load cycle

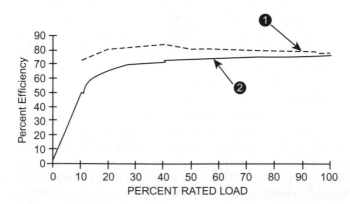

Figure 191 Specimen boiler efficiency vs per cent load (courtesy Clayton Boilers, Cheshire).

needs to be appraised before selection of the boiler system but ❶ has good low loading compared to ❷.

Since boilers operate most of the time at less than 100 per cent load rating, FUEL costs cannot readily be compared unless this information is available and in comparable terms.

It is important to remember when comparing THERMAL EFFICIENCY claims that the percentage increase in fuel costs will be greater than that nominal difference in efficiency. For example, 80 per cent versus 75 per cent efficiency at partial load, a 5 per cent difference in efficiency, translates to a 6.25 per cent saving in fuel usage.

$$1 - \frac{75}{80} \times 100 = 6.25 \text{ per cent savings in fuel usage.}$$

Figure 192 shows how 'useful heat' can be lost through soot build-up on boiler tubes and lead to a drop in efficiency manifested as an increase in fuel consumption. (◊ SOOTBLOWING)

Drax coal-fired power station which supplies 7 per cent of the UK's electricity needs, is undergoing an upgrade (after 33 years service), which will raise its thermal efficiency from 38 to 40 per cent. In 2006, the plant emitted 22.7 Million tonnes of CO_2. After the retrofit, this will drop by 15 per cent by a combination of upgrading and co-firing with biofuels. Coal is still vital for the UK's power industry, and even after CARBON CAPTURE AND STORAGE is implemented, is predicted to have the edge over gas. Recall 'peak gas' occurs in 2012. Peak coal is over a hundred years away. Continuity of energy supply is an absolute must. (◊ WIND POWER, NUCLEAR POWER)

Thermal oxide reprocessing plant (THORP). The plant run by British Nuclear Fuels Limited at Windscale, Cumbria, for reprocessing oxide fuel from thermal reactors. (◊ NUCLEAR POWER)

Thermal pollution. The heat released from the combustion of fossil fuels or the dissipation of energy from prime movers, such as electric motors, which is eventually converted to heat. All such releases end up as waste heat in the sinks of air and water.

Direct thermal pollution of water usually occurs at power stations where 60 per cent or more of the heat content of a fuel ends up as waste heat which must be removed by cooling water which is then discharged to rivers or coastal waters. This form of thermal pollution depletes DISSOLVED OXYGEN and can change aquatic ecosystems. It also increases any existing BIOCHEMI-CAL OXYGEN DEMAND and *reduces* the capacity of a stream to assimilate ORGANIC wastes. The waste heat from power stations could be used for glasshouse or DISTRICT HEATING, but the heating load and power station capacity and duty require careful matching. In 1989, low river conditions in the River Ouse in Yorkshire, which is used for cooling the Drax power

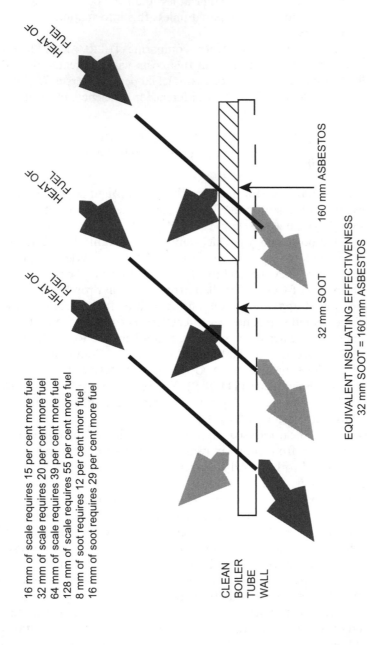

16 mm of scale requires 15 per cent more fuel
32 mm of scale requires 20 per cent more fuel
64 mm of scale requires 39 per cent more fuel
128 mm of scale requires 55 per cent more fuel
8 mm of soot requires 12 per cent more fuel
16 mm of soot requires 29 per cent more fuel

HEAT OF FUEL

HEAT OF FUEL

HEAT OF FUEL

160 mm ASBESTOS

32 mm SOOT

EQUIVALENT INSULATING EFFECTIVENESS
32 mm SOOT = 160 mm ASBESTOS

CLEAN
BOILER
TUBE
WALL

Figure 192
Source: The Clayton Report, Clayton Boilers of Belgium, NV, 1995 (UK, Woolstoon, Warrington).

station, were such that diluted sewage effluent passed into the cooling towers with subsequent deposition of dried particulate matter in the vicinity.

Thermal pollution of the atmosphere can cause local instabilities and in North America there is reason to believe that the industrialized Atlantic Seaboard has its own climate induced by the energy of air and pollutants released.

Thermal processing. A collective term for the disposal or conversion of domestic refuse or wastes by INCINERATION, PYROLYSIS OR GASIFICATION.

Thermal wheel. A slowly rotating brick-lined heat exchanger which recovers heat from (say) furnace exhaust gases on one half, while the other half is giving up its recovered heat to incoming gases such as air for combustion in boilers or blast furnaces. Recovery of 30–50 per cent of the heat in exhaust gases can be obtained.

Thermionic converter. A device for DIRECT ENERGY CONVERSION which gives an electrical current from the electrons emitted by the heating of a suitable metal. This thermionic emission produces a net flow of electrons from a high-temperature electrode (the cathode) to the lower temperature electrode (the anode), both being enclosed in an evacuated enclosure.

However, the efficiency of the device is low unless the gap between the electrodes is very small (less than 25 μm). In addition, as radiation takes place from the hot surface to the cold surface, large temperature differences cannot be used.

Thermodynamics. The analytical methodologies which enable the conversion of heat into work to be analysed and improved. (◊ AVAILABILITY, LAWS OF THERMODYNAMICS)

Thermodynamic heating. If a COMBINED CYCLE/COMBINED HEAT AND POWER plant is used in conjunction with HEAT PUMPS for DISTRICT HEATING then it is possible to obtain from 100 per cent FUEL ENERGY, heat totalling 220 per cent, where the heat pump extracts heat at ambient temperature.

The relevant ENERGY balance is given in Figure 193 for a COEFFICIENT OF PERFORMANCE of 3.5.

Even if 10 per cent transmission losses are assumed for the electricity and the heat, the useful heat is still 198 per cent, which is more than double the heat produced by a good boiler. It should be noted that this figure refers to the maximum heat load; averaged over the year, conditions are even more favourable.

Thermodynamic temperature scale. ◊ TEMPERATURE, KELVIN.

Thermoelectric converter. A device for DIRECT ENERGY CONVERSION which gives an electric current when the two junctions of a loop of two wires of different metals are kept at different temperatures. This is the principle of the thermocouple for TEMPERATURE measurement. However, for ordinary metals the voltage produced per unit of temperature difference is extremely

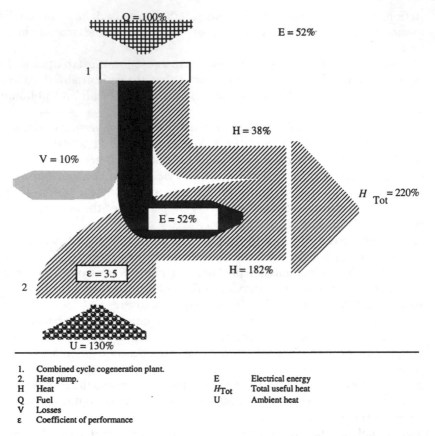

1. Combined cycle cogeneration plant.
2. Heat pump.

H	Heat	E	Electrical energy
Q	Fuel	H_{Tot}	Total useful heat
V	Losses	U	Ambient heat
ε	Coefficient of performance		

Figure 193 Principle of thermodynamic heating. The combination cycle cogeneration plant with sequential combustion and heat pumps yields useful heat equivalent to 220 per cent of the primary fuel energy (100 per cent).
Source: *ABB Review*, March 1995. (Reproduced by permission of ABB Limited)

low and it is only with the advent of semiconductors that the voltage obtainable rose from microvolts per degree kelvin to millivolts per degree kelvin (a factor of 1000). Semi-conductor technology also allows vast numbers to be connected in series, so that useful outputs can be obtained. The electrical charge carriers can be electrons as in metals (or in semiconductor terminology 'n-type' materials) or they can be positive (or 'p-type' materials). Thus an 'n-pair' with low resistance to heat flow and low electrical resistance would make the ideal unit for thermoelectric converter construction. The semiconductor is indeed such a device and just as for the THERMIONIC CONVERTER the 'working fluid' is the flow of charge carriers, so that the Carnot restrictions also apply. (◊ CARNOT EFFICIENCY)

The thermoelectric device operates at lower temperatures than the thermionic converter and is therefore suited for waste-heat recovery applications as in, say, the exhaust gas stream of a gas turbine. Practical efficiencies are 10 per cent or less due to the temperature restrictions but this can be a useful addition in a total energy situation. The USSR manufactures a thermoelectric generator powered by a paraffin heater for radio receivers in rural areas.

Thermoplastic. A plastic which deforms on heating and which can be heated repeatedly and reformed. Hence, source-separated plastics can be recycled, e.g. POLYVINYL CHLORIDE and POLYETHYLENE.

Thermoselect process. A proprietary process for GASIFICATION and direct smelting of MUNICIPAL SOLID WASTE (MSW) in which it is stated that organic components are totally destroyed and inorganic compounds smelted at 2000 °C.

The products are a combustible synthesis gas, mineral products and metals. The sequence of operations is given in Figure 194.

An energy balance is given in Figure 195 for two production lines totalling 20 tonnes per hour. This implies that 1 tonne of MSW provides 350 kWh electricity. (◊ INCINERATION)

Thorium cycle. A nuclear fuel cycle in which fertile thorium-232 is converted to fissile uranium-233. It is regarded as a possible alternative to the uranium-238/plutonium-239 cycle.

Three Mile Island. The site of a major (1979) US nuclear reactor incident in which a circulating pump failure plus a stuck valve on the pressurizer allowed the pressurized water coolant to escape. The reactor core overheated in minutes and core meltdown was narrowly avoided. Total cost of clean-up was in excess of $US 1 billion. The US NUCLEAR POWER industry was severely set back by this avoidable event which was compounded by operator error and mechanical failures.

The lesson is, nothing is certain except mishap, and the most careful planning and emergency procedures must be openly agreed with the communities who host these facilities. (◊ TOKAIMURA)

Three-minute mean concentration. The maximum permissible concentration of pollutant at ground level from a stationary source, i.e. the concentration of pollutant averaged over three minutes. For example, SULPHUR DIOXIDE from a factory chimney is often specified as a three-minute mean value. Yearly averages are likely to be between 5 and 15 per cent of this value, and monthly averages up to 25 per cent owing to wind changes and PLUME-dispersal effects.

In addition to the three-minute mean concentration an absolute mass emission limit may also be placed on the total amount of a pollutant that may be emitted from a stationary source per hour. Thus, the mass emission limit for a smelter discharging less than 85 cubic metres per minute of gases up

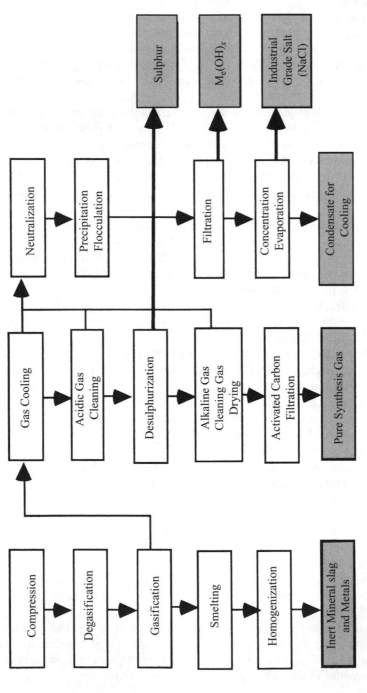

Figure 194 Thermoselect sequence steps.

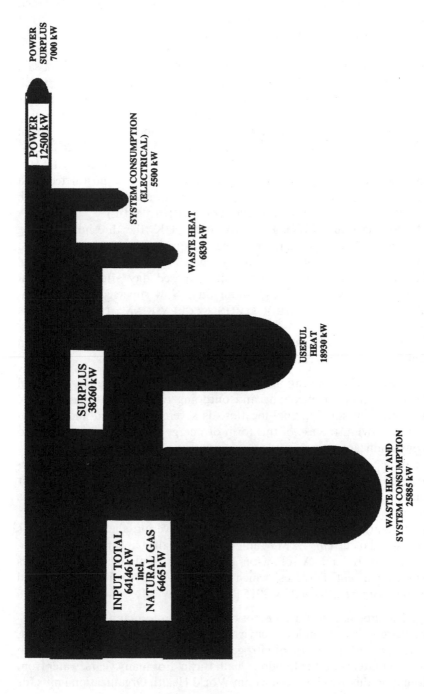

Figure 195 Energy balance for thermoselect process (two production lines, throughput of 10 te/h each). *Source*: R. Stahlberg, 'Thermoselect recovery of energy and raw materials from waste', *Recycling 95 Congress Proceedings*, Chalmers University, 1–3 February 1995.

the stack is set at 45 kilograms of lead per 168 hours which may not approach anywhere near the three-minute mean value, but is required to prevent despoilation of land through build-up of lead with its dangers to animals. (◊ EMISSION STANDARD)

Threshold. In environmental terms, the dividing line between exposure to, say, a noise level above which physical damage may take place and below which damage will probably not take place. The threshold is a concept that is applied in many areas of industrial exposure to airborne pollutants. However, the basis is that below the threshold none but the most sensitive or infirm would suffer. There are substances for which many believe there is no threshold effect whatsoever, e.g. VINYL CHLORIDE and ASBESTOS. IONIZING RADIATION is another area where a threshold may in effect be postulated by a recommended dose limit but the only safe exposure is no exposure at all. (◊ CARCINOGEN; MUTAGEN; THRESHOLD LIMITING VALUE)

Threshold limiting value (TLV). Formerly in the UK, the threshold limiting value of an airborne pollutant was the maximum concentration of that pollutant (or mixture of pollutants) to which it was believed 'healthy' workers in industry could be repeatedly exposed day after day without adverse health effects, on an eight-hour day. Now superseded by occupational control limits. (◊ CONTROL LIMITS, OCCUPATIONAL; TIME-WEIGHTED AVERAGE)

Tidal power. Generation of power by the ebb and flow of the tides. This is an example of an 'income' source of energy and takes two forms:

1. BARRAGES with conventional hydroelectric turbines installed in special sluices, so that the incoming and outgoing tides generate power. The availability of suitable tidal locations is severely restricted and, as with WATER POWER, the use of this form of energy source is not expected to grow significantly.
2. Floating power stations which rise and fall with the tide and have wave-powered floats or paddles which can drive water turbines for power generation. This method is not restricted to the availability of sites as for barrages.
3. A substantial development is now (2008) powering ahead. The world's longest 'tidal current' 1.2 MW unit is to be installed at Strangford Lough, Co. Down, Ireland. A follow-on 10 MW unit is planned by promoters 'Marine Current Turbines' which claims that up to 500 MW of tidal capacity can be achieved by 2015 [◊ WAVE POWER]

Time lag. The time needed for an event or disruption to manifest itself.

There are many examples in environmental matters of a time-lag effect. One such example is the use of nitrogen fertilizers on agricultural land from which water filters into the London Chalk Basin. For many years, water from the aquifer in this area has been within World Health Organization limits for nitrates; now after 20 years or more of fertilizer application, some wells are showing higher than normal nitrate levels.

The latency period for the development of cancer is well known. This can take tens of years dating from exposure to the carcinogenic agent in the first instant. It is for this reason that before the introduction of new health products, food additives, hair dyes, etc., the most exhaustive tests must be carried out for mutagenic and carcinogenic properties. Hair dyes are a good case in point; they have been implicated as being carcinogenic but confirmation will have to wait for at least a decade in all probability.

The time-lag effect often obscures the cause of the problem encountered and leads to treatment of the symptoms only. (◊ CARCINOGEN; MUTAGEN)

Time-weighted average (TWA). Used in the control of occupational exposure to hazardous substances or harmful emissions such as NOISE or IONIZING RADI-ATION. This is done by controlling the DOSE that the body is permitted to receive over a period of time. A long-term exposure limit (LTEL) is the sum of the concentrations averaged over 8 hours. If, say, for a compound this is 100 ppm, then exposure to a concentration of 200 ppm would only be permitted to 4 hours, which is the same as a TWA of 100 ppm for 8 hours, i.e. time-weighted.

Note that there is often an STEL or short-term exposure limit (concentration averaged over 10 min) for chemical or radiation exposure. In the above example, if the STEL were 600 ppm then exposure to 400 ppm for 4 hours is permissible. If the STEL were 300 ppm it would not be allowed. (◊ CONTROL LIMITS, OCCUPATIONAL)

Tioxide. Titanium dioxide (TiO_2) is used as a white pigment in paints, plastics and ceramics. On 17 February 1999, concentrated hydrochloric ACID contaminated 70 000 square metres of protected land at Seal Sands in Tesside. It is unclear how it leaked from a faulty drain at Tioxide Europe's Greatham Works, into a stormwater drain and onto the site. The ENVIRONMENT AGENCY served the company with a prohibition notice to ban it using the faulty part of the drain.

Sea water was pumped into the affected region of marshland to dilute the acid to a pH of 2.5 from 1.1.

Environment protection officers described the acid leak as one of the worst incidents they had ever attended. About half the 70 000 square metres of marshland was thought to have been affected and within a day or two, dead invertebrates were found coming to the surface of the water.

Biologists assessed the impact on wildlife in the marshland. The Environment Agency worked with the company to prevent a similar leak happening again.

The Environment Agency has prosecuted ICI on five occasions since 1997.

- 20 March 1997 – fined £15 000 over spillage of ethylene dichloride at Runcorn plant
- 2 July 1997 – fined £34 000 for chemical leak at Runcorn that polluted Weston Canal
- 14 October 1997 – fined £7000 for discharging trade effluent into Tees Estuary from North Tees Plant
- 12 March 1998 – fined £3 200 000 for chloroform leak at Runcorn plant
- 26 June 1998 – fined £80 000 for leak of metal cleaning chemical into Weston Canal at Runcorn.

Tioxide did not inform the public, or even the local fire brigade, for two days. Local MP, Frank Cook, was outraged at the company. 'It is almost CHERNOBYL-like in its reluctance to come into the light of day.'

Sources: *The Chemical Engineer*, 25 February 1999, p. 3; 'Minister and Agency condemn ICI acid leak', *Environmental Action*, April/May 1999, issue 19, p. 3.

Tisza. River in Hungary devastated by cyanide spill from a gold mining operation on 30 January 2000. One of Europe's greater ecological disasters. Concentrations measured >30 mg litre.

Titanium (Ti). The ninth commonest element in the earth's crust. Over 90 per cent of titanium produced is used in the form of the oxide, titania, as a white pigment or mineral filler in the paint, paper and plastic industries. Titanium metal and its alloys have a favourable weight-to-strength ratio, combined with excellent resistance to corrosion. (◊ TIOXIDE)

Titration. A method for the quantitative determination of a substance in solution by the addition in measured amounts of a reagent that reacts with the substance until the reaction is complete. This is indicated by a colour change in the solution, by precipitation, by the colour change of an added indicator, or by electrical measurement which indicates the end point of the reaction.

Tokaimura. A major NUCLEAR POWER-related mishap (30 September 1999) which stands comparison with those of CHERNOBYL and THREE MILE ISLAND.

The incident took place in a URANIUM processing plant where enriched fissile U_{235} (20 per cent) fuel elements were manufactured. The U_{235} mass in the process is deemed to have gone critical due to (a) too much U_{235} allowed to accumulate in the processing and (b) the presence of water which slows down (moderates) NEUTRONS which then causes further ATOMS to split, i.e. a nuclear reaction took place momentarily.

Radiation levels 1 mile from the site were said to be ca. 15 000 times 'normal' (at least 31 000 residents affected, 19 workers hospitalized).

A concerted exercise to reassure the public took place in both Japan and the West. The fact remains that human error is ever present and the fortunate lack of serious consequences should not detract from the human error implications attached to matters nuclear. The *Daily Telegraph* editorial on Takaimura on 2 October 1999 stated: 'To take Thursday's accident as a pretext for blanket condemnation is unjustified.'

On the IAEA scale of 0 to 7 for nuclear accidents, Tokaimura rated 4–5 (Chernobyl was 5).

Tolerable daily intake (TDI). A threshold value set by the EC Scientific Committee on Food for individual additives or trace chemicals below which no adverse health effects may be expected, e.g. the value for dioctyladipate (DOA) is 30 mg/kg body weight. ($\Diamond$ PHTHALATES)

Ton/tonne. The British (UK) ton is an old unit of weight equivalent to 20 hundredweights or 2240 pounds, avoirdupois measure. In the USA it is called the long ton or gross ton; the American equivalent is the short ton or net ton, 0.893 × UK ton. The metric equivalent, the tonne (tonneau) or metric ton, is equal to 1000 kilograms.

Tonne kilometres. A measure used by the rail industry to scale freight transport quantities, i.e. number of tonnes transported and distance (km) hauled. There is scope for improvement. In 1995, approximately 40 billion tonne-kilometres of freight were hauled; in 2006, approximately 22 billion tonnes – km. (*'Times'* report more freight trains will free up roads but disrupt rail journeys, B Webster, 6 Sept 06).

Total Daily Ingestion (TDI). A TDI is the amount of contaminant, expressed on a body weight basis, that can be ingested daily over a lifetime without appreciable risk.

Source: *MAFF Food Safety Information Bulletin*, No. 8, March 1998.

Total dissolved solids (TDS). The solids residue expressed in mg/litre after evaporating a sample of water or effluent.

Total energy. The integrated use of all or most of the heat generated by the combustion of fossil fuels or use of nuclear fuels. Thus, instead of just generating electricity, a total energy scheme would generate electricity and sell heat for both DISTRICT HEATING and factory processes. ($\Diamond$ COMBINED HEAT AND POWER)

Toxic action of pollutants. There are three main mechanisms by which the human organism is affected by toxic pollutants.

1. They influence enzymatic action by, for example, combining with an enzyme so that it cannot function. ($\Diamond$ ENZYMES)
2. They can combine chemically with the constituents of cells, as, for example, carbon monoxide combining with blood haemoglobin so that oxygen transport to the brain is affected.

3. Secondary action because of their presence. Hay fever is brought about by pollen and the system reacts to produce histamine.

The factors of importance are the concentration of the pollutant, the length of exposure, the age, the activity – whether slight or heavy exertion – and the health of the exposed person/population.

Toxic wastes. ◊ WASTES, HAZARDOUS (I); WASTES, OPTIONS (IV); DOSE.

Toxicity-based criteria for the regulatory control of wastewater discharges.
Toxicity-based conditions in discharge licences (UK) are based on the following stages:

1. discharge prioritization (effluent complexity and environmental impact);
2. discharge characterization (toxicity of discharge and predicted no effect concentrations (PNEC) which is compared to the actual concentration);
3. toxicity reduction and licensing (toxicity reduction based on 2 (above) and/or toxicity-based condition in licence);
4. compliance monitoring (self-monitoring by discharger, with EA audit and independent monitoring).

Figure 196 gives the outline protocol for setting toxicity-based conditions for discharge licences.

Source: Support paper for the Environmental Agency Consultation Document, 1996.

Toxicity characteristics leaching procedure. US test used as an indication of the leaching elements in incinerator ash. The acidic solutions used exaggerate the leachability of the ash. (US ash practice is to use monofills where the ash cannot come into contact with acids). Hence, there is minimal LEACHATE production from such sites and great confidence can be placed in ash monofills.

The alternative of LANDFILL of crude MSW can produce highly-polluting LEACHATES and LANDFILL GAS, both of which can require decades of after-treatment following deposition. Horses for courses.

Trace elements. Elements which occur in minute quantities as natural constituents of living organisms and tissues. They are necessary for the maintenance of growth and development; the shortage of any one may result in reduced growth, physiological troubles and eventually death. However, in large quantities they are generally harmful.

Trace elements include copper, silicon, cobalt, iron, zinc, iodine and manganese.

Tracer. Commonly used to determine the movement of groundwater through AQUIFERS (or LEACHATE in LANDFILL SITES) in order to determine both the direction and velocity of groundwater flow. They are also used to study dispersion in the atmosphere.

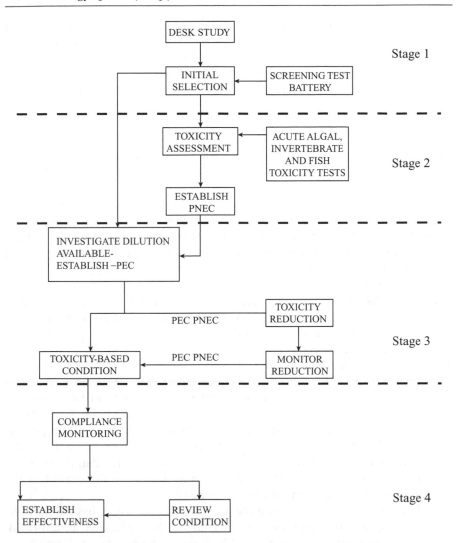

Figure 196 Protocol for setting toxicity-based conditions for discharge licences.

Tracers may be classified by method of detection, namely, colorimetry, chemical determination, radioactivity, electrical conductivity, etc. Certain radioisotopic tracers, e.g. tritium, can be used as field tracers without danger of contamination, but others must be carefully controlled because of dangerous radiation levels.

Tradable Energy Quotas (TEQs). TEQs are an electronic system for rationing energy. TEQs are measured in TEQ units. Every adult is given an equal number of TEQ units. Industry and Government bid for their units at a weekly tender. At a start of the scheme, a full year's supply is placed on the market. Then every week, the number of units in the market is topped up with a week's supply. Units can be traded. If you use less than your entitlement, you can sell your surplus. If you need more you can buy them.

Source: David Fleming, of the Lean Economy Connection Think Tank, *Your Environment*, published by the Environment Agency, April 2006.

Trade effluent. Any wastewaters from an industrial process. For direct discharge into a watercourse, strict consent conditions have to be met and expensive treatment plant may have to be installed. In England and Wales the Environment Agency generally determines consent conditions and maintains a register. It is often cheaper to discharge to a sewer and have the wastewater treated at the local sewage works. The trader pays a charge for the services rendered. The control of trade effluents to sewers is controlled by the Control of Pollution Act (1974).

Tramp elements. ◊ METALS.

Transfer station. A depot from where waste from local municipal waste collection vehicles is unloaded into larger road vehicles, rail wagons or barges for shipment to a treatment or disposal site.

Transferable drug resistance. ◊ ANTIBIOTIC.

Transfrontier [Waste] Shipment Regulations. The legal vehicle that waste may be shipped abroad for RECYCLING but not disposal. This means that UK household waste can end up in environmentally unsound Far East countries for recycling. Scope for concern exists in one case: Grosvenor Waste Management pleaded guilty to six out of 19 charges concerning the export of household materials abroad. The company agreed to pay £85 000 towards the Environment Agency's legal costs. The company stated that the prosecution were not able to say that the shipments that were the subject of the indictment, were destined for overseas landfill.

The waste management industry has a viewpoint which has been put as follows by a major company. Shanks Group has criticized 'ill-informed' national press coverage of recyclable material being exported to China. Director of Materials Management, Paul Dumpleton, said two million tonnes of UK material supposedly 'dumped' in China was in fact being purchased by re-processors there. 'This economic transaction by its very nature makes the dumping of recycled material in its delivered form totally uneconomic', he insisted.

Source: Legal Column, *Waste Management and Recovery*, 2007, 9 February, p4.

Transmissivity. The transmissivity of an aquifer is the product of the HYDRAULIC CONDUCTIVITY (permeability) and the aquifer thickness. It is an essential measure of the movement of groundwater through an aquifer, particularly in the vicinity of production wells. Measured in gallons per day per foot or square metres per day.

Transpiration. Water is transferred from the soil to the leaves of plants by capillary action and osmosis. At the leaf surface, the water transpires or evaporates and the vapour diffuses to the atmosphere.

As evaporation also takes place from the surfaces of lakes and rivers, it is common to use the term evapo-transpiration to account for the land-based water vapour component of the HYDROLOGICAL CYCLE.

Transport emissions. The Tyndall Centre for Climate Change Research estimated carbon dioxide emissions from transport to be as follows:

Car (petrol) – 298 g per passenger mile
Car (diesel) – 225 g per passenger mile
Rail – 116 g per passenger mile
Coach – 90 g per passenger mile

Source: *The Daily Telegraph*, 2 December 2006.

Transport emissions. Travel contribution of CO_2 – it all depends.
Amsterdam – Rome (805) miles

Coach	31 kg CO_2
Train	87 kg CO_2
Plane	153 kg CO_2
Car	154 kg CO_2

London–Manchester 300 miles

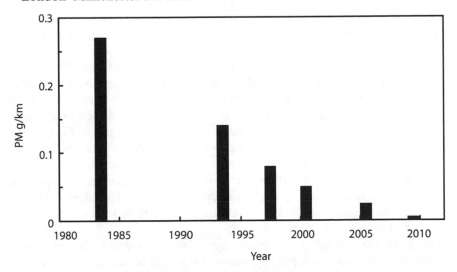

Figure 197 The decrease in European legislated PM passenger car emissions. Since 1983 the emissions have reduced by an order of magnitude and future stringent legislation will demand the fitment of filters.

Train	$13\,kg\ CO_2$	⎫ Commission for
Plane	$99\,kg\ CO_2$	⎬ integrated transport.
Car	$53\,kg\ CO_2$	⎭

On short distances, aircraft emissions are disproportionately greater due to take-off and landing fuel use.

(Dutch) Climate Action Network – Europe.

Check your car's CO_2 emissions. Are they 400 g/km (SUV) or 150 g/km (compact)? You can make a difference. How you drive is significant: e.g. 0–62 mph: 1.6 l engine (Mitsubishi Lancer) in 11.8 sec, equates to 163 g/km; 2 l engine (10 sec, 0–62 mph) 207 g/km.

Source: Mitsubishi Lancer brochure, 2006.

BP has launched a website WWW.TARGETNEUTRAL.COM to allow vehicle drivers to calculate the cost of making their CO_2 emissions carbon neutral, e.g. via renewable energy schemes or biomass gasification. The average UK car driven 10 000 miles/year produces 4 tonnes CO_2.

Tributyltin (TBT). A marine anti-fouling agent which is on the RED LIST. It is extremely toxic; concentrations as low as two parts per trillion can affect marine life. Banned in the UK and France for use on boats greater than 25 m in length.

Trichlorofluoromethane ($CFCl_3$) (Freon 11). A member of the group known as halogenated fluorocarbons, used as an AEROSOL PROPELLANT. It breaks down chemically, releasing free chlorine which combines with ozone to deplete the earth's OZONE SHIELD. However, due to TIME LAG, the effects are only now becoming apparent. If all $CFCl_3$ production were stopped immediately, the decomposition will still continue for decades. The EC instituted a total sales ban on all CFCs in 2000. (◊ CHLOROFLUOROCARBONS)

Trichoderma viride. A fungus which is capable of breaking down crystalline cellulose by the production of an ENZYME of the cellulase group which can accomplish the decomposition in a few days. Mutant strains of *Trichoderma viride* have been developed which accelerate the decomposition. *Trichoderma viride* is the basis of the Natick process for the production of glucose from cellulose. (◊ ENZYME TECHNOLOGY)

Triple bottom line. Shorthand for the simultaneous provision of economic, environmental and social gains by responsible industrial and service organizations. Social accountability will feature even more in the future and may be a highly subjective topic.

Triple point. The temperature, pressure (and volume) at which the solid, liquid and gaseous phases of a pure substance are in equilibrium. The triple point of water is used as a fixed reference point in physics and thermodynamics, and is 0.01 °C (273.16 K) at an almost total vacuum of 0.006112 bar. This enables practical and consistent temperature scales to be defined. ($\lozenge$ CRITICAL POINT; TEMPERATURE; PRESSURE)

Trippage. (1) The number of trips that a returnable bottle makes in its lifetime. In the UK, average trippage is 25 times for a milk bottle. In the USA, 10 times for a soft drink bottle. ($\lozenge$ RECYCLING)

(2) The 'shutting down' of, for example, a steam power plant, due to malfunction or overload.

(3) The number of round trips that haulage vehicles can make in a given period.

Tritium (H$_3$). An ISOTOPE of hydrogen, atomic mass number 3. It is used as a fuel in fusion power research.

Trophic levels. $\lozenge$ FOOD CHAIN.

Tropopause. The upper limit of the TROPOSPHERE.

Troposphere. The 'lower' portion of the atmosphere about 8 kilometres high at the poles and 16 kilometres high at the equator. In the troposphere the earth's temperature usually decreases with height. It is marked by considerable turbulence and contains the bulk of the atmospheric mass. ($\lozenge$ STRATOSPHERE)

Turbidity. Reduced transparency of the atmosphere, caused by absorption and scattering or radiation by solid or liquid particles, other than clouds, held there in suspension. In a liquid, scattering of light due to suspended particulate matter.

Turbidity monitoring through the water treatment process is a vital element in checking that the treatment operations are working properly.

> The unifying factor in all CRYPTOSPORIDIUM outbreak situations is the potential for peaks in turbidity to be present in the treated water leaving the works. The fact that turbidity events were not recognised in all cases could be a reflection of inadequacy in the continuity of turbidity monitoring, the interpretation of results, or in the calibration and control of the equipment.

Source: DETR Expert Views Report 1998.

Turbulence. Movements which are uncoordinated and in a state of continuous change in liquids and gases.

Turndown ratio. The ratio of maximum output to minimum output of a plant or boiler based on continuous operation. ($\lozenge$ THERMAL EFFICIENCY)

TWA. $\lozenge$ TIME-WEIGHTED AVERAGE.

Typhoid. Water-borne disease caused by drinking sewage-contaminated water. ($\lozenge$ PUBLIC HEALTH)

TX Active. TX Active is a photocatalytic cement which when used in building materials and coverings, absorbs and eliminates 20 to 80 per cent of air pollutants, depending on atmospheric conditions and the level of sunlight available to trigger the process. Titanium dioxide is combined into wall facings and road surfaces which can remove pollutants highly effectively. Sunlight starts a chemical reaction between the CO_2 and the titanium dioxide that removes the gas from the air and crystallizes it into a salt that sits on the surface of buildings and roads until washed away by rain. Laboratory tests on photoactive cements indicated that three minutes of sunlight are sufficient to reduce pollutants by up to 75 per cent.

Tyres. There are approximately 0.5 million tonnes per year (tpy) of scrap tyres (2006) produced in the UK. These have a gross energy value of 0.5 tpy of COAL equivalent and burning them for ENERGY recovery is one way of both effecting their disposal and recovering something of value. Now banned from LANDFILL under the EU LANDFILL DIRECTIVE, they can be used for landfill engineering purposes in leachate drainage.

Methods of reuse include rubber crumb manufacture for shock-resistant surfaces in playgrounds, or as an asphalt additive, but both of these uses are costly. Another method is to bale them for reinforcing access roads on moors and woodlands, etc.

One prototype US approach is bioprocessing to remove the SULPHUR in the tyres by grinding the rubber to MICRON size at a cost of $1.4/kg which compares with virgin rubber compounds at $2.2/kg.

The Pacific Northwest Laboratory has successfully demonstrated the technology (shown schematically in Figure 198), using micro-organisms which selectively attach to the carbon-sulphur cross-links.

Another approach is to use PYROLYSIS to recover the ENERGY content, plus the steel reinforcement in the tyre casings. The mass and energy balance for one such scheme is given in Figure 199.

The Environment Agency has made the following recommendations:

- Tyre manufacturers should develop longer-lasting, quieter and more energy-efficient tyres.
- Drivers should take better care of tyres, cutting out unnecessary journeys and driving more carefully to increase their lifespan. Correct tyre pressures are not only essential for safety but save energy and increase the life of a tyre.
- The life of tyre casings can be increased by retreading them. Drivers should more readily consider using retreaded tyres, and manufacturers be more ready to make them.
- Greater use should be made of granulated rubber, for example in road and playground surfaces. Less than 11 per cent of used tyres are currently recycled in this way in the UK. Other countries such as France have brought in legislation to increase this form of recycling.
- More ways should be found to recover the energy value of worn or scrap tyres. (Cement kilns already use them as a fuel.) (◊ INCINERATION; GAS-IFICATION; PYROLYSIS)

Source: Tyres in the Environment, Environment Agency, February/March 1999.

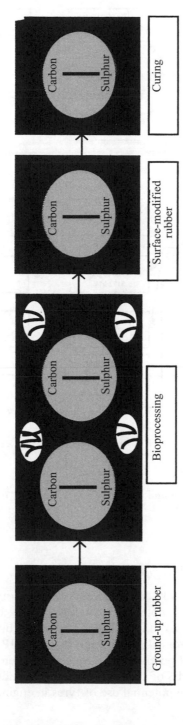

Figure 198 Bioprocessing of waste tyre rubber.
Source: Bioprocessing helps make waste tyres recyclable, *World Wastes*, March 1995.

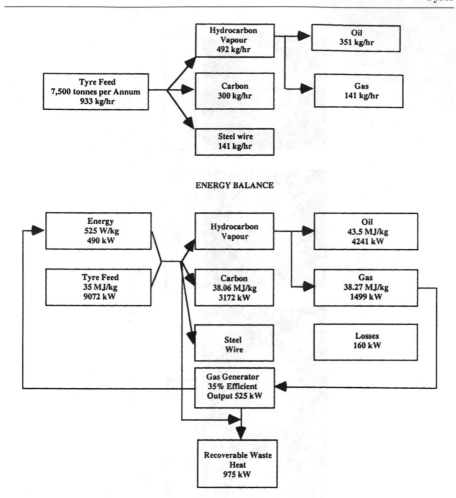

Figure 199 Tyre pyrolysis mass and energy balances (courtesy of BRC Environmental Services Limited, February 1994).

Case Study on sustainable re-use of tyres

H R Wallingford led a study on the potential reuse of tyres in port, coastal and river engineering, prompted by a European-wide ban in 2003, on the disposal of whole tyres to landfill sites. The study built on earlier work undertaken at Southampton University and at TRL-VIRIDIS. Research had identified a wide range of potential uses, especially when the tyres were compressed and bound in bales of about 100. The study addressed the full range of issues arising from the potential use of tyres in engineering works and in the water environment. Sites were monitored for physical movement, stability, and chemical leaching.

The scale of the used tyre problem is shown by the data in the table below, together with figures on the recovery options:

Tyre arisings and use of recovery options for various EU countries in 1998

Country	Tyre Arisings (tonne)	Overall Recovery Rate (%)	Reuse (%)	Retreading (%)	Materials Recycling (%)	Energy Recovery (%)	Export (%)
Belgium	45 000	94		22	11	33	28
Finland	3000	80		6	60	2.5	11.5
France	370 000	39		20	9	7	3
Germany	596 000	92	2	24	15	45	16
Italy	333 000	60		25	9	33	3
Netherlands	45 000	100	16	29	8	47	
Spain	241 000	19		13.5	0.5	3.5	1.5
Sweden	58 000	98	19	8.5	6.5	54	10
UK	468 000	70	16	18.5	10.5	18	7.5

Tyres in use on vehicles contribute to water pollution via road run-off. Inappropriate disposal can lead to tyre fires causing pollution of the air, and contamination of land and vegetation. Illegally-dumped tyres are unaesthetic and a fire risk. Dumping of tyres can occur on a massive scale, such as in the Hampole Quarry in Yorkshire, where it is estimated that around 1.5 million tyres were dumped illegally between the late 1970s and the early 1990s. A fire will release acrid smoke and mobile contaminants such as zinc and phenol.

Used tyres have been utilized in several ways. These include:

The use of whole tyres in:
Embankments / retaining walls using loose tyres
Embankments / retaining walls using linked tyres
Post-erosion protection
Anti-scour mattresses
Floating breakwaters
Submerged reefs
Fenders for shipping

The use of tyre bales in:
Hearting to concrete structures
Hearting to coastal rock structures
Hearting to embankments
Beach nourishment substitute

Armouring to banks of small rivers and streams
The use of tyre chips in:
Lightweight backfill behind retaining walls
Aggregate in asphalt concrete

The use of tyres in port, coastal and river engineering, has several benefits:

- they can offer distinct engineering benefits over traditional aggregates;
- they can be used as an alternative to primary materials, thereby reducing an environmental burden on extraction;
- their use can help to reduce the burden of waste disposal (including illegal stockpiling and disposal, such as flytipping, with their associated risks);
- the impacts on the environment associated with some other uses of old tyres can be reduced.

Source: H R Wallingford, Sustainable re-use of tyres in Port, Coastal and River Engineering: Guidance for planning, implementation and maintenance, Report SR 669, H R Wallingford, Wallingford, Oxfordshire, 2005.

U

UK Ambient Air Quality Standards and Objectives. The framework for improving air quality in the UK is set by Part IV of the Environment Act 1995. Two main elements of the Act are:

(i) a requirement to produce a National Air Quality Strategy containing standards, objectives and measures to achieve objectives.

(ii) a Local Air Quality Management system giving statutory duties for local authorities to review their air quality, assess it against air quality objectives set in regulations and, where it is likely the objectives will not be met, declare air quality management areas and draw up and implement appropriate action plans.

The National Air Quality Strategy was published in March 1997. The Strategy contains standards and policy objectives for eight pollutants, based on advice from the independent Expert Panel on Air Quality Standards (EPAQS).

Ultimate analysis. The proportion by weight of the chemical and inert constituents of a FUEL. Usually done as an elemental analysis by percentage of CARBON, OXYGEN, HYDROGEN, NITROGEN, SULPHUR, water and inerts (ASH).

The table (overleaf) gives analyses for a range of solid fuels from anthracite to wood. It illustrates the great care and accuracy which combustion engineers put into their work. It is not a black art but a robust scientific and engineering-based industry.

Ultrafiltration. A MEMBRANE separation process which uses larger pore size membranes than those of REVERSE OSMOSIS. Used in the tertiary treatment of biologically treated sewage effluent.

Selected Ultimate Analyses for Solid Fuels.

Fuel	Composition (per cent)							Calorific value (MJ/kg) as sampled	
	Carbon	Hydrogen	Nitrogen	Sulphur	Oxygen	Moisture	Ash	Gross	Net
Coals									
Anthracite	78.2	2.4	0.9	1.0	1.5	8.0	8.0	29.66	28.94
Coking steam	77.1	3.5	1.2	1.0	2.2	7.0	8.0	30.70	29.75
Medium volatile coking	75.8	4.1	1.3	1.2	2.6	7.0	8.0	30.82	29.75
Non-coking	59.0	3.7	1.2	1.7	8.4	18.0	8.0	23.84	22.59
Wood	42.5	5.1	0.9*		36.5	15.0	Trace	15.82	14.35
Peat	43.7	4.2	1.5*		26.6	20.0	4.0	15.91	14.49
Lignite	56.0	4.0	1.6*		18.4	15.0	5.0	21.45	20.19
Charcoal	90.2	2.4	1.5*		2.9	2.0	1.0	33.70	33.12
Coke	82.0	0.4	1.7*		0.9	8.0	7.0	28.63	28.35
Low temperature coke	79.0	2.6	1.7*		1.7	8.0	7.0	29.19	28.42

* Nitrogen and sulphur.

Source: Adapted from Kemps Engineers Yearbook 1998, 'Solid Fuels'.

Ultrasound. SOUND waves which are above the normal level of hearing (above a frequency of 20 kHz or 20 000 cycles per second).

Ultrasound is used in cleaning of fine parts, measuring fluids, and for medical purposes. Also, chromic acid mist, which is generated by electroplating processes (because of the high current densities used), can be reduced by ultrasound application.

Ultrasound is also used in silver recycling. The slowest stage in silver electrolysis is the silver ion's journey across the diffusion layer, a static layer of fluid that normally clings to solid surfaces, including cathodes. This holds up the whole process. Ultrasound disrupts this layer by creating tiny bubbles in the fluid that send out high-pressure jets as they collapse. These jets squirt through the diffusion layer and give the ions an easier passage.

Sources: *HSE Newletter*, No. 109, October 1996; *New Scientist*, 24 August 1996.

Ultraviolet disinfection. In sewage effluents, even after secondary treatment there are large populations of micro-organisms of enteric origin present. These comprise bacteria and viruses, many capable of causing disease. Microbial cells contain nucleic acids which store genetic data and to survive and reproduce, a cell must be able to replicate the biochemical data in these nucleic acids. As these acids can absorb light of different wavelengths, it has been shown that ultraviolet irradiation can be absorbed by them, thereby damaging or rearranging their genetic information. That damage renders a cell unable to replicate and results in its death. It is therefore possible to arrange for secondary sewage effluent to be irradiated by a suitable array of ultraviolet lamps in order to render the final effluent disinfected. The table below shows the results before and after UV treatment of the Bellozanne treatment plant effluent in the State of Jersey.

Example of results from small-scale trial.

Intensity (mW/cm^2)	T%	S.S. (mg/l)	Total coliforms/100 ml			Faecal coliforms/100 ml		
			Before	After	Reduction	Before	After	Reduction
4.8	56	—	56 000	30	99.95	5000	<10	99.80
1.8	10	71	14 000 000	290 000	97.93	2 200 000	85 000	94.14
			10 000 000	290 000	97.10	2 000 000	69 000	96.5
4.8	50	7	90 000	50	99.94	4000	<10	99.75
			79 000	40	99.95	2000	<10	99.50

SS = Suspended Solids.
T = Transmissivity.
Sources: D. O. Lloyd and J. Clark, 'Waste Management in Jersey', Paper 5, States of Jersey Seminar, March 1993.
G. W. Shepherd, Paper 20, States of Jersey Seminar, March 1993.

Ultraviolet radiation (UV). Radiation which falls between visible light waves and X-rays. The longest UV waves have wavelengths slightly shorter than those of ultraviolet light (the limits of the human eye). UV radiation stimulates the body to produce vitamin D.

Unburnt hydrocarbons. Airborne particles of hydrocarbon fuels not consumed in combustion and which are emitted in exhaust or flue gases particularly from an INTERNAL COMBUSTION ENGINE. Main components are CARBON particles, PARAFFINS, olefins, and aromatic compounds. (◊ AUTOMOBILE EMISSIONS; PHOTOCHEMICAL SMOG)

Uncertainty principle. For pairs of related particle mechanics quantities (e.g. position and momentum/time and energy), it is impossible to determine both variables accurately, e.g. position can normally be determined very accurately and momentum cannot. The product of the errors cannot be determined to an accuracy less than Planck's constant (which has the value of energy × time and is 6.62×10^{-34} Js) divided by 2π.

UN Convention on the Law of the Sea (UNCLOS) 1982. This ground-breaking Convention had its foundations in the run-up to the Third UN Conference on the Law of the Sea in 1973. It stipulates: limits of territorial waters to be 12 nautical miles from the coast; no weapons of mass destruction to be on or under the sea bed of the high seas; Economic Exclusion Zones to be 200 nautical miles from land; a procedure for settling disputes (International Tribunal on the Law of the Sea). It was finally adopted by the UN in 1994. Fisheries and pollution are two important topics still to be tackled, but UNCLOS is a tribute to the UN's persistence.

Unit cost. The cost per unit of production or service, e.g. the cost per tonne of waste incinerated.

Unsaturated zone. That part of an aquifer where water only partially fills the interstices, i.e. the unsaturated zone exists everywhere above the WATER TABLE.

Uranium (U). The principal radioactive element used in the production of nuclear (or radioactive) energy. Uranium has several isotopes but only the natural isotope known as ^{235}U is 'radioactive' or unstable. Uranium-235 comprises only about 0.7 per cent of all natural uranium, the rest being largely ^{238}U which can only be used as a nuclear fuel after treatment.

Uranium-235 breaks down into lighter elements, neutrons and heat during nuclear fission. The 238 form can be converted into the active element plutonium-239 by the process known as breeding, and the form of the element thorium, known as thorium-232, can be converted into fissionable ^{233}U in a breeder reactor.

1 kilogram of uranium produces the thermal energy equivalent to about 11 800 barrels of crude oil. Conversion efficiencies are less than fossil-fuelled power stations as the maximum steam temperature of nuclear reactors is lower. (◊ NUCLEAR POWER; NUCLEAR REACTOR DESIGNS)

Urea. A nitrogenous component for FERTILIZER manufacture. Also used as an animal feed additive and in plastics manufacture (urea formaldehyde). Manufactured from CARBON DIOXIDE and AMMONIA reacted at high temperature and pressure.

Formula $H_2N \cdot CO \cdot NH_2$

Also used in NITROGEN OXIDES control in fossil fuelled boilers or waste incinerators.

V

Vacuum filter. Used in SLUDGE dewatering. Usually consists of a cloth-covered drum or belt which is partially submerged in the sludge. The vacuum applied to the other side of the cloth effects dewatering.

Valdez. Name of a supertanker owned by the US Exxon group which ran aground on a reef in Prince William Sound, Alaska (March 1989), spilling 11 million gallons of crude oil and causing one of the largest anthropological ecological disasters. The oil slick covered over 3000 square miles.

Ten years on: the recovery of the species below was scientifically investigated.

Not recovering	Recovering	Recovered
Common loon	Black oystercatcher	Bald eagle
Cormorant	Clams	River otter
Harbour seal	Common murres	
Harlequin duck	Intertidal communities	
Killer whale	Marbled murrelets	
Pigeon guillemot	Mussels	
	Pacific herring	
	Pink salmon	
	Sea otter	
	Sediments	
	Sockeye salmon	
	Subtidal communities	

Investigators noted that the heavier components, the polynuclear aromatic hydrocarbons (PAHs), were a serious problem. Experiments showed that a suspension of less than 0.5 ppb was enough to burst the cell membranes of the fish studied. This suggests that no coastal region can ever be protected from oil pollution; offshore oil platform and ballast water discharge practices still need review as these can increase PAH concentrations up to 15 ppb.

Source: *The Chemical Engineer*, 25 March 1999.

Valley wind. Daytime air flow up a valley caused by the upwards convection of heated air from surrounding hills.

Valorization. The recovery of value from waste. This can (and does) embrace RECYCLING, COMPOSTING and energy recovery. It embodies a pragmatism which, in the case of PACKAGING wastes after RECYCLING, has been carried out (or judged inappropriate in the case of contaminated PLASTICS); INCINERATION with energy recovery can be practised. The name of the game is added value and not dogmatic adherence to one solution only.

Valpak. The body set up by the UK packaging industry to recover real value from PACKAGING waste.

Its functions are as follows:

- Valpak will agree with material-specific organizations (MOs) acting on behalf of businesses in that material sector what funds will be required to execute the plan. Valpak will also agree with waste management companies and local authorities their funding needs.
- Valpak will register appropriate companies (including importers of packaged goods) and inform them of the formula to be applied in calculating the packaging levy that they should pay to Valpak. The Producer Responsibility Group (PRG) together with the relevant Government Departments has agreed to appoint an independent expert to recommend the point(s) within the packaging chain at which the levy should be raised. Companies who ship packaged goods directly to industrial or commercial organizations will be treated like packers.
- Valpak will manage the collection and disbursement of funds.
- Valpak will co-ordinate the setting up of additional domestic collection and sorting schemes to meet re-processing and end-use market demand.
- Valpak will monitor value recovery achievements and regularly report on progress.

LANDFILL diversion and value recovery are greater for commercial (and industrial) packaging waste than for domestic waste. This makes environmental and economic sense because it is easier to collect and re-process commercial material owing to its relative non-contamination, and concentration/amount available often at a single source of supply (e.g. supermarkets) vis-à-vis contaminated, dispersed materials from households. (◊ VALORIZATION; PACKAGING)

Source: *Report on Public Consultation*, Producer Responsibility Group, Committee Response to the PRG 'Real Value from Packaging Waste', June 1994.

Value. The worth of something to its owner. There are two varieties. Value in use is the pleasure a commodity actually generates for its owner, or the public at large. Value in exchange is the quantity of other commodities (or more usually money) a commodity can be swapped for. Fresh air, for example, has high value in use, but low value in exchange, ditto peace and quiet. (◊ COST-BENEFIT ANALYSIS)

Vanadium (V). Hard, white metal, used as a steel alloying element and as a chemical industry catalyst. The THRESHOLD LIMITING VALUE is 500 micrograms per cubic metre for dusts and 50 micrograms per cubic metre for fumes of vanadium pentoxide (V_2O_5).

Vanadium affects most metabolic processes in the human organism. The lethal dose is between 60 and 120 milligrams. Chronic exposure to environmental air concentrations of vanadium can lead to bronchitis.

Van Allen radiation belts. Two regions of charged atomic particles surrounding the earth, formed by the earth's magnetic field. The inner belt is ca. 1.6 earth radii and the outer ca. 3.7 earth radii. Discovered by US physicist J. Van Allen in 1958.

Vapour pressure. The pressure exerted by a vapour. Usually saturated vapour pressure, i.e. the pressure of the vapour in contact with the parent liquid, e.g. water vapour PRESSURE. Saturated vapour pressure increases with TEMPERATURE hence STEAM can be generated if the parent liquid (water) is contained at high pressure in boiler tubes.

Variable. A number that may take different values in different situations. For instance, CARBON DIOXIDE produced by a car varies according to the amount of oil-derived FUEL consumed and engine EFFICIENCY.

Variable costs. Costs which vary directly with the rate of output, e.g. labour, FUEL and raw material costs. Also known as operating costs.

Variance. A measure of the degree of dispersion of a series of numbers around their mean. The greater the variance, the larger the spread of the series around its mean. (◊ AVERAGE)

VCM (vinyl chloride monomer). ◊ POLYVINYL CHLORIDE.

Vector. An organism (e.g. animal, fungus) which transmits or acts as a carrier of parasites, e.g. the *Anopheles* mosquito is the vector for the malaria parasite, the *Aedes* mosquito the vector for yellow fever, the rat flea the vector for the plague, and the tsetse fly the vector for sleeping sickness.

The use of insecticides to eliminate such vectors, together with generally improved standards of medical practice, have resulted in significant reductions in death rates in developing countries.

Vehicle emissions.

The problem

Holiday smog. Britain suffered its worst winter smog for years over Christmas 1994 but the Government failed to issue a health warning.

High levels of nitrogen dioxide, which brings on respiratory disease, and is one of the main factors in the asthma epidemic, settled over four of the country's biggest cities for many hours on 23 December and Christmas Eve.

The holiday smog, which followed the most polluted summer this decade, reached dangerous levels when nitrogen dioxide emitted by a rush of cars on last-minute Christmas errands failed to disperse because of a TEMPERATURE INVERSION (when cold air near ground level is trapped by warmer air higher up).

The nitrogen dioxide gas reached 198 ppb (parts per billion) in east Birmingham on 23 December, nearly treble the previous record for the year, and 257 ppb in Walsall, about two and a half times as bad as previously.

High levels were also recorded in Leeds and Manchester. London suffered worst, with two days of pollution twice as bad as anything that had been encountered all year. Over 23 December and Christmas Eve there were 27 hours of serious nitrogen dioxide pollution over much of the centre of the city. The highest level of all, 288 ppb, was recorded near Victoria.

The crisis passed on Christmas Day as the traffic died down and winds dispersed the pollution.

At no time during the crisis did the Government issue health warnings to advise people with respiratory trouble to stay indoors or ask drivers not to take out their cars, even though it had issued them during the polluted summer.

The Department of the Environment said that it had not warned the public because it thought that the weather was going to change earlier than it did.

The amounts

The table opposite gives the amount of airborne pollutants from road transport, by class of vehicle, for 2004.

How gaseous emissions are measured. Gaseous emissions can be measured directly from either undiluted hot exhaust or from cool exhaust after dilution with air.

UK Transport Emissions in 2004 (all in kilotonnes)

Type of Vehicle	Carbon Dioxide	NO_x	PM_{10}	VOCs	SO_2	CO
Buses and Coaches	966	39.40	1.27	2.71	0.09	8.27
Cars	19 444	246.23	12.89	103.41	1.64	1171.80
HGVs	7 577	251	7.87	19.60	0.63	47
LGVs	4 360	59.19	12.79	9.57	0.39	68.96
Mopeds and Motorcycles	117.2	1.19	0.69	9.38	0.06	71.50

Source: Accessed 26 January 2007.

HC	Heated flame ionization detector
NO_x	chemiluminescence analyser
CO	non-dispersive infrared
Particulates	These are usually measured after diluting the exhaust with air to simulate the condition at the outlet of the exhaust system. A sample is drawn through a filter and weighed. Particulates are sampled from a dilution tunnel, either a full flow constant volume sampling system, or a mini-dilution tunnel in which only a sample of the exhaust is diluted.
Smoke	Can be measured in the undiluted exhaust either by passing a sample through a white filter and measuring the darkening effect, or by passing a beam of light through the exhaust and measuring the opacity.
Sulphur dioxide	UV fluorescence

Reduction of vehicle emissions

Volvo Trucks has cut emissions from its transport operations in and around its manufacturing bases in Gothenburg, Sweden. Traffic-generated pollution has been cut by 50 per cent in just 5 years, by using cleaner fuels, improving vehicle scheduling and productivity, and from developments in truck design.

Particulate emissions have been reduced by 25 per cent and the level of hydrocarbons and nitrogen oxide has decreased by approximately 10 per cent each by using diesel fuel with lower levels of sulphur and aromatics both in suppliers' vehicles and in Volvo's own trucks.

Relocating loading and unloading terminals for goods and finished products, improving road signs, driver information points and equipment have produced reductions in regional transport of around 2800 km/day. And by co-ordinating incoming deliveries to Volvo and making more use of information technology, long haul transport volume has decreased by around 1600 km/day. Further reductions of up to 300 km/day are envisaged if plans for combined road/rail transport are implemented.

The advanced technology in Volvo's new 12-litre engine introduced in its FH range in the autumn of 1993 brought further benefits, albeit at a late stage in the programme. As Volvo's older vehicles are replaced, the D12A engine in the FH series will make a significant contribution towards reducing emission levels still further.

Source: Issued by Volvo Truck and Bus Ltd, December 1994.

Legislation. The emission legislation in the European Community is set out in the following pages.

Another option is to switch to rail or water transport.

Electric cars. One should be aware that for electric vehicles, generating electricity for recharging batteries can cause considerable environmental harm.

Reduction of noise nuisance from transport

The Royal Commission on Environmental Pollution 18th Report proposed the following:

To reduce daytime exposure to road and rail noise to not more than 65 dB $L_{Aeq.16h}$ at the external walls of housing;
To reduce night-time exposure to road and rail NOISE to not more than 59 dB $L_{Aeq.8h}$ at the external walls of housing.

Noise measurements relating to transport. In assessing human exposure to noise we also need to take account of variation in noise levels over time. This can be expressed in a number of different ways, for example:

dB L_{Aeq} the mean level of the sound
dB L_{A90} the level of sound exceeded for 90 per cent of the time (represents the average low level or background noise)
dB L_{A10} the level of sound exceeded for 10 per cent of the time. (represents the average peak level)

Other suffixes are used to represent the level of exposure over all or part of a 24-hour period, for example:

dB $L_{Aeq.8h}$
dB $L_{Aeq.16h}$
dB $L_{Aeq.18h}$

Vehicle-kilometres ('Food miles'). The aggregate distance travelled by (say) HGVs transporting food for supermarkets. This figure rose by 5 per cent in 2003–2004 (Defra, 2006). If measured in tonne-kilometres, the increase would

Passenger car exhaust emissions legislation (EEC Directives 91/441/EEC, 94/12/EEC etc.).

Applicability	Effective date											
	1992 Stage I				1994		1996 Stage II*				1999 Stage III***	
	IDI		DI		DI		IDI		DI		IDI & DI	
	TA	CoP	TA	CoP	TA	CoP	TA	CoP	TA	CoP	TA	CoP
HC & NO$_x$ (g/km)	0.97	1.13	1.36	1.58	0.97	1.13		0.7		0.9	0.5	
CO (g/km)	2.72	3.16	2.72	4.42	2.72	3.16		1		1	0.5	
Pm (g/km)	0.14	0.18	0.2	0.25	1.14	0.18		0.08		0.1	0.04	

Notes: *, Proposal; ***, Stage III at discussion stage; TA, type approval; CoP, conformity of production; DI, direct injection; IDI, indirect injection.
Source: Automotive Diesel Engines and the Future, Ricardo Consulting Engineers Ltd, Shoreham-by-Sea, West Sussex, 1994.

Heavy duty engine exhaust emissions legislation (EEC Directive 91/542/EEC).

Applicability	Effective date							
	Until June 1992		July 1992 Stage I		October 1995 Stage II		1999? Stage III (Discussion)	
	TA	CoP	TA	CoP	TA	CoP	TA	CoP
HC	2.4	2.6	1.1	1.23	1.1		0.7	
CO	11.2	12.3	4.5	4.9	4.0		2.5	
NO$_x$ (g/kWh)	14.4	15.8	8.0	9.0	7.0		<5.0	
PT								
<85kW			0.61	0.68	0.15		<0.12	
<85kW			0.36	0.4				

Source: Automotive Diesel Engines and the Future, Ricardo Consulting Engineers Ltd, Shoreham-by-Sea, West Sussex, 1994.

Air quality standards: WHO guidelines, EC standards (bold type) and EC guide values (italics).

	Less than 1 hour	1 hour	8 hours	24 hours	Units
Carbon monoxide	100 (15 mins) 60 (30 mins)	30	10	1	mg/m^3
Nitrogen dioxide		400 **200**		150	μg/m^3
Sulphur dioxide	500 (10 mins)	350		*100–150* *40–60*	μg/m^3
Combined exposure to sulphur dioxide and suspended particulates (black smoke)				125 SO$_2$ plus one of the following: 125 black smoke 120 total suspended particulates 70 thoracic particles	50 SO$_2$ and 50 black smoke
Suspended particulates (black smoke)				**80** **130** **250** *100–150* *40–60*	
Ozone (health)		150–200	100–120 **110**		μ g/m^3
Ozone (vegetation)		200 **200**	65 **65**		**60** (April to September, over the growing season)
Lead					0.5–1.0
Formaldehyde	10 (30 mins)				**2.0**

Source: Royal Commission on Environmental Pollution, 18th Report, 1994.

Freight transport modes: energy use and emissions.

	Rail	Water transport	Road	Pipeline	Air
Specific primary energy consumption (kJ/tonne-km)	677	423	2890	168	15839
Specific total emissions (g/tonne-km)					
Carbon dioxide	41	30	207	10	1206
Methane	0.06	0.04	0.3	0.02	2.0
Volatile organic compounds	0.08	0.1	1.1	0.02	3.0
Nitrogen oxides	0.2	0.4	3.6	0.02	5.5
Carbon monoxide	0.05	0.12	2.4	0.0	1.4

Source: J. Whitelegg, Traffic Congestion – Is there a way out?, Leading Edge Press, 1992.

have been greater. The difference is due to the use of longer HGVs. Once again, an industry will choose its norms for comparison which put its impacts in the best light. What can be said is that food industry CO_2 vehicle emissions have increased (by 5 per cent) and hence the impacts are increasing. A strong case for sound ecological purchasing and food miles labelling perhaps? Do we really need roses grown in Kenya for the UK market?

Venturi effect. A local decrease of PRESSURE, e.g. caused when the wind blows through a narrow mountain pass or between buildings. The effect can be put to use in pipes by inserting a contracting length, a throat and then an expanding length and measuring the pressure drop between the throat and upstream. This can then be correlated with the flow rate of the fluid in the pipe. Also used in a high efficiency SCRUBBER for air pollution control. (⧫ VENTURI TUBE)

Venturi tube. A tube whose internal diameter gradually decreases to a throat and then gradually increases again to its original value. Such tubes are used in flowmeters and also in venturi SCRUBBERS.

Vermicomposting. Vermicomposting is the breaking down of organic matter by some species of earthworm. Vermicompost is a nutrient-rich, natural fertilizer and soil conditioner. The earthworm species (or composting worms) most often used are Red Wigglers (*Eisenia foetida*) or Red Earthworms (*Lumbricus rubellus*). These species are only rarely found in soil and are adapted to the special conditions in rotting vegetation, compost and manure piles. Small-scale vermicomposting is well suited to turning kitchen waste into high-quality soil, where space is limited. In addition to worms, a healthy vermicomposting system hosts many other organisms such as insects, mould and bacteria. Although these all play a role in the composting process, the earthworm is the major catalyst for the composting process.

Vinyl chloride (CH_2CHCl). A colourless gas used in the manufacture of POLYVINYL CHLORIDE (PVC), a PLASTIC. The vapour has been found to be highly carcinogenic and has been linked with the deaths of 20 workers globally from angioma, a rare form of liver cancer. In the UK there is evidence that the vapour has caused impotence, stiffening of the joints, bad circulation and shortness of breath. In the USA very strong standards set exposures at a maximum of 5 parts per million and TIME-WEIGHTED AVERAGED exposures as low as 1 part per million. This has also been adopted by Sweden.

There is a great concern at the potential health effects of this substance and no threshold effect is postulated for it, as for other pollutants, i.e. exposure to any concentration may cause damage. (⧫ CARCINOGEN; MUTAGEN; THRESHOLD LIMITING VALUE)

Viscosity. A measure of the internal flow resistance of a liquid or gas. Viscosity for liquids decreases as temperature rises.

Visual impact appraisal. ZVIs, or zones of visual intervisibility (also intrusion or influence), are maps which display the areas from which an object or a number of objects may be expected to be seen. (An area under study is usually modelled by the computer.)

Other techniques used include photomontage, created by either photographing a perspective which has been very accurately modelled into a photographic background, or by the direct manipulation of the photo image in the computer. The technique is extremely useful for wind turbine locations in areas of scientific interest, and for tall chimney stack siting.

Vitrification. The process of making a glassy non-crystalline solid from materials capable of vitrifying when heated into their vitrification range, e.g. clays vitrify when heated to high temperatures, hence ceramics and brickmaking. Vitrification is claimed to be one of the 'ultimate solutions' for the safe containment of NUCLEAR REACTOR WASTES but has not been put to the test for the hundreds (or thousands) of years needed for it to be shown as being totally reliable.

VOC. ◊ VOLATILE ORGANIC COMPOUNDS.

Voest-Alpine process. An Austrian high-temperature gasification process which can use contaminated scrap plastic (e.g. car scrap) as a feedstock. It is particularly useful for thermosetting plastics which cannot be recycled (like a THERMOPLASTIC). The process comprises a reactor supplied with preheated air and a primary fuel to attain a temperature of 1600 °C which decomposes the plastics feedstock. The resulting combustible gas is purified and used as an industrial fuel. The residues consist of glassy granules which can be land-filled or, it is claimed, used as a construction material.

Void ratio. The ratio of VOID volume in a filter medium to the total volume occupied. (◊ SEWAGE TREATMENT)

Voids. The open spaces between solid material in a porous medium. Voids, PORES and interstices are terms which are closely interrelated and in part synonymous.

Volatile acids. Mainly acetic, propionic and butyric acids which are produced during anaerobic decomposition of sewage sludge and organic wastes.

Volatile matter[1]. The ratio of the weight of dry matter lost on heating a SLUDGE sample to 600 °C to the initial weight of sludge. This is often used as an approximation of its organic content.

Volatile matter[2]. In FUEL analysis, the loss of weight (corrected for moisture) when a solid fuel is heated to 900 °C in the absence of air. The greater the volatile matter, the more secondary air is needed to avoid SMOKE products during COMBUSTION.

Volatile organic compounds (VOCs). ORGANIC compounds (e.g. ethylene, propylene, BENZENE, styrene, acetone) which evaporate readily and contribute to air pollution directly or through chemical or photochemical reactions to produce secondary air pollutants, principally OZONE and PEROXYACETYL NITRATE. VOCs embrace both HYDROCARBONS and compounds of carbon and hydrogen containing other elements such as oxygen, nitrogen or chlorine. The degree to which a VOC contributes to ozone formation depends on:

(a) total mass of the VOC emitted
(b) molecular mass of the VOC
(c) reactivity of the hydrocarbon with OH RADICALS
(d) the VOC chemical structure.

Common artificial sources of VOCs include paint thinner, dry cleaning solvent, and some constituents of petroleum fuels. Trees are also an important biological source of VOC. It is also known that trees emit large amounts of VOCs, especially isoprene and terpenes. Significant biological sources of methane are termites, cows (ruminants) and cultivation (estimated emissions 15, 75 and 100 million tonnes per year, respectively). Another significant source of VOC emission is crude oil tanking. Both during offloading and loading of crude oil tankers, VOC are released to the atmosphere.

VOCs are an important outdoor air pollutant. In this field they are often divided up into the separate categories of methane (CH_4) and non-methane (NMVOCs). Methane is an extremely efficient greenhouse gas which contributes to enhanced global warming. Other hydrocarbon VOCs are also significant greenhouse gases via their role in creating ozone and in prolonging the life of methane in the atmosphere, although the effect varies depending on local air quality. Within the NMVOCs, the aromatic compounds benzene, toluene and xylene are suspected carcinogens and may lead to leukaemia through prolonged exposure. 1,3-butadiene is another dangerous compound which is often associated with industrial uses.

Some VOCs also react with nitrogen oxide in the air in the presence of sunlight to form ozone. Although ozone is beneficial in the upper atmosphere because it absorbs UV thus protecting humans, plants and animals from exposure to dangerous solar radiation, it poses a health threat in the lower atmosphere by causing respiratory problems. In addition high concentrations of low level ozone can damage crops and buildings.

Many VOCs found around the house, such as paint strippers and wood preservative, contribute to sick building syndrome because of their high vapour pressure. VOCs are often used in paint, plastics and cosmetics. The US Environmental Protection Agency (EPA) has found concentrations of VOCs in indoor air to be two to five times greater than in outdoor air. (⬦ SECONDARY POLLUTION)

Volatilization. The driving off of VOLATILE MATTER as a vapour or liquid by heating.

Volumetric measurements. Methods of measuring gaseous pollutants (usually in air) that make use of accurately measured volumes of liquid reagents and volume(s) of air reacted with them.

VOC emissions.

Chemical	Emission (tonnes per year)		Maximum annual average (site)[a]	2-year network average (ppbv)[b]	Annual ESL (ppbv)	ACGIH TLV-TWA 420 (ppbv)
	1987	1988				
Ethylene	2148	2715	20.9 (7)	13.2	—[c]	—[d]
Propylene	2324	1379	19.2 (8)	11.2	—[c]	—[d]
Toluene	1359	992	7.7 (15)	5.7	98	238
Benzene	489	804	5.1 (8)	3.4	1	24
Xylene (mixed isomers)	551	622	5.3 (15)	2.5	100	238
Styrene	710	600	1.8 (3)	0.3	50	119
Acetone	450	497	8.9 (8)	6.2	245	1786
Chlorobenzene	379	477	0.4 (11)	<0.2	75	179 (24)[e]
Dichloromethane	465	371	2.9 (11 & 13)	3.0	7	119
Methyl tert-butyl ether	285	312	1.7 (15)	0.4	—[f]	—[g]

[a] The maximum annual (average) concentration measured during the two-year period at the site is indicated in parentheses.
[b] Average concentration for all sites during the two-year monitoring period (parts per billion volume).
[c] 30-minute average screening level is less than the annual average screening level.
[d] Simple asphyxiant. No TLV-TWA has been established.
[e] On TLV-TWA Notice of Intended Changes list (Threshold Limit Value-Time Weighted Average).
[f] No effects screening level has been established.
[g] No TLV-TWA has been established.
ESL = Health Effect Screening Level.
ACGIH = American Congress of Government and Industrial Hygienists.

Voluntary separation. The separation from domestic refuse of glass bottles, food and beverage cans or newspapers by individuals or groups, at home or in local collection centres. (◊ RECYCLING)

Vortex separator. A vertical cylindrical tank into which air- or liquid-borne particulate matter is introduced tangentially. Separation is effected by centrifugal force with the particulate matter falling downwards for removal from the conical base.

W

Waste. The current definition of waste in force in the UK is the definition given in Article 1(a) of the amended EC Framework Directive on Waste, which states that: 'waste shall mean any substance or object in the categories set out in annex 1 which the holder discards or intends or is required to discard'.

There are currently 16 waste categories in annex 1 to the Directive:

- production or consumption residues not otherwise specified below;
- off-specification products;
- products whose date for appropriate use has expired;
- materials spilled, lost or having undergone other mishap, including any material, equipment, etc. contaminated as a result of that mishap;
- materials contaminated or soiled as a result of planned actions (e.g. residues from cleaning operations, packaging materials, containers, etc.);
- unusable parts (e.g. reject batteries, exhausted catalysts, etc.);
- substances that no longer perform satisfactorily (e.g. contaminated solvents, exhausted tempering salts, etc.);
- residues of industrial processes (e.g. slags, still bottoms, etc.);
- residues from pollution abatement processes (e.g. scrubber sludges, baghouse dusts, spent filters, etc.);
- machining/finishing residues (e.g. lathe turnings, mill scales, etc.);
- residues from raw materials extraction and processing (e.g. mining operations, oil field slops, etc.);
- adulterated materials (e.g. oils contaminated with PCBs, etc.);
- any materials, substances or products whose use has been banned by law;
- products for which the holder has no further use (e.g. agricultural, household, office, commercial and shop discards, etc.);
- contaminated materials, substances or products resulting from remedial action with respect to land;

- any materials, substances or products which are not contained in the above categories.

(◊ WASTES (SOLID AND HAZARDOUS))

What is waste?

As an outcome of every production process, waste is an inescapable consequence of a consumer society. As such, waste is a 'product' like any other.

Waste is also a paradoxical product – it demands contrary thinking. Companies need to minimize its production and pay 'buyers' to take it away.

Estimated UK waste production in 2004 is shown below.

Estimated total annual waste arisings by sector: 2004

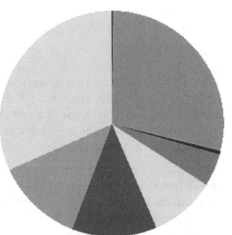

■ Agriculture (inc. Fishing) <1%

■ Mining and Quarrying 29%

■ Sewage sludge <1%

■ Dredged materials 5%

□ Household 9%

■ Commercial 12%

▨ Industrial 13%

▫ Construction and Demolition 32%

Total = 335 million tonnes

Source: Defra, ODPM, Environment Agency, Water UK. http://www.defra.gov.uk/environment/statistics/waste/kf/wrkf02/.htm. Accessed 26 January 2007.

Waste Collection Authority (WCA). Outside Greater London, the Council or a District Council charged with the responsibility for the collection of household waste.

The Environmental Protection Act 1990 confers the following duties and powers on the District Councils as Waste Collection Authorities:

- A duty to arrange for the collection of household waste and, if requested, from commercial and industrial premises.
- A duty to deliver for disposal all waste collected by the Authority, other than waste for which arrangements for recycling have been made, to places directed by the Waste Disposal Authority.
- A duty to inform the Waste Disposal Authority of any new arrangements it proposes to make for recycling.
- Powers to provide plant and equipment for the sorting and baling of waste retained by the Authority for recycling.

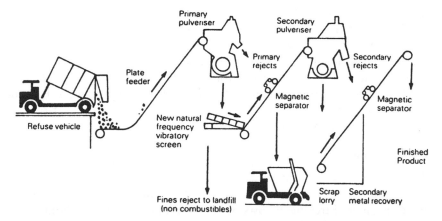

Figure 199   Blue Circle refuse processing flow chart.

- Powers to require household waste to be placed in receptacles of a specified type and number, including for the separation of waste which is to be recycled and that which is not.
- A duty to carry out investigations and prepare a recycling plan in respect of household and commercial waste arising in its area.
- Powers to make payments based on net savings in collection costs (collection credits) to third parties who collect waste, which would not otherwise be collected by the Authority, for recycling.
- Powers to buy or otherwise acquire waste with a view to recycling it and use, sell or otherwise dispose of waste (or anything produced from it) belonging to the Authority.

Waste derived fuel. Fuel made from wastes which can be loose, e.g. shredded paper and plastics from MUNICIPAL SOLID WASTE (MSW) or compressed ('densified') into pellets. Fuel from MSW can have up to half the energy value of UK industrial coal. It is typically used as a solid fuel supplement in ratios of up to 50:50 coal and waste derived fuel on a heat input basis, or it can be blown in the loose state into specially adapted boilers or cement kilns. The Blue Circle process is shown in Figure 199.

    Other waste derived fuels are based on agricultural residues such as rice hulls, sawdust, logging residues and STRAW. Most are characterized by energy values roughly 30 per cent that of an industrial coal. They can also pose handling and storage problems.

    Waste oil and spent SOLVENTS can also be used as fuels, preferably in specially adapted boilers. However, contamination with PCBs can lead to DIOXIN emissions. (◊ INCINERATION; PELLETED FUEL RDF, MRF)

Waste Disposal Authority (WDA). Authorities established by the Local Government Act (England & Wales) 1972 and in 1973 for Scotland.

    In England, they are basically the County Councils and statutory authorities such as the London Waste Regulatory Authority. In Wales and Scotland, they are the District Councils. Their functions are now changed under the Environmental Protection Act 1990 which specifies Waste Regulation Authorities in their place.

Waste Electrical and Electronic Equipment (WEEE). From 1 July 2007 the responsibility to ensure the environmentally sound recycling of Waste Electrical and Electronic Equipment (WEEE) lies with producers. Extended producer responsibility will come into force for WEEE that is collected from private households at specific collection facilities.

Role of producers

Producers of Electrical and Electronic Equipment (EEE) are obligated to join a Producer Compliance Scheme (PCS), which must finance the collection, reprocessing and treatment costs of WEEE from private households.

Role of distributors

Distributors of EEE must participate in one of the following:
Join the National Distributor Take-back Scheme – DTS (finances a national network of Designated Collection Facilities – DCFs).
Offer in store take-back.
Distributors of household EEE must also provide consumers with information about the WEEE directive and arrangements for disposal.

Role of local authorities

A National Distributor Take-back Scheme (DTS) has been set up to create a National network of Designated Collection Facilities (DCFs).
Local Authorities may register current waste management facilities to become Designated Collection Facilities.
In these circumstances the Local Authority enters an agreement with a Producer Compliance Scheme, to arrange the collection and recycling of WEEE materials
DCFs are expected to collect WEEE into the following five streams:

A – large household appliances other than cooling appliances
B – cooling appliances
C – TVs and monitors
D – gas discharge lamps
E – all other WEEE

The Producer Compliance Scheme then makes arrangements for the collection and final reprocessing of the WEEE at an Approved Authorised Treatment Facility (AATF).
The AATF generates evidence notes based on the amount of WEEE it processes from DCFs. This evidence is passed back to the producer compliance schemes in order to meet producer obligations.

In Normal Events

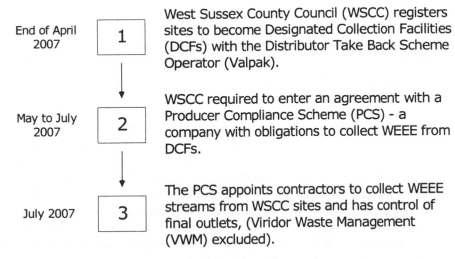

End of April 2007 — 1 — West Sussex County Council (WSCC) registers sites to become Designated Collection Facilities (DCFs) with the Distributor Take Back Scheme Operator (Valpak).

May to July 2007 — 2 — WSCC required to enter an agreement with a Producer Compliance Scheme (PCS) - a company with obligations to collect WEEE from DCFs.

July 2007 — 3 — The PCS appoints contractors to collect WEEE streams from WSCC sites and has control of final outlets, (Viridor Waste Management (VWM) excluded).

Proposed Arrangements

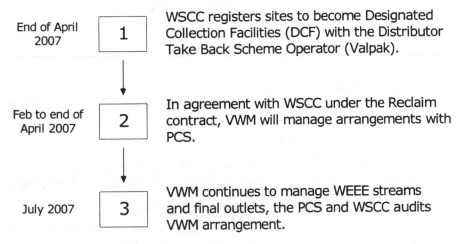

End of April 2007 — 1 — WSCC registers sites to become Designated Collection Facilities (DCF) with the Distributor Take Back Scheme Operator (Valpak).

Feb to end of April 2007 — 2 — In agreement with WSCC under the Reclaim contract, VWM will manage arrangements with PCS.

July 2007 — 3 — VWM continues to manage WEEE streams and final outlets, the PCS and WSCC audits VWM arrangement.

Implementation of the WEEE Directive.

Case Study: The Likely Scenario For Managing Waste Electrical and Electronic Equipment (WEEE) in West Sussex

The flowchart shown above sets out two scenarios for the management of WEEE in West Sussex, from 1 July 2007. In both events West Sussex County Council (WSCC) would register its Household Waste Recycling Sites and Transfer Stations to become Designated Collection Facilities

(DCFs), which will form part of the national network of collection facilities for accepting Household WEEE.

Under Scenario 1 – In normal events WSCC is required to enter into an agreement directly with a Producer Compliance Scheme (PCS) – which has obligations to collect and reprocess WEEE from its DCFs. In this scenario, the PCS would appoint its own subcontractors to collect the five WEEE streams and would control the final reprocessing outlets. The problem with this scenario is that it does not take into account WSCC long-term contractual arrangements with its waste management contractor – Viridor Waste Management. Viridor currently manage the collection arrangements and reprocessing routes for four of the five WEEE streams. Under this scenario there are likely to be significant financial and operational issues to WSCC.

Under Scenario 2 – Proposed Arrangements, WSCC transfers the operational responsibility and a proportion of the financial risk to Viridor to manage arrangements with the PCS. In this scenario, Viridor are subcontracted by the PCS to manage the collection and reprocessing routes for all five categories of WEEE. This overcomes any significant operational and financial issues that would occur under Scenario 1.

Waste export. Grosvenor Waste Management (GWM) Ltd admitted at Maidstone Crown Court six breaches of the Transfrontier Shipments of Waste Regulations. The prosecution came after 15 containers were detained at Southampton Port. The accompanying documentation claimed they contained waste paper for recycling, but Environment Agency officers noticed a strong smell of decaying domestic and household waste, and found flies in many of the containers. In addition to a fine of £55 000, GWM Ltd was ordered to pay £85 000 in costs, and suffered financial losses as result of the case amounting to about £400 000.

Source: *The Times*, 6 April 2007.

Dr Liz Goodwin, the new chief executive of WRAP, has stated 'export is always going to have a role' in dealing with recyclables. (◊ RECYCLING, STATISTICS)

Source: *Resource Management and Recovery*, 20 April 2007.

Waste factor. A term used in ENERGY ANALYSIS to measure the departure from ideal energy needed to effect a transformation. It is defined as:

$$\frac{(\text{actual energy required to effect transformation}) - (\text{ideal energy})}{(\text{actual energy})}$$

Ideal energy is obtained from physical or chemical calculations which show the theoretical energy consumption required to manufacture a given component or product.

A waste factor close to 1 represents a very wasteful and inefficient process. Thus, it can be used to show scope for improvement in energy utilization.

Product	Actual energy (1968) (MJ/kg)	Ideal energy (MJ/kg)	Waste factor
Iron	25	6	0.76
Petrol	4.2	0.4	0.90
Paper	38	0.2	0.99
Aluminium	190	25	0.87
Cement	7.8	0.8	0.90

From E. Gyftopoulos *et al.*, Free Energy Use, Actual and Ideal, Thermo-Electron Corp., Waltham, Mass., 1974.

Waste heat. Term used to denote heat that is normally rejected to the environment, e.g. the hot water discharge from power-station cooling circuits which may be 10–15 °C above ambient temperature. Other examples include the jacket cooling water from large industrial residual oil fuelled engines which may be at a temperature of ca. 90–100 °C. The higher the temperature of the waste heat source, the greater the prospect of using it for an additional source of energy for heating purposes such as DISTRICT HEATING where a modified steam power station cooling circuit can provide both power and high-temperature hot water.

Waste Management Hierarchy. The hierarchy of waste management begins with waste minimization before proceeding to actual disposal (see Figure 200). Dr Henrik Wenzel, of the Technical University of Denmark, has suggested that increased recycling creates extra capacity in waste incinerators, so that more waste can be burned. He argues that since this 'cascade effect' (see Figure below) results in less landfilling of wastes, it therefore strengthens the case for

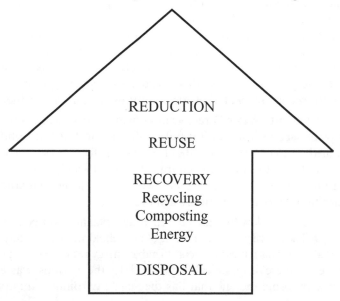

Figure 200 The waste hierarchy.

Landfill   Incineration   Recycling

Figure 201   The Cascade Effect.
Source: Resource Management & Recovery, 1 December 2006, p. 3.

recycling. Dr Wenzel said the approach is important because 'the use of electricity derived from fossil fuels can be a negative in the recycling process, especially for cardboard. Adding in the benefit of creating space in energy-from-waste
plants, helps make it better to recycle in terms of life-cycle analysis'.

A new five-stage hierarchy of waste management has been suggested:

1) Prevention, 2) re-use, 3) recycling, 4) recovery, and then 5) landfill.

This allows energy recovery to raise its profile, and is long overdue!

1. Do not create the waste product in the first place.
2. If there is a waste product, reuse it.
3. If it cannot be reused, recover or reclaim the primary material for new
   manufactured products, if there is an environmental benefit.
4. If primary materials recovery is not practicable (and this includes excessive recovery costs *vis-à-vis* other waste management options) recover the
   waste for secondary materials, or if combustible, use it for fuel.

   Under a 2006 EU Waste Directive modification, waste-to-energy recovery
   may be classed as disposal if it fails a 60% EFFICIENCY threshold, which
   is framed so that COMBINED HEAT & POWER for DISTRICT HEATING or
   industrial heat utilisation is encouraged. This is widely done in Scandinavia, France, Germany, Austria, etc. where overall thermal efficiencies of
   greater than 80% are commonly achieved.

   The UK has been slow to exploit the environmental and cost benefits of waste-
   to-energy. The calorific value in 1 tonne of household waste is roughly 1/3 that of
   industrial coal. This represents a considerable energy resource saving if utilised.
5. If none of these is practicable proceed to the various waste disposal
   options, choosing the one that has the least environmental impact.

Note that the hierarchy is for guidance only, it is not God-given, or to be
followed in an unquestioning manner which sadly often elevates recycling as

an absolute good, whereas it can easily make environmental and resource demands out of all proportion to any environmental benefits with the exception of metals recycling and glass remelting.

Waste management. The management of wastes at all stages from production, handling, storage, transport, processing and ultimate disposal, which includes the Duty of Care.

Revised EU Waste Management Strategy

The European Commission (concerned by the continuing increase in waste arisings) adopted the revision of the Community's 1989 Waste Management Strategy in 1996.

The revised strategy affirms the waste management hierarchy of prevention, recovery and final disposal as a guide to preferred waste management options but also introduces producer responsibility as another key guiding principle.

The Commission will reinforce these principles by encouraging the:

- promotion of clean technologies and products;
- improvement of the environmental dimension of technical standards drafted by CEN (European Standardization Committee);
- prevention of the production of hazardous waste by limiting or banning certain heavy metals or dangerous substances in products and processes;
- use of economic instruments to influence waste prevention;
- development of the eco-audit and eco-label schemes.

In addition, the Commission has undertaken to:

- present a new proposal for a landfill directive;
- further reduce illegal exports of waste and related criminal activities;
- promote consumer information and education to contribute to a gradual change in consumption patterns.

(◊ WASTES – DUTY OF CARE; WASTES (VII) WASTE MINIMIZATION)

Source: *CBI Environmental Newsletter*, September 1996.

Extract from *Summary of Conclusions and Recommendations*, House of Commons Report Inquiry into Sustainable Waste Management, Vol. I, 17 June 1998.

It is important to stress from the beginning of our Report our profound disappointment, on the basis of the evidence we have received, that waste management in this country is still characterised by inertia, careless administration and *ad hoc*, rather than science based, decisions. Lip-service alone, in far too many instances, has been paid to the principles of reducing waste and diverting it from disposal. Central government has lacked the commitment, and local government the resources, to put a sustainable waste management strategy into practice.

The continuing lack of information in Government about waste is extraordinary: it would appear to be common sense that one first identifies the nature and scale of the problem before attempting to sort it out. The production of accurate statistics on waste arising, the composition of waste at the point of arising and on the demographic structure of households (which affects that composition) must be a Government priority.

The uncertainty over the significance of the targets in *Making Waste Work*, and the continuing failure to provide either scientific justification or material support for their achievement has seriously damaged their credibility. We do not believe that the targets were ever aspirational, but if they were, they have provoked a negative response. We believe that the targets must now be re-affirmed with vigour.

We acknowledge that the principle of the Best Practicable Environmental Option is the key to the development of locally sustainable solutions for waste management; but waste reduction should always be considered as a precursor to implementation of the Best Practicable Environmental Option in any specific instance.

A revolution is underway in the manner in which the UK manages its wastes – and not before time too.

Source: *Sustainable Waste Management*, Environment, Transport and Regional Affairs Committee, HV484-I (1997–98), House of Commons.

Hampshire has taken an integrated approach towards waste management, and consultation with the public has confirmed the 'Integra' strategy (see Figure 202):

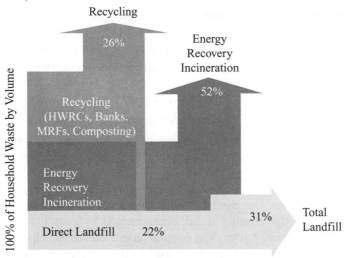

If anaerobic digestion is implemented in Southampton the total landfill demand will be reduced by 4%

Figure 202 'Integra' strategy for reducing household waste.
Source Integra News, Issue 3, Autumn 1998. (Reproduced with permission from Onyx Environmental Group)

- keep dustbin waste to 1995 levels;
- 50 per cent RECYCLING (including COMPOSTING) by the year 2010, with 40 per cent in the longer term;
- the use of ENERGY recovery systems for residual waste;
- an ANAEROBIC digestion project in Southampton;
- three energy recovery INCINERATORS (in the north, southeast and southwest of Hampshire);
- landfill the remainder.

The recycling rate achieved in 2004/05 was 27%, with over 95% of Hampstire households having access to kerbside collection of recyclables.

Waste pretreatment. The landfilling of non-hazardous wastes now requires pretreatment. This can be physical, thermal, chemical or biological processing (including sorting) which changes the characteristics of waste in order to reduce its volume or hazardous nature, facilitate its handing or enhance recovery. Naturally, this is not quite so simple in practice, and EA guidance is available. Waste producers must ensure that they meet the Three Point Test, shown in Figure 203.

Wastes – Duty of Care. The Royal Commission on Environmental Pollution 11th Report, 1985, recommended in para. 13.74:

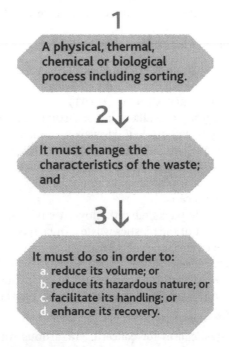

Figure 203   The Three Point Test.
Source: *Resource Management and Recovery*, 9 February 2007, p. 16.

(a) that producers of controlled wastes have a duty of care to take all reasonable steps, having regard to the hazards presented by their wastes, to ensure that their wastes are subsequently managed and disposed of without harm to the environment;

(b) that the steps that it is reasonable for the waste producer to take in different circumstances to discharge his duty of care shall be contained in a code of practice issued by the Secretary of State; and

(c) that producers of commercial, industrial and certain domestic wastes who engage contractors to transport their wastes remain liable for the proper disposal of those wastes unless they use a contractor registered as a waste transporter in accordance with our later recommendation and provide him in writing with an unambiguous indication of the nature of the wastes and clear instructions for their disposal.

EC framework directive on waste

This has four main mandatory elements:

(a) each member state must have a competent authority or authorities with responsibility for waste;

(b) the competent authorities must prepare waste management plans;

(c) undertakings handling waste must have a permit from the competent authority;

(d) the polluter-pays principle must be applied.

Waste is defined as any substance or object which the holder disposes of or is required to dispose of pursuant to the provisions of national law in force. (◊ DUTY OF CARE; ENVIRONMENTAL AUDIT)

Wastes (plastics). If recycling results in net environmental and economic benefits, it should clearly be undertaken. However, in the case of plastics packaging *vis a vis* other readily recyclable materials a cradle (production) to grave approach is needed. Various studies have demonstrated that the overall environmental impacts of plastics packaging can be very much less than that of aluminium/tin plate cans.

   Many non-recyclable packs have the lowest environmental impact. After use, the energy content of the plastics waste can be recovered in energy-from-waste plants whose environmental performance is of the highest order.

Wastes (solid and hazardous). This extended entry is provided to outline most matters relevant to both public concern and the environment resulting from historical solid and hazardous waste disposal practices (◊ LANDFILL DIRECTIVE):

(i) Hazardous wastes, causes for concern. Hazardous wastes disposal in the UK has long been a cause for concern. Starting with *The Report of the Technical Committee on the Disposal of Solid Toxic Wastes*, March 1970 (the

Committee was appointed in July 1964), through the House of Lords Select Committee on Science and Technology *Hazardous Wastes Disposal*, July 1981, the Royal Commission on Environmental Pollution (RCEP) *Managing Wastes: The Duty of Care*, December 1985, *Three Hazardous Wastes Inspectorate Reports* (HWI), 1985–1988, the House of Commons Environment Committee *Toxic Waste*, February 1989, and finishing with another House of Lords Report, *Hazardous Waste Disposal*, April 1989.

The RCEP 11th Report, *Managing Wastes: The Duty of Care* (1985), stated that the main areas of public concern about wastes were:

- Toxic wastes which have mutagenic, teratogenic or carcinogenic effects, or other latent, slow effects on health;
- The tipping of hazardous wastes and the risk of acute poisoning to humans and other living organisms;
- The spread of toxic or noxious materials placed on the ground to gardens, homes, play areas, etc.;
- Irreversible damage to the environment by irretrievable disposal of persistent substances to land or the marine environment;
- Contamination of groundwater;
- The lowering of the amenity value of an area through the presence of wastes – especially litter, but also the noise, smell and other nuisances associated with waste collection and disposal;
- The apparent waste of resources which could be recovered from discarded materials.

The 1989 House of Lords Report said in paragraphs 3 and 7:

3. Eight years after making their first report on Hazardous Waste Disposal, the committee are dismayed at how little has changed. Legislation to improve the control of waste disposal has still not been introduced; the standards of control by local authorities still vary widely; many of those authorities have not drawn up the plans for waste disposal required of them by the Control of Pollution Act 1974; incidents of pollution by waste continue.

7. In 1981 the Committee considered 'the present degree of control exercised by waste disposal authorities to be defective and urgently in need of strengthening . . . It must be made possible for the public to feel confident that real control is taking place and that disposal is geared to the best practicable, not the cheapest tolerable, means.' The Committee still believe this to be the key to acceptable waste disposal.

Source: Hazardous Waste Disposal, House of Lords 1st Report, Session 1980–81.

This extract gives a flavour of the concerns felt by the House of Lords. It is fair to say that most major disposal contractors wish to see standards raised but they have to compete with those whose practices leave much to be desired. Also, thanks to a deadline issued by the Department of the Environment most WASTE DISPOSAL PLANS (V) have been filed – 15 years after the passage of the Control of Pollution Act 1974. Another example of what can happen is given below from the second *Hazardous Waste Inspectorate* (HWI) *Report*, appropriately titled *Hazardous Waste Management – Ramshackle and Antediluvian?*, July 1986:

2.11.2   The second potential incident involved tonneage quantities of solid waste materials which react violently with water to produce a poisonous gas which, under the appropriate conditions of humidity, is spontaneously flammable in air. The disposal of these materials had been the subject of an invitation to tender, and one of the proposals, fortunately intercepted by the HWI, had been for a landfill disposal route in sealed drums. Investigations showed that this did not reflect the disposal philosophy of the environmental adviser of the producing company, but that the invitation to tender had originated from a particular manufacturing site, with no reference to HQ. Again, a proposal of potentially disastrous consequences. Again, without intervention by the Inspectorate this landfill disposal could have gone ahead and without, we believe, breaching the landfill site licence.

2.12   This is an example (and there are others) of what can go wrong at co-disposal sites. The HWI commented – 'It is small wonder that the UK's championship of co-disposal landfill is looked at askance by other nations if these examples were to be typical of current practice.'

The HWI 2nd Report gave further examples of malpractice including:

2.16   The HWI is also aware of two cases of the deliberate landfill disposal of low flash-point solvent wastes at sites in two different WDA areas in England. These wastes, which were water-miscible solvents and which arose in bulk, were deliberately diluted with water (and in one case, with leachate from the site at which their disposal was intended) in order to bring the aqueous solvent mixture within the landfill site licence conditions in terms of flash point or fire point. This is entirely contrary to DOE advice which recommends that bulk flammable solvents should be recovered, used as a fuel or incinerated, and that large volumes of aqueous waste should not, in general, be landfilled.

The HWI stated that the preceding examples were evidence of a continuing trend by waste generators to seek the cheapest rather than the environmentally most suitable disposal option.

The Inspectorate gave further examples in the report. The RCEP 11th Report has commented:

NIMBY (not in my back yard) cannot be dismissed as irrational: it is the expression of concern about a risk as perceived by the local inhabitants. But the fact there is public concern does not necessarily mean that there is a real environmental hazard. A proper evaluation of the risk requires access to the relevant information and its interpretation, and the public will not be reassured by the interpretations provided by the putative polluter, who has an interest. In the absence of public confidence in the role of bodies that are both authoritative and independent, interpretations by the press or pressure groups are often accepted, even if they go beyond what an informed expert would regard as justified by the evidence.

(ii) Waste classifications. The common definition for waste is that for which there is no further use, the implication being that there is nothing that can be done, hence they must be disposed of. However, the UK Environmental Protection Act (EPA) 1990 (Section 75) defines waste as:

(a) any substance which constitutes a scrap material or an effluent or other unwanted surplus substance arising from the application of any process; and
(b) any substance or article which requires to be disposed of as being broken, worn out, contaminated or otherwise spoiled, but does not include a substance which is an explosive within the meaning of the Explosives Act 1875.

There are innumerable classifications for waste. However, they can be grouped by: (i) *origin*, e.g. clinical wastes, household or urban solid wastes, industrial wastes, nuclear wastes, agriculture; (ii) *form*, e.g. liquid, solid, gaseous, slurries, powders; (iii) *properties*, e.g. toxic, reactive, acidic, alkaline, inert, volatile, carcinogenic; (iv) *legal definition*, e.g. special, controlled, household and industrial, where specific definitions or criteria are employed as follows:

Controlled wastes: are defined by the UK EPA 1990 Section 75 as household; industrial and commercial waste and any such waste; agricultural wastes apart from crop residues and animal manures.

Commercial wastes: are defined in the EPA 1990 Section 75(7). Waste from premises used wholly or mainly for the purposes of a trade or business or the purposes of sport, recreation or entertainment excluding (i) household and industrial waste; (ii) mining, quarrying and agricultural waste; (iii) waste of any other description prescribed by regulations made by the Secretary of State for the purposes of the above.

Demolition wastes: Masonry and rubble wastes arising from the demolition or reconstruction of buildings or other civil engineering structures.

Difficult wastes: Another UK term (not legally defined) which includes wastes that could be harmful to the environment or whose physical properties present handling problems. A list of potentially 'difficult' wastes is given below. Note these are not necessarily *special wastes*, but rather fall loosely into 'hazardous' wastes definitions or classes used in other countries. 'Hazard' relates to the situation and circumstances as well as the properties of the waste materials. Toxicity is also not conducive to rigorous definition unless target organisms and levels and duration of exposure are stated. Hence special wastes is a subset of hazardous wastes that has been more precisely defined leaving authorities free to devise their own classification and criteria for hazardous wastes which has led (in the UK) to variability of standards between authorities.

Classification of Difficult Wastes include:

(a) Inorganic acids
(b) Organic acids and related compounds
(c) Alkalis
(d) Toxic metal compounds
(e) Non-toxic compounds
(f) Metal (elemental)
(g) Metal oxides
(h) Inorganic compounds
(j) Other inorganic materials
(k) Organic compounds
(l) Polymetallic materials and precursors
(m) Fuel, oils and greases
(n) Fine chemicals and biocides
(p) Miscellaneous chemical waste
(q) Filter materials, treatment sludge and contaminated rubbish
(r) Interceptor wastes, tars, paint, dyes and pigments
(s) Miscellaneous wastes
(t) Animal and food wastes

Hazardous wastes – definition: There are major problems in finding an acceptable definition of hazardous wastes, because of the following:

(i) The hazards created depend on the waste disposal method used.
(ii) The quantity of toxic material present, e.g. household waste, may contain small quantities of toxic compounds and be innocuous. On the other hand, larger concentrations of the same toxic material could pose very serious problems. However, small disposable concentrations of some toxic compounds can contaminate large bodies of water, e.g. phenol. (The taste threshold for phenol in water can be as low as 0.001 ppm.)
(iii) The hazards posed by other industrial materials so that hazardous wastes are not unjustly singled out for opprobrium.

The 1981 House of Lords Report considered these matters and recommended that the WHO definition should be used. Namely:

(a) short-term hazards, such as acute toxicity by ingestion, inhalation or skin absorption, corrosivity or other skin or eye contact hazards or the risk of fire or explosion; or
(b) long-term environmental hazards including chronic toxicity upon repeated exposure, carcinogenicity (which may in some cases result from acute exposure but with a long latent period), resistance to detoxification processes such as biodegradation, the potential to pollute underground or surface waters, or aesthetically objectionable properties such as offensive smells.

This recommendation does not have official acceptance in the UK.

Household wastes: Defined in the UK EPA 1990 Section 75(5) as waste from:

(a) domestic property, that is to say, a building or self-contained part of a building which is used wholly for the purposes of living accommodation;
(b) a caravan (as defined in section 29(1) of the Caravan Sites and Control of Development Act 1960) which usually and for the time being is situated on a caravan site (within the meaning of that Act);
(c) a residential home;
(d) premises forming part of a university or school or other educational establishment;
(e) premises forming part of a hospital or nursing home.

Industrial wastes: 'Industrial waste' is defined in the UK EPA 1990 Section 75(6) as waste from:

(a) any factory (within the meaning of the Factories Act 1961);
(b) any premises used for the purposes of, or in connection with, the provision to the public of transport services by land, water or air;
(c) any premises used for the purposes of, or in connection with, the supply to the public of gas, water or electricity or the provision of sewerage services; or
(d) any premises used for the purposes of, or in connection with, the provision to the public of postal or telecommunications services.

Generally taken to include waste from any industrial undertaking or organization.

Inert wastes: Wastes that will not react physically, chemically or biologically and are non-polluting under normal conditions, e.g. demolition waste, although this could contain organic matter and give rise to LEACHATE generation.

Poisonous wastes (*UK*): This was used under the now defunct UK Deposit of Poisonous Waste Act 1972 (passed incidentally by Parliament in 10 days after several well-publicized incidents of drums of industrial cyanide wastes being dumped in children's playgrounds, ditches, etc. by unscrupulous transport companies). The scope of this Act was very broad, in effect requiring 'the giving of notices in connection with the removal and deposit of waste and for connected purposes'.

The definition of poisonous was given as

> to subject persons or animals to material risk of death, injury or impairment of health or as to threaten the pollution or contamination (whether on the surface or underground) of any water supply; and where waste is deposited in containers, this shall not of itself be taken to exclude any risk which might be expected to arise if the waste were not in containers.

This law had the effect of ensuring very strict controls on all manner of poisonous, noxious or polluting waste whose presence on the land is liable to give rise to an environmental hazard. However, the Deposit of Poisonous Waste Act was repealed (being adjudged to be too all-embracing and liable to misinterpretation) and replaced by the Control of Pollution (Special Waste) Regulations 1980.

Special wastes (*UK*): The Special Waste Regulations 1996 came into effect on 1 September 1996 and replaced the Special Waste Regulations 1980. They implement the 1991 EU Hazardous Waste Directive (91/689/EEC) and the associated EU List of Hazardous Waste. Special wastes are defined by reference to:

(a) a list of substances which, if they are present in wastes, may cause the waste to have characteristic properties which make it dangerous or difficult to dispose of. These substances are set out in a list, e.g. B502 – barium compounds, B521 – zinc compounds;

(b) a list of the characteristic properties which cause waste to be dangerous or difficult to dispose of. These are given below:
Explosive
Flammable (liquid)
Flammable (solid)
Emission of flammable gas
Oxidizing
Toxic (poisonous; harmful)
Infectious
Corrosive/chemical irritant
Reactive
Toxic (delayed)
Ecotoxic

The inclusion of ecotoxicity remedies the important omission in the original 1980 regulations. It is defined as follows:

(a) A substance which may present an immediate, delayed or accumulative risk for one or more sectors of the environment; and
(b) in determining whether a substance presents a risk to the environment particular account shall be taken of:

  (i) its effect on animals and other living organisms, including aquatic organisms;
  (ii) its effect on plants, water and land; and
  (iii) the likelihood of its entering the food chain and its persistence in the food chain.

The EC's Hazardous Waste List (1996) entries are in bold in the European Waste Catalogue (DoE Circular 6/96). Waste producers must perform a proper assessment of all the potential hazards associated with a waste so that they are able to describe it thoroughly on the waste consignment note.

(iii) Wastes, clinical. Clinical waste is defined in regulation 1 (2) of the Controlled Waste Regulations 1992 (SI 1992/588) as meaning:

(a) any waste which consists wholly or partly of human or animal tissue, blood, other body fluids, excretions, drugs or other pharmaceutical products, swabs, or dressings, or syringe needles or other sharp instruments, being waste which unless rendered safe may prove hazardous to any person coming into contact with it; and
(b) any other waste arisings from medical, nursing, dental, veterinary, pharmaceutical or similar practice, investigation, treatment, care, teaching or research, or the collection of blood for transfusion, being waste which may cause infection to any person coming into contact with it.

The need for very strict control and secure hygienic disposal of clinical wastes cannot be over-emphasized and is discussed in depth in Waste Management Paper No. 25 *Clinical Wastes*. These papers are a series numbering from 1 to 28 on important waste management topics published in the UK by HMSO and regularly updated. However unprofessional contractors can give the industry a bad name.

In February 2003, the Environment Agency secured fines totalling £100 000 against Eurocare Environmental Services, one of Britain's leading clinical waste disposal contractors, for a string of waste and pollution offences. Deliberate and reckless criminality, gross mismanagement, shortcomings and flouting of the law at all levels of the company, were the hallmarks of the case.

Source: ENDS Report 337, February 2003.

Group A Wastes:

(a) All human tissue, including blood (whether infected or not), animal carcasses and tissue from veterinary centres, hospitals and laboratories, and all related swabs and dressings.
(b) Waste materials where the assessment indicates a risk arising from, for example, infectious disease cases.
(c) Soiled surgical dressings, swabs and other soiled waste from treatment areas.

Group B Wastes: Discarded syringe needles, cartridges, broken glass and any other contaminated disposable sharp instruments or items.

Group C Wastes: Microbiological cultures and potentially affected waste from pathology departments (laboratory and post-mortem rooms) and other clinical or research laboratories.

Group D Wastes: Certain pharmaceutical products and chemical wastes.

Group E Wastes: Items used to dispose of urine, faeces and other bodily secretions or excretions assessed as not falling within Group A. This includes used disposable bed pans, incontinence pads, stoma bags and urine containers.

The shape of careful clinical wastes management is exemplified by the City of Copenhagen which has established seven individual collection schemes for clinical wastes:

• Collection of special hospital waste from all hospitals and from other sources generating more than 50 kg of special hospital waste per week.
• Collection of special hospital waste from doctors' and dentists' practices, nursing homes, visiting nurses, etc., generating less than 50 kg of special waste per week.
• Collection of chemical waste from doctors' and dentists' practices, nursing homes, visiting nurses, etc.
• Collection of dead animals from veterinary hospitals, laboratories, etc.
• Collection from pharmacies of discharged medicine, disposable syringes, etc.
• Emptying of syringe boxes in public places.
• Clean-up of syringes, etc., from certain troubled back-yards, playgrounds, etc.

ALL THE ABOVE MUST BE SEPARATED FROM HOUSEHOLD WASTE.

The collection and transportation system is designed with one main objective: safe handling of the waste from the user via collection to its ultimate disposal.

This ensures two purposes:

(a) rational and safe handling which protects the staff in all steps of the collection and treatment system from direct contact to the waste, and
(b) rendering the collection system visible by the waste handlers as they must never be unaware of which waste product they are handling.

The above precautions are essential in today's society.

Sources: The UK National Association of Waste Disposal has also published *Clinical Waste Guidelines*, November 1994.
C. Vennicke, 'Incineration of Hospital Waste', presented at Wastes Management Meeting WM95, Copenhagen, 1995.

(iv) Waste disposal options for industrial wastes. There is a wide range of options available and given that a strategy for WASTE MINIMIZATION (vii) is followed, and that recovery, reuse or recycling options are exhausted or not feasible, then the choice boils down to:

(a) *Disposal on site*. This is used by many companies who LANDFILL their waste arisings as the cheapest option where permissible. Such sites must have a WASTE MANAGEMENT LICENCE (vi) and are of course subject to the Special Waste Regulations (UK).
(b) A variety of absorption and filtering processes take place. The substances that may be deposited and their manner of deposition is laid down in the site licence. Landfill disposal has been viewed as a cheap option and is often not a totally environmentally secure means of disposal. But, in strictly controlled sealed landfill sites with thorough checks on the WATER BALANCE of the site, backed up by rigorous LEACHATE treatment and LANDFILL GAS controls, and sound aftercare programmes, landfill disposal can be very acceptable for restricted low toxicity waste. (◊ LANDFILL)

The scope for abuse of landfill is considerable.

The 1981 House of Lords Report on *Hazardous Waste Disposal* identified the following requirements for the 'sensible' landfill of hazardous waste:

1. Proper selection of landfill site;
2. sampling, analysis and specification of the waste;

3. skilled operation of the selected site for the disposal of the waste specified (including any necessary treatment); and

4. external monitoring of the site both during operation and after the site has been closed.

Unfortunately, very few UK landfill sites are equipped to analyse wastes received 'at the gate' and reliance has to be placed on the manufacturer's description. Reputable companies will sample but standards and practices varied as paragraph 2.17.1 from the HWI 2nd Report of 1986 illustrates:

2.17.1 The HWI is aware of a major waste disposal contractor who was forced either to lose a longstanding and very substantial disposal contract by a treatment route to a landfill competitor, or himself to offer a cheaper landfill route. The contractor chose the latter course and some 2,500 tonnes per annum of acid at pH 1 and alkali at pH 8 have now been diverted from treatment to a landfill disposal route.

(c) *Treatment*. This falls into four categories: biological, physical, chemical and fixation.

Biological treatment can only be used on organic aqueous wastes and may be either an AEROBIC or an ANAEROBIC process.

Chemical treatment embraces a spectrum of techniques to, for example, make soluble wastes insoluble, destroy toxicity (oxidation of cyanides) or neutralization of ACIDS or ALKALIS.

Physical treatment techniques include settling, sedimentation, filtration, flotation, evaporation, distillation, e.g. the separation of oil and water using settlement lagoons or the recovery of SOLVENTS by distillation are widely used.

Fixation is the encapsulation or sealing of the hazardous wastes in a variety of matrices such as organic polymers, resins, fly ash/cement mixtures. The matrices are unreactive and are claimed to encapsulate the wastes although controlled release might be a better term. The long-term integrity of some of these processes has been questioned, but if properly conducted, they are undoubtedly a much better environmental option than straight landfill.

(d) *Deep underground disposal*. This is practised in the UK in a few deep underground mineshafts which discharge the liquid low or non-toxic ('nuisance') wastes into old mine workings where there is minimal risk to water supplies. Worked out salt mines or brine deposits can also be used. This is used in West Germany for '*difficult wastes*' which are very securely packaged.

(e) *Incineration*. This is really the only environmentally secure option if properly conducted, for pathogenic materials, inflammable liquids, carcinogenic substances such as PCBs or materials containing DIOXINS. Organic wastes are destroyed and there is the possibility of energy recovery from spent SOLVENTS. The highly aggressive operating conditions and intermittent loading of waste incinerators means high maintenance costs. (◊ INCINERATION)

(v) Waste disposal plan. Required under Section 50 of the Environmental Protection Act 1990. The plan of the Waste Regulation Authority (WRA) must include information on:

(a) the kinds and quantities of controlled waste which the authority expects to be situated in its area during the period specified in the plan;
(b) the kinds and quantities of controlled waste which the authority expects to be brought into or taken for disposal out of its area during that period;
(c) the kinds and quantities of controlled waste which the authority expects to be disposed of within its area during that period;
(d) the methods and the respective priorities for the methods by which, in the opinion of the authority, controlled waste in its area should be disposed of or treated during that period;
(e) the policy of the authority as respects the discharge of its functions in relation to licences and any relevant guidance issued by the Secretary of State;
(f) the sites and equipment which persons are providing and which during that period are expected to provide for disposing of controlled waste; and
(g) the estimated costs of the methods of disposal or treatment provided for in the plan.

A waste recycling plan must also be produced.

In preparing the plan the WRA is also required to consult any relevant water companies, collection authorities (District Councils in England), any WDAs affected by the export of controlled wastes, and relevant other persons engaged in the disposal of controlled waste. Members of the public can make representations but are not statutory consultees, as the WDA may make alterations which it 'considers appropriate in consequence of the representations'.

(vi) Waste management licence. A waste management licence is defined under Section 35 of the Environmental Protection Act 1990:

A waste management licence is a licence granted by a waste regulation authority (now undertaken by EA under the powers of the ENVIRONMENT ACT 1995) authorising the treatment, keeping or disposal of any specified description of controlled waste in or on specified land or the treatment or disposal of any specified description of controlled waste by means of specified mobile plant.

The licence is granted to the following persons, that is to say:

(a) in the case of a licence relating to the treatment, keeping or disposal of waste in or on land, to the person who is in occupation of the land; and

(b) in the case of a licence relating to the treatment or disposal of waste by means of mobile plant, to the person who operates the plant.

on the terms and conditions as appear to the waste regulation authority to be appropriate; the conditions may relate:

(a) to the activities which the licence authorizes, and

(b) to the precautions to be taken and works to be carried out in connection with or in consequence of those activities;

and accordingly requirements may be imposed in the licence which are to be complied with before the activities which the licence authorizes have begun, or after the activities which the licence authorizes have ceased.

Under the EPA, the suitability of a person to be a licence holder will be taken into account when an application is made. The Environment Agency must be satisfied that the applicant is a 'fit and proper person'. Whether a person is or is not a fit and proper person to hold a licence will be determined based on whether he or another relevant person has been convicted of a relevant offence, whether he is technically competent to manage the licensed activities, and whether he is able to make adequate financial provision to discharge the obligations arising from the licence. This is an important check, as too often in the past unsuitable licensees have either not discharged their responsibilities or have simply walked away from them by handing their licence back to the then Waste Disposal Authority.

The grounds for refusing a licence include prevention of pollution of the environment, harm to human health or severe detriment to amenities of the locality. The environment is defined in Section 29 as:

all, or any, of the following media, namely land, water and the air. Pollution of the environment means pollution of the environment due to the release or escape (into any environmental medium) from:

(a) the land on which controlled waste is treated,

(b) the land on which controlled waste is kept,

(c) the land in or on which controlled waste is deposited,

(d) fixed plant by means of which controlled waste is treated, kept or disposed of,

of substances or articles constituting or resulting from the waste and capable (by reason of the quantity or concentrations involved) of causing harm to human or any other living organisms supported by the environment.

A licence can no longer be handed back at any time. On an application to surrender a licence, the Agency will inspect the site and may request additional information. The Agency shall determine whether it is likely or unlikely that the condition of the land will cause pollution of the environment or harm to human health. Where the surrender of a licence is accepted the Agency will issue the applicant with a Certificate of Completion. This is a major advance on the Control of Pollution Act 1974, where the conditions of the licence ceased to be enforceable once the licence was handed back regardless of the state in which the land was left. This loophole has left several local authorities with massive problems due to past site mismanagement.

Contravention of the conditions of the waste management licence is an offence under the Act. The current situation regarding conviction for contravention of licence conditions depends on the offence being committed while waste is being deposited. This has proved very difficult to establish when action has been taken for contravention of licence conditions under the Control of Pollution Act. A licence may be revoked by the Environment Agency, under Section 38, if it appears to the Agency:

(a) that the holder of the licence has ceased to be a fit and proper person by reason of his having been convicted of a relevant offence; or
(b) that the continuation of the activities authorized by the licence would cause pollution of the environment or harm to human health or would be seriously detrimental to the amenities of the locality affected; and
(c) that the pollution, harm or detriment cannot be avoided by modifying the conditions of the licence.

Licence conditions may, for example, include:

(a) duration of licence;
(b) supervision by the licence holder of the licensed activities;
(c) the kind and quantities of waste which may be dealt with;
(d) precautions to be taken on land to which the licence relates;
(e) compliance with planning permission conditions;
(f) hours during which waste may be dealt with;
(g) the works to be carried out relating to the land, plant or equipment to which the licence relates before the activities authorized by the licence are begun or after the activities which the licence authorizes have ceased.

Site licences are available for public inspection at the relevant Environment Agency office.

On 6 April 2008, new Environmental Permitting Regulations will come into force in England and Wales. Initially they will replace the Waste Management Licensing and Pollution Prevention and Control (PPC) systems. All current waste management licences and PPC permits will automatically become 'Environmental Permits'. Risk assessment and compliance will increasingly move to a site approach.

Source: Environment Agency, Bristol, UK, *Your Environment*, Feb-April 2008.

Case Study

An investigation after a major fire at a waste storage and treatment site found that it had been accepting dangerous waste forbidden under its waste management licence, a court has heard. (Two firemen required hospital treatment). The company (Shanks) was fined a total of £45 000 and ordered to pay costs of nearly £21 640 by Hartlepool Magistrate's Court. The court heard that the Shanks site was licensed to accept waste but that the licence did not cover lithium. A lithium fire can burn at temperatures of up to 2000 °C and cannot be extinguished with water or many types of powder extinguisher. The case against Shanks was brought jointly by the Environment Agency and the Health and Safety Executive (HSE).

Source: Environment Agency, *Your Environment*, April 2006, p12.

(vii) Waste minimization.

Waste reduction is not just an option: it is an imperative. We need to create, through education and example, a new social spirit in which Government, industry and citizens positively want to see waste reduced at all stages. In encouraging the Government to be more active in promoting waste prevention and recycling and in creating the right market conditions for them, we recognise that decisions sometimes have to be made for reasons other than those of economics: there can be other perceived benefits in the form of public education, enhanced amenity, employment opportunities and so on.

Source: House of Lords Select Committee Report on the Proposal for a Council Directive on the landfill of waste, September 1998.

As an example of a waste minimization strategy for the chemical industry, the following stages are suggested:

1. Design for maximum chemical conversion.
2. Design for maximum energy efficiency.
3. Use low hazard solvents.
4. Make minimum use of solvents.
5. Make minimum use of process water.
6. Ensure minimum dilution of any carrier liquid (otherwise there will be greater clean-up costs).
7. Ensure low inventories of materials and product.

Source: Discussion at Institution of Chemical Engineers meeting on Pollution Control, I.C.I. Pharmaceuticals, Alderley Park, Cheshire, 2 November 1989.

In addition, through applying the POLLUTION PREVENTION PAYS approach, many wastes can be reduced in volume or altered to be less environmentally damaging. (◊ WASTE MANAGEMENT; WASTE DISPOSAL OPTIONS FOR INDUSTRIAL WASTE)

Case study 1. detergent reclamation

This is an excellent example of the appliance of science (with acknowledgements to T. U. Deft).

Detergent effectiveness is determined by four factors: chemistry, mechanical energy, heat and time. First, the chemistry. Washing is a separation process in which detergent is used to detach microscopically small particles and molecules from the fabric. In addition to surfactants, detergent may also contain other components, such as bleach or disinfectants. (Household washing powders contain even more additives including enzymes, scents, brighteners and water softeners not used industrially.) The surfactants play a critical role in the laundering process. They adsorb onto the particles and (grease) molecules, which owing to their sheath of surface-active components are transformed into stable colloids (charged particles suspended in solution) which are then carried off by the water.

Clean laundry requires mechanical efforts as well, e.g. via the rotating drum of a washing machine.

A third important factor in the laundering process is heat. Soaking off the dirt particles and molecules tends to work better at higher temperatures. One of the reasons is that at higher temperatures, particles of grease will liquefy, which renders them easier to remove. At washing temperatures of 60 °C and over, the heat also acts as a disinfectant.

Chemicals, mechanical energy and heat affect the fourth factor determining the efficacy of the laundering process, i.e. time. The longer the process takes, the cleaner the result will be. There are, of course, diminishing returns and an optimum is struck.

The quantity of detergent added is a compromise between laundry result, the cost of energy, water and detergent, and environmental considerations. More detergent per kilogram of laundry can be added, so the laundering process can be speeded up, which will save on energy and water, but this is offset by the extra cost of the detergent.

Recovering used detergent from industrial laundry water will enable higher concentrations of detergent to be used without increasing the cost of detergent or wastewater treatment. To reclaim the detergent, or rather the surfactants, from the laundry water, innovative work by Paul Brasser at Delft Technical University has shown that a process based on crystallization by cooling can effect recovery.

Individually, surfactants do not dissolve at all well, whereas collectively they do. Above a certain temperature, separate molecules converge to form micelles (groups of approximately 50 molecules arranged in such a way that the apolar tails point inwards, with the polar heads facing out; this makes a micelle easy to dissolve in water). The forming of micelles is affected by the temperature and the concentration of molecules. As the temperature rises above a certain value, the solubility increases to the point where micelles can be formed. The concentration at which micelles start to form is referred to as the critical micelle concentration, or CMC. The temperature at which the phenomenon occurs is called the Krafft point (after the scientist Krafft). Above the Krafft point, the concentration of single molecules in the solution remains constant, but the number of micelles will keep growing unchecked, i.e. the total solubility increases enormously.

The reverse process also occurs, and thus forms the basis for the recovery process developed by Brasser. By lowering the temperature of a solution of micelles to below the Krafft point, the micelles will disintegrate into single molecules. As a result, the molecule concentration will rise far above the point of solubility, causing them to be deposited in the form of crystals.

The net result for industrial laundries is a cost effective recycling of surfactants provided heat and detergent recovery are included in the reckoning.

Source: J. V. Kastern, 'Detergent Recovery Offers both Economical and Ecological Advantages', *Delft Outlook*, No. 4, 1998, TU Delft.

Case study 2. Reducing environmental impacts of car paint repair

Figures 204 and 205 compare the flow of lacquer, solvent and thinner through a typical and a 'perfect' painting shop, respectively, at a car repair firm.

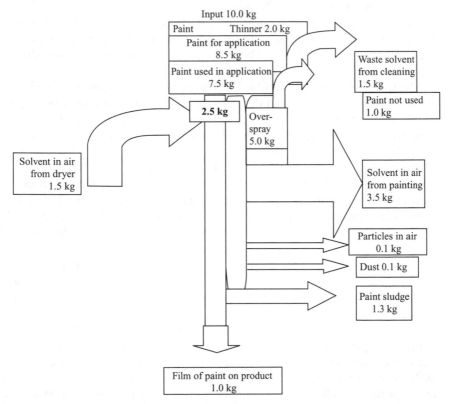

Figure 204   Flow scheme for a typical painting shop.

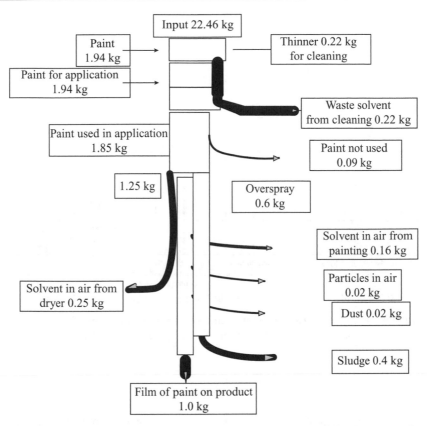

Figure 205 Flow scheme of a 'perfect' painting shop using Best Available Technique. *Source*: H. Heitzinger and P. Schnitzer, 'Reducing environmental impacts of car paint repair', *UNEP Industry and Environment*, July–September 1998, p. 33. (Reproduced by permission of the United Nations Environment Programme)

(viii) Waste quantities. Until relatively recently there were no reliable sources of information on the quantities of waste produced in the UK. However, the results of the Department for Environment, Food & Rural Affairs surveys of municipal waste generation in England and Wales are now published annually and the Environment Agency's National Waste Production Survey should provide better information on industrial and commercial waste.

Figure 206 shows volumes of waste in Switzerland. The table on page 741 shows the average composition of municipal solid waste in some Latin American cities, and Figure 207 shows the composition of household waste in England in 2004.

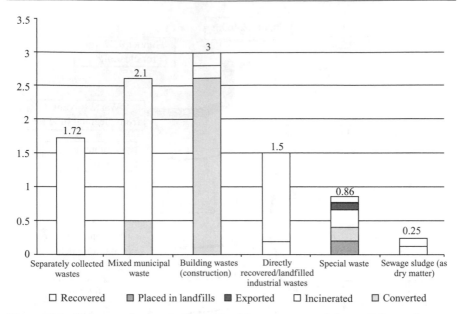

Figure 206 Volumes of waste in Switzerland by category and form of disposal.
Source: *ISWA Times*, Issue 4, 1998. (Reproduced by permission of ISWA)

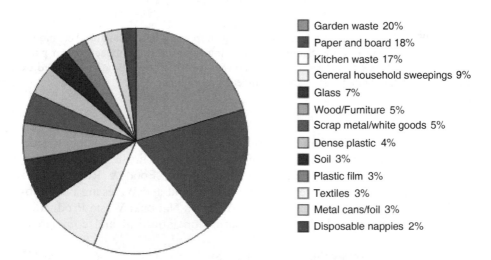

Figure 207 Composition of household waste in England in 2004.
Source: www.defra.gov.uk/environment/statistics/waste/kf/wrkf18.htm
Accessed 2 January 2008.

(ix) Waste recycling plan. Required under the Environmental Protection Act 1990 (Section 49). Each waste collection authority, as respects household and commercial waste arising in its area, is required:

(a) to carry out an investigation with a view to deciding what arrangements are appropriate for dealing with the waste by separating, baling or otherwise packaging it for the purpose of recycling;
(b) to decide what arrangements are in the opinion of the authority needed for that purpose;
(c) to prepare a statement ('the plan') of the arrangements made and proposed to be made by the authority and other persons for dealing with waste in those ways;
(d) to carry out from time to time further investigations with a view to deciding what changes in the plan are needed; and
(e) to make any modification of the plan which the authority thinks appropriate in consequence of any such further investigation.

The information in the plan should include:

(a) the kinds and quantities of controlled waste which the authority expects to collect during the period specified in the plan;
(b) the kinds and quantities of controlled waste which the authority expects to purchase during that period;
(c) the kinds and quantities of controlled waste which the authority expects to deal with in the ways specified in subsection (a), during that period;

Average composition (%) of municipal solid waste in some Latin American cities.

Material	Mexico City, Mexico	Caracas, Venezuela	Asuncion, Paraguay	Bogota, Columbia	Lima, Peru	Guatemala City, Guatemala
Paper	16.7	34.9	12.2	18.3	24.3	13.9
Putrescible matter	56.4	40.4	60.8	57.1	34.3	65.0
Metals	5.7	6.0	2.3	1.7	3.4	1.8
Glass	3.7	6.6	4.6	4.6	1.7	3.2
Plastics	5.8	7.8	4.4	14.2	2.9	8.1
Textiles	6.0	2.0	2.5	3.8	1.7	3.6
Ceramics, dust, stones	5.7	2.3	13.2	0.3	31.7	4.4
Total	100.0	100.0	100.0	100.0	100.0	100.0

Source: L. F. Diaz, G. M. Savage, L. L. Eggerth and C. G. Golueke, *Solid Waste Management for Economically Developing Countries*, International Solid Waste Association, Denmark, 1996.

(d) the arrangements which the authority expects to make during that period with waste disposal contractors or, in Scotland, waste disposal authorities and waste disposal contractors, for them to deal with waste in those ways;

(e) the plant and equipment which the authority expects to provide; and

(f) the estimated costs or savings attributable to the methods of dealing with the waste in the ways provided for in the plan.

Recycle or compost 25 per cent of household waste by 2005; 30 per cent by 2010 and 33 per cent by 2015.

Source: Waste Management Strategy for England and Wales DETR, May 2000.

Central Government policy towards waste management is based upon the following general principles:

A hierarchy (under EU review at the moment) of:

Waste reduction
Reuse
Recovery (including materials recycling, energy recovery and composting)
Safe disposal

The 'proximity principle' under which waste should be disposed of (or otherwise managed) close to the point at which it is generated. Where waste has to be transported, consideration should be given to use of rail or water transport, if economically feasible.

'Regional self-sufficiency' whereby each region should expect to provide sufficient facilities to treat or dispose of all the waste it produces and development plans should reflect this need. (◊ WASTE EXPORT)

Waste Strategy for England 2007. The following is a Summary of the 'new' Waste Strategy:

Incentives for individuals and businesses to recycle waste, leading to 40 per cent household waste recycling rate by 2010, and 50 per cent by 2020.

More action from businesses to cut waste, including packaging waste.

Greater emphasis on waste reduction – home composting, food waste reduction, businesses cutting down on packaging use.

Target to reduce the amount of household waste not re-used, recycled or composted to 12.2 million tonnes by 2020 – a reduction of 45 per cent from 2000.

The end of free single use bags to be encouraged.

Recycling points in public areas like shopping malls, train stations and cinema multiplexes, via a voluntary code of practice.

Consultation on banning biodegradable and recyclable waste from landfill.

Increasing the amount of energy produced by a variety of energy from waste schemes, using waste that cannot be reused or recycled.

Author's Comment: Stand by for obfuscation by environmentalists (e.g. claims that 'Food packaging that isn't recycled ends up going to landfill or being incinerated, both of which produce greenhouse gases', while true, are misleading as 'incineration' produces ca. $550\,kWh_e$/tonne of waste, and produces it at nett lower CO_2 emission than coal or gas-fired generation.

Source of Quote: Sarah Collard, *The Daily Telegraph Magazine*, 7 July 2007, 'Is it worth it?, p73).

Waste to energy. *Case Study – Uppsala Energi, Sweden*
Uppsala Waste to Energy plant is a good example of the Swedish philosophy regarding recovery from waste and minimizing the use of landfill.

- Municipal solid waste (MSW) is a resource. After hazardous and economically valuable material have been sorted out in households, etc., the residual municipal solid waste (RMSW) is a biofuel which shall be used in waste to energy plants.
- Optimal design of the waste to energy plant minimizes the impact on the environment from the plant and keeps the cost at an acceptable level.
- Effective utilization of energy in the waste by steam/hot water production maximizes the income for the plant and minimizes the use of alternative fossil fuels.
- The plant has a very high reliability mainly due to interested and well-educated personnel.
- The company is open and is accepted by public, politicians and authorities.

Introduction

Uppsala has Sweden's oldest university and is situated about 60 km north of Stockholm. The municipality of Uppsala has 120 000 inhabitants, the fourth biggest city in Sweden. The Uppsala region is developing very fast mainly because of its excellent location and the university.

Uppsala Energi AB is a complete energy company which produces and distributes electricity, heat and steam to approximately 75 000 customers. In 1980 Uppsala Energi was 92 per cent dependent on oil. Through expansion of waste to energy, conversion of boilers fired with oil to peat and other biofuels, and installation of heat pumps, dependence on oil was reduced to only 5 per cent in 1995. Uppsala Energi produces about 1900 GWh steam/hot water and 300 GWh electricity per year. The heat produced covers about 95 per cent of the heating requirements for the city of Uppsala.

Residual municipal solid waste (i.e. post-recycling residuals)

Residual municipal solid waste (RMSW) is wastes from households, shops and small industries after hazardous material, glass and paper have been sorted out. About 250 000 tons of RMSW from 20 municipalities around Uppsala are burnt in four boilers at Uppsala Energi's plant in Uppsala. Net calorific value of the RMSW = 10–12 MJ/kg.

Boilers

RMSW is burnt in four stoker boilers with the following capacities:

- 15 ton/h
- 10 ton/h
- 3 ton/h
- 3 ton/h

About 70 MW steam with a pressure of 16 bar and temperature of 200 °C is generated; some of this is sold directly to nearby industries but the main part is condensed and converted to hot water for district heating. Combustion conditions are very good with high temperatures (approx. 100 °C), intensive turbulence, long residence time and excess of oxygen. Flue gas temperature after the boilers is 220 °C. Urea is added into the 900 °C region in the boilers to reduce NO_x, through a selective non-catalytic reduction (SNCR) system.

From the bottom ash, magnetic material is separated out and sold to steel plants.

Air pollution control (APC)

All the units were equipped with electrostatic precipitators when they were built. In 1986 a common scrubber/condenser for all the four units was installed which reduced the emission of water-soluble gases, mainly hydrochloric acid and mercury chloride. Simultaneously it recovered about 17 MW low temperature waste heat from condensing water in the flue gas.

In the year 1990/91 the flue gas cleaning, system was renovated and extended with a 'Filsorption' stage after the condenser. The 'Filsorption' process uses a fabric filter as a fixed bed reactor for the flue gas and powder sorbent which is blown into the gas. Impurities in the flue gas are then removed by absorption, filtration and chemical reaction.

The Air Pollution Control system in Uppsala now comprises:

- an SNCR system for NO_x reduction
- electrostatic precipitators, one to each boiler
- a common waste heat boiler (WHB) which decreases the temperature to 140 °C with a production of hot water of about 8.5 MW
- a combined wet scrubber/condenser for flue gas cleaning and waste heat recovery

- 'Filsorption' with lime and carbon acts as an almost absolute filter for the impurities in the flue gas.

As a wet scrubber transfers contamination in the flue gas to water, extensive water treatment is necessary.

The APC system is also designed to operate with the scrubber/condenser in by-pass. In this case the unit operates at 140 °C.

Emission to the atmosphere

Cleaning gases in stages is very efficient; electrostatic precipitators remove 99 per cent of the dust in the first stage and the wet scrubber removes water-soluble gases to a high degree. The 'Filsorption' stage with additives has a universal function: it removes almost all impurities in particulate as well as gaseous form, and reduces the emissions to very low levels, tending to zero.

The table on the next page shows emissions to the atmosphere.

Emission to water

Water produced in the condenser must, of course, be cleaned before it is discharged. The water treatment plant comprises:

- neutralization with limestone and hydrated lime
- precipitation of heavy metals and organic sulphide
- flocculation
- sedimentation
- filtration.

Emission to land

The slurry from the sedimentation in the water treatment is mixed with the fly ash from electrostatic precipitators and the fabric filter.

The excess sulphide in the slurry reacts with and binds the heavy metals in the fly ash. In this way the leaching of heavy metals in the landfill is greatly reduced.

The mixture of slurry/fly ash is placed in a monofill area of a controlled landfill.

Energy balance

Input: 90–100 MW
Output:

Steam	70 MW
Hot water	
Waste heat boiler:	8.5 MW
Condenser/heat pump:	17 MW

Component	Measured emissions (concentration in mg/m³ dry gas but dioxins in ng/m³ dry gas)				
	Scrubber in by-pass 10% CO₂	Total APC in operation 10% CO₂	Guaranteed emission 10% CO₂	Swedish Regulation 10% CO₂	German Regulation 10% CO₂
Dust	<1	<1	10	20	10
HCl	<15	<1	30	100	10
HF	<0.1	<0.1			1
SO₂	<20	<1	50		50
Cd + Ti	—	=0.001			0.05
Hg tot	<0.01	<0.0055	0.03	0.03	0.05
Σ 7HM	—	<0.05			0.5
TEQ Dioxin	<0.05	<0.005	0.1	0.1	0.1
CO		20			50
THC		<0.02			20
NO$_x$		86 mg/J			200

TEQ Dioxin according to Eadon. German Regulations according to NATO.
Σ 7HM = sum of seven heavy metals. German Regulations sum of 10 HM.

The following table summarizes the performance of the water treatment plant.

	Limit	Actual emission
Condense (m³/year)	<150 000	139 649
pH value	>7	>8.1
Dioxin (ng/l)	<0.1	0.0045*
Hg (μg/l)	<10	<3.0
Pb (μg/l)	<50	<15.0
Cd (μg/l)	<10	<2.9
Cr (μg/l)	<50	<6.6
Ni (μg/l)	<50	<16.9
Zn (μg/l)	<600	76.4
Co (μg/l)	<10	<5.0

* Result from three measurements.

Economy

The total cost for burning one ton of RMSW is about 360 SEK which includes capital (@ 50 per cent) operation, maintenance, treatment and disposal of ashes. This corresponds to approx. 120 SEK/MWh.

The yearly maximum production of steam/hot water is 700 GWh.

All energy produced from the waste to energy plant is sold because part of the steam is delivered to process industries all the year round. (◊ INCINERATION)

Source: Uppsala Energi, Sweden, (1998).

Wasteplex. A concept of a highly integrated centre for the reception of domestic and industrial wastes from population catchment areas of around 500 000 people. This allows economic sorting of the wastes into directly and indirectly reusable fractions. (◊ RECYCLING)

Wastes separation. The use of mechanical, manual or other means of segregating specific materials from a mixed waste stream, normally a municipal or domestic waste stream but commercial wastes that are not of uniform composition and origin may also be separated. (◊ RECYCLING)

Water. One of the prime resources for life which must be preserved from contamination for the public WATER SUPPLY. It has many industrial uses: a conveying medium for wastes, slurries, woodpulp, etc.; a heat-exchange medium; it is used in steam raising and also as a solvent. Its basic chemical formula is H_2O, yet it has other molecular combinations such as H_8O_4, which gives it many unique properties such as expansion on freezing.

For the purpose of this entry the chemical, physical and biological attributes are considered.

1. *Chemical properties.* These have a great influence on the use for drinking or industrial purposes or the support of aquatic life. The main parameters are:

   (i) *pH* – a measure of the acidity or alkalinity. Natural water supplies are usually in the pH range 6–8, whereas industrially contaminated waters can have any value in the range pH 1–13 and require neutralization to pH 7 before discharge. (◊ pH)

   (ii) *Hardness* – this prevents soap lathering properly. Also when the water is heated, scale may be deposited on the heating surfaces, which can ruin the efficiency of heat transfer and in extreme instances, cause burn-out of boiler tubes or heat exchangers. The hardness is measured by the concentration of calcium ions (Ca^{2+}) and magnesium ions (Mg^{2+}), present usually as calcium carbonate ($CaCO_3$) and magnesium sulphate ($MgSO_4$).

   (iii) DISSOLVED OXYGEN and, related to this, the chemical and biological oxygen demand. These values are of great importance to biological systems. (◊ BIOCHEMICAL OXYGEN DEMAND; CHEMICAL OXYGEN DEMAND)

2. *Physical characteristics.* The colour, taste and appearance of water are of great importance in its suitability for consumption. The solids in solution

and in suspension play an important part. The following characteristics are measured:

   (i) *Turbidity* – an indication of the presence of colloidal ('gluey' suspensions of fine particles) particles such as silt or bacteria from, say, sewage treatment.
  (ii) *Colour* – usually due to dissolved substances such as peat acids which give highland streams their brown appearance.
 (iii) *Taste* and *odour* due to the presence of dissolved solids or gases, e.g. 0.001 milligrams per litre (1 part in 100 million) of phenolic liquors can taint water and render it unpalatable. The gases can be of biological origin from algae or from bacterial action on wastes or impurities.
  (iv) *Temperature* – this has a direct bearing on the saturation concentration of dissolved oxygen in the water; the greater the temperature, the less oxygen the water can hold in solution and therefore temperature has important biological consequences.

3. *Biological properties*. The variety and type of water organisms and aquatic life give a very good indication of its biological state of health and the BMWP Score is used as such a measure. Other tests are made to indicate the presence of various classes of micro-organism algae and bacteria, and especially the bacterium, *Escherichia coli*, which indicates the possible presence of faecal matter.

(◊ COLIFORM COUNT; WATER SUPPLY; WATER QUALITY STANDARDS; PUBLIC HEALTH)

Water balance. Method used to estimate industrial water consumption or the quantities of LEACHATE produced in a LANDFILL site.

This is best understood by considering the landfill hydrological cycle (which can be regarded as a local sub-system of the hydrological cycle). This local sub-system may be complex, with important liquid inputs to, and outputs from, the landfill storage system, not normally encountered or considered in regional hydrologic cycles. An idealized picture of the main components is shown in Figure 208. The water balance for the site is determined by the difference between these rates of inflow and outflow, viz.

$$\text{(Rate of inflow)} - \text{(Rate of outflow)} = \Delta S$$

where ΔS is the rate of change in the quantity of liquid, S, stored within the landfill. Thus, the rate of change in storage is given by

$$\Delta S = (P + R_1 + G + L) - (E + T + R_2 + R_3 + I + Q)$$

Of these, the liquid abstracted from storage (Q), the surface springs and seepages (R_3), and any recharge to groundwater (I), can be considered as

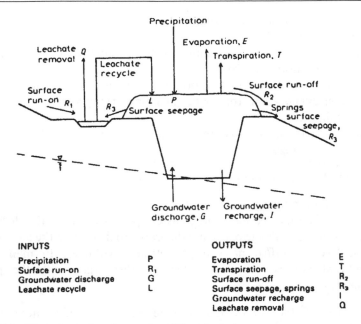

Figure 208   Hydrological cycle at a landfill water balance.

leachate and require to be managed as such. In some cases, surface run-off R_2, which would normally be clean, becomes mixed with leachate and increases the volume requiring attention.

The equation above reflects the reality that most of the component factors vary continuously with time. However, in order to calculate a water balance for a particular period, say one year, it is normally only feasible to divide the period into a limited number of discrete smaller periods (e.g. days, weeks or months) and to compute values for each period. The size of the sub-period chosen can make a difference to the calculated values for the overall rate and pattern of leachate production.

Water conservation. The following example is based on *Practical Water Management in Paper and Board Mills* (DETR, 1998).

However, the methodology is applicable to any process which requires water as part of the production process.

Preparing water balances

Before a mill can determine the scope of any water conservation programme, it will need to establish its current level of performance by preparing WATER BALANCES both for the site as a whole and for individual items of equipment:

- metered fresh water intake volumes;
- metered effluent volumes;
- estimates of any other water inputs to the production process, e.g. from chemical additions and from pulp;
- estimates of water outputs that do not form part of the metered effluent, e.g. water evaporation from paper machine drying sections and vacuum pump exhausts;
- estimates of the effect that stormwater discharges will have on metered effluent flow rates.

To obtain the relevant data, the following may be required:

- additional monitoring of effluent flow rates;
- a review of design specifications to assess possible evaporation losses;
- a review of the drainage plans and diagrams to determine the contribution of surface run-off/storm water to effluent volumes. In many cases, significant volumes of stormwater are discharged with the process effluent.

The mill can use the data obtained to determine its specific water consumption. This provides a measure of the efficiency of water use.

The same approach can be applied to individual items of equipment. This requires:

- metering the fresh water intake of individual machines;
- measuring the effluent flow rates from each machine, using appropriate flow measurement structures (e.g. weirs or flumes) in the effluent channels;
- a review of design specifications to identify the range of flow rates that can be expected.

The need for additional metering can be minimized by using mass balances to estimate flow rates where appropriate.

The information obtained can be used:

- to construct a water balance diagram showing water use in different parts of the process;
- to provide a detailed breakdown of site water use, allowing staff to identify priorities for action in a water conservation programme;
- to facilitate checks on drain losses. This can be done on a shift basis and, provided the information is reported promptly, will allow the rapid identification of trends and unusually high losses, and ensure that corrective action is taken before significant increases in running costs (or worse, major pollution events) occur. (⟡ TIOXIDE; WATER BALANCE)

Case Study: UK Millennium Dome

The Dome toilets and urinals (606 in total, each using 6 litres water per flush instead of 9), were flushed with non-potable water. $120 \, m^3/d$ of grey water needed for the scheme was reclaimed. $100 \, m^3/d$ water was collected from the Dome roof. After preliminary storage in a lagoon, it was filtered through two $250 \, m^2$ surface area REED BEDS and pumped to a holding tank before repurification. The reed beds' lifespan is expected to be 10–20 years.

Case Study: London Transport

London Transport has 14 depots, each offering up to $30\,000 \, m^2$ of roof surface with an average of 560 mm of rain falling on London each year, a significant resource ($16\,800 \, m^3$/depot) that could be harnessed.

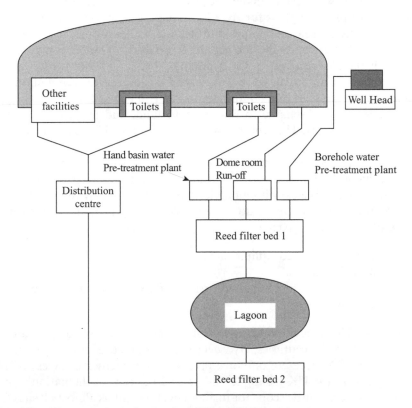

Figure 209 Water conservation at the Millennium Dome.
Source: The Chemical Engineer, 10 September 1998. (Reproduced by permission)

This captured water could be used for train washing and other non-potable uses.

Source: L. Hibbert, 'Mind the Energy Gap', *Professional Engineering*, January 1999.

Water consumption. The quantity of water abstracted and used for any purpose, irrespective of the state or time in which it is returned to the source or to the atmosphere. Per capita consumption is the amount of water supplied to an area divided by the size of the resident population. The UK average per capita consumption is 45 000 litres or 10 000 gallons per year.

Water cycle. Also called the hydrological cycle. The circulation of water from the oceans through the atmosphere to the land and ultimately back to the oceans.

Water Framework Directive. A European Directive intended to provide a co-ordinated approach to water management within the EU by bringing together strands of EU water policy under one piece of framework legislation. Member states must produce plans for river basin management districts that set out a programme of measures aimed at protecting bodies of surface and ground-water. Each plan must include economic analysis of water use and move towards full cost recovery in water pricing.

Water pollution. Can be caused by a variety of means, e.g. farm pollution from animal wastes and SILAGE liquor, LEACHATE from LANDFILL sites, and spoil heaps, SOLVENT discharge to sewers or to land, and inadequate SEWAGE TREATMENT works. The typical range of pollution incidents is shown on the facing page.

Her Majesty's Inspectorate of Pollution, First Annual Report (1987–88) stated that in 1986, there were 4333 water authority sewage treatment works (in England and Wales) with numerical consents, of which 1000 (23%) failed to comply with their DISCHARGE CONSENTS. The principal reasons for this state of affairs were given as:

(a) Old and overloaded equipment.
(b) Poor maintenance.
(c) Possible overloading because of trade effluent discharges to sewer.
(d) Under-investment.

Water power. The free renewable source of energy provided by falling water has been put to use for centuries. Hydroelectricity is practised in many countries, yet in the USA there are only five plants with capacities in excess of 1000 megawatts. In the UK it supplies less than 2 per cent of the nation's power needs and there is little scope for further development as almost all sites have been used. It is capital intensive, requiring costly dams and earthworks which

Total pollution incidents.

Year	Total	%	Serious	%	Prosecutions	%
1987						
Farm	3890	19	990	51	225	67
Industrial	7575	37	617	32	98	29
Sewage	4177	20	221	11	6	2
Other	5017	24	117	6	8	2
Total	20659	100	1945	100	337	100
1980						
Farm	1671	15				
Total	10797	100				

Source: Water Authorities Association, *Water Pollution from Farm Waste*, 1988 *England & Wales*, 1989.

may have a finite lifetime due to silting up behind the dam. This will restrict its use at any one site to 100–200 years unless silt-laden floodwater is diverted to lengthen the life of the reservoir. Globally there are locations that can, in total, supply energy comparable to the present rate of energy consumption. South America, Africa, Asia and the USSR account for 80 per cent of the potential capacity. Because of the high capital investment per megawatt capacity, its contribution to future energy supplies is not expected to be significant in the West. (◊ BARRAGE)

Water quality monitoring. Rivers are routinely surveyed to assess their water quality. Measurements are taken by inspectors from the regulatory authority, in the UK, these being the Environment Agency in England and Wales, the Scottish Environment Protection Agency in Scotland and the Environment and Heritage Service in Northern Ireland. The parameters measured regularly are the dissolved oxygen level, pH, total ammonia and the Biochemical Oxygen Demand (BOD). Biological assessment, typically through the determination of the BMWP Score, is also be carried out.

Water quality objectives (WQOs). WQOs specify formal minimum quality standards, and have already been set to give effect to the EC Dangerous Substances legislation arising from Directive 76/464/EEC, and subsequent daughter Directives (via the Surface Waters (Dangerous Substances) (Classification) Regulations 1989 (LS 2286) and 1992 (SI 337)), and the EC Bathing Waters legislation arising from Directive 76/160/EEC (via the Bathing Waters (Classification) Regulations 1991 (SI 1597)).

The requirements of these Directives, and the methodologies necessary for compliance assessment, are contained within the Directives and the domestic Regulations and Notices arising from them.

Details of WQOs assigned to river stretches, compliance with WQOs, and the monitoring data upon which compliance assessment is based, together with any of the 'exceptional circumstances' which have been identified by the Environment Agency as applying, will be included on the public register.

The quality of water for domestic use should comply with purity standards which are internationally acceptable. The World Health Organization has published its Guidelines for Drinking-Water Quality (Third Edition, 2006). In the EU the quality of water supplied to the public is controlled by the EU Directive on the Quality of Water Intended for Human Consumption (98/83/ EC). This directive covers all water for domestic use and water used by the food industry where this affects the final product (and thus consumers' health). It does not apply to water used for agricultural purposes, natural mineral water or medicinal waters. The parameters in the directive will be reviewed at least every five years. The new directive splits the quality parameters into two categories: Mandatory microbiological and chemical parameters, and Indicator parameters.

Mandatory parameters

For tap water, there are two microbiological and 26 chemical parameters (Tables a and b). Instead of 'guide levels' and 'maximum admissible concentration' (as in the old directive), there are now 'parametric values' (PVs).

(a) Microbiological parameters

Parameter	Parametric value (*number per 100 ml*)
Escherichia coli (E. coli)	0
Enterococci	0

The following applies to water offered for sale in bottles or containers:

Parameter	Parametric value
Escherichia coli (E. coli)	0 per 250 ml
Enterococci	0 per 250 ml
Pseudomonas aeruginosa	0 per 250 ml
Colony count at 22 °C	10 per ml
Colony count at 37 °C	20 per ml

(b) Chemical parameters

Parameter	Parametric value	Unit	Notes
Acrylamide	0.10	$\mu g\,l^{-1}$	Note 1
Antimony	5.0	$\mu g\,l^{-1}$	
Arsenic	10	$\mu g\,l^{-1}$	
Benzene	1.0	$\mu g\,l^{-1}$	
Benzo(a)pyrene	0.010	$\mu g\,l^{-1}$	
Boron	1.0	$mg\,l^{-1}$	
Bromate	10	$\mu g\,l^{-1}$	Note 2
Cadmium	5.0	$\mu g\,l^{-1}$	
Chromium	50	$\mu g\,l^{-1}$	
Copper	2.0	$mg\,l^{-1}$	Note 3
Cyanide	50	$\mu g\,l^{-1}$	
1,2-dichloroethane	3.0	$\mu g\,l^{-1}$	
Epichlorohydrin	0.10	$\mu g\,l^{-1}$	Note 1
Fluoride	1.5	$mg\,l^{-1}$	
Lead	10	$\mu g\,l^{-1}$	Notes 3 and 4
Mercury	1.0	$\mu g\,l^{-1}$	
Nickel	20	$\mu g\,l^{-1}$	Note 3
Nitrate	50	$mg\,l^{-1}$	Note 5
Nitrite	0.50	$mg\,l^{-1}$	Note 5
Pesticides	0.10	$\mu g\,l^{-1}$	Notes 6 and 7
Pesticides – total	0.50	$\mu g\,l^{-1}$	Notes 6 and 8
Polycyclic aromatic hydrocarbons	0.10	$\mu g\,l^{-1}$	Sum of concentrations of specified compounds; Note 9
Selenium	10	$\mu g\,l^{-1}$	
Tetrachloroethene and trichloroethene	10	$\mu g\,l^{-1}$	Sum of concentrations of specified parameters
Trihalomethanes – total	100	$\mu g\,l^{-1}$	Sum of concentrations of specified compounds; Note 10
Vinyl chloride	0.50	$\mu g\,l^{-1}$	Note 1

Note 1: The parametric value refers to the residual monomer concentration in the water as calculated according to specifications of the maximum release from the corresponding polymer in contact with the water.

Note 2: Where possible, without compromising disinfection, Member States should strive for a lower value. For the water referred to in Article 6(1)(a), (b) and (d) [see below], the value must be met, at the latest, 10 calendar years after the entry into force of the directive. The parametric value for bromate from five years after the entry into force of the directive until 10 years after its entry into force is $25\,\mu g\,l^{-1}$.

Article 6

Point of compliance

The parametric values set in accordance with Article 5 shall be complied with: in the case of water supplied from a distribution network, at the point, (a) within premises or an establishment, at which it emerges from the taps that are normally used for human consumption; in the case of water supplied from a tanker, at the point at which it (b) emerges from the tanker; in the case of water put into bottles or containers intended for sale, at (c) the point at which the water is put into bottles or containers; in the case of water used in food production undertaking, at the point (d) where the water is used in the undertaking.

Note 3: The value applies to a sample of water intended for human consumption obtained by an adequate sampling method at the tap and taken so as to be representative of a weekly average value ingested by consumers. Where appropriate, the sampling and monitoring methods must be applied in a harmonized fashion to be drawn up in accordance with Article 7(4). Member States must take account of the occurrence of peak levels that may cause adverse effects on human health.

Note 4: For water referred to in Article 6(1)(a), (b) and (d), the value must be met, at the latest, 15 calendar years after the entry into force of this directive. The parametric value for lead from five years after the entry into force of this directive, until 15 years after its entry into force is $25\,\mu g l^{-1}$.

Member States must ensure that all appropriate measures are taken to reduce the concentration of lead in water intended for human consumption as much as possible during the period needed to achieve compliance with the parametric value.

When implementing the measures to achieve compliance with that value, Member States must progressively give priority where lead concentrations in water intended for human consumption are highest.

Note 5: Member States must ensure that the condition that [nitrate]/50+[nitrate]/3 ≤ 1, the square brackets signifying the concentrations in $mg l^{-1}$ for nitrate (NO_3) and nitrite (NO_2), is complied with and that the value of $0.10\,mg l^{-1}$ for nitrites is complied with ex-water treatment works.

'Pesticides' means:
organic insecticides,
organic herbicides,
organic fungicides,
organic nematocides,
Note 6: organic acaricides,
organic algicides,
organic rodenticides,
organic slimicides,
related products, (*inter alia*, growth regulators)
and their relevant metabolites, degradation and reaction products.

Only those pesticides which are likely to be present in a given supply need be monitored.

Note 7: The parametric value applies to each individual pesticide. In the case of aldrin, dieldrin, heptachlor and heptachlor epoxide the parametric value is $0.030\,\mu g l^{-1}$.

Note 8: 'Pesticides–total' means the sum of all individual pesticides detected and quantified in the monitoring procedure.

Note 9: The specified compounds are:
benzo(b)fluoranthene,
benzo(k)fluoranthene,
benzo(ghi)perylene,
indeno(1,2,3-cd)pyrene

Note 10: Where possible, without compromising disinfection, Member States should strive for a lower value.

The specified compounds are: chloroform, bromoform, dibromochloromethane, bromodichloromethane.

For the water referred to in Article 6(1)(a), (b) and (d), the value must be met, at the latest, 10 calendar years after the entry into force of this directive. The parametric value for total THMs from five years after the entry into force of this directive until 10 years after its entry into force is $150\,\mu g l^{-1}$.

Member States must ensure that all appropriate measures are taken to reduce the concentration of THMs in water intended for human consumption as much as possible during the period needed to achieve compliance with the parametric value.

When implementing the measures to achieve this value, Member States must progressively give priority to those areas where THM concentrations in water intended for human consumption are highest.

Indicator parameters

These parameters were set for monitoring purposes. This is a new concept. Generally speaking, the PVs for these parameters are not based on health considerations and are non-binding. When the PVs are exceeded, remedial action is only necessary if a Member State judges there to be risk to health. Most of the Indicator parameters are included in the list of parameters to be analysed in Check monitoring. This has the aim of providing information on the organoleptic and microbiological quality of supplied water, as well as on the effectiveness of treatment. The Indicator parameters are listed in the table below. A fair number of the existing EC standards have been turned into Indicator parameters in the new directive, or omitted altogether. The Department for Environment, Food and Rural Affairs, however, proposes to retain most of the existing EC standards as binding limits. This means, for instance, that standards for aluminium, iron, manganese, colour and turbidity intended to prevent the supply of discoloured water, will be retained.

Indicator parameters

Parameter	Parametric value	Unit	Notes
Aluminium	200	$\mu g\,l^{-1}$	
Ammonium	0.50	$mg\,l^{-1}$	
Chloride	250	$mg\,l^{-1}$	Note 1
Clostridium perfringens (including spores)	0	number per 100 ml	Note 2
Colour	Acceptable to consumers and no abnormal change		
Conductivity	2500	$\mu S\,cm^{-1}$ at 20 °C	Note 1
Hydrogen ion concentration	6.5 and 9.5	pH units	Notes 1 and 3
Iron	200	$\mu g\,l^{-1}$	
Manganese	50	$\mu g\,l^{-1}$	
Odour	Acceptable to consumers and no abnormal change		
Oxidizability	5.0	$mg\,l^{-1}$ O_2	Note 4
Sulphate	250	$mg\,l^{-1}$	Note 1
Sodium	200	$mg\,l^{-1}$	
Taste	Acceptable to consumers and no abnormal change		
Colony count at 22 °C	No abnormal change		
Coliform bacteria	0	number per 100 ml	Note 5
Total organic carbon (TOC)	No abnormal change		Note 6

Parameter	Parametric value	Unit	Notes
Turbidity	Acceptable to consumers and no abnormal change		Note 7
Radioactivity			
Tritium	100	Bq 1^{-1}	Notes 8 and 10
Total indicative dose	0.10	mSv year^{-1}	Notes 9 and 10

Note 1: The water should not be aggressive.

Note 2: This parameter need not be measured unless the water originates from or is influenced by surface water. In the event of non-compliance with this parametric value, the Member State concerned must investigate the supply to ensure that there is no potential danger to human health arising from the presence of pathogenic micro-organisms, e.g. *Cryptosporidium*. Member States must include the results of all such investigations in the reports they must submit under Article 13(2).

Note 3: For still water put into bottles or containers, the minimum value may be reduced to 4.5 pH units.

For water put into bottles or containers which is naturally rich in or artificially enriched with carbon dioxide, the minimum value may be lower.

Note 4: This parameter need not be measured if the parameter TOC is analysed.

Note 5: For water put into bottles or containers the unit is number per 250 ml.

Note 6: This parameter need not be measured for supplies of less than 10 000 m³ a day.

Note 7: In the case of surface water treatment, Member States should strive for a parametric value not exceeding 1.0 NTU (nephelometric turbidity unit) in the water ex-treatment works.

Note 8: Monitoring frequencies to be set later in Annex II.

Note 9: Excluding tritium, potassium-40, radon and radon decay products; monitoring frequencies, monitoring methods and the most relevant locations for monitoring points to be set later in Annex II.

Note 10: 1The proposals required by Note 8 on monitoring frequencies, and Note 9 on monitoring frequencies, monitoring methods and the most relevant locations for monitoring points in Annex II shall be adopted in accordance with the procedure laid down in Article 12. When elaborating these proposals the Commission shall take into account *inter alia* the relevant provisions under existing legislation or appropriate monitoring programmes including monitoring results as derived from them. The Commission shall submit these proposals at the latest within 18 months following the date referred to in Article 18 of the Directive.

2A Member State is not required to monitor drinking water for tritium or radioactivity to establish total indicative dose where it is satisfied that, on the basis of other monitoring carried out, the levels of tritium of the calculated total indicated dose are well below the parametric value. In that case, it shall communicate the grounds for its decision to the Commission, including the results of this other monitoring carried out.

The EEC now has several Directives on water quality. Discharges to the aquatic environment are controlled by Directives which protect surface and groundwater against pollution. Each of these directives has two lists (List 1 and List 2) and the aim is to eliminate pollution by substances quoted in List 1 and reduce pollution by those in List 2. These Directives are 'enabling'

Directives which will come fully into force as daughter Directives are enacted. Directives on cadmium, mercury and hexachlorocyclohexane have been passed, while there are proposed directives for aldrin, dieldrin, endrin, DDT, carbon tetrachloride and pentachlorophenol. (◊ DIRECTIVES; EC)

The Directive on the *Quality of Surface Water Intended for Drinking Water* proposes standards for surface water to be used in the public supply and defines three categories of water treatment (A1, A2, A3) from simple physical treatment and disinfection to intensive physical and chemical treatment. The treatment to be used depends on the quality of the water abstracted. The Directive uses I (Imperative) values for parameters known to have an adverse effect on health and G (Guide) values for those which are less adverse. There is also a Directive which complements the 'surface water abstraction' Directive by indicating the methods of measurement and the frequency of sampling and analysis required.

The quality of coastal and estuarial waters is considered in Directives on the quality of bathing water, the quality required for shellfish water and the quality of water for freshwater fish.

The Bathing Water Directive sets out quality requirements for bathing water, defines a monitoring programme, and gives reference methods of analysis.

The Freshwater Fish and Shellfish Directives apply to designated waters which require protection to make them suitable for fish life or to ensure that the quality of shellfish is suitable for human consumption.

The Directive on waste from the titanium oxide industry has as its aim the prevention and progressive reduction of pollution from the industry until eventually the pollution is eliminated. This Directive applies to all sections of the environment and not just the aquatic environment.

There has been control on water pollution in the UK since 1951. This control has been enacted in the Rivers (Prevention of Pollution) Acts 1951 and 1961 along with the Control of Pollution Act 1974. Section 31 of the latter Act controls pollution to the 'three-mile offshore limit' and also certain underground waters.

Water supply. A complex subject illustrated in Figure 210 with the processes used to supply London with 'wholesome' water from the River Thames.

The presence of the coliform bacteria, *Escherichia coli*, is monitored in the source as an indicator of faecal pollution. In Figure 210 the count is originally 6680 per 284 cubic centimetres. The abstracted water is pumped to a reservoir for storage and settlement where faecal bacteria are reduced through natural processes and a large portion of the suspended solids settles out. (◊ COLIFORM COUNT)

The water is taken from the storage reservoir via the draw-off arrangement under the dam base to an aeration basin so that it is fully oxygenated, and then sent to the primary filter and micro-strainers where most plankton

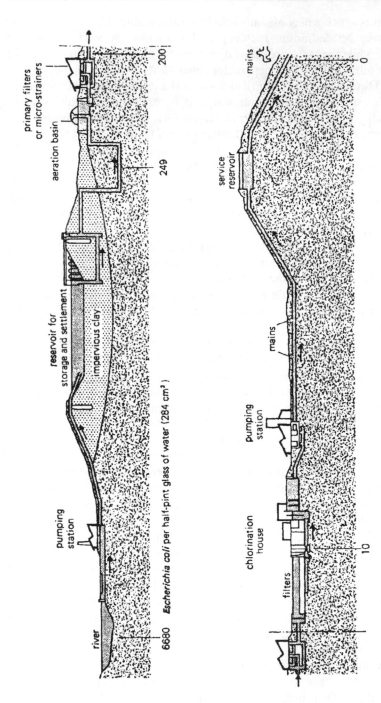

Figure 210 *Escherichia coli* count in London's water supply.

and algae are removed. (Prior to this stage, coagulation is often practised whereby colloidal particles which carry negative charges and thus will not naturally come together are coagulated by special agents and removed by sedimentation.)

Sand or other filtration follows and this is a most important stage as the sand is very much more than a fine strainer. The sandfilter is a large shallow basin on which there is 1 metre of sand on a gravel base. The water is introduced at the top of the bed and drawn off at the bottom through collector pipes and channels. Filtration in this case takes place under gravity although pressure filters are also available.

Two separate layers develop in the bed: in slow sand filtration the top 2 millimetres comprise micro-organisms which decompose organic matter and consume nitrates and phosphates and release oxygen. This is the autotrophic layer which has a strong purifying action, on whose surface there is a dirt film which can impede the flow of water considerably. Below the autotrophic layer is the heterotrophic zone (30 centimetres) where non-pathogenic bacteria continue the decomposition of the organic matter and render the water in a very pure condition. In the London example the *E. coli* count is 200 before filtering and 10 after. The sand filter thus performs a complex range of duties. Its performance depends on the physical and chemical characteristics of the suspended solids and the chemical composition of the water.

Filtration can be by either the *slow sand filter* or the *rapid sand filter* method. The slow method requires a lot of space and is not now often used as the dirt layer previously referred to must be skimmed off by draining the filter, and new sand spread on top every two or three months. The flow of water per unit area is low. The rapid gravity sand filter is 20 times faster and this is accomplished by backwashing with compressed air. The rapid filter is not nearly so efficient as the slow filter so coagulation is required as a pre-treatment process. The rapid filter does not oxidize organic matter and nitrogen compounds so these may require removal at a later stage, or anthracite sand filters can be used. These are a layer of large anthracite grains on top of smaller sand grains. These filters clog less rapidly as the particles adhere to the larger grains and thus higher filtration rates are possible.

The water with an *E. coli* count of 10 is now disinfected by chlorine injection at a rate proportional to the water flow. The *E. coli* count is now zero and the presumption is that the water is now wholesome. It is then pumped to the mains where it is distributed to the consumers. Note that service reservoirs and/or water towers are required to maintain adequate pressure in the supply mains and to iron out fluctuations in demand.

A form of sterilization (not shown in Figure 210) is the use of OZONE, which is a powerful oxidizing agent and which also reduces colour, taste and odour in water. A dose as low as 0.0001 per cent or 1 part per million destroys all bacteria within 10 minutes. The ozone must be manufactured by electric-arc discharge on site and the process is thus more expensive than chlorination. No matter what method is adopted, extensive testing and analysis is carried out by all water companies.

Water table. The upper surface of the SATURATION ZONE below which all void spaces are filled with water.

Water usage (domestic). The average family of four uses 600 litres of water a day:

200 litres for showers and drinking
200 litres for flushing the lavatory
200 litres for washing up, washing machines and other uses.
A bath uses 80 litres of water
A power shower can use 80 litres of water, an ordinary shower 35 litres
A hosepipe can use 600 litres of water an hour
Water consumption by appliances:
Washing machine 30–100 litres per wash programme
     (based on wash load capacity of 5 kg)
Washer dryer 60–200 litres per wash and drying programme
     (based on a wash load capacity of 5 kg)
Dishwasher 10–30 litres per wash
     (based on a 8-place setting)
     10–50 litres per wash
     (based on a 12-place setting)

Watt (W). Unit of POWER defined as 1 JOULE per second. Conversion factor 1 horse power = 745.7 W.

A megawatt (MW) is equal to 1000 kilowatts (kW) or 1 000 000 watts (W). It is used as a measurement of electrical generating capacity. (⇨ KILOWATT-HOUR)

Watt-hour (Wh). Unit of ENERGY equal to 1 watt acting for 1 hour, i.e. 3600 joules.

Wavelength. The characteristic length (λ) over which a wave repeats itself, i.e. the distance between two successive wave crests or troughs. Different types of electromagnetic radiation are characterized by a particular range of wavelengths. The wavelength and frequency (f) of electromagnetic radiation are related by the equation.

$$c = f\lambda$$

where c (= $3.00 \times 10^{8}\,\mathrm{m\,s^{-1}}$) is the speed of light.

Wave power. The harnessing of the energy in the waves to generate electricity. Engineering assessments show that to get the equivalent amount of energy as from 1 kilogram of coal requires 1 tonne of sea water falling through 3 kilometres. Thus, the capital costs for the harnessing of wave power may be high.

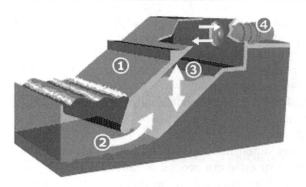

Wave power station

The world's first commercial wave power station, on the island of Islay, is the product of years of research into how to effectively harvest energy from the world's oceans.

The Islay wave power generator was designed and built by Wavegen and researchers from Queen's University in Belfast and has financial backing from the European Union. Known as Limpet 500 (Land Installed Marine Powered Energy Transformer), it feeds 500 kilowatts of electricity into the island's grid. It started operation in November 2000. Limpet was born out of a 10-year research project on the island where the team had built a demonstration plant capable of generating 75 Kilowatts of electricity.

The power generator consists of two basic elements:

1) A wave energy collector
2) A generator to turn this into electricity

The energy collector (Figure above) comprises a sloping reinforced shell built into the rock face on the shoreline with an inlet big enough to allow seawater to freely enter and leave a central chamber (1). As waves enter the shell chamber (2), the level of water rises, compressing the air in the top of the chamber (3). This air is then forced through a 'blowhole' and into the 'Wells Turbine' (4), designed by Professor Alan Wells of Queen's University. The turbine has been designed to continue turning the same way irrespective of the direction of the airflow. As the water inside the chamber recedes as the waves outside draw back, the air is sucked back under pressure into the chamber, keeping the turbine moving. This constant stream of air in both directions, created by the oscillating water column, produces enough movement in the turbine to drive a generator which converts the energy into electricity.

Wavegen says that there could be sufficient recoverable wave power around the UK to generate enough power to exceed domestic electricity demands. Furthermore, renewable energy supporters say some research suggests that less than 0.1 per cent of the renewable energy within the world's oceans could supply more than five times the global demand for energy – if it could be economically harvested. That would probably involve large-scale wave plants in near-shore or off-shore environments, a technology still being developed. However, large-scale on-shore wave power generating stations could face similar problems to those encountered by some wind farm projects, where opposition has focused on the aesthetic and noise impact of the machinery on the environment.

Wave power supporters say that the answer lies not in huge plants but in a combination of on-shore generation and near-shore generation (using a different technology) focused on meeting local or regional needs. On-shore or near-shore plants, they argue, could also be designed as part of harbour walls or water-breaks, performing a dual role for a community.

Source: www.wavegen.co.uk/news_cliffbased.htm. Accessed 15 January 2007.

Weathering. The chemical and mechanical breakdown of rocks and minerals under the action of atmospheric agencies.

Wells and boreholes. A hole drilled into the ground to tap an aquifer for water supplies (or an oil or gas field for oil or gas). Shallow wells in soft formations can be dug by hand or with power augers. Deeper and/or resistant rock formations are drilled by machine. Once the well has been drilled it must be completed, i.e. the hole cased if necessary to prevent collapse, with a slotted casing to allow water to enter. The well is then developed by explosives, chemicals or compressed air to shatter the rock and increase the yield.

Once pumping commences, the safe yield is established. This can be restricted by salt-water INFILTRATION in some areas. This means that a well's yield is not necessarily the maximum rate at which water can be extracted, but the maximum rate at which the quality of the water or the supply of other nearby wells is not affected. *Note*: Wells can have other uses, e.g. as observation wells to monitor the abstraction of water from an aquifer or the migration of LEACHATE or LANDFILL GAS from a LANDFILL SITE.

Whale harvests. Whale fisheries serve as an archetypal model of over-exploitation. As the large whales (blue and fin) were hunted to near or total extinction, the industry shifted to harvesting not only the young of the larger species but also smaller species.

After the Second World War the International Whaling Commission (IWC) was established, with the brief of regulating harvests and protecting certain endangered species. In practice the IWC has turned out to be a relatively impotent organization with inadequate power of inspection and non-existent powers of enforcement. The IWC made an early fundamental error in establishing quotas on the basis of 'blue whale units' (bwu), a unit being one blue whale, or its equivalent (2 fin whales, 2½ hump-back whales, or 6 sei whales). The total permitted quota was initially set at 16000 units with the blue whale the preferred species. In the 1950s the number of blue whales caught declined sharply and the whalers turned their attention to the fin and hump-back whales. During the 1960s the blue whale and fin whale catch continued to decline, and in 1963 a Commission committee report warned that blues and hump-backs were in serious danger of extinction. It recommended that these species be totally protected and the fin whale catch limited. It also urged that the 'blue whale unit' be dropped in favour of limits on individual species. Predictably, these warnings and recommendations were ignored and the total catch quota was limited to 10000 bwu. By this time the fin whale harvest was estimated to be three times the MAXIMUM SUSTAIN-ABLE YIELD. Subsequently the committee recommended a bwu total for 1964–5 of 4000; for 1965–6, 3000; and only 2000 for 1966–7, in order to allow recovery of the whale stocks. Again it was ignored. All four countries then engaged in Antarctic whaling – Japan, the Netherlands, Norway and Russia – voted against accepting the recommendation and continued to practise their trade according to the sound economic principle of obtaining a maximum return on invested capital. (◊ ECONOMICS)

Norwegian and Japanese whalers continue to kill minke whales despite a 1986 moratorium.

In 1988, only 22 blue and fin whales were recorded, where once the blue whales numbered 250000.

There is now, hopefully, some evidence that a long process of recovery is under way and that population growth rates of 5–10 per cent per year are now taking place from a very low base. The fear is that the practitioners of whaling will use this increase as justification for the continuous extension of their activities. One positive aspect of the whale population increase is that the effects of toxic compounds in the marine environment may not be as severe as was feared.

Fish removed are also over-exploited and have remained static globally at around 100 million tonnes per year. The FAO has stated that: '70 per cent of fish stocks are over-fished, depleted or rebuilding from earlier over-fishing. The most exploited areas are around the North Atlantic, around the American and European land masses, the North Sea, the central Baltic, the Mediterranean and Black Seas, and the central-west Pacific.'

Fish farming measures to stop over-fishing and a reduction in the waste of fish discarded at sea are all urged. However, the use of antibiotics and the intensive faecal production from farmed fish mean that strict management is required.

Whey. Organic liquid left from cheese manufacture which is a highly polluting material, one cubic metre of whey being equivalent to the daily pollution production from approximately 600 people. In wastewater terminology, the composition of a typical whey is:

Total solids	66 000 mg/l
Volatile solids	59 400 mg/l
Chemical oxygen demand	65 000 g/l
Biological oxygen demand	45 000 mg/l
Total nitrogen	1500 mg/l

A waste of this strength is uneconomic to treat by conventional AEROBIC PROCESSES, due to the high power requirements for aeration, and for this reason whey has not traditionally been treated in aerobic biological systems.

In ANAEROBIC digestion, however, whey is broken down to produce a gas rich in METHANE, giving the treatment process a large positive ENERGY balance. (◊ ANAEROBIC DIGESTION)

Wind. The movement of air parallel to the earth's surface caused by differences in atmospheric pressure and the earth's rotation. Above a certain height the wind blows at constant velocity in a direction parallel to the isobars – lines of constant pressure on the weather chart. The velocity is proportional to the isobar spacing – the pressure gradient – and is known as the *gradient wind*. This occurs above the gradient height, approximately 600 metres above ground. Below the gradient height, the wind is slowed down by ground effects, mainly friction, and is also stirred near ground level by obstructions so that eddies are set up which are very useful in diluting and dispersing air pollutants emitted at or near ground level.

Wind energy. Energy (normally electrical power) obtained from the wind using wind turbines. The exploitation of wind energy should be regarded in the context of UK and EU policy on SUSTAINABLE DEVELOPMENT. Wind energy as with other forms of renewable energy occurs naturally and repeatedly in the environment. For the UK, commercial development of wind energy is still relatively new. It began in 1990 with the introduction of the NON-FOSSIL FUEL OBLIGATION (NFFO) and by November 1994 over 35 wind energy projects had been built including several with individual turbines and 26 wind farms comprising either a large number of turbines or a cluster of a few.

In order to provide some numbers, the power available from the Finnish Phyätunturi Arctic Test Wind Turbine site are given in the table below:

Rated power	220 kW
at wind speed	13 m/s
Rotor diameter	25 m
Rotor swept area	490 m^2
Weights	
machinery	6800 kg
rotor	5700 kg
tower	13 500 kg
Manufacturer	Wind World AS, Denmark
Estimated annual production on Phyätunturi	600 MWh

Source: CADET Renewable Energy Newsletter, January 1995.

However there can be legitimate concerns over inappropriate siting of wind turbines. A development chart is given in Figure 211 to help responsible development.

How Much Does Wind Energy Cost?

The price of wind energy has fallen considerably over the last 10 years and compares favourably with conventional energy generation. Power generation costs are determined by the installed costs of the plant (including interest during construction), operation and maintenance costs, fuel costs, energy productivity, cost of capital and the capital repayment period. In the case of wind energy, the fuel – the wind itself – is free. The available wind speed determines the final cost of wind energy from specific wind farms sites.

The table below, taken from the Sustainable Development Commission's report 'Wind Power in the UK', shows a summary outlining the average costs for wind power. The average generation costs of onshore wind power are around 3.2 pence per kilowatt hour (p/kWh) and around 5.5 p/kWh for off-shore. Once the cost of carbon to society and the environment is included in electricity generation costs, the price of wind power will be even lower since wind energy is a clean and renewable source of electricity generation, producing no harmful emissions.

Every modern wind turbine will save over 4000 tonnes of CO_2 emissions annually.

Source: npower advertisement, 2007.

A study by the Renewable Energy Foundation (a charity which aims to evaluate renewable energy systems) has shown that England and Wales are not windy enough to allow large turbines to work at the rates claimed for them. Wind turbines in Cornwall, expected to be the most efficient, operated at only 24.1 per cent of capacity, on average, while turbines in mid-Wales on average achieved 23.8 per cent, those on the Yorkshire Dales 24.9 per cent, and those in Cumbria 25.9 per cent. Turbines in Scotland fared much better: Southern Scotland 31.5 per cent, Caithness, Orkney and Shetland 32.9 per cent, and offshore (North Hoyle and Scroby Sands) 32.6 per cent.

Source: The Daily Telegraph, 9 December 2006.

The world's biggest offshore wind farm, set to produce 1G W of electricity, is to be built 20 km off the Kent and Essex coast, in the Thames Estuary.

Wind power. Factors affecting performance are:

(1) *The windiness of the site.* The power available from the wind is a function of the cube of the wind speed. Therefore, if the wind blows at twice the speed, its energy content will increase eight-fold. In practice, turbines at a site where the wind speed averages 8 metres per second (m/s) will

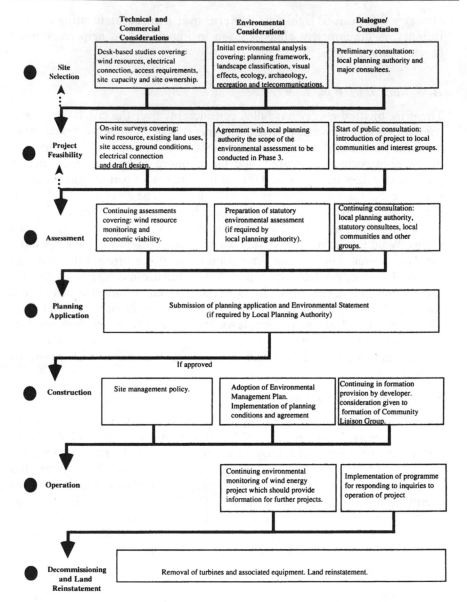

Figure 211 Development chart for wind energy, taken from Best Practice Guidelines for Wind Energy Development.
Source: British Wind Energy Association, November 1994. (Reproduced by permission)

Summary of wind generation costs.

Source	Capital cost, £/kW	O&M, £kW	Capacity Factor, %	Tdr %	Life	Gencost, p/kWh	Comments
Onshore							
NFF05	-	-	-	8	15	2.7	Average price
Oxera	605–800	15	30	?	20	3.1	
WPM	800	n.q.	36	6	15	3.3	'High' cost, 8.5 m/s site
	550	n.q.	27	6	15	3.0	'Low' cost, 7.2 m/s site
IEA/DK	585	16	27	5	20	2.65	
Offshore							
Oxera	1100–1430	35–42	35	n.q.	20	5.5	
WPM	1200	n.q.	38	6	15	5.7	'High' cost, 8.8 m/s site
	970	n.q.	31	6	15	4.9	'Low' cost, 7.8 m/s site
IEA/DK	1130	36	27	5	20	3.2	

Note: n.q. Not quoted. NFF05: Fifth Round of the Non-Fossil Fuel Obligation Scheme, a subsidy for renewable electricity. Oxera: an analysis carried out for the DTI by Oxford Economic Research Associates. WPM: analysis which examined coast data from over 3300 MW of wind around the world. Two figures are quoted; per 'high' and 'low' installed costs. IEA/DK: International Energy Agency data from Denmark, which has a wealth of wind energy experience.
Source: 'Wind Power in the UK', the Sustainable Development Commission, London, 2005.

Wind energy in European countries.

Country	Megawatts installed at the end of 1996
Denmark	835
France	5.7
Germany	1552
Greece	29
Italy	70.5
Ireland	11
Netherlands	299
Portugal	19.1
Spain	249
Sweden	103
UK	273
Other countries	50.2
Total	3496

Source: European Wind Energy Association, *Wind Energy Technology*, 1998.

Electricity generating costs from different fuels.

Technology	Plant cost (£/kW)	Fuel cost (p/kWh)	O & M* (p/kWh)	Total generating cost (p/kWh)
Coal	850–1090	1.3	0.7	3.5–4.5
Nuclear	1100–1350	0.4	0.6	4.5–6.0
Gas	380–560	1.1	0.3	2.3–2.9
Wind	650–900	0	0.8–1.0	2.7–4.9

* Operation and maintenance costs include utility overheads and profit.
Source: British Wind Energy Association, *Wind Energy Economics*, 1998.

produce around 80 per cent more electricity than those where the wind speed is 6 m/s.

(2) *Availability*. This is the capability to operate when the wind is available – an indication of the turbine's reliability. Availability is typically 98 per cent or more for modern European machines.

(3) *Turbine arrangement*. Turbines in wind farms must be carefully arranged to gain the maximum energy from the prevailing wind.

Harnessing wind as a renewable energy source involves converting the power within a moving air mass (wind) into rotating shaft power, via aerodynamic blades, to generate electricity. Wind power is proportional to the cube of the wind's speed, so relatively minor increases in speed result in large changes in potential output. Individual turbines vary in size and power output from a few hundred watts to two or three megawatts (as a guide, a typical domestic system would be 2.5–6 kilowatts, depending on the location and size of the system). Uses range from very small turbines supplying energy for battery charging systems (e.g. on boats or in homes), to turbines grouped on wind farms supplying electricity to the grid. Wind speed increases with height so it is best to have the turbine high on a mast or tower. Generally speaking the ideal siting is a smooth-top hill with a flat, clear exposure, free from excessive turbulence and obstructions such as large trees, houses or other buildings.

Small-scale wind power is particularly suitable for remote off-grid locations where conventional methods of supply are expensive or impractical. Most small wind turbines generate direct current (DC) electricity. Off-grid systems require battery storage and an inverter to convert DC electricity to AC (alternating current – mains electricity). Household wind energy systems are typically sized up to 6 kW but there are larger turbines of up to 50 kW available for larger community-scale Projects. Small-scale applications range from individual battery charging systems to those that provide power for homes, schools or community halls. A small system of 600 watts could be used for charging batteries for caravans and boats. A larger system of 5–6 kW could be used to provide power to a community hall or other public building. The optimum size for the average household would be 1.5–3.0 kW.

Technology has moved on apace and the world's largest offshore wind turbine (first of two) is being installed in the Moray Firth, Scotland. It is 170 m high with 63 m long blades and is designed to generate 5 MW$_e$. This is part of an appraisal for 200 such turbines which could when operational, supply 20 per cent of Scotland's energy demand.

Regarding Wind Turbines in the Thames Estuary: 'If, is as suggested, the installed capacity of the 400-plus turbines is 1.3 GW (1300 MW) then even with a generous load factor of 30 per cent the average output will only be 390 MW. This would in fact be enough to provide 5 kW to 78 000 homes, about enough to power an electric kettle and a toaster. If, as there frequently is, a high pressure system sitting over south-east England, then there will be zero output from these wind farms. The claims about carbon dioxide savings are equally dishonest. Using widely-accepted data, the annual, theoretical savings of CO_2 for these turbines would be approximately 1.46 Mt and would reduce global levels by a farcical 0.005 per cent'.

Source: Graham, R. *The Daily Telegraph*, Letters, 23 December 2006.

Notes:

1. Nuclear reactor can generate 1.000 MW$_e$ continuously.
2. Drax coal-fired power station generates 4000 MW$_e$ continuously.
3. The Thames Estuary wind farm received approximately £160 m/year in subsidies.

Wind profile. A graphical representation of the variation of wind speed as a function of height or distance.

Wind rose. A star-shaped diagram indicating the relative frequencies of different wind directions for a given location and period of time. The most common form consists of a circle from whose centre are drawn a number of radii to indicate different points of the compass, the length of each radius being proportional to the percentage of time during the given period that the wind blew from that direction.

Wind speed. The wind speed of a potential WIND ENERGY site is a crucial factor in determining the economic viability of a project. Since energy yield is a function of wind speed (proportional to the cube of the wind speed), the higher the wind speed, the greater the energy yield.

Wobbe Index. Also called the Wobbe Number (WN) – the calorific value divided by the square root of the specific gravity (i.e. the density of the gas relative to air). Thus $WN = CV/\sqrt{SG}$. In the SI system it has units of MJ/m^3. Two gases of differing composition but having the same WN will deliver the same amount of energy for any given injector under the same injector pressure.

Wood (Fuel)–Case Study. A CARBON NEUTRAL fuel, if new trees are planted to replace those felled, and/or managed coppicing is practised. Its use to heat ca. $30\,m^2$ of offices in converted farm buildings at Godington House in Kent shows that 75 t of useable wood per acre of coppice can be harvested once every 12 years (60 t at 30 per cent moisture content). The boiler uses 20 t per year. Hence 4 acres of established woodland are required for coppicing 1/3 acre on a 12-year cycle. Costs are quoted as '45–50 t woodchips which have the heating value of 40 litres of fuel oil which costs £165'.

Source: Adam Nicholson, 'Log onto the future of heating', *The Daily Telegraph*, Country Section, 24 February 20007.

Woodmark label. This label tells buyers that their wood product has come from well-managed forests, and that by purchasing the product they are helping to support responsible timber production, rather than poor forestry practice or destructive logging.

Wood alcohol. ◊ METHANOL.

Work. The product of FORCE × distance moved. Fundamental unit is the JOULE.

Working plan. Document submitted in support of waste disposal licence and planning application detailing the engineering and restoration proposals and the conduct of operations. It should include a technical drawing or drawings for each stage of the development and a detailed statement of the way the operations are to be carried out, including provisions for LANDFILL GAS control, leachate treatment and any aftercare programme.

WRAP. The Waste & Resource Action Plan, WRAP, is a not-for-profit company created in 2000 as part of the UK Government's waste strategy. WRAP's mission is to help develop markets for material resources that would otherwise have become waste. WRAP also provides advisory services to local authorities and helps influence public behaviour through national level communication programmes. WRAP works in partnership to encourage and enable businesses and consumers to be more efficient in their use of materials and recycle more things more often. This helps to minimize landfill, reduce carbon emissions and improve our environment.

X

Xenon effect. The rapid but temporary poisoning of a reactor by the build-up of xenon-135 from the radioactive decay of the fission product iodine-135. (Xenon-135 is a strong absorber of neutrons and until it (and its parent) have largely decayed away, reactor start-up can be difficult.) (⟡ POISON, NUCLEAR)

X-ray. Ionizing or electromagnetic radiations of the same type as light but with much shorter wavelength. The absorption of the rays depends on the density of the material; as bone is denser than flesh this allows X-ray photographs to be taken. All radiation exposure carries risks. However, in medical irradiation the benefits incurred by the diagnostic X-rays far outweigh the risks. Having said that, pregnant women and/or fertile women are not normally exposed to X-rays as the foetus is particularly susceptible. (⟡ IONIZING RADIATION)

Y

Yeast. Unicellular fungi which are able to multiply asexually by a budding process. Yeasts are of great importance in FERMENTATION, the manufacture of pharmaceuticals and SINGLE-CELL PROTEIN. One major strain used for producing single-cell protein from molasses and the sugar in the spent liquors from sulphite pulp manufacture is *Candida utilis*, which has an approximate 50–60 per cent protein content.

Yellow-cake. Concentrated crude uranium oxide, the form in which most uranium is shipped from the mining areas to the fuel manufacturers. (⟡ NUCLEAR POWER)

Z

Zeolites. Comprise aluminium–silicates which contain sodium, calcium, potassium used in ION EXCHANGE water softening where calcium/magnesium IONS are exchanged for sodium.

Zero-carbon homes. 2016 is the date for HMG's target for all new-build homes to have zero carbon emissions. Currently (2008) conventional UK homes emit on average 6t CO_2 annually, approximating to 25 per cent of total CO_2 output.

Note: The greenhouse effect of aircraft CO_2 emissions can be four times greater than terrestrial emissions. Better care is needed in interpreting environmental data and, unfortunately, this caveat applies to politicians' deeds, as the example below shows:

Climate change should be left for scientists to sort out, Tony Blair has announced. It will not be fought by making small sacrifices such as giving up holidays in far-flung corners of the world, the Prime Minister said.

'These things are a bit impractical actually, to expect people to do that', he added of giving up long-haul flights. 'It is like telling people you should not drive anywhere'.

'What we need to do is to look at how you make air travel more energy-efficient, how you develop the new fuels that will allow us to burn less energy and emit less'.

Mr. Blair, who often holidays in the Caribbean, all but exonerated Britain from its duty to cut its greenhouse gas emissions. 'Britain is two percent of the world's emissions', he added. 'We shut down all of Britain's emissions tomorrow – the growth in China will make up the difference within two years'.

It is understood that a climate offset levy has been paid on the long haul holidays referred to. Again trees need time to grow, thus the 'offset' is deferred and there is no like for like mitigation.

Source: *Metro*, 9 January 2007, p5.

Zero landfill[1]. A goal now being adopted by large construction and food manufacturing firms, which requires that all 'waste' be either recycled or reused (e.g. construction and demolition debris to be crushed and reused for car park foundations). Food wastes will require composting and/or anaerobic digestion with their associated environmental impacts.

Zero landfill[2]. The ultimate goal of integrated waste management. Asda (food retailer) and Wates (construction) have both set themselves the goal of zero landfill by 2010 (2006, Asda sent 86 000 t of food and other waste to landfill, Wates 70 000 t mainly bricks, plaster board and timber).

Food wastes can be treated in closed 'in vessel' composting systems leading to their elimination from the environment plus a useful product – a win-win situation. (⟡ CONSTRUCTION & DEMOLITION WASTE)

Zero-sum game. Whatever strategy is chosen, one player's gain is equal to other players' losses. The sum of gains will always equal the sum of losses. (The whole summing to zero.)

Zinc (Zn). Metallic element, atomic number 30, atomic mass 65.38. Principal ore, zinc sulphide (ZnS).

Zinc equivalent. The sum of the copper, zinc and nickel contents of a sludge, in milligrams per kilogram, after multiplying the copper content by 2 and the nickel content by 8. Used to indicate the rate of addition of metals and the cumulative concentration in the soil with reference to the effect on the growth of crops. A maximum safe limit of 250 mg/kg zinc equivalent in dry top soil has been suggested. (⟡ SLUDGE (SEWAGE))

Zircaloy. An alloy of zirconium and tin used for fuel cladding in water reactors.

Zone of visual influence. A zone of visual influence provides a representation (usually presented as a map with markings or colourings) of the area over which a site and/or a proposed development, such as a wind turbine or chimney stack, may be visible.

Zoonoses. Infections that can be transmitted from animals to humans such as cryptosporidiosis, a diarrhoeal disease which can be life-threatening in individuals with suppressed immune systems. Other zoonoses with the potential to cause ill health include chlamydia (also known as enzootic abortion, which can cause abortion in pregnant women and is a risk during lambing and open days), leptospirosis, and ringworm.

Zooplankton. Floating and drifting aquatic animal life. (⟡ PHYTOPLANKTON)

Pollution and the environment-organizations

General

Departments of State and other Public Bodies

Communities and Local Government, Eland House, Bressenden Place, London, SW1E 5DU, UK (020 7944 4400) Fax: 020 7944 4101 www.communities.gov.uk

Countryside Council for Wales, Maes y Ffynnon, Penrhosgarnedd, Bangor, Gwynedd, LL57 2DW, UK (0845 1306 229) Fax: 01248 355782 www.ccw.gov.uk

Department for Environment, Food & Rural Affairs, Nobel House, 17 Smith Square, London, SW1P 3JR, UK (020 7238 6000) www.defra.gov.uk

Department of the Environment for Northern Ireland, Clarence Court, 10-18 Adelaide Street, Belfast, BT2 8GB, Northern Ireland, UK (028 9054 0540) www.doeni.gov.uk

Drinking Water Inspectorate, Floor 2/A1, Ashdown House, 123 Victoria Street, London, SW1E 6DE, UK (020 7082 8024) Fax: 020 7082 8028 www.dwi.gov.uk

Environment Agency (in England), General Enquiries: 08708 506506 www.environment-agency.gov.uk

> Agricultural Waste Registration: 0845 603 3113
> Floodline: 0845 988 1188
> Hazardous Waste Registration: 08708 502 858
> Incident hotline: 0800 807060
>> Head Office, Rio House, Waterside Drive, Aztec West, Almondsbury, Bristol, BS32 4UD, UK (08708 506506)
>> Anglian Regional Office, Kingfisher House, Goldhay Way, Orton Goldhay, Peterborough, Cambridgeshire, PE2 5ZR, UK (08708 506506)
>> Midlands Regional Office, Sapphire East, 550 Streetsbrook Road, Solihull, West Midlands, B91 1QT, UK (08708 506506)
>> North East Regional Office, Rivers House, 21 Park Square South, Leeds, West Yorkshire, LS1 2QG, UK (08708 506506)

North West Regional Office, PO Box 12, Richard Fairclough House, Knutford Road, Latchford, Warrington, Cheshire, WA4 1HT, UK (08708 506506)

Southern Regional Office, Guildbourne House, Chatsworth Road, Worthing, Sussex, BN11 1LD, UK (08708 506506)

South West Regional Office, Manley House, Kestrel Way, Exeter, Devon, EX2 7LQ, UK (08708 506506)

Thames Regional Office, Kings Meadow House, Kings Meadow Road, Reading, Berkshire, RG1 8DQ, UK (08708 506506)

Environment Agency (in Wales), Cambria House, 29 Newport Road, Cardiff, CF24 0TP, Wales, UK (08708 506506) www.environment-agency.gov.uk

Environment and Heritage Service, Northern Ireland www.ehsni.gov.uk

Natural England, Northminster House, Peterborough, PE1 1UA, UK (0845 600 3078) Fax: 01733 455103 www.naturalengland.org.uk

Natural Environment Research Council, Polaris House, North Star Avenue, Swindon, SN2 1EU (01793 411500) Fax: 01793 411501 www.nerc.ac.uk

The Royal Commission on Environmental Pollution, Third Floor, The Sanctuary, Westminster, London, SW1P 3JS, UK (0207 7998970) Fax: 0207 799 8971 www.rcep. org.uk

Scottish Environment Protection Agency, Erskine Court, Castle Business Park, Stirling, FK9 4TR, Scotland, UK (01786 457700) Fax: 01786 446885 www.sepa.org.uk

Scottish Natural Heritage, Great Glen House, Leachkin Road, Inverness, IV3 8NW, UK (+44 (0)1463 725000) Fax: +44 (0)1463 725067 www.snh.org.uk

Sustainable Development Commission, Ground Floor, Ergon House, Horseferry Road, London, SW1P 2AL, UK (020 7238 4995) Fax: 020 7238 4981 www.sd-commission. org.uk

The Sustainable Development Unit, Defra, 4E, 9 Millbank, c/o Nobel House, 17 Smith Square, London, SW1P 3JR, UK (+44 (0)20 7238 5811) www.sustainable-development.gov.uk

Non-Governmental Organizations

Association for the Protection of Rural Scotland, Gladstone's Land 3rd Floor, 483 Lawnmarket, Edinburgh, EH1 2NT, Scotland, UK ((0131) 225 7012) www. ruralscotland.org

Campaign for the Protection of Rural England, 128 Southwark Street, London, SE1 0SW, UK (020 7981 2800) Fax: 020 7981 2899 www.cpre.org.uk

Campaign for the Protection of Rural Wales, Ty Gwyn, 31 High Street, Welshpool, Powys, SY21 7YD, Wales, UK (01938 552525 or 01938 556212) Fax: 01938 552741 www.cprw.org.uk

Centre for Alternative Technology, Machynlleth, Powys, SY20 9AZ, UK (+44 (0)1654 705950) Fax: +44 (0)1654 702782 www.cat.org.uk

Chartered Institute of Environmental Health, Chadwick Court, 15 Hatfields, London, SE1 8DJ, UK (020 7928 6006) Fax: 020 7827 5862 www.cieh.org

Chartered Institution of Wastes Management, 9 Saxon Court, St. Peter's Gardens, Marefair, Northampton, NN1 1SX, UK (+ 44 (0) 1604 620426) Fax: + 44 (0) 1604 621339 www.ciwm.co.uk

Chartered Institution of Water and Environmental Management (CIWEM), 15 John
 Street, London, WC1N 2EB, UK (020 7831 3110) Fax: 020 7405 4967 www.ciwem.
 org
Confederation of British Industry, Centre Point, 103 New Oxford Street, London,
 WC1A 1DU, UK (0207 379 7400) www.cbi.org.uk
The Conservation Foundation, 1 Kensington Gore, London, SW7 2AR, UK (0207 591
 3111) Fax: 0207 591 3110 www.conservationfoundation.co.uk
Consumer Council for Water (0845 039 2837 or 0121 345 1000) www.ccwater.org.uk
Consumers' Association, 2 Marylebone Road, London, NW1 4DF, UK (020 7770 7000)
 Fax: 020 7770 7600 www.which.co.uk
Electronic Development and Environment Information System http://www.eldis.org
Environmental Data Services (ENDS) http://www.ends.co.uk
Federation of City Farms and Community Gardens, The GreenHouse, Hereford Street,
 Bristol, BS3 4NA, UK (0117 923 1800) Fax: 0117 923 1900 www.farmgarden.org.uk
Friends of the Earth, 26-28 Underwood Street, London, N1 7JQ, UK (020 7490 1555)
 Fax: 020 7490 0881 www.foe.co.uk
Green Party, 1a Waterlow Road, London, N19 5NJ, UK (020 7272 4474) Fax: 020 7272
 6653 www.greenparty.org.uk
Greenpeace, Canonbury Villas, London, N1 2PN, UK (020 7865 8100) Fax: 020 7865
 8200 www.greenpeace.org.uk
National Society for Clean Air and Environmental Protection, 44 Grand Parade, Brigh-
 ton, East Sussex, BN2 9QA, UK (01273 878 770) Fax: 01273 606 626 www.nsca.org.
 uk
The National Trust, PO Box 39, Warrington, WA5 7WD, UK (0870 458 4000) Fax: 020
 8466 6824 www.nationaltrust.org.uk
Society for the Environment (SocEnv), PO Box 860, Lincoln, LN1 3WW, UK (0845 226
 3625) www.socenv.org.uk
Town & Country Planning Association, 17 Carlton House Terrace, London, SW1Y 5AS,
 UK (020 7930 8903) Fax: 020 7930 3280 www.tcpa.org.uk
UK Co-housing Network www.cohousing.org.uk
Ulster Society For the Preservation of the Countryside, 22 Donegall Road, Belfast,
 Northern Ireland, UK (028 9071 0024)
The Wildlife Trusts, The Kiln, Waterside, Mather Road, Newark, Nottinghamshire,
 NG24 1WT, UK (0870 036 7711) Fax: 0870 036 0101 www.wildlifetrusts.org

Air Pollution

Departments of State and Other Public Bodies

Air Quality in London www.londonair.org.uk
Association for the Conservation of Energy, Westgate House, 2a Prebend Street, London,
 N1 8PT, UK (+44 (020) 7359 8000) Fax: +44 (020) 7359 0863 www.ukace.org
The Conservation Foundation, 1 Kensington Gore, London, SW7 2AR, UK (0207 591
 3111) Fax: 0207 591 3110 www.conservationfoundation.co.uk
Consumers' Association, 2 Marylebone Road, London, NW1 4DF, UK (020 7770 7000)
 Fax: 020 7770 7600 www.which.co.uk
Convention on Long-Range Transboundary Air Pollution http://www.unece.org/env/lrtap/

Corinair 90 Emissions Data http://www.eea.europa.eu/

Met Office, FitzRoy Road, Exeter, Devon, EX1 3PB, UK (0870 900 0100) Fax: 0870 900 5050 www.metoffice.gov.uk

Motor Industry Research Association, Watling Street, Nuneaton, Warwickshire, CV10 0TU (+44 (0)24 7635 5000) Fax: +44 (0)24 7635 5355 www.mira.co.uk

UK Air Quality Archive www.airquality.co.uk

Countryside, Farming, Wildlife

Departments of State and Other Public Bodies

Department for Environment, Food & Rural Affairs, Nobel House, 17 Smith Square, London, SW1P 3JR (020 7238 6000) www.defra.gov.uk

Health and Safety Executive, Caerphilly Business Park, Caerphilly, CF83 3GG, Wales, UK (0845 345 0055) Fax: 0845 408 9566 www.hse.gov.uk

Scottish Executive www.scotland.gov.uk

Non-Governmental Organizations

British Trust for Ornithology, The Nunnery, Thetford, IP24 2PU, UK (01842 750 050) www.bto.org

Chemical Industries Association, Kings Buildings, Smith Square, London, SW1P 3JJ, UK (020 7834 3399) Fax: 020 7834 4469 www.cia.org.uk

Garden Organic (formerly the Henry Doubleday Research Association), Ryton Organic Gardens, Coventry, Warwickshire, CV8 3LG, UK (+44 (0) 24 7630 3517) Fax: +44 (0) 24 7663 9229 www.gardenorganic.org.uk

National Farmers' Union, Agriculture House, Stoneleigh Park, Stoneleigh, Warwickshire, CV8 2TZ, UK (024 7685 8500) Fax: 024 7685 8501 www.nfuonline.com

National Society for Clean Air and Environmental Protection, 44 Grand Parade, Brighton, East Sussex, BN2 9QA (01273 878 770) Fax: 01273 606 626 www.nsca.org.uk

Royal Society for the Protection of Birds, The Lodge, Sandy, Bedfordshire, SG19 2DL, UK (01767 680 551) www.rspb.org.uk

Soil Association, South Plaza, Marlborough Street, Bristol, BS1 3NX (0117 314 5000) Fax: 0117 314 5001 www.soilassociation.org

WWF-UK, Panda House, Weyside Park, Godalming, Surrey, GU7 1XR (01483 426444) Fax: 01483 426409 www.wwf.org.uk

Marine Pollution

Departments of State and Other Public Bodies

Centre for Environment, Fisheries and Aquaculture Science (CEFAS), Pakefield Road, Lowestoft, Suffolk, NR33 0HT, UK (+44 (0)1502 524430) Fax: +44 (0)1502 524569 http://www.cefas.co.uk

Maritime and Coastguard Agency, Tutt Head, Mumbles, Swansea, West Glamorgan, SA3 4HW, Wales, UK (0870 6006505) www.mcga.gov.uk

The Scottish Association for Marine Science, Dunstaffnage Marine Laboratory, Oban, Argyll, PA37 1QA, Scotland, UK ((+44) (0)1631 559000) Fax: (+44) (0)1631 559001 www.sams.ac.uk

Non-Governmental Organizations

Advisory Committee on Protection of the Sea, 11 Dartmouth Street, London, SW1H 9BN, UK (+44 207 7993033) Fax: +44 207 7992933 http://www.acops.org

British Oil Spill Control Association (BOSCA), 4th Floor, 30 Great Guildford Street, London, SE1 0HS, UK (+44 (0)20 7928 9199) Fax: +44 (0)20 7928 6599 www.maritimeindustries.org/bosca

The Energy Institute, 61 New Cavendish Street, London, W1G 7AR, UK (+44 (0) 20 7467 7100) Fax: +44 (0) 20 7255 1472 www.energyinst.org.uk

Institute of Fisheries Management, 22 Rushworth Avenue, West Bridgford, Nottingham, NG2 7LF, UK (01159 822317) www.ifm.org.uk

The Institute of Marine Engineering, Science and Technology, 76 Mark Lane, London, EC3R 7JN, UK (020 7488 1854) www.imarest.org.uk

Marine Conservation Society, Unit 3, Wolf Business Park, Alton Road, Ross-on-Wye, Herefordshire, HR9 5NB, UK (01989 566017) Fax: 01989 567815 www.mcsuk.org

Noise

BRE (formerly the Building Research Establishment), Bucknalls Lane, Watford, WD25 9XX, UK (01923 664 000) www.bre.co.uk

Civil Aviation Authority, CAA House, 45-59 Kingsway, London, WC2B 6TE, UK (020 7379 7311) www.caa.co.uk

Department for Environment, Food & Rural Affairs, Nobel House, 17 Smith Square, London, SW1P 3JR, UK (020 7238 6000) www.defra.gov.uk

Non-Governmental Organizations

Association of Noise Consultants, 105 St Peter's Street, St Albans, Herts, AL1 3EJ, UK (01727 896092) Fax: 01727 896026 http://www.association-of-noise-consultants.co.uk

Chartered Institute of Environmental Health, Chadwick Court, 15 Hatfields, London, SE1 8DJ, UK (020 7928 6006) Fax: 020 7827 5862 www.cieh.org

Institute of Acoustics, 77A St Peter's Street, St Albans, Herts, AL1 3BN, UK (01727 848195) www.ioa.org.uk

Institute of Sound and Vibration Research, University Road, Highfield, Southampton, S017 1BJ, UK (+44 (0) 23 8059 2294) Fax: +44 (0) 23 8059 3190 www.isvr.soton.ac.uk

Nuclear

Department for Business, Enterprise and Regulatory Reform, 1 Victoria Street, London, SW1H 0ET, UK (020 7215 5000) Fax: 020 7215 0105 www.berr.gov.uk

Department for Environment, Food & Rural Affairs, Nobel House, 17 Smith Square, London, SW1P 3JR, UK (020 7238 6000) www.defra.gov.uk

HSE – Nuclear Directorate, CASE Team, Desk 25, 4N.1 Redgrave Court, Merton Road, Bootle, L20 7HS, UK (0151 951 3484 / 3290) www.hse.gov.uk

Health Protection Agency, Centre for Radiation, Chemical and Environmental Hazards, Radiation Protection Division, Chilton, Didcot, Oxon, OX11 0RQ, UK (01235 831600) Fax: 01235 833891 www.hpa.org.uk

UKAEA, Culham Science Centre, Abingdon, Oxon, OX14 3DB, UK (01235 528822) Fax: 01235 466675 www.ukaea.org.uk

Non-Governmental Organizations

British Nuclear Energy Society, 1-7 Great George Street, London, SW1P 3AA, UK (020 7665 2241 (direct line)) Fax: 020 7799 1325 (local) www.bnes.com

Institution of Nuclear Engineers, Ellen House, 1 Penerley Road, London, SE6 2LQ (0207 717 6000) www.inuce.org.uk/

Occupational Health and Safety

Departments of state, research establishments, and other public bodies

Health and Safety Executive, Caerphilly Business Park, Caerphilly, CF83 3GG, Wales, UK (0845 345 0055) Fax: 0845 408 9566 www.hse.gov.uk

Medical Research Council, 20 Park Crescent, London, W1B 1AL, UK (+44 (0)20 7636 5422) Fax: +44 (0)20 7436 6179 www.mrc.ac.uk

Non-Governmental Organizations

The British Occupational Hygiene Society, 5/6 Melbourne Business Court, Millennium Way, Pride Park, Derby, DE24 8LZ, UK (01332 298101) Fax: 01332 298099 www. bohs.org

Institution of Occupational Health and Safety, The Grange, Highfield Drive, Wigston, Leicester, LE18 1NN, UK (0116 257 3100) Fax: 0116 257 3101 www.iosh.co.uk

Occupational and Environmental Diseases Association (OEDA), (formerly the Society for the Prevention of Asbestosis and Industrial Diseases), PO Box 26, Enfield, Middlesex, EN1 2NT, UK www.oeda.demon.co.uk

The Society of Occupational Medicine, 6 St Andrews Place, Regents Park, London, NW1 4LB, UK (+44 (0) 20 7486 2641) Fax: +44 (0) 20 7486 0028 www.som.org.uk

Trades Union Congress, Congress House, Great Russell Street, London, WC1B 3LS, UK (020 7636 4030) Fax: 020 7636 0632 www.tuc.org.uk

Public Health, Toxicology

The Department of Health, Richmond House, 79 Whitehall, London, SW1A 2NS, UK (020 7210 4850) www.dh.gov.uk

Health Protection Agency, 7th Floor, Holborn Gate, 330 High Holborn, London, WC1V 7PP, UK (020 7759 2700 / 2701) Fax: 020 7759 2733 www.hpa.org.uk

Medical Research Council, 20 Park Crescent, London, W1B 1AL (+44 (0)20 7636 5422) Fax: +44 (0)20 7436 6179 www.mrc.ac.uk

UKAEA, Culham Science Centre, Abingdon, Oxon, OX14 3DB, UK (01235 528822) Fax: 01235 466675 www.ukaea.org.uk

Non-Governmental Organizations

British International Biological Research Association, Westmead House, 123 Westmead Road, Sutton, Surrey, SM1 4JH, UK (+44 (0)20 8722 4701) Fax: +44 (0)20 8722 4706 www.bibra.co.uk

Chartered Institute of Environmental Health, Chadwick Court, 15 Hatfields, London, SE1 8DJ, UK (020 7928 6006) Fax: 020 7827 5862 www.cieh.org

Chemical Industries Association, Kings Buildings, Smith Square, London, SW1P 3JJ (020 7834 3399) Fax: 020 7834 4469 www.cia.org.uk

Health Protection Scotland, Clifton House, Clifton Place, Glasgow, G3 7LN, Scotland, UK (0141 300 1100) Fax: 0141 300 1170 www.hps.scot.nhs.uk

The Royal Institute of Public Health, 28 Portland Place, London, W1B 1DE, UK (020 7580 2731) Fax: 020 7580 6157 www.riph.org.uk

Royal Society of Chemistry, London, Burlington House, Piccadilly, London, W1J 0BA (+44 (0)20 7437 8656) Fax: +44 (0)20 7437 8883 www.rsc.org

The Royal Society for the Promotion of Health, 38A St. George's Drive, London, SW1V 4BH, UK ((+44) (0) 20 7630 0121) Fax: (+44) (0) 20 7976 6847 www.rsph.org

Society of Chemical Industry, 14/15 Belgrave Square, London, SW1X 8PS (+44 (0) 20 7598 1500) Fax: +44 (0) 20 7598 1545 www.soci.org

UK Public Health Association, 2nd Floor, 28 Portland Place, London, W1B 1DE, UK (020 7291 8351) Fax: 020 7436 9525 www.ukpha.org.uk

Wastes, Waste Treatment and Disposal

Departments of State, Research Establishments, and Other Public Bodies

Department for Environment, Food & Rural Affairs, Nobel House, 17 Smith Square, London, SW1P 3JR, UK (020 7238 6000) www.defra.gov.uk

Department of the Environment for Northern Ireland, Clarence Court, 10-18 Adelaide Street, Belfast, BT2 8GB, UK (028 9054 0540) www.doeni.gov.uk

Environment Agency (in England) www.environment-agency.gov.uk

General Enquiries: 08708 506 506

Agricultural Waste Registration: 0845 603 3113

Floodline: 0845 988 1188

Hazardous Waste Registration: 08708 502 858

Incident hotline: 0800 807060

Head Office: 08708 506506

Rio House, Waterside Drive, Aztec West, Almondsbury, Bristol, BS32 4UD, UK (08708 506506)

Anglian Regional Office, Kingfisher House, Goldhay Way, Orton Goldhay, Peterborough, Cambridgeshire, PE2 5ZR, UK (08708 506506)

Midlands Regional Office, Sapphire East, 550 Streetsbrook Road, Solihull, West Midlands, B91 1QT, UK (08708 506506)

North East Regional Office, Rivers House, 21 Park Square South, Leeds, West Yorkshire, LS1 2QG, UK (08708 506506)

North West Regional Office, PO Box 12, Richard Fairclough House, Knutford Road, Latchford, Warrington, Cheshire, WA4 1HT, UK (08708 506506)

Southern Regional Office, Guildbourne House, Chatsworth Road, Worthing, Sussex, BN11 1LD, UK (08708 506506)

South West Regional Office, Manley House, Kestrel Way, Exeter, Devon, EX2 7LQ, UK (08708 506506)

Thames Regional Office, Kings Meadow House, Kings Meadow Road, Reading, Berkshire, RG1 8DQ, UK (08708 506506)

Environment Agency (in Wales), Cambria House, 29 Newport Road, Cardiff, CF24 0TP, UK (08708 506506) www.environment-agency.gov.uk

Environment and Heritage Service, Northern Ireland www.ehsni.gov.uk

Scottish Environment Protection Agency, Erskine Court, Castle Business Park, Stirling, FK9 4TR, UK (01786 457700) Fax: 01786 446885 www.sepa.org.uk

UKAEA, Culham Science Centre, Abingdon, Oxon, OX14 3DB, UK (01235 528822) Fax: 01235 466675 www.ukaea.org.uk

Non-Governmental Organizations

British Metals Recycling Association, 16 High Street, Brampton, Huntingdon, Cambs, PE28 4TU, UK (01480 455249) Fax: 01480 453680 www.recyclemetals.org

Chartered Institute of Public Finance and Accountancy, 3 Robert Street, London, WC 2N 6 RL, UK (0207 543 5600) Fax: 0207 543 5700

The Chartered Institution of Wastes Management, 9 Saxon Court, St. Peter's Gardens, Marefair, Northampton, NN1 1SX, UK (+ 44 (0) 1604 620426) Fax: + 44 (0) 1604 621339 www.ciwm.co.uk

ENCAMS (the charity which runs the Tidy Britain Campaign), Elizabeth House, The Pier, Wigan, WN3 4EX, UK (01942 612621) Fax: 01942 824778 www.encams.org

Environmental Services Association, 154 Buckingham Palace Road, London, SW 1W 9TR, UK (0207 824 8882) Fax: 0207 824 8753 www.esauk.org

The Institute of Materials, Minerals and Mining, 1 Carlton House Terrace, London, SW1Y 5DB, UK (020 7451 7300) www.materials.org.uk

Society of Chemical Industry, 14/15 Belgrave Square, London, SW1X 8PS, UK (+44 (0) 20 7598 1500) Fax: +44 (0) 20 7598 1545 www.soci.org

Waste Watch, 56-64 Leonard Street, London, EC2A 4LT, UK (0207 549 0300) Fax: 0207 549 0301 www.wastewatch.org.uk

Water and Water Pollution

Departments of State, Research Establishments, and Other Public Bodies

British Waterways, Willow Grange, Church Road, Watford, Herts, WD17 4QA, UK (01923 201120) Fax: 01923 201400 www.britishwaterways.co.uk

Department for Environment, Food & Rural Affairs, Nobel House, 17 Smith Square, London, SW1P 3JR, UK (020 7238 6000) www.defra.gov.uk

Department of the Environment for Northern Ireland, Clarence Court, 10-18 Adelaide Street, Belfast, BT2 8GB, UK (028 9054 0540) www.doeni.gov.uk

Drinking Water Inspectorate, Floor 2/A1, Ashdown House, 123 Victoria Street, London, SW1E 6DE (020 7082 8024) Fax: 020 7082 8028 www.dwi.gov.uk

Environment Agency (in England) www.environment-agency.gov.uk

　　General Enquiries: 08708 506 506

　　Agricultural Waste Registration: 0845 603 3113

　　Floodline: 0845 988 1188

　　Hazardous Waste Registration: 08708 502 858

　　Incident hotline: 0800 807060

　　Head Office, Rio House, Waterside Drive, Aztec West, Almondsbury, Bristol, BS32 4UD, UK (08708 506506)

Anglian Regional Office, Kingfisher House, Goldhay Way, Orton Goldhay, Peterborough, Cambridgeshire, PE2 5ZR, UK (08708 506506)

Midlands Regional Office, Sapphire East, 550 Streetsbrook Road, Solihull, West Midlands, B91 1QT, UK (08708 506506)

North East Regional Office, Rivers House, 21 Park Square South, Leeds, West Yorkshire, LS1 2QG, UK (08708 506506)

North West Regional Office, PO Box 12, Richard Fairclough House, Knutford Road, Latchford, Warrington, Cheshire, WA4 1HT, UK (08708 506506)

Southern Regional Office, Guildbourne House, Chatsworth Road, Worthing, Sussex, BN11 1LD, UK (08708 506506)

South West Regional Office, Manley House, Kestrel Way, Exeter, Devon, EX2 7LQ, UK (08708 506506)

Thames Regional Office, Kings Meadow House, Kings Meadow Road, Reading, Berkshire, RG1 8DQ, UK (08708 506506)

Environment Agency (in Wales), Cambria House, 29 Newport Road, Cardiff, CF24 0TP, UK (08708 506506) www.environment-agency.gov.uk

Environment and Heritage Service, Northern Ireland www.ehsni.gov.uk

Ofwat, Centre City Tower, 7 Hill Street, Birmingham, B5 4UA, UK (0121 625 1300 / 1373) Fax: 0121 625 1400 www.ofwat.gov.uk

Scottish Environment Protection Agency, Erskine Court, Castle Business Park, Stirling, FK9 4TR, UK (01786 457700) Fax: 01786 446885 www.sepa.org.uk

Non-Governmental Organizations

Centre for Ecology and Hydrology, Polaris House, North Star Avenue, Swindon, Wiltshire, SN2 1EU, UK (01793 442516) Fax: 01793 442528 www.ceh.ac.uk

Chartered Institution of Water and Environmental Management (CIWEM), 15 John Street, London, WC1N 2EB, UK (020 7831 3110) Fax: 020 7405 4967 www.ciwem.org

The Institution of Water Officers, 4 Carlton Court, Team Valley, Gateshead, NE11 0AZ, UK (0191 422 0088) www.iwohq.demon.co.uk

UK Water Industry Research Limited, 1 Queen Anne's Gate, London, SW1H 9BT, UK (+44(0)20 7344 1807) Fax: +44(0)20 7344 1859 www.ukwir.org.uk

Water UK, 1 Queen Anne's Gate, London, SW1H 9BT (020 7344 1844) Fax: 020 7344 1866 www.water.org.uk

International

CADDET – Energy Efficiency and Renewable Energy at your fingertips http://www.caddet.org

Foundation for International Environmental Law and Development, 3 Endsleigh Street, London, WC1H 0DD, UK (0207 73882117) Fax: 020 73882826 www.field.org.uk

Global Ecolabelling Network, Green Labelling Unit, Department for Environment, Food and Rural Affairs, Zone 4 / G15 E Ashdown House, 123 Victoria Street, London SW1E 6DE (+44 20 7082 8672) http://www.defra.gov.uk/environment/consumerprod/ecolabel/index.htm

Institute for European Environmental Policy, 28 Queen Anne's Gate, London, SW1H 9AB, UK (+44 (0) 20 7799 2244) Fax: +44 (0) 20 7799 2600 www.ieep.eu

Intergovernmental Panel on Climate Change, c/o World Meteorological Organization, 7bis Avenue de la Paix, C.P. 2300, CH- 1211 Geneva 2, Switzerland (+41-22-730-8208 / 84) Fax +41-22-730-8025 / 13 www.ipcc.ch

International Association for Impact Assessment, 1330, 23rd Street South, Suite C, Fargo, ND 58103 USA (+ 1 701 297 7908) Fax: +1 701 297 7917 www.iaia.org

International Council for Local Environmental Initiatives, City Hall, West Tower, 16th Floor, 100 Queen St. West, Toronto, Ontario, M5H 2N2 Canada (+1-416/392-1462) Fax +1-416/392-1478 www.iclei.org

International Energy Agency, 9, rue de la Fédération, 75739 Paris Cedex 15, France (33 1) 40 57 65 00/01 fax: (33 1) 40 57 65 59 www.iea.org

International Institute for Sustainable Development, 161 Portage Avenue East, 6th Floor, Winnipeg, Manitoba, Canada, R3B 0Y4 (+1 204 958-7700) Fax: +1 204 958-7710 www.iisd.org

International Union of Air Pollution Prevention and Environmental Protection Associations, 44, Grand Parade, Brighton, BN2 9QA, UK (+44 1273 878770) Fax: +44 1273 606626 www.iuappa.com

International Water Association (IWA), Alliance House, 12 Caxton Street, London, SW1H 0QS, UK (+44 207 654 5500) Fax: +44 207 654 5555 www.iwahq.org.uk

UN Economic Commission for Europe, Palais des Nations, CH – 1211 Geneva 10, Switzerland (+41 (0) 22 917 12 34) Fax: +41 (0) 22 917 05 05 www.unece.org

United Nations Development Programme, One United Nations Plaza, New York, NY 10017 USA (+1 (212) 906-5000) Fax: +1 (212) 906-5364 www.undp.org

United Nations Environment Programme, United Nations Avenue, Gigiri, PO Box 30552, 00100, Nairobi, Kenya ((254-20) 7621234) Fax: (254-20) 7624489/90 www.unep.org

World Business Council for Sustainable Development, 4, chemin de Conches, 1231 Conches-Geneva, Switzerland (+41 (22) 839 3100) Fax: +41 (22) 839 3131 www.wbcsd.org

World Health Organization, Avenue Appia 20, 1211 Geneva 27, Switzerland ((+ 41 22) 791 21 11) fax: (+ 41 22) 791 3111 www.who.int

Europe

CORDIS (Community Research & Development Information Service) www.cordis.europa.eu/en/home.html

European Centre for Nature Conservation, Visiting address: Reitseplein 3, 5037 AA Tilburg, Postal address: PO Box 90154, 5000 LG Tilburg, The Netherlands (+ 31-13-5944944) Fax: +31-13-5944945 http://www.ecnc.org

European Commission Directorate-General on Energy and Transport, Rue J.-A. Demot, 24-28, B – 1040 Brussels www.ec.europa.eu/dgs/energy_transport

European Environment Agency, Kongens Nytorv 6, DK-1050 Copenhagen K, Denmark ((+45) 33 36 71 00) Fax: (+45) 33 36 71 99 www.eea.europa.eu

European Integrated Pollution Prevention and Control Bureau, IPTS-European Commission, WTC, Isla de la Cartuja s/n, E-1092 Sevilla, Spain Fax: 0034 95 448 84 26 http://eippcb.jrc.es

European Partners for the Environment, Av. de la Toison d'Or 67, B-1060 Brussels, Belgium (++32 2 771 15 34) Fax: ++32 2 539 48 15 www.epe.be

United States of America

Governmental Organizations

National Renewable Energy Laboratory, 1617 Cole Blvd, Golden, CO 80401-3393, Main Phone Number (303) 275-3000 www.nrel.gov

Nuclear Information and Resource Service, 6930 Carroll Avenue, Suite 340, Takoma Park, MD 20912 (+301-270-6477) Fax: +301-270-4291 www.nirs.gov

Office of Fusion Energy Sciences, U.S. Department of Energy, SC-24, 19901 Germantown Road, Germantown, Maryland 20874-1290 (+ 301 903 4941) Fax: +301 903 8584 www.ofes.fusion.doe.gov

U.S. Department of Energy, Office of Environmental Management, 1000 Independence Ave., SW, Washington, DC 20585 (+202-586-5000) www.em.doe.gov

U.S. Environmental Protection Agency, Ariel Rios Building, 1200 Pennsylvania Avenue, N.W, Washington, DC 20460 (+ (202) 272-0167) www.epa.gov

Non-Governmental Organizations

Green Seal, 1001 Connecticut Avenue, NW, Suite 827, Washington, DC 20036-5525, USA (202-872-6400) Fax: 202-872-4324 www.greenseal.org

Resource Renewal Institute, Fort Mason Centre, Building D, San Francisco, CA 94123 (+415.928.3774) Fax: +415.928.4050 www.rri.org

Pacific Institute, 654 13th Street, Preservation Park, Oakland, CA 94612, U.S.A (+1-510-251-1600) Fax: +1-510-251-2203 www.pacinst.org

Union of Concerned Scientists, 2 Brattle Square, Cambridge, MA 02238-9105 (617-547-5552) Fax: 617-864-9405 www.ucsusa.org

U.S. Climate Change Science Program / U.S. Global Change Research Program, Suite 250, 1717 Pennsylvania Ave, NW, Washington, DC 20006 (+1 202 223 6262) Fax: +1 202 223 3065 www.usgcrp.gov

APPENDIX II

The Periodic Table

giving atomic number and chemical symbol for each element

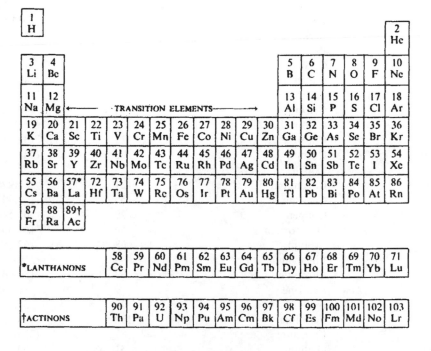

1 H																	2 He
3 Li	4 Be											5 B	6 C	7 N	8 O	9 F	10 Ne
11 Na	12 Mg	←————— TRANSITION ELEMENTS —————→										13 Al	14 Si	15 P	16 S	17 Cl	18 Ar
19 K	20 Ca	21 Sc	22 Ti	23 V	24 Cr	25 Mn	26 Fe	27 Co	28 Ni	29 Cu	30 Zn	31 Ga	32 Ge	33 As	34 Se	35 Br	36 Kr
37 Rb	38 Sr	39 Y	40 Zr	41 Nb	42 Mo	43 Tc	44 Ru	45 Rh	46 Pd	47 Ag	48 Cd	49 In	50 Sn	51 Sb	52 Te	53 I	54 Xe
55 Cs	56 Ba	57* La	72 Hf	73 Ta	74 W	75 Re	76 Os	77 Ir	78 Pt	79 Au	80 Hg	81 Tl	82 Pb	83 Bi	84 Po	85 At	86 Rn
87 Fr	88 Ra	89† Ac															

*LANTHANONS	58 Ce	59 Pr	60 Nd	61 Pm	62 Sm	63 Eu	64 Gd	65 Tb	66 Dy	67 Ho	68 Er	69 Tm	70 Yb	71 Lu
†ACTINONS	90 Th	91 Pa	92 U	93 Np	94 Pu	95 Am	96 Cm	97 Bk	98 Cf	99 Es	100 Fm	101 Md	102 No	103 Lr

├———→ ARTIFICIAL ELEMENTS

Table of Chemical Elements

Atomic masses are based on the 1969 international agreed values of the International Union of Pure and Applied Chemistry, the basis being the carbon-12 isotope. Values in brackets indicate the mass numbers of the most stable isotopes. See also the Periodic Table.

Symbol	Name	Atomic	
		No.	Mass
Ac	Actinium	89	(227)
Ag	Silver	47	107.868
Al	Aluminium	13	26.9815
Am	Americium	95	(243)
Ar	Argon	18	39.948
As	Arsenic	33	74.9216
At	Astatine	85	(210)
Au	Gold	79	196.9665
B	Boron	5	10.81
Ba	Barium	56	137.34
Be	Beryllium	4	9.0122
Bi	Bismuth	83	208.9806
Bk	Berkelium	97	(249)
Br	Bromine	35	79.904
C	Carbon	6	12.011
Ca	Calcium	20	40.08
Cd	Cadmium	48	112.40
Ce	Cerium	58	140.12
Cf	Californium	98	(251)
Cl	Chlorine	17	35.453
Cm	Curium	96	(247)
Co	Cobalt	27	58.9332
Cr	Chromium	24	51.996
Cs	Caesium	55	132.9055
Cu	Copper	29	63.546
Dy	Dysprosium	66	162.50
Er	Erbium	68	167.26
Es	Einsteinium	99	(254)
Eu	Europium	63	151.96
F	Fluorine	9	18.9984
Fe	Iron (ferrum)	26	55.847
Fm	Fermium	100	(253)
Fr	Francium	87	(223)
Ga	Gallium	31	69.72
Gd	Gadolinum	64	157.25

Symbol	Name	Atomic	
		No.	*Mass*
Ge	Germanium	32	72.59
H	Hydrogen	1	1.0080
Ha	Hahnium	105	—
He	Helium	2	4.0026
Hf	Hafnium	72	178.49
Hg	Mercury	80	200.59
Ho	Holmium	67	164.9303
I	Iodine	53	126.9045
In	Indium	49	114.82
Ir	Iridium	77	192.22
K	Potassium	19	39.102
Kr	Krypton	36	83.80
La	Lanthanum	57	138.9055
Li	Lithium	3	6.941
Lu	Lutetium	71	174.97
Lr	Lawrencium	103	—
Md	Mendelevium	101	(256)
Mg	Magnesium	12	24.305
Mn	Manganese	25	54.9380
Mo	Molybdenum	42	95.94
N	Nitrogen	7	14.0067
Na	Sodium (natrium)	11	22.9898
Nb	Niobrium (columbium)	41	92.9064
Nd	Neodymium	60	144.24
Ne	Neon	10	20.179
Ni	Nickel	28	58.71
No	Nobelium	102	(254)
Np	Neptunium	93	(237)
O	Oxygen	8	15.9994
Os	Osmium	76	190.2
P	Phosphorus	15	30.9738
Pa	Protactinium	91	231.0359
Pb	Lead (plumbum)	82	207.2
Pd	Palladium	46	106.4
Pm	Promethium	61	(145)
Po	Polonium	84	(210)
Pr	Praseodymium	59	140.9077
Pt	Platinum	78	195.09
Pu	Plutonium	94	(242)
Ra	Radium	88	226.0254
Rb	Rubidium	37	85.4678
Re	Rhenium	75	186.2

Symbol	Name	Atomic	
		No.	Mass
Rf	Rutherfordium	104	—
Rh	Rhodium	45	102.9055
Rn	Radon (niton)	86	(222)
Ru	Ruthenium	44	101.07
S	Sulphur	16	32.06
Sb	Antimony	51	121.75
Sc	Scandium	21	44.9559
Se	Selenium	34	78.96
Si	Silicon	14	28.086
Sm	Samarium	62	150.4
Sn	Tin (stannum)	50	118.69
Sr	Strontium	38	87.62
Ta	Tantalum	73	180.9479
Tb	Terbium	65	158.8254
Tc	Technetium	43	(99)
Te	Tellurium	52	127.60
Th	Thorium	90	232.0381
Ti	Titanium	22	47.90
Tl	Thallium	81	204.37
Tm	Thulium	69	168.9342
U	Uranium	92	238.029
V	Vanadium	23	50.9414
W	Tungsten (wolfram)	74	183.85
Xe	Xenon	54	131.30
Y	Yttrium	39	88.9059
Yb	Ytterbium	70	173.04
Zn	Zinc	30	65.37
Zr	Zirconium	40	91.22

Table of prefixes for SI units

Prefix	Symbol	Factor
tera	T	$10^{12} = 1\,000\,000\,000\,000$
giga	G	$10^{9} = 1\,000\,000\,000$
mega	M	$10^{6} = 1\,000\,000$
kilo	k	$10^{3} = 1000$
hecto	h	$10^{2} = 100$
deca	da	$10^{1} = 10$
deci	d	$10^{-1} = 0.1$
centi	c	$10^{-2} = 0.01$
milli	m	$10^{-3} = 0.001$
micro	μ	$10^{-6} = 0.000\,001$
nano	n	$10^{-9} = 0.000\,000\,001$
pico	p	$10^{-12} = 0.000\,000\,000\,001$
femto	f	$10^{-15} = 0.000\,000\,000\,000\,001$
atto	a	$10^{-18} = 0.000\,000\,000\,000\,000\,001$

Conversion table for SI and British units

Physical property	British unit	SI unit	SI unit	British unit	Cgs unit
length	1 ft	0.305 m	1 m	3.28 ft	100 cm
	1 mile	1.61 km	1 km	0.621 mile	10^5 cm
area	1 ft^2	0.0929 m^2	1 m^2	10.76 ft^2	10^4 cm^2
	1 acre	4047.0 m^2	1 km^2	2.471×10^2 acres	
	1 acres	0.405 ha	1 ha	2.471 acres	10^8 cm^2
volume	1 ft^3	0.0283 m^3	1 m^3	35.31 ft^3	10^6 cm^3
	1 gall (UK)	4.55 litres	1 litre	0.220 gallon (UK)	10^3 cm^3
mass	1 lb	0.454 kg	1 kg	2.204 lb	10^3 g
density	1 lb/ft^3	16.02 kg/m^3	1 kg/m^3	0.0624 lb/ft^3	10^3 g/cm^3
time	1 sec	1 s	1 s	1 second	1 s
			a defined fraction of a solar day		
			3600 s = 1 h		
velocity	1 ft/s	0.305 m/s	1 m/s	3.28 ft/s	100 cm/s
			1 km/h	0.911 ft/s	
	1 mile/h	1.61 km/h	1 km/h	0.621 mile/h	27.8 cm/s
force	1 lbf	4.45 N	1 N(kg m/s^2)	0.225 lbf	10^5 fyn (dynes)
(mass × cceleration)					(1 dyn = 1 g cm/s^2)

pressure (force per unit area)	1 lbf/ft²	47.5 N/m²	1 Nn/m² (1 pascal – Pa)	0.0207 lbf/ft²	10 dyn/cm²
pressure in meteorology and acoustics work			1 bar	29.53 in Hg	750 mm Hg (10⁶ dyn/cm²)
			10⁵ N/m²		
			1.013 bar	29.9 in Hg	760 mm Hg
energy (force × distance)	1 ft lbf	1.352 J	1 J (Nm) (joule)	0.738 ft lbf	10⁷ ergs
heat equivalent	1 Btu	1.055 kJ	1 kJ	0.948 Btu	10¹⁰ ergs
power (rate of doing work)	1 ft lbf/s	1.356 W	1 W (J/s)	0.738 ft lbf/s	10⁷ ergs/s
	1 hp	0.746 kW	1 kW	1.34 hp	10¹⁰ erg/s
flow rate	1 million gall/d	0.0526 m³/s (Cumecs)	1 m³ s⁻¹ (Cumecs)	19.01 million gall/d	10⁶ cm³/s
			1 × 10⁶ m³/d	220 million gall/d	
	1 gall/min	0.0761 s⁻¹	1 l/s	13.20 gall/min	10³ cm³/s